ENCYCLOPEDIA OF 20TH-CENTURY
TECHNOLOGY

ENCYCLOPEDIA OF 20TH-CENTURY
TECHNOLOGY

Volume 2
M–Z
INDEX

Colin A. Hempstead, Editor
William E. Worthington, Jr., Associate Editor

ROUTLEDGE
NEW YORK AND LONDON

Published in 2004 by
Routledge
29 West 35th Street
New York, NY 10001
www.routledge-ny.com

Published in Great Britain by
Routledge
2 Park Square
Milton Park, Abingdon
Oxon OX14 4RN, UK
www.routledge.co.uk

Routledge is an imprint of the Taylor & Francis Group.

10 9 8 7 6 5 4 3 2 1

Library of Congress Cataloging-in-Publication Data

Encyclopedia of 20th-century technology / Colin A. Hempstead, editor; William E. Worthington, associate editor.
p. cm.
Includes bibliographical references and index.
ISBN 1-57958-386-5 (set : alk. paper)—ISBN 1-57958-463-2 (vol. 1 : alk. paper)—
ISBN 1-57958-464-0 (vol. 2 alk. paper)
1. Technology—Encyclopedias. I. Hempstead, Colin. II. Worthington, William E., 1948–
T9.E462 2005
603—dc22

Advisers

Dr. Jon Agar, Department of History and Philosophy of Science, University of Cambridge

Professor Janet Bainbridge, Chief Executive, EPICC (European Process Industries Competitiveness Centre), Teesside

Dr. Hans Joachim Braun, Universität der Bundeswehr Hamburg

Dr. Robert Bud, Principal Curator of Science, Science Museum, London

Dr. Michael Duffy, formerly Department of Engineering, University of Sunderland

Dr. Slava Gerovitch, Dibner Institute for the History of Science and Technology, MIT

Dr. Ernst Homburg, Department of History, University of Maastricht

Dr. Sally Horrocks, Department of Economic and Social History, University of Leicester

R. Douglas Hurt, Professor and Director, Graduate Program in Agricultural History and Rural Studies, Department of History, Iowa State University

Dr. Peter Morris, Science Museum, London

Professor John Pickstone, Centre for the History of Science, Technology and Medicine, University of Manchester

Keith Thrower, former Technical Director at Racal Electronics, UK

Contents

List of Entries

Acknowledgments

A host of workers and authors contributed to this encyclopedia, and I wish to extend my thanks to every person without whom these volumes would be stillborn. My particular thanks are offered to Gillian Lindsey of Routledge. Gillian conceived the idea of an encyclopedia of 20th-century technology, and appointed me the editor of the work in 2000. Her energy and ideas were legion, although she glossed over the amount of work for me! However, the editorship was rewarding, offering the possibility of producing a worthwhile publication with academic colleagues from around the globe. The selection of technologies and of particular subjects suggested by Gillian and me were critiqued and extended by our advisers. Their contributions, drawn from their specialist knowledge and scholarship, were invaluable. When circumstances forced my withdrawal from the active editorship, William Worthington, then with the National Museum of American History in Washington, stepped into the hot seat. To William I give my heartfelt thanks.

Finally I acknowledge the publishers and the 20th century which presented all of us with the opportunity to examine and extol some of the content and effects of modern technology. Nevertheless, the encyclopedia is partial, and any omissions and shortcomings are mine.

Colin Hempstead

My thanks go to Gillian Lindsey for presenting me with the challenge of filling the void left by Colin's departure. However, the prospect of assuming a role in a project already well under way and natural differences in approach and style were concerns. Nonetheless, the final third of the encyclopedia was crafted in such a way that it blends seamlessly with the sections completed under Colin's careful guidance. This was due in no small part to the untiring efforts of Sally Barhydt, and to her I extend sincere thanks.

William E. Worthington, Jr.

M

Mass Spectrometry

Mass spectrometry is the separation by mass-to-charge ratios (m/z) and the measurement of m/z and abundances of mono- and polyatomic ions in the gas phase. A graphical display of ion abundance as a function of m/z is called a mass spectrum. All mass spectrometers operate under greatly reduced pressure in order to produce and sustain ions in the gas phase. The British physicist J.J. Thomson observed the first mass spectrum in 1912, which consisted of differences in the trajectories of two isotopic ions of the inert gas neon, subjected to electric and magnetic fields. Near the end of the decade, Thomson's student at Cambridge University, Francis Aston, developed the first practical mass spectrograph, which was based on the apparatus used by Thomson and employed an ion-sensitive photographic plate as the detector. During the 1920s and 1930s, other investigators including A.J. Dempster at the University of Chicago and J. Mattauch and R.Z. Herzog in Austria developed mass spectrographs, and Dempster invented the electron ionization technique that is widely used today to produce ions in the gas phase. These early instruments used magnetic and electric fields to separate ions by m/z, and the discoveries made with them had a profound impact on the chemical, physical, and life sciences. The Nobel Prize in Chemistry for 1922 was awarded to Francis Aston "...for his discovery, by means of his mass spectrograph, of isotopes, in a large number of nonradioactive elements..." The atomic weights of all the elements were determined by mass spectrometric measurements of the exact masses and relative abundances of the naturally occurring and man-made isotopes.

Alfred O.C. Nier at The University of Minnesota designed and constructed improved magnetic deflection mass spectrometers during the 1930s and 1940s, and used these instruments to discover new isotopes and to measure the isotopic compositions of many elements including the uranium isotopes. Mass spectrometers were used for analyses of isotopic compositions, measurements of isotopic enrichment, and the production of uranium-235 during the World War II project to develop the atomic bomb. The Consolidated Engineering Corporation introduced the first commercial mass spectrometer in the U.S. in 1940. This instrument used magnetic deflection to separate ions and was widely employed in the petroleum industry for qualitative and quantitative analyses of gas mixtures during and after World War II. During the early 1950s Nier and his student E.G. Johnson designed a double-focusing high-resolving power mass spectrometer consisting of tandem electrostatic and magnetic sectors and the electrical detection of ions rather than photographic plates. The Nier–Johnson design was used in many commercial spectrometers, and became a standard instrument for exact measurements of the m/z of elemental and polyatomic inorganic and organic ions.

During the late 1940s and 1950s, mass spectrometer designs began to appear that were not dependent on heavy, power-consuming electromagnets. In 1955 W.C. Wiley and I.H. McLaren of the Bendix Aviation Corporation described a time-of-flight (TOF) mass spectrometer, and the Bendix Corporation marketed instruments based on their design throughout the 1960s. In a TOF mass spectrometer, ions are accelerated in batches into an evacuated flight tube by a rapid series of

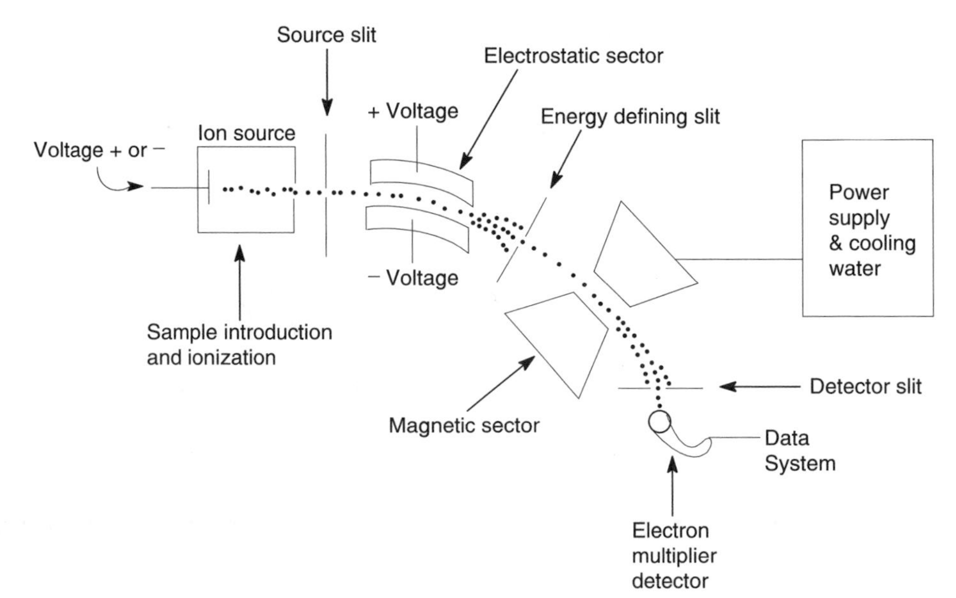

Figure 1. A mass spectrometer with electrostatic and magnetic sectors in the Nier-Johnson double focusing configuration.

high potential pulses at the acceleration electrode. The time of flight of ions from acceleration to detection is a function of m/z. However, the resolving power of vintage 1960s instruments was still too low for most applications. During the 1980s several design innovations, including the electrostatic ion mirror developed by B.A. Mamyrin and D.V. Shmikk in Russia, and very fast semiconductor electronics gave significantly improved TOF mass spectrometer performance. The TOF mass spectrometer is widely used, especially for the characterization of large molecules of biological origin and in experiments where very high-speed data acquisition is required.

Another innovation was the linear quadrupole mass spectrometer, or mass filter, described by W. Paul and H. Steinwedel of The University of Bonn in 1953. Radio frequency and direct current potentials are applied to four parallel electrodes positioned at the apices of a square and, depending on the potentials, ions of specific m/z pass through the center of the quadrupolar field to the detector while all other ions are deflected and discharged on the rods. The linear quadrupole is the most widely used type of mass spectrometer, especially when mass spectrometry is combined with gas and liquid chromatography (GC and LC) sample introduction. It is compact in size, low in weight, relatively low in cost, and has a tolerance to higher pressures—about 10^{-5} Torr—than most other types of mass spectrometers. However, it does not have high resolving power or exact m/z measurement capability. The ion trap mass spectrometer is a three-dimensional quadrupole in which ions are stored for milliseconds by a radio frequency potential applied to a ring electrode. Ions are ejected from the trap according to their m/z by increasing the amplitude of the radio frequency potential. Wolfgang Paul was awarded the Nobel Prize in Physics in 1989 for the inventions of the linear quadrupole and the quadrupole ion trap. The ion trap mass spectrometer is widely used in GC/MS and LC/MS applications.

The Fourier transform mass spectrometer (FTMS) was derived from the ion cyclotron resonance spectrometer that was introduced commercially by Varian Associates in 1966. In 1974 M.B. Comisarow and A.G. Marshall of the University of British Columbia developed the FTMS which uses a strong magnetic field, typically from a superconducting magnet, to trap ions and cause them to spiral towards the detector at circular frequencies that are related to their m/z. Unlike other mass spectrometers, ions are not physically separated, but are detected by measuring an image current induced in the walls of the trap. The m/z of the ions are determined by Fourier transforms of the image current data. The FTMS requires a very low operating pressure; that is, 10^{-8} Torr or lower, but is capable of very high resolving power and exact mass measurements.

Samples for mass spectrometry may be gases, liquids, or solids but their components must be converted into gas-phase positive or negative ions before mass analysis. Many types of sample introduction systems are used, and sample introduction and ionization may be separate or combined in a single process. Ions are injected from the sample introduction/ionization source into the spectrometer for very rapid analysis that is typically on a microsecond or shorter time scale. Ions may be formed and injected continuously, pulsed into the spectrometer in batches, or introduced and stored in batches. Two mass spectrometers are often arranged in tandem configurations to facilitate special experiments, for example, the separation of an ion of specific m/z in the first spectrometer, collision-induced dissociation of the ion, and analysis of the decomposition products in the second mass spectrometer. All mass spectrometers are very sensitive and produce measurable signals from quantities of sample in the range of micrograms (10^{-6}) to femtograms (10^{-15}) or less. Nearly all commercial mass spectrometers produced during the last third of the century incorporate dedicated digital computers that control the operations of the mass spectrometer, acquire and store mass spectrometric data in digital form, and process or display the m/z and abundance data. Operations of most types of mass spectrometers, especially the TOF, quadrupole ion trap, and FTMS, would be impossible without very fast digital computer technology. The major application of mass spectrometry is the determination of chemical composition and structure in many fields of investigation including agriculture, beverages and foods, biological systems and processes, the environment, geology, industrial materials, petroleum exploration and processing, pharmaceutical discovery, natural products, and space exploration.

WILLIAM L. BUDDE

Further Reading

Amster, I.J. Fourier transform mass spectrometry. *J. Mass Spectrom.*, 31, 1325–1337, 1996.

Barshcik, C.M., Duckworth, D.C. and Smith, D.H., Eds. *Inorganic Mass Spectrometry: Fundamentals and Applications.* Marcel Dekker, New York, 2000.

Beynon, J.H. *Mass Spectrometry and its Applications to Organic Chemistry*. Elsevier, Amsterdam, 1960.

Budde, W.L. *Analytical Mass Spectrometry: Strategies for Environmental and Related Applications.* Oxford University Press, New York, 2001.

Cotter, R.J. *Time-of-Flight Mass Spectrometry: Instrumentation and Applications in Biological Research.* American Chemical Society, Washington D.C., 1997.

March, R.E. and Todd, J.F.J., Eds. *Practical Aspects of Ion Trap Mass Spectrometry*; vols 1–3. CRC Press, Boca Raton, FL, 1995.

Montaser, A., Ed. *Inductively Coupled Plasma Mass Spectrometry*. Wiley, New York, 1998.

Nier, K.A. A history of the mass spectrometer, in *Instruments of Science: An Historical Encyclopedia*, 1998.

Watson, J.T. *Introduction to Mass Spectrometry*, 3rd edn. Lippincott–Raven Publishers, Hagerstown, MD, 1997.

Materials and Industrial Processes

The history of the development of materials during the twentieth century, as revolutionary as these developments may seem, has been part of an evolutionary past. The human race is unique among animals on this planet (except for some apes) in that we continually strive to make life easier for ourselves by inventing tools. Some essential features of all developments include the avoidance of excessive effort or extreme conditions and the achievement of improved properties as part of a war against the problems of durability, wear, rot, and corrosion. This is probably true for the development of nonferrous metallurgy since the Bronze Age. The firing of certain clays yielded pots, but many of these were porous. It was the development of glazes and particularly fluxes for glazes that led to containers that could hold fluids and avoided colonization by harmful bacteria. Fluxes could also be added to sand to reduce the melting temperature, and this led to glass manufacture. Ferrous metallurgy since the Iron Age represents a curious development. Iron is durable and malleable but not resistant to corrosion, and cast iron is brittle. However, the advantages outweighed the disadvantages and over time, techniques evolved and skills were handed down. The work of the blacksmith and specialized products such as Toledo steel have a long history.

Industrial Revolution and the Rise of Materials Science

There was a rapid acceleration of materials technology during the Industrial Revolution. Iron and steel were needed for transport (ships, trains) and in the textile industry. Iron itself became a marketable commodity; ex-catalog, made to order cast-iron shop-fronts dating from the nineteenth century can be found in Halifax, Nova Scotia. There was a need to control composition, just as there was a movement from small-batch processes to large-batch and continuous processes; the blast furnace, reverbatory furnaces, the Bessemer process and its modern derivatives. Techniques had to be developed to work the material into usable products. Hammer mills, drop forges, and rolling mills were constructed. There were improvements to lathes, drilling machines, boring and rifling machines so that they were capable of tight tolerances over large areas of large work-pieces. But before this could be achieved in a reproducible way there was a need for standards and a uniform set of units as well as standard measures of flatness, hardness and tensile strength.

If the Industrial Revolution had major long-term effects, other developments in the Victorian era could be viewed as revolutionary in their own right, whether they were independent or spawned by the demands of increased industrialization and commercialization. The advent of the electric telegraph could be interpreted as merely the integration of the ideas of many people who were working on electricity at that time. However, it was adopted because it was needed. The provision of uniform time by which railways could be run would have been reason enough. Its value in railway signaling provided safety for the participants in the first generation of mass transport, unused to large-scale fatalities such as could occur in a train crash. The benefits for business and for the dissemination of news ensured that it was here to stay. However, the new technology demanded new materials for new applications. Subaqueous and submarine telegraphy required that the conductor be insulated and gutta percha (*cis*-polyisoprene) was found to be suitable. It is isomeric with latex rubber (*trans*-polyisoprene) and is likewise a natural product, extracted from trees that only grow in certain parts of the world. Thus it is possible to interpret an element of British Imperial strategy in the late nineteenth century as a move to protect (and control) the supply of this valuable natural resource. The demands of mass communications extended further. Improved ceramics were developed for the humble insulators that were used to support wires on overhead poles. One of the positive outcomes of the unsuccessful attempt to lay a transatlantic cable in 1857 was the unequivocal demonstration of the need for a copper conductor of the highest possible purity. Quality control was paramount and it could be argued that the ultimate success of the transatlantic venture in 1866 was largely due to the introduction of a strict system of externally monitored quality control during both the manufacture and subsequent laying of the cable.

Just as we could argue about telegraphy as a product or a mover in global developments, we could engage in a similar debate about electricity itself. High-quality copper that had been developed for telegraphy was available to provide low-loss conductors. Electric motors replaced steam and other motive sources. Energy costs for manufacturing were reduced and the demand created a new science as well as a new industry for the generation and distribution of electrical power.

Other aspects of the response to the demands that were created during the Industrial Revolution include the entire history of the chemical industry, particularly the replacement of natural dyes with synthetic dyes and the impact that this had on the textile industry. There is also the development of explosives as a means of easing the effort of mining, quarrying, tunneling and providing a strategic edge in warfare. Nitrogylcerin is easy to manufacture, but it is notoriously unstable. Blasting charges did not always undergo uniform detonation and many miners and quarry-men were killed or injured by secondary explosions. Nobel's response by integrating the explosive in an absorbent earth (kielelghur) was a timely invention that solved a problem that was seriously hampering industrial development.

Twentieth Century Technological and Consumer Needs

Materials technology in the twentieth century has been enhanced by understanding the underlying science, and tools that characterize microstructure, such as electron microscopy. As technologies required ever more stringent performance requirements of synthetic materials, novel synthesis and processing approaches have been developed. Since the 1970s, control of structural growth layer by layer (epitaxial growth in semiconductors), microstructure of composites, and targeted molecular design of polymers have made it possible to manipulate the "microstructure" to produce classes of materials with unique properties.

Fundamental developments in materials processing have helped to shape world history. There was a time when sodium and potassium nitrate, prime components in gunpowder and many explosives were in relatively short supply. During World War I, Germany did not have ready access to the natural deposits of nitrates in Chile and would have been unable to manufacture gunpowder were it not for the Haber process by which nitrogen in the air can be "cracked" and turned into ammonia (see Nitrogen Fixation). When this was converted into nitric acid, it was possible to manufacture nitroglycerin and ammonium nitrate as blasting explosives, trinitrotoluene (TNT) as a high explosive, gun-cotton as projectile explosive, as well as lead azide and mercury fulminate as detonating explosives.

The use of materials and processes to exploit latent properties to provide an easier life for more people at reduced cost continued to operate during the twentieth century, although there has been a much greater incidence of developments for nothing more than leisure. The automobile can be seen as a device for reducing effort, but it can also be seen as a means for facilitating leisure. While the derivation of a multitude of chemicals from coal-tar (a byproduct of the coal gas industry) was a nineteenth century achievement, in the twentieth century the chemical industry has responded to the demands of a fuel-hungry mobile society by developing ever more efficient methods of maximizing the yield of fuels extracted from crude oil. This was made possible due to advances in our understanding of the mechanisms of heterogeneous catalysis (see Cracking). However, as the twenty-first century approached there was also an awareness of environmental issues. Cars are now subject to environmental restrictions. Leaded petrol is gone. There are limitations on exhaust emissions, and full recycling seems imminent.

The food industry required a protective medium for packaging and for many years cellophane (invented in 1912) fulfilled this role. Similarly, as the diffusion of electricity into domestic markets gathered pace there was a requirement for a cheap and reliable insulating plastic for plugs, plug sockets, and switches. In 1909 the response to this was bakelite, a thermosetting plastic based on phenol and formaldehyde (see Plastic, Thermosetting).

Polythene, one of the earliest plastics to come into common use, was first discovered serendipitously in 1933. Of immense importance for cable insulation during World War II, polythene is a quality insulator with a very low dielectric loss and was an ideal material for use in radar. Since then, there has been a progression from low molecular weight material obtained using extreme conditions of temperature and pressure to high molecular weight, stereo-regular polythene produced in solution using a Ziegler–Natta catalyst. The development and ubiquitous deployment of polyvinyl chloride (PVC) has almost universally replaced rubber as the insulating medium in domestic electrical installations. There have been other significant developments in the polymer industry. Polyvinyl alcohol (PVA) has replaced natural gum-arabic as a domestic adhesive and is widely used as an additive in concrete. Urea-formaldehyde and polystyrene, when manufactured as a foam, provide excellent thermal insulation used for insulating buildings. The use of epoxy resin with glass and carbon fibers to make composite materials has opened up entirely new markets. Car bodies, boat hulls and boat superstructures are regularly made using fiberglass. Carbon fiber has even better mechanical properties and is used in yacht masts, aircraft fuselages, golf club shafts, and other stressful applications. One of the more recent polymeric products that has helped to change our lives is the cyanoacrylate or superglue group of adhesives.

The twentieth century has seen the development of specialist steels at reasonable prices: chrome–vanadium steel, high-speed steel and tungsten carbide tipped tools. The cost of stainless steel has come down to a level where it is being used in applications that would have been unthinkable 50 years ago. There have also been significant developments in processing of specialist metals: titanium, zirconium, tungsten, and uranium. The addition of small amounts of thorium to tungsten was a major factor in the manufacture of low-cost, reliable electric light bulbs. The development of high-quality, low-cost aluminum and its lightweight alloys have been essential to the aviation industry.

After World War II there was a significant research effort to develop application-specific ferrite materials (ceramic ferromagnets). These were normally manufactured using the sinter process, in which a fine powder of the base material (a mixture of iron oxide with other oxides) is pressed into a mould using a wax or PVA binder. Once removed from the mould, the unit is raised to a dull-red heat and kept at this temperature for several days, during which time microdiffusion bonds the material together. Ferrites were used for rod antennas and were a key component in magnetic core memories for old mainframe computers, beginning with the Whirlwind computer in the 1950s. Ferrite films have been used as the prime data storage medium from the start of the modern computer era. They have always been used in hard disks and have continued in use as floppy disks moved from 8.5 inches to 5.25 inches to 3.5-inch high-density disks, subsequently displaced by rewriteable CDs. Layers of ferrite material deposited on thin plastic films have been used as the storage medium for tape-recorders, cassette recorders, and video tapes. The dark strip on the back of most bank, credit, and debit cards is a ferrite film that contains a magnetic record of the data relevant to the card. Specialist ferrites are now used in high-frequency transformers within switched mode power supplies, These provide more power within a smaller volume than has ever been possible before and are universally used in desktop computers as well as in the charging units for portable computers. In a curious exchange of expertise, sinter technology developed for magnetic ceramics is now used in the manufacture of complex steel components such as gears for the car industry. There is an enormous reduction in the energy requirement when compared with cast units. The time and energy required for postprocessing is also significantly reduced.

The early part of the twentieth century witnessed an explosion in electronics. Thermionic valves (diode, triode, tetrode and pentode), or vacuum tubes, were developed to act as rectifiers, detectors, and amplifiers. They changed the nature of radio reception and made modern television a reality. Of themselves, these would have been noteworthy in world history, but they were progressively replaced after World War II by the much more reliable transistor. The twentieth century could be called the "semiconductor century." Although the selenium photoelectric cell was used in the late nineteenth century by early television scanning disk cameras (Nipkow disk) and crystalline lead sulfide (galena) was important as a crystal detector in the early days of wireless, it has been the controlled manufacture of germanium, silicon, gallium arsenide, and its relatives that has been significant. Dislocation-free crystals are required, and large diameter crystals must be grown under precisely tailored thermal gradients. Never before has any material been made in such large volumes at such high levels of purity. One of the first steps in the production of high-purity silicon involves the van Arkel process. This uses an important property in heterogeneous chemistry, namely thermal dissociation. Purified silane (SiH_4) gas is passed over a very hot tungsten wire causing it to break up into hydrogen and silicon, which deposits on the hot filament. Once this deposition has been run for some time, the rods are removed from the reactor, the tungsten is etched away and the product is ready for conversion into single crystal silicon. The situation proved to be much more difficult when efforts were made to grow single crystal gallium phosphide, the basic material for light emitting diodes (LEDs). The high temperatures needed for growth caused the more volatile phosphorous to evaporate. This presented a problem until the advent of the liquid encapsulation Czochralski (LEC) technique. Before that, the breach-end of a large naval gun was used as a pressurized reactor in the first experimental-scale production of this material.

Integration of Technologies

A key feature of recent developments has been the integration of technologies; new technology and materials frequently developed for other areas of industry were often applied elsewhere. The conversion of a semiconductor such as silicon into a useful device such as a microprocessor chip requires a wide range of technologies in the fabrication steps. Chief among these is microlithography, which provides the geometric definition (current limit for line widths is about 0.2 micrometers) of the different areas that may be the subject of a particular process step.

A slightly earlier example of the integration of technologies is the cathode ray tube in a color television receiver—the one thermionic valve that has survived when all other valves have been supplanted by discrete and integrated semiconductors. This miracle of engineering contains three electron guns, which together with their control components are connected to the outside by means of glass-metal seals. Research into the chemistry of phosphors has meant that we now have durable screens that deliver realistic colors. The ubiquitous fluorescent light is a mercury vapor discharge tube that emits ultraviolet light. The inner surface of the glass enclosure is coated with a phosphor that converts ultraviolet light to visible light. The experience gained with television screen phosphors has been incorporated into fluorescent light technology, so that the color temperature can be adjusted to the requirements of the customer.

Discharge tubes involving sodium have transformed street lighting although there is much talk of light pollution. The orange output from early sodium lights made them unpopular in shopping areas where the colors of articles on display could not be distinguished, and high-pressure mercury discharge tubes were preferred. Recent developments in high-pressure sodium tubes have provided an efficient lamp with a good color spectrum. With improvements in the design of street lighting, there has been a significant decrease in light pollution. The dissociation of gaseous compounds on hot surfaces has also been exploited to increase the light output in what are called tungsten–halogen bulbs. These are operated at much higher temperatures than normal bulbs, and they require quartz rather than the normal glass enclosures. At their normal operating temperature there is a significant "boil-off" of tungsten from the filament. This would lead to early failure except that the tungsten vapor reacts with a halogen such as iodine or bromine, which is present in the ambient gas. Since tungsten halide undergoes thermal dissociation, this mechanism ensures that the metal is quickly redeposited on the filament.

Ferrite films in magnetic storage media have increased in storage density as the methods of locating the read/write heads on the platter of hard disk have improved, but it was the integration of micro-optics for positioning that brought about the jump to multigigabyte hard drives, which by the end of the twentieth century were available at extremely low prices. In a separate application of ferrites, the integration of high-frequency electronics involving ferrites with fluorescent tube technology has resulted in low-wattage, energy-saving light bulbs. We have also seen the integration of long-chain molecules with special optical properties in array circuits where individual locations (segments) are controlled by means of transparent (tin oxide) electrical conductors in the liquid crystal display (LCD). Almost as soon as they were introduced, they caused a revolution in digital watch technology; as the twentieth century drew to a close, large-area LCDs were displacing the CRT as the display medium for television and computer monitors. As late as the mid-1970s optical fibers were a laboratory curiosity. Since then techniques have been developed to manufacture long fibers with very low losses. The integration with electronic circuits involving semiconductor lasers and high-speed photo-detectors provides the basics for a high-density communications link, the heart of the Internet era.

Just as few could have dreamt of the ubiquitous cell phone even in the late 1980s, future developments of materials and processes are anyone's guess. It is likely that tailor-made semiconductors will be prominent, but a challenge that remains is the development of improved methods of energy storage. The semiconductor manufacturers have gone a long way to reducing power requirement, but any user of a portable computer knows how limited the useful battery is. Unless there is a fundamental limit, then material scientists need to respond to the challenge of a cheap, safe and environmentally acceptable battery or battery replacement.

See also **Absorbent Materials; Adhesives; Alloys; Ceramic Materials; Chemicals; Chemicals Process Engineering; Composite Materials; Electrochemistry; Fibers, Synthetic and Semi-Synthetic; Iron and Steel Manufacture; Liquid Crystals; Nanotechnology; Optical Materials; Plastics; Semiconductors; Smart and Biomimetic Materials; Synthetic Resins; Synthetic Rubber; Thin Film Materials and Technology; Timber Engineering**

DONALD DE COGAN

Further Reading

Aftalion, F. *A History of the International Chemical Industry From the "Early Days" to 2000*, 2nd edn. Translated by Otto Benfey T. Chemical Heritage Foundation, Philadelphia, 2001.

Braun, H.-J. and Herlea, A., Eds. *Materials: Research, Development and Applications*. Brepols, Turnhout, Belgium, 2002.

Buschow, K.H.J., Cahn, R.W., Flemings, M.C., Ilschner, B., Kramer, E.J. and Mahajan, S. The science and technology of materials: an introduction, in *Encyclopedia of Materials: Science and Technology*, Buschow, K.H., Ed. Elsevier, Amsterdam, 2001.

Cahn, R.W. *The Coming of Materials Science*. Pergamon Press, Oxford, 2001.

Hummel, R.E. *Understanding Materials Science: History, Properties, Applications*. Springer Verlag, Berlin, 1998.

Mowery, D.C. and Rosenberg, N. *Paths of Innovation: Technological Change in 20th-Century America*. Cambridge University Press, Cambridge, 1998.

Medicine

Traditionally, the practice of medicine starts at the interface between the patient (or client, in late twentieth century phraseology) and the medical advisor, and that relationship should be one of mutual trust and respect. In this overview the dramatic developments in medical technology during the twentieth century are counterpointed with the changes that have occurred in the public perception of doctors.

At the end of the nineteenth century, William Osler's *Principles and Practice of Medicine* was the standard reference, and an attitude of therapeutic nihilism permeated medical thought. Cures simply could not be expected. By the end of the twentieth century, the situation was not only different, but attitudes were nearly reversed. Pharmaceutical research had resulted in powerful drugs, and complex surgery and highly technical diagnostic procedures were available. Despite the efforts of governments and individual practitioners however, the gap between the rich and the poor still existed, and not only in the poor countries. Attitudes had changed; the social structure of society had changed; patients had changed; and doctors had to change, too.

A woman born in the U.S. in 1900 had a life expectancy of 50.7 years. However, in neighboring Mexico a woman's life expectancy was only 24 years. By the end of the century female life expectancy in the U.S. had risen to 79.4 years, but in Mexico the improvement was dramatically more impressive, leading to a life expectancy of 76.5 years. Obviously something very dramatic had happened, particularly if you lived in a less affluent country. Health care had improved in both the wealthy and less well-endowed countries, resulting in significant convergence of "outcome."

Tables listing the causes of death in a given year give an indication of the areas in which twentieth century medical technology exercised its influence. In North America in 1900 the top four causes of death were tuberculosis, pneumonia, heart disease, and enteritis (the latter being largely infantile gastroenteritis, or infections in the gastrointestinal tracts of children). By 1998 the table was headed by diseases of the heart, cancer of all forms, stroke, and chronic lung disease.

The late nineteenth century has been regarded as the golden age of microbiology, but it was nearly 40 years before there was any effective therapy for infectious diseases. Mortality from these conditions had, however, been falling for the first three decades of the twentieth century, and this can be attributed to the improvements in nutrition, public health, and a more general acceptance of the dangers of cross infection in hospital wards. In the U.K. a great debt is owed to the public works engineers of the Victorian era. Vaccination had been introduced in the second part of the nineteenth century and eventually became compulsory after the last smallpox epidemic in London in 1901. So successful was the vaccination program worldwide that in 1980 the World Health Organization (WHO) declared that smallpox was extinct.

The first antimicrobial agents had been in use as long ago as the sixteenth century when mercury was used in the treatment of syphilis. It was early in the twentieth century that German bacteriologist Paul Ehrlich introduced an organic arsenical for the treatment of that condition and noted that aniline dyes killed bacteria. The first antimicrobial agent, Prontosil (a sulfonamide) was developed in 1935 as a result of research originating in the German dye industry. This was the start of a revolution in the treatment of bacterial diseases.

In the U.K. in 1928, Alexander Fleming had observed a mold (apocryphally, blown in through an open window) that partially cleared a dish where a colony of staphylococci was growing. This "mold" was to become penicillin. Eventually the active agent was isolated and used in humans for the first time in 1941. Ernst Chain and Howard Florey started their research into what they were the first to call antibiotics, substances produced by microorganisms that kill other bacteria. By the end of the 1950s, streptomycin sulfate had been developed and tuberculosis was then curable with drugs. At the close of the twentieth century there are antibiotics to treat most bacterial infections, but progress in the development of antiviral agents has been painfully slow.

Ether, nitrous oxide, and chloroform had been introduced in Boston by Wells, and by Simpson in Edinburgh in 1846 to 1848, but it was in the twentieth century that advances in anesthesia permitted safer, longer operations. New primary general anesthetics such as halothane were safer and less toxic, but it was the introduction of muscle relaxants (developed from the curare employed by South American Indians on their arrows), mechanical ventilators, and increased understanding of pathophysiology (and the ability to both monitor and treat metabolic abnormalities) that freed the surgeons from the shackles of operating only on the younger, low-risk patients. Synthetic materials, frequently developed as a byproduct of space research at the National Aeronautic and Space Administration (NASA) led to dramatic developments in vascular surgery.

The impact of heart surgery, notably coronary artery bypass surgery, as well as the development of the artificial heart and valves are discussed elsewhere. The importance and relevance of such surgery to the prognosis of coronary heart disease cannot be overemphasized. Dramatic changes in the treatment of so-called heart attacks with drugs, angioplasty to open blocked blood vessels, and implants of stents and pacemakers have helped to increase survival rates.

Death rates from the common forms of cancer did not improve greatly in the second half of the century, and the incidence of smoking induced lung cancer had risen significantly. Cure rates for lung cancer, breast cancer, ovarian cancer, and esophageal/gastric malignancies remained distressingly low and have hardly changed although some patients had their lives extended with treatments of radiation and/or highly toxic chemotherapeutic agents. There were dramatic results in some of the less common malignancies such as lymphoma or some types of acute leukemia, notably Hodgkin's disease and some childhood leukemia. We can but hope that with further research these triumphs are the template for similar success in the more common malignancies.

After World War I an extended epidemic of influenza (Spanish flu) from 1918 to 1920 killed 25 million people, three times as many as had died in the war. In the year 2000 it was estimated that in sub-Saharan Africa alone there were 28.1 million people with HIV/AIDS infection. There were an additional 12 million people infected in Asia and the rest of the world. Without access to the retroviral drugs discovered and made available in the last decade of the twentieth century, most of these people will die, producing a devastating demographic effect. The 1918–1920 influenza pandemic had to be faced in a state of total medical impotence, but there is, or could be, an answer to the HIV/AIDS pandemic. There are drugs to control the disease or drugs that can be given to pregnant women to prevent another generation being infected at birth. But drugs cost money, the afflicted are predominately in the less affluent world, and there are controversial issues of pharmaceutical patents and preventive methods, thus a limited impetus for action. In terms of potential global catastrophe, HIV/AIDS is the greatest challenge faced by doctors and governments. Some years earlier, tuberculosis presented a similar risk to the third world as the disease-causing organism became drug resistant. By a significant reduction in the cost of antituberculosis drugs, the threat was substantially reduced

By the end of the twentieth century organ transplantation was commonplace. The first successful renal transplant was carried out in Boston in 1954 between identical twins, so there were no immunological problems. Studies into the immunological basis of organ rejection had been carried out in the 1940s by Medawar. Once antirejection drugs (most notably azathiaprine and cyclosporin) and laboratory tissue typing had been developed, transplantation became a routine procedure. The first heart transplant was carried out in 1967 by South African surgeon Christiaan Barnard. The list of transplantable organs is now extensive, including lungs, pancreas, liver, small bowel, bone marrow, and even limbs. (Historically, corneal transplants antedate all these, but the cornea is avascular and so rejection does not occur, and tissue matching is not necessary) The problem with transplants is the supply of donor organs, which is limited for obvious reasons. As cloning became feasible there were efforts to create transgenic, cloned animals, (e.g., pigs) which could be used for human transplantation. As the century ended this work was all experimental, but it seems probable that this will be the answer to the provision of some donor tissues.

Reproductive physiology and contraception were also important areas of advance that impinged on the healthy population. Early in the century mechanical barrier methods of contraception (condoms had been used for hundreds of years) or topical spermicides were freely available. The oral contraceptives, a mixture of hormones that inhibit ovulation, were introduced in 1959 by Pincus and Sanger. They revolutionized sexual behavior and liberated women from the fear of unwanted pregnancy. Fertility was also being studied. Artificial, external insemination either by partner or anonymous donor became routine. *In vitro* fertilization using harvested donor eggs, the storage of eggs and sperm, the use of surrogate mothers, and the ability, with the use of hormone therapy, to impregnate postmenopausal women all were commonplace by the end of the century.

Psychiatric disorders are also largely amenable to treatment thanks to the very powerful psychotropic drugs developed in the last 40 years of the century. What are now considered barbaric "treatments" such as insulin comas, ice baths, and inducing malaria to cause high fever to exorcise the demons were used in psychiatric hospitals and asylums through the middle of the twentieth century and have now been relegated to the history books. There remained some level of discomfort with psychiatric disorders, which is seen in the reaction to a substratum of illness that people are reluctant to consider to be of partly psychiatric origin. This group of illnesses includes chronic fatigue syndrome, Gulf War syndrome, and several others. Attempts by physicians to suggest a non-organic component or to fail to have a specific diagnosis and treatment plan cannot be accepted by a public accustomed to knowing the causes and finding "cures." These disabling conditions often prevent the sufferers from leading a normal life, and researchers continue to explore psychoneurological and immunological responses in an attempt to find successful treatments.

It may seem paradoxical that at a time when physicians had powerful drugs and technological treatments at their fingertips, patients should flock to consult nonmedical alternative practitioners. It has been reported that in 1990 Americans made 425 million visits to alternative practitioners (chiropractors, acupuncturists, massage therapists, etc.) but only 388 million to orthodox primary care physicians. In the latter part of the century "orthodox" physicians have, in some very basic way, been perceived to have let their patients down. There is a widespread perception that physicians have become mechanistic, do not have enough time to talk with their patients, talk down to them, and ignore what they say. Doctors could, however, learn something from alternative medicine colleagues. Some unorthodox treatments, most notably manipulative therapy, have proved as effective as high-tech neuro- or orthopedic surgery. A good orthodox physician is well aware of the valuable therapeutic effect of an unhurried consultation with time allotted for a full explanation to the patient.

As patients expectations have mounted so has the demand for screening. There is a mistaken belief that with detailed routine screening, including a multitude of blood tests, radiological or ultrasound scans, electrocardiograms and so on, the Holy Grail of very early diagnosis will be achieved and cures guaranteed. At the end of the 1990s there were controversies related to some screening programs, notably the value of mammography screening for breast cancer and PSA (prostate specific antigen) screening for prostate cancer. The opponents of these procedures state that the very early diagnosis of breast cancer makes no difference to outcome and that PSA screening is unreliable and results in many men having prostate surgery or radiotherapy quite unnecessarily. On the other hand, screening for carcinoma of the cervix has been both successful and cost effective.

An important change in the 1990s has been the concept of evidence-based medicine; that is, treatment based on research findings of effective outcomes. Many therapeutic procedures have never been submitted to the discipline of a controlled trial. (The first controlled clinical trial was as long ago as 1754, on the treatment of scurvy with lemon juice.) Teaching medical students and other healthcare practitioners in an evidence-based curriculum provides the new generation with the science and technology and the art of medicine.

The list of medical advances in the twentieth century is formidable, and many of the technologies are described in separate entries in this encyclopedia. By the close of the century medical "breakthroughs" had become a staple component of daily news. However, people born in the second half of the century who neither knew nor cared about the paucity of effective therapy before 1940 remained obstinately obsessed with the degenerative disorders such as arthritis, Alzheimer's disease, Parkinson's disease and the care of the elderly. By the end of the twentieth century, the demographics of aging populations in the western industrial countries with expectations that medical science could solve these problems meant that even these less exciting fields of research could attract grant dollars.

Memories are short, and if asked about medical major achievements of the twentieth century the questioner will be bombarded with a variety of answers mainly relating to the hot news of the 1990s: the unraveling of DNA, the Human Genome Project, the latest chemotherapeutic agent for cancer and so on. However, it is a sad fact that while the genome project is of incalculable scientific importance and interest, only a tiny number of patients with rare immune system disorders have, so far, been treated with gene therapy, with mixed results. It may be that in the future gene therapy will be helpful in the management of the wide spectrum of disorders that the genome project has already shown to have a genetic basis. In terms of direct interventional treatment any attempt to draw up a table of the most important medical advances in the last 100 years would, inevitably, be controversial and highly subjective. The development of antibiotics, insulin, and oral contraceptives would be very close to the top of most people's drug list. The science of DNA, the genome project, and the unraveling of the immune system hold much promise for the future. Coronary bypass surgery, hip replacement, and renal transplantation have benefited more people than some of the more exotic surgical introductions. Treatment of pneumonia and osteoporosis may have extended more lives than cancer chemotherapy. Consideration to both quantity and quality of life as a consequence of medical care is the norm.

See also **Cancer, Chemotherapies; Dialysis; Immunological Technology; Implants, Joints and Stents; Health; Hormone Therapy; Neurology; Organ Transplantation; Psychiatry, Pharmaceutical Treatment**

JOHN J. HAMBLIN

Further Reading

Balint, M. *The Doctor, His Patient and the Illness*. Pitman, London, 1957. Seminal work on doctor/patient interface.

Black, D. *Inequalities in Health Care*. Her Majesty's Stationary Office, London, 1980. Commissioned by one government, ignored by the next.

Dutton, D. *Worse than the Disease: Pitfalls of Medical Progress*. Cambridge University Press, Cambridge, 1988.

Ernst, E., Ed. *Desktop Guide to Complementary Medicine and Alternative Medicine*. Mosby, London, 2001.

Osler, W. *Principals and Practice of Medicine*. 1889. Classic textbook. How things were at beginning of the century.

Porter, R. *The Greatest Benefit to Mankind*. Harper Collins, London, 1997. Excellent history of medicine.

Showalter, E. *Hystories*. Picador, London, 1992. Controversial discussion of Gulf War Syndrome, ME etc.

Various. Editorials. *Br. Med. J.*, 324, 181–186, 2002. Five editorials on the HIV/AIDS epidemic.

Metals, Ferrous, *see* **Iron and Steel Manufacture**

Methods in the History of Technology

The introduction of the artifacts discussed in this encyclopedia, ranging from the automobile and the radio to the spacecraft and the computer, from wired electric lines to wireless electronic nets, and from the atomic bomb to the nuclear reactor, took place in a century more accustomed to rapid technical change than the nineteenth century—a century marked by the creation of the "machinery question" and the development of political economy to answer it. By the twentieth century, the question no longer concerned the introduction of one wave of novel machines, but had to be reformulated to analyze the historical pattern of succeeding waves of novel machines. The question could no longer be answered by political economy alone, birthed to make sense of the world of steam engines. Over the course of the twentieth century, as steam engines were already passé while new machines kept appearing, the study of the past had to be enlisted to help society understand its own relationship to technology. The twentieth century, then, witnessed the emergence and establishment of the history of technology as a distinct historical subdiscipline.

As new machines were coming at an accelerating pace, a mass faith in the equivalence between technical and social progress reposed as the twentieth century's ideological analog to medieval religious dogma. In response, the methods of the history of technology have been overdetermined by the challenge to interpret the appeal of the so-called ideology of "technological determinism." Noticeably, history of technology as such was logically impossible before the twentieth century because the modern use of the word "technology" was not established before the first decades of the twentieth century. Satisfied by how much their subdiscipline has advanced by the wise agreement to not force one definition of technology upon its members, the community of professional technology historians now agrees that the vagueness of the word "technology" is only consistent with the protean persistence of technological determinism. In the face of a definitional openness of technology, perceptions of the object of the history of technology, propositions about what its key concepts ought to be, and, correspondingly, suggestions over practices and methods have remained pluralistic and defy any easy act of subsumption under a single theoretical framework.

For convenience, we may distinguish between two historiographical periods, separated by the foundation of the Society for the History of Technology (SHOT) in 1958, with *Technology and Culture* serving as its journal of record. Melvin Kranzberg is unanimously recognized as the single most important founder of SHOT. By focusing on technological change as the outcome of "invention," individual contributions before 1958 tended to choose methods that privileged the study of individual agency over social structure. Yet, debates investigating the proper relationship between the two concerning technical change were also present in schools of thought like the "sociology of invention," known by the works of S. Colum Gilfillan, W. Fielding Ogburn, and Abbott Payson Usher. Culture was certainly the starting point for Usher, an economic historian (*A History of Mechanical Inventions*, 1929), the literary and social critic Lewis Mumford (*Technics and Civilization*, 1934), and the art historian Sigfried Giedion (*Mechanization Takes Command*, 1948); the three individuals commonly credited for being among the most distinguished contributors to history of technology's pre-SHOT period. Moreover, the interest on invention survives to the present, quite clearly as an interest in the study of the transformation of inventiveness as a socially situated manifestation of human creativity during the transition from the individual inventor's workbench to the expansive settings of industrial research.

We now know the importance of history of technology was acknowledged by several other early historiographical currents emanating from outside the U.S. Known for its sensitivity to the history of material civilization, the French *Annales* school invited historians to favorably regard the history of technology. As early as 1935, Lucien Febvre called for a tripartite methodological synthesis of a competent understanding of the technology under consideration, of a proper placement of this technology to a series, and of the appropriate move from these series to total history. Subsequent interpretations of Febvre's manifesto frequently placed the accent on the first or the third of its ingredients, thereby privileging technical and social history respectively. Technical history has been a strength of the long-lived British *Transaction of the Newcomen Society*, founded in 1920 by the Newcomen Society for the Study of the History of Engineering and Technology. The publication of an analogously focused German history of technology journal as early as in 1909 by the Association of German Engineers (now called *Technikgeschichte*) suggests that the interest in enriching engineering studies was a second strong motivation behind the foundation of the history of technology. Evidently, a concern with proper engineering education loomed large even in SHOT's narrow constitution as a community of professional historians. Bringing along a rather spontaneous interest in the history of technology as a depository of ideas for solving contemporary technical problems and as a testimony to the scientific nature of engineering knowledge, engineers have tended to support the methodological emphasis on the history of the "heroic" moment of technological change, namely, invention. History of technology is still marked by occasional instances of unbridled antagonisms between scholars with engineering education and with those with a humanities background.

Finally, the development of the history of technology could be stimulated (or blocked) by an overarching ideological orientation. In trying to make sense of the limited development of the history of technology in the socialist societies of the twentieth century, scholars argue that underlying the crude evolutionary interpretation of Marxism emphasizing the primacy of "forces of production" was usually a hardened variant of technological determinism that left little room for historical interpretation of the society–technology relationship. Technology historians sought to advance the history of technology against the Cold War difficulties in scholarly communication through the 1968 formation of the International Committee for the History of Technology (ICOHTEC), a Scientific Section of the International Union of the History and Philosophy of Science (IUHPS), which is part of UNESCO. The agreed methodological choice of those participating in ICOHTEC was to table issues most likely to reproduce political divisions. Hence, the result was a heightened attention to subjects that appeared more technical, and, as such, less directly subjected to nationalistic interpretations.

In the 1990s, ICOHTEC published its own journal, *ICON*. In addition to the journals already mentioned, an historian wishing to publish his or her research on the history of technology can now consider two more international journals, *History and Technology* and *History of Technology*. The history of technology has by now matured enough to support special interest groups and their publications. For example, an historian interested in the history of computing technology may consider the *IEEE Annals for the History of the History of Computing*, or, for example, assuming a special interest in the history of computing with the slide rule, the American *Journal of the Oughtred Society* or the British *Slide Rule Gazette* will prove satisfactory. Another sign of the maturity of the history of technology is its sustained support by specialized museums and, with increasing frequency, even the general museums will offer particular thematic displays of interest to those in the history of technology. Also, the availability of textbooks on the history of technology, with several focused on twentieth century technology, either from a national or an international perspective, are finding their way into print. Thomas Hughes' 1989 *American Genesis: A Century of Invention and Technological Enthusiasm, 1870–1970* is already considered a classic in the field.

From the perspective of the methodologies tried, the post-1958 period can be conveniently split into two subperiods, one that ended with the mid-1980s acknowledgment of the victory of variants of "contextualism" over the couple "internalism"–"externalism," and a second, lasting since then, which started with the cautious experimentation with versions of "social constructivism." In his analysis of the methodological profile of *Technology and Culture* between 1959 and 1980, John Staudenmaier, the third (and present) editor of SHOT's journal after Melvin Kranzberg and Robert Post, found that "methodological style" already favored was "contextual" (50 percent), followed by almost identical shares of the "internalist" (17 percent) and the "externalist" (14 percent) approach, and by "nonhistorical analysis" and metastylistic "historiographical reflection" representing the remaining percentage (12 percent and 7 percent respectively). Internalism and externalism, represented the emphasis on agency and structure respectively, or, as Staudenmaier put it, technological "design and ambience." Contextualism stood for a synthetic consideration of both. If contextualism prefigured in the works of Usher, Mumford, and Giedion, internalism was dominant in the multivolume histories of technology published in Britain, France, and the Soviet Union. *The History of Technology*, edited between 1954 and 1958 by Charles Singer, E. J. Holmyard, and A. Rupert Hall, just before the emergence of SHOT, was the most influential of all.

All available historiographical balance sheets point toward an extreme variation of contextualism. Having set as its methodological standard to retrieve historical correspondence between technical relationships and social relationships, contextualism frequently stopped short after starting from either end. As the successful opening up the black box of technology proved difficult, the contribution of contextualism, according to scholars sympathetic to such thought, consisted in setting a standard that allowed the community of professional historians to evaluate what exactly was achieved when the black box remained closed or was unpacked only partially. The acknowledged difficulty to get close to that standard was proposed as the reason for the distinct and peculiar existence of the history of technology. In his 1996 SHOT presidential address, Alex Roland stated regarding the difficulty of penetrating deep into the black box, technology historians were unique in that they were at least trying to get inside.

This difficulty generated second guesses about the value of the history of technology as a distinct subdiscipline. First came the challenge from historians who argued that the isolation of the historical study of technology from the rest of history perpetuated the risk of the spontaneous reproduction of technological determinism within the history of technology, constantly blocking the history of technology as such from providing broader historical interpretations. Far more important was the opposite challenge, that of seeking to avoid the risk of reproducing technological determinism by abandoning the attempt at broader historical interpretation altogether. As it should be expected, this challenge took the form of the invasion of the history of technology by sociology, now known as "social constructivism." Following things to their logical conclusion, social constructivists came to question the necessity of assuming a border between the inside and the outside of the black box. In the Social Construction of Technology framework (SCOT), Trevor Pinch and Wiebe Bijker's variant of social constructivism, such a border is missing as long as an artifact is subject to "interpretative flexibility" by various "relevant social groups," before "stabilization" and "closure" is achieved. In Michel Callon, Bruno Latour, and John Law's actor–network theory (ANT) variant of social constructivism, the very demarcation between nature and society is turned into a question. To SCOT's attention toward technological success and failure, ANT adds a symmetrical treatment of human actors and natural phenomena as "actants," thereby making the historian's interest in social causality altogether irrelevant.

While contextualism starts with methods to understand causes, social constructivism sees the method as an end-in-itself. While both agree on a symmetrical treatment of success and failure from a synchronic perspective, they differ in their interest for patterns. For example, the contextualist historian of the bicycle is interested in the connection between the history of the bicycle and that of the Fordist automobile whereas the social constructivist sociologist of the bicycle is interested only in the high–low wheel bicycle connection within the history of the bicycle (contrast, for example, the history of the bicycle in David Hounshell's 1984 *From the American System to Mass Production, 1800–1932* and in Wiebe Bijker's 1995 *Of Bicycles, Bakellites, and Bulbs: Toward a Theory of Sociotechnical Change*). Social constructivism has then been blamed for a constitutional inability to consider the long run. Its focus on the history of the short run has been charged with restricting the study of the transformation of technology in use. Accordingly, social constructivism has been accused by David Edgerton for conflating the history of technology with the history of invention, thereby reproducing in effect the pre-SHOT methodological emphases of internalism that cuts the technical from the social to re-invite technological determinism. In his 2003 update of the historiography of technology, Staudenmaier interprets the outcome of the contextualism–constructivism encounter positively by arguing that it has shown that critiques of technological determinism are now more forced to:

> "... take a laborious route, seeking historical contingency deep in the gears and circuits of technological design itself." *[Staudenmaier, 1992, p.170.]*

Social constructivist methods of studying technological change may be comfortably placed under the inter–trans–cross disciplinary methodologies of science and technology studies, which grew on philosophical critiques of positivism to differentiate between historical, sociological, and anthropological studies of science and technology. By contrast, authors of the contextualist school are concerned with pursuing methodologies that will facilitate the firm recognition of the discipline by the historical profession in large. Accordingly, contextualist studies of historical change tend to merge with cultural and intellectual histories of technology.

Having endorsed pluralism as their methodological principle, technology historians have experimented with sharing and borrowing from other historical subdisciplines. Their methodological focus on opening technical black boxes is now shared by business history, influenced now by schools competing with neoclassical economics, known as "evolutionary" or "institutional" economics. In pointing to habituation rather than rationality as a source of economic action, these schools of thought favor the substitution of historical study of technological "trajectories" and firm "routines" for the assumption that technological change is the automatic result of profit-maximizing decisions on the basis of perfect market information. At the other end of the spectrum, technology historians have also developed methodological ties with labor historians, especially towards a shared focus on the puzzling issue of the relationship between mechanization, employment and unemployment, and the skilling (deskilling or reskilling) processes. The methodological ideal of this mutually beneficial interaction has been a history of workers not stripped from machines and of machines not stripped from workers.

Problematic for as long as technology was assumed to be the mere application of science, the relationship between the history of technology and the history of science has dramatically improved after the most recent generation of science historians added the study of scientific practice to scientific theory, and after the corresponding generation of technology historians started to retrieve the importance of the whole of technical knowledge instead of exhausting themselves at establishing the epistemic status of engineering knowledge. Arching toward the other end of the spectrum, where the study for the knowledge of artifacts is replaced by the study of artifacts for knowledge, history of technology has enriched its methodological apparatus by a new relationship to material culture and similarly included fields (e.g., industrial archaeology). Technology historians have been helped by their encounter with material culture to detect a contradiction in that they themselves have been methodologically studying the history of artifacts by relying on documents rather than artifacts. Against the programmatic consensus on opening the black box, the majority of technology historians continue to privilege texts over experiential sources, thereby hesitating to agree that documents are only one of the many classes of artifacts.

Joseph Corn's own reading of the *Technology and Culture* articles that Staudenmaier read for his review of the methodological transition is suggestive; preferring to write about ideas or institutions related to technical change, slightly more than half of the authors did not write about objects at all; of those who focused on the history of technical artifacts slightly more than 70 percent relied exclusively on primary and secondary textual sources, only occasionally supplemented by oral interviews; and, more importantly, only 15 percent of those considered employing some reference to material evidence. Authors focused on objects were more likely to disappear as one moves from ancient, medieval, and early modern history to the twentieth century. Scholars who focused on objects usually worked in museums and disproportionate numbers of them were trained in archaeology. As far as the methods of those who attempted to learn from things goes, Corn identified and classified five of them: ordinary looking, technical analysis, simulation, testing through use, and archaeological science.

Finally, technological determinism in the recent decades was challenged by the findings of the study of previously invisible experiences with technology: women, children, handicapped, non-whites, and non-Westerners. The list also includes the study of the relationship between technological change and ecological destruction, and the study of technological change from the perspective of use in war by a state (as opposed to its invention within an enterprising firm). Critical here is the general methodological invitation to consider technological change in consumption rather than in production, technology as changed in use rather than through invention. The issue is not one of neglecting the study of technical change at the laboratory and the factory in favor of studying the reconfiguration of technology at the world fair and the house, but on integrating the study of the two. Novel as this methodology may seem, it was suggested by Lewis Mumford, who, as Carroll Pursell has reminded us, argued against the exclusive identification of technology with tools and machines, which, in Mumford's count, had left out hearths, pits, houses, pots, sacks, clothes, traps, bins, byres, baskets, bags, ditches, reservoirs, canals, and cities. To which Pursell adds cupboards, packing cases, ship containers, trailers, and suitcases before noting that these "static containers" (Mumford) or "containers" (Pursell) are associated with domestic and agricultural work, historically the domain of women.

Pursell's list of technologies rendered invisible includes the seismic engineering that is designed into the built environment in California, the highly mechanized kitchen adjacent to the swank hotel dining room, and tableware, chairs and everything that is usually not seen as technology. A variant of the methodology aiming at seeing the invisible has been tried by looking at actors mediating between production and consumption of technologies that we would all recognize as such. Included in these actors are technological enthusiasts, from audio outlaws to computer hobbyists. The orientation to the mismatch between what a technology was thought to be and what it turned out to be in use has produced great insights on some of the technologies that defined the twentieth century—the gap between the imagined and the real uses of nuclear energy and the rest of the crucial twentieth century technologies included in Joseph Corn's 1986 collection in *Imagining Tomorrow: History, Technology and the American Future* is a case in point.

ARISTOTLE TYMPAS

Further Reading

Braun, H.J. Current research in the history of technology in Europe. *Hist. Technol.*, 21, 167–188, 1999.

Corn, J.J. Object lessons/object myths? what historians of technology learn from things, in *Learning from Things: Method and Theory of Material Culture Studies*, Kingery, D., Ed. Smithsonian Institution Press, Washington D.C., 1996.

Cutcliffe, S.H. and Post, R.C., Eds. *In Context: History and the History of Technology—Essays in Honor of Melvin Kranzberg*. Lehigh University Press, Bethlehem, 1989, 133–149.

Edgerton, D. From innovation to use: ten eclectic theses on the historiography of technology. *Hist. Technol.*, 16, 1999, 111–136.

Fox, R. *Methods and Themes in the History of Technology*. Hardwood Academic, Amsterdam, 1996.

Hounshell, D.A. Hughesian history of technology and Chandlerian business history: parallels, departures, and critics. *Hist. Technol.*, 12, 3, 205–224, 1995.

Laudan, R. Natural alliance or forced marriage? changing relations between the histories of science and technology. *Technol. Cult.*, 36, s17–28, 1995.

Lerman, N.E., Mohun, A.P. and Oldenziel, R. *Gender and Technology: A Reader*. John Hopkins University Press, Baltimore, MA, 2003.

Levin, M.R. What the French have to say about the history of technology. *Technol. Cult.*, 37, 1, 158–168, 1996.

Lubar, S. Representation and power. *Technol. Cult.*, 36, s54–82, 1995.

McGee, D. Making up mind: the early sociology of invention. *Technol. Cult.*, 36, 4, 773–801, 1995.

Pfaffenberger, B. Social anthropology of technology. *Ann. Rev. Anthropol.*, 21 491–516, 1992.

Pursell, C. Seeing the invisible: new perceptions in the history of technology. *ICON: J. Int. Comm. Hist. Technol.*, 1, 9–15, 1995.

Roe Smith, M., Ed. *Military Enterprise and Technological Change: Perspectives on the American Experience*. MIT Press, Cambridge, Massachusetts, 1985.

Roe Smith, M. and Marx, L., Eds. *Does Technology Drive History: The Dilemma of Technological Determinism*. MIT Press, Cambridge, MA, 1994.

Roland, A. What hath Kranzberg wrought? Or, does the history of technology matter. *Technol. Cult.*, 38, 3 697–713, 1997.

Rurup, R. Historians and modern technology: reflections on the development and current problems of the history of technology. *Technol. Cult.*, 15, 161–193, 1974.

Scranton, P. None-too-porous boundaries: labor history and the history of technology. *Technol. Cult.*, 29, 722–743, 1988.

Seely, B. SHOT, the history of technology and engineering education. *Technol. Cult.*, 4, 36, 739–772, 1995.

Sinclair, B., Ed. *New Perspectives on Technology and American Culture*. American Philosophical Society, Philadelphia, 1986.

Staudenmaier, J.M.S.J. *Technology's Storytellers: Reweaving the Human Fabric*. MIT Press, Cambridge, MA, 1985.

Staudenmaier, J.M.S.J. Rationality, agency, contingency: recent trends in the history of technology. *Rev. Am. Hist.*, 30, 168–181, 2002.

Stine J.K. and Tarr, J.A. At the intersections of histories: technology and the environment. *Technol. Cult.*, 39, 601–640, 1998.

Weber, W. History of technology in Germany after 1945: institutions, methods, fields of interest. *ICON: J. Int. Comm. Hist. Technol.*, 1, 148–171, 1995.

Weber, R.J. and Perkins, D.N. *Inventive Minds: Creativity in Technology*. Oxford University Press, New York, 1992.

Wright, J.L., Ed. *Possible Dreams: Enthusiasm for Technology in America*. Henry Ford Museum & Greenfield Village, Dearborn, Michigan, 1992.

Microscopy, Electron Scanning

The scanning electron microscope (SEM) enables the imaging of surfaces at very high resolution. The SEM was a development of the earlier transmission electron microscope (TEM), and is a close analog of the conventional optical microscope but uses high-energy electrons instead of photons. This enables it to have a very much higher resolution because, for example, the wavelength of a high-energy (c.100 kilovolt) electron is less than 1/10000 that of a photon of blue light. The specimen in a TEM has to be very thin so that the electrons can pass through and be focused by electron lenses on to a fluorescent screen to produce the magnified visible image. Surfaces can be imaged directly only with difficulty with a TEM, at a glancing angle. In a SEM, an electron beam probe interacts with the surface producing backscattering and emission of secondary electrons as the high-energy electrons strike the surface; the electron beam is scanned over each part of the surface in turn (pixels), and scattered and secondary electrons for each position in the scan are detected, usually by a cathode ray oscilloscope.

The inventors of the TEM were Max Knoll and Ernst Ruska at the Technische Hochschule, Berlin in the early 1930s, while the development of the SEM started a few years later, also by Knoll after he moved to the Telefunken Company and was working on the development of television iconoscope-type camera tubes. In the course of this work he built an electron beam scanner (1935) for studying the secondary-electron emitting properties of the iconoscope targets. This apparatus (Figure 2) had many of the features of a modern SEM—the electron beam was scanned to produce a 200-line, 50-fields per second horizontal line scanning (raster), and a cathode-ray tube, deflected

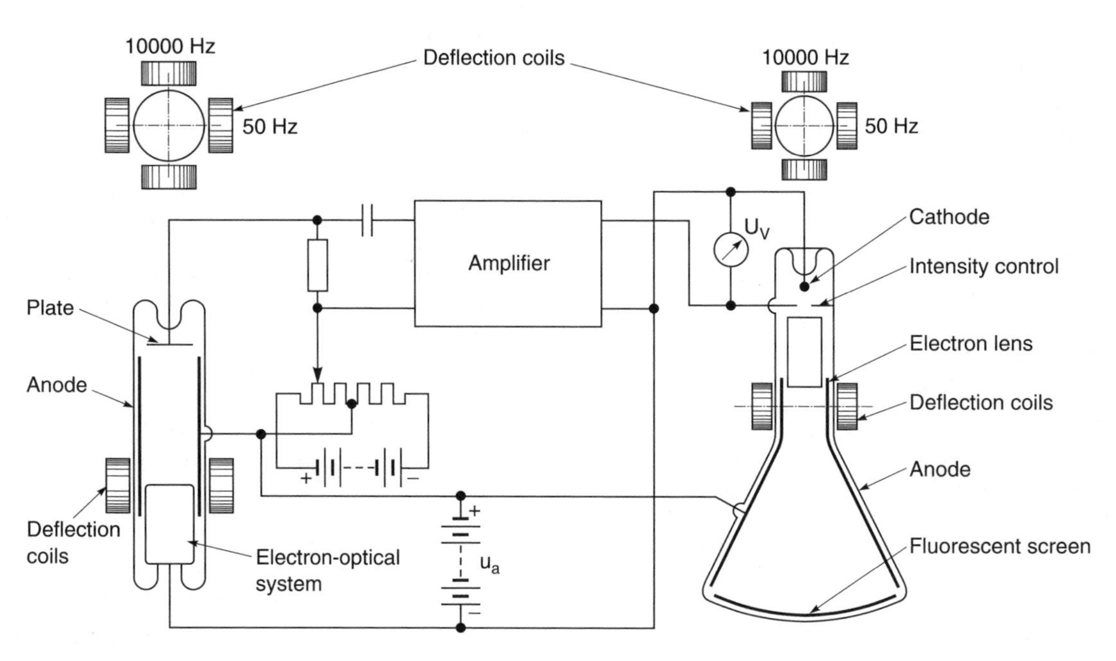

Figure 2. Diagram of Knoll's 1935 electron beam scanner.

in synchronism with scanning of the specimen and modulated in brightness by the secondary electron current from the specimen (labeled "plate" in Figure 2), displayed a visible image. This current depended on the elemental constituents of the specimen surface and, if etched, its topography; but the scanner lacked one essential feature of an SEM, a highly "demagnified" probe. In the SEM, magnification is simply the ratio of the different scanned areas at the sample and at the image. Higher magnification comes from scanning a smaller area with a sharply focused beam, made by the SEM electromagnetic lenses. Knoll was working mainly at unity magnification (or sometimes up to ten times) using a single lens to focus the beam from the electron gun on to the specimen.

The idea was extended in 1937 by Manfred von Ardenne in his private laboratory at Berlin-Lichterfelde using magnetic lenses to produce an electron probe of less than 10 nanometers (nm) diameter for a scanning transmission electron microscope (STEM), but it was unsuccessful as an SEM because he did not have a suitable detector.

A further attempt was begun at the Radio Corporation of America (RCA) in 1939 by a team led by Vladimir Zworykin (the inventor of the iconoscope). They were the first to construct a working SEM but the project was abandoned because its resolution was much inferior to that of a TEM and because of a development in surface imaging, the replica technique, which was reported in 1940 by Hans Mahl in Germany. These replicas were membranes of plastic or oxide, which bore the imprint of the etched surface being studied, and which were thin enough to be imaged in a TEM. The use of the TEM in metallurgy began from this date and for many years replicas were regarded as a satisfactory procedure for surface imaging although they were tedious to make and prone to have artifacts (false structure); even so, the consensus among most electron microscopists in the 1940s was that any further attempt to develop an SEM would be a waste of time. It required someone outside the field, Charles Oatley of the Cambridge University Engineering Department, to start the development of a microscope that could image surfaces directly and which led to the modern type of SEM.

In 1948 Oatley gave a research student, Dennis McMullan, the task of building an SEM, which resulted in a microscope with two electrostatic lenses and used an electron multiplier as the detector. Figure 3 shows a very simplified schematic diagram of an SEM from the 1950s and it is still applicable today. The components and construction of TEMs and SEMs are similar with the fundamental difference that in a TEM the lenses are used to produce a magnified image of the specimen on a viewing screen, while in an SEM they demagnify the electron source (in the gun) to produce a very small diameter scanning spot focused on the specimen. The collector receives the electrons that are reflected from the specimen

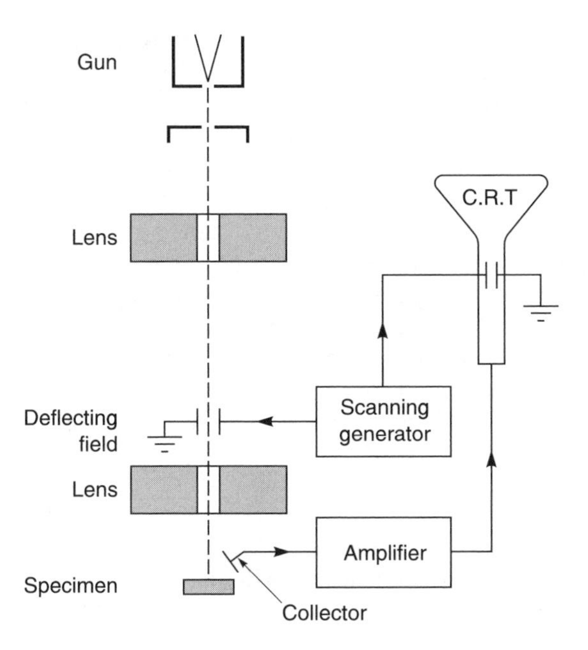

Figure 3. Simplified schematic drawing of a scanning electron microscope (1952).

Figure 4. An early visual image (of etched aluminum surface) produced with the first Cambridge SEM; 5-second scan; angle of incidence of 16 kilovolt electrons = 25 degrees; horizontal field width = 37 micrometers.

surface, and produces an electrical signal that can used to modulate the brightness of the cathode-ray tube used to display the image, as in Knoll's original electron beam scanner (Figure 2). One of the first (1952) images, of an etched aluminum surface placed at 25 degrees to the beam, shows the three-dimensional effect that is the hallmark of an SEM (Figure 4).

By the early 1960s a succession of Oatley's students had developed several SEMs that produced high-quality images, although still with substantially lower resolution than TEMs, and consequently were of little interest to the electron microscope community. However, the Cambridge Instrument Company in 1965 marketed an SEM based on Oatley's work and within a few years had sold several hundred "Stereoscans" (so-named because of the three-dimensional effect); the virtues of the SEM had quickly become apparent to electron microscopists who had actually seen one. The Stereoscan set the pattern for SEMs marketed by other manufacturers and present instruments are recognizable descendants but with many improvements due to advances in electronics, improved electron-optical components (magnetic lenses, high-intensity field-emission electron guns, etc.), and computer control.

The resolution of an SEM depends not only on the size of the focused electron-scanning spot but also, very importantly, on the scattering of electrons that penetrate the specimen and escape from a larger area of the surface, to be collected by the detector. The depth of penetration of the electrons is a function of their energy and therefore the lowest acceptable energy is used, say 1 kiloelectron volt compared with several 100 kiloelectron volts in a TEM ("acceptable" is specified because the diameter of the electron spot increases at low energies due to lens aberrations, and the current in the beam is also reduced). Advances in electron microscope technology, particularly the field-emission electron gun have enabled very low-energy electrons to be used in an SEM while still achieving good resolution. Typical spot sizes are 1 nanometer with 20 kiloelectron volt energy electrons, 2 nanometers with 1 kiloelectron volt, and 5 nanometers with 200 electron volts.

In an SEM, when the scanning beam hits the specimen not only are secondary electrons with a range of energies emitted but other radiations are as well, in particular x-rays. The energies (or wavelengths) of these depend on the constituents of the specimen surface and if a suitable detector is used, a map of the elements at the surface can be generated, which can be compared with the electron topographical image. The simplest detector is a cooled silicon diode, doped with lithium, which produces pulses proportional to the energy of the x-ray photons. Better energy resolution can be obtained with a diffracting x-ray spectrometer if

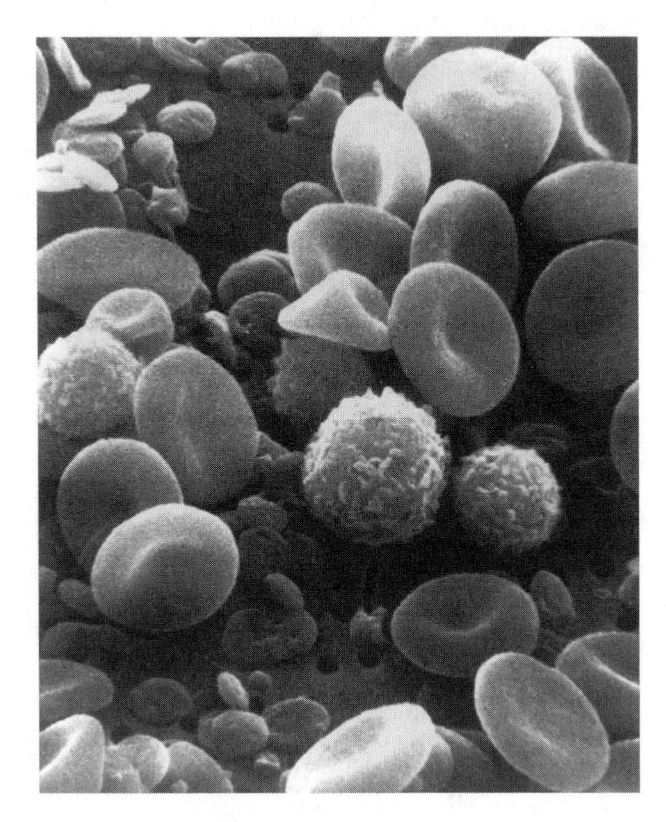

Figure 5. Scanning electron microscope image of human red blood cells and lymphocytes.
[*Source: National Cancer Institute.*]

one can be accommodated on the SEM near to the specimen. Such microanalyzers were first developed in the 1950s by Raymond Castaing in France and by Ellis Cosslett and Peter Duncumb at the Cavendish Laboratory in Cambridge: these were stand-alone instruments but later this facility was incorporated into many SEMs.

Light is also emitted by cathodoluminescent specimens and can be analyzed in an optical spectrometer; such specimens include biological substances that can be stained with suitable luminescent compounds.

Special types of SEM are used for particular classes of specimen, such as "wet" biological samples that would impair the high vacuum of a conventional SEM with water vapor. The ESEM (environmental SEM), which was first developed by Gerry Danilatos (around 1985) in Australia, operates with a specimen chamber at a relatively high pressure (around 500 pascals, still at low vacuum) so that water does not evaporate from the specimen. The column of the SEM (the region between the electron gun and the specimen) is maintained at high vacuum to enable the electron gun and lenses to operate normally.

Since the introduction of commercial SEMs in 1965, many uses have been found for them, both in advanced research and in routine analysis and imaging for science and industry. Imaging of biological cells, rock specimens, microelectronic components, and metallurgical defects can all be carried out at micrometer and nanometer scale. In fabrication of nanoscale devices and microelectro mechanical systems (MEMS), SEM imaging is an important inspection and metrological technique.

The basic components of electron microscopes continue to be improved; in particular the correction of the aberrations of the electron lenses, and no doubt the SEM will benefit from these developments.

See also **Electronics; Microscopy, Electron Transmission; Iconoscope; Microscopy, Optical**

DENNIS MCMULLAN

Further Reading

Goldstein, J., Newbury, D., Kehlin, P., Joy, D.C., Lyman, C.C., Echlin, P., Lifshin, E., Sawyer, L. and Michael, J.R., *Scanning Electron Microscopy & X-ray Microanalysis*. Kluwer, Dordrecht, 2003. The 3rd edition of a standard work on the SEM.

Oatley, C. *The Scanning Electron Microscope: Part 1, The Instrument*. Cambridge University Press, London, 1972. An early authoritative account by a pioneer.

Rasmussen, N. *Picture Control: The Electron Microscope and the Transformation of Biology in America, 1940–1960*. Stanford University Press, Stanford, CA, 1997.

Rasmussen, N. and Hawkes, P. The electron microscope, in *Instruments of Science: An Historical Encyclopedia*, Bud, R., Johnston, S. and Warner, D. Eds. Garland, London, 1997, pp. 386–389.

Reimer, L. *Scanning Electron Microscopy: Physics of Image Formation and Microanalysis*. Springer Verlag, Berlin, 1998.

Useful Websites

McMullan, D. Scanning electron microscopy 1928–1965. *Scanning*, 17, 175–185, 1995. An account of the early history of the SEM. This article is available on the web address of the Cambridge University Engineering Department: http://www.g-eng.cam.ac.uk/125/achievements/mcmullan/mcm.htm, June 2000

Microscopy, Electron Transmission

The study of electron lenses began in 1927, when Hans Busch showed that the effect of the magnetic field of a long coil on electrons can be described by the same optical law as that of a glass lens on a light beam. Soon after, Ernst Ruska in Berlin tested the familiar lens formula on a coil enclosed in an iron yoke, the first electron lens. From there, it was a short step to combine two such lenses to form a primitive electron microscope. The first images were published by Max Knoll and Ruska in 1931, and the resolution of the light microscope was surpassed in 1933. The 1930s saw rapid developments in theoretical electron optics, notably by Otto Scherzer and Walter Glaser, and toward the end of the decade, the Siemens company began serial production. Thirty-eight instruments were built before production was interrupted by World War II.

Why should an electron microscope be better than the familiar light microscope? In the 1920s, Louis de Broglie showed that a beam of charged particles, such as electrons, has a wavelength just like a beam of light. The resolution of any microscope is fixed by the value of this wavelength, which is typically about 0.5 micrometers for light. For electrons, however, the wavelength is of the order of a few picometers (1 micrometer = 1,000,000 picometers) only, and we can thus expect to see much smaller objects with electrons.

What does an electron microscope look like? From the outside, a long metal column with various connections is all that is visible. At the head of the column is the "gun," a simple triode structure consisting of an emitter (cathode), a control electrode (known as the wehnelt), and an anode, held at 100 or more kilovolts relative to the cathode, in the case of a thermionic gun. In field-emission guns, the first electrode beyond the filament creates a high electric field at the emitter. Several lenses follow, each a rotationally symmetric electromagnet, pierced by a central canal through which the electrons pass, the yoke being terminated by a pair of high-quality iron parts known as polepieces, which create the true lens field. In one of these lenses, the objective, the specimen is introduced via an airlock. At the bottom of the column is a window though which the operator, or microscopist, can examine the fluorescent screen on which the image is formed. Outside the column are power supplies for the gun and the lenses and vacuum connections designed to reduce the pressure inside the column to about 100 micropascals (a pascal is a unit of pressure equal to 1 newton per square meter) in routine operation or 10 nanopacals if a field-emission gun is employed. The overall design is shown schematically in Figure 6.

Today, a transmission electron microscope (TEM) consists of a source, condenser lenses to direct the electron beam onto the specimen, an objective lens to provide the first stage of magnification, and several intermediate lenses and a final projector. The image is formed on a fluorescent screen for visual appraisal or on a photographic emulsion or a mosaic "charge-coupled" device (known as a CCD plate) for permanent recording. In addition, an energy analyzer is frequently incorporated in the microscope column for a type of microscopy that gives information about the chemical composition of the specimen (analytical electron microscopy); a biprism may be included for electron holography, which is a valuable technique that allows us to "see" magnetic fields directly, for example. The latest generation of microscopes may include an aberration corrector and a monochromator (see below). The power supplies to all these components are under computer control.

The "conventional" TEM forms a reasonably sharp image of the specimen (or of the diffraction pattern of the specimen in the diffraction mode), and the entire image is formed simultaneously.

In the 1960s, a new type of transmission instrument, the scanning transmission electron microscope (STEM), was developed by Albert Crewe. Here, the specimen is explored by a very fine electron probe, and the information imprinted on the probe by the specimen is retrieved by a set of collectors downstream. This usefully complements the TEM but at the cost of additional technological complexity. In particular, a very bright source is needed to keep the exposure time (the time needed to record a good image) within reasonable limits; such sources (field-emission guns) require a substantially higher vacuum than the thermionic guns that are used for routine work in the TEM.

Electron guns typically accelerate the electrons to energies between 100 and 400 kiloelectron volts, although a few very high voltage instruments have been built (3 magavolts in the Toulouse instrument, for example). At 100 kilovolts, the corresponding wavelength is around 4 picometers (1 picometer = 10^{-12} meters), but it is not possible to attain a resolving power of this order with the TEM. Electron lenses have very high aberrations and must hence be operated at a small numerical aperture, which limits the resolution to a few angströms (1 Å = 0.1 nanometers = 100 picometers). They are also very sensitive to variations in wavelength, and the energy spread of the beam emitted by the source must not exceed a few electron volts. For these reasons, the latest generation of instruments is equipped with aberration correctors and monochromators. Aberration correctors consist of sets of elements not possessing rotational symmetry. For the STEM, the device consists of a series of quadrupoles and octopoles,

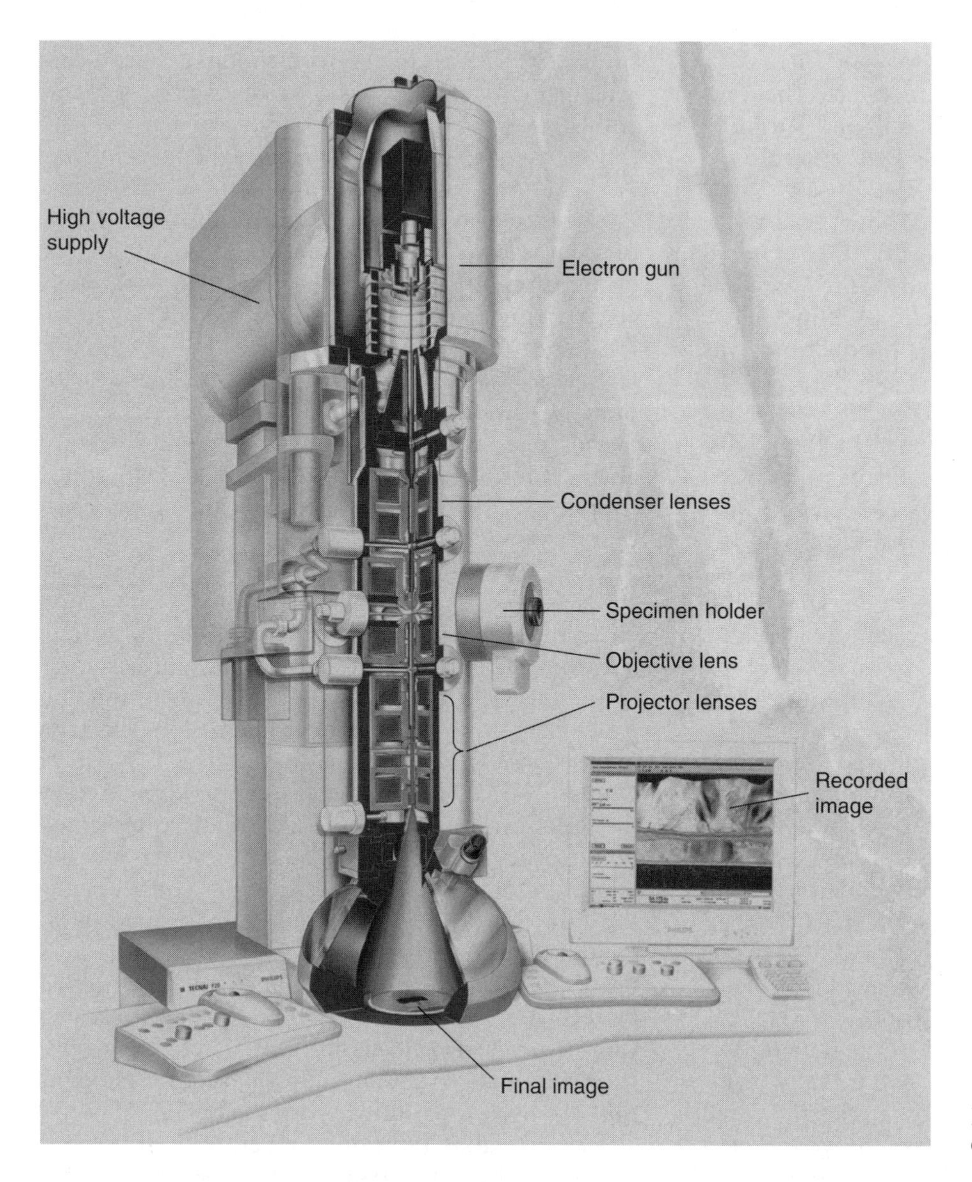

Figure 6. Transmission electron microscope.

which are capable of eliminating both the principal geometrical aberrations and the "parasitic" aberrations caused by imperfect construction or alignment. A quadrupole is the direct analog of the glass lenses used to correct the astigmatism of the human eye; there is no simple analog of the octopole, the role of which is to cancel the principal defect of electron lenses. Many STEMs are being equipped with this device today. The idea of using such optical elements was first proposed in the 1940s by Otto Scherzer, and numerous attempts were made to put it into practice before Ondrej Krivanek built a successful corrector in Cambridge in the 1990s. For the TEM, a device based on sextupoles, developed by Max Haider of CEOS on the basis of designs proposed by Harald Rose (Darmstadt) and Albert Crewe (Chicago), is preferred. As their name suggests, sextupoles consist of six magnetic poles that surround the electron beam and, once again, cancel the main defect of the microscope objective. Monochromators are devices for selecting electrons with an energy very close to the average value of the beam energy. The others are excluded so that the beam that is used for image formation is very nearly "monochromatic."

In the early years, electron microscopy developed slowly owing to the difficulty of preparing suitable specimens and the belief that the specimen would be rapidly burnt to a cinder by the electron bombardment. Gradually, however, it was realized that an electron image is formed by deflection of the electrons inside the specimen and not by absorption. In the 1950s, methods of preparing specimens thin enough to give a good image for both the life and physical sciences were found; from then on, electron microscopy has provided invaluable information in a host of fields that could have been obtained in no other way. In the physical sciences, the relationship between the macroscopic properties of materials and their microstructure has been elucidated. The structure of magnetic materials can be seen directly. The motion of dislocations in metals and alloys can be followed in real time. Similar observations are equally valuable in the earth sciences, in polymer studies, and in such areas of applied science as catalysis and aerospace. In the life sciences, the record is just as impressive. The cell division cycle has been described in detail, and the structure of bacteria and, more importantly, of viruses has been elucidated. Thanks to the extraordinary progress of electron tomography, the three-dimensional (3-D) architecture of many biological structures has been established at resolutions on the nanometer scale. This involves the collection of hundreds of different views of the same object. After digitization, these images are processed and combined in a computer to yield 3-D images allowing the biologist to relate structure to function. There are few areas of human knowledge that have not benefited from the electron microscope.

See also **Microscopy, Electron Scanning; Microscopy, Optical**

PETER HAWKES

Further Reading

Ernst, F. and Rühle, M., Eds. *High-Resolution Imaging and Spectroscopy of Materials*. Springer Verlag, Berlin, 2003.

Haguenau, F., Hawkes, P., Hutchison, J., Satiat-Jeunemaître, B., Simon, G. and Williams, D. Key events in the history of electron microscopy. *Microsc. Microanal.*, 9, 96–138, 2003.

Hawkes, P. and Kasper, E. *Principles of Electron Optics*, 3 vols. Academic Press, London, 1996. A full, modern treatise on the optics of electron microscopes and their accessories.

Hawkes, P., Ed. *Selected Papers on Electron Optics*. SPIE Optical Engineering Press, Bellingham, 1994.

Hawkes, P. The beginnings of electron microscopy. *Adv. Electron. Electron Phys.*, 16 (s), 1985. A collection of history and reminiscence by the pioneers of the subject in many countries.

Marton, L. *Early History of the Electron Microscope*. San Francisco Press, San Francisco, 1994. An engaging personal account of the development of electron microscopy by one of the pioneers.

Mulvey, T. The growth of electron microscopy. *Advances in Imaging and Electron Physics*, 96, 1996.

Reimer, L. *Transmission Electron Microscopy: Physics of image formation and microanalysis*. Springer Verlag, Berlin, 1997.

Reimer, L. *Scanning Electron Microscopy: Physics of image formation and microanalysis*. Springer Verlag, Berlin, 1998. These two volumes provide an authoritative and well-written account of most aspects of electron microscopy.

Reimer, L., Ed. *Energy-Filtering Transmission Electron Microscopy*. Springer Verlag, Berlin, 1995.

Ruska, E. *The Early Development of Electron Lenses and Electron Microscopy*. Hirzel, Stuttgart, 1980. Original German edition, 1979. The authoritative history of the first years of the electron microscope by the builder of the first instrument.

Tonomura, A. *Electron Holography*. Springer Verlag, Berlin, 1999. The standard text on the subject by one of its greatest practitioners.

Tonomura, A. *The Quantum World Unveiled by Electron Waves*. World Scientific, Singapore, 1998. A beautifully illustrated general introduction to electron interference and holography.

Williams, D.B. and Carter, C.B. *Transmission Electron Microscopy, A Textbook for Materials Science*. Plenum Press, New York, 1996. An extremely complete and clear teaching text, covering most aspects of the subject in great detail.

Useful Websites

Royal Microscopical Society Handbooks. Many of these deal in simple language with some aspect of electron microscopy. For full details, consult info@rms.org.uk or sales@bios.co.uk

Microscopy, Optical

A microscope is an optical instrument consisting of a sequence of lenses, the purpose of which is to overcome the limit imposed by the least distance of distinct vision of the human eye, typically 25 centimeters. At this distance, the eye cannot see objects smaller than about 0.1 millimeters; 0.2 millimeters can be seen without effort. By placing a single lens, a magnifying glass, in front of the eye, the object being examined can be brought much closer and the eye then sees a distant virtual image, which is an enlarged representation of the object. The magnification M is given approximately by $25/f$, where f is the focal length of the lens (in centimeters). It was with such a primitive microscope that Anton van Leeuwenhoek made his celebrated observations of organisms in water and spermatozoa in the seventeenth century. When higher magnifications are required, a second lens is added (Figure 7); this two-lens design, with an objective lens close to the specimen and an eyepiece, is the basic form of the compound microscope, invented around 1610 by Galileo. The magnification is now given by $-(25/f)(v/u)$, where the distances v and u are defined in the figure (the minus sign indicates that the image is inverted). The design of the objective lens requires considerable care and skill since a simple glass lens suffers from defects known as aberrations, asso-

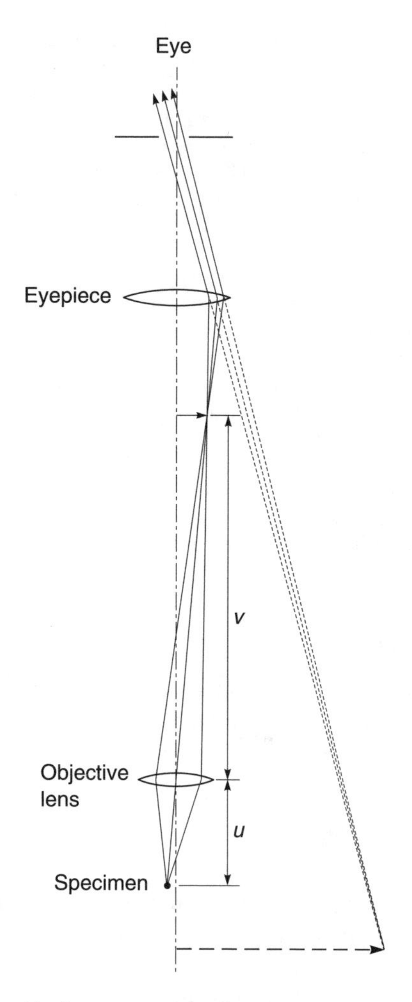

Figure 7. Basic compound microscope.

ciated with the steep inclination of some of the rays (spherical aberration, giving spatial distortion) and the spread of colors in white light (chromatic aberration). These defects can be eliminated or kept acceptably small by a suitable choice of the shapes of the lens surfaces and of the type of glass employed.

The specimen is placed on a glass slide and may be protected by a glass cover slip. The specimen stage can usually be rotated and translated. For opaque specimens, the microscope may be equipped with an epi-illuminator, a source that shines light from above the specimen; light reflected from the surface then generates the image.

It might seem from the above geometrical arguments that there is no limit to the magnification of a microscope. Although it is true that extra lenses could be added to attain very high values of the magnification, it was shown by Ernst Abbe of Zeiss in the nineteenth century that there is a limit beyond which no further fine detail can be discerned in the image. The microscope has a limit of resolution, a consequence of the wave nature of light. This limit is given by $k \tilde{\lambda}$ NA), in which k is a constant ($k \approx 0.6$), λ denotes the wavelength of light (about 0.6 micrometers in the middle of the visible spectrum), and NA is the "numerical aperture," a quantity that measures the light-gathering ability of the objective lens and is determined by the largest angle "seen" by the objective (the so-called angle of acceptance) and the refractive index of the medium between the specimen and the lens surface; in order to increase this refractive index, oil may introduced between the lens surface and the specimen. The numerical aperture is at best about 1.45. In practice, the resolution can reach about 0.2 micrometers with an oil-immersion objective and about 0.3 micrometers without oil. The useful magnification is hence about 1200 and 800 times, respectively; increasing the magnification will reveal no further detail.

The contrast seen in the image usually arises from absorption of the light incident on the specimen by opaque features of the latter. The amplitude of reflected light is thus reduced by varying amounts, depending on the opacity, and these variations in amplitude are perceived by the eye as the image of the specimen. However, some biological specimens are largely transparent and generate no amplitude contrast. They may nevertheless be of interest because of thickness variations or changes in the refractive index of the material of which the object is composed. Such specimens are said to alter the phase of the illuminating beam but not its amplitude and special arrangements have been devised to create "phase contrast" from such objects. In order to explain how this is achieved, we must first introduce the notion of coherence. Light sources are typically large and produce white light, which contains a broad range of visible wavelengths. Such sources are said to be incoherent because the light emitted form any point on the source is unrelated to that from any other point. At the other extreme are point sources emitting a very narrow range of wavelengths; these are both spatially (small emitting area) and temporally (narrow wavelength spread) coherent and it is such sources that are required for phase-contrast microscopy, discovered by Frits Zernike in the 1930s. Here, part of the light from the specimen passes through a phase plate, a thin layer of glass, the thickness of which is chosen in such a way that the phase of the light passing through it is altered by a suitable amount. This phase-shifted beam is recombined with the other part of the light beam and the result is that the phase differences, invisible to the unaided eye, are converted into amplitude (or brightness) variations. For this invention, Zernike was awarded the Nobel Prize in 1953.

The basic form of the microscope may be modified and extended in many ways. Frequently, a binocular eyepiece is added, and the microscope tube may be duplicated to give stereoscopic effects. From the 1960s it became common for lenses to be coated, for example, to reduce glare. It is usual today to add a recording medium, film or a numerical device, on which individual images or sequences can be captured. In the 1970s it became possible to capture microscopical images with a television camera, and then to digitize them. Images recorded in numerical form are often transferred to a computer for subsequent analysis or processing. For the study of crystals, particularly minerals in thinly cut rock sections, the incident illumination may be polarized by means of a Nicol prism, the polarizer (a natural rhomb of calcite, cut in two parts, which are then cemented together with Canada balsam). The light emerging from the specimen is analyzed by means of a second Nicol prism (the analyzer). Polarized light microscopy (often called petrographic microscopy) gives valuable information about the boundaries between mineral grains and can be used for identification because the refractive index of many crystals is not the same in all directions (anisotropy). Polarized light selects a particular direction and, on rotating the specimen, different amounts of light will be transmitted, depending on the relative orientations of the crystal and the direction of polarization.

In a confocal microscope the optics are designed in such a way that information is gathered only from a very thin slice of the object, rejecting out of focus light from other points of the specimen. A screen with a pinhole at the other side of the lens removes all rays of light not initially aimed at the focal point. The specimen is illuminated by this spot of light and the specimen scanned point by point. There is never a complete image of the sample—at any given instant, only one point of the sample is observed and the detector builds up the image one pixel at a time. Scanning optical microscopes (SOM) had been proposed in 1928 by Edward H. Synge, and first built in 1951 by John Z. Young and F. Roberts. The performance of a SOM was improved when in 1955 Marvin Minsky invented the confocal scanning microscope. The resolution of Minsky's confocal microscope was up to twice that of a simple SOM.

The phenomenon of fluorescence has given rise to the family of confocal fluorescence microscopes. A few specimens emit light (or fluoresce) naturally but the fluorescence microscope exploits the fact that other specimens can be "stained" with fluorescent dyes, the colors of which correspond to different features of the object. Fluorescence microscopes in use today follow the basic design of Johan S. Ploem from the 1970s. Light from a laser (i.e., very high intensity) is directed onto the specimen by means of a dichroic mirror, a plate that reflects light of all wavelengths below a threshold value and transmits all wavelengths above this threshold. This incident light causes the dye in the specimen to fluoresce at a color with a longer wavelength; a small pinhole in the image plane selects the fluorescence signal, which is then recorded. Such an optical arrangement is very sensitive to the exact level in the specimen from which the fluorescence emanates. In order to form a full image of the specimen, scanning mirrors are added (Figure 8), with which the beam scans the specimen in a raster pattern (like the spot on a television screen); these mirrors also cancel the scanning effect on the return beam. The latter is intercepted by the pinhole and the signal that passes through the opening falls on the detector, which records an image of a particular layer of the specimen. By focusing on successive layers, a full three-dimensional image can be built up.

Another new member of the microscope family is the scanning near-field optical microscope (SNOM), a close relative of the atomic-force microscope, which provides information far beyond the traditional resolution limit defined earlier. The SNOM was independently developed by Dieter W. Pohl and others at IBM in Zurich and Aron Lewis at Cornell University. Here, a sharp conical optical fiber tip is placed very close to the surface of the specimen. The diameter of the tip and the distance between tip and surface are much smaller than the wavelength of light, typically 50 nanometers or less. The surface is then displaced systematically and the tip is raised or lowered to ensure that the tip–surface distance remains constant. Two signals are recorded: the intensity of the light reflected from the surface (or fluorescence excited in the sample) and the vertical movement of the tip. The first is used to form the SNOM image while the second provides topographic information about the surface. Very many variants on this basic design have been developed, notably a transmission version for transparent specimens.

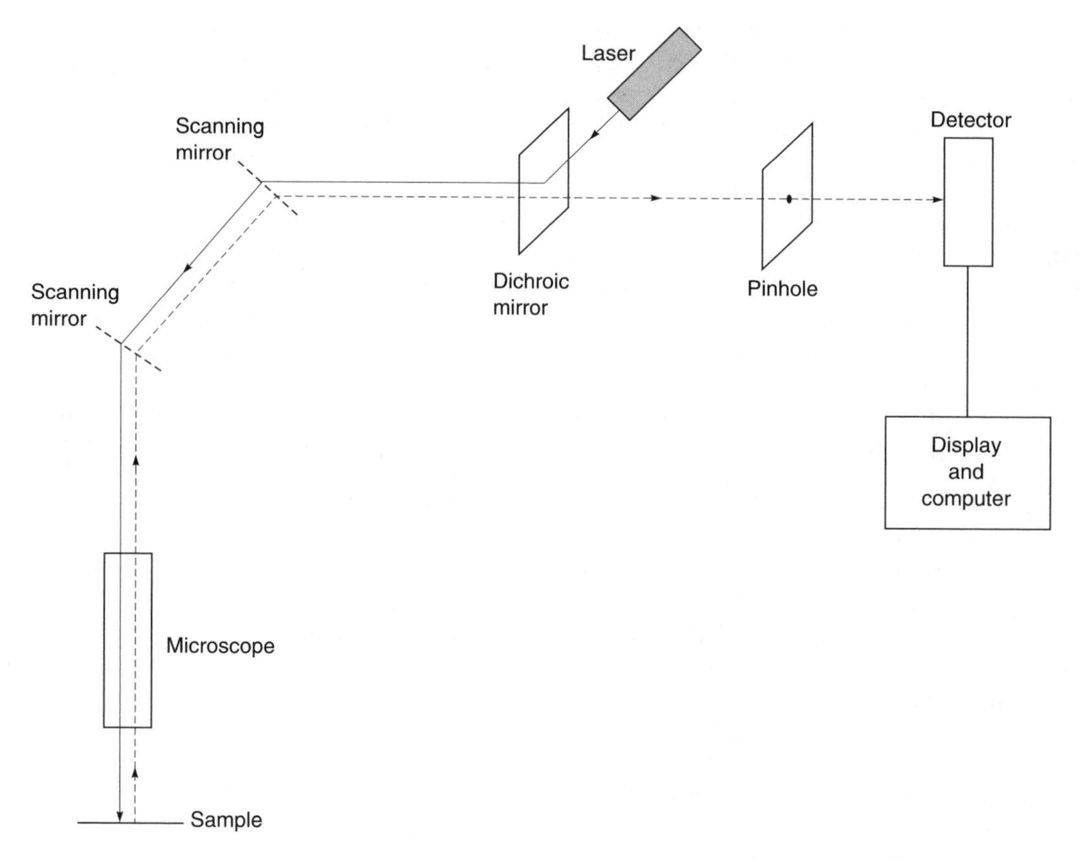

Figure 8. Confocal microscope.

See also **Microscopy, Electron Scanning; Microscopy, Electron Transmission**

PETER HAWKES

Further Reading

The series of Royal Microscopical Society handbooks published by Bios give simple introductions to most types of microscope. The most relevant volumes as well as a combined book and CD-ROM on microscopy and other titles are listed here.

Bonnell, D.A., Ed. *Scanning Probe Microscopy and Spectroscopy*. Wiley, Chichester, 2001.
Bradbury, S. and Bracegirdle, B. *Introduction to Light Microscopy*. Bios, Oxford, 1998.
Bradbury, S. and Bracegirdle, B. *Contrast Techniques in Light Microscopy*. Bios, Oxford, 1996.
Goldstein, D.J. *Understanding the Light Microscope: a Computer-Aided Introduction with CD-ROM*. Academic Press, London, 1999.
Herman, B. *Fluorescence Microscopy*. Bios, Oxford, 1997.
Murphy, D.B. *Fundamentals of Light Microscopy*. Wiley, Chichester, 2001.
Oldfield, R. *Light Microscopy: An Illustrated Guide*. Wolfe, Aylesbury, 1994.
Robinson, P.C. and Bradbury, S. *Qualitative Polarized Light Microscopy*. Bios, Oxford, 1992.
Sheppard, C.J.R. and Shotton, D.M. *Confocal Laser Scanning Microscopy*. Bios, Oxford, 1997.
Slayter, E.M. and Slayter, H.S. *Light and Electron Microscopy*. Cambridge University Press, Cambridge, 1993.

Microwave Ovens

A byproduct of the development of radar during World War II, the microwave oven is now a common part of the modern kitchen.

The microwave oven uses high-frequency radio waves generated by a magnetron tube to heat food by molecular friction. To put it another way, it causes the water molecules in the food to vibrate vigorously, and this vibration creates heat. Unlike conventional methods of cooking, where food is cooked from the outside in, the microwave oven cooks from the inside out. The high-frequency waves penetrate to the interior of the food. This results in considerably shorter preparation times than for other means of cooking. A culinary shortcoming, however, is that the microwave oven cannot ordinarily make foods, notably meats, brown and crisp.

Because of its derivation from radar, the microwave oven is an exception in the history of kitchen technology, the components of which generally evolved from simple origins to their present level of technology; the microwave oven arrived on the scene at essentially its present state of development.

Although the microwave oven appeared after World War II, the concept of cooking with radio waves predated its development. Even before the end of the war, *Science Digest* (October 1944) predicted postwar development of a "special electronic oven" that would "employ high-frequency radio waves" for cooking. It foresaw furthermore that

> "A post-war innovation of the frozen food processors will be the completely prepared dinner. The shopper will choose between menus oered by competing companies . . . Then, one minute before dinnertime, she will place the precooked frozen meal, in its sectioned, plastic container, into a special electronic oven. This oven will employ high-frequency radio waves, which penetrate all foods equally, warming a whole chicken as fast as a portion of peas. In a few seconds a bell will ring and the whole dinner will pop up like a piece of toast—ready to serve and eat."

The prediction in *Science Digest* was apparently based on a report on the development of frozen dinners for the armed forces by the W.L. Maxson Company of New York.

Radar (radio detecting and ranging) was used for the first time by Great Britain during the Battle of Britain in 1940 and was then shared by Britain with the U.S. by then-secret agreement. The American firm that was designated to collaborate with Britain was the Raytheon Company, working in cooperation with the Radiation Laboratory of the Massachusetts Institute of Technology. Raytheon produced roughly 500,000 magnetron tubes during the war.

After the war, with demand for magnetron tubes sharply reduced, Raytheon proceeded with development of the electronic oven—what would come to be called the microwave oven—and introduced the first one to the public in 1946. Called the Radarange, it was demonstrated publicly for the first time at a press conference at New York City's Waldorf-Astoria Hotel in October 1946. By now the public was familiar with radar, and a *New York Times* headline on a story about the press conference had merely to say, "Stove Operating with Basic Radar Tube Will Cook Household Meals in Seconds."

The technology reflected in the Radarange was entirely new and, until the end of the war, highly restricted. Nearly overnight, it was not only finding its way into the public domain but being heralded as something for the most ordinary of places—the household kitchen. It was a remarkably abrupt about face. Equally remarkable was that the average housewife was thought ready to accept a device technologically so much more sophisticated than a kitchen range or any of a number of other kitchen appliances she was used to operating.

As a practical matter, however, the average housewife would not have a microwave oven in her kitchen for a number of years. The Raytheon Radarange, for the time being, was intended only for commercial use. It was roughly the size of a refrigerator, about the equivalent of five modern microwave ovens placed one atop the other. The principal reason for its size was that it required water-cooling, and hence a connection to plumbing, to keep the magnetron tube from overheating.

While it thus lacked the compactness of something for countertop use, in another respect it was easily recognizable as the progenitor of its modern counterpart: "a screened oven door, so that foods may be watched as they cook" (quoting a contemporary account). Though in full production by 1947, Radaranges were used only in restaurants, on trains, and aboard ocean liners.

By 1949 a prototype for household use was ready for demonstration, and was shown to the press at a test kitchen at Columbia University in New York City. But this second-generation Radarange was still almost the size of a refrigerator; and since the magnetron tube alone cost some $500 to produce, the household version promised to be well beyond the budget of the average home.

The household Radarange never went into production. It was not until 1955 that there was a microwave oven suitable for household use. The manufacturer was Tappan, which bought the rights to Raytheon's patent. The Model RL-1 was designed for in-the-wall installation. At 61 × 61 × 69 centimeters, it was not as compact as the average modern microwave oven, but it could be used on the countertop as well as in the wall.

The key to making it appreciably smaller than the Radarange was Tappan's development of an air-cooling system to keep the magnetron tube from overheating, a clear improvement over the Radarange's water-cooling system. The 1955 Tappan microwave nevertheless remained expensive at $1200. Only 54 were sold the first year.

Raytheon and Amana Refrigeration Inc. of Amana, Iowa, announced plans for production of microwave ovens in 1967 and other companies followed. The price was down to less than $500 by the early 1970s. It is estimated that more than

100,000 ovens were being sold annually in the early 1970s.

Improvements in methods for producing the magnetron resulted in a continual reduction in the price. As the microwave oven gained popularity in Japan in the 1970s, competition was further stimulated, driving down the price still more. By the turn of the twenty-first century, microwave ovens could be purchased for well under $100.

See also **Electronics; Radar, High Frequency and High Power; Radio-Frequency Electronics**

MERRITT IERLEY

Further Reading

Amana Refrigeration Inc. *The Amana Radarange Microwave Oven Cook Book*. Amana Refrigeration, Amana, IA, 1975.

Block, S.S. New foods to tempt your palate. *Science Digest*, October, 1944.

Ierley, M. *The Comforts of Home: The American House and the Evolution of Modern Convenience*. Clarkson Potter, New York, 1999.

New York Times. News of food: stove operating with basic radar tube will cook household meals in seconds. *New York Times*, 8 October, 1946.

Van Zante, H.J. *The Microwave Oven*. Houghton Mifflin, Boston, 1973.

Military versus Civil Technologies

The exchange of technical ideas between the military world and the civilian world can be found throughout the history of technology, from the defensive machines of Archimedes in Syracuse about 250 BC, through the first application of the telescope by Galileo in military and commercial intelligence, to the application of nuclear fission to both weaponry and power production. In the twentieth century, as the military establishments of the great powers sought to harness inventive capabilities, they turned to precedents in the commercial and academic world, seeking new ways to organize research and development. By the 1960s, the phrase "technology transfer" described the exchange of technique and device between civilian and military cultures, as well as between one nation and another, and provided a name for the phenomenon that had always characterized the development of tools, technique, process, and application.

Until late in the nineteenth century, the process of invention itself was traditionally viewed in much the same light as the process of scientific discovery. That is, writers on the topic focused on the inspired individual who, by application of intuition and intellect, solved a particular mechanical or developmental problem in his (or her) head and then worked to implement the innovation through iterative trial and error until a perfected resolution was achieved. The literature tended to present such tales as moral lessons, showing that virtues of dedication, hard work, and persistence in the face of skepticism and tradition eventually conquered obstacles, leading either to financial reward and fame or to belated recognition. More often than not, well into the twentieth century, such tales were also presented as a gloss on national virtues, with British precedents stressed in works authored by Britons, American genius dominating the works of authors in the U.S., and similar echoes of national pride found in the recorded achievements of Italians, Russians, Germans, and others.

Furthermore, the process of invention was viewed as an extension of the "great man" school of historical writing that dominated historiography of politics and statehood. Rarely did scholars look behind the biographical narrative of the scientist or technological inventive genius to try to uncover the cultural, social, or psychological roots of invention and their cross application in military and civilian spheres.

However, as technological challenges and opportunities became more complex in the late nineteenth century, the locus of achievement subtly changed from the lone genius to the team and to the accumulation of component innovations into ever more complex systems. Academics, businesses, and the military alike began to establish institutions in which programmed problems would be solved and the applications worked out in group settings. The "invention factory" established in 1876 by Thomas Edison in Menlo Park, New Jersey, although often cited in American literature as the precedent for such an approach, was part of a much larger international movement. In Britain, the Admiralty supported the construction of a model basin or towing tank by William Froude, allowing him to move from his self-financed experiments at Torquay to a staffed center at Haslar. The U.S. Navy established a smokeless powder research program at a torpedo station in Newport, Rhode Island. The German technical school at Charlottenberg established an engineering station that served as a model for an American engine laboratory built by the Navy. American land grant colleges supported agricultural experiment stations in the 1880s. Russian, Italian, British, and German institutions connected with naval research or academic institutions and sometimes funded by industrialists like Nobel, Diesel, and the Du Ponts were all in place around the turn of the twentieth century, proliferating widely in the decade from 1900 to 1910.

Although writers and the general public may have persisted in the perception that invention was the action of the lone genius, businesses and governments alike recognized that the growing complexity of technological progress often required the application of skills from a wide variety of disciplines. They set up shops, laboratories, institutes, and project offices to foster technical creativity. Even when inventions were produced by laboratories or team efforts, however, the quest for heroes in the public mind required that Americans believe that Edison invented the light bulb while the British attributed it to Joseph Swan, that Italians believe that Guglielmo Marconi invented wireless telegraphy, and that the French, British, Americans, and Germans believe that their own scientists had single-handedly invented such component-rich devices as the photographic camera, the automobile, and the dirigible balloon. In fact, all such complex inventions, with both military and civilian applications, were the product of team invention, the exchange and cross-licensing of patents, and the international flow of purloined, imitated, and sometimes legitimately purchased technology transfer.

The world of weaponry had grown by similar combinations of individual innovation, accretion of parts, and exchange of ideas. A major improvement to naval guns came in 1851 with a design developed by John Dahlgren, an American naval officer. Reasoning that the greatest force of the burning charge was at the breech of the weapon, Dahlgren designed a gun with a thick breech around the bore and a thinner barrel further toward the muzzle. Yet to bring his device to completion required skilled craftsmen, shop workers, and metallurgists. A process developed in the 1850s by T.J. Rodman at the West Point Foundry cast the guns on a hollow core. Rodman and his crew allowed the guns to cool from the inside, greatly strengthening the inner side of the barrel.

A secure, screw-breech system for loading artillery was first patented by B. Chambers in 1849 in the U.S. With the development of smokeless powder in the 1890s and with the discovery of the ideal formula for the material by Dmitri Mendeleev, designers sought to improve the breech loading system to allow for rapid firing and reloading, and several rapid-fire designs came into use after 1895. At least five different rapid-fire breech-loading designs were developed in Germany, Britain, Russia, and the U.S. in the 1890s. Although some were named after their individual inventors, all were the product of group efforts both in machining and in testing. Only in the twentieth century, however, did most armies and navies begin to designate their weapons by "Mark" numbers, sometimes retaining an individual name along with it. Thus, the U.S. introduced the M-1 rifle in the 1930s, but also designated it the "Garand."

Some authors have continued to argue that most inventions come from the individual or at most, the small firm. In a study produced in 1969, John Jewkes of Oxford University detailed the case histories of 75 twentieth century inventions, ranging from acrylic fibers through to the zipper, and showed that a large proportion of them were conceived and developed by individuals or small firms, not by industrial or military facilities. The focus of Jewkes' work tended to be civilian technologies such as the antibiotic penicillin and the gasoline additive tetraethyl lead, rather than devices conceived for military purposes.

Behind the persistent public mythology of the lone heroic inventor, the research and development laboratories of the technology corporations, the naval establishments, the armories, and the government-funded experiment stations sought means to program or schedule the process of invention. For some it seemed unlikely that the head of an organization could "order" progress. Alexander Fleming, the British researcher who accidentally discovered penicillin in 1928, believed that "a team is the worst possible way" of conducting research.

Thomas Midgely, however, who tracked down tetraethyl lead to add to gasoline as a means to eliminate engine "knock," was always ready to describe how his discovery, although accidental, came out of a funded, directed, and tedious process in which a team employed by C.F. Kettering of Delco explored alternate chemicals over the period 1919 to 1921. Kettering recognized that the knocking of early four-stroke internal combustion engines varied with fuel and that some chemical additive might reduce or eliminate the problem. In effect, Midgely produced a discovery (and an invention, in the process of making tetraethyl lead), on order. Kettering himself was a firm believer in putting experts together and tying the efforts of academically trained specialists to the practical experience of mechanics, tinkerers, and experienced craftsmen. Setting up such a group and then giving them a problem to solve was the essence of the new invention-on-order system emerging in twentieth century laboratories and workshops.

Penicillin, although discovered by Fleming individually, was not produced as a medicine until nearly 15 years after its discovery, and this was only through the hard work of a team working in the period from 1939 to 1943. That group included Australian-born Howard Florey and German-born Ernst B. Chain who found a way to produce purified penicillin, the active ingredient in the mold that Fleming had identified. Florey and Chain worked initially at Oxford University. Chain identified the chemical structure of crystalline penicillin and identified four different types. From 1941 to 1943, Florey worked with staff of the Research Laboratory of the U.S. Department of Agriculture in Peoria, Illinois, developing methods for production. Fleming, Florey, and Chain shared the 1945 Nobel Prize for Physiology or Medicine for the complete process of discovery and isolation of the antibiotic. In this case, the interaction of the military and civilian spheres was demonstrated in the vastly increased need for a drug to combat infection and disease brought on by World War II. Florey himself first worked on field tests among wounded combat victims in Sicily and Tunisia.

The identification of talent, establishing of research direction (or definition of a research problem), and management of the team effort have all presented difficult issues for industrial, academic, and military laboratories and their research and development (R&D) activities. Since interdisciplinary work is often required among people with training in varied fields such as chemistry, mechanics, computer technology, materials sciences, and others; merely structuring a team and managing it may present difficulties. Establishing a shared vocabulary sometimes requires team members to break out of their disciplines and learn to convey specialized data and concepts in the language of generalists.

One common solution to the organizational issue has been to house the specialists in departments with a disciplinary focus and then to assign them to projects on a temporary basis, in a so-called matrix organization that exists for the length of the project or development. An extension of this principle in the late twentieth century was to draw specialists from entirely different organizations in academia, the military, and in industry to work on the same project while they collected their salaries from the home organizations. In such situations, "integrated project teams" (IPTs) were created for the duration of a developmental project. IPTs proliferated in military technology development efforts in the 1990s.

Although cases of technology transfer from civilian applications to the military and vice-versa can be identified, the creation of new technology specifically for a military application is very rarely a simple matter. Converting the civilian discovery of nuclear fission to a workable nuclear weapon occurred fairly promptly. Fission was identified in December of 1938, and the first test of a device that could be fitted into a weapon case was held in July 1945. The Manhattan Engineer District, formed in 1942, consolidated work at numerous civilian and military facilities into a single project, managed as a large enterprise, all conducted behind a screen of security. Although penetrated by Soviet agents, the American project remained unknown to the Germans and to the Japanese until the first weapon was detonated over Hiroshima in August 1945. Even so, it involved tens of thousands of construction workers and hundreds of engineers and scientists, and it was conducted at research and production facilities scattered across the U.S. and Canada.

The Manhattan Project has often been regarded as the first case of "big science" in the U.S., although prior projects to build a cyclotron at the University of California in the 1930s, to construct a 200-inch (5-meter) telescope for Mt. Palomar in Pasadena, the Soviet and German efforts to construct a nuclear weapon, and the German project to build the V-2 rocket at Peenemunde were precedents or contemporary in nature. What characterized the so-called big science projects was massive funding, the organization of hundreds if not thousands of specialists, and the pursuit of a specific technological goal, all on a massive budget.

Despite such vast projects involving many different specialists, the persistence of the "great man" mythology elevated the administrators and lead scientists of such projects to the rank of historical figures. Thus J. Robert Oppenheimer became known as the "father of the bomb," while similar roles were attributed to such science and technology administrators as Werner Heisenberg, Igor Kurchatov, and Wernher von Braun.

Even the concept that such massive projects represented big science was itself a matter of debate in later years. Were such applications of scientific method actually scientific endeavors, or were they the work of engineers? In fact, the task of building the nuclear weapon represented a case of the application, not the discovery, of scientific theory, and the design of machinery and processes to build a working weapon. Although there was no such field as "nuclear engineering" in 1938, the work done by the teams at the Metallurgical Laboratory at the University of Chicago, at Hanford and Oak Ridge in the U.S., and at Montreal and Chalk River in Canada were all engineering tasks. Scientists working as engineers and cooperating with industrial engineers and chemical engineers from such firms as DuPont collaborated with civil engineers from the U.S. Army Corps of Engineers. Together they designed production reactors to make plutonium and separation plants to isolate fissionable uranium-235 from the more plentiful isotope uranium-238, and they built the "gadgets" themselves. Regarded as a triumph of science and the work of notable physicists (Oppenheimer, Leo Szilard, Enrico Fermi, Niels Bohr, Neddermeyer, Eugene Paul Wigner, and others), in fact the weapon and the work involved in its creation was an immense engineering task.

In the U.S. following World War II, the administrator of the Office of Scientific Research and Development (OSRD), Dr. Vannevar Bush, published a work that was highly influential in capturing the sense that "science had won the war." *Science, The Endless Frontier*, published both as a report on the work of the OSRD and then as a popular work to build political support for a continued effort to fund science, made the clear argument that science had to precede application and that research had to precede development. For a generation in the U.S., Britain, and Canada, funding for military advances was directed to "research and development" projects. In fact, most of those projects represented the application of existing technology and established science rather than efforts to fund pure, basic, or abstract science, as advocated by Bush. Even so, engineers were baffled by the emphasis in the popular press and in the minds of government administrators on "rocket science" and on the vocabulary that insisted that research preceded science.

In the late 1960s and into the 1970s in both the U.S. and Britain, a lively discussion emerged over the sources of invention that challenged the Bush paradigm of science leading technology toward innovation. The *Project Hindsight* report in 1969 by Chalmers Sherwin and Raymond Isenson captured many criticisms of the Bush paradigm. The report by Sherwin, a preliminary version of which was published in the journal *Science* in 1967, stirred up a hornet's nest of responses and letters. A follow-up study by the Illinois Institute of Technology, known as the *TRACES Report* in 1969, demonstrated the ultimate scientific basis for many technological advances. However, the debate continued in both the U.S. and Britain, leading to close studies by I.C.R. Byatt and A.V. Cohen in 1969, M. Gibbons and R.D. Johnston in 1972, J. Langrish in 1972, and F.R. Jevons in 1976, all of which concluded that very few important recent inventions could be attributed to advances in science.

The issue was no sterile debate between the two professions; it had serious implications for the funding of military R&D in both Britain and the U.S. In the U.S., the compromise was typically bureaucratic. In the military appropriation budgets, the Defense Department established a continuum from Basic Research, funded in a budget category or "budget element" 6.1 in the military appropriation budgets, through Applied Research (6.2), Advanced Technology Development (6.3), Demonstration and Validation (6.4), Engineering and Manufacturing Development (6.5), Management Support (6.6), and Operational Systems Development (6.7). Although the categories evolved over the period from the 1960s through the 1990s, by 1993 the pattern persisted. However, the 6.1 category of Basic Research received a very low proportion of defense budgeting, with far greater amounts proposed (and funded) in each budget cycle for categories 6.3 through 6.5. Although science was given priority of place in the intellectual scheme, in the practical world the dollars went into the costly work involved in actually building weapons and devices rather than maintaining the scientist at his bench in the laboratory.

This budgetary scheme reflected the historical reality. When Lise Meitner and her nephew Otto Frisch conceived of the idea of nuclear fission in 1938, they did so over the telephone between Stockholm and Copenhagen, while on Christmas vacation away from any expensive facilities. Seven years later, the Manhattan Project had spent 2 billion dollars to construct an industry and a device implementing the concept. Despite the fact that pure science tends to be less expensive than the construction of weapons or weapons platforms, a continuing debate remains over the degree to which basic research budget categories within the military appropriation request should be expanded.

See also **Engineering, Cultural, Methodologicial, and Definitional Issues; Engineering, Production and Economic Growth; Globalization; Organization of Technology and Science; Research and Development in the Twentieth Century; Social and Political Determinants of Technological Change; Warfare**

Rodney P. Carlisle

Further Reading

Bush, V. *Science, The Endless Frontier*. Washington D.C., 1945.

Bush, V. *Modern Arms and Free Men: A Discussion of the Role of Science in Preserving Democracy*. Simon & Schuster, New York, 1949.

Byatt, I.C.R. and Cohen, A.V. *An Attempt to Quantify the Economic Benefits of Science Research*. Department of Education and Science, Science Policy Studies no. 4, London, 1969.

Gibbons, M. and Johnston, R.D. *The Interaction of Science and Technology*. Report to the Council for Scientific Policy (U.K), December 1977.

Illinois Institute of Technology. *Technology in Retrospect and Critical Events in Science* ("TRACES"). Illinois Institute of Technology, Chicago, 1968.

Jevons, F.R. The interaction of science and technology today, or, is science the mother of invention? *Technol. Cult.*, 729–42, 1976.

Jewkes, J., Sawers, D. and Stillerman, R. *The Sources of Invention*. 2nd edn. W.W. Norton, New York, 1969.

Langrish, J. and Gibbons, M., Evans, W.G. and Jevons, F.R. *Wealth from Knowledge*. Wiley, New York, 1972.

Rhodes, R. *The Making of the Atomic Bomb*. Simon & Schuster, New York, 1986.

Sherwin, C.W. and Isenson, R.S. *Project Hindsight: Final Report*. U.S. Department of Defense, Office of the Director of Research and Engineering, Washington D.C., October 1969.

Missiles, Air-to-Air

Interest in air-to-air missiles (AAMs, also known as air intercept missiles or AIMs) was initially prompted by the need to defend against heavy bombers in World War II. Unguided rockets were deployed for the purpose during the war, but the firing aircraft had to get dangerously close, and even so the rockets' probability of approaching within killing range of their targets was poor. Nazi Germany developed two types of rocket-propelled missiles employing command guidance and produced some examples, but neither saw service use.

Wartime work on fire control for aerial gunnery had clarified many of the guidance issues. It rapidly became apparent that there were two principal avenues of promise:

1. Semiactive radar
2. Passive infra-red (IR).

In semiactive homing the interceptor aircraft that had fired the missile kept its radar trained on the target and the missile homing head incorporated a radar receiver that received the reflected radar energy and made its own computations of best course to intercept. The radar was normally in x-band (i.e., near 9 gigahertz frequency) for good precision. IR homing heads, operating in the 1–2 micrometer band, detected the very hot jet efflux and thus were effective only for attack from astern.

AAMs were generally low-aspect vehicles with cruciform wing and tail or canard surfaces. As thermionic valves (vacuum tubes) had to be used, electronics were avoided or drastically simplified and electromechanical systems were preferred where possible. Early AAMs were bulky, expensive, and unreliable; interceptors generally carried mixed armaments of missiles together with unguided rockets or cannon.

A major milestone was the development of the IR-guided AIM-9A Sidewinder missile by the U.S. Naval Ordnance Test Station in China Lake, California. Every effort was made to find clever and simple solutions in order to reduce complexity, cost, and size. Moreover, the Sidewinder was the first AAM to show a capability in fighter-to-fighter combat. It was limited to firing positions in a narrow cone no more than about 3 kilometers behind the target fighter, but this nevertheless represented a major advance over what was possible with cannon. It received its first combat test in fighting between Taiwanese and Chinese forces in 1958, when the American-equipped Taiwanese showed significant superiority. In a succession of variants the Sidewinder became the standard short-range missile not only for the U.S. forces but those of many American allies as well. Highly evolved versions remained in wide service at century's end.

By the 1960s, many fighters were dispensing with cannon altogether and relying entirely on AAMs. This approach came under test in the conflict between the U.S. and North Vietnam (which relied on Soviet-supplied arms) in the late 1960s. In addition to Sidewinder, the American planes carried the larger semiactive radar AIM-7 Sparrow which, in principle, had a wider engagement envelope, and of course offered the attraction of operation in clouds. In early innings, however, only about 5 percent of the Sparrows found their targets. There were a number of reasons, of which inadequacies in Sparrow performance and reliability played a significant role.

These early combat experiences led to a reversion to cannon armament for fighters, but at the same time prompted intensive efforts to develop improved missiles. It was found that IR seekers were vulnerable to locking onto the sun or its reflections from water, glass, metal, or even clouds. Vietnam experience also served to heighten awareness of the difficulty in attaining an ideal firing position in combat with an agile fighter and the consequent need for the AAM to have a much wider engagement envelope. Two other issues that had not thus far been particularly troublesome in combat emerged as a result of tests. First was the need for the fighter to keep its radar and thus its nose pointed toward its adversary throughout the flight of its semiactive homing AAMs. In a head-on engagement, this forced it to close to dangerously close ranges to ensure a hit, even with relatively long-ranged AAMs. It also was appreciated that both radar and IR missiles were potentially vulnerable to various sorts of countermeasures and decoys.

The increasing use of solid-state and ultimately integrated digital electronics made it possible at once to cut the weight and cost of guidance and control functions and increase their reliability, resulting in substantial opportunities to improve AAMs for those who could afford the necessary technology. These and other new technologies were applied to making improved versions of Sidewinder and Sparrow, which gave a good account of themselves in the Gulf War of 1991.

A new generation of AAMs was also developed, of which the American AIM-120 AMRAAM (advanced medium range AAM) is among the best known and most widely employed. AMRAAM navigates to a preset terminal point, which may be updated during flight by coded radio command from the launching fighter. On reaching the terminal point it turns on a self-contained radar, acquires the target, and homes to an intercept autonomously. In the final eight years of the century, this weapon was used over Iraq and the former Republic of Yugoslavia, reportedly with good results.

Only American AAMs have seen wide combat service, which accounts for their prominence in this account. A number of other nations have been active in AAM development, however. Reportedly, the Soviet Union was particularly vigorous in developing so-called "high off-boresight" weapons, meaning AAMs that can be successfully employed even when the nose of the firing fighter is pointed far away from the target. Advanced versions of the Sidewinder and comparable missiles of other nations were planned to have similar capabilities.

The development of stealthy aircraft posed special challenges for AAMs: how is an AAM to home (and fuse) on a very stealthy target? And how is a stealthy aircraft to employ AAMs without compromising its stealth? The technology and techniques being applied to these and other issues remain shrouded in secrecy, ensuring that the full story of twentieth century AAMs will not be known for many years to come.

William D. O'Neil

Further reading

Dow, R.B. *Fundamentals of Advanced Missiles.* Wiley, New York, 1958. Many fundamentals remain relevant, in addition to its value in documenting earlier technologies.

Fleeman, E.L. *Tactical Missile Design.* AIAA, Washington D.C., 2001. Comprehensive engineering textbook at introductory level.

Frieden, D.R., Ed. *Principles of Naval Weapons Systems.* Naval Institute Press, Annapolis, 1985. Extensive material relating to AAMs at modest level of technical complexity.

Gunston, B. *The Illustrated Encyclopedia of the World's Rockets and Missiles.* Crescent Books, New York, 1979. Capsule histories of developments from World War II through 1970s.

Ivanov, A. Radar guidance of missiles, in *Radar Handbook*, Merrill I. Skolnik, Ed., 2nd edn. McGraw-Hill, New York, 1990.

Lennox, D., Ed. *Jane's Air-Launched Weapons.* Jane's Information Group, Coulsdon, 2000. Earlier editions provide added information on earlier developments.

Pocock, R.F. *German Guided Missiles of the Second World War.* Ian Allan, London, 1967.

Whitford, R. *Fundamentals of Fighter Design.* Airlife Publishing, Shrewsbury, 2000. Useful brief summary of AAMs in pp. 189–203.

Missiles, Air-to-Surface

Precision attack of ground targets was envisioned as a major mission of air forces from their first conception, even before the advent of practicable airplanes. Until the 1970s most air forces believed that this could be best accomplished through exact aiming of cannon, unguided rockets, or freely-falling bombs, at least for most targets. But although impressive results were sometimes achieved through these methods in tests and exercises, combat performance was generally disappointing, with average miss distances on the order of scores, hundreds, or even thousands of meters.

Even moderate accuracy required the aircraft to approach the target closely, exposing it to defensive fire. In World War I, Germany developed a family of gliders guided by commands transmitted by unreeling electrical wires and intended to deliver a torpedo while the aircraft remained several kilometers distant. The war ended before these saw service.

Late in the 1930s, Germany again took up development, now focused on precision attack of ships, bridges, tunnel entrances, and other difficult targets. Glide and rocket-propelled bombs guided by radio command were employed successfully from mid-1943. Comparable weapons were also employed by U.S. services. The need for command guidance limited the launching aircraft's standoff range and exposed it to defensive fire. Moreover, the U.S. and the U.K. deployed jammers to interfere with the radio command links of the German missiles. Just prior to the end of the war, the U.S. Navy introduced the Bat, a missile with autonomous radar guidance that successfully attacked ships and bridges at night at ranges of tens of kilometers, providing high levels of safety for the launch aircraft.

In the 1950s and early 1960s, however, interest in air-to-surface missiles (ASMs) was limited owing to concerns about cost, complexity, and reliability. The most widely used weapon was the U.S. Bullpup, rocket-propelled and guided by radio command. Several types of nuclear-armed ASMs were introduced, however, to permit strategic attacks to penetrate heavy air defenses. Due to the large effective radius of the nuclear warhead these missiles could employ preset inertial guidance, and some had ranges of more than 1000 kilometers.

The Vietnam War in the 1960s and early 1970s combined with advances in seeking and guidance technology to stimulate development of many new weapons. Again the weapons seeing widest use were American: Walleye and Maverick (TV guidance), Shrike and Standard ARM (homing on enemy radar emissions), and Paveway (homing on reflection from illumination by laser—so-called laser-guided bomb or LGB). These weapons combined miss distances of the order of a few meters with standoffs of around 5 kilometers or more. ASMs accounted for very few of the weapons employed in the Vietnam War, but achieved disproportionate results.

Armed helicopters intended for anti-tank attack were armed with adapted versions of ground anti-tank missiles. For anti-ship duties, specialized weapons were developed, of which the most widely employed has been the French Exocet, using autonomous radar homing. Similar to World War II's Bat in broad principle, Exocet was far more advanced, reliable, and effective. In the 1982 Falklands War, Argentine air forces destroyed or damaged a number of British ships with Exocets.

The Gulf War of 1991, in which a U.S.-led coalition rolled back Iraq's 1990 invasion of Kuwait, laid any remaining doubts concerning the value of modern ASMs. While many more conventional unguided bombs were dropped, ASMs (to include guided bombs, chiefly LGBs) achieved much the most critical and dramatic results. Iraqi radars, command posts, transportation and logistical facilities, bunkers, and tanks were picked off in great numbers with precisely guided ASMs. Iraqi ground forces, deprived of mobility and much vital matériel and quite demoralized, collapsed swiftly in the face of a powerful allied armored ground thrust, ensuring very low allied casualties.

Even in the U.S., the majority of air forces had not been equipped to make effective use of ASMs prior to the 1990s. The experience of the Gulf War led all who could afford to do so to rush to adopt ASMs. LGBs in particular, combining high precision with relatively low cost, quickly became all but universal. However, the Americans were not entirely satisfied with laser-guided weapons, which involved carriage of an expensive and bulky laser pod on the aircraft, only worked in reasonably clear weather, and required the aircraft to remain near the target throughout the LGB's fall. As the century closed, the U.S. introduced a new family of weapons that navigated to a prespecified point on the ground by reference to signals from the global positioning system (GPS) of satellites. The GPS-guided weapons were slightly less accurate than the LGBs (which remained in service) but more flexible. They require some means of determining the GPS coordinates of targets, implying a considerable increase in the technical sophistication of the collection, processing, and dissemination of intelligence about targets. Nevertheless, they were seen as an important step ahead and in a few engagements just at century's close performed well. Several variants were planned, ranging from two types of glide weapons allowing standoffs of around 50 kilometers (when released from high altitude) to powered weapons ranging hundreds of kilometers.

Hybrid weapons also were under development, employing GPS to bring them close enough to the target to permit transition to a autonomous precision homing system employing a laser radar, millimeter-wave radar, or passive imaging seeker. These gave promise of miss distances of a fraction of a meter against moving targets. Improvements in GPS were expected to cut miss distances of GPS-only guidance to around 1 meter.

As has been demonstrated fairly frequently, most guidance systems are vulnerable to some extent or another to countermeasures, whether technical or tactical. Many weapons have declined in effectiveness with use, as enemies took their measure. Nevertheless, by the close of the century it seemed that a powerful air force armed with sophisticated ASMs must enjoy a decisive advantage over an opponent not so favored. Thus, much more strongly than ever before, the ASM had transferred dominance in war from the side with the biggest battalions to that with the most sophisticated air forces, a triumph of wealth and technology over mass.

WILLIAM D. O'NEIL

Further Reading

Dow, R.B. *Fundamentals of Advanced Missiles*. Wiley, New York, 1958. Many fundamentals remain relevant, in addition to its value in documenting earlier technologies.

Frieden, D.R., Ed. *Principles of Naval Weapons Systems*. Naval Institute Press, Annapolis, 1985. Extensive material relating to ASMs at modest level of technical complexity.

Gunston, B. *The Illustrated Encyclopedia of the World's Rockets and Missiles,* Crescent Books, New York, 1979. Capsule histories of developments from World War II through 1970s.

Ivanov, A. Radar guidance of missiles, in *Radar Handbook,* Skolnik, M.I., Ed., 2nd edn. McGraw-Hill, New York, 1990.

Lennox, D., Ed. *Jane's Air-Launched Weapons*. Jane's Information Group, Coulsdon, 2000. Earlier editions provide added information on earlier developments.

Pocock, R.F. *German Guided Missiles of the Second World War*. Ian Allan, London, 1967.

Zarchan, P. *Tactical and Strategic Missile Guidance*. AIAA, Washington D.C., 1990.

Missiles, Defensive

Missile defenses are complex systems composed of three major components: sensors to detect the launch of missiles and track them as they advance toward their targets, weapon systems to destroy the attacking missiles, and a command and control system that interconnects sensors and weapons. As a result of technological advances, these three components have evolved over the years since World War II, producing two major periods in the history of missile defense and suggesting the advent of a third by about 2025.

Between 1945 and the early 1980s, the Soviet Union and the U.S. developed and deployed missile defenses that used nuclear-tipped interceptors. The choice of nuclear weapons was dictated by technical limitations—interceptors of this period had to be guided from the ground using tracking information generated by ground-based radars. Given the limited accuracies that could be achieved with this approach to guidance, a nuclear warhead was required to achieve a reasonable probability of destroying an attacking warhead. However, there were drawbacks associated with nuclear-armed interceptors, including the stringent controls associated with nuclear weapons, the danger of stationing these interceptors near defended cities, and the fact that nuclear detonations would blind missile defense radars.

By late 1975, the U.S. and the Soviet Union had deployed operational missile defense systems. Under the 1972 ABM treaty and a 1974 protocol to the treaty, both systems were restricted to one hundred interceptors at a single site. While Russia continues to operate the Moscow site established by the Soviet Union, the U.S. closed its one Safeguard site at Grand Forks, North Dakota, in early 1976.

After deactivating Safeguard, the U.S. focused on research and development that would eliminate the need for defensive missiles with nuclear weapons. By this time, computers were becoming more powerful (operating faster with greater memory capacity) and decreasing in size. Phased array radars with electronic beam steering had continued to improve. Infrared sensors were becoming more and more sensitive, advancing from the simple one-detector devices of World War II to arrays with thousands of detectors. Furthermore, the outputs of these detectors were now being integrated and enhanced by the constantly improving computers.

Capitalizing on these advances, the U.S. Army developed and tested an experimental vehicle that destroyed its target through the kinetic energy generated when objects collide at high speeds. This hit-to-kill (HTK) interceptor was boosted outside the atmosphere where its own infrared sensor located the target and provided data to on-board computers that guided the interceptor to a collision with its target. In the early 1980s, the Army tested its new interceptor in a series of four flight tests known collectively as the Homing Overlay Experiment (HOE). After three failures, the test vehicle destroyed its target during the fourth test on 10 June 1984, demonstrating the feasibility of HTK interceptors.

As the HOE flights were taking place, the U.S. started the Strategic Defense Initiative (SDI) program. Focused on non-nuclear defense concepts, SDI started by surveying the missile defense technology base to see if advances since Safeguard would justify a new effort to develop missile defenses. Having determined that such an effort was justified, SDI leaders developed a new system architecture that included both ground- and space-based components that could attack and destroy a missile in all phases of its flight. At the same time, SDI produced significant advances in a broad array of missile defense technologies. As a result, virtually every missile defense component was made smaller, faster, and more powerful.

Illustrative of these developments are the following:

- The HOE kill vehicle of 1984 weighed 1100 kilograms. The EKV interceptor included in the midcourse defense system pursued under Presidents Clinton and George W. Bush performs the same mission but weighs only 55 kilograms.
- The infrared focal plane arrays used in surveillance satellites of the Defense Support Program during the mid-1980s, the most advanced IR sensors of that day, included 6000 IR detectors. Focal plane arrays developed under the SDI program included over 65,000 detectors, yet these devices were small enough to fit on the end of one's finger.
- The x-band radar that helped guide the EKV interceptor to the vicinity of its target was capable of "seeing" a golf ball at a range of over 3860 kilometers.
- The second stage engine on a 1980s Delta rocket weighed 45 kilograms and produced 4424 kilograms of thrust for a thrust-to-weigh ratio of 60 to 1. In one of its programs in the mid-1980s, the SDI program set out to develop motors that could produce 454 kilograms of thrust for each pound of engine. One result was the HEDI lateral thruster, which weighed 5.1 kilograms and produced 4800 pounds of thrust for a thrust-to-weight ratio of 930 to 1.

These advances fed into the missile defense programs pursued under the administrations of Presidents George H. W. Bush, William J. Clinton, and George W. Bush. They included progress in directed energy weapons (DEW) that promised to spawn a third phase in missile defense history, the DEW era. The applicability of lasers to missile defense had been recognized in the early 1960s, shortly after the laser's invention. By the 1970s a number of laser programs were under development within the U.S. Defense Department, and these programs continued into the 1980s when they were consolidated and expanded under the SDI program. Also included here was work on particle beam systems.

DEW systems possess two advantages over interceptor missiles. First and foremost is the velocity of their "projectiles," bursts of energy, which are propagated at the speed of light. Second, directed energy weapons can be fired repeatedly. These two qualities allow these weapons to destroy multiple targets over long ranges.

At the beginning of the twenty-first century, the U.S. was developing the airborne laser (ABL) system that would mount a chemical oxygen–iodine laser in a modified Boeing 747-400 freighter aircraft. This system was expected to achieve initial operational status as early as 2009, when it would offer the ability to destroy ballistic missiles during their boost phase before they can deploy their multiple warheads and release decoys to confuse defenses. The ABL and other developments on the horizon suggest a third period in missile defense history, the DEW era, which could begin as early as 2025.

See also **Radar Aboard Aircraft; Radar, Defensive Systems in World War II; Radar, High Frequency and High Power; Radar, Long Range Early Warning Systems**

DONALD R. BAUCOM

Further Reading

Baucom, D.R. *The Origins of SDI: 1944–1983*. University Press of Kansas, Lawrence, KA, 1992.

Bruce-Briggs, B. *The Shield of Faith: A Chronicle of Strategic Defense from Zeppelins to Star Wars*. Simon & Schuster, New York, 1988.

Graham, B. *Hit-to-Kill: The New Battle over Shielding America from Missile Attack*. Public Affairs, New York, 2001.

Missiles, Long Range and Ballistic

During the 1960s, the U.S. and the Soviet Union began to develop and deploy long-range ballistic missiles, both intercontinental ballistic missiles (ICBMs) and intermediate-range ballistic missiles (IRBMs). The former would have ranges over 8000 kilometers, and the latter would be limited to about 2400 kilometers. The German V-2 rocket built during World War II represented a short- or medium-range ballistic missile. The efficiency and long range of these missiles derived from the fact that they required fuel only to be launched up through the atmosphere and directed towards the target. They used virtually no fuel traveling through near outer space. They were "ballistic" rather than guided in that they fell at their target after a ballistic arc, like a bullet.

Over the period of the Cold War, the U.S. Air Force deployed more than ten different ICBMs, considering the various models and modifications of the missiles. Sometimes the same warhead would be employed on different missiles, leading to some confusion or conflict among sources. The U.S. Air

Force ICBMs, in order of introduction, were as follows.

Thor: 1957–1975
Atlas: 1950s–1975
Jupiter C: 1950s–1960s
Titan I: 1950s–1960s
Minuteman I: 1962–1969
Titan II: 1963–1987
Minuteman II: 1965–time of writing
Minuteman III: 1970–time of writing
Peacekeeper: 1981–time of writing
Midgetman: cancelled

The Thor, Atlas, Jupiter C, and Titan missiles were liquid-fueled, presenting serious hazards for handlers and requiring considerable advance notice for fueling time. The Minuteman and Peacekeeper missiles, by contrast, were solid-fueled, safer, and could be kept on alert at all times. The Titan II, although liquid-fueled, could be maintained in a ready state. The explosion of a Titan II missile in its silo near Damascus, Arkansas in 1980 demonstrated the hazards associated with that type of missile and contributed to the decision to retire it.

By the mid-1980s, both the U.S. and the Soviet Union each had over 1000 such missiles, some with multiple independently targetable re-entry vehicles (MIRVs) on them. A 1985 estimate indicated that the U.S. had 2130 warheads on ICBMs, while the Soviet Union had 6420 warheads on similar missiles. At the peak of the Cold War in 1985, before reductions under arms-control agreements, the balance of long-range ICBMs was as shown in Table 1.

The Soviet SS20 was an intermediate-range nuclear missile introduced in 1978 that could threaten targets all across Western Europe. The U.S. responded by the deployment to Europe of Pershing II (an intermediate-range, or IRBM) and ground-launched cruise missiles (GLCMs) in 1983–1984 that could reach targets well inside the Soviet Union. These "Euromissiles," although they could hold at risk strategic targets, were regarded by the U.S. as long-range theater nuclear forces, rather than as strategic weapons. With ranges of 1600 to 2400 kilometers, they were not intercontinental in range, but since they could target facilities within Soviet borders, they altered the Soviet's perception of the nuclear balance of power. The Pershing II

Table 1 The balance of long-range ICBMs in 1985, before reductions under arms-control agreements.

	Number of launchers	Warheads per missile	Total warheads
U.S.			
Titan II	30	1	30
Minuteman II	450	1	450
Minuteman III Mark 12	250	3	750
Mark 12a	300	3	900
Totals	1030		2130
U.S.S.R.			
SS11	520	1	520
SS13	60	1	60
SS17	150	4	600
SS18	308	10	3080
SS19	360	6	2160
Totals	1398		6420

and the SS20 were sometimes called medium-range ballistic missiles (MRBMs) to distinguish them from ICBMs.

The Soviets argued that the Pershing II missiles and the Europe-based GLCMs should be considered in any treaty limiting the deployment of ICBMs. This issue, among others, became a stumbling block in negotiations toward SALT II, the second strategic arms limitation treaty. The issue was finally resolved by treating it under a separate intermediate-range nuclear force (or INF) treaty, signed in 1988, under which the GLCMs and the Pershing II missiles were removed from Europe.

On July 31, 1991, Presidents Mikhail Gorbachev of the Soviet Union and George H.W. Bush of the U.S. signed in Moscow the strategic arms reduction treaty (START I). The treaty reflected the trend in improved U.S.–Soviet relations that had been built over the prior five years. When President Ronald Reagan first proposed deep cuts in nuclear arsenals in May 1982, the concept had been called SALT III, but negotiations towards the eventual strategic arms reduction treaty did not begin in a serious fashion until 1990–1991.

The START agreement imposed equal ceilings or totals on megatonnage and on warheads on each side. The treaty also set up a complicated list of sublimits on the total number of delivery vehicles, including intercontinental ballistic missiles (ICBMs), submarine-launched ballistic missiles (SLBMs), and heavy bombers. The reductions were carried out in three phases over a period of seven years after the treaty came into force. After that seven-year period of implementation, each country would be allowed 1600 strategic nuclear delivery vehicles and no more than 6000 accountable warheads. Even though the treaty was not ratified by the Soviet Union before its dissolution in December 1991, Russia and the other three republics of the former Soviet Union then holding nuclear arms agreed to the terms of the agreement. Russia confirmed its adherence to START I in several legal steps, confirmed in agreement between President Bush and President Boris Yeltsin on June 16, 1992. The START agreement allowed each country to make up its allowed total of 6000 warheads by different combinations of weapons carried by bombers, ICBMs, and SLBMs, an arrangement that allowed the U.S. to take advantage of her long-range aircraft, and the Soviet Union to rely on her heavy ICBMs.

Both countries soon adopted a series of confidence-building measures to increase transparency and to provide early experience with verification techniques. For example, both countries opened ICBM silo hatches and displayed submarine missiles so that they could be counted by satellite. Later, under START II, both nations would witness on the ground the actual destruction of launchers, missiles, and other delivery systems.

The development of intermediate-range and tactical missiles by North Korea, Pakistan, India, and China and the modification of theater missiles by Iraq (such as the Soviet "Scud" or SS-1), all capable of carrying weapons of mass destruction (chemical, biological or nuclear warheads) continue to present threats to international stability in the twenty-first century.

RODNEY P. CARLISLE

Further Reading

Berman, R.P. and Baker, J.C. *Soviet Strategic Forces: Requirements and Responses*. Brookings Institution, Washington D.C., 1982.

Collins, J.M. *U.S.–Soviet Military Balance, 1980–1985*. Pergamon-Brassey's, Washington D.C., 1985.

Garthoff, R. *Détente and Confrontation: American Soviet Relations from Nixon to Reagan*. Brookings Institution, Washington D.C., 1994.

Gottemoeller, R. *Strategic Arms Control in the Post-Start Era*. Brassey's, London, 1992.

Pretty, R. *Jane's Pocket Book of Missiles*. Macdonald and Jane's Publishers, London, 1978.

Soviet Military Power. U.S. Government Printing Office, Washington D.C., 1981.

Tammen, R.I. *MIRV and the Arms Race*. Praeger, New York, 1973.

Tsipis, K. Arsenal: Understanding Weapons in the Nuclear Age. Simon & Schuster, New York, 1983.

Missiles, Long Range and Cruise

A cruise missile is an air-breathing missile that can carry a high-explosive warhead or a weapon of mass destruction such as a nuclear warhead for an intermediate range of up to several hundred kilometers. When launched from the ground, such missiles are known as ground-launched cruise missiles (GLCMs). Some historians of weapons technology regard the German V-1 or "buzz-bomb" operated in World War II, propelled with a ram-jet, air-breathing engine, as the first GLCM. The weapons do not require remote guidance, but automatically home in on pre-assigned targets, acting autonomously.

Modern American cruise missile development had its origin in a $20 million program initiated in the 1973 budget. Reliance on long-range ballistic missiles during the 1960s had led to the retirement of earlier cruise-missile models such as the

Regulus, Matador, Mace, and Hound Dog. From the 1970s onward, cruise missile development followed two paths, one for air-launched weapons (by the Air Force) and the other for ship-launched weapons for the Navy.

When modern cruise missiles were first under development in the 1970s and 1980s, there was considerable uncertainty about their circular error probable (CEP), with estimates predicted to run from under 30 meters to over 180 meters. The circular error probable is a measure of accuracy referring to a radius from the target point in which 50 percent or more of the missiles would strike. Thus a CEP of 30 meters represented a 60-meter circle with the target at its center in which one half or more of the missiles would strike.

Such a weapon, carrying a nuclear device, would be far more accurate than an intercontinental submarine-launched ballistic missile or than most intercontinental ballistic missiles (ICBM). A smaller nuclear device could be scheduled for a particular target if carried on a low-CEP GLCM. In fact, the development of low-CEP missiles and more powerful high explosives made it quite possible to develop non-nuclear war-fighting strategies that would be as effective in destroying enemy targets as earlier nuclear strategies. Accurate cruise missiles had the added advantage of reducing collateral (unintended) damage.

Among the best-known types of modern cruise missiles are the French-built air-launched cruise missile (ALCM), the Exocet; the American-built ship-launched cruise missile (SLCM), the Tomahawk; and the Chinese-built anti-ship ALCM, the Silkworm. GLCMs are distinguished from intermediate-range ballistic missiles (IRBMs) and ICBMs in that the cruise missiles have an internal system of targeting, and are air-breathing, while IRBMs and ICBMs are fired in a ballistic arc, utilizing either a solid or liquid rocket fuel together with oxygen to sustain the rocket at extremely high altitude and for exo-atmospheric flight. Since cruise missiles fly low and are air-breathing, they can be built lighter than IRBMs, which must carry an oxidizer, taking up weight that in a cruise missile can be devoted to warhead. The issue of the range of GLCMs led to controversies over how to classify them in arms control discussions between the Soviets and the Americans.

Following a policy announced in 1979, 464 American-built cruise missiles were stationed in Germany, Britain, Belgium, Holland, and Italy, in the period 1983–1985. Some of these highly accurate and long-range GLCMs, could reach targets deep in the Soviet Union. The press and weapons commentators called the GLCMs "Euromissiles." When the Soviets discussed strategic arms limitation or reduction in the mid and late 1980s, they wished to include the GLCMs, since they could reach strategic targets within their own borders. However, the U.S. insisted that the GLCMs were long-range theater nuclear force weapons, and should be covered in the treaty dealing with intermediate-range nuclear forces (the INF treaty) rather than in a strategic arms reduction treaty. The issue was difficult to resolve, but the GLCMs were finally included in the INF treaty negotiated in 1987 and signed in 1988. The GLCMs, together with 108 intermediate-range Pershing II ballistic missiles, were removed from Europe under that treaty.

Since cruise missiles are capable of carrying nuclear weapons, experts in nuclear proliferation include in their evaluations of nuclear-armed or nuclear proliferating states, such as Israel, Iraq, India or Pakistan, the weapons delivery systems. Domestically constructed cruise missiles or imported ones from suppliers in France, Russia, China, or North Korea add to the danger to neighbors represented by such nations. For example, China has provided Silkworm missiles and SLCMs to Iran.

The American Tomahawk land-attack missile (TLAM) was originally designed to be able to fit in a submarine's torpedo tube. In effect, by arming TLAM missiles with nuclear warheads, attack submarines could be converted into nuclear missile-launching platforms. The Tomahawk anti-ship missile (TASM) is another type of cruise missile. Since such a missile would carry 450 kilograms of high explosive, it would take only one such missile to disable a major warship. The TASM had a much longer reach than a torpedo.

Since cruise missiles can be launched from aircraft or ships many miles from their targets, they are classed as "stand-off" weapons. The launching platform, either ship or airplane, need not come within range of the defenses of the target. With the development in the 1980s of terrain-following and then laser-guided air-launched cruise missiles (ALCMs), extremely accurate missiles could be aimed at a particular airshaft, window, or door of a target building. Terrain-contour following missiles require that an accurate map of the ground over which the missile flies be loaded into an onboard computer. Such maps can be downloaded from satellite images. Cruise missiles have been employed with high-explosive warheads in several wars since the early 1980s. Particularly effective was the use by the Argentines of the

French Exocet during the Argentine–British war over the Malvinas (Falkland) Islands in 1982, and the use by Americans of ALCMs and SLCMs in operations over Iraq in 1991, Kosovo in 1999, and Afghanistan in2001.

RODNEY P. CARLISLE

Further Reading

Betts, R.K. Ed. *Cruise Missiles: Technology, Strategy, and Politics*. Brookings Institution, Washington D.C., 1981.
Cockburn, A. The air-launched version of the cruise missile: building on a foundation of failure? *Defense Week*, 23 June 1980.
Gray, C.S. *The Future of Land-Based Missile Forces*. Adelphi Paper no. 140, International Institute for Strategic Studies, London, 1978.
Huisken, R. The Origin of the Strategic Cruise Missile. Praeger, New York, 1981.
Jones, R. et al. *Tracking Nuclear Proliferation*. Carnegie Endowment, Washington D.C., 1998.
Tsipis, K. Cruise missiles. *Scientific American*, February 1977.

Missiles, Short Range and Guided

Early development of U.S. guided missiles can be seen as a direct case of technology transfer from the German V-2 program to the U.S. The Redstone missile, operational in 1956, was developed under the leadership of Werner von Braun and other German scientists who emigrated to the U.S. immediately after World War II. The project had initially been named "URSA," changed to "Major," and finally "Redstone." The guidance system was developed by the U.S. Army Ballistic Missile Agency at Redstone Arsenal, Huntsville, Alabama. The rocket motors were built by North American Aviation. The liquid-fueled rocket burned alcohol and liquid alcohol, and had a maximum range of the order of 280 to 320 kilometers, although later multistage missiles with the Redstone as the first stage reached ranges up to 4800 kilometers.

The upper limit of an intermediate range missile was arbitrarily set in the U.S. at about 2400 kilometers, and the upper limit of short-range missiles varied in official literature from 950 to 1300 kilometers. Other early short- and medium-range missiles developed by the U.S. are listed in Table 2.

In November 1956, U.S. Secretary of Defense Charles Wilson announced that the Army would have jurisdiction over ground-launched guided missiles with ranges under 320 kilometers, and that the Air Force would have control over those with higher ranges. Accordingly, the Army withdrew from the Jupiter program, and supported the Vanguard program to launch Explorer satellites. The Jupiter-C and prior work on Redstone missiles became part of that program. Juno I carried the first successful U.S. satellite into orbit on January 31, 1958.

One of the most difficult issues in early intermediate- and short-range missile work was perfecting a guidance system. Although ballistic missiles

Table 2 Early short- and medium range missiles developed by the U.S.

Missile	Service	Year	Manufacturer	Guidance Type	Range: km
Corporal	Army	1951	Firestone/Gilfillan	Radar	120
Sergeant	Army	1958	Sperry/Thiokol	Unguided	Artillery
Lance	Army	1972	Vought	Light beam, inertial	120
Honest John	Army	1953	Douglas/Emerson	Unguided	140
Jupiter	Army	1955	Chrysler/North American	Radar	
Matador	Air Force	1954	Martin	Ground controlled	900
Mace	Air Force	1958	Martin	Self-contained	900
Regulus I	Naval Bu Aer	1950	Chance Vought	Sperry intertial	800
Regulus II	Naval Bu Aer	1958	Chance Vought	Self-contained	1600
Thor	Air Force	1958	Douglas/North American	Inertial	2400

are generally regarded as unguided, of course they were guided in their initial stage; that is, pointed at the target as one would point artillery. True guided missiles, like cruise missiles, rely on systems that control the flight from launch to target, whereas ballistic missiles, whether long- or medium-range, were usually only guided or aimed during the initial lift-off and flight-path selection phases. Some confusion in the popular literature derived from this fact, and many ballistic missiles, guided only in lift-off, were known as guided missiles.

Early guidance systems for this initial stage were usually inertial, self-contained guidance systems. Inertial guidance systems relied on Newton's Second Law of Motion, and incorporated a computer (fairly primitive in early systems) to integrate velocity and distance with direction to target and accordingly control thruster angles and stability of the missile during ascent. Other systems included celestial guidance systems that corrected the basic inertial guidance with supplementary position and velocity information gathered from celestial star tracking, either optical or radio. A beam-rider system utilized a radar or light beam that the missile would follow in initial stages.

Short-range or artillery-type rockets surface-to-surface, or SS, missiles such as the Honest John (with ranges up to 37 kilometers) were simply aimed at the target. Mounted on a mobile transporter-erector-launcher (TEL), or on a truck, the vehicle would be pointed in the right direction, the launch rack elevated to the proper angle, and the missile simply launched in a parabolic arc. Such surface-to-surface missiles were guided by the aiming and elevation of the launching rack. The Lance, with a range to 120 kilometers, had a simple inertial guidance system, as did the 140-kilometer range Sargeant. The Pershing I, with a range up to 740 kilometers, had a more sophisticated inertial guidance system.

Short- and medium-range systems developed before the 1970s by other countries with a nuclear capability are listed in Table 3.

The most famous of the intermediate range missiles listed above is the SS-1 "Scud." Manufactured in several models with increasing range, they were sold abroad and used extensively by Iraq in the 1990–1991 Gulf War. Scud A had a range of 130 kilometers, Scud B a range 270 kilometers, and Scud C up to 450 kilometers. The Iraqis modified the Scuds further, by linking two together and achieving a range up to a reputed 900 kilometers, striking targets in Israel and deep in Saudi Arabia. In addition to initial aiming with the TEL, the Scud had a system of inertial guidance with fixed external vanes and movable auxiliary vanes positioned in the motor thrust efflux.

With most nations surrounding their military capabilities with considerable secrecy, different published range figures and guidance types sometimes contradicted each other. Furthermore, the distinction between an intermediate-range missile (up to about 2400 kilometers) and an intercontinental- or long-range missile was a matter of definition over which there was never complete agreement. Some publications would include submarine-launched missiles up to intermediate range in short- and medium-range ballistic missile listings. Although the missiles listed here were generally capable of carrying a nuclear warhead, most were also loaded with conventional explosive warheads.

RODNEY P. CARLISLE

Table 3 Short- and medium-range systems developed before the 1970s.

Country	Missile	Range: km	Guidance
France	Pluton	120	inertial
France	SSBS	3000	inertial, gimballed thrust motors
USSR	Scud (SS-1)	130–450	inertial
USSR	Frog	32	spin stabilized
USSR	Scaleboard	800	inertial
USSR	Sandal (SS-4)	1770	radio command and inertial
China	CSS-1	NA	inertial

NA: not available.

Further Reading

Collins, J.M. *U.S.-Soviet Military Balance, 1980–1985*. Pergamon-Brassey's, Washington D.C., 1985.

Merrill, G., Ed. *Dictionary of Guided Missiles and Space Flight*. Van Nostrand, New York, 1959.

Pretty, R. *Jane's Pocket Book of Missiles*. Macdonald and Jane's Publishers, London, 1978.

Soviet Military Power. U.S. Government Printing Office, Washington D.C., 1981.

Tsipis, K. *Arsenal: Understanding Weapons in the Nuclear Age*. Simon & Schuster, New York, 1983.

Walker, P.W. Precision guided weapons, *Scientific American*, August 1981.

Missiles, Surface-to-Air and Anti-Ballistic Missiles

In World War II, when Japanese kamikaze aircraft showed the amount of damage that could be inflicted with a single explosive-laden plane, it became apparent that machine gun and anti-aircraft fire were insufficient protection against current and future weapons. The answer was to combine radar detection, guided rockets, and the proximity fuse into surface-to-air missiles or SAMs. Intensive development in the postwar years produced the Sea Sparrow in the 1950s as one of the first, successful SAMS. When identifying a Warsaw Pact weapon as a surface-to-air missile, NATO forces would assign it an "SA" or surface–air, number.

Propelled by solid fuel-rockets, with radar guidance and or heat-seeking systems, and proximity fuses, there have been dozens of models and modifications of SAMs developed over the last half of the twentieth century. They have ranged from small shoulder-mounted devices effective against helicopters and low-flying aircraft to ship- and ground-based missile systems capable of destroying extremely high-flying aircraft or even ballistic missiles. In addition to SAMs that were loaded with high explosive, both the U.S. and the Soviet Union began to develop missiles, armed with nuclear warheads, which would be capable of destroying incoming intercontinental ballistic missiles (ICBMs). Such anti-ballistic missiles (ABMs) would be set to detonate outside the atmosphere, destroying incoming missiles with intense radiation or electromagnetic pulse. The distinction between a high explosive-loaded SAM with high-altitude capability and a nuclear-armed ABM was not always made clear in the press, nor in strategic literature.

The development of nuclear-armed ABM systems proceeded in the U.S. in the late 1960s, resulting in the creation of two missiles: Sprint and Spartan. Under the anti-ballistic missile agreement signed at the same time as the strategic arms limitation treaty (SALT-I) in 1974, both the U.S. and the Soviet Union agreed not to deploy ABMs at more than two sites in each country, later reduced to one site each. In the case of the U.S., a Spartan defense was built near ICBM silos at Grand Forks, North Dakota. However, the system was installed, tested, and then turned off. The Sprint and Spartan missiles were deactivated and stored.

The ABM treaty (which the U.S. renounced in 2002) prohibited ABMs capable of intercepting ICBMs, but it did not prohibit the development of surface-to-air missiles capable of destroying tactical or medium-range ballistic missiles, armed either with nuclear or high-explosive warheads. The U.S., the Soviet Union, and Britain proceeded to develop surface-to-air missiles with extremely high altitude capabilities. These SAMs that could serve as part of a ballistic missile defense (BMD) systems have sometimes been regarded as types of ABMs.

The Nike-Hercules, although intended as an anti-aircraft missile, could be loaded with either high explosive or nuclear warheads, and proved its efficacy against some ballistic missiles, when tested against a Corporal ballistic missile and against another Nike-Hercules. The announced altitude limit of the Nike-Hercules was 46,000 meters. Nike-Hercules missiles were deployed to NATO nations in the 1970s. Since NATO targets could be reached by intermediate-range, rather than intercontinental-range ballistic missiles, such deployment of the Nike-Hercules as a BMD system did not represent a violation of the ABM treaty.

The Nike-Hercules was followed by the Patriot primarily designed as an anti-aircraft ground-to-air missile, loaded with a high-explosive warhead. Development of the Patriot, begun in 1972, was delayed in 1977 by a decision to replace the missile's onboard computer system with a digital, rather than an analog system. Test firings in 1984 proved out the computer system. The core of the Patriot system was the Raythen MPQ-53 radar. Whereas the earlier Nike-Hercules required four separate radar types, the Patriot had a single system for surveillance, target acquisition, tracking, ranging and range rate, as well as missile tracking, command guidance and target illumination. The Patriot was carried on a two-axle semitailer and then leveled with jacks, and propelled by a solid-fuel Thiokol rocket motor.

With an announced altitude limit of 24,000 meters, the Patriot was only moderately successful as a defense against the Scud in the Gulf War. When the U.S. deployed the Patriot in 1987, and then used it as a defense against the Soviet-made surface-to-surface Scud missiles during the Gulf War, 1990–1991, the Patriot, properly classified as a SAM, was often loosely referred to as a form of ABM. When used as a close-in defense system, the detonation of the Patriot could cause ground casualties or damage since it simply blew apart the incoming missile, letting large debris, engines, and fuel tanks rain down at or near the intended target.

The Soviets developed in the early 1960s a surface-to-air missile they designated the S-200 Volga, referred to in the West as the SA-5 Gammon missile. Western observers assumed that it was the Gammon that brought down the U-2 spy plane over the Soviet Union on May 1, 1960. The SA 5B, deployed in 1970, had a nuclear warhead; the SA 5C, deployed in 1975, had the option of either a nuclear or high-explosive warhead. With a potential altitude of 29,000 meters, the Gammon was thought to be capable of destroying incoming ballistic missiles. Western observers were divided over whether the SA 5B and 5C represented violations of the ABM treaty.

The Soviet SA-12B Giant surface-to-air missile was also capable of extremely high altitudes, over 30,000 meters. Like the SA 5s, it appeared to be in violation of the ABM agreements when introduced in 1986.

Many other anti-aircraft missiles designed and built in the U.S., France, the Soviet Union, and the U.K. were capable of high-altitude performance and many could have been capable of intercepting some ballistic missiles. Sweden, Japan, Israel, Italy and other countries developed a wide variety of mobile SAMs with altitude ceilings up to 3000 meters—quite effective against combat aircraft.

RODNEY P. CARLISLE

Further Reading

Chant, C. *Air Defense Systems and Weapons: World AAA and SAM Systems in the 1990s*. Brassey's Defence Publishers, London, 1989.

Pretty, R. *Jane's Pocket Book of Missiles*. Macdonald and Jane's Publishers, London, 1978.

Mobile (Cell) Telephones

In the last two decades of the twentieth century, mobile or cell phones developed from a minority communication tool, characterized by its prevalence in the 1980s among young professionals, to a pervasive cultural object. In many developed countries, more than three quarters of the population owned a cell phone by the end of the century.

Cell phone technology is a highly evolved form of the personal radio systems used by truck drivers (citizens band, or CB, radio) and police forces in which receiver/transmitter units communicate with one another or a base antenna. Such systems work adequately over short distances with a low volume of traffic but cannot be expanded to cope with mass communication due to the limited space (bandwidth) available in the electromagnetic spectrum. Transmitting and receiving on one frequency, they allow for talking or listening but not both simultaneously.

For mobile radio systems to make the step up to effective telephony, a large number of two-way conversations needed to be accommodated, requiring a duplex channel (two separate frequencies, taking up double the bandwidth). In order to establish national mobile phone networks without limiting capacity or the range of travel of handsets, a number of technological improvements had to occur.

The major problem with radio-based communications is the limited bandwidth available to transmit all the data of the calls that people wish to make. The high-frequency region of the electromagnetic spectrum has little space left in which to transmit and receive calls. To efficiently use the allocated bandwidth, mobile phone operators began by splitting it up into smaller segments (channels). Before networks were established, a single, high-power base antenna would cover a large area, but within the radius of the transmitter (normally greater than 32 kilometers), the number of users was strictly limited by the small number of channels available within the bandwidth.

An early step toward establishing a network was introduced in St. Louis, Missouri, in 1946, with a further 25 American cities adopting similar services in the next year. These were the first "zoned" systems, with multiple receiving stations, allowing users to travel around the city transmitting calls from car-based units to the nearest available receiver (this was controlled automatically in a mobile telephone switching office (MTSO) based on the relative signal/noise ratios detected). However, all transmissions from the mobile radiotelephones were picked up by a central station, making the systems again limited by the small number of channels available. This lack of space also required calls to be "half-duplex" only, without simultaneous talking and listening.

Research by Bell labs in the 1940s and 1950s showed that channels could be reused across the network if the areas using the same channels could be separated sufficiently to remove cochannel interference. This was a huge breakthrough, but it would still be three decades before mass-communication systems were up and running. During this time, transmitters and receivers became more efficient, so channels required less bandwidth and more could be squeezed into the same space. In the 1980s, governments began to allocate bands of the electromagnetic spectrum exclusively for nationwide mobile communication. These developments allowed for the implementation of modern cellular networks, which removed the long waiting lists caused by years of excess demand.

Cellular networks require small base stations to communicate on a duplex channel with the mobile phones within their cell (area of coverage). The system is modeled around a hexagonal honeycomb, with a base station at the center of each hexagon (see Figure 9). Each cell has a limited number of channels assigned to it, which can be reused without interference in nonadjacent cells. Crucially, as more users are added to each network, the number of channels can stay the same and the cells can be split into smaller cells by reducing the power and range of base stations. This allows for almost unlimited increases in capacity. As a phone travels from cell to cell, the MTSO allocates the phone to the antenna with the strongest signal, switching the frequency channel to continue the call without interruption. This is known as a "handoff." The phone can travel across the whole network by being regularly reassigned channels (i.e., frequencies).

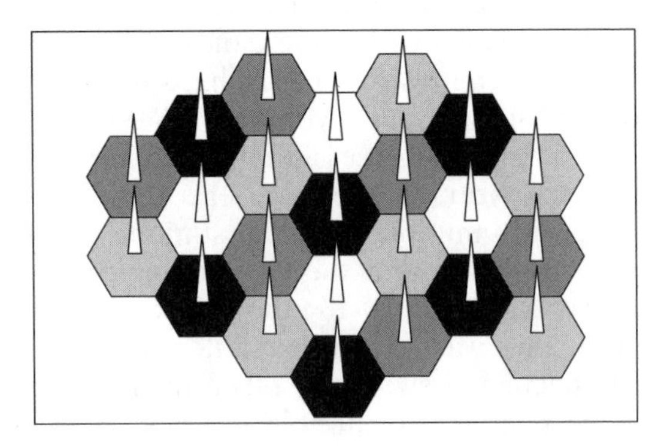

Figure 9. A representation of a network with a four-cell cluster arrangement, which allows four lots of channels to be reused in non-adjacent cells. In the diagram, the four shades represent cells that can re-use the same frequencies without interference if the system works perfectly. In reality, cells more closely resemble overlapping circles rather than hexagons, so a seven-cell cluster is normally used, increasing the distance between cells using the same channels and reducing the likelihood of interference.

The network system obviously demands a large number of base stations in busy areas, but these allow clear communication at low power, which in turn allows for smaller batteries, reducing the size and cost of the phones. In the 1980s, improved network coverage along with advances in battery efficiency, computer hardware (microchips and integrated circuits), and software allowed mobile phones to become truly mobile. What was once exclusively a business tool, requiring a cumbersome separate battery unit, could be easily slipped into a pocket and used relatively cheaply, often completely replacing a traditional phone line.

Mobile phone networks cover most urban areas of the world and continue to expand. With increased coverage and greatly reduced cost, the number of users increased more than ten-fold in the 1990s. In 2000 alone, around 400 million handsets were sold. It is estimated that by 2005, there will be more than 1 billion mobile phone subscribers worldwide.

In the early 1990s, there was a worldwide move to a digital standard for mobile phones. Digital compression techniques allow for a greater volume of information to be transmitted within the same frequency band, increasing the capacity of each cell. In Europe, GSM (global standard for mobile communications) operating on the 900 megahertz band, became the digital standard (although many countries, including the U.K., now have a separate 1800 megahertz band used by newer networks). In the U.S., mobile phone systems operate at 800 and 1900 megahertz on the PCS (personal communication system) standard. Although the systems are incompatible, both use a technique called time division multiple access (TDMA). This works by splitting the available transmission time on each channel into a number of time slots (typically eight), allowing more users to simultaneously use each channel.

Digital mobile telephony also allows for the reliable transmission of data. One development is the addition of WAP (wireless application protocol) facilities to handsets, which permits rudimentary Internet access. However, slow downloads and a paucity of available information have prevented WAP from being adopted as quickly as other less advanced features such as SMS (short message service) text-messaging, which has revolutionized the way many users interact with their phones. A third generation of mobile phone technology (the analog standard is the first and digital the second) promises to further change the way we use our handsets, offering multimedia and respectable Internet access, but this will involve the construction of new networks in the twenty-first century.

See also **Telecommunications; Telephony, Digital**

JACK STILGOE

Further Reading

Akaiwa, Y. *Introduction to Digital Mobile Communication*. Wiley, New York, 1997.

Lee, W.C.Y., *Mobile Cellular Telecommunications: Analog and Digital Systems*, 2nd edn. McGraw-Hill, New York, 1995.

Lee, William C.Y. *Lee's Essentials of Wireless Communications*. McGraw-Hill, New York, 2000. A less technical book by the same author, introducing the relevant issues.

Lewis, W.D. Bell Telephone Laboratories, coordinated broadband mobile telephone system. *IRE Transactions*, May 1960, p. 43.

Garg, V. and Wilkes, J. *Wireless and Personal Communications Systems*. Prentice-Hall, Englewood Cliffs, NJ, 1996.

Rahnema, M. Overview of the GSM system and protocol architecture. *IEEE Communications Magazine*, April 1993.

Schulte, H.J. Jr. and Cornell, W.A. Bell Telephone Laboratories, multi-area mobile telephone system. *IRE Transactions*, May 1960, p. 49.

Schwartz, M. *Telecommunication Networks*. Addison-Wesley, Reading, MA, 1987.

Modems, *see* **Electronic Communications**

Motorcycles

Without the bicycle there was nothing to motorize. Originating in France in the late 1770s, the hobbyhorse became the safety bicycle over 110 years through many people's efforts, including Thomas Humber, Dan Rudge, George Singer, James Starley, and Harry Lawson. Lawson's patent of 30 September 1879 shows a pedal-powered, rear-wheel chain-driven bicycle, something perfected by John Kemp Starley and William Sutton in 1885. The technology to motorize a bicycle had existed for years. A German-built steam-driven vélocipède is shown in a drawing dated 5 April 1818. Sylvester Roper of Roxbury, Massachusetts, exhibited a bicycle with a charcoal-fired two-cylinder engine at fairs in the eastern U.S. in 1867, and in France, the Michaux-Perreaux was built in 1868 by attaching a small commercial steam engine to a bicycle.

These attempts suffered from the disadvantageous power to weight ratio of steam engines. The solution came in 1883, when Gottlieb Daimler and Wilhelm Maybach perfected the high-speed internal combustion engine. Applications of the technology to motorize bicycles came quickly, and included work by two English pioneers: Edward Butler, who built his "Petrol-Cycle" in 1887; and J.D. Roots, who developed a motor-tricycle in 1892.

Brothers Henry and Wilhelm Hildebrand, Alois Wolfmuller and Hans Greisenhof made the first commercial motorcycle in Germany. They developed their machine from 1892 and manufactured it from 1894 as the Hildebrand and Wolfmuller "Motorrad," meaning motorcycle, the first use of the word. A number of Hildebrand & Wolfmullers were imported into Britain, where Colonel H. Capel Holden also produced motorcycles from 1897.

In 1895 the French DeDion–Buton Company built a small, light, high-revving four-stroke single engine that used battery- and coil-ignition. It was 138 cc, developed 0.5 horsepower, and made the mass production of motorcycles possible. DeDion-Buton used it in tricycles, but it was widely copied, including by Indian and Harley-Davidson in the U.S.

By 1900, at least 14 companies had tried their hand at motorcycle production in Britain. The twentieth century added to their number, and in all 639 makes of motorcycle would be produced in the U.K., the most of any country in the world. The first motorcycles retained the bicycle's "diamond" frame, into which the mechanical parts were fitted, the majority going between the main down tube and front fork, the engine driving the rear wheel through a belt. They had no gearbox or clutch, being fixed speed. Riders pushed the bike forward, ran alongside to start it, and then jumped aboard. Some manufacturers fitted clutches, which permitted stationary starting and allowed the engine to run without the bike moving and was favored by amateur mechanics.

Most early motorcycles had a four-stroke engine, usually with a single cylinder. This had inherent drawbacks. Only the third of the four strokes was powered, and on the second, or compression stroke, the piston was furthest from its last power stroke and encountering its greatest resistance. Without careful setting of valve timings, single-cylinder four-stroke engines could run erratically.

Two ways were found around these drawbacks. By 1905 V-twin engines had been produced in Europe and the U.S. These have two cylinders, usually angled at between 26 degrees and 30 degrees apart. By having one cylinder working two strokes ahead or behind the other, the engine runs on more power strokes, more smoothly, with more power. The other way around the drawbacks was the two-stroke engine. Sir Dugald Clerk developed an engine working to a two-stroke principle in 1881. This used two-cylinders, one to compress the mixture and another in which it was exploded. Work to produce a single-cylinder two-stroke engine was undertaken by Alfred Scott between 1900 and 1908, which he used in his company's motorcycles from 1909; followed by Levis from 1910, and Velocette from 1912.

Sporting experience showed the value of gears. Some of the first were adjustable, like those on the Zenith Gradua in 1909, and the Rudge Multi in 1911. A rider-operated lever varied the diameter of an engine and rear wheel drive pulley, giving a variable ratio. More conventional two-speed gears were more widely fitted from around 1910, and by 1914 three- and four-speed gearboxes were fitted by some manufacturers.

Early motorcycles used belt-drive, the use of a v-section one being widely adopted. Belts were wont to stretch, slip, or break; but they were still used on single-gear, sports and lightweight machines into 1920s. Some manufacturers favored drive chains; especially on twin-cylinder models where the extra power could cause slippage. A compromise was "chain-cum-belt" drive. Here the drive from the engine to the gearbox was by chain, and that from the gearbox to the rear wheel by belt.

The British government suspended motorcycle production late in 1916 for aircraft and munitions work. It resumed by 1919. Many servicemen had ridden motorcycles and wanted to own one, providing a boom time for new makers. The 1920s also saw changes to motorcycle design and specification. Frames became elongated and low-built, the saddle was set further back, nearer the ground, and petrol tanks were mounted on top of the frame, in front of the saddle, rather than suspended beneath the top tube. Transmissions progressed too. Veloce introduced the first foot-operated gear change in 1925, the decade also seeing the progressive adoption of all-chain drive by most makers.

The marketing of motorcycles changed too. From around 1928 manufacturers named their machines. Thus Model 7s, in "Standard" or "De Luxe" versions, became a "Big Twin Export," "Speed Chief," or "Super Sports." Replicas of motorcycles successful in races and trials were also produced, as were machines for the imported Australian sport speedway.

With recession in the late 1920s, fewer motorcycle firms were formed, and established ones experienced difficulties. A number went under, while others had to cut costs and prices to survive. By the early 1930s serious and ultimately fatal damage had been done to the motorcycle industry in Britain and elsewhere. The 1930s was a decade of consolidation, characterized by the formation of Associated Motor Cycles Ltd (AMC), which eventually owned some of the greatest names such as Sunbeam and Norton.

European governments generally had a more enlightened view of motorcycles and their use, and through a combination of exemption from tax, based either on weight or engine capacity, or from a riding test, they fostered a market for lighter weight machines and a climate in which manufacturers looked for ways to make frames ever lighter and small engines ever more powerful.

Technical innovation came through the test bed of racing, where, in the 1930s, the German firms BMW and DKW, and the Italian Guzzis challenged British dominance. They produced lighter machines with higher revving engines, which, by 1936, enjoyed great success in competition. The threat this posed to the traditionally conservative British motorcycle industry was compounded by the fact that both of these countries had totalitarian governments.

World War II stifled motorcycle production, but the postwar period looked brighter. Prewar racing experience using sprung frames, introduced by Guzzi in 1935, and by Norton in 1936, influenced the design of road machines after 1945. Pressed steel frames were also increasingly common, offering both an economy in manufacture and weight, with the added advantage of better weatherproofing and enclosed engines. The use of new lightweight alloys allowed engine development to proceed apace, with single- and multicylindered versions appearing which offered power outputs that equaled or exceeded those of traditional designs, from units weighing considerably less and offering greater fuel economy.

The use of these new materials allowed manufacturers to meet the growing demand for personal motor transport seen throughout Europe and the Far East after the war. Two new designs of motorcycle appeared—the cyclemotor, or moped, and the motor scooter—and smaller motorcycles were remodeled as miniature or lightweight machines. Mopeds retained pedals and generally had small engines in the 49 to 98 cc range. One of the earliest was the NSU Quickly in Germany, but Japanese manufacturers soon joined the market, most notably Honda, who began makeshift production in October 1946. The ultimate development of the form came in July 1958 with the introduction by Honda of its C series Super Cub machines. Known colloquially as the "Honda 50," its 30 million sales to date make it the world's best-selling motor vehicle.

The motor scooter was an Italian concept that owed much to the aircraft engineering experience of its manufacturer Piaggio & Co. Their Vespa design, introduced in April 1946, was an instant success, and was followed in 1948 by Innocenti's Lambretta. Imported Italian scooters proved popular, and some British firms introduced their own versions in response, but whereas the former had style and were lightweight, British scooters were heavy and far from handsome.

As motorcycle weights were reduced and their power increased, improved brakes were needed, with hydraulic systems replacing cables and then the introduction of disk brakes, first on the front wheel and then on the rear. Racing experience continued to influence the design of road machines, notably in the use of streamlining and fairings, especially on Italian motorcycles.

The Ariel Leader of 1958 represented the first major departure from traditional motorcycle design by a British manufacturer since the war. It used unit construction, with all its mechanicals enclosed behind steel panels, and power coming from a 247-cc two-cylinder, two-stroke engine mounted on to an integral crankcase and gearbox casting. Externally, fairings and a windshield offered a previously unheard of level of protection to the rider on a machine that was very handleable. Despite its innovation, at a shade over £200, the Leader was expensive; scooters were cheaper to buy and run, and £50 more would buy the more powerful Triumph 500 Twin. Competition was also on the horizon, and 1959 saw the launch of both the Honda 250 cc C72 Dream motorcycle, with a pressed steel frame, swing arm and front-leading link forks, sophisticated OHC all-aluminum engine, electric starter and indicators, and the Austin Seven and Morris Mini Minor motorcar, which offered a family four wheels for £400.

Honda led the onslaught on the British and U.S. markets by Japanese manufacturers. Interest in the machines was generated by their higher technical specification; electric starters were a standard on Hondas Yamahas and Suzukis by 1960. Japanese firms also built their reputation on racing success, and in 1961 Honda stunned the racing world with Mike Hailwood's twin victories at the Isle of Man, the first of an unprecedented string of wins. In the U.S. Honda made their impact through advertising, beginning in 1962 with their clever "You meet the nicest people on a Honda" campaign that directly addressed the myth that motorcycles were only for tough guys and rebels. It reached out and made Honda and motorcycling in general, appealing to everyone, also opening up a vast new market to other Japanese firms, such as Kawasaki, who began volume motorcycle production in 1962.

The 1960s saw the virtual collapse of the established motorcycle industry in Britain, especially after the failure of AMC in 1966. In contrast European and Japanese manufacturers responded to the new markets in the U.S. and Australia, where high-performance motorcycles continued to woo people out of cars and on to two wheels. A luxury high performance motorcycle market began in 1965 with the launch of the Honda CB450, which offered a 163 kilometers per hour top speed, and the Italian Guzzi V7. This set a trend that continued, and led to the development of ever more powerful motorcycles, and the coining of a new word—superbike—first applied to the Honda CB750F on its launch at the 1968 Tokyo Motorcycle Show. A 750-cc machine, with four cylinders and disk brakes, the CB750F was the biggest motorcycle out of Japan to date, and it proved that a high-performance motorcycle could also be very reliable.

See also **Automobiles, Internal Combustion**

PAUL COLLINS

Further Reading

Collins, P. *British Motorcycles since 1900*. Ian Allan, Shepperton, 1998. A technological and sociological analysis of the motorcycle, plus a survey of all 639 makes of British motorcycle.

Caunter, C.F. *Motor Cycles: A Historical Survey*. Her Majesty's Stationary Office, London, 1982. Exhaustive technological detail on the development of motorcycles.

Hough R. and Setright, L.J.K. *A History of the World's Motorcycles*. George Allen & Unwin, London, 1973. World view of motorcycle history.

Motorways, *see* **Highways**

N

Nanotechnology, Materials and Applications

The *Shorter Oxford English Dictionary* defines nanotechnology as "the branch of technology that deals with dimensions and tolerances of 0.1 to 100 nanometers." Thus nanotechnology is defined not by discipline but by size, and chemists, medical scientists, and electronic engineers working on this scale may all call themselves nanotechnologists. Nanotechnology is at the interface between physics, chemistry, engineering, and biology; the fundamental processes of living matter occur on the nanometer scale. A nanometer is 10^{-9} meters or about 4 atoms wide; a human hair is about 70,000 nanometers thick.

The impetus for nanotechnology came from a famous talk by the Nobel physicist Richard Feynman in 1959. Feynman observed that there seemed to be no natural lower limit of size that would constrain the design and manufacture of very small devices. His argument was pursued in the 1980s by the American Eric Drexler, who devised a conceptual framework for nanotechnology. He envisaged "assemblers," devices controlled by computer that would be able to manipulate individual molecules or even atoms, and which would be

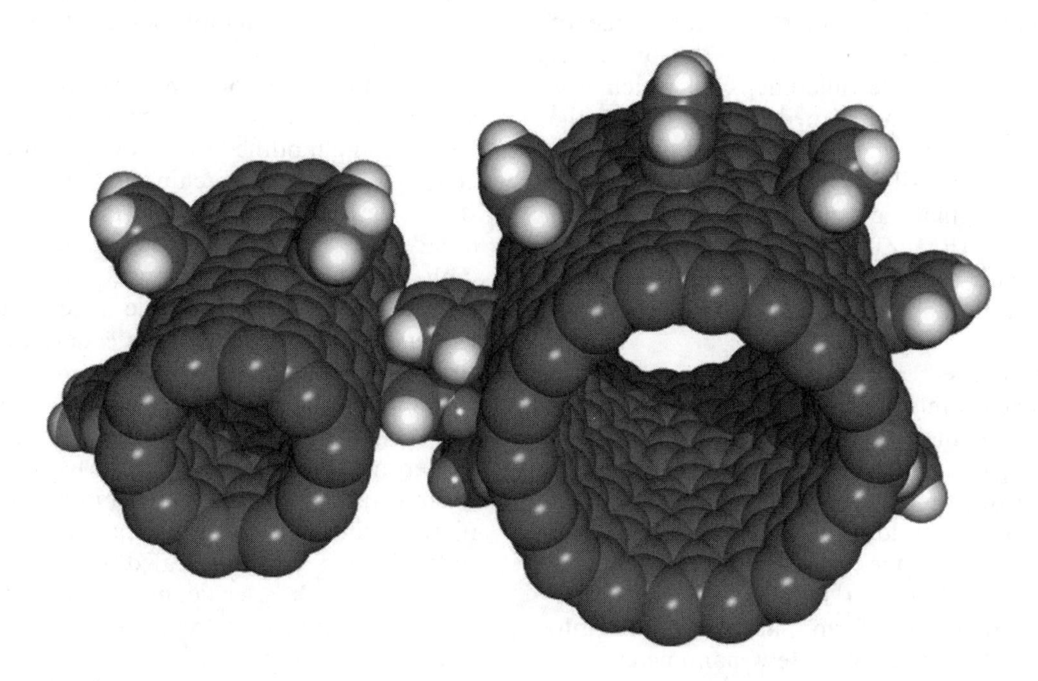

Figure 1. Simulation of fullerene gear train.
[*Source: http://science.nas.nasa.gov/Groups/Nanotechnology/gallery/*]

able to build replicas of themselves. Drexler made large claims for nanotechnology as an enabling technology that would change the human future. Its most exotic applications were to be in medicine: sensing systems and repairing machines would be implanted within individual cells; tiny robots would roam the bloodstream, seeking out and destroying viruses, delivering chemotherapeutic drugs directly to cancer cells, reaming out clogged arteries. Eventually self-repair would permit people to live forever. Such ideas initially attracted more attention from speculative writers than from the scientific community. However, technology has now begun to catch up with his visions, and some important milestones of nanotechnology have already been passed in the laboratory.

"Bottom-up" nanotechnology refers to the construction or assembly of devices or structures molecule by molecule from their constituent atoms, using techniques that have no conventional counterpart such as the scanning tunneling microscope (STM) and electron beam lithography. The STM, developed by Gerd Karl Binnig and Heinrich Rohrer in IBM's Zurich Research Laboratory in 1981, can actually pick up individual molecules on the tip of its probe and reposition them under control, demonstrated in 1990 when the American researcher Don Eigler was able to write the letters IBM on a nickel substrate with 35 individual xenon atoms. "Motors" have been constructed of less than 100 atoms in size, powered by biochemical reactions or by incident light. At the moment these have only a limited range of movement but have demonstrated the principle. The spherical molecule fullerene, C_{60}, which was discovered in 1985 and earned its discoverers the Nobel Prize in 1996, has a diameter of 1 nanometer and offers the possibility of molecular-scale ball bearings for small machines. In 1996 James Gimzewski at IBM Zurich manipulated fullerene molecules with an STM to build a 10×10 abacus with an overall diameter of less than 1 nanometer (see Figure 2).

Figure 2. Abacus constructed from fullerene molecules. [*Source: IBM Zurich http://www.zurich.ibm.com/pub/hug/PR/Abacus/abacus.gif*]

Materials based on carbon chemistry appear to be the most promising building blocks for nanotechnology. Diamond was suggested by Drexler. In practice, rigid nanostructures have successfully been made out of DNA molecules, but currently the most popular building element for nanodevices is the carbon nanotube. Nanotubes, discovered in 1991 by Sumio Iijima of Japan, are single molecules of carbon in sheet form that can be rolled into tubes with diameters of a few nanometers but lengths of up to 1 millimeter. Nanotubes have already been used to make weighing devices and have successfully been rolled across a flat surface, suggesting their suitability as machine elements. Nanotube bearings appear to be wear-free. Nanotube circuit elements fabricated from the mid-1990s offer the possibility of molecular computing using single-molecule switches to process data. Carbon nanotubes have been commercially available since 1999, but cost is still a problem; weight-for-weight, nanotubes are more expensive than gold.

"Top-down" nanotechnology refers to the miniaturization of existing mechanical and electrical devices using modified forms of existing fabrication techniques such as etching or micromachining. Microelectromechanical systems or MEMS are already in use in a wide range of technical applications. In medicine, micrometerscale MEMS devices are available to control internal delivery of drugs. Microfluidic devices also exist capable of electrically and chemically interfacing with single cells. Other MEMS devices in production in the early twenty-first century include tiny accelerometers for activating airbags and optical sensing and projection chips mounting a million micromirrors. The hope is that at least some of these devices can be scaled down to nanosize NEMS, though a critic has drawn attention to the dangers of merely "reinventing the Swiss watch industry."

An everyday example of a MEMS device, part of which already operates in the nanoscale region, is the read–write head on a computer hard drive. Although the active area of the head itself is some tens of micrometers in diameter, it "flies" above the disk platter on an air cushion that is only a few nanometers thick on the latest laptops. Because the operating clearance is smaller than the mean free path of the air molecules, the ideal gas laws can no longer predict the behavior of the air cushion, and the "wings" of the head must be elaborately contoured to stabilize its flight. So flat must the surfaces of platter and flier be to avoid moving contact that in static contact molecular forces will weld them together. Therefore special areas textured with nanoscale pimples must be provided on the platter to prevent intimate contact when the head is parked. The consequent metrological problems are likely to apply to other NEMS devices.

Nanotechnology has yet to make any significant impact on society, and has been attacked as a "cargo cult" pseudoscience that promises more than it can ever deliver. Even its supporters admit that medical robots small enough to operate in the bloodstream are at least a generation away. So far most of the progress has been top-down, but in the long term the bottom-up approach is likely to yield more important and fundamental results. However, several of Feynman's suggestions have already been realized, and large financial and technical resources are currently being invested in bringing these to market; the U.S. government earmarked $675 million for nanotechnology research in 2002. It seems likely that at least some of Drexler's predictions will become reality within a few years.

T. R. Thomas

Further Reading

Allegrini, M., Garcia, N., Marti, O., Eds. *Nanometer Scale Science and Technology*. IOS Press, Amsterdam, 2001.

Bhushan, B., Ed. *Handbook of Micro/Nanotribology*. CRC Press, Boca Raton, FL, 1998.

Drexler, E.K. *Engines of Creation*. Fourth Estate, London, 1996.

Fujimasa, I. *Micromachines*. Oxford University Press, Oxford, 1996.

Gross, M. *Travels to the Nanoworld: Miniature Machinery in Nature and Technology*. Perseus, Cambridge, MA, 2001.

Lyshevski, S.E. *Nano- and Micro-electromechanical Systems*. CRC Press, Boca Raton, FL, 2000.

Regis, E. *Nano: The Emerging Science of Nanotechnology*. Little, Brown & Company, London, 1996.

Stix, G. Waiting for breakthroughs. *Scientific American*, April 1996.

Timp, G.L., Ed. *Nanotechnology*. Springer Verlag, New York, 1998.

Navigation, *see* **Global Positioning System (GPS); Gyro-Compass and Inertial Guidance; Radionavigation**

Neurology

Following the neuroanatomical discoveries of Englishman Thomas Willis in the seventeenth century and Scotsman Alexander Monro "Secundus" in the eighteenth, and the neurophysiological work of Frenchmen Pierre Flourens, Guillaume Duchenne, and Jean-Martin Charcot in the nineteenth, knowledge of the brain and nervous system advanced significantly. But rapid progress in the clinical neurological sciences occurred only after discovery of x-rays by the German Wilhelm Roentgen in 1895 and subsequent developments made it possible to create images of living brain and nerve tissue. The diagnostic science of neurological imaging is broadly called neuroradiology.

In chronological order of development, six general neuroradiological techniques have dominated: plain x-ray films, pneumography, radiopaque myelography, cerebral angiography, computed tomography (CT), and magnetic resonance imaging (MRI). The development of the first four techniques is generally complete, but the fifth and sixth are still being refined. Other techniques of neurological examination include echoencephalography, electroencephalography, various ultrasound applications, and removal of cerebrospinal fluid (CSF) from subarachnoid space. The era from plain films to cerebral angiography took roughly 75 years, from 1898 to 1973.

Plain X-Rays

Plain x-ray films, available after 1896, were first used extensively for skull pictures in the Spanish–American War. Austrian Artur Schüller, the acknowledged father of neuroradiology, investigated their application to intracranial diagnosis and published his groundbreaking work in 1912. Swede Erik Lysholm refined the photographic method in the 1920s and 1930s. His manual of "skull tables" allowed precise identification of intracranial anatomical relations. Even as recently as the 1960s, physicians preferred plain films for some kinds of neurological diagnosis.

Pneumography/ventriculography

Pneumography, invented by American Walter Dandy in 1918, is the x-ray photography of the skull (pneumoencephalography and pneumoventriculography) or spine (pneumomyelography) after air has been introduced. Dandy observed that abdominal surgical x-rays showed gas as black against the gray of soft tissues and the white of bones, and he had read American William Henry Luckett's surgical case report of air from the sinuses entering a fractured skull and showing the ventricles and subarachnoid space in an x-ray. He discovered that if air was introduced through a lumbar puncture needle, a pneumoencephalogram could be produced. His pneumographic procedure was dangerous in cases of brain tumors or papilledema because the increased pressure could herniate the brain stem. For such cases, making holes in the skull and putting the needles directly into the ventricles (ventriculography) was safer.

Formaldehyde, first used as an anatomical fixative in 1893, enabled the pioneer work of Swiss psychiatrist Adolf Meyer in the postmortem confirmation of neuroradiological diagnoses. Meyer discovered that the precise anatomical relations of the brain can be preserved for inspection if formaldehyde is injected into the CSF and allowed to stand 24 hours before autopsy. By this method, Meyer learned that what often killed brain cancer patients was not the tumors themselves but the herniation of brain tissue through the tentorial notch or into the foramen magnum. Meyer described sideward and downward shifts of the brain tissue in 1920; American Arthur Ecker described upward shifts in 1948. Transtentorial herniation became a common topic of neurosurgical research in the 1950s.

By 1934 Dandy had refined his diagnostic procedure into a standard routine. His patients at Johns Hopkins University would first undergo a general and neurological exam by a surgical resident, a neuro-ophthalmological exam by Frank B. Walsh, and an otological exam by Benjamin M. Volk. The next day Dandy himself would perform ventriculography and make his judgment on the basis of all these reports. Errors were often made, and the mortality rate was high, even for benign intracranial tumors; but at the time this method was the best available in America.

Ventriculographic technique was refined to a higher level in Sweden in the 1930s, primarily by Lysholm's team directed by Herbert Olivecrona. Pneumography, especially pneumoencephalography, was improved from the 1920s through the 1950s by Germans Otfrid Foerster and Erich Fischer-Brügge, Americans Leo M. Davidoff and Cornelius G. Dyke, Briton Henry Head, Australian E. Graeme Robertson, and Italian Giovanni Ruggiero.

Radiopaque myelography

Radiopaque myelography, x-rays of the spine with a variety of contrast media, was begun in 1921 by Frenchmen Jean Athanase Sicard and Jacques Forestier. Since then, the quest for safer, more effective, more easily applied contrast media is a major part of the history of neuroradiology.

Cerebral angiography

In 1926 or 1927 Portuguese physician and statesman Antonio Caetano de Egas Moniz invented cerebral angiography, x-rays of the skull after introducing a contrast medium into both carotid arteries. Egas Moniz mounted his camera on the "radio-carousel" invented by his colleague José Pereira Caldas to get a large series of angiograms in rapid succession. His earliest contrast media for cerebral angiography were very dangerous substances such as strontium iodide. A colloidal thorium dioxide solution, Thorotrast, was com-

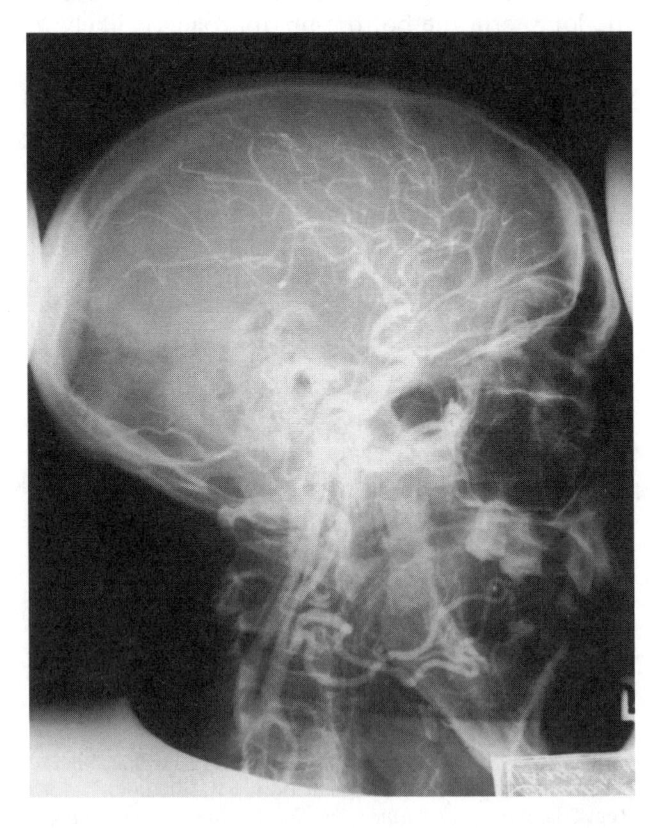

Figure 3. Cerebral angiogram made by Arthur D. Ecker, M.D., in 1949.
[*Courtesy of the Department of Historical Collections, Health Sciences Library, SUNY Upstate Medical University.*]

monly used after 1929 because it provided better contrast and was somewhat safer.

Egas Moniz's method discouraged patients and physicians alike because it required incisions to expose both carotid arteries and left unsightly scars on the neck. The development in 1936 of percutaneous carotid injections of Thorotrast alleviated that difficulty. But Thorotrast is radioactive, and if any of the injected solution fell outside the carotid artery it would cause proliferation of neck tissue with disastrous results. In the 1940s some organic iodides, such as Perabrodil, were found to be safer than Thorotrast. Yet even into the 1950s some physicians argued in favor of Thorotrast and against percutaneous carotid injections.

The U.S. lagged behind Europe, especially Sweden, in the acceptance and development of cerebral angiography. In 1951 Ecker, who had translated Egas Moniz's method for his own use in the 1930s, published the first American monograph on the topic. The Anglophone world to its disadvantage neglected the German work of Fischer-Brügge and Hermann Coenen in the early 1950s.

Computed tomography

Mathematical equations may sit idle long after their discovery until someone finds practical uses for them. Such was the case with the equations necessary for the development of computed tomography (CT) scans, the greatest single advance in radiology since x-rays. Austrian Johann Radon discovered equations for determining plane functions from line integrals in 1917, but South African physicist Allan Macleod Cormack only learned of them in 1972, nine years after he independently solved that problem and successfully applied it to computerized composite imaging. When British physicist Godfrey Newbold Hounsfield led the team of radiologists that built the first clinical CT scanner in 1971, he was unaware of either Radon or Cormack. Many improvements of CT have appeared since the 1970s, including positron emission tomography (PET) and single photon emission computed tomography (SPECT).

Magnetic resonance imaging

In the 1990s magnetic resonance imaging (MRI) and PET enabled ultrasophisticated, computer-driven brain mapping projects, such as that directed by Arthur Toga at the University of California at Los Angeles. MRI is safer than PET because it does not expose patients to radiation. MRI was originally called nuclear magnetic resonance (NMR) because it resonates hydrogen nuclei in the body, but in 1983 the American College of Radiology voted to change the name to avoid misunderstandings within the burgeoning "No Nukes" movement. A refinement of MRI, magnetoencephalography (MEG), measures brain activity directly rather than inferring it from relative oxygen levels, electrical impulses, and other intracranial data.

See also **Angiography; Electroencephalogram (EEG); Nuclear Magnetic Resonance (NMR, MRI); Positron Emission Tomography (PET); Psychiatry, Diagnosis and Non-Drug Treatments; Tomography in Medicine; X-Rays in Diagnostic Medicine**

ERIC V.D. LUFT

Further Reading

Bucy, P.C., Ed. *Modern Neurosurgical Giants*. Elsevier, New York, 1986.

Clarke, E. and Dewhurst, K. *An Illustrated History of Brain Function: Imaging the Brain from Antiquity to the Present*. Norman, San Francisco, 1996.

Clarke, E. and O'Malley, C.D. *The Human Brain and Spinal Cord: A Historical Study*. University of California Press, Berkeley, 1968.

Clifford, R.F. and Bynum, W.F., Eds. *Historical Aspects of the Neurosciences: A Festschrift for Macdonald Critchley*. Raven, New York, 1982.

Corsi, P., Ed. *The Enchanted Loom: Chapters in the History of Neuroscience*. Oxford University Press, New York, 1991.

Ecker, A.D. *The Normal Cerebral Angiogram*. Charles C. Thomas, Springfield, 1951.

Fox, W.L. *Dandy of Johns Hopkins*. Williams & Wilkins, Baltimore, 1984.

Krayenbühl, H.A. and Yasargil, M.G. *Cerebral Angiography*. J.B. Lippincott, Philadelphia, 1968.

Taveras, J.M. Diamond Jubilee Lecture: Neuroradiology, Past, Present, and Future. *Radiology*, 175, 593–602, 1990.

Nitrogen Fixation

In 1898, the British scientist William Crookes in his presidential address to the British Association for the Advancement of Science warned of an impending fertilizer crisis. The answer lay in the fixation of atmospheric nitrogen. Around 1900, industrial fixation with calcium carbide to produce cyanamide, the process of the German chemists Nikodemus Caro and Adolf Frank, was introduced. This process relied on inexpensive hydroelectricity, which is why the American Cyanamid Company was set up at Ontario, Canada, in 1907 to exploit the power of Niagara Falls. Electrochemical fixing of nitrogen as its monoxide was first realized in Norway, with the electric arc process of Kristian Birkeland and Samuel Eyde in 1903. The nitrogen monoxide formed nitrogen dioxide, which reacted with water to give nitric acid, which was then converted into the fertilizer calcium nitrate. The yield was low, and as with the Caro–Frank process, the method could be worked commercially only because of the availability of hydroelectricity.

In Germany, BASF of Ludwigshafen was interested in diversification into nitrogen fixation. From 1908, the company funded research into nitrogen fixation by Fritz Haber at the Karlsruhe Technische Hochschule. Haber specialized in the physical chemistry of gas reactions and drew on earlier studies started in 1903 on the catalytic formation of ammonia from its elements, nitrogen and hydrogen. He attacked the problem with high pressures, catalysts, and elevated temperatures. Even under optimum conditions the yield was low, around 5 percent, but Haber arranged for unreacted hydrogen and nitrogen to be recirculated. Though exothermic, the reaction was carried out at 600°C in order to increase the rate. The preferred catalyst was either osmium or uranium. The main part of the apparatus was the furnace (later known as a converter) in which the gases were preheated by the outgoing reaction mixture. At a pressure of 200 atmospheres the gases were forced to react in the presence of the catalyst. Cooling moved the equilibrium in the direction of producing ammonia, which was liquified and separated from unreacted hydrogen and nitrogen.

On July 2, 1909, BASF catalyst expert Alwin Mittasch was convinced of the potential when he observed the benchtop reactor at work. Patents were filed in Germany and elsewhere, and Haber came to an agreement with BASF over royalties. BASF chemist Carl Bosch confronted the difficulties of scaling up Haber's 0.75-meter-high converter to a pilot plant. BASF had to seek cheaper catalysts than osmium and uranium, but with similar levels of activity, to build reactors to withstand high temperatures and pressures, and to establish inexpensive sources of nitrogen and hydrogen.

Mittasch undertook catalyst experiments in miniature high-pressure tubes. Suitable catalysts based on iron compounds, such as a Swedish magnetite, were available just by chance. In 1910, an iron–aluminum catalyst with activity close to that of osmium and uranium was chosen. A stable iron catalyst with aluminum and potassium used as the promoter was found to be successful in 1911, and Mittasch soon added calcium as a third promoter. In this way, he came up with the catalyst that would be favored in industrial synthesis of ammonia.

The main step forward was achieved in 1911, with Bosch's double-wall converter. The inner wall was made of low-carbon soft steel; the outer wall was of ordinary steel. Hydrogen diffusing through the inner wall underwent loss of pressure and then came into contact with the outer wall, which though hot from external heating did not become brittle. Stress on the outer wall was reduced considerably, particularly when a little hydrogen was allowed to escape into the atmosphere through small holes, called Bosch holes. Internal heating of the converter was by combustion of gas, later replaced by internal electrical heating.

In 1910, hydrogen became available from the reaction between steam and red-hot coke, which produced hydrogen and carbon monoxide (water gas). At first, nitrogen was available from the Linde process for liquefaction of air. The most important part of the physical processing of the gases was their compression. To achieve this under previously untried conditions, Bosch had to seek out powerful leak-proof gas compressors. The recycling of unreacted gases introduced novel approaches to optimization and control loops. Monitoring of the various physical processes and the chemical reaction that took place in the reactor required instruments for measuring gas flow at high pressures, gas density, and product mixture composition. Fast-acting magnetic shut-off valves were also required.

The BASF ammonia factory, at Oppau, near Ludwigshafen, was opened in September 1913. The converter was 8 meters high, and weighed 8.5 tons. A much cheaper source of nitrogen was available from the action of air on hot coke that gave producer gas. Later, a mixture of nitrogen and hydrogen in the proportions required by the chemical equation (1:3) was made from producer gas, water gas and steam. By 1915, ammonia converters of 12 meters in height and 75 tons in weight were in operation. In the spring of that year, synthetic ammonia was converted into nitric acid for munitions production in Germany. The process was soon named Haber–Bosch (see Figure 4).

Luigi Casale and Giacomo Fauser in Italy and Georges Claude in France independently invented high-pressure ammonia processes. The Italian and French processes differed from the Haber–Bosch process in that they employed higher pressures and different catalysts. Casale's process of 1924 operated at 650 to 750 atmospheres and incorporated a circulating system similar to that used in the

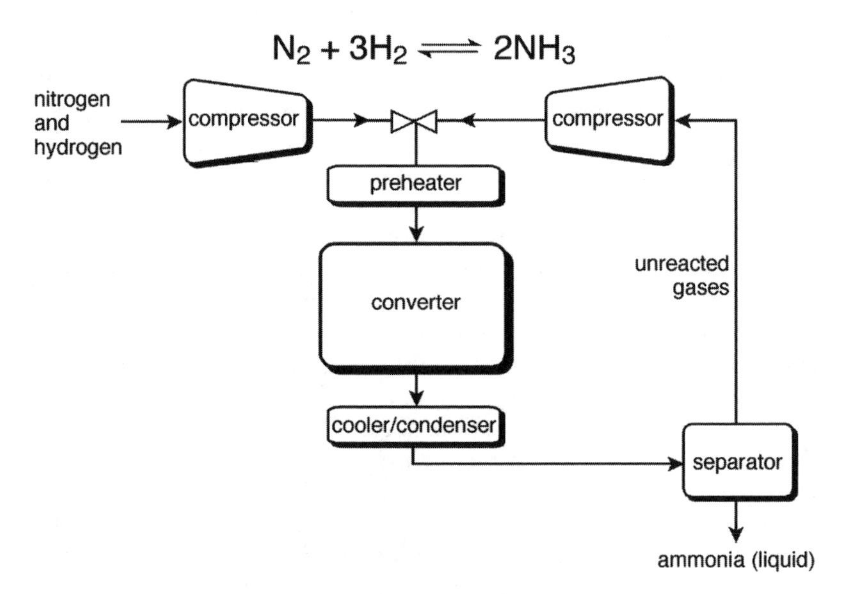

Figure 4. Flow chart showing the Haber–Bosch process for producing ammonia.

Haber–Bosch process. Fauser, a consulting engineer, used a 250-millimeter cannon as reactor in the garden of his home. From 1925, his studies were backed by Montecatini, and a viable ammonia process operating at 300 atmospheres was established. Georges Claude investigated high-pressure ammonia synthesis from 1917 using pressures of around 1000 atmospheres. Because of the high pressure, only 10 percent of unreacted nitrogen and hydrogen remained after passage through four convertors in series.

An estimate of world ammonia capacity for 1932–1933 showed that the Haber–Bosch process represented 53 percent of the total. Haber's reward was directorship of the new Kaiser Wilhelm-Institut für Physikalische Chemie und Elektrochemie in Berlin in 1912 and the Nobel Prize in 1918. Bosch's role in high-pressure synthesis was acknowledged with the Nobel Prize, received jointly with Friedrich Bergius, in 1931.

See also **Fertilizers**

Anthony S. Travis

Further Reading

Bosch, C. *Nobel Lectures: Chemistry, 1922–1941*. Elsevier, Amsterdam, 1966, pp. 326–340.

Haber, L.F. *The Chemical Industry 1900-1930: International Growth and Technological Change*. Clarendon, Oxford, 1971.

Reader, W.J. *Imperial Chemical Industries: A History*, vol. 1: *The Forerunners, 1870–1926*. Oxford University Press, London, 1970–75. Includes difficulties in imitating the Haber–Bosch process at ICI, and how they were overcome.

Reinhardt, C. Über Wissenschaft und Wirtschaft. Fritz Habers Zusammenarbeit mit der BASF 1908 bis 1911, in Naturwissenschaft und Technik in der Geschichte, 25 Jahre Lehrstuhl für der Naturwissenschaft und Technik am Historischen Institut der Univesität Stuttgart, Albrecht, H., Ed. Verlag für Geschichte der Naturwissenchaften und der Technik, Stuttgart, 1993, pp. 287–315. A source of references to the German literature on Haber's ammonia work, including historical and historiographical accounts.

Smil, V. *Enriching the Earth: Fritz Haber, Carl Bosch, and the Transformation of World Food Production.* MIT Press, Cambridge, MA, 2001.

Stoltzenberg, D. *Fritz Haber: Chemiker, Nobelpreisträger, Deutscher, Jude* VCH, Weinheim, 1994. Nitrogen fixation is covered in chapter 5, "Die Fixierung des Stickstoffs," pp. 133–197.

Haber, L. F. *The Poisonous Cloud: Chemical Warfare in the First World War*, Clarendon, Oxford, 1986. Studies on Haber normally place great emphasis on his controversial gas warfare work.

Travis, A.S. High pressure industrial chemistry: the first steps, 1909–1913, and the impact, in *Determinants in the Development of the European Chemical Industry, 1900–1939: New Technologies, Political Frameworks, Markets and Companies*, Travis, A.S., Homburg, E., Schröter, H. and Morris, P.J.T., Eds. Kluwer, Dordrecht, 1998, pp. 1–21.

Nuclear Fuels

Nuclear fuels are the propellants of nuclear reactors. In the context of the nuclear industry, they are divided into fertile materials and fissile materials. The former are treated radioactive minerals, while the latter are produced in nuclear reactors or through chemical separation with the fertile materials. In nature there are two types of fertile materials, natural uranium (uranium-238) and thorium (thorium-232). The first is used to produce the fissile isotopes uranium-235 and plutonium-239. The second is used to obtain the fissile uranium isotope uranium-233. It is also possible to reprocess plutonium to obtain another type of fissile plutonium isotope, plutonium-241.

In 1789, the German Martin H. Klaproth, professor of chemistry at the University of Berlin, isolated natural uranium in the form of uranium dioxide (UO_2). Klaproth succeeded in extracting the mineral from a sample of pitchblende in the Joachimsthal ores (Czechoslovakia). This was named uranium in honor of William Herschel's discovery of the planet Uranus eight years before. While Henri Bequerel's experiments with uranium salts in 1896 demonstrated that uranium was naturally radioactive, the fissile properties of uranium were not recognized until 1938. The German physical chemists Otto Hahn and Fritz Strassmann demonstrated that neutron-bombarded uranium produced minimal quantities of lighter elements. Hence, uranium nuclei could undergo fission with the liberation of neutrons, radiation, and energy. In certain conditions, fission-produced neutrons could fission other uranium nuclei and thereby activate a self-sustaining reaction. But in February 1939, a joint paper by the Danish physicist Niels Bohr and his American colleague John Wheeler clarified that only a small fraction (1/139) of natural uranium (i.e., uranium-235), was responsible for the fission. The uranium-235 nucleus contains an even number of protons and an odd number of neutrons (even–odd nucleus). This asymmetry makes its binding energy smaller than the uranium-238 isotope (odd–odd nucleus). Thus Bohr and Wheeler concluded that uranium-238 captures neutrons, while uranium-235 undergoes fission.

From 1940 onward, several processes of separation were conceived in order to isolate the fissile uranium-235 from natural uranium and produce it at industrial scale. In 1940, the physical chemist Francis Simon designed the gaseous diffusion process at the University of Oxford in the U.K. Simon proposed to transform uranium oxide into a gaseous form, uranium hexafluoride (UF_6) and then pass it through a metallic membrane punctured with millions of microscopic holes. The lighter $^{235}UF_6$ would pass through the membranes with higher speed and so be isolated. The American physicists John Dunning and Eugene Booth of Columbia University adopted the process in 1941 and designed the first plant for separation based on the gaseous diffusion process. Construction of the plant began in 1943 at the Clinton Engineering Works in Tennessee in the U.S., and by the summer of 1945, it was successfully in operation.

Another process of enrichment was electromagnetic isotope separation, first analyzed by the American physicist Ernest O. Lawrence at the Radiation Laboratory of Berkeley. In the electromagnetic process, ions of gaseous uranium compound move in a circular ring due to the action of a strong magnetic field. Then the ions separate into two beams, the lighter uranium-235 ions following a narrower arc than the uranium-238 ions. In 1942, Lawrence designed the CALUTRON (California University cyclotron) that was eventually built in many models in the U.S. after 1943.

Finally, a third process of enrichment was the thermal separation of uranium, analyzed in detail by the American physicist Philip Abelson, who designed the first plant for thermal separation in 1943. In thermal separation the volatile uranium hexafluoride compound $^{235}UF_6$ is separated from the heavier UF_6 by heating.

Although these industrial methods for production of enriched uranium were aimed at the production of fissile material for atomic weapons, their concept was adopted by the nuclear industry in the 1950s when uranium became a fuel for commercial generation of electricity. The gaseous diffusion process proved by far the most successful, and 98 percent of enriched uranium is currently produced this way.

Plutonium does not exist in nature. It was artificially produced for the first time in January 1940 by bombarding uranyl nitrate hexahydrate (UNH_6) with an intense emitter of α-rays in a 9-inch (230-millimeter) cyclotron. The American chemist Glenn T. Seaborg at the Radiation Laboratory at Berkeley named the unknown chemical element plutonium after the planet Pluto, discovered in 1930.

The potentialities of plutonium (Pu) as a fissile material were thoroughly investigated between 1940 and 1943 and believed to be even higher than those of enriched uranium. Seaborg noted that plutonium could be obtained from natural uranium, which captures bombarded neutrons (^{239}U), decays into neptunium-239 (^{239}Np, 23.5 minutes), and then to plutonium-239 (2.3 days). In 1943, Seaborg also developed the chemical method of separating plutonium from irradiated uranium in which the uranium would be put in a chemical carrier and then undergo several processes of oxidation and reduction. The Clinton laboratories were the first research facility in which a new pile and a new chemical separation plant were built for the experimental production of plutonium from natural uranium. In 1944 a new complex of reactors and chemical extraction plants were erected in Hanford, Washington, by Pasco for the first large-scale production of plutonium. Similarly to uranium, plutonium was initially produced for military purposes. From the 1950s onward, plutonium was used also as fissile material in nuclear reactors for electricity production.

Thorium was first discovered and extracted as a metal by the Swedish chemist Jöns J. Berzelius (1828), who named it after Thor, the Nordic God of thunder. The main source of thorium is the mineral monazite. Thorium's fissile properties were dismissed by the 1939 Bohr–Wheeler paper, but in December 1940 the Swiss physical chemist Egon Bretscher predicted that thorium would produce the fissile isotope uranium-233. In 1942, Seaborg observed that thorium-233, similarly to natural uranium, captures neutrons (^{234}Th) and decays to protactinium-233 (^{233}Pa, 22.2 minutes) to uranium-233 (27.4 days). Uranium-233 is a less valuable fissile material compared touranium-235 and plutonium-239. This ruled out its use as nuclear bomb material in World War II, but the abundance of natural thorium compared to natural uranium fostered its use in nuclear reactors for peaceful purposes beginning in the 1950s.

See also **Fission and Fusion Bombs; Particle Accelerators: Cyclotrons, Synchrotrons, and Colliders; Nuclear Reactors**

SIMONE TURCHETTI

Further Reading

Finch, W.I. *Uranium, Its Impact on the National and Global Energy Mix and Its History, Distribution, Production, Nuclear Fuel-Cycle, Future, Relation to the Environment*. U.S. Geological Survey, Washington, 1997.

Gowing, M. *Britain and Atomic Energy, 1939–1945*. Macmillan, London, 1964.

Kathren, R.L., Gough J.B. and Benefiel G.T., Eds. *The Plutonium Story, The Journals of Professor Glenn T. Seaborg, 1939–1946*. Battelle Press, Columbus, 1994.

Rhodes, R. *The Making of the Atomic Bomb*. Penguin, New York, 1986.

Smith, E., Fox, A.H., Sawyer, T. and Austin, H.R. *Applied Atomic Power*. Blackie, London, 1946.

Ursu, I. *Physics and Technology of Nuclear Materials*. Pergamon Press, Oxford, 1985.

Nuclear Magnetic Resonance (NMR) and Magnetic Resonance Imaging (MRI)

The phenomenon of nuclear magnetic resonance (NMR) was discovered independently in 1946 by Felix Bloch at Stanford University, and by Edward Purcell at Harvard University. For this they were jointly awarded the 1952 Nobel Prize in Physics. Chemists were quick to see the potential of NMR, as the NMR signal gives valuable information about the structure of molecules. This is because the atomic electrons, which determine the chemical properties of materials, interact with the nuclei which give rise to the NMR signals. Much later, in 1973, Paul C. Lauterbur outlined how NMR could be used to form images for medical diagnosis. Raymond Damadian's 1971 finding that cancerous tissue could have different NMR properties from normal tissue had prompted his work. This eventually led to a major new application of NMR in medical imaging, which we now call magnetic resonance imaging (MRI). Although Damadian and Lauterbur were both working in the U.S., much of the pioneering work needed to turn the revolutionary idea into a practical reality was done in the U.K. in the 1970s and 1980s, at the Universities of Nottingham and Aberdeen and at the EMI Company's London research laboratories. In 2003, the Nobel Prize in Medicine was awarded to Lauterbur, University of Illinois at Urbana, and Sir Peter Mansfield, University of Nottingham, for their discoveries, emphasizing the diagnostic importance and widespread use of NMR.

Atomic nuclei are positively charged, and some (but not all) have the quantum mechanical property termed spin. If an object is both charged and spinning, it will generate a magnetic field in the same way that a current circulating round a loop will generate a magnetic field. Thus, a nucleus that has spin can be thought of as being a tiny bar magnet. Normally, nuclear spins do not have any preferred direction of alignment. However, if they are placed in a strong magnetic field they will tend to align with it, in much the same way as a set of compass needles align with the Earth's magnetic field. The alignment brought about by the strong magnetic field produces an observable nuclear magnetism. Until the nuclei are placed in the strong magnetic field their magnetic dipoles are randomly oriented and their average effect is zero, so that the phenomenon of NMR cannot occur. In fact, even in a very strong magnetic field their alignment is only weak, as the nuclear magnetic moments are randomly disturbed by thermal agitation as if they were compass needles being violently shaken.

The gyroscope, or spinning top, is a good analogy for the behavior of the nuclear magnetization in a strong magnetic field. If the spinning top is perturbed from its initial alignment with the Earth's gravitational field while supported by a table top, then it precesses around a vertical axis through its point of contact with the table. The precession is a much slower motion (in terms of revolutions per second) than the spin of the top around its own axis. Similarly, if the nuclear magnetic dipoles are perturbed from their alignment with the magnetic field, they precess around it, as shown in Figure 5. The precession frequency f_0 is often referred to as the Larmor frequency (after Joseph Larmor, an Irish physicist who investigated the behavior of electrons in magnetic fields at the end of the nineteenth century) and is given by the equation $f_0 = \gamma B_0$. Different nuclei have different values of γ. Protons have a gyromagnetic ratio of 42.6 megahertz per tesla, while the figure for sodium is 11.3 megahertz per tesla. Thus protons (hydrogen nuclei) precess at 42.6 megahertz in a magnetic field of strength 1 tesla, while sodium nuclei (specifically of the sodium isotope with atomic weight 23) precess at only 11.3 megahertz. It is possible to determine the frequency of a signal very accurately, by comparing it to a stable reference signal of known frequency. Thus NMR can be used to make a highly precise magnetometer—a device for measuring magnetic field strengths very accurately.

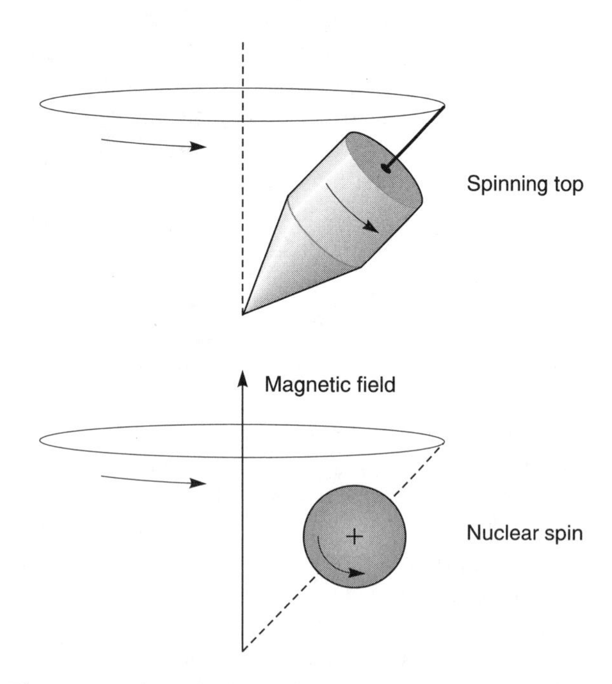

Figure 5. The spinning top is a good analogy to the behavior of the nuclear magnetism in a strong magnetic field. Once the top is pushed away from the vertical, it precesses around the gravitational field of the Earth. Similarly, once the spinning nuclei are pushed away from their alignment with the magnetic field, they also precess. The rate of precession is proportional to the strength of the magnetic field.

Whereas the spinning top can be pushed from the vertical by a tap of the finger, the nuclear spins have to be pushed by an oscillating magnetic field applied at right angles to the main magnetic field. The oscillation frequency has to equal the Larmor precession frequency f_0, or nothing happens. This is what is meant by resonance. The field strength used in a typical NMR or MRI machine is a few tesla, so that the Larmor frequency is in the radio-frequency (RF) range. The RF magnetic field is applied by means of a tuned RF coil surrounding the patient's body or head, with power supplied by a radio-frequency power amplifier.

The first NMR experiments in the late 1940s used the "continuous wave" technique, in which the frequency of the RF field is steadily increased (or decreased) while passing through resonance. In 1950 Erwin Hahn proposed the "pulsed" method of NMR in which the entire frequency response is obtained following a short powerful burst of transmitted RF energy called an RF magnetic field pulse. The difference in methods can be understood by considering two possible methods of testing a church bell. In the analogy for the continuous wave method, a loudspeaker is used to produce a pure note, the frequency of which is steadily increased. When the natural tone of the bell is reached, the bell will begin to vibrate in sympathy with the applied sound. The pulsed method is faster and more direct, and can be likened to striking the bell, and listening to its note as the sound dies away. The note contains a mixture of all the natural frequencies of the bell. Once the nuclei are precessing, their magnetic fields are also precessing, so that a tiny voltage is induced in the RF coil by the principle of electromagnetic induction. The frequency of this oscillating voltage equals the precession frequency. If a number of different frequencies are present in the signal, because the nuclei are in a number of different electronic environments, then the different frequency components have to be separated by a mathematical process called Fourier transformation, carried out by a computer.

The use of NMR to investigate the atomic and molecular structure of materials is called NMR spectroscopy. During the 1950s and 1960s, NMR spectroscopy became a widely used technique for the nondestructive analysis of small samples, particularly of liquids. The electron cloud surrounding the nuclei within individual molecules modifies the strength of the magnetic fields sensed by the NMR sensitive nuclei, and hence changes the frequency of the NMR signal that the nuclei emit. In addition, neighboring nuclei can also influence each other. Pulses of RF energy are used to perturb the NMR sensitive nuclei, and sensitive RF receivers are used to pick up the signals they give out. The pulsed method allied with computerized Fourier transformation revolutionized NMR spectroscopy in the 1970s. The "spectrum" is a plot of signal strength versus NMR frequency and contains much useful information about the chemical structure of the material under testing. NMR spectroscopy is now widely used in the fields of biomaterials, polymer chemistry, and solid-state physics.

Pulsed NMR techniques can also be used to form medical images. MRI normally uses signals arising from hydrogen nuclei, because they are so much more abundant in the body than any other NMR-sensitive nucleus and therefore give a measurable signal even from small volumes of tissue. However, phosphorus and sodium MRI is also possible. Hydrogen nuclei that are MRI visible occur predominantly in water in the tissues and in body fat. Figure 6 shows a patient lying on a table, about to be moved into an MRI scanner, with the head resting inside the RF head coil. The electromagnet has a large tunnel in which the patient lies and uses a special wire immersed in liquid helium so that the wire superconducts. Thus the magnet does not need any electrical power to generate the field. To form an image, it is essential to have a method of determining the position of the nuclear spins within the magnet. This is accomplished by field gradients, a method proposed by Paul Lauterbur of the State University of New York at Stony Brook in 1973. A field gradient coil modifies the strength of the main magnetic field along a particular direction so that it varies in a linear way. Three independent field gradient coils are used in order to generate the gradients in x, y, or z directions (along the magnet tunnel, left to right across the tunnel, and vertically). When a field gradient is on, therefore, the Larmor frequency will also vary in a linear fashion along the direction of the field gradient.

The first step in imaging is to tip the nuclear spins away from their alignment along the main magnetic field using a short-pulsed RF magnetic field oscillating at the Larmor frequency, as explained above. If the pulse is applied in the presence of a field gradient along the patient (along z), then only one transverse plane across the body will respond, as only one plane has a Larmor frequency that matches the RF frequency. The imaging process is now simplified to imaging a two-dimensional slice in an x–y plane, rather than imaging the entire three-dimensional body. The gradient along the patient is switched off, and a gradient switched on across the body (along x). This makes the nuclei precess faster on one side of the body, and more slowly on the other. Thus the frequency of the NMR signals varies in a linear way across the body, so that a particular NMR signal frequency corresponds to a particular left to

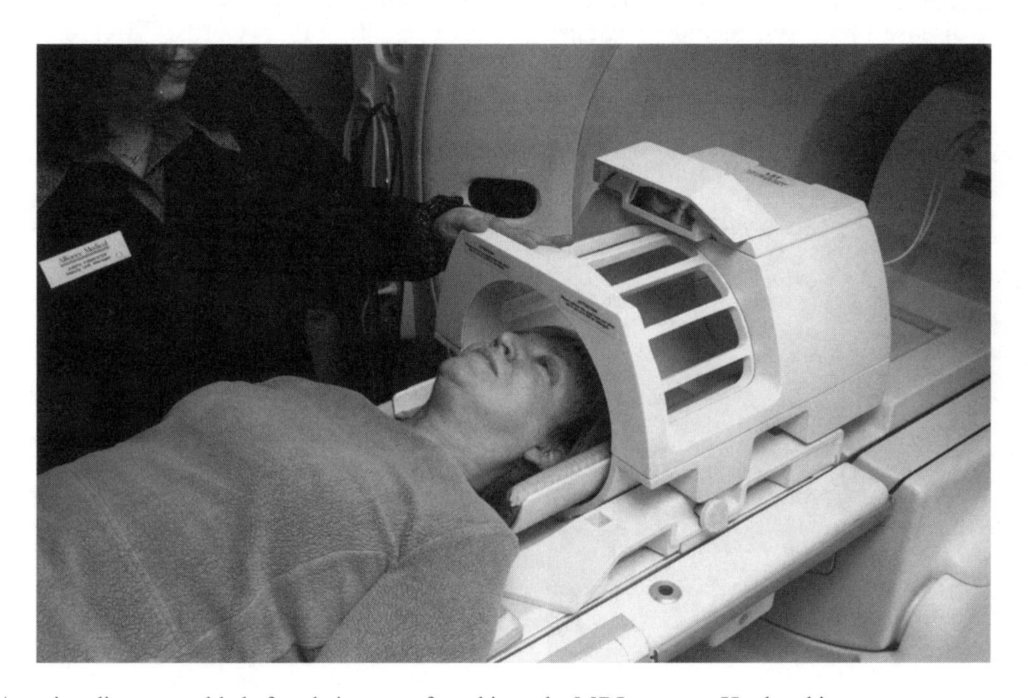

Figure 6. A patient lies on a table before being transferred into the MRI scanner. Her head is being imaged and has to be placed inside the radio-frequency head coil beforehand.

right (x) position. In order to form the image, the vertical (y) position also has to be encoded onto the NMR signal. Although a full discussion is impossible here, the technique involves repeating the above process many times, with a small pulse applied to the vertical y gradient just before the signal is collected and with the amplitude of the pulse being stepped up between each repetition. This is the "spin-warp" imaging method, invented by William Edelstein and James Hutchison of Aberdeen University in 1980. It is still the most widely used imaging method today. Modern MRI scanners rely on fast computers to carry out the large number of digital Fourier transforms necessary to form the images and to control very precisely the timing of the pulsed currents flowing in the gradient and RF coils.

Figure 7 shows a midline slice through the center of a human head on an MRI image. Note the excellent anatomical detail achieved. Areas with no protons, such as the air-filled sinuses behind the nose, give no signal and appear dark. Dense hard bone in the skull also has few protons and gives no signal. Watery fluid such as the cerebrospinal fluid around the spinal cord gives a different signal (brighter in this case) from that of the soft tissue of the brain because the NMR properties of the water protons are affected by their biophysical environment. In other words, image detail arises from tissue structure on a microscopic level, as well as from differences in proton density. This is the particular strength of MRI in medical diagnosis because many disease processes, including cancer and multiple sclerosis, and conditions such as infection and hemorrhage alter tissue structure and can therefore be visualized.

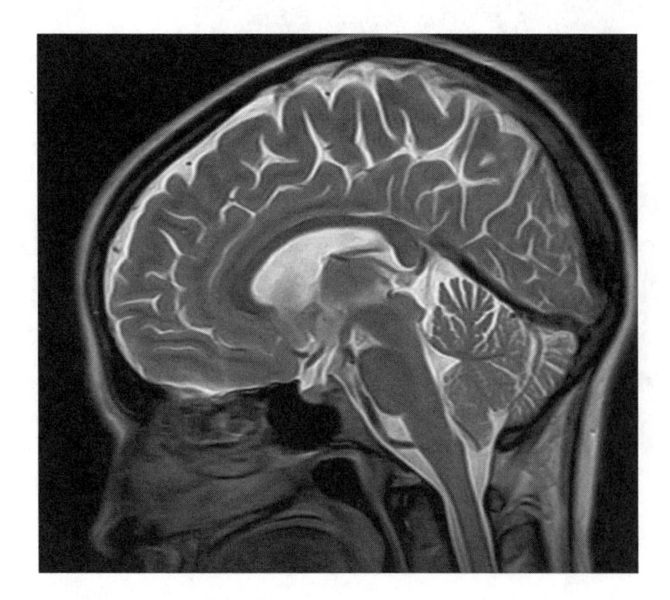

Figure 7. A section of the brain imaged by MRI. Note the excellent anatomical detail the scan gives.

A recent development in MRI is its use to repeatedly image the brain every 2 or 3 seconds as it performs different tasks or responds to various stimuli, a technique called functional MRI (fMRI), developed independently by Seiji Ogawa of AT&T Bell Laboratories in New Jersey and John Belliveau at Massachusetts General Hospital in Boston in the early 1990s. This requires a high-performance scanner capable of performing the very high-speed imaging technique of "echo-planar" imaging, invented by Peter Mansfield at Nottingham University in 1977. The fMRI technique relies on the oxygen-dependent magnetic effect of the iron atoms contained in the hemoglobin molecules of the blood. When part of the brain is active, its oxygen consumption increases, thus stimulating a large increase in the local blood supply, and the oxygen concentration in the tiny blood-filled capillary vessels is raised. For the particular rapid-imaging method used in fMRI, this causes a small increase in the signal detected. In its simplest form, fMRI alternates periods of rest, lasting 30 seconds or so, with equal periods of activation. Relatively large signal changes of a few percent can be obtained with visual activation using flashing lights or rapidly alternating checkerboard patterns (black squares turning to white and vice-versa), while more subtle cognitive tasks involving memory or reasoning give smaller changes. In other words, MRI can image your thoughts, an application never dreamt of by Felix Bloch and Edward Purcell in 1946.

See also **Cardiovascular Disease, Diagnostic Methods; Neurology**

THOMAS W. REDPATH

Further Reading

Callaghan, P.T. *Principles of Nuclear Magnetic Resonance Microscopy*. Oxford University Press, Oxford, 1991.

Farrar, T.C. and Becker, E.D. *Pulse and Fourier Transform NMR*. Academic Press, Orlando, 1971.

Gadian, D. *NMR and its Applications to Living Systems*, 2nd edn. Oxford University Press, Oxford, 1995.

Hore P.J. *Nuclear Magnetic Resonance*. Oxford University Press, Oxford, 1995.

Redpath, T.W. MRI developments in perspective (review). *Br. J Radiol.*, 70, s70–80, 1997.

Rinck, P.A. *Magnetic Resonance in Medicine*. Blackwell Wissenschafts-Verlag, Berlin, 2001.

Stark, D.D. and Bradley, W.G. *Magnetic Resonance Imaging*, 3rd edn. Mosby, St. Louis, 1999.

Nuclear Reactor Materials

Over the latter half of the twentieth century, several materials were used by the nuclear industry in the construction and functioning of nuclear reactors. Those materials have been divided into the following categories: moderators, structural materials, coolants, and shielding materials.

The moderator is used in nuclear reactors to slow neutrons down to fission energy. Slowed down or "thermal" neutrons have a far higher probability to cause nuclear fission than fast neutrons; the use of a moderator is therefore essential in nuclear reactors that exploit thermal neutrons (thermal reactors). Moderation is obtained through repeated elastic collisions of the neutrons with the moderator's nuclei and the transfer of kinetic energy from the first to the second.

Despite initial attempts to use light water as a moderator in early experiments with nuclear piles, graphite appeared far more reliable and constituted the main type of moderator for a long time. The use of graphite is widespread. Its manufacturing process was first outlined in 1896 by the American industrialist Edward G. Acheson. He used petroleum coke and coal-tar pitch in a large furnace capable of reaching temperatures of 3000°C. When in 1942 the Italian physicist Enrico Fermi designed the first experimental nuclear reactor, CP-1, at the University of Chicago, he was advised by his Hungarian colleague Leo Szilard to try using graphite as a moderator. In December 1942, Fermi succeeded in reaching criticality with the graphite pile CP-I. This success influenced future designs of nuclear piles, although the low moderating power of graphite implied the design of big power plants with large cores and therefore deployment of large quantities of structural materials. Graphite is still in use today in nuclear reactors of the graphite core reactor (GCR) type.

In 1943, graphite's properties were thoroughly examined by the theoretical physicist Eugene Wigner at the University of Chicago. He argued that radioactivity heavily affects graphite and modifies its strength and thermal conductivity. Eventually several scientists and industrialists within the nuclear establishment considered the use of other types of moderator such as heavy water (D_2O) and light water (H_2O). Heavy water was produced for the first time in 1934 by the British Imperial Chemical Industries (ICI) through a process of water electrolysis. Previously, the American physical chemist Harold Urey had isolated the isotope deuterium (^2H or D). The first experimental heavy water pile CP-3 was built in 1944 at the U.S. Argonne National Laboratory. It consisted of a tank filled with 6.5 tons of heavy water containing 121 rods of uranium. Although the Argonne CP-3 was the first heavy water reactor, the technology associated with this type of moderator was mainly developed in Canada. From 1943 British, Canadian, and French scientists gathered in Montreal as members of the Anglo–Canadian project code-named "Tube Alloys." Their purpose was to build a heterogeneous pile in which uranium rods were immersed in heavy water. In 1945 they built the prototype zero energy experimental pile (ZEEP) at the new Chalk River Nuclear Laboratories (Canada), and in 1947 the heavy-water nuclear reactor X (NRX) was brought into operation. Heavy water is still used in the Canadian-designed commercial reactors of the CANDU type. Heavy water reactors have a larger output compared with graphite reactors, and they have reduced dimensions. Their main drawback lies in the production of heavy water, which is still a very expensive and cumbersome process.

After being initially dismissed as a moderator, light water was reconsidered because its moderating power is higher than both graphite and heavy water. Light water was used in reactors of the light water reactor (LWR) type from the 1950s onward.

Structural materials are used to make the core and several other essential parts of a nuclear reactor. Thus, they must possess a high resistance to mechanical stress, stability to radiation and high temperature, and low neutron absorption. Materials used for structural purposes are aluminum, beryllium, carbon (graphite), chromium, iron, magnesium, nickel, vanadium and zirconium. Stainless steels, alloys of iron, chromium, nickel, and other metals, are by far the most used structural materials since the beginning of the nuclear industry.

In the nineteenth century, the experiments conducted by the British natural philosopher Michael Faraday with iron and nitric acid had shown that iron alloys had a very high corrosion resistance (or passive state). The characteristics of iron alloys were thoroughly examined at the beginning of the twentieth century. In 1911, P. Monnartz investigated and classified the Fe–Cr (ferritic) alloys. He also developed a number of methods (e.g., introduction of molybdenum) to enhance their passive state. In 1912, the Germans Eduard Maurer and Benno Strauss, chemists of the Krupp research laboratory, patented a method for the treatment of Fe–Cr–Ni alloys. From 1914 those alloys became the main structural materials for chemical plants, such as for ammonia or nitric acid production. In the same year, the English metallurgist Henry Brearley used iron alloys to make "rustless" cutlery, hence the name "stainless" steels. In 1931, Brearley synthesized a new type of stainless steel—the austenitic—using chromium and nickel. From the end of World War II, several new stainless steels were designed, introducing new elements such as titanium, silicon, and nitrogen. Since then, stainless steels were applied as structural materials in the petrochemical industry, thus their use was suggested for the nuclear industry as well. Stainless steels are the main off-core structural materials for thermal reactors and are essential materials for fast neutron reactors. Their main drawback is that they absorb neutrons, which limits their in-core applications. Carbon and ferritic steels are used for nuclear pressure vessels in LWR and heavy water reactors (HWR). Lightly alloyed steels are used in PWR reactors. Ferritic steels were also used in the commercial reactors of the CANDU–PHRW type for the manufacture of pressure tubes and fittings.

Coolants are used in nuclear reactors to evacuate the thermal energy produced from the core. Materials used for cooling purposes are water (pressurized or boiling), heavy water, liquid metals (mainly sodium), or gases (carbon dioxide, helium, or nitrogen oxide). The process of energy transfer may imply a phase transition in the coolant (i.e., water storing heat through vaporization). The containment of radiation and heat within nuclear power plants is obtained through the shielding materials. Nuclear reactors are provided with an internal thermal screen made of water or heavy water and an external biological screen made of air, water, or concrete. In fast reactors, stainless steels are used for a thermal screen while sodium, iron, or boron is used as the biological screen.

See also **Nuclear Fuels; Nuclear Reactors, Thermal, Graphite-Moderated; Nuclear Reactors, Thermal, Water-Moderated; Nuclear Reactors, Fast Breeders**

SIMONE TURCHETTI

Further Reading

Dahl, P. F., *Heavy Water and the Wartime Race for Nuclear Energy*. Institute of Physics Publishing, Bristol, 1999.

Fleming, R.P.H. Design and development, structure and properties of some new stainless steels for high-temperature use, in *Mechanical Behaviour and Nuclear Applications of Stainless Steel at Elevated Temperatures: Proceeding of the International Conference, Varese, Italy, 1981*. Metals Society, London, 1982.

Gowing, M. *Britain and Atomic Energy, 1939–1945*. Macmillan, London, 1964.

Nightingale, R. E. *Nuclear Graphite*. Academic Press, New York, 1962.

Rhodes, R. *The Making of the Atomic Bomb*. Penguin, New York, 1986.

Streicher, M. A. Stainless steels: past, present and future, in *The Metallurgical Evolution of Stainless Steels*, Pickering, F.B., Ed. Metals Society, London, 1979.

Ursu, I. *Physics and Technology of Nuclear Materials*. Pergamon Press, Oxford, 1985.

Nuclear Reactors: Fast Breeders

The idea of a fast breeder reactor (FBR) was first conceived in 1946 by the Canadian physicist Walter H. Zinn at the Argonne National Laboratory in the U.S. On the basis of wartime developments in nuclear reactor research, Zinn thought a combination of two options within reactor technologies was feasible: fast neutron nuclear fission and the breeding principle. Fast reactors produce nuclear fission with fast neutrons rather than thermal neutrons. Fast neutrons prompt critical reactions with a large energy release in a short time and without a moderator operating in the core. Breeder reactors have a core of fissile material (i.e., uranium-235 or plutonium-239) produced in nuclear reactors or through chemical separation, and a blanket of fertile material (i.e., uranium-238), the treated radioactive mineral. Once in operation, breeder reactors incinerate the fissile material in the core and emit neutrons as fission products. Thus the blanket is neutron bombarded, the fertile material is irradiated, and afterward transformed into fissile material (i.e., plutonium-239) by neutron capture and following decay. In optimal operation conditions, the fissile material produced through breeding equals the fissile material incinerated in the core, so that the reactor perpetuates indefinitely the production of its fuel.

In 1948, Zinn designed a FBR provided with a good coolant, a liquid alloy of sodium and potassium (NaK), to promptly transfer the enormous heat produced in the core. This project soon received priority within the U.S. Atomic Energy Commission (AEC). The construction of EBR-I (experimental breeder reactor) started in June 1949 at the Idaho reactor research station and was completed in 1951. In December 1951 the reactor went critical, and EBR-I was the first nuclear reactor capable of producing electricity at the initial rate of 45 megawatts. In 1952, EBR-I proved to be a breeder reactor as part of the natural uranium in the blanket was irradiated and then transformed into plutonium. In the same year, a leak in the nuclear reactor heat exchanger caused its temporary interruption. The leak highlighted the fact that the corrosive nature of sodium would be a major concern for FBR technology. In 1955 a combination of causes (operator error and a temperature effect) caused the meltdown of the EBR-I core. The core was eventually replaced, but the reactor was shut down in 1963. In that year, the Enrico Fermi I—a full-scale FBR cooled with liquid sodium—started operations in Lagoona Beach, Michigan, in the U.S. In 1966 during a normal test, the instrumentation registered an abnormal temperature in the core and an erratic rate of change in the neutron population. After the plant was shut down, it was discovered that two of the fuel assemblies had melted. Enrico Fermi I reached full power again only in 1970, and it was shut down afterwards. The two nuclear accidents, the fourth level of the International Atomic Energy Administration (IAEA) scale of nuclear events, represented a major drawback for the FBR American program.

The U.S. slowed down investments in FBR technology in 1970. However, FBRs were still considered very reliable in Europe, Japan, and the former Soviet Union, where they were regarded as the "nuclear system of the future." From 1969, the first French prototype Rapsodie (20 megawatts) was built by the Commissariat à l'Energie Atomique (CEA) at the Saclay research laboratory. In 1974, the FBR Phénix was built at Avignon (250 megawatts), and in 1984, the full-scale FBR Super Phénix was built (1200 megawatts) at Malville on the Rhone River. From the 1970s, the Soviet Union heavily invested in FBR technologies with two experimental FBRs (15 megawatts); an intermediate FBR (BN350, 150 megawatts, 1973); and a full-scale FBR (BN600, 600 megawatts). In 1977 the Dounreay PFR (270 megawatts) was the first FBR operating in Britain. In Germany in 1978 the prototype KNK Karlsrhue (18 megawatts) was operative, while a new FBR, the SNR 300 was built in 1987 (327 megawatts).

However, in thirty years the opinion of experts and the public on FBR technology changed as FBR did not prove to be efficient nuclear systems. The Soviet FBRs suffered serious sodium leaks and raised concern among nuclear reactor experts. The French Super Phénix operated at full power for just six months, and in 1997 its closure was announced as part of the new French government's plans. The British Dounreay PFR was shut down in 1994 after several small accidents occurred, and Britain generally opted for investing in FBR development at the European rather than British level. In Europe the development of a common program for a European Fast Reactor (EFR) capable of producing 1450 megawatts was evaluated but never fully accomplished.

The development of FBR technology was seriously evaluated after several accidents occurred in existing prototypes and commercial reactors. Those accidents have also slowed down the process of commercialization of FBR technology. The critics of these types of reactors point out that liquid sodium is a highly corrosive chemical compound that causes leaks, tube vibrations, and flow instabilities. Sodium also reacts with air and water and interacts with the fuel in emergency conditions. This is considered a major concern as the explosive nature of the interaction between fuel and coolant can lead to uncontrolled critical conditions in the rods. Moreover, technical problems that do not emerge at the level of prototypical FBRs affect full-scale commercial reactors because sodium instability increases when larger quantities of coolant are deployed. This implies corrections to original designs and escalating costs. For example, the French Super Phénix needed an expenditure of six times its original estimated cost to be completed.

Supporters of FBR technology argue that there is no reason to be concerned about liquid sodium's properties and behavior in nuclear reactors. There are no fundamental engineering problems connected with the construction and operation of FBRs. The main problem is simply the balance between incinerated fissile material and fissile material to replace it. In other words, FBRs incinerate fairly well but breed poorly, making the incineration/breeding balance very difficult to obtain. Among several positive aspects, FBR supporters stress that the breeding principle helps in confining radioactive waste. FBRs allow treatment of radioactive fissile material in the reactor establishment within 12 months rather than the transportation of radioactive material in and out from the reactor and its treatment in five years. Thus they provide a good way to confine nuclear fuel and radioactive materials. Furthermore, once the optimal incineration/breeding balance is obtained, the efficient incineration of plutonium is another favorable element of FBR technology as this helps to diminish the amount of plutonium, one of the most toxic chemical compounds in the world.

See also **Nuclear Reactors, Thermal Graphite-Moderated; Nuclear Reactors, Thermal Water-Moderated; Nuclear Fuels**

SIMONE TURCHETTI

Further Reading

Dombey, N. Fast reactors and problems in their development, in *The Fast Breeder Reactor. Need? Cost? Risk?* Sweet, C., Ed. Macmillan, London, 1980.

Hecht, G. *The Radiance of France, Nuclear Power and National Identity after WW2*. MIT Press, Cambridge, MA, 1998.

Hewitt, G.F. and Collier, J.G. *Introduction to Nuclear Power*. Taylor & Francis, New York, 2000.

Hewlett, R. and Duncan, F. *Atomic Shield: A History of the United States Atomic Energy Commission, vol. II, 1947–1952*. University of California Press, Berkeley, 1990.

Marshall, W. The UK fast breeder programme, in *The Fast Breeder Reactor. Need? Cost? Risk?* Sweet, C., Ed. Macmillan, London, 1980.

Mounfield, P. *World Nuclear Power*. Routledge, London, 1991.

Nuclear Reactors: Fusion, Early Designs

The production of nuclear energy through the fusion of two light chemical elements is better known as a controlled thermonuclear reaction (CTR). In the 1950s, explosive or uncontrolled thermonuclear reaction was achieved with the manufacture of hydrogen bombs, but CTR was never successfully accomplished. Among the several types of reactions considered, the following CTR has been attempted in fusion reactors:

$$^{2}H + {}^{3}H \rightarrow {}^{4}He + n + 5{,}2 \times 10^{-13}\ J$$

In the above reaction, deuterium (^{2}H or D) and tritium (^{3}H or T) fuse to release helium (He), neutrons, and energy (calculated in joules, J). Deuterium is contained in a small percentage in water, while tritium can be artificially produced from lithium by neutron bombardment in a breeding reactor.

In order to reach the fusion point, a gaseous mixture containing deuterium and tritium should be heated to 100,000,000°C and hold that temperature for enough time to activate a self-sustaining reaction. At elevated temperatures, a gaseous mixture becomes plasma, a state in which electrons and ions are no longer physically bonded. (The term plasma was first used in 1922 by the American physical chemist Irving Langmuir because the properties of a super-heated gas reminded him of blood plasma.)

Heating and confinement of plasma are the two main features of any fusion reactor. Plasma must avoid any contact with the walls of the vessel containing it in order to avoid the loss of temperature and subsequent instability that makes a controlled thermonuclear reaction impossible to achieve. Early designs of fusion reactors focused on confinement of plasma using magnetic fields.

In 1951, the American astrophysicist Lyman Spitzer at Princeton University designed the Stellarator, so called because thermonuclear reactions occur in stars and the device was aimed at reproducing the energy released when such reactions occur. The Stellarator was designed in three models—A, B, and C—between 1953 and 1961. Spitzer argued that "To keep the ions from hitting the wall, some type of force is required that will act at a distance. A magnetic field seems to offer the only promise." He designed a closed tube in the shape of a circular ring (torus) surrounded by external magnetic coils. In the device, the plasma was heated by electric current or radio waves and then introduced into the torus in vacuum conditions. Ideally, the coils should hold the plasma and move it in the ring for enough time to generate a CTR. The 1961 Stellarator C was the first prototype of a fusion reactor. This device reached temperatures of thousands of degrees Celsius and held the plasma for a few milliseconds, but it did not achieve CTR.

After 1952, another type of design was considered for fusion reactors: the pinch effect machine. Exploiting an idea originally developed in 1946 by his British colleagues George P. Thomson and Moses Blackman, the British physicist James Tuck devised the Perhapstron (to ridicule the Stellarator's "grandeur") at the U.S. Los Alamos National Laboratory. Perhapstron was a pinch effect machine in which an electric current is discharged in the plasma. In this way the current shapes a circular magnetic field that holds the plasma and pinches it into a thin filament. The problem of Tuck's device was the plasma's instability; that is, its capacity to hold the plasma by magnetic means. In 1955 Edward Teller addressed plasma instability as a major concern for CTR research. For this reason in 1957 a new Los Alamos pinch machine, Scylla, was designed. This was an "azimuthal" machine in which longitudinal wires around the pinch tube excited longitudinal currents in the plasma.

In August 1957 another experimental pinch prototype, the Zero Energy Thermonuclear Assembly (ZETA) started operation at the Harwell Atomic Research Establishment in Britain. This was the largest stabilized pinch machine ever built, and it was initially believed that CTR was finally achieved as neutrons were detected pouring out of ZETA. However, more accurate studies eventually clarified that the detected neutrons were not fusion products.

In the same year, Edward Gartner, leader of the Fusion Group at the U.S. Oak Ridge National Laboratory, established that a beam of ions trapped in a carbon arc and eventually sent into a pinch machine would enhance plasma stability. On this premise, two fusion reactor prototypes named Direct Current X (DCX) were successfully developed in the early 1960s. On the whole, pinch machines provided useful experimental data on CTR, but they never achieved fusion.

If pinch machines and stellarators focused on toroidal structures, the magnetic mirror machines were designed to attempt plasma confinement in straight tubes. In 1952, the American physicist Herbert York, head of the project for the new U.S. Lawrence Livermore National Laboratory, directed a research program for building a fusion device based on a linear structure. A young American PhD graduate, Richard F. Post, designed TableTop, a device in which the coils surrounding the straight tube would produce different magnetic fields. In that way the plasma would travel from the center to the ends of the tube and then back again, creating a mirroring, resonance effect. In 1954, the American Frederic H. Coengsen designed ToyTop, a machine to test plasma heating by magnetic compression. In 1960, ToyTop and TableTop were combined in the design of the large-scale mirror machine, ALICE (adiabatic low-energy injection and confinement experiment). The major drawback for mirror machines was "flute instability," the tendency of the plasma to avoid confinement in the straight tube. In order to avoid flute instability, the Russian physicist Abraham Ioffe proposed introducing magnetic wells at the ends of the tube. In 1957 another Russian physicist, Len Andreevich Artsimovich, designed OGRA, the first Soviet mirror machine manufactured at the Institute of Physics of Moscow and fitted with Ioffe's magnetic wells. Meanwhile, the British physicist Stephen Pease at the Culham Laboratory designed magnetic wells shaped as a tennis ball seam. The wells were eventually attached to the second Los Alamos ALICE prototype built in 1966.

During the 1960s, many fusion reactor projects went under critical review. This was largely because of the lack of success in harnessing fusion energy. Several projects were therefore withdrawn, such as the Livermore mirror machine Astron, designed by the Greek-born American physicist Nicholas C. Christofilos. In 1968, Artsimovich presented a new type of machine—the tokamak—whose innovative design would restore hope (and funding) in fusion reactors research.

See also **Fission and Fusion Bombs; Nuclear Reactors: Fast Breeders; Nuclear Reactors: Fusion, Later Designs (Tokamak)**

Simone Turchetti

Further Reading

Bromberg, J.L. *Fusion. Science, Politics and the Invention of a New Energy Source.* MIT Press, Cambridge, MA, 1982.

Hendry, J. and Lawson, J. *Fusion Research in the UK, 1945–1960.* AEA Technology, Harwell, 1993.

Herman, R. Fusion, *The Search for Endless Energy.* Cambridge University Press, Cambridge, 1990.

Hewitt, G. and Collier, J. *Introduction to Nuclear Power.* Taylor & Francis, New York, 2000.

Mounfield, P.R. *World Nuclear Power.* Routledge, London, 1991.

Stacey, W.M. Fusion: *An Introduction to the Physics and Technology of Magnetic Confinement Fusion.* Wiley, New York, 1981.

Nuclear Reactors: Fusion, Later Designs (Tokamak)

The tokamak reactor design aims to replicate in a terrestrial environment the conditions for fusion that exist in stars. There, the forces of high temperature and compression that prevent electrons from adhering to hydrogen nuclei and fusing the positively charged atoms (ions) into helium ions (a process that releases massive energy), are balanced by immense gravitational forces. In simulating these conditions, scientists use magnetic fields to stabilize, confine, and suspend ionized fuel (plasma) in a reactor vessel. This prevents the plasma from contacting the vessel walls, which would cool, and thus terminate, the reaction.

In the 1940s and 1950s, scientists developed two methods of magnetic plasma confinement. A magnetic field could be created within the plasma itself by running a current through it, producing a "pinch" effect. While initial "pinch" devices were tube-like, it was discovered that the plasma column cooled when it touched the ends of the cylinder. This led researchers to adopt the toroidal (hollow doughnut) reactor, within which the plasma formed a floating ring. The American physicist Lyman Spitzer fitted an electromagnet to such a reactor (the "Stellerator"), producing a toroidal magnetic field.

In the early 1950s, the Soviet physicists Andrei Sakharov and Igor Tamm proposed a reactor that generated both internal plasma and external toroidal magnetic fields. This concept was adopted by their colleague Lev Artsimovich in his T-3 reactor, the first "tokamak" (the Russian acronym for toroidal chamber and magnetic coil), unveiled in 1968. The tokamak magnetic field is thus the combination of two magnetic fields: the stronger horizontal, toroidal field interacts with the weaker vertical, poloidal plasma field to produce a helical magnetic field. In confining its plasma for 0.01 to 0.02 seconds and heating it to 10,000,000°C, the T-3 produced results that suggested fusion energy was feasible.

The tokamak reactor subsequently became the standard tool for fusion research. The energy crisis of the 1970s resulted in state support for major projects in a number of industrialized countries including France, Japan, the U.K., and the U.S. The largest and most notable were the American Tokamak Fusion Test Reactor (TFTR), approved by the Atomic Energy Commission in 1974 and completed in 1982 at Princeton University, and the British–European Joint European Torus (JET), which began operations in 1983 in Culham, Oxfordshire, U.K. Other important tokamaks include Japan's JT-60 and General Atomics' DIII-D.

Plasma heating in fusion reactors had previously been achieved via the resistance produced when electrical current or radio waves were run through the plasma. The latest tokamaks employed a new technology, the neutral beam injector, which augmented heating by accelerating particles into the plasma. The device converts moving ions (which in transit would be influenced by the magnetic field such that they could only shallowly penetrate the plasma) into neutral atoms. These particles are then shot deep into the plasma, where they collide with electrons and ions, reionize, and become confined within the magnetic field, producing heat in the process.

Scientists have used tokamaks as research tools to explore a number of fusion problems, including energy confinement and the behavior of neutrons and helium ions in the fusion fuel reaction. Scientists consider a 50:50 deuterium–tritium (DT) combination the optimal fuel for two reasons. First, it has a fusion temperature of 100,000,000°C, the lowest of all fuel configurations. Second, because the chances of high-speed particles colliding and fusing in the superheated plasma increase with larger quantities of neutrons in the fuel mix, deuterium by itself is inefficient as a fuel because its nuclei have only one neutron. Deuterium must be combined with tritium, the nuclei of which carry two neutrons, for sustained, efficient fusion to occur in a tokamak. Though most of the energy in a DT reaction is produced by the release of fast neutrons, these particles are not affected by the tokamak magnetic field and escape with their plasma-heating energy. However, helium ions, along with DT ions, are subject to magnetic attraction, and their prolonged confinement is indispensable for plasma heating and, ultimately, ignition, the point at which a fusion reaction becomes self-sustaining.

The course of fusion experimentation has been governed by economic and environmental conditions. During the energy crisis of the 1970s, U.S. scientists focused on DT fuel dynamics in approaching the problem of ignition. The easing of the crisis in the 1980s and concern over the use of mildly radioactive tritium impelled researchers to examine the energy confinement question, resulting in experiments employing only deuterium. Not until the early 1990s were tests staged using DT. The JET produced the world's first significant controlled fusion power in 1991, using a 90:10 DT mixture to generate nearly two megawatts. In 1993, the TFTR produced the world's first 50:50 DT reaction, yielding three megawatts. The TFTR and the JET have since achieved peak outputs of 10.7 and 16 megawatts, respectively. However, tokamak reactors currently consume much more power than they yield. During a typical experiment, the JET requires up to 500 megawatts to supply its transformer, toroidal/poloidal field coils, and various heating appliances.

The TFTR ceased operation in 1997 and was replaced at Princeton in 1999 by the National Spherical Torus Experiment (NSTX), a small tokamak designed to explore plasma physics using only deuterium. Today, the JET is the only tokamak equipped to operate with tritium. The legacy of the tokamak reactor is an increased understanding of fusion theory and experience in reactor operations such that the scientific community currently believes ignition and "break-even" operations (where power input equals power output) may be possible with the planned International Thermonuclear Experimental Reactor (ITER). If built, ITER would be the largest and most advanced tokamak, with a projected output of 410 megawatts. The U.S. joined the European Union, Japan, and the Soviet Union as a formal ITER partner in 1988, but American participation has since fluctuated. In 1998, Congress directed the Department of Energy to withdraw from the project, although the U.S. rejoined talks with international participants in early 2003. Although negotiations continue today over the siting of ITER, with France and Japan as the leading candidates, the future of the project is not certain.

See also **Nuclear Reactor Materials; Nuclear Reactors, Weapons Material; Nuclear Reactors: Fast Breeders; Nuclear Reactors: Fusion, Early Designs; Nuclear Reactors: Thermal, Graphite-Moderated; Nuclear Reactors: Thermal, Water-Moderated**

MATTHEW EISLER

Further Reading

Bromberg, J.L. *Fusion: Science, Politics, and the Invention of a New Energy Source*. MIT Press, Cambridge, MA, 1982.
Butler, D. France and Japan lock horns in battle to host research reactor. *Nature*, 426, 4 December, 483, 2003.
Feng, K.M., Zhang, G.S. and Deng, M.G. Transmutation of minor actinides in a spherical torus tokamak fusion reactor, FDTR. *Fusion Eng. Design*, 63/64, 1, 127–133, 2002.
Fowler, T.K. *The Fusion Quest*. The Johns Hopkins University Press, Baltimore, 1997.
Gross, R.A. *Fusion Energy*. Wiley, New York, 1984.
Harms, A.A., et al., *Principles of Fusion Energy: An Introduction to Fusion Energy for Students of Science and Engineering*. World Scientific, River Edge, NJ, 2000.
Herman, R. *Fusion: The Search for Endless Energy*. Cambridge University Press, New York, 1990.
Mitrishkin, Y.V., Kurachi, K. and Kimura, H. Plasma multivariable robust control system design and simulation for a thermonuclear tokamak-reactor. *Int. J. Control*, 76, 13, 1358–1374, 2003.
Nishikawa, K. and Wakatani, M. *Plasma Physics: Basic Theory with Fusion Applications*. Springer Verlag, New York, 2000.
Souers, P.C. *Hydrogen Properties for Fusion Energy*. University of California Press, Berkeley, 1986.

Useful Websites

European Fusion Development Agreement-JET: http://www.jet.efda.org/pages/history-of-jet.html
International Thermonuclear Experimental Reactor: http://wwwofe.er.doe.gov/iter.html
Princeton Plasma Physics Laboratory: http://www.pppl.gov/news/pages/tftr_removal.html

Nuclear Reactors: Thermal, Graphite-Moderated

In a nuclear reactor, an element low on the atomic scale such as carbon or hydrogen is used to absorb kinetic energy to slow down naturally emitted neutrons from the radioactive fuel. In most power reactors, refined but unenriched natural uranium (^{238}U or uranium-238) is the preferred fuel over 99 percent of the time. When the neutrons move more slowly or at a "moderated" speed, the chances of collision between the neutrons and other uranium nuclei, leading to fission and a chain reaction, are increased. Reactor designs are often named for the type of moderator used.

The first reactors, including the experimental pile built in 1942 at Chicago during World War II and the early production reactors built in 1943 at Hanford in Washington state, used graphite as a moderator. Later reactors used water, heavy water, sodium, or other materials as moderators. In the U.S., almost all power reactors and all submarine and ship propulsion reactors relied on pressurized water systems or boiling water systems, first installed in the late 1950s. Acronyms for all these systems have become conventional, with the most common being the boiling water reactor (BWR), pressurized water reactor (PWR), and light water-cooled graphite-moderated reactor (LWGR).

The former Soviet Union developed three power reactor designs in the 1950s and 1960s. The *reactory bolshoi moshchnosti kanalynye* (RBMK) or "channelized large power reactor" was one of their most common and, equivalent to a LWGR, very similar to the first production reactors built in the U.S. at Hanford. A second Soviet type was the *vodo-vodyannoy energeticheskiy* reactor (VVER) or water-cooled and water-moderated reactor, rather similar to the American and European PWRs. The Soviets exported VVERs to several eastern European countries. The third Soviet type was a fast reactor designated "BN" as a breeder reactor.

Britain developed the Magnox reactor, a gas-cooled, graphite-moderated reactor, using the system both for plutonium production for weapons and for power production.

Accidents involving graphite reactors are particularly dangerous, since graphite is flammable. A release of radioactivity in 1957 at the British Windscale Reactor near Sellafield, Cumbria, a graphite production reactor, was not immediately disclosed. Even accidents with water-cooled reactors, such as that at Three Mile Island in Pennsylvania on March 28, 1979, cause national and international concern. However, far more serious was the Chernobyl fire of April 26 1986, in a 1000-megawatt rated RBMK graphite-moderated reactor. That fire spread radioactive contamination across not only the Ukraine but also much of eastern and northern Europe as well. As at Windscale, details of the Chernobyl accident were temporarily suppressed. Gas-cooled graphite reactors are prevented from burning by the fact that they are cooled with carbon dioxide. However, if oxygen-containing air leaks into the system and the cooling system fails, the graphite can ignite.

In the period from 1964 to 1987, a large LWGR used for both plutonium production and power generation remained in operation on the Hanford reservation in the U.S. Designated "N" reactor, it was the only graphite-moderated reactor still in operation in the U.S. during the Chernobyl accident. After a safety review conducted by the National Academy of Sciences, N reactor was closed permanently.

The first British power reactor went into operation in 1956 at Calder Hall, in Seascale, Cumbria, at 50 megawatts electric (MWe). The early British design was called the Magnox type and was a gas-cooled reactor (GCR). Several of these GCRs were exported to France, Italy, Japan, and Spain. The Magnox reactor uses magnesium oxide as canning for the fuel slugs, and is cooled with carbon dioxide. The Magnox design had been worked out at Harwell, and early models were constructed at both Calder Hall and Chapelcross. A later design, first constructed at Hinkley Point B in Somerset is an advanced gas-cooled reactor (AGR). Whereas the GCRs operate at about 245°C, the AGRs operate at about 630°C. In some descriptions, the AGRs are known as high-temperature gas-cooled reactors.

Over the decades, Britain built more than 30 power reactors, providing nearly 20 gigawatts of power by the 1990s (about 12.5 gigawatts net). All are of the Magnox GCR type or the advanced gas-cooled reactor (AGR) type, except for one PWR, at Sizewell, Suffolk, built in the 1990s. The Magnox and AGR reactors differ in several aspects, but both are graphite-moderated.

At the beginning of the twenty-first century, the U.K. remained one of the largest producers of nuclear power in the world, ranking after the U.S., France, Japan, Germany, Russia, the Ukraine, and Canada. Nuclear power, nearly all of it from graphite-moderated reactors (see Table 1), made up slightly more than one fourth of the electricity supplied in the U.K.

See also **Nuclear Reactors: Fast Breeders; Nuclear Reactors: Fusion, Early Designs; Nuclear Reactors: Fusion, Tokomak; Nuclear Reactors: Thermal, Water-Moderated**

RODNEY CARLISLE

Further Reading

Carlisle, R., Ed. *Encyclopedia of the Atomic Age*. Facts on File, New York, 2001.
Carlisle, R. and Zenzen, J. *Supplying the Nuclear Arsenal*. Johns Hopkins University Press, Baltimore, 1996.
Cochrane, T.B., *et al. Making the Russian Bomb: From Stalin to Yeltsin*. Westview Press, Boulder, CO, 1995.
Hewlett, R.G. and Holl, J.M. *Atoms for Peace and War, 1953–1961*. University of California Press, Berkeley, 1989.
Patterson, W.C. *Nuclear Power*. Penguin, Harmondsworth, 1976
Rhodes, R. *The Making of the Atomic Bomb*. Simon & Schuster, New York, 1986.

Table 1 British Graphite Reactors.

Characteristic	Magnox or graphite core reactor	Advanced gas-cooled reactor (AGR)
Thermal output	800+ MWt	1400+MWt
Electrical output	50–475 megawatt electric	200–700 megawatt electric
Efficiency	c.33%	c. 40%
Moderator	Graphite	Graphite
Fuel	Natural uranium	Uranium oxide 2% enriched
Cladding	Clad in magnox alloy	Stainless steel
Coolant	Carbon dioxide gas	Carbon dioxide gas
Coolant temperature	245°C	634°C.
Vessel	Welded steel	Prestressed concrete
Hazard	Magnox alloy low melting point	Cladding has higher melting point

*Megawatt electric output as percentage of megawatt thermal output.

Nuclear Reactors: Thermal, Water-Moderated

Nuclear reactors are usually classified by their coolant and their moderators. The moderator is a material, low in the atomic scale, whose atomic nucleus has the effect of slowing down or moderating the speed of fast neutrons emitted during nuclear fission. By slowing the speed of neutrons, the moderator increases the chance of collision of neutrons with the nuclei of fissionable nuclear fuel atoms. The original reactor designed by Enrico Fermi during the Manhattan Project at Chicago, known as Chicago Pile One, or CP-1, was a graphite-moderated, air-cooled reactor. Many British and French nuclear reactors for the generation of electrical power use carbon in the form of graphite, and they are cooled with carbon dioxide gas. These types are known as Magnox reactors. However, the common designs for power generation developed in the U.S. used water both as coolant and as a moderator.

Water-cooled reactors fall into two large families. Heavy water reactors contain water in which the hydrogen atom is replaced with the hydrogen isotope deuterium. This type of reactor is manufactured for export by Canada. The pressurized heavy water reactor (PHWR) has been exported and installed in India, Romania, and elsewhere. The U.S. built five heavy water reactors at Savannah River, South Carolina, in the 1950s to serve as production reactors for the manufacture of plutonium and tritium for nuclear weapons. By the late 1980s, all the Savannah River production reactors had been closed. After some experimentation with graphite-moderated gas-cooled designs and with heavy-water moderation during the 1950s, the U.S. followed the "light water" path.

In the U.S., two reactor designs use regular or light water both as coolant and moderator. The light water reactor (LWR) has been designed in two forms. The smaller, pressurized water reactor (PWR), originally designed for use aboard submarines as a propulsion reactor for generating electricity to power the drive motors of the vessel, was the first adapted for use in the U.S. for power purposes. The first reactor of this type on land for power generation was installed at Shippingport, Pennsylvania, in 1957. The Shippingport reactor, installed under the supervision of the naval reactor chief, Admiral Hyman Rickover, was modeled on one designed for the propulsion of surface ships such as cruisers and aircraft carriers.

Most PWRs in the U.S. were manufactured by Westinghouse Corporation, while most boiling water reactors (BWRs) were manufactured by General Electric. In the pressurized water reactor type, the water is circulated through the reactor and super-heated, then piped through a steam generator where a separate water system is heated to produce steam. That steam is used to drive the turbine generators. In the BWR, a single loop takes water through the reactor, raising it to steam temperature that is then channeled through pipes and valves to the turbine. It is then condensed back to water to be pumped back into the reactor. In both systems, the steam is cooled by an independent system of outside water circulated through a characteristic cooling tower. For this reason, both types of light water reactors are often sited near a body of natural fresh water. The effluent from the cooling tower, although not radioactive in the slightest, is usually released back to the natural water supply at a somewhat elevated temperature. This "thermal pollution" has become a concern of some environmentalists, who decry its effect on local ecosystems.

By the beginning of the twenty-first century, all power reactors in the U.S. were light water types, with about two-thirds of the reactors as pressurized water reactors and about one-third following the boiling water reactor design. Advocates of gas-cooled designs in the U.S. argued that the inherent safety of the gas-cooled design, in which a runaway reaction automatically causes the reactor to close down, is superior to the water-cooled and moderated designs.

An advantage of the Canadian-designed PHWRs is that they operate on natural, unenriched uranium. One disadvantage is that the heavy water utilized in the reactors may be diverted for use in reactors to produce weapons-grade plutonium, as has been suspected in the case of the PHWR program in India.

One type of reactor made in the Soviet Union was the *vodo-vodyannoy energeticheskiy* reactor (VVER), or water-cooled and water-moderated reactor, equivalent to the American PWRs. The VVERs were built not only for use in the Soviet Union but also for export to satellite nations. The VVER-440, developed before 1970, was the most common. The later VVER-440s and a VVER-1000 that was developed in 1975 had added safety features.

The 1979 Three Mile Island accident, in which a Babcock and Wilcox-manufactured 906-megawatt-electrical PWR failed, was responsible more than any other factor for changing U.S. attitudes toward nuclear power. The accident began with a failure in the cooling system, resulting in an interruption in the flow of water to the steam generator. A series of errors, including valves accidentally left in the closed position, unobserved warning lights, and a faulty pressure relief valve, all led to a spike in heat and reactivity.

Operators grew concerned that hydrogen produced by reaction of steam with the fuel cladding at high temperatures might explode when mixed with oxygen resulting from the breakup of water under radiation. The operators and outside emergency personnel disagreed over whether a hydrogen explosion could occur, but word of a possible explosion was released to the public. The Pennsylvania authorities ordered an evacuation of children and pregnant women from the immediate area surrounding the reactor. The after-accident appraisals were contradictory and depended to an extent on the orientation of the appraisal authors. Fortunately, there had been very little radioactive release. Even so, the damage to the reactor core was severe. The concrete containment vessel performed very well, and there were no injuries or deaths. The reactor was shut down and entombed in concrete. Despite problems, engineers concluded that the accident demonstrated that some existing safety systems worked very well.

Following the Three Mile Island accident, reactor orders in the U.S. greatly declined. Although new reactors have been built, and the U.S. remains the leader in electrical power generation from nuclear sources with about 100 power reactors in total, most specialists predict that the proportion of nuclear power production versus other sources will decrease in the U.S. in the coming decades.

See also **Nuclear Reactors: Fast Breeders; Nuclear Reactors: Fusion, Early Designs; Nuclear Reactors: Fusion, Later Designs (Tokomak); Nuclear Reactors: Thermal, Graphite-Moderated**

RODNEY P. CARLISLE

Further Reading

Balogh, B. *Chain Reaction: Expert Debate and Public Participation in American Commercial Power, 1945–1975*. Cambridge University Press, New York, 1991.

Bodansky, D. *Nuclear Energy: Principles, Practices, and Prospects*. American Institute of Physics, Woodbury, NY, 1996.

Cantelon, P. and Williams, R. *Crisis Contained: The Department of Energy at Three Mile Island*. University of Southern Illinois Press, Carbondale, 1984.

Duderstadt, J.J. and Kikuchi, C. *Nuclear Power: Technology on Trial*. University of Michigan Press, Ann Arbor, 1979.

Holl, J., Anders, R. and Buck, A. *The United States Civilian Nuclear Power Policy, 1954–1984: A Summary History*. Department of Energy, Washington D.C., 1986.

Kemeny, J.G. *Report of the President's Commission on the Accident at Three Mile Island*. Pergamon Press, New York, 1979.

Ragovin, M. *Three Mile Island: A Report to the Commissioners and to the Public*. Nuclear Regulatory Commission, Washington D.C., 1980.

Nuclear Reactors, Weapons Material

The first successful nuclear reactor, called an "atomic pile" because of its structure of graphite bricks, was completed and operational on December 2, 1942, in Chicago in the U.S. Although originally built to demonstrate a controlled nuclear reaction, the reactor was later dismantled and the depleted uranium removed in order to recover minute amounts of plutonium for use in a nuclear weapon. In effect, Chicago Pile-One (CP-1) was not only the world's first nuclear reactor but also the world's first reactor used to produce material for a nuclear weapon.

CP-1 and a few other experimental reactors built at Chicago were soon followed by a second generation of five reactors designed solely for large-scale production of plutonium, constructed by DuPont Corporation at Hanford, Washington. Dedicated to production of weapons material, they are designated "production reactors" to distinguish them from later power-generating reactors and ship or submarine propulsion reactors. Later production reactors also made the rare isotope of hydrogen, tritium, for use in weapons. Tritium causes a partial fusion reaction, boosting the neutron supply in a fission reaction.

Production reactors were first built in the U.S. in 1943 at Hanford to take advantage of the constant flow of cold water from the Columbia River for cooling the reactors. The isolation from centers of population made security easier and reduced the number of people who might suffer radiation exposure in case of an accident. The Hanford reactors consisted of 12-meter-high cores of graphite, through which horizontal tubes fed cold water from the river. Operators inserted slugs of tin-encased natural uranium into the tubes. The slugs were designed with fins to hold them in place in the tubes, allowing the cooling water to flow directly over the slugs (and later, through longitudinal holes, in the slugs themselves). Elements common to later reactors were included: moderated by graphite, controlled with cadmium rods, and cooled by natural water.

The uranium-238 in the natural uranium slugs was transformed by a several-step reaction into plutonium. The uranium-235 split into fission fragments and generated two or three neutrons to sustain the reaction. Operators pushed the fuel slugs out of the back of the reactor and took them to a processing plant to separate plutonium. At the beginning of November 1944, the first reactor produced plutonium that was later used in the Trinity test at Alamogordo, New Mexico. The atomic weapon dropped on Nagasaki, Japan, in August 1945 also contained plutonium from the Hanford reactors.

Following World War II, an additional five reactors were built at Hanford. Responding to the Soviet test of a nuclear weapon in August 1949, and reacting to the outbreak of the Korean War in June 1950, the U.S. Congress funded a second production reactor complex at Savannah River, South Carolina. The five Savannah River Site reactors used heavy water as a moderator; that is, water based on the deuterium isotope of hydrogen. In the U.S., a few reactors were kept open throughout the Cold War, supplying both plutonium and the tritium isotope of hydrogen. The tritium was produced by the insertion of lithium deuteride pellets or targets in the reactors and bombarding them with neutrons.

All the nuclear-armed nations built reactors to make plutonium. The British facility at Calder Hall in Cumbria, the Soviet facilities at Chelyabinsk-40 and later at Krasnoyarsk-26 and at Tomsk-7, and the French Marcoule facility all housed production reactors. N reactor, built by the U.S. at Hanford in 1964, was a dual-purpose reactor designed both to serve as a production reactor for making plutonium and to generate electricity. Eventually other dual-purpose reactors included some of the Soviet reactors at Tomsk-7 and Krasnoyarsk-26, the British reactors at Calder Hall, and the French Marcoule complex. The U.S. Congress debated the wisdom of mixing the weapons material production function with electrical power production for the civilian power grid. In the U.S., the dilemma was resolved at N reactor by producing steam that was piped off site and used there to produce electricity. Apparently Russian experts disagreed over the propriety of the Russian dual-purpose program at Tomsk-7. The fact that some British dual-purpose reactors produced both electric power for commercial use and plutonium for weapons remained a matter of continuing controversy in the U.K. as well.

India maintains several production reactors, including Cirus and a larger one at Dhruva. One Indian reactor, Kakrapar II, opened in 1995 as a power reactor but could be used to generate material for weapons use. Pakistan maintains a small plutonium production reactor, rated at 50 megawatts at Khusab. Israel has a production reactor with a classified level of production in the Negev at Dimona. China is known to have two graphite-moderated production reactors. A 1000-megawatt reactor is located in Guangyuan, Sichuan province, and a smaller one, estimated at between 400 to 500 megawatts, is reputed to be at the Jiuquan Atomic Energy Complex at Subei, in Gansu province.

In the U.S., the last production reactors closed in 1988. In Russia, three production reactors remained in operation until the year 2000, including underground graphite-moderated reactors at Krasnoyarsk-26 (renamed Zheleznogorsk) and at Tomsk-7. In 1998, U.S. officials quietly announced that a power reactor in the Tennessee Valley Authority complex would be utilized for both power and weapons material production.

See also **Nuclear Fuels; Nuclear Reactor**

Rodney P. Carlisle

Further Reading

Carlisle, R., Ed. *Encyclopedia of the Atomic Age*. Facts on File, New York, 2001.

Carlisle, R. and Zenzen, J. *Supplying the Nuclear Arsenal*. Johns Hopkins University Press, Baltimore, 1996.

Cochrane, T.B., et al. *Making the Russian Bomb: From Stalin to Yeltsin*. Westview Press, Boulder, CO, 1995.

Gowing, M. *Britain and Atomic Energy 1939–1945*. Macmillan, London, 1964.

Gowing, M. Independence and Deterrence: Britain and Atomic Energy, 1945–1952. Macmillan, London, 1974.

Hewlett, R.G. and Anderson, O. *The New World, 1939–1946, vol.1: History of the United States Atomic Energy Commission*. Atomic Energy Commission, Washington D.C., 1972.

Hewlett, R.G. and Duncan, F. *Atomic Shield, 1947–1952, vol. 2: History of the United States Atomic Energy Commission*. Atomic Energy Commission, Washington D.C., 1972.

Hewlett, R.G. and Holl, J.M. *Atoms for Peace and War, 1953–1961*. University of California Press, Berkeley, 1989.

Holloway, D. *Stalin and the Bomb*. Yale University Press, New Haven, 1994.

Patterson, W.C. *Nuclear Power*. Cox & Wyman, London, 1976.

Rhodes, R. *The Making of the Atomic Bomb*. Simon & Schuster, New York, 1986.

Nuclear Waste Processing and Storage

Radioactive waste consists of liquid, solid, or gaseous materials containing high, medium, or low levels of radioactivity. There are two alternative policies underlying technologies for processing and storage of radioactive waste. If the level of radioactivity in the waste is low or medium, the materials are diluted in liquid carriers and then dispersed when the concentration of radioactivity is considered not hazardous for the environment. If the level of radioactivity in the waste is medium or high, then the radioactive waste is concentrated, contained, and isolated in proper repositories.

Low- and medium-level radioactive waste includes materials that have been contaminated in nuclear reactors during operation and that may contain fissile products or a low percentage of fuel. It also includes contaminated protective clothing and equipment. Some of the medium gaseous radioactive waste is due to fuel reprocessing. In this process, the dissolution of spent fuel in nitric acid causes the production of toxic gases containing radioactive isotopes such as krypton (^{85}Kr), iodine (^{129}I), carbon (^{14}C), and tritium (^{3}H). While tritium and carbon can be collected in liquid carriers, there is no solution currently available for krypton and iodine. These elements are therefore released into the atmosphere in diluted gaseous mixtures. Their containment consists of controlling the time, place and manner of dispersion.

Dispersion in the sea (or ocean dumping) was considered, in the past, a viable option for low-level radioactive waste. Waste was packed in steel or concrete drums and dumped at an average depth of 4000 meters in the sea. From 1946 to 1970, the U.S. dumped radioactive waste in 50 locations in the Pacific and Atlantic Oceans. The U.K. dumped radioactive waste in the Atlantic from 1949 to 1983. In the 1980s, rising concern was expressed for the marine biological environment and also for the possible irradiation of seafood with consequent hazard for human beings. In 1985 the United Nations imposed a moratorium on sea dumping of radioactive waste.

Processing and storage of high- and medium-level radioactive waste is considered the "Achilles heel" of the nuclear industry, as the technologies available have not yet provided viable solutions. Fission products such as iodine, technetium, neodymium, zirconium, molybdenum, and cesium produced by the fissile materials plutonium and uranium in nuclear reactors are by far the most toxic and radioactive elements known in nature, and their storage presents problems for the natural and human environment. Moreover, the problem faced by the scientific community is that a number of fission products are long-lived and will emit radiation and heat for a period varying from 10 to 100,000 years.

In the short term, high-level liquid and solid waste is stored in the proximity of the nuclear plants using water ponds. Subsequently, the high-level liquid waste is solidified through a process of vitrification with borosilicate glass. Vitrification was successfully achieved in France by the Commissariat À l'Energie Atomique (CEA) in 1978, and the Atelier de Vitrification de Marcule (AVM) was the first plant designed for this purpose. It represents an improvement in radioactive waste containment as the radioactive waste is chemically bonded into the atomic structure of an inert and stable solid. From 1988, the vitrification process was also adopted by Cogema in La Hague (France), by the U.K. Atomic Energy Agency at Windscale (Britain), and by the Vortec Corporation in the U.S. Borosilicate glass is also used for the production of canisters in which spent fuel and solid high waste is stored. Canisters will eventually be put into deep repositories capable of containing radioactivity and heat without causing major hazard to the environment, but no repositories of this kind have yet been accomplished.

Between the 1950s and the 1980s, several sites were used for short-term waste using shallow burial in trenches, tumuli, tunnels, or concrete bunkers. The oldest site of this type is at Savannah River (Idaho Desert, U.S.), which stores the contaminated material from Los Alamos National Laboratory in New Mexico. The U.K.'s short-term waste repository is located at Drigg in Cumbria. More modern facilities for short-term repositories use concrete pipes and concrete bases sunk vertically to prevent leakage as at Hanford in Britain, Tokai in Japan, and La Manche in France.

Since 1960, three types of geological sites have been considered for long-term radioactive waste disposal: clay-rich rocks, hard crystalline igneous and metamorphic rocks (granite, gabbro, and basalt), and evaporites (bedded salts or salt domes). In 1960, the Project Salt Vault for the creation of a facility in the former Lyons salt mine in Kansas (U.S.) was discussed, but in 1971 the project was finally dropped. A new project for repositories in deep salt beds was considered in 1988, but it never led to final construction. From 1982 onward, the scientific community stressed that deep geological disposal should be considered the final target for any high-level waste strategy. It also provided details of how containment should be pursued. Facilities should be provided with barriers capable of dividing the depository from the surrounding area. Repositories should contain the canisters, ventilate them with air, and survey them with CCTV systems. They should be built in concrete and manufactured according to a geological analysis of the rocks, the water, and seismic activity. At the moment the only repository that may be used for long-term disposal is the undersea plant of Sipra in Sweden. Various projects for long-term disposal have not been developed because they have met severe public protests by local communities unwilling to accept radioactive waste in the proximity of their living areas (as highlighted by the acronym NIMBY for Not-In-My-Back-Yard).

In recent years, several scientific establishments have considered a different approach to long-term radioactive waste, focusing on the destruction of fission products. In 1997, the European Organisation for Nuclear Research presented a project for an energy amplifier capable of incinerating waste through fast neutron bombardment. The project aimed to substitute in 29 years the geological disposal of the high waste produced by nine nuclear reactors operating in Spain. In 2000, the Nuclear Energy Agency (OECD-NEA) published the results of an international meeting on the process of partitioning and transmutation (P&T). The goal is to shorten the life of some fission products through incineration in nuclear reactors or treatment in nuclear accelerators. P&T research programs have been developed in Japan (Omega, 1988) and in France (Spin, 1991).

See also **Nuclear Reactors, Weapons Material**

SIMONE TURCHETTI

Further Reading

Annex A: The Spin Programme—Assets and Prospects, in Assessment Report on Actinide and Fission Product Partitioning and Transmutation. Sixth International Information Exchange Meeting, Madrid, Spain, 11–13 December, 2000.

Annex B: The Omega Programme—Partitioning and Transmutation R&D Programme of Japan, in Assessment Report on Actinide and Fission Product Partitioning and Transmutation. Sixth International Information Exchange Meeting, Madrid, Spain, 11–13 December 2000.

Blowers, A., Lowry, D. and Solomon, B.D. The International Politics of Nuclear Waste. Macmillan, London, 1991.

Mounfield, P. World Nuclear Power. Routledge, London, 1991.

Rubbia C., Buono S., Kadi Y. and Rubio J.A. *Fast Neutron Incineration in the Energy Amplifier as Alternative to Geological Storage: the Case of Spain.* European Organisation for Nuclear Research, Geneva, 1997.

Ursu, I. *Physics and Technology of Nuclear Materials.* Pergamon Press, Oxford, 1985.

Oil from Coal Process

The twentieth-century coal-to-petroleum or synthetic fuel industry evolved in three stages:

1. Invention and early development of Bergius coal liquefaction (hydrogenation) and Fischer–Tropsch (F–T) gas synthesis from 1910 to 1926.
2. Germany's industrialization of the Bergius and F–T processes from 1927 to 1945.
3. Global transfer of the German technology to Britain, France, Japan, Canada, the U.S., and other nations from the 1930s to the 1990s.

Petroleum had become essential to the economies of industrialized nations by the 1920s. The mass production of automobiles, the introduction of airplanes and petroleum-powered ships, and the recognition of petroleum's high energy content compared to wood and coal, required a shift from solid to liquid fuels as a major energy source. Industrialized nations responded in different ways. Germany, Britain, Canada, France, Japan, Italy, and other nations, having little or no domestic petroleum, continued to import it. Germany, Japan, and Italy also acquired by force the petroleum resources of other nations during their 1930s–1940s World War II occupations in Europe and the Far East. In addition to sources of naturally occurring petroleum, Germany, Britain, France, and Canada in the 1920s–1940s synthesized petroleum from their domestic coal or bitumen resources, and during the 1930s–1940s war years Germany and Japan synthesized petroleum from the coal resources they seized from occupied nations. A much more favorable energy situation existed in the U.S., and it experienced few problems in making an energy shift from solid to liquid fuels because it possessed large resources of both petroleum and coal.

Germany was the first of the industrialized nations to synthesize petroleum when Friedrich Bergius in Rheinau-Mannheim in 1913 and Franz Fischer and Hans Tropsch at the Kaiser Wilhelm Institute for Coal Research (KWI) in Mülheim in 1926 invented processes for converting coal to petroleum. Bergius crushed and dissolved a coal containing less than 85 percent carbon in a heavy oil to form a paste. He reacted the coal–oil paste with hydrogen gas at high pressure (200 atmospheres at 400°C) and obtained petroleum-like liquids. Bergius sold his patents to BASF in July 1925, and from 1925 to 1930 Matthias Pier at BASF (IG Farben in December 1925) made major advancements that significantly improved product yield and quality. Pier developed sulfur-resistant catalysts, such as tungsten sulfide (WS_2), and separated the conversion into two stages, a liquid stage and a vapor stage. Fischer and Tropsch reacted coal with steam to give a gaseous mixture of carbon monoxide and hydrogen and then converted the mixture at low pressure (1–10 atmospheres at 180–200°C) to petroleum-like liquids. Fischer and his co-workers in the 1920s–1930s developed the cobalt catalysts that were critical to the F–T's success, and in 1934 Ruhrchemie acquired the patent rights to the synthesis. These pioneering researches enabled Germany to develop a technologically successful synthetic fuel industry that grew from a single commercial-size coal liquefaction plant in 1927 to twelve coal liquefaction and nine F–T commercial-size plants that in 1944 reached a peak production of 26 million barrels of synthetic fuel.

Britain's synthetic fuel program evolved from post-World War I laboratory and pilot-plant studies that began at the University of Birmingham in 1920 on F–T synthesis and in 1923 on coal liquefaction. The Fuel Research Station in East Greenwich also began research on coal liquefaction in 1923, and the program reached its zenith in 1935 when Imperial Chemical Industries (ICI) constructed a coal liquefaction plant at Billingham that had the capacity to synthesize annually 1.28 million barrels of petroleum. British research and development matched Germany's, but because of liquefaction's high cost and the government's decision to rely on petroleum imports rather than price supports for an expanded domestic industry, Billingham remained the only British commercial-size synthetic fuel plant. F–T synthesis in the 1930s–1940s never advanced beyond the construction of four small experimental plants: Birmingham, the Fuel Research Station's two plants that operated from 1935 to 1939, and Synthetic Oils Ltd. near Glasgow.

Britain and Germany had the most successful synthetic fuel programs. The others were either smaller-scale operations such as France's three demonstration plants (two coal liquefaction and one F–T), Canada's bitumen liquefaction pilot plants, and Italy's two crude petroleum hydrogenating (refining) plants; or technological failures such as Japan's five commercial-size plants (two coal liquefaction and three F–Ts) that produced only about 360,000 barrels of liquid fuel during the World War II years.

The U.S. Bureau of Mines had begun small-scale research on F–T synthesis in 1927 and coal liquefaction in 1936, but did no serious work on them until the government expressed considerable concern about the country's rapidly increasing petroleum consumption in the immediate post-World War II years. At that time the Bureau began a demonstration program, and from 1949 to 1953 when government funding ended, it operated a small 200–300 barrel per day coal liquefaction plant and a smaller 50 barrel per day F–T plant at Louisiana, Missouri. In addition to the Bureau's program, American industrialists constructed four synthetic fuel plants in the late 1940s and mid-1950s, none of which achieved full capacity before shutdown in the 1950s for economic and technical reasons. Three were F–T plants located in Garden City, Kansas; Brownsville, Texas; and Liberty, Pennsylvania. The fourth plant was a coal liquefaction plant in Institute, West Virginia.

Following the plant shutdowns in the U.S. and until the global energy crises of 1970–1974 and 1979–1981, all major synthetic fuel research and development ceased except for the construction in 1955 of the South African Coal, Oil, and Gas Corporation's (SASOL) F–T plant in Sasolburg, south of Johannesburg. South Africa's desire for energy independence and the low quality of its coal dictated the choice of F–T synthesis rather than coal liquefaction. Its Johannesburg plant remained the only operational commercial-size synthetic fuel plant until the 1970s energy crises and South Africa's concern about hostile world reaction to its apartheid policy prompted SASOL to construct two more F–T plants in 1973 and 1976 in Secunda.

The 1970s energy crises also revitalized synthetic fuel research and development in the U.S. and Germany and led to joint government–industry programs that quickly disappeared once the crises had passed. Gulf Oil, Atlantic Richfield, and Exxon in the U.S., Saarbergwerke AG in Saarbrüken, Ruhrkohle AG in Essen, and Veba Chemie in Gelsenkirchen, Germany, constructed F–T and coal liquefaction pilot plants in the 1970s and early 1980s only to end their operation with the collapse of petroleum prices a few years later.

In the mid-1990s two developments triggered another synthetic fuel revival in the U.S.:

1. Petroleum imports again reached 50 percent of total consumption or what they were during the 1973–1974 Arab petroleum embargo.
2. An abundance of natural gas, equivalent to 800,000 million barrels of petroleum, but largely inaccessible by pipeline, existed.

Syntroleum in Tulsa, Oklahoma; Exxon in Baytown, Texas; and Atlantic Richfield in Plano, Texas, developed modified F–T syntheses that produced liquid fuels from natural gas and thereby offered a way of reducing the U.S.'s dependence on petroleum imports. The Department of Energy (DOE) at its Pittsburgh Energy Technology Center through the 1980s–1990s also continued small-scale research on improved versions of coal liquefaction. DOE pointed out that global coal reserves greatly exceeded petroleum reserves, anywhere from 5 to 24 times, and that it expected petroleum reserves to decline significantly in 2010–2030. Syntroleum, Shell in Malaysia, and SASOL and Chevron in Qatar have continued F–T research, whereas DOE switched its coal liquefaction research to *standby*. The only ongoing coal liquefaction research is a pilot plant study by Hydrocarbon Technologies Inc. in Lawrenceville, New Jersey, now Headwaters Inc. in Draper, Utah.

A combination of four factors, therefore, has led industrialized nations at various times during the twentieth century to conclude that synthetic fuel could contribute to their growing liquid fuel requirements:

1. The shift from solid to liquid fuel as a major energy source.
2. The invention of the Bergius and F–T coal-to-petroleum conversion or synthetic fuel processes.
3. Recognition that global petroleum reserves were finite and much less than global coal reserves and that petroleum's days as a plentiful energy source were limited.
4. The desire for energy independence.

With the exception of South Africa's three F–T plants, the synthetic fuel industry, like most alternative energies, has endured a series of fits and starts that has plagued its history. The historical record has demonstrated that after nearly 90 years of research and development, synthetic liquid fuel has not emerged as an important alternative energy source. Despite the technological success of synthesizing petroleum from coal, its lack of progress and cyclical history are the result of a lack of government and industry interest in making a firm and a long-term commitment to synthetic fuel research and development. The synthetic fuel industry experienced intermittent periods of intense activity internationally in times of crises, only to face quick dismissal as unnecessary or uneconomical upon disappearance of the crises. Even its argument that synthetic liquid fuels are much cleaner burning than coal, and if substituted for coal they would reduce the emissions that have contributed to acid rain formation, greenhouse effect, and to an overall deterioration of air quality has failed to silence its critics. The hope of transforming its accomplishments at the demonstration stage into commercial-size production has not yet materialized.

See also **Feedstocks**

ANTHONY N. STRANGES

Further Reading

Much of the information on the German, Japanese, British, and American synthetic fuel programs has come from the 300,000 pages of documents and the 305 microfilm reels that the Technical Oil Mission and other Allied investigative teams in Europe collected on the German synthetic fuel program, and the 185 microfilmed reports that the U.S. Naval Technical Mission to Japan compiled. The reports on the Japanese synthetic fuel industry are the same as those in the National Diet Library in Tokyo and in the National Archives in Washington. Complete sets of both microfilm collections and documents are in the Texas A&M University Archives.

Donath, E. Hydrogenation of coal and tar, in *Chemistry of Coal Utilization,* suppl. vol., Lowry, H.H., Ed. Wiley, New York, 1963.

Report on the Petroleum and Synthetic Oil Industry of Germany. The Ministry of Fuel and Power, His Majesty's Stationery Office, London, 1947.

Storch, H.H. Golumbic, N. and Anderson, R. *The Fischer–Tropsch and Related Syntheses*. Wiley, New York, 1951.

Stranges, A.N. Canada's mines branch and its synthetic fuel program for energy independence. *Technol. Cult.*, 32, 521–554, 1991.

Stranges, A.N. From Birmingham to Billingham: high-pressure coal hydrogenation in Great Britian. *Technol. Cult.*, 20, 726–757, 1985.

Stranges, A.N. Germany's synthetic fuel industry 1927–45, in *The German Chemical Industry in the Twentieth Century*, Lesch, J.E., Ed. Kluwer, Dordrecht, 2000.

Stranges, A.N. Synthetic fuel production in Japan: a case study in technological failure. *Ann. Sci.*, 50, 229–265, 1993.

Stranges, A.N. Synthetic petroleum from coal hydrogenation: its history and present state of development in the United States. *J. Chem. Ed.*, 60, 617–625, 1983.

Stranges, A.N. The US Bureau of Mines' synthetic fuel programme. *Ann. Sci.*, 54, 29–68, 1997.

Useful Websites

Fischer-Tropsch Archive: http://www.fischer-tropsch.org

Oil Rigs

Although historical accounts exist that describe oil and natural gas drilling techniques in ancient Mesopotamia and China, modern oil rig drilling has its roots primarily in salt-boring technology. By AD 350, China was constructing salt drilling wells that ran as deep as 900 meters into the ground. In the nineteenth century, Europe and the U.S. began importing this salt drilling technology from China. George Bissell, an American entrepreneur, realized that salt-boring techniques could be applied to the drilling for oil. Bissell and other investors hired Edwin Drake to construct and oversee rigs designed for oil drilling. Their venture proved successful when on 27 August 1859, Drake struck oil in Titusville, Pennsylvania.

Drake utilized cable tool drilling, or percussion drilling, which consists of a cable that raises and drops a heavy metal bit capped with a chisel end into the ground. The bit punches a hole into the earth by breaking through rock with regularly repeated blows. Although cable tool drilling remains in use today in shallow, low-pressure oil and gas wells, rotary drilling has become the industry standard.

The transition from cable tool drilling to rotary drilling was the most significant advancement in rig-drilling technology during the twentieth century. Anthony Lucas and Patillo Higgins popularized rotary drilling when they used it in 1901 on their Spindeltop well located near Beaumont, Texas. Rotary drilling involves the use of a sharp, rotating drill bit which can cut through the hardest rock and dig deeper and higher pressured wells than is possible with cable tool drilling. Rotary drilling technology is composed of four basic components:

1. The prime mover, which provides power to the drilling rig.
2. The hoisting equipment, which includes the derrick that raises and lowers necessary equipment into the well hole.
3. The circulating system, which controls the well's pressure, cools and lubricates the drill bits, removes debris, and helps reduce the chance for a blowout.
4. The rotating equipment, which enables the rotation of the drill bit.

Offshore drilling dates back to the mid-nineteenth century when American T.F. Rowland was granted a patent for his design of a four-legged offshore drilling rig. However, it was not until the 1940s that North American construction of offshore wells began to flourish. Although onshore and offshore drilling technologies are similar, offshore drilling poses unique challenges. For example, onshore drilling uses the ground as a platform, but in the offshore environment, the floor can be located thousands of feet below sea level thus requiring the construction of an artificial drilling platform.

Offshore drilling rigs are located on either moveable or permanent sites. Moveable rigs are designed for mobility, which enable exploratory drilling opportunities in numerous places. The drilling barge, the jack-up rig, and the submersible rig are designed for use in shallow water. The semisubmersible rig and the drillship are able to withstand the often harsher conditions present in deeper water.

Both the drilling barge and the jack-up rig must be towed to their drilling site in shallow water. The drilling barge is a large, floating platform that drills in inland, calm water. The jack-up rig also drills in calm water. It is equipped with "legs" that can be lowered to the sea bottom, which, unlike the drilling barge, enables its working platform to remain above the water's surface.

Like jack-up rigs, submersible rigs are designed for direct contact with the lake or ocean floor. These rigs contain a platform that has two vertically connected hulls. The crew and the drilling platform reside in the top hull. The lower hull is equipped with a circulatory air system. When the hull is filled with air, the entire rig becomes buoyant which provides it with the ability to move from site to site. When the air is released, the rig drops to the floor. However jack-up rigs are not sturdy enough for use in deep water.

The most common offshore drilling rig is the semisubmersible rig. The semisubmersible rig is similar to the submersible rig in that it also possesses a lower hull that can inflate and deflate. However, when the semisubmersible rig's lower hull releases its air, the rig floats above the drill site instead of submerging to the floor. The lower hull is then filled with water, which provides the rig with the needed stability to drill in deep water. In addition, these rigs are held in place by anchors that weigh up to 10 tons each. This design provides semisubmersible rigs with the necessary reliability for drilling in offshore turbulent waters.

Drillships are used for drilling in very deep waters. These ships carry a drilling platform and a derrick, and are designed with a hole, called a "moonpool," which extends through the ship's hull. Drilling is done through the moonpool with the assistance of sensors and satellite positioning systems that ensure that the drillship is directly above the desired site.

Once a moveable drilling rig locates an oil or natural gas reservoir, a permanent platform can be built over the site to continue the drilling. These offshore rigs are among the largest man-made artifacts on the planet. Their height ranges in size from twice that of the Hoover Dam upward to that of the Empire State Building. The type of permanent platforms utilized for drilling is dependent on the depth range of the water. Fixed platforms are used in waters measuring up to around 450 meters. Compliant towers are used for drilling in sites where the floor is 450 to 900 meters below sea level. Seastar platforms are used where the drilling sites are located at 150 to 1060 meters below sea level. For deeper water levels, floating production systems can be used at up to 1.8 kilometers below sea level, tension leg platforms at up to 2.1 kilometers below sea level, and the SPAR system at up to 3 kilometers below sea level.

Although petroleum and natural gas provides much of the globe with its energy, drilling technology has proven controversial. Citizen and public interest groups have voiced their concerns about the negative impact drilling has on the environment. In addition, countries are increasingly confronting the challenges posed by decommissioned offshore rigs.

See also **Prospecting, Minerals**

Julia Chenot Goodfox

Further Reading

Gatlin, C. *Petroleum Engineering: Drilling and Well Completions*. Prentice Hall, Englewood Cliffs, NJ, 1960.

Hall, R. S., Ed. *Drilling and Producing Offshore*. PennWell Books, Tulsa, 1984.

Hyne, N.J. *Nontechnical Guide to Petroleum Geology, Exploration, Drilling and Production*, 2nd edn. PennWell Books, Tulsa, 2001.

Uren, L. *Petroleum Production Engineering*, 4th edn. McGraw-Hill, New York, 1956.

Yergin, D.Y. *The Prize: The Epic Quest for Oil, Money and Power*. Free Press, New York, 1991.

Useful Websites

American Petroleum Institute: http://api-ec.api.org
Greenpeace: http://archive.greenpeace.org
Natural Gas Supply Association: http://www.naturalgas.org

Ophthalmology

Ophthalmology is the medical specialty concerned with eye diseases and, in contrast to optometry, it deals with the pathologies of anatomy that require surgery or medical treatment. Ophthalmologists use more technology because they operate on the eye, but they share many of the technologies used to test and assess vision. Training for ophthalmology requires an MD degree followed by four years of specialization.

Prior to the nineteenth century, ophthalmology was included in the ear, nose, and throat (ENT) specialty. The first professor of ophthalmology in Europe was Joseph Baer (1812). Albrecht von Graefe founded the first archive for ophthalmological instruments and established the Ophthalmological Society in Heidelberg. In 1929 that name was changed to The German Ophthalmological Society. In the U.S., the American Board of Ophthalmology was established in 1916 to certify practitioners and maintain standards for the profession. At the time that the Canadian Ophthalmological Society was formed in 1937, ophthalmologists were part of the Eye, Ear, Nose and Throat (EENT) section of the Canadian Medical Association. In the United Kingdom, The European Ophthalmic Pathology Club was officially established at the Royal College of Surgeons in 1962. Worldwide there are over 300 ophthalmological societies, networks and associations.

Astronomers and physicists contributed to the early development of ophthalmology with their knowledge of lenses and the behavior of light. Famed astronomer Johannes Kepler wrote about the role of the retina and the lens in vision. Most surgery was for removal of foreign objects in the eye or cataract couching (or extraction), a fearsome procedure prior to the eighteenth century. Although this procedure to remove cataracts, a clouding of the lens that blocked the passage of light and caused poor vision or blindness, had been known since ancient times (as described in Sanskrit documents), it was dangerous. Both Bach and Handel died from postsurgical infections after successful removal of their cataracts in the eighteenth century. At that time, only the couching knife and needle, crude surgical instruments, were available. The technique involved perforating the sclera and pushing the lens backward with a blunt instrument. Jacques Daviel, born in Normandy, France, educated in Rouen and professor at Hotel Dieu in Marseille (1693–1752), considered the father of ophthalmology, changed the technique to perforate the cornea and remove the whole lens, which was known as extracapsular extraction.

Until the invention of the ophthalmoscope in 1851 by Hermann von Helmholtz, the terms glaucoma (a condition of increased pressure in the eye that can damage the optic disk and cause loss of vision) and cataract were often used interchangeably because no one had been able to see inside the eye of a living person. The first ophthalmoscope used a concave mirror mounted on a handle. In the center of the mirror was a hole through which light was shone into the patient's eye. The viewer saw the reflected image of the retina, the most posterior structure of the eye. It is there that abnormalities of the fundus such as glaucoma could be visualized. Cataracts, however, are found on the lens, a more anterior structure. The physician could finally distinguish the two conditions using the ophthalmoscope

Nineteenth century ophthalmic instruments were elegant-looking brass or copper polished metal materials mounted on solid ivory, ebony, or other kinds of wooden bases. They varied in their design and were operated by reflected light sources or natural light. In the early 1900s, the basic diagnostic tools were the slit-lamp, ophthalmoscope, and retinoscope, which were already in use by both optometrists and ophthalmologists. The tonometer was also used to measure the intraocular pressure of the eye fluid, significant for the diagnosis of glaucoma. As the advent of plastics and other metals flooded the technology market in the 1940s, instruments became lighter, smaller, and more easily operated. The use of a battery-operated light source allowed the clinician more mobility. This was useful when examining the elderly or very young children who were unable to follow directions and position themselves correctly during examinations. The late twentieth century tonometer available was hand-held, used a 1.5-volt battery, and could be manipulated with only one hand.

In the 1950s a new type of light using fused quartz instead of a glass bulb was developed by General Electric. By the early 1960s, a tungsten halogen lamp was developed, and by the 1990s, a very bright halogen light source was used for ophthalmic instruments.

The invention of the slit lamp and its subsequent improvement through the final decades of the nineteenth century and first two decades of the twentieth century made it possible to study the cornea and anterior segment of the living human eye in a thorough and meaningful manner. Alfred Vogt in Switzerland and Allvar Gullstrand in Sweden are credited with its development. In the early 1930s, Vogt published *Lehrbuch und Atlas der Spaltlampenmikroskopie des Leibenden Auges,* which provided the foundation for a new branch of ophthalmology called slit-lamp biomicroscopy.

With the use of topical anesthetics, originally by Carl Koller for trachoma and iritis, and the advent of antibiotics in the mid-twentieth century, great strides were made in eye surgery. Not only could cataracts be removed safely but all kinds of microsurgical procedures were introduced.

After lens removal in cataract surgery, the patient had to wear glasses with extremely thick and heavy lenses and often had problems with depth perception and peripheral vision. The application of a contact lens solved this problem. However, since cataract is a condition more common in elderly patients, there was a need for a more permanent lens, one that would not require the fine motor manipulation required for frequent removal, replacement, and cleaning. During World War II, a serendipitous observation among pilots whose eyes were injured but not damaged by plastic windshield fragments led to the development of the intraocular lens implant. In the ensuing years, a physician to Royal Air Force pilots also observed that pilots who had sustained the intrusion of slivers of cockpit glass did not appear to have a foreign body reaction. He hypothesized that a small lens could be fashioned from similar material. This initial idea resulted in a successful plastic lens design for an intraocular lens implant to use in cataract surgery.

Radial keratotomy was a procedure developed in Japan in the 1940s and enthusiastically continued and perfected in Russia in the late 1950s. In the mid-1970s, Svyatoslav Fyodorov (Moscow) began performing refractive keratotomy with rather primitive instruments: a small razor blade in a holder and a gauge to check the depth of the incision. This revolutionary surgery changed the curvature of the cornea so that a person with myopia (nearsightedness) no longer needed glasses. Ophthalmologists from the U.S. traveled to Russia to learn the technique and used it until the 1990s, eventually replacing the blade with a laser. At the end of the twentieth century, *lasik* surgery to correct vision was performed at outpatient centers and valued because of its convenience and cosmetic result.

Many tests for vision and pathologies of the eye were developed during the twentieth century. The Amsler grid was the simplest and least expensive. It consists of a rectangle divided into small squares with a dot in the middle. The patient stares at the dot and then notes if any of the lines appear distorted. If so, the distortion indicates an underlying problem with the retina, and the patient would seek a thorough ophthalmological examination. More expensive and complex tests such as angiography, computed tomography (CT) scan, and magnetic resonance imagery (MRI) scan require the use of contrast materials and hospital equipment.

Digital technology is now used in cameras such as the fundus camera, which connects to the slit lamp or a combination of imaging devices used in tandem. The image is displayed on a monitor and can be manipulated, stored, and added to a database. A portable slit lamp is now available that is cordless, has a rechargeable battery, uses a halogen light source, and can support a digital video camera (made by Kowa Optimed).

The rise in ophthalmic technology during the twentieth century became so specialized that new educational programs were started in the U.S. to accommodate the rising need for workers within the field. One program at Eastern Virginia Medical School/Old Dominion University opened in 1985 and offered the following subspecialties: ophthalmic photography, ophthalmic ultrasonography, contact lens, ophthalmic surgical technology, electrophysiology, and low vision optics.

See also **Lasers, Applications; Optometry**

LANA THOMPSON

Further Reading

Garrison, F.H. *An Introduction to the History of Medicine*. W.B. Saunders, Philadelphia, 1929.
Lyons, A. and Petrucelli, J. *Medicine: An Illustrated History*. Harry Abrams, New York, 1987.
Magner, L. *A History of Medicine*. Marcel Dekker, New York, 1992.
Medow, N. and Ravin, R. Slit lamp provided closer glimpse at eye. *Ophthalmology Times*. November 1, 1999.

Useful Websites/Further Information

http://www.eyemdlink.com
http://www.eyesite.com
Haag-Streit USA. 5500 Courseview Drive, Mason, OH 45040-2303
Keeler Instruments Inc., 456 Parkway, Broomall, PA 19008
Kowa Optimed, Inc. 20001 S. Vermont Avenue. Torrance, CA 90502
National Optronics, 100 Avon Street, Charlottesville, VA 22902

For information about ophthalmologic technician programs, contact: Lions Center for Sight, 600 Gresham Drive, Norfolk, Virginia 23507,(757) 668-3747, E-mail: optech@-visi.net

Optical Amplifiers

The history of optical telecommunications has been the story of the overcoming of successive technical roadblocks. With the removal of each roadblock came a leap in possible transmission distances or data transmission rates.

The concept of transmitting light down "glassy lightpipes" (optical fibers) having taken root in some minds in the late 1960s, the obvious first objective was to reduce the attenuation (the weakening of the light beam with distance). Once this was achieved, and power loss of less than a factor of 10 in 20 kilometers was feasible, the next fundamental limitation was dispersion—the smearing of the pulses of light. A major improvement in dispersion was achieved with the development of single-mode fibers, which replaced multimode fibers and their intermodal dispersion problems. With this achievement, researchers once more focused on attenuation, since the new high data rate communications were limited to fiber spans of 40 to 60 kilometers between expensive electronic regeneration equipment. Long-distance telecommunications companies, which routinely carried traffic much further than 100 kilometers, saw the economic advantages of systems capable of transmitting high optical data rates for hundreds of kilometers. Proponents of optical communications promoted a vision of WDM (wavelength division multiplexing) that made this value even greater, but also made critical the development of a transparent signal regeneration technology, since no scenario in which wavelengths were separated at each regenerator and then recombined made business sense.

While there remained a prospect for the further reduction of fiber attenuation, theoretical limits seemed to indicate that only about a factor of 2 in reach extension could be achieved. The required reach extension was 10 times. Fortunately there was an alternative approach—optical amplification—in which the power of a light beam was increased (amplified) directly. Using stimulated emission, the same physics that underlies lasers, it was theoretically possible to prepare an "excited" medium through which light would pass, being amplified as it went. This amplification would preserve the phase relationships of the photons, so that any signal imposed on the light stream would be preserved. The theoretical cost is an admixture of random, or spontaneous, emission—background light that causes noise on the detector. This signal degradation is more than offset by the massive reduction in shot noise achieved by maintaining the signal power.

There were at least two obvious techniques for achieving optical signal gain, Raman amplification and amplification in semiconductor materials (semiconductor optical amplifiers, SOAs). Raman amplification occurs in all materials with its details depending on the material, and requires an intense "pump" beam of light at a wavelength shorter than that of the signal to drive the process, but will provide gain at a wavelength that can be controlled through the pump wavelength. SOAs were a clear extension of semiconductor lasers, and would share their advantages of bandgap engineering, allowing the material to be designed for operation in the required wavelength range, and direct electrical pumping. Both approaches would naturally lend themselves to a guided wave approach, Raman in optical fibers and SOAs in the waveguide structures already well developed for semiconductor lasers.

In fact, the practical realization of optical amplification came from quite a different direction, from a technology that did not offer the benefits of flexible operating wavelength. Researchers showed that to get to the threshold for Raman amplification required pump powers beyond the reach of cost-effective and practical implementations of the day. Engineers also failed to overcome the limitations in SOA attributes, ranging from functional impairments like limited bandwidth and extra underlying attenuation to more fundamental processes such as cross-talk (interference) between channels of different wavelengths, and intersymbol interference (ISI) within a channel.

Optical fiber made of very dry silica has a wavelength "window" of low attenuation, roughly from 1500 to 1600 nanometers in the near-infrared range (known for historical reasons as the third telecommunications window). While this window would have difficulties with intramodal dispersion at very high bit rates, it was otherwise an attractive region in which to operate. Therefore, any amplification scheme that could work in this window would be technically advantaged, even if it were not applicable to other wavelength bands. At the end of the 1980s, the amplification technology that leaped to prominence was the EDFA, the erbium-doped fiber amplifier. In some ways derived from ideas of the early 1960s, the approach was similar to well-understood laser systems. The fundamental problem was to fabricate an optical fiber with the light-guiding core doped with a small quantity of a rare-earth element. The rare-earth elements have rich spectroscopy in the visible and near-infrared region, and neodymium and erbium turned out to be particularly interesting to telecommunications. Once researchers in fiber fabrication invented processes to manufacture the fiber, erbium ions in a silica–germania host glass proved to be an almost ideal medium to provide gain in the short wavelength end of the third telecommunications window. The first EDFA was realized by researchers at the University of Southampton in the U.K. in 1986. These researchers and their counterparts worldwide quickly demonstrated effective performance, with gains of up to 30 decibels (a power increase of 1000 times) for relatively low pump powers, which scaled with the signal power. The pump light could be derived from existing fiber-coupled laser diodes, and there was the prospect of using different pump wavelengths with even more favorable properties if the laser diode technology could be developed.

In retrospect, we can see that the development of EDFAs in the late 1980s, with early commercialization at the end of the decade and in the early 1990s, set the stage for over a decade of continuous rapid improvement in optical telecommunications systems and drove rapid and major enhancements in semiconductor laser technology, dispersion compensation techniques and optical componentry. From the very first, practical EDFAs required pump/signal multiplexers, diode pumps and fiber-pigtailed isolators, to prevent these extremely high-gain devices from being turned into lasers by inevitable stray back-reflections. The EDFA combined almost immediate economic dominance by its ability to replace electrical regenerators, and its great potential value, which was clear to many, could be unlocked by developments in sister technologies. Therefore as pump powers increased, EDFAs could move gracefully from enhancing single-wavelength systems to enabling WDM (wavelength division multiplexed systems) carrying many wavelength channels, which multiplied the effective value of each installed amplifier manifold times. In fact, EDFAs were expensive enough to help justify a migration to WDM. As reliable pump lasers at 980 nanometers were developed, EDFAs could move from reliance on pumping at 1480 nanometers, and achieve fundamentally better performance. As 1480-nanometer lasers were increased in power, longer wavelength sub-bands within the third window could be used. Finally, designers configured EDFAs to allow midstage access; that is, a preamp and postamp could be sandwiched around a lossy element without significant degradation in end-to-end system performance. This platform supported the use of dispersion-compensating modules, allowing higher data rates per channel (10 gigabits per second), and of optical add-drop multiplexers, which turned point-to-point systems into optical networks.

Ironically, these technology trends, which were supported by the commercial application of EDFAs, have led to the reinvigoration of a competing technology, Raman amplification. The threshold pump power requirements, which once appeared prohibitive, are now achievable because available laser pump powers have increased by ten times to meet the requirements of WDM systems with over 100 channels. In fact, Raman amplification may enable the next telecommunications revolution, ultra-longhaul (ULH) systems with transmitter-to-receiver distances of more than 1500 kilometers, even up to the 5000 kilometers of a coast-to-coast transmission in the U.S. To achieve this, engineers can realize distributed Raman amplification by injecting pump power into the fiber span to give gain (or loss reduction) along a large fraction of the span. They can thus maintain optical signal-to-noise ratios several times higher than in a pure EDFA system with the same span lengths, where the amplification occurs in a module at the end of the span. While most ULH proposals involve a mixture of EDFAs and Raman amps, at least one supplier has proposed an all-Raman solution using distributed Raman amplification as well as discrete modules as direct substitutes for EDFAs.

The WDM backbone, carrying enormous amounts of data and voice traffic between more localized metropolitan and access communications networks, no longer represents a bottleneck in data transmission, and attention is shifting to the access network with the expectation that users will ultimately enjoy the advantages of bandwidth much greater than that at the time of writing. Business models for this market emphasize inexpensive technologies, and devices that cost little to manufacture and operate may have an overall advantage even if their technical attributes are less attractive than those of costlier alternatives. Thus SOAs are currently generating significant interest for their potential low-cost production and low operating power.

EDFAs, as the first practical implementation of optical amplification, drove the technological development of telecommunications in the last decade of the twentieth century, but their very success and influence may have sown seeds that will lead to their fall from prominence in the first decade of the twenty-first. Optical amplification is, however, definitely here to stay—without it, modern telecommunications systems would not function.

See also **Lasers in Optoelectronics; Optical Materials; Optoelectronics, Dense Wavelength Division Multiplexing; Telecommunications**

MARTIN HEMPSTEAD

Further Reading

Becker, P.C., Olsen, N.A. and Simpson, J.R. *Erbium-Doped Fiber Amplifiers*. Academic Press, New York, 1999.

Hecht, J. City of Light: *The Story of Fiber Optics*. Oxford University Press, Oxford, 1999.

Optical Fibers, *see* **Optical Materials**

Optical Materials

Optical materials were essential for many of the twentieth century's most significant technological accomplishments. Humans have been intrigued with the development of optical materials and light behavior for centuries. Thomas Alva Edison, Max Planck, Albert Einstein, Max Born, and Niels Bohr presented significant optical theories and innovations that provided a foundation for research, development, and application of twentieth century optical materials. Significant optical milestones included the introduction of lens coating in the 1930s and lasers in 1960; fiber optics emerged by mid-century, and thousands of scientists innovated and adapted uses for optical materials in the following decades. Engineers established the International Society for Optical Engineering (SPIE) in 1955 to coordinate professional efforts.

During the 1870s, German chemist Dr. Otto Schott sought to make glass with optical qualities, devising a lithium formula. By 1884, he produced optical glass specifically for lens and prism applications such as microscopes. With colleague Dr. Ernst Abbe and industrialist Carl Zeiss, Schott established the pioneering optical glass-manufacturing site at Jena, Germany. As a result, Germany was the main manufacturer of optical glass prior to World War I.

Initially, scientists were most concerned with the composition of optical glass. The production of optical glass and removal of impurities is crucial to achieve optimal performance. The homogeneity of optical glass differentiates it from most glass. Its consistent, predictable quality; strength and durability despite physical stresses and temperatures; precise images; minimization of distortion; and versatility for desired applications all make it ideal for optical usages. In the early twentieth century, German technologists advanced optical glass technology with such innovations as Schott's borosilicate and barium glass. By 1939, George W. Morey of the U.S. devised optical glass composed of tantalum, lanthanum, and thorium that increased refraction.

Before World War II, most optical glass manufacture resulted from clay pot melting processes. After the war, technologists focused on manufacturing aspects to improve optical glass. Some technicians used platinum pots and rare earth elements to create optical glass with minimal flaws (such as bubbles, weaknesses, and striae) where refraction differed. Japanese technologists at Hoya Corporation initiated a continuous melting technique in 1965 that produced flaw-free optical glass. Engineers worldwide devised better materials and melting processes, including electric melting, to create optical glass for electronics and cameras that were becoming lighter, smaller, and more precise because of optical glass lens advancements. Most modern optical glass is made from fused silica that is synthetically manufactured to enhance purity and quality.

In 1938, American Katherine J. Blodgett patented a layered soap film to produce transparent glass that did not reflect light rays and distort images in lenses because the soap and light waves neutralized each other. Her findings inspired optical coating work to apply transparent thin films to lenses, glasses, and electronic devices. Scientists developed materials especially for those purposes. Most modern optical coating materials are metals, semiconductors, or insulators.

In the late twentieth century, plastic lenses replaced eyeglass lenses. In addition to being more shatterproof, plastic material such as polycarbonate enables technicians to apply or incorporate a wider variety of optical technology, primarily in coatings, to enhance vision. Each side of a plastic lens is layered with thin films that aid in making the lens invulnerable to water, glares, and scratches. A physical vapor deposition technique is used for the antireflective coating in which such metal compounds as aluminum oxide undergo vaporization in a vacuum chamber and several layers of atoms are deposited on a lens. A hydrophobic coating is layered on the lens by chemical vapor deposition or submersion in a solution. Engineers improved coatings that protect eyes from ultraviolet light.

The transparent vinyl thermoplastic polymer polymethylmethacrylate (PMMA) was first industrially produced in the mid-1930s after chemists initiated research the previous decade. PMMA is commonly referred to as acrylic, a word derived from the name acrylates, the polymer family that includes PMMA. This hard synthetic plastic cannot be shattered like glass and is used for various optical needs. In 1928, the company Rohm & Haas (based in Philadelphia and Darmstadt) began to sell acrylics based on resins made by Rudolph Fittig in the nineteenth century and adapted by German chemist Otto Rohm, for coating use as a celluloid substitute. At Imperial Chemical Industries, English chemists John Crawford and Rowland Hill made a harder acrylic in 1934 in a heating, pouring, baking, and cooling process to create acrylic sheets that were marketed as Perspex. Competition between acrylic manufacturers resulted in improved acrylic production methods and applications. Acrylic is more vulnerable to impact damage than polycarbonate in lenses, but PMMA is considered a valuable optical material for its superb clarity, electrical characteristics, and resistance to scratching. Engineers developed computer-guided looms to weave acrylic optical fibers. Acrylic is also used for LED and other optical displays. Lucite (DuPont) and Plexiglass (Rohm & Haas) are well-known trade names for acrylic optical products.

Because of its lightness, flexibility, affordability, and ability to be tinted with color, plastic is often shaped to make contact lenses. Leonardo da Vinci first suggested contact lenses in the early sixteenth century, and scientists developed similar glass lenses in the following centuries. In the mid-1930s, American Dr. William Feinbloom created hard plastic contact lenses. Californian Dr. Kevin Tuohy received a PMMA corneal contact lens patent in 1950. Industries such as Bausch and Lomb produced lenses, and researchers improved designs to resolve problems patients reported. English ophthalmologist Harold Ridley innovated the intraocular lens (IOL) in 1949 for lens replacement in cataract surgery. Although PMMA is used for IOLs, chemists developed softer acrylics and silicones that can be folded to insert in smaller eye incisions than required for PMMA IOLs.

Softer plastics gradually replaced PMMA as a lens material except for specific uses such as treating astigmatism. By the 1960s, contact lenses became more appealing when Czechoslovakian optometrists Otto Wicherle and Drahoslav Lim used the polymer hydroxyethylmethacrylate (HEMA) to make soft contact lenses. They selected materials that can absorb moisture to become flexible. By 1979, rigid gas-permeable (RGP) contact lenses containing primarily silicone (which permits oxygen to reach the eyes) were introduced. RGPs often correct vision problems that soft lenses cannot, last longer, and do not collect as much debris as soft lenses. Technological advancements in contact lenses included bifocals, disposable contacts, and improved material formulas that resulted in thinner lenses and increased oxygen access to the eyes.

Fiber optics enable telecommunications to span the globe swiftly and clearly. Glass fibers transmit infrared light pulses that are sounds or digital information transformed into light by semiconductor lasers. Optical fibers, made of a core inside a separate glass cladding, consist primarily of extremely pure silica. Scientists add fluorine and boron to the cladding and phosphorous and germanium to the core to manipulate those components' refractive index as needed for efficient light movement through fibers. The cladding prevents light from leaking out of the core by reflecting light within the boundaries of the core (total internal reflection).

Visionaries who proposed fiber optic technology in the 1920s and 1930s but did not follow through with their ideas included American Clarence Hansell, who received a patent for bundling glass fibers, and German Heinrich Lamm, who experimented with glass fibers for surgery. Narinder S. Kapany is often credited as fiber optics' inventor, but his Massachusetts Institute of Technology doctoral advisor, Harold H. Hopkins, suggested Kapany's topic. Hopkins had been investigating glass fibers and wanted to bundle them to transmit images when he received a grant which he used to hire Kapany. The initial letter describing their work was published in *Nature* on 2 January 1954 beneath a letter about fiber bundling that Dutch scientist Abraham C.S. van Heel had submitted months before. Kapany wrote a *Scientific American* feature in 1960 and the first book on that topic.

Dr. Charles K. Kao is also cited as the innovator of fiber optics because of his research at the English ITT Standard Telecommunications Laboratories in the 1960s. Kao and Charles Hockham wrote about fiber optic possibilities. Kao was particularly interested in purifying optical materials such as silica compounds to remove metal impurities and improve transmissions. During the early 1970s, Corning Glass Works researchers Robert Maurer, Don Keck, and Peter Schultz developed a heat process to make extremely clear glass fibers from fused silica doped with germanium. Their achievement resulted in fibers with much lower attenuation (power loss with distance along the fiber), and fiber optic communications became feasible. Losses in optical fibers were much lower than in copper cables, and fewer repeaters meant lower cost systems. At the same time, researchers developed semiconductor lasers for use as fiber-optic light sources.

In addition to these pioneers, since the 1960s, many people and corporations, often fiercely competing to secure patents, have contributed to the advancement and distribution of fiber optics worldwide. Some researchers focused on multimode optic fibers instead of single-mode fibers, delaying the technology in some regions because of interference and noise caused by conflicting modes and waves that can disrupt transmissions. Toni Karbowiak's 1964 waveguide aided acceptance of optical fibers when they first became publicly available in the 1970s. On 19 May 1971, Queen Elizabeth II observed a fiber-optic video presentation, and other trials were held in the U.S. By 1975, a pioneering fiber-optic system was in use for communications by police in Great Britain. Engineers at British Telecom, Bell, and GTE introduced fiber optics into their systems.

Researchers advanced fiber optics quickly in the 1980s, improving services particularly for long-distance telephone calls. International service advanced in 1988 with the introduction of TAT-8, the initial transatlantic fiber-optic cable that provided increased circuits, greater capacity, and clearer signals than satellite and wire connections. More fiber-optic cables connected continents, with each fiber capable of transmitting several hundred million bits every second. During that decade, the University of Southampton's Dave Payne realized erbium would be the most useful amplifier material in fibers for clear, uninterrupted signals transmitted via ocean cables.

Fiber optics reduced the need for amplifiers due to attenuation. Unless water vapor is present, most silica in optic fibers does not absorb material sufficiently to interrupt signal movement. At higher wavelengths, silica is prone to absorb material, and substitute fiber sources such as fluorozirconate are often used. Engineers focus on solving dispersion problems related to silica interaction with frequencies when various parts of a signal move at different speeds in fibers toward receivers, as this often creates pulse interference.

Researchers are constantly advancing fiber-optic materials, systems, and applications to achieve greater speed, amount of information transmitted, and practical uses. For example, fiber optics enables surgeons to access and examine internal organs without performing surgery. Fiber optics help guide missiles, find earthquake victims, provide transportation signals, and act as sensors. Although many communities' telephones, Internet, and cable television services relied on fiber-optic networks, because such technology is expensive, few private homes had optic fibers by the end of the twentieth century and were connected to those systems by wires.

See also **Electronic Communications; Lasers in Optoelectronics; Materials and Industrial Processes; Optometry; Telephony, Long Distance**

Elizabeth D. Schafer

Further Reading

Agrawal, G.P. *Fiber-Optic Communication Systems.* Wiley, New York, 1997.

Buck, J.A. *Fundamentals of Optical Fibers.* Wiley, New York, 1995.

Chaffee, C. D. *The Rewiring of America: The Fiber Optics Revolution.* Academic Press, Boston, 1999.

Fenichell, S. *Plastic: The Making of a Synthetic Century.* Harper Business, New York, 1996.

Hecht, J. *City of Light: The Story of Fiber Optics.* Oxford University Press, New York, 1999.

Izumitani, T.S. *Optical Glass.* American Institute of Physics, New York, 1986.

Kao, C.K. *Optical Fiber Systems: Technology, Design, and Applications.* McGraw-Hill, New York, 1982.

Kao, C.K. *Optical Fibre.* IEE Materials & Devices Series vol. 6, Peregrinus/Institution of Electrical Engineers, London, 1988.

Kao, C.K., Ed. *Optical Fiber Technology II*. Wiley/Institute of Electrical and Electronics Engineers, New York, 1981.
Kapany, N.S. *Fiber Optics: Principles and Applications*. Academic Press, New York, 1967.
Marker, AJ. III, Ed. Properties and characteristics of optical glass. *Proceedings of SPIE—The International Society for Optical Engineering*, 970, SPIE, Bellingham, 1988.
Meikle, J.L. *American Plastic: A Cultural History*. Rutgers University Press, New Brunswick, NJ, 1995.
Morey, G.W. *The Properties of Glass*. Reinhold, New York, 1938.
Mossman, S., Ed. *Early Plastics: Perspectives, 1850–1950*. Leicester University Press, London, 1997.
Mossman, S.T.I. and Morris, P.J.T., Eds. *The Development of Plastics*. Royal Society of Chemistry, Cambridge, 1994.
Park, D. *Fire Within the Eye: A Historical Essay on the Nature and Meaning of Light*. Princeton University Press, Princeton, 1997.
Seymour, R.B., Ed. *History of Polymer Science and Technology*. Dekker, New York, 1982.
Seymour, R.B. *Pioneers in Polymer Science*. Kluwer, Dordrecht, 1989.
Su, F., Ed. *Technology of Our Times: People and Innovation in Optics and Optoelectronics*. SPIE Optical Engineering Press, Bellingham, 1990.
Willey, R.R. *Practical Design and Production of Optical Thin Films*. Dekker, New York, 1996.
Yariv, A. *Optical Electronics*, 5th edn. Oxford University Press, New York, 1997.

Useful Websites

International Society for Optical Engineering: http://www.spie.org

Optoelectronics, Dense Wavelength Division Multiplexing

The cluster of technologies that are dense wavelength division multiplexing (DWDM) emerged during the last decade of the twentieth century. Multiplexing is the sending of separate signals with one transmitter in one optical signal, to increase data capacity. The pioneers of optical fiber communications had very early understood the potential of telecommunication fibers to carry more than one communication channel if the channel were on different optical wavelengths ("colors") but it took an army of researchers many years to realize the necessary technical solutions.

The advantages of DWDM are very clear—the technology offers a relatively straightforward path to exploiting the enormous inherent bandwidth of optical fiber. DWDM involves an array of modulated light sources at discrete wavelengths, which are combined ("multiplexed") onto a single transmission fiber. The signal may travel many hundreds of kilometers, being amplified every 80 to 100 kilometers by erbium-doped fiber amplifiers (EDFAs), which are continuations of the transparent path and require no separation of the channels. Ultimately, the channels are demultiplexed at the terminal equipment and each channel is individually detected and its modulated data converted to digital electronic format.

Until EDFAs emerged at the beginning of the 1990s, displacing electronic regenerators, DWDM was an expensive luxury. Electronic regeneration required the demultiplexing of channels at each node, so DWDM had little or no advantage over the use of a different fiber for each channel. The single channel-one fiber approach at least avoided expensive optical multiplexing, and allowed the use of any signal laser source, without tight wavelength control, for each channel, thus saving on the cost of maintaining a large inventory of expensive spare sources. The EDFA made it most economical to put all channels on the same fiber, saving cost at the nodes at the expense of terminal multiplexing and demultiplexing.

A number of technologies were required to make DWDM successful, and we will touch on each in turn.

- Since the channels are defined by the filter passbands in the optical multiplexers, light sources—semiconductor lasers—had to be stable in wavelength and available on a well-defined grid of wavelengths. They also had to be capable of modulation in a format that would be stable over the transparent links. In particular, they had to have acceptable "chirp" (phase distortion over the pulse) so that dispersion in the optical fiber would not degrade the pulse. The initial solutions were directly modulated diodes with integral grating-based distributed feedback (DFB) to lock the lasing wavelength.
- The multiplexers (which in reverse also served as demultiplexers) were of course key to the entire DWDM approach. Their function was to take a number of channels of different wavelengths each on a separate fiber and put them out on a single fiber. While this could be done with simple couplers such as fused biconic taper (FBT) couplers, the loss at each coupler is a factor of 2 in power, so high channel count DWDM would require very high laser powers. Since beams of light at different wavelengths can in principle be combined without loss, a more elegant and more scalable approach was to

use some kind of resonant coupling structure, for example, grating-assisted couplers or dichroic thin film filters.

- EDFAs suited for multichannel amplification had to be designed for high power, which required the development of more powerful pump lasers. The uniformity of gain across the wavelength band was also a challenge. While this was initially solved by careful positioning of the signal wavelengths in the flattest part of the erbium gain spectrum, the inevitable demand for more and more channels forced the use of gain-flattening filters in the amplifiers to tailor the gain spectrum. GFFs are optical filters with spectral loss curves engineered to match the erbium spectrum. Initially researchers manufactured them from carefully designed multilayer thin film filters, and later they used fiber Bragg grating (FBG) technology, when the grating community had developed techniques for fabricating complex chirped gratings.

The first DWDM systems were installed in submarine cables in 1990. Terrestrial systems followed later—they initially sported just four channels, each channel operating at 2.5 gigabits per second (Gbps), and were installed in about 1994. However, the expectations of long-distance telecommunications carriers quickly exceeded this unprecedented single fiber bandwidth of 10 Gbps, and an explosion of activity in the next few years pushed channel data rates quickly up to the next level—OC192 or 10 Gbps—and channel counts from 4 to 16 to 32 to 96, while the separation between adjacent channel wavelengths decreased accordingly. The capabilities of EDFA pump, multiplexer and GFF technologies had to be continuously upgraded to support this revolutionary increase in communication capacity. At OC192 rates, fiber dispersion became a problem and the new technology of dispersion compensation was also introduced, based on dispersion-compensating fibers that had been invented earlier in the decade. Dispersion compensation modules (DCMs) are typically sandwiched between EDFAs to compensate for their loss and thus minimize their impact on signal-to-noise ratio. Finally, the design of the fiber itself was modified to deliver better performance for the high-density high data-rate communications on the horizon. By the end of this explosion the potential capacity of new systems was about 1 terabit per second (10^{12} bits per second) per fiber, and the rediscovery of Raman amplification was set to extend system reach from 600 to 5000 kilometers.

With this breathtaking success under its belt, the telecommunications engineering community in 2000 confidently anticipated the continued rapid evolution of channel data rates to 40 Gbps, a doubling or more of the useable fiber bandwidth and the imminent fulfillment of the Raman promise. New challenging problems would have to be tackled—optical nonlinearities in the fiber (self-phase modulation (SPM), four-wave mixing (FWM), cross-phase modulation (XPM), etc.) and polarization mode dispersion (PMD) were the next hurdles to be overcome. New signal modulation formats and enhanced forward error-correction schemes would help to overcome these challenges. Unfortunately, the technological revolution had been accompanied by a massive expansion in fiber plant and high-performance system installation, fueled by the same expectation of continually exploding demand for bandwidth and solid revenue streams that had driven the investments in technology. While bandwidth continued to grow, albeit more slowly than predicted, the revenue did not materialize, and the fiber buildout turned out to have delivered a glut in capacity that devastated the telecommunications market as the bubble burst in the early years of the twenty-first century.

See also **Electronic Communications; Lasers in Optoelectronics; Optical Materials; Optoelectronics, Frequency Changing; Telephony, Digital**

Martin Hempstead

Further Reading

Laude, J.-P. *DWDM Fundamentals, Components, and Applications*. Artech House, 2002.

Ramaswami, R. and Sivarajan, K. *Optical Networks: A Practical Perspective*, 2nd edn. Morgan Kaufman, 2001.

Optoelectronics, Frequency Changing

Lasers produce monochromatic light; that is, light with a single frequency or wavelength. Many laser applications, such as atomic spectroscopy, depend on the ability of a particular laser to be frequency tunable. Fine tuning (small shifts in output frequency) can be achieved by adjustments of operating characteristics of dye and semiconductor lasers. In the area of fiber-optic communications, a large number of output wavelengths is desirable for wavelength division multiplexed (WDM) multichannel optical communication systems that utilize many wavelengths to increase data capacity. This can be achieved by nonlinear effects that create new optical frequencies from fundamental frequencies (frequency doubling), or with tunable laser sources.

A nonlinear laser-induced effect, discovered by Peter Franken and co-workers in 1961, occurs when an intense laser beam propagates through a nonlinear optical medium (quartz, for example). Light at double the frequency of the input beam is produced, an effect known as frequency doubling or second harmonic generation. Second harmonic generation can usefully convert the coherent output of a fixed-frequency laser to a different spectral region. For example, the infrared radiation of a Nd:YAG laser operating at 1064 nanometers (infrared) can be converted into 532-nanometer visible radiation with a conversion efficiency of more than 50 percent. Novel laser sources produced by frequency doubling may offer advantages over existing laser sources that are bulky or inefficient; ultrashort laser pulses produced by frequency doubling or tripling may also have applications in the early twenty-first century for high-density optical storage, increased data transmission rates, and ultrafast spectroscopy of biochemical processes.

As the intensity of the incident light exceeds a certain threshold value, a nonlinear effect known as stimulated Brillouin scattering, which scatters light to different wavelengths, becomes important in fibers. Power is lost and the frequency-shifted wave may "cross-talk" with wavelengths in neighboring channels. Both of these effects degrade the optical signals and are undesirable.

Tunable dye lasers were discovered in 1966 by Peter Sorokin and John Lankard in the U.S. and Fritz Schäfer in Germany. Liquid lasers of this type can be made for almost any wavelength from the ultraviolet to the infrared, dependent upon the fluorescence of the dye. Dye lasers may be tuned, through 100-nanometer ranges or more, by either changing the concentration of the dye or by adding and turning a diffraction grating in place of one of the cavity mirrors.

The class of laser used in the majority of today's telecommunication systems is the semiconductor laser. This class of laser was first made from gallium arsenide (GaAs), but now they are commonly formed by a compound of elements from groups III and V of the periodic table (see Semiconductors, Compound). The end faces of the crystal (0.1 to 1 millimeter thickness) form mirrors that create the necessary cavity to trap the light and sustain stimulated emission.

Semiconductor lasers operate as either "fixed" (a single wavelength, or frequency) or "tunable" (offering coarse or fine tuning of many frequencies across a specific frequency band). The wavelengths of tunable semiconductor lasers depend upon both the properties of the III–V gain medium and the physical structure of the laser cavity surrounding the gain medium. Specifically, the length of the cavity (i.e., the distance between the mirrors) and the speed of light within the gain medium within the cavity determine a laser's wavelength. A semiconductor laser can therefore be tuned by mechanically adjusting the cavity length or by changing the refractive index (the speed) of the gain medium. Alternatively, light can be adjusted externally to the laser source, using micromachined elements such as micromirrors or actuators.

Tunable semiconductor lasers can be grouped into four types:

1. The "edge emitting" triad of distributed feedback (DFB)
2. Distributed Bragg reflector (DBR)
3. External cavity diode lasers (ECDL)
4. The "surface emitting" type known as vertical cavity surface-emitting lasers (VCSEL)

As one might suspect, edge-emitting devices emit light at the substrate edges, whereas VCSELs emit light at the surface of the laser diode chip.

Rather than placing the resonator mirrors at the edges of the device, the mirrors in a VCSEL are located on the top and bottom of the semiconductor material. Somewhat confusingly, these mirrors are typically DBR devices. This arrangement causes light to "bounce" vertically in a laser chip, so that the light emerges through the top of the device, rather than the edge. As a result, VCSELs produce beams of a more circular nature than their cousins and beams that do not diverge as rapidly. These characteristics enable a more efficient coupling of VCSELs to optical fibers. VCSELs, furthermore, benefit from single-process manufacturing and a relatively straightforward tuning process involving a microelectromechanical-systems (MEMs) cantilever arm placed directly above the optical cavity. Moving the arm a matter of a few micrometers up or down changes the frequency of the device by up to 5 percent.

The edge-emitter family features diffraction gratings etched on a single chip, as in DFB lasers; gratings placed near the active region of the laser cavity, in the case of DBRs; and one or two mirrors combined with a conventional laser chip to reflect light back into the cavity, as found in ECDLs.

DFB lasers can be tuned by controlling the temperature of the laser diode cavity. Because this technique requires large temperature differences, a single DFB has a small tuning range. However, it is possible to link multiple DFBs together to create multiple cavities and therefore wider tuning outputs. DBRs are actually variations of the DFB. In addition to having their grating (or mirror) section in a separate portion of the chip, a DBR has a gain section and a phase section. Tuning occurs when current is injected into the phase and mirror sections to change the carrier density of those two sections and, as a result, the wavelength of light they refract. As with DFBs, DBRs have a somewhat limited tuning range, but techniques have been developed to expand that capability; for example specialized gratings known as "sampled gratings" and grating and bidirectional coupler combinations. The ECDL achieves tunability by physically moving a wavelength selective element, such as a grating or prism, to tune the laser output. One method involves moving a reflector up and down relative to the surface of a diffraction grating, with the varying distances determining specific wavelengths. This particular tuning method gives it a wide tuning range, but a slow tuning speed compared to the other methods.

Because of their inherent design, specifically characteristics of the gain medium and cavity, VCSELs have lower power than the other three types and as a result are used principally for local or metro (metropolitan area) wavelength applications. ECDLs have the highest power output of the tunables discussed here and are used for long distance networks. The others lie somewhere in the "middle space" between VCSELs and ECDLs and are used for metro and regional applications.

As noted above, tunable lasers have emerged for use in optical communication systems and specifically "wave division multiplexing" (WDM) or dense WDM (DWDM) applications, which entail the process of sending many different wavelengths carrying information down a single fiber optic strand to increase data capacity. WDM emerged as a solution to the bandwidth crunch imposed on telecommunications by the ever-increasing growth of the Internet and other broadband applications and it presents an exciting and new enabling application for tunable lasers in local, regional and long-distance networks.

In future all-optical networks (i.e., those without conversions to electronic signals), fiber optic systems will require tunable lasers that can provide a signal into any WDM channel and that can switch among channels, tuning to a new output wavelength in nanoseconds. In 2001 and 2002, some of the first true tunable semiconductor lasers moved from prototypes to early production.

See also **Lasers in Optoelectronics; Optoelectronics, Dense Wavelength Division Multiplexing**

CHARLES DUVALL

Further Reading

Bruce, E. Tunable lasers, *IEEE Spectrum,* 35–39, 2002.

Chang-Hasnain, C.J. Tunable VCSEL. *IEEE Journal on Selected Topics in Quantum Electronics*, 6, 6, 978–987, 2000.

Schäfer, F. P., Schmidt, W. and Volze, J. Organic dye solution laser, *Appl. Phys. Lett.* 9, 306–309, 1966.

Sorokin, P. P. and Lankard, J.R. Stimulated emission observed from an organic dye, choloalurinum phthalocyanine. *IBM J. Res. Develop*., 10, 162–163, 1966.

Optometry

The word, "optometry" was introduced to ophthalmology in 1904 and is important as an adjunct to that medical specialty because it focuses on examination of the eyes, analysis of vision, and the prescription of corrective or preventive measures for any deficits or problems. An optometrist will refer a patient to an ophthalmologist when pathologies of the eye are found during examination of visual function. Most of the technology used in optometry involves measurement and lenses. The only pharmaceuticals used are pupil dilators, drugs used to make the pupils larger so that the posterior parts of the eye can be examined.

Optometry developed from a split in the optical profession that took place in the nineteenth century that resulted two kinds of opticians, refracting and dispensing. Refracting opticians became optometrists. In Canada and the U.S., an optometrist must obtain an undergraduate degree and then undertake four years of specialized graduate education plus clinical and resident training to become an OD, or doctor of optometry. By contrast, a dispensing optician's training is from six months to two years (which varies by location), and does not necessarily require formal education. The optician can apprentice with an ophthalmologist, optometrist, or another optician. An optician makes or dispenses lenses and eyeglasses; an optometrist performs eye examinations.

Technologies of optometry can be subdivided into two categories: development of eyeglasses, contact lenses, and coatings; and instruments of measurement.

The idea that a convex lens could be used to magnify was known to the ancients. The use of a lens as a magnifying glass to aid vision is attributed to Roger Bacon in Opus Majus (major work), written in 1268. Soon after, Italian monks D'Armate (1284), da Rivalto (1306), or possibly a competitive contemporary, da Spina, invented eyeglasses. This early device was hinged and held two magnifying spheres that rested uncomfortably on the nose without support.

In 1303, Bernard of Gordon (Montpellier, France) fashioned a pair of spectacles with a fixed bar, but it took 400 years of various experiments, from ribbons looped around the ears, to weights that balanced glasses on the nose, until a design was engineered to allow the glasses enough stability to free the reader's hands. Corrective lenses were developed in the early seventeenth century.

The popularity of spectacles grew throughout Europe, but because they were promoted by itinerant peddlers with little or no education, medical professionals were hesitant to endorse such products. Ironically, the growth of optometry was not because of oculists or physicians but in spite of them. Each time these unlicensed vendors and peddlers were able to fit spectacles to someone visually impaired, it increased their knowledge and skill.

By the nineteenth century, telescopes, microscopes, and cameras, and optical instruments required lenses for their use. However it was the adaptation of these lenses to that gave optometry its refined technology. An early ophthalmometer that used a small telescope was developed by E. Javal and H. Schiötz in Germany in 1881 and later modified into an astigmometer. The ophthalmoscope was invented by Hermann Von Helmholtz (Germany) in 1851; the retinoscope was introduced by Cuignet in 1873 in France; and Placido's keratoscope appeared in 1882. By the beginning of the twentieth century these instruments were in place to fit the established practices of optometrists.

The greatest changes to optometry in the twentieth century were in modifications of the instruments for diagnosis and the proliferation of materials that rendered lenses—for both glasses and instruments of measurement—thinner, lighter, more durable, safer, and more accurate. The four major instruments used by optometrists are the retinoscope, or skiascope, the slit lamp, the phoropter (introduced in 1938), and the ophthalmoscope.

The retinoscope illuminates the retina. It is used to test astigmatism, farsightedness, and nearsightedness. Early machines used a light source directed from a mirror. In the early 1900s, the technique was refined and termed spot retinoscopy. The operator recorded the direction of movement of the light on the retinal surface and the angle of the light rays that emerged as a reflex from the patient's eye, a phenomenon known as refraction. In 1926, ophthalmologist Jack Copeland inadvertently dropped the spot retinoscope he was using. In what might be considered a serendipitous event, he recognized that the damaged equipment could still be used and that it produced a clearer image of the patient's eye. Thus the streak retinoscope was invented. He patented it in 1927, and it remained unchanged until 1968 when a cordless retinoscope was developed by Optec (the Optec 360). In the twentieth century, independent light sources—an electric bulb or battery-operated machine—were employed instead of mirrors. In 1992, a new retinoscope was developed which did not require measurement by the operator at all. Instead, the patient's eye was measured against computerized predetermined calibrations and then transposed into meaningful data that was used by the optometrist to determine the patient's prescription if correction were needed and the lenses that would be required.

The slit lamp looks like a very elaborate microscope. It sits on a table and has a chin and headrest attached. The light source with controls is on a movable arm and the operator is able to adjust brightness, the size of the beam of light, and insert different filters. It uses a set of hand-held lights directed at the front of the patient's eye and examines the eyelid, the sclera (white of the eye), conjunctiva (mucous membranes), iris (colored part of the eye), lens, and the cornea (outer covering). It is called a slit lamp because its light source shines as a slit. This specialized magnifying microscope-type device was invented by Allvar Gullstrand in 1911 in Sweden. Later, attachments were added to include a camera for taking photographs, a tonometer for measurement of intraocular pressure (significant in glaucoma), a pachymeter to measure corneal thickness, and a laser treatment module for the ophthalmologist. It also had changeable lenses.

The phoropter is imposing in appearance. It is suspended from a bar and looks like a giant pair of glasses with five lenses on each side. The machine is used to test vision by moving sets of lenses in front of the patient's eye. Traditional phoropters are used to both measure and correct vision in order to derive an optical prescription for the patient. In the last decade of the twentieth century, a MEMS (microelectromechanical semiconductor)-based adaptive optics phoropter (MAOP) was developed that used adaptive-optics technologies and a deformable mirror. Interestingly, this technology was originally developed for astronomy applications. Of benefit to the optometrist is that the newer machine requires less space and automatically calculates the numbers required for the vision correction prescription.

The von Helmholtz ophthalmoscope consisted of a handle and a group of lenses that could be interchanged. It is an instrument that allows the doctor to look inside a person's eye and view the optic disc. Modern ophthalmoscopes are either direct or indirect. The direct is a hand-held instrument with a battery-powered light source. It has a series of lenses that are dialed in to focus the doctor's view of the central retina. The indirect ophthalmoscope is used to examine the entire retina. This instrument is worn on the doctor's head and another lens is placed in front of the patient's eye. The ophthalmoscope is used by both optometrists and ophthalmologists.

Materials

Eyeglasses were made exclusively from glass until the 1950s when the revolution in plastics began. Two companies pioneered the development of a thinner, lighter, and flatter lens made from plastic: American Optical (AO) and Columbia Southern Chemical Company. In 1937, AO produced a polymethyl methacrylate (PMMA), a hard plastic material that could be molded for lenses, but it was not scratch resistant and often distorted under high temperatures. During World War II, Columbia Southern Chemical Company produced a series of 200 polymers. From these, number 39, the chemical composition of which is allyl diglycol carbonate (ADC), had ideal qualities for a lens. It did not soften or distort at high temperatures and was scratch resistant. It was cast rather than molded. Named CR-39, this became the industry standard for eyeglass and contact lenses.

In 1971, a flexible plastic contact lens containing a gel was developed that allowed the wearer to change the lens. The advantage was cosmetic and economical. Eye color could be changed with the lens and the price was greatly reduced, thus concerns about loss or breakage were less of an issue. In 1978, a rigid gas permeable lens (RGP) was developed that could be custom fitted to the cornea. By 1983, they were available for commercial distribution. Although they were hard, they allowed for the exchange of oxygen on the surface of the eye and were particularly advantageous to people who had undergone surgery for cataract removal. An extended-wear plastic, developed in 1981 allowed the wearer to keep the lens on the eye for longer than 24 hours and even sleep with it. These could remain as long as one week without change. A few years later, the disposable contact lens was on the market. By the late 1990s, the rigid gas-permeable contact lenses formula of fluorosilicone acrylate was modified. At the turn of the century, contact lens possibilities included disposable tinted lenses that could be used for two weeks; disposable lenses with ultraviolet ray protection; and multifocal disposable soft lenses that made a contact "bifocal" possible, thereby moving from eyeglasses and contacts to contacts alone.

See also **Ophthalmology; Optical Materials**

ELIZABETH D. SCHAFER

Further Reading

Garrison, F.H. *An Introduction to the History of Medicine*, 4th edn. Saunders, Philadelphia, 1929.
Gregg, J. *The Story of Optometry*. Ronald Press, New York, 1965.
Lyons and Petrucelli. *Medicine, an Illustrated History*. Abrams, New York, 1987.
Stein, H., Slatt, B. and Stein, R. *The Ophthalmic Assistant: A Guide for Ophthalmic Personnel*. Mosby, 2000.

Useful Websites/Further Information

http://web.grinnell.edu/groups/sca/glasses/glasses.htm
http://www.contactlens.co.uk/education/public/history.htm
http://www.contactlenses.org/rgpvsoft.htm
http://www.mse.utah.edu/students/MSE5471/ContGlas/Glasses%20Processing.htm
http://spectacle.berkeley.edu/class/opt10/lect2.shtm

Many institutions and companies can be consulted for information on specific machines. A few are: The Lawrence Livermore National Laboratory in California; Bausch & Lomb in Rochester, NY; Boston Micromachines Corp., Watertown, MA; Sandia National Laboratories, Livermore, CA; and Wavefront Sciences, Albuquerque, NM.

Organ Transplantation

Organ transplantation is a specialized case of the transplantation of living body tissue. In the 1880s scientists and surgeons started developing a concept of organ replacement that formed the basis of the technique. According to this concept it is possible to treat complex internal diseases by replacing the lost function of a particular organ. The roots of organ transplantation lay in the ever more sophisticated surgical strategy of removing diseased tissues. Surgeons had noticed that the lack of particular organ tissue led to the development of specific disease phenomena, but reinserting the tissue into the body reversed disease development.

For the general acceptance of this concept it was significant that surgeons and physiologists were able to create and stop disease symptoms at will, using experimental animals under the controlled conditions of the laboratory. The first organ to be examined in this way was the thyroid gland. It was also the subject of the very first organ transplant by the Swiss surgeon and Nobel laureate Theodor Kocher in 1883. The principle was then applied to other organs, starting with other endocrine glands—pancreas, testes, ovaries, and adrenal glands. In 1905 Alexis Carrel and C.C. Guthrie in New York City carried out the first heart transplant in a dog. In 1906 Mathieu Jaboulay in Lyons performed the first kidney transplant in a human being, using an animal kidney.

The technical aspects of organ transplantation were mastered in the first decade of the twentieth century when Carrel developed a reliable and effective technique for suturing blood vessels. Carrel was awarded the Nobel Prize in 1912 for his work on vascular surgery and organ transplantation. With growing technical perfection it became clear that organ transplants between different individuals (allotransplantation) normally failed because of some specific factor associated with the biological identity of individuals. Some researchers ascribed the rejection of foreign tissue to the same mechanism that was also responsible for the body's defense against infectious agents. In the 1920s the German surgeon Georg Schöne introduced the notion of transplant immunity to describe the phenomenon. However, all attempts to prevent the rejection of allotransplants by suppressing the immune reaction or by selecting suitable donors failed. As a result of the inability to overcome these difficulties, organ transplantation was gradually abandoned in the course of the 1920s.

In 1945 a new phase in the history of organ transplants was initiated when surgeons at the Peter Bent Brigham Hospital in Boston transplanted a kidney from a dead donor to a woman suffering from renal failure. Even though this and subsequent transplants failed, the American surgeons did not abandon their efforts. In 1954, again in Boston, a patient with renal failure was given a kidney from his identical twin brother. The transplantation was successful, and in 1990 the operating surgeon Joseph E. Murray was awarded the Nobel Prize.

In the latter case it was possible to avoid transplant rejection by the selection of an appropriate donor. In order to make allotransplanation applicable on a broader scale, however, surgeons had to pursue a different strategy, which consisted in suppressing the recipient's immunological reaction against the transplant. In the 1940s, Peter Medawar and Macfarlane Burnet had described the principles of immunological rejection, following wartime work on tissue grafting. Initial trials using x-ray radiation for immunological suppression proved too damaging for the recipient, so further attempts concentrated on chemical immune suppression. In 1962 the first successful kidney transplantation from a nonrelated donor was performed in Boston. Immune suppression had been achieved by the antimetabolic agent azathioprine. This approach was subsequently perfected, one of its milestones being the introduction of the immune suppressor cyclosporine in 1982, which enabled more effective but simultaneously more selective suppression of tissue rejection. At the same time, efforts to select suitable organs from nonrelated donors were being made with the help of tissue typing using the human leucocyte antigen (HLA) system as a marker of compatibility. The allocation of transplants was organized by special organizations such as Eurotransplant, founded in 1967 to enable distribution in Austria, Belgium, Luxemburg and Germany.

Apart from kidneys, other organs were also soon transplanted in humans. Most spectacular was the first successful heart transplant in a human performed by Christiaan Barnard in Cape Town, South Africa in 1967. Because of technical and biological difficulties, however, heart transplantation was almost stopped during the 1970s, only to be resumed after the introduction of cyclosporine in the 1980s. At that time the transplantation of other organs, such as the liver, pancreas, and lungs were becoming more successful and therefore a more popular therapeutic option.

However, despite the development of new drugs for immune suppression and the introduction of immunological means to influence rejection (e.g., antilymphocyte globulines), the maintenance of long-term function of transplanted tissue is still considered an unresolved issue in the field. The other main problem concerns the procurement of organs. In the late 1960s, organ procurement from dead donors was regulated by formalizing criteria for the diagnosis of brain death. Brain death is a state in which the brain has died, but the rest of the body is kept alive with intensive care measures such as artificial ventilation. Regulations were issued in different countries. For instance in the U.S. the so-called Ad Hoc Committee of the Harvard Medical School published a set of guidelines in 1968. Nonetheless, donation rates in no way kept up with the demand for organ tissue. Living donor transplants, which would be a viable alternative, are largely restricted to the kidneys, though split liver transplants have been performed with hepatic tissue from living donors. Another strategy to relieve organ shortage is the use of animal tissue. The procedure is called xenotransplantation and was tried from the very beginning of transplant medicine in the late nineteenth century. Results, however, were poor, and despite a number of research and development programs with pig organs in the 1980s and 1990s, xenotransplants do not seem to be a realistic option for the near future. The idea of growing tissues or even whole organs, which had been pursued by Alexis Carrel together with the engineer and aviator Charles Lindbergh in the 1930s, became popular again in the 1990s and led to the development of the field of tissue engineering. However, even if surgery should overcome all technical obstacles, transplantation will continue to raise a number of relevant cultural and bioethical issues concerning personal identity and the definition of human life. Organ transplantation is, after all, a technology that transcends boundaries of the individual body that had previously been taken for granted.

See also **Blood Transfusion and Blood Products; Immunological Technology**

THOMAS SCHLICH

Further Reading

Brent, L. *A History of Transplantation Immunology*. Academic Press, San Diego, 1997.
Fox, R.C. and Swazey, J.P. *The Courage to Fail. A Social View of Organ Transplants and Dialysis*. University of Chicago Press, Chicago, 1974.
Fox, R.C. and Swazey, J.P. *Spare Parts. Organ Replacement in American Society*. Oxford University Press, New York, 1992.
Hakim, N.S. and Papalois, V.E., Eds. *History of Organ and Cell Transplantation*. World Scientific, 2003.
Küss, R. and Bourget, P. *An Illustrated History of Organ Transplantation. The Great Adventure of the Century*. Laboratoires Sandoz, Rueil-Malmaison, 1992.
Lock, M. *Twice Dead. Organ Transplants and the Reinvention of Death*. University of California Press, Berkeley, 2002.
Moore, F.D. Give and Take: the Development of Tissue Transplantation. W.B. Saunders, Philadelphia, 1964.
Schlich, T. *Die Erfindung der Organtransplantation. Erfolg und Scheitern des chirurgischen Organersatzes (1880–1930)* [The Invention of Organ Transplantation: Success and Failure of Surgical Organ Replacement]. Campus, Frankfurt, 1998.
Woodruff, M.F.A. *The Transplantation of Tissues and Organs*. Charles C. Thomas, Springfield, 1960.

Organization of Technology and Science

The composition and configuration of institutions supporting technological advance changed dramatically over the course of the twentieth century. While national patterns differed substantially in the organization of science and technology, most industrialized nations experienced a few common trends. In the first half of the century, corporate research laboratories supplanted the workshops of independent inventor–entrepreneurs as the focal point of innovative activity. The establishment of corporate research laboratories and the increasing employment of engineers and scientists in industry coincided with, and to some extent facilitated, the expansion of institutions contributing to scientific and technological advance, including universities, government bureaus, and private research institutes. Following the end of World War II in 1945, institutions supporting science and technology expanded and new patterns of funding and coordination emerged. National governments, even those without significant military research and development (R&D) programs, exercised greater influence over the network of research institutions within their borders. Large-scale projects and collaborations across disciplines and institutions became common. In the third phase, beginning in the 1970s, industry reduced its support for basic research, shifted technical resources to divisional research programs, and increasingly entered into national and international strategic alliances. The influence of governments over the content of research in private industry diminished somewhat. Corporations, however, began to exert greater influence over research in universities. Market mechanisms influenced the organization of technology and science at the end of the twentieth century to a greater extent than during the previous period.

Differences in funding patterns, institutional roles and responsibilities, and relationships among institutions have accounted for national variations in the organization of technology and science. Questions relevant to discerning national variations include:

- Where is the locus of innovative activity?
- To what extent and through what mechanisms have markets, interest groups, and gov-

ernment agencies influenced the content and direction of private research activity?
- What distinctions exist in the motivations and goals of research in universities, firms, and government laboratories?
- What is the character of relationships between and among universities, firms, and government agencies?

While a systematic categorization of nations according to their organizational patterns is beyond the scope of this essay, these questions provide a basis for limited comparisons on discreet issues discussed here. In this entry, major changes in the organization of technology and science during the twentieth century will be explored, with sensitivity to national variations.

The Emergence of Institutions Supporting Technological Advance

Formal institutions supporting technological advance were in their infancy at the start of the twentieth century. In the late nineteenth century in the U.S., new technologies emerged primarily from the activities of independent inventors, such as Thomas Edison, and mechanics working in machine shops. Testing laboratories, such as the Pennsylvania Railroads, supported research to standardize equipment and improve materials, and universities remained almost completely disconnected from the process of technological change. The situation in Germany was different. The German chemical and electrical industries first began to exploit scientific knowledge systematically for technological purposes around 1880. German companies had the advantage of a national system of universities and state-sponsored research institutes with unparalleled capabilities in scientific education and research. Large firms, such as Bayer and Hoechst, funded individual professors in universities who, along with their graduate students, conducted studies relevant to the concerns of their sponsors. Most importantly, German firms hired graduate students with doctorates in chemistry and physics and established the first science-based industrial research laboratories. The German research model thus consisted of science-based research as a distinct corporate function and a network of scientific institutions that provided corporations access to the latest advances in scientific knowledge and a steady supply of new recruits for their industrial laboratories.

Leading corporations in the U.S., including General Electric, American Telephone and Telegraph, DuPont, and Eastman Kodak, established science-based research as a distinct organizational function in the first two decades of the twentieth century. Competitive threats from national and international rivals and fear of government antitrust action were among the reasons these companies established centralized research laboratories. To recruit and retain talented scientists, many of whom were trained in European universities, the managers of these pioneering laboratories often found it beneficial to provide their researchers with a modicum of freedom to choose their own research topics and to publish the results of their work. This tension between professional commitments and the goals of the corporation remained a major issue within industrial laboratories throughout the twentieth century. Nonetheless, the physicists and chemists that staffed the early laboratories generally conducted basic research related to their respective companies' core technologies.

While some scientists in American universities looked askance at their colleagues and students who sought employment in industry, applied science and academic engineering programs developed explicitly to meet the technical needs of industry. The chemical and chemical engineering departments at the Massachusetts Institute of Technology (MIT) grew large during the 1920s on consulting fees and corporate-sponsored research. Engineering programs at state universities such as Purdue University, the University of Illinois, and the University of Michigan, actively engaged in research relevant to the needs of industry. The chemical, electrical, and aeronautical engineering programs of such schools trained the legions of engineers who populated the production, development, testing, and even the research organizations of industrial corporations. Furthermore, the role of American universities in training scientists who found employment in corporate laboratories increased steadily throughout the first half of the twentieth century, and the pattern of collaboration between university scientists and industry spread to other fields such as the biomedical sciences.

In the American context, therefore, a large and diverse network of universities with strong training and research capabilities served as a foundation for the growth of industrial research. The British educational system, by contrast, did little to promote the expansion of research in industry. Limited government financial support for engineering education meant that Britain produced far fewer university-trained engineers than Germany and the U.S. Furthermore, university scientists lacked incentives to develop relationships with corporations, and the linkages between the two remained tenuous throughout the first half of the century. Instead of developing the engineering and applied research capabilities of universities, the British government established cooperative research associations, which conducted routine studies to improve processes, but developed few new technologies. Under the British government's Department of Scientific and Industrial Research, the associations proved to be poor substitutes for corporate research. While some companies in cutting-edge industries developed in-house research capabilities, the focus of government policy on the promotion of cooperative associations and the isolation of science in the British context prevented the development of strong links between education, research, and inventive activity.

A handful of government agencies in the U.S. provided technical services similar to those of the British cooperative research associations. In 1901, Congress created the most prominent of these agencies, the Bureau of Standards, to establish, test, and maintain standards for industry, government, educational institutions, and scientific interests. The Bureau maintained programs in a wide range of fields including electricity, metallurgy, weights and measures, chemistry, and instruments. It generally limited its research program to solving technical problems related to the development of standards, but in dispersed industries that lacked research capabilities, such as the building trades, the Bureau supported research that supported technological advance more broadly. The Bureau of Mines conducted research to improve safety conditions and to improve efficiency and reduce waste in the mining and petroleum industries. The wind tunnels and laboratories of the National Advisory Committee for Aeronautics (NACA) served as centers of research and testing for the budding aviation industry. Although these organizations provided crucial services to industry in the years prior to World War II, their influence on the pace of technological change and the relations among research institutions was modest.

Other institutions that contributed to the advance of science and technology included independent research institutes and private foundations. Institutes such as Arthur D. Little, the Mellon Institute, and the Battelle Institute provided technical assistance to private corporations on a contract basis. Although the institutes were created in many instances to serve small firms, large firms turned to them for process research and help with routine technical problems, such as analyses of the properties of metals and chemicals. Private foundations operated independently of industry and supported basic research and education. Built on the wealth of prominent industrialists, these organizations possessed the means to set research agendas and to influence the national research environment. The Carnegie Institute of Washington, for example, supported individual and collaborative studies in the fields of genetics and terrestrial magnetism. The Rockefeller Foundation of New York supported basic research in the biomedical sciences and worked to spread medical cures and public health measures throughout the world. Following World War I, the Carnegie Corporation established a large endowment to support the National Academy of Sciences, and the Rockefeller Foundation began funding postgraduate education in physics and chemistry through the National Research Council.

Prior to World War II, support for science isolated from industrial concerns grew steadily in the U.S. but remained small in comparison with Europe. The lack of an institute for basic scientific research supported with government funds comparable to the Royal Institution of London or the French Academy of Sciences troubled the leaders of the scientific community who sought to raise the esteem of their profession. Failed efforts at achieving such ends, including the creation of the National Research Council during World War I and the Scientific Advisory Board during the Great Depression, did little to dampen the scientific community's quest to increase federal support for basic research. Whereas professional engineers in the U.S. had found accommodation with industry by the early twentieth century, scientists, even those working for and within industrial laboratories, remained ambivalent about their relationship with business. Despite such ambivalence, the American institutional context allowed for the institutionalization of science in industry and the formation of alliances between universities and industrial laboratories.

World War II and the Cold War

World War II represented a watershed in the organization of technology and science. The unprecedented technological achievements of the wartime mobilization, including the proximity fuse, radar, and the atom bomb, erased all doubts about the potential technological benefits of increased collaboration between industry, the universities, and government. Following the war, Vannevar Bush's famous report to the president, entitled *Science, the Endless Frontier* provided a formal justification for the establishment of a National Science Foundation in the U.S. and for increased federal funding of basic research in universities. Throughout much of the world, national military establishments came to occupy an important role in shaping the national scientific and technological infrastructure. Military demand for new weapons and government funding for science drove the expansion of existing institutions, facilitated the development of new methods for organizing research projects, and inspired the creation of new institutions for promoting, funding, managing, and conducting research.

Massive government expenditures for R&D and procurement provided the context for the expansion of corporate research following the war. A great many corporations hired teams of scientists to pursue basic scientific research and established new laboratories separate from engineering and production operations. Underlying this trend was a general misunderstanding, perpetuated by Vannevar Bush, of the innovation process. At the core of Bush's "linear model" of research was the notion that scientific research alone provided the basis for the development of new technologies. Innovation, according to the model, proceeds in a stepwise fashion, from research to development to production and then marketing. Unimpeded by market concerns and potential applications, scientists left on their own would produce fundamental ideas that applied scientists would then effortlessly transform into new technologies. Corporations that adhered to this model experienced great difficulty producing innovations, and many abandoned their research laboratories altogether by the early 1970s.

Although it had earlier precedents, "big science" emerged as a major theme in the organization of technology and science following World War II. Many activities with distinct organizational features have been lumped together under the term "big science," and the term is often used loosely to refer to any scientific or technological endeavor that involves vast financial resources and large numbers of researchers. Geographically concentrated projects centered on massive instruments and dedicated to advancing scientific knowledge, in this view, fall into the same category as research programs involving multiple laboratories dispersed over great distances, or even projects with a distinctly technological objective. In considering the organizational characteristics of large-scale science and technology, it is useful to distinguish such projects by their objectives (e.g., scientific or technological advance), geographic characteristics (e.g., concentrated or dispersed), and managerial structure (e.g., hierarchical or diffuse). Such categories, to be sure, may not fully capture the complexity of some projects, which, for example, may be centered in a single location but involve networks of individuals that stretch across continents. Nevertheless, distinguishing such projects by these dimensions allows a more complete sense of the diversity of approaches to the organization of large-scale technological and scientific projects in the postwar period.

The Manhattan Project is widely regarded as the wartime exemplar of big science. Under the leadership of General Leslie Groves, the U.S. Army's Manhattan Engineering District assembled and coordinated the activities of academics scientists, industrial firms, and construction contractors for the development of the atomic bomb. Even though a wide range of academic and industrial institutions spread throughout the country contributed to the project, Groves maintained tight control by compartmentalizing different aspects of the project. Groves ultimately had authority to control the distribution of funds and to transfer tasks to other groups if he did not get the results he wanted when he wanted them. Therefore, despite the role of certain sites as central nodes in the network of institutions contributing to the project, the sense of teamwork and community that evolved in certain compartments, and the scientific content of some of the work, the project was geographically dispersed, hierarchically controlled, and technologically oriented.

Big science projects in the postwar era, especially in the fields of astronomy and physics, were often organized around large instruments and directed toward scientific goals. In the case of nuclear physics, scientific entrepreneurs located in universities and government laboratories in the U.S. harnessed the resources of the state to fund the creation of increasingly larger particle accelerators. The construction and maintenance of accelerators, at places like Stanford, Berkeley, and Brookhaven, required advanced engineering knowledge and close collaboration between theoreticians, experimentalists, and engineers. Some accelerators were available as a shared resource for use by multiple research groups—a pattern also common in the field of astronomy. Particle physics research in Europe differed from the U.S. in a number of ways. Most important was the concentration of resources at a single institution, the

European Center for Nuclear Research (CERN). The multinational effort avoided military influence experienced in some U.S. laboratories, but it also experienced greater operational setbacks due to a much more rigid separation between scientists and engineers. During the 1950s, the Joint Institute of Nuclear Research in Dubna, Russia, served as the center of research for physicists in communist countries. As in the U.S., the Russians maintained particle accelerators at a number of locations and used them for military-related research.

The instruments produced to answer scientific questions or to coordinate scientific activities sometimes found applications beyond the laboratory. Nuclear magnetic resonance imaging, which physicists developed to measure the movement of nuclei, has been used in analytical chemistry and diagnosis of medical patients. The design of ARPAnet reflected the decentralized structure of the community of computer scientists and engineers that created and used it. In a few decades, the open-ended computer network moved from a tool for sharing data among researchers to a platform for economic transactions, personal communication, and entertainment, called the Internet.

National endeavors for the creation of large-scale technologies, such as Project Apollo, more closely resembled the Manhattan Project in organization and orientation than traditional big science projects. In the case of Project Apollo, the civilian administrators of the National Aeronautics and Space Administration (NASA) had responsibility for mobilizing and coordinating a geographically dispersed network of contractors, advisory committees, university researchers, in-house research centers, and government bureaucracies. A well-defined mandate backed by tax dollars and political support gave NASA administrators the authority they needed to protect researchers from external disruptions to prevent competition from within their network of bureaucracies and contractors from undermining the project. The Soviet space program, like the American program, was characterized by both centralization of resources and authority and high levels of competition among different interests within the space bureaucracy. Proposals for new spacecraft usually emerged from design bureaus that competed against one another in a formal review process. A scientific-technical council reviewed the proposals and made recommendations to a military–industrial commission for the final decision. Party politics played a role in the process. Subsumed under the Soviet military bureaucracy, overwhelmed by competition among design bureaus for limited resources, and lacking the narrow focus of the U.S. space program in the 1960s, the Soviet program lost the race to the moon.

Managers of large-scale development projects developed new techniques to cope with the organizational complexities they faced. From the U.S. Air Force's Atlas missile program emerged a number of concepts, including concurrency, which involved the pursuit of research, design, and production engineering in parallel on large-scale projects. Concurrency required the development of systems analysis capabilities in order to specify technical details at the start of a project. Concurrency and systems analysis were among the techniques incorporated into U.S. Air Force regulations in the early 1960s under the name configuration management. With configuration management, the U.S. Air Force required contractors to share cost estimates and designs with program managers and document changes that occurred during the course of a project, all of which were subject to approval by a configuration control board. The centralized, hierarchical reporting system imposed additional layers of bureaucracy on projects, although some observers claim that it controlled costs and forced contractors to abide by schedules and performance standards.

Collaboration, Internationalization, and University–Industry Relations

The decision of the U.S. Congress in 1993 to terminate the Superconducting Supercollider Project, a particle accelerator of unprecedented power, served as evidence of physicists' declining influence on science policy, but it did not mark the end of big science. Even though it relied on networks and multiple research sites rather than hierarchical communities amassed around large-scale instruments, the biogenetics community's Human Genome Project represented a continuation of the big science tradition. A short-lived defense conversion program following the fall of the Soviet Union in 1991 did little to halt the long-term growth of military budgets. With high levels of spending on research and costly weapons projects such as the Strategic Defense Initiative and a nanotechnology initiative, the U.S. Department of Defense assured its continued influence over the national research agenda. However, new patterns in the organization of science and technology began to emerge in the final decades of the twentieth century.

The failure of isolated research laboratories to make good on their promises to transform scientific ideas into marketable technologies, as well as a general decline in funding for military research and basic science in the 1970s, laid the foundation for new patterns of research to emerge. Disillusioned companies either eliminated their centralized research laboratories or reorganized their research programs by creating closer links between research and product development and engineering and shifting research to corporate divisions. By the 1980s, decentralization and collaboration with external partners replaced isolated science as the central organizing principles of industrial research. International competitive threats served as the justification for the creation of industry-wide consortia. With significant financial backing from the military, the U.S. semiconductors industry established the Semiconductor Manufacturing and Technology Institute (SEMATECH) to improve commercial semiconductor manufacturing capabilities. Research consortia proliferated throughout the 1980s and 1990s. Great Britain established its LINK program in 1997 to support the establishment of consortia in precommercial research across the spectrum of advanced technologies. The European Union's Brite–Eura program supported consortia dedicated to the improvement of manufacturing technologies in a broad range of industries. Not all consortia were government creations but most received financial support from governments. More importantly, collaborative efforts took many different forms. Joint ventures and strategic alliances among companies in the same industry and with suppliers and subcontractors also became common near the end of the century.

While companies engaged in international alliances for generations, collaborative research efforts between overseas competitors, such as Toyota and General Motors, became increasingly common toward the end of the century. Often these international collaborations involved research to create entirely new technologies. Another trend that had earlier precedents but became increasingly common in the 1980s and 1990s was the establishment of research laboratories in foreign countries. Firms from Europe, Japan, and the U.S. established laboratories outside their countries to recruit researchers and to exploit the scientific and technological capabilities found in these markets and universities.

Increased corporate spending on university research was perhaps the defining characteristic in the organization of science and technology in the late twentieth century. Rather than hire a smattering of university professors as consultants or let small contracts to university departments, major corporations awarded multi-million dollar grants to gain access to the fruits of university research. Monsanto's $23 million grant to Harvard University in 1974 started the trend. Companies in the chemical and biotechnology industries, fields in which scientific breakthroughs are quickly translated into new products, were among the most active supporters of academic research. The trend was not limited to U.S. firms. Hoeschst of West Germany gave $50 million for research in molecular biology to Harvard's Massachusetts General Hospital. Although some university scientists rejected private funding, claiming it perverts basic research, declining state and federal support for universities had all but made industry support essential for maintaining the health and well-being of university research programs.

The shift from fundamental to product-related research and the decentralization of research within firms, the growth of national and international partnerships, the overseas expansion of R&D, and increasingly close relations between universities and corporations represented the emergence of a market-driven approach to R&D in which personal and institutional networks, rather than centralized hierarchies, defined the context of innovation to an increasing extent. The role of universities in preserving and advancing national scientific and technological capabilities appears to have increased in importance in this new organizational environment.

GLEN ASNER

Further Reading

Branscomb, L.M., Kodama, F. and Florida, R., Eds. *Industrializing Knowledge: University–Industry Linkages in Japan and the United States*. MIT Press, Cambridge, MA, 1999.

Bush, V. *Science, the Endless Frontier*. U.S. Government Printing Office, Washington D.C., 1945.

Capshaw, J.H. and Rader, K.A. Big science: price to the present. *OSIRIS*, 2nd series, 7, 3–25, 1992.

Dunning, J. Multinational enterprises and globalization of innovatory capacity. *Res. Policy*, 23, 67–88, 1994.

Dupree, A.H. *Science in the Federal Government: A History of Policies and Activities to 1940*. Harvard University Press, Cambridge, MA, 1957.

Galison, P. and Hevly, B., Eds. *Big Science: The Growth of Large-Scale Research*. Stanford University Press, Stanford, 1992.

Geiger, R.L. *Research and Relevant Knowledge: American Research Universities Since World War II*. Oxford University Press, New York, 1993.

Graham, L.R. Big science in the last years of the big Soviet Union. *OSIRIS*, 2nd series, 7, 49–71, 1992.

Hounshell, D.A. The evolution of industrial research in the United States, in *Engines of Innovation: U.S. Industrial Research at the End of an Era*, Rosenbloom, R.S. and Spencer, W.J., Eds. Harvard Business School Press, Boston, 1996.

Kevles, D.J. *The Physicists: The History of a Scientific Community in Modern America*, 3rd edn. Harvard University Press, Cambridge, MA, 1995. Originally published by Knopf, New York, 1977.

Mack, P., Ed. *From Engineering Science to Big Science: The NACA and NASA Collier Trophy Research Project Winners*. NASA History Office, Washington D.C., 1998.

Marsch, U. Strategies for success: research organization in German chemical companies and IG Farben until 1936. *Hist. Technol.*, 12, 23–77, 1994.

Meyer-Thurow, G. The Industrialization of invention: a case study from the German chemical industry. *ISIS*, 73, 268, 363–381, 1982.

Millard, A. *Edison and the Business of Innovation*. Johns Hopkins University Press, Baltimore, 1990.

Mowery, D.C. and Rosenberg, N. *Technology and the Pursuit of Economic Growth*. Cambridge University Press, New York, 1989.

Mowery, D.C. and Teece, D.J. Strategic alliances and industrial research, in *Engines of Innovation: U.S. Industrial Research at the End of an Era*, Rosenbloom, R.S. and Spencer, W.J, Eds. Harvard Business School Press, Boston, 1996.

Niosi, J. The internationalization of industrial R&D: from technology transfer to the learning organization. *Res. Policy*, 28, 107–117, 1999.

Pearce, R.D. *The Internationalization of Research and Development by Multinational Enterprises*. St. Martin's Press, New York, 1989.

Reich, L.S. *The Making of American Industrial Research: Science and Business at GE and Bell, 1876–1926*. Cambridge University Press, New York, 1985.

Rosenberg, N. and Nelson, R.R. The roles of universities in the advance of industrial technology, in *Engines of Innovation: U.S. Industrial Research at the End of an Era*, Rosenbloom, R.S. and Spencer, W.J, Eds. Harvard Business School Press, Boston, 1996.

Seidel, R.W. Accelerators and national security: the evolution of science policy for high-energy physics, 1947–1967. *Hist. Technol.*, 11, 361–391, 1994.

Servos, J.W. The industrial relations of science: chemical engineering at MIT, 1900–1939. *ISIS*, 71, 259, 531–549, 1980.

Servos, J.W. Changing partners: the Mellon Institute, private industry, and the federal patron. *Technol. Cult.*, 35, 221–257, 1994.

Siddiqi, A. *Challenge to Apollo: The Soviet Union and Space Race, 1945–1974*. NASA History Office, Washington D.C., 1998.

Smith, B.L.R. *American Science Policy Since World War II*, Brookings Institution, Washington D.C., 1990.

Swann, J.P. *Academic Scientists and the Pharmaceutical Industry: Cooperative Research in Twentieth-Century America*. Johns Hopkins University Press, Baltimore, 1988.

Varcoe, I. Cooperative research associations in British industry, 1918–1934. *Minerva*, 19, 433–463, 1981.

Westwick, P.J. *The National Labs: Science in an American System, 1947–1974*. Harvard University Press, Cambridge, MA, 2003.

Wise, G. and Whitney, W.R. *General Electric, and the Origins of U.S. Industrial Research*. Columbia University Press, New York, 1985.

P

Packet Switching

Historically the first communications networks were telegraphic—the electrical telegraph replacing the mechanical semaphore stations in the mid-nineteenth century. The network consisted of strategically located stations linked by telegraph lines. Using alphabetic codes such the five-unit Baudot code (still used in telex today), messages were relayed from station to station. This approach is known as message store-and-forward or message switching. Messages could be relayed coast-to-coast in a matter of hours or even minutes. One of the problems with message switching is that messages are highly variable in length. A short message may have to queue behind one or more long messages from other senders, and the delivery time is therefore rather variable.

Telegraph networks were largely eclipsed by the advent of the voice (telephone) network, which first appeared in the late nineteenth century, and provided the immediacy of voice conversation. The Public Switched Telephone Network allows a subscriber to dial a connection to another subscriber, with the connection being a series of telephone lines connected together through switches at the telephone exchanges along the route. This technique is known as circuit switching, as a circuit is set up between the subscribers, and is held until the call is cleared (see Telephony, Automatic Systems).

One of the disadvantages of circuit switching is the fact that the capacity of the link is often significantly underused due to silences in the conversation, but the spare capacity cannot be shared with other traffic. Another disadvantage is the time it takes to establish the connection before the conversation can begin. One could liken this to sending a railway engine from London to Edinburgh to set the points before returning to pick up the carriages. What is required is a compromise between the immediacy of conversation on an established circuit-switched connection, with the ad hoc delivery of a store-and-forward message system. This is what packet switching is designed to provide.

Packet switching is the data transmission technique whereby messages are cut up into small fixed-length pieces for routing through a data communication network, and then the pieces are reassembled into the original message at the receiving end. The term packet switching was coined in 1965 by Donald Davies, at the U.K.'s National Physical Laboratory (NPL). He proposed the approach as a solution to the burgeoning requirements for computer networks and data communication. At about the same time, another researcher, Paul Baran of the Rand Corporation in the U.S., proposed a very similar scheme, but for a highly reliable, digital, nationwide military voice network that could switch to another route when individual nodes or exchanges were destroyed by enemy action. Neither author was aware of the work of the other until a year or so later.

The technique can be applied to a wide variety of traffic types including interactive computing (as in client/server systems), file transfer, telemetry, digital telephone, or even delivery of video or audio streams. By dividing up large messages into small, fixed-length messages (packets of say, 1000 bits) and sending these like telegrams, the message handling exchanges can route each packet very simply. Being small, packets can be held in random access memory (RAM) and routed dynamically to the next best link. Using a line of 1 megabit per second, a packet of 1000 bits takes 1 millisecond to transmit. If a packet switch can route a packet in 1 millisecond and if there are five hops between subscribers, the typical cross-network delay would be 10 milliseconds. At 50 percent loading, a 1-megabit per second line could deliver 500 packets per second and a multiprocessor switch could process thousands of packets per second. Doing this makes the packet-switched network appear as a set of direct connections between subscribers.

Message reassembly is handled at the receiving end, as is the detection of any lost packets. The technique here is for the sender to number the packets, and the receiver to request the retransmission of missing items. This happens infrequently and does not turn out to be such an onerous task, but in any case, the software for doing this is now readily available.

During 1966, Davies's team at the NPL produced a design for an actual communication network. This work was published in the proceedings of the ACM (Association for Computing Machinery) conference, held in Gatlinburg, U.S. in 1967. At that meeting a paper was presented by Larry Roberts of the Advanced Research Projects Agency (ARPA) project, describing proposals for a computer network that facilitated the sharing of resources between university and research campuses. It was a seminal meeting as the NPL proposal illustrated how the communications for such a resource-sharing computer network could be realized. During 1968 and 1969, development work proceeded in both U.S. and the U.K. The ARPA network or ARPAnet communications system was built by Bolt, Beranek & Newman Inc., and the first exchanges—interface message processors (IMPs)—were delivered in 1969. During the 1970s the network grew to some 30-plus "nodes" covering mainland U.S., with outposts in Hawaii and Europe. At the same time, a campus network was built by the NPL team to meet the laboratory's local data processing needs. Elements of the pilot NPL network first operated in 1969 and this was later expanded to link some 20 computers and 300 user terminals. The U.K. and ARPAnet networks were later connected together, and then to other experimental networks in Europe, such as the European Informatics Network, which linked France, Germany, Italy, Sweden, Switzerland and U.K.

Theoretical work was being undertaken in parallel, involving mathematical modeling and simulation into the performance of such systems. The ARPA development was underpinned by the work of Leonard Kleinrock at University of California, Los Angeles (UCLA), who had been developing such modeling techniques as a graduate student at MIT in the mid-1960s. Simulation work on packet networks was also undertaken by the NPL group.

The network research community formed the Inter-Network Working Group (INWG)—chaired by Vint Cerf—and out of this came the inter-network protocol transmission control protocol/Internet protocol (TCP/IP, the now de facto Internet standard). The ARPA network rapidly became the focus of attention for both the computer industry and the regulated telecommunication network providers. Pressure mounted for such services in the public domain. With the momentum gathering for computer networking during the 1970s and 1980s, effort was put into public standards for such networks. In the telecommunications arena, the "regulated carriers" introduced packet services, based on the X25 standard.

Public wide area networks were characterized by the low data rates that could be supported by the telephone network infrastructure. However, the NPL work in particular had demonstrated both the benefits and the feasibility of very high data rates (around 1 megabit per second). Despite the development of digital telephone lines (2 megabits per second) the benefits of fast digital networks probably first came to the public attention in local area networks (LANs) with the design for Ethernet (10 megabits per second) by Bob Metcalf of Xerox. This design, and the competing IBM Token-ring LAN, are also packet-organized network technologies.

In the 1990s, the International Public Record Carriers started to deploy even faster digital communication systems based on packet technology (frame relay and asynchronous transfer mode (ATM), the latter at 155 megabits per second). The term "Internet" was coined and Tim Berners-Lee conceived the notion of the World Wide Web. This popular technology outstripped the ability of the regulated carriers to satisfy demand, and the essentially unregulated Internet spread like wildfire, fueled by the burgeoning home PC market. At the turn of the century, packet switching was embedded in so much that it was accepted as daily experience. Home users could connect to the Internet at near-megabit rates using ADSL (asynchronous digital subscriber line), and gigabit Ethernet was introduced. In optical communication fiber networks, optoelectronic switches route optical signals. Current optical–electronic–optical

(O–E–O) conversion will not be able to support the terabit-per-second capacities that will be needed in the near future. Future high-speed packet switching will only be realized by all-optical (photonic) networks.

See also **Computer Networks; Electronic Communications; Internet; Telephony, Automatic Systems; Telephony, Digital; World Wide Web**

ROGER SCANTLEBURY

Further Reading

Abbate, J. *Inventing the Internet*. MIT Press, Cambridge, Massachusetts, 2000.

Campbell-Kelly, M. Data communications at the National Physical Laboratory (1965–1975). *Ann. Hist. Comput.*, 9, 3/4, 221–247, 1988.

Gillies, J. and Cailliau, R. *How the Web was Born*. Oxford University Press, Oxford, 2000.

Norberg, A.L. and O'Neill, J.E. Improving connections among researchers: the development of packet-switching computer networks, in *Transforming Computer Technology*. Johns Hopkins University Press, Baltimore, 1986.

Yates, D.M. *Turing's Legacy: A History of Computing at the NPL, 1945–1995*. Science Museum, London.

Particle Accelerators: Cyclotrons, Synchrotrons, and Colliders

Particle accelerators, or "atom smashers" as they are popularly known, are devices that produce concentrated beams of charged particles of very high energy. These beams have been used to study nuclear and atomic structure (x-ray crystallography), to create radioactive isotopes, and to irradiate cancerous tumors with x-rays. After World War II they constituted the essential infrastructure for the new field of high-energy physics. As accelerator power climbed to ever-higher energies, the laboratories where these microscopes probed to the heart of matter were transformed into huge industrial-scale centers of "big science." Particle accelerators became particle factories supporting multidisciplinary teams of researchers whose size has increased from less than a dozen in the 1960s to as many as 1500 by the end of the twentieth century.

In the early 1930s a number of attempts were made to produce an energetic beam of charged particles under controlled conditions. One of the simplest devices was that devised by John Cockroft and Ernest Walton working in Ernest Rutherford's famous Cavendish Laboratory in Cambridge. By charging a bank of capacitors in parallel at low potential and then discharging them through a load resistor to develop a high potential, they could multiply voltage by a factor of 4. The British scientists attained a steady output of about 500 kilovolts. This was used to accelerate protons (positively charged hydrogen ions) that were then smashed into atoms in a metallic target. Another scheme was the electrostatic generator developed by Robert J. Van de Graaff working at Princeton University. Ions produced by a corona discharge from needle points were transported by motor-driven insulating belts to two spherical conductors mounted on insulating rods. There the belts were discharged into the terminals. A potential difference of up to 1.5 million volts could be accumulated on the spheres, limited only by voltage breakdown, and sparking between them. This approach was so successful that Van de Graff and others formed their own company after the war to commercialize their device.

In these generators, charged particles picked up energy by falling through a large voltage difference. An alternative idea, also considered at the time, was to have the charged particles gain energy in several small steps. In this way, "the high-voltage energy would be accumulated on the particles, not on the apparatus" (Heilbron and Seidel, 1989). It was Ernest O. Lawrence, working at the Radiation Laboratory at the University of California in Berkeley, along with his graduate student M. Stanley Livingston, who first successfully applied this concept in the autumn of 1931, accelerating protons to over 1 million volts in a cyclotron. Lawrence won the 1939 Nobel Prize in Physics for his invention.

The cyclotron comprised two hollow half-cylinders or D's, with a small gap between them. A magnetic field was applied perpendicularly to the D's. Ions were injected at the center, and were given a small kick by a radio oscillator as they crossed the gap. They described a circular trajectory in the magnetic field, incrementally increasing their energy (say, by 10 kilovolts), and the radius of their path, each time they crossed the gap (say, 100 times in all). Tracing a spiral as they moved from the center of the D's to the circumference they thus emerged with 1 million volts of energy (in this case).

The key to increasing energy lay in the size of the D's. The first experimental setup on which Livingston demonstrated the feasibility of the idea could be fitted in the palm of one's hand. Within the decade it was followed by the 27-inch (69-centimeter) D, the 60-inch (152-centimeter) D, and then the huge 184-inch (467-centimeter) D; this measure being the diameter of the magnet face.

This machine, largely funded by the Rockefeller Foundation at a cost of $1.4 million in 1940, was designed to reach energies higher than 100 million volts, and required its own building to house it.

Cyclotron energies were restricted by the fact that, as the velocity of the particles approached that of the speed of light, the mass changed in accordance with Einstein's relativistic principles, and the orbital frequency of the particles changed with it. The principles on which the cyclotron was based thus broke down. This limitation on achievable energy was removed with the discovery of phase stability in 1945. The implementation of this technique depended on the whether it was applied to a proton or an electron accelerator. In the case of protons, the decreasing orbital frequency was compensated for by increasing both the strength of the magnetic field and the frequency of the accelerating voltage. Phase stability allowed engineers to do away with the huge pole faces required in a high-energy cyclotron. The particle beams circulated in evacuated beam pipes surrounded by magnets, radio-frequency generators and power supplies. Economic considerations were the only remaining constraint on particle energy, and the field of high-energy physics was born.

Many innovations have been exploited to increase accelerator energy while containing costs. Strong focusing, discovered in the U.S. in 1952, was an ingenious technique for limiting the cross-section of the particle beam, and therefore of the beam tube in which it circulated. The size (and therefore the cost) of the magnets confining the beam was thus sharply reduced. Colliding beams of particles traveling in opposite directions is another important way of increasing the energy available for doing physics. The first practical demonstration of this principle occurred at the European Laboratory for Particle Physics (CERN, near Geneva) in 1971. The beams were produced in two different intersecting storage rings (the ISR machine). Drawing on this experience, CERN scientists won the Nobel Prize for Physics in 1984 for colliding beams of opposite charge circulating in the same ring. The use of superconducting magnets became practicable in the 1980s, and these have opened yet another cost-reducing path to higher energy.

Most laboratories use circular accelerators. Linear accelerators have also been built, notably the 3.2-kilometer-long machine authorized in 1959 near Stanford University in California. Practicable length is, however, a limitation on energy, and linacs are generally used as injectors into ring systems. Today these are gigantic and very expensive. The new machine under construction at CERN, the large hadron collider, will be housed in a tunnel 27 kilometers in circumference. In 2001 its cost to completion was estimated to be about $1,800 million. The superconducting super collider (SSC), cancelled by the Clinton administration in 1993 when its costs began to creep above $10,000 million, was designed to reach 20 million million volts in a 85-kilometer-long subterranean tunnel in Texas.

Only some governments have been able to afford tools of this kind for such esoteric research, and they have done so for reasons of state. CERN, established in 1954 by 12 European governments who pooled their resources, combined a determination to keep nuclear scientists in Europe with foreign policies sympathetic to the construction of a European community. In the U.S., particle accelerators were identified with world scientific and technological leadership, national prestige, and the "peaceful atom," and used to promote international scientific exchange behind the Iron Curtain. Correlatively, some would argue, the demise of the SSC is simply one more response to the collapse of the Soviet bloc.

See also **Cancer: Radiation Therapy; Particle Accelerators, Linear**

JOHN KRIGE

Further Reading

Galison, P. and Hevly, B. Eds. *Big Science. The Growth of Large-Scale Research.* Stanford University Press, Stanford, 1992.

Heilbron, J.L., Seidel, R.W. and Wheaton, B.R. Lawrence and his laboratory: nuclear science at Berkeley. *LBL News Magazine*, 6, 3, 1981.

Heilbron, J.L. and Seidel, R.L. *Lawrence and His Laboratory. A History of the Lawrence Berkeley Laboratory*, vol. I. University of California Press, Berkeley, 1989.

Hermann, A., Krige, J., Mersits, U. and Pestre, D. *History of CERN. Vol. I: Launching the European Organization for Nuclear Research.* North Holland, Amsterdam, 1987.

Livingston, M. S. *Particle Accelerators. A Brief History.* Harvard University Press, Cambridge, MA, 1969.

Riordan, M. The demise of the superconducting super collider. *Phys. Perspect.*, 2, 411–425, 2000.

Westfall, C. and Hoddeson, L. Thinking small in big science: the founding of Fermilab. *Technol. Cult.*, 37, 457–492, 1996.

Useful Websites

CERN, European Organization of Nuclear Research: http://www.cern.ch

Particle Accelerators, Linear

A linear accelerator or linac, as it is commonly called, is a device that uses an oscillating electric field to accelerate charged particles (atomic or subatomic particles) in a straight line. The final energy of the charged particles is achieved by repeated acceleration on a linear path through energy steps.

The high-energy particle beams produced are used in elementary particle physics research, as neutron sources for research and materials inspection, and in radiation therapy for treatment of deep cancers (either as direct electron treatment, or from x-rays produced when high-energy electrons strike a target). Accelerated electrons or x-rays, both of which also cause ionization, can also be used to irradiate food against insect pests and pathogenic bacteria.

The 1920s saw not only the first formal proposal of a linear accelerator by the Swedish scientist G. Ising, but also the first report of a working one by the Norwegian R. Wideröe. It was in 1928 when Wideröe described his experiments with a machine consisting of three consecutive cylindrical electrodes, whereby an alternating electric field was applied between the central electrode and the two side ones. The frequency of alternation was such that, during the time in which the potassium (K^+) and sodium (Na^+) ions he used traversed the central electrode, the electrode potentials were reversed. Actual acceleration took place when particles traveled from one electrode to the next. This is how, with the central electrode shielding the particles for the length of time that the field would be decelerating, ions reached a final energy twice the energy obtained from a single transversal of the field. The radio-frequency voltage of 25 kilovolts used sped ions up to energies of 50 kiloelectron volts of energy. During the following two decades, several proposals and experimental devices came to light in the attempt to accelerate particles to higher energies. From what we have said, it is clear that high-voltage machines can be considered to be early linear accelerators. In these devices, a high-voltage generator provides a potential difference into an accelerating chamber, and the particle beam acquires energies of the order of the voltage delivered by the generator. In 1932 the British John D. Cockcroft and the Irish Ernest T. S. Walton used a linear machine of this kind to accelerate protons to 500 kiloelectron volts and disintegrate lithium nuclei at the Cavendish Laboratory at Cambridge University.

During the 1930s, a group of American physicists worked in Berkeley, California, developing and improving Wideröe's idea. The work of Ernest O. Lawrence and D. H. Sloan culminated in a machine that accelerated mercury (Hg^+) ions to energies of up to 1.26 megaelectron volts. The design was based on 30 consecutive cylindrical metallic electrodes of increasing length—note that particle velocity increases as it travels through the accelerator—placed along the axis of a vacuum chamber. These electrodes were connected alternately to the positive and negative polarities of a 42-kilovolt alternating current voltage generator. Obviously, the longer the accelerator, the higher the energies obtained. The later design of 36 drift tubes, by Sloan and W. M. Coates, achieved final energies of 2.85 megaelectron volts. Nevertheless, heavy ion linear accelerators seemed less useful in the field of nuclear physics than circular accelerators. The investment in the field of radar and radio communication during World War II allowed the development of power generators in the megawatt power range and in the gigahertz frequency range. This was what physicists needed for successfully accelerating lighter particles and building practical linear accelerators with the energies required for nuclear physics studies. At Stanford University, American William W. Hansen and his research group proceeded toward the construction of an electron linac, a project conceived before the war, based on Russell and Sigurd Varian's klystron (which generated the microwave power necessary to accelerate electrons, and was used in World War II in radar aboard aircraft), and completed in 1947. The machine, later known as Stanford Mark I, was 2.7 meters long, powered by a single magnetron that yielded 0.9 megawatts, and provided electrons with an energy of around 4.5 megaelectron volts. A short time later, it was extended to 4.3 meters long and energies of 6 megaelectron volts were obtained. At the same time, at the University of California, design studies under the direction of American Luis W. Alvarez led to the construction of a proton linear accelerator. The surplus radar components available from the war were adopted to power a machine 12.2 meters long. The Berkeley 32-megaelectron-volt proton linear accelerator was in full operation by 1947, two years after its construction began.

D.W. Fry's group at the Telecommunications Research Establishment (TRE), Great Malvern, U.K, worked more or less simultaneously with Hansen on a linear electron accelerator, but with limited knowledge of Hansen's work until mid-

1947. Fry's accelerator was driven by a magnetron, and produced a beam of 0.5 megaelectron volt electrons towards the end of 1946. The British linear accelerator was soon adopted by hospitals for cancer therapy, with the first patient treated in August 1953 at Hammersmith Hospital in London. In the U.S., Henry Kaplan, working with Edward Ginzton of the Stanford Physics Department, worked on medical applications. The Stanford medical linear accelerator was completed in 1955.

Several developments and improvements made work at higher energies in modern linear accelerators possible. In each case, different problems appeared to be related to the shape of electromagnetic fields, particle dynamics, power requirements, cavities, and defocusing. Well-known examples are the 91-meter Stanford Mark III, operating in 1964 at 1.2 gigaelectron volts energy; and the largest linac built, the 3.2-kilometer machine at the Stanford Linear Accelerator Center (SLAC), with emerging electron energies of more than 20 gigavolt electrons in 1967. During the 1960s and the 1970s, ideas such as the linear induction accelerator, where inductive electric fields are used, contributed to progress in the pursuit of higher energies and new challenges. The invention of the low-energy radio-frequency quadrupole (RFQ) linear accelerator, where the electric fields are produced by the radio-frequency fields applied to four electrodes collinear with the beam axis, was proposed by the Russians I. M. Kapchiski and V. A. Teplyakov and became one of the most important improvements.

The construction and development of linear accelerators has continued today, not only because of interest in electron and ion optics, space science, industrial applications, and thermonuclear fusion, but also for medical purposes. An example of this is the Los Alamos Meson Physics Facility (LAMPF), an 800-megaelectron-volt proton linear accelerator sponsored by the National Cancer Institute. There is no doubt that the excellent collimation of the emergent beam, as opposed to the spreading that results from circular accelerators, is one of the most attractive advantages that linear accelerators offer for such purposes. Nevertheless, the most common use of proton linear accelerators is as injectors for high-energy machines. Electron linear accelerators have an advantage over circular accelerators however in that they present no practical limitation. A good example of future machines of this kind is the worldwide collaboration compact linear collider (CLIC), a linear accelerator planned by CERN. Nearly 40 kilometers long, future lepton machines will probably be linear colliders like the one proposed by CERN. Powered from a drive beam with a high frequency of 30 gigahertz, the CLIC is thought to cover a center-of-mass energy range for electron–positron collisions of 0.5 to 5 teraelectron volts.

See also **Cancer, Radiation Therapy; Particle Accelerators: Cyclotrons, Synchrotrons, Colliders**

Pedro Ruiz Castell

Further Reading

Ginzton, E.L. and Nunan, C.S. History of microwave electron linear accelerators for radiotherapy. *Int. J. Radiat. Oncol. Biol. Phys.*, 11, 205–216.
Lapostolle, P.M. and Septier, A.L., Eds. *Linear Accelerators*. North Holland, Amsterdam, 1970.
Livingston, M.S. and Blewett, J.P. *Particle Accelerators*. McGraw-Hill, New York, 1962.
Livingston, M.S. *Particle Accelerators: A Brief History*. Harvard University Press, Cambridge, MA, 1969.
Pierce, J.R. *Theory and Design of Electron Beams*. Van Nostrand, New York, 1949.
Reiser, M. *Theory and Design of Charged Particle Beams*. Wiley, New York, 1994.
Scharf, W.H. *Particle Accelerators: Applications in Technology and Research*. Research Studies Press, Taunton, 1989.
Wiedeman, H. *Particle Accelerator Physics: Basic Principles and Linear Beam Dynamics*. Springer Verlag, Berlin, 1993.

Useful Websites

AccSys Technology, Inc.: http://www.accsys.com
CERN, European Organization of Nuclear Research: http://www.cern.ch
Stanford Linear Accelerator Center: http://www.slac.stanford.edu

Personal Computer, *see* **Computers, Personal**

Personal Stereo

The scaled-down cassette tape player represents not only one of the most successful audio products of the twentieth century but also a drastic change in the way we listen to music. The personal stereo has become a universal product that can be found in every corner of the globe. It has brought high-quality reproduction of sound into every field of human activity.

The concept of the personal stereo can be traced to two great electrical manufacturers: Sony of Japan and Philips of the Netherlands. Both of these companies created a great new market for electrical goods by reducing the size of their products. Key employees in both companies shared a passion for music and a goal of perfect reproduction of sound.

Akio Morita and Masuru Ibuka formed a company in 1946 to make a variety of electrical testing devices and instruments, but their real interests lay in music and they decided to concentrate on audio products. They renamed their company Sony (derived from "sonic") to emphasize this strategy. A pioneer in developing the pocket-sized transistor radio, Sony penetrated the American market with small, fully transistorized television receivers.

The original marketing strategy for manufacturers of all mechanical entertainers had been to put one into every home. This was the goal for Edison's phonograph, the player piano, the Victor Talking Machine Company's Victrola and the radio receiver. However, Sony and other Japanese manufacturers found out that if a product was small enough and cheap enough, two or three might be purchased for home use. This was the marketing lesson of the transistor radio that was successfully applied to televisions and tape players.

The personal stereo was the result of the convergence of two technologies: the transistor, which enabled miniaturization of electronic components, and the compact cassette—a worldwide standard for magnetic recording tape. The latter was devised by Philips as a replacement for the cumbersome reels used in tape recorders. Users found threading tape around the reels troublesome. The size of the reels (and the power requirements to turn them) made it difficult to reduce the size of tape recorders. Philips' cassette was one of many similar innovations in the 1960s based on the tape cartridge concept already in use in film cameras. At this point there was no idea that a smaller tape recorder would have applications in musical entertainment; the goal was to develop small, portable recorders to be used as dictating machines. Philips' executive Lou Ottens played an important part in this project, applying the descriptive adjective "compact" to the cassette.

Masuru Ibuka of Sony initiated the research project that led to the Walkman personal stereo. He wanted to be able to listen to high-fidelity recorded sound wherever he went and instructed his team to produce a player small enough to fit inside his pocket. (Another group of Sony engineers was working on a video recording cassette that could also fit into Ibuka's pocket.) The Walkman was based on a systems approach that made use of advances in several unrelated areas, including improvements in magnetic tape, integrated circuits, and new types of batteries (notably the nickel–cadmium combination, which offered higher output in smaller sizes). The problem of reducing the size of the loudspeaker without serious deterioration of sound quality blocked the path to very small cassette players. Sony's engineers produced a very small dynamic loudspeaker using plastic diaphragms and lighter materials for the magnets. These were incorporated into tiny stereo headphones.

The Sony Soundabout portable cassette player was introduced in 1979. It was initially treated as a novelty in the audio equipment industry. Sony's engineers, working under the direction of Kozo Ohsone, reduced the size and cost of the machine. In 1981 the Walkman II was introduced. It was 25 percent smaller than the original version and had 50 percent less moving parts. It took about two years for Sony's Japanese competitors, including Matsushita, Toshiba and Aiwa, to bring out portable personal stereos. Such was the popularity of the device that any miniature cassette player was called a Walkman, irrespective of the manufacturer. In 1986 the term entered the *Oxford English Dictionary*.

Constant innovation added new features to the personal stereo: Dolby noise reduction circuits were added in 1982 and rechargeable batteries were introduced in 1985. The machine grew smaller and smaller until it was hardly bigger than the compact cassette it played.

Within two years of the introduction of the compact disk in 1982, Sony had brought out a portable player named the Discman.

Ohsone led the development team that had to overcome the considerable challenges of reducing vibration in the unit (which disturbed the optical reading of microscopic lines of data) and reducing the size of the laser reader. Like its cassette counterpart, the Walkman technology was systematically improved while its size and price was continually reduced.

The size of the market for personal stereo systems ensured that many new recording technologies developed in the 1990s would be reduced in size and offered with earphones. This was the case for digital audio tape (DAT), Philips' digital compact cassette (DCC), and Sony's minidisc (MD). All these technologies came with the vital advantage of a recording capability—the major commercial consideration in competing with magnetic tape units.

The minidisc has been the most successful digital version of the personal stereo recorder. It employs the same optical technology as the compact disk but in a smaller size. It contains enough buffer memory to overcome the skipping of tracks that caused problems with the Discman.

The ubiquitous Walkman has had a noticeable effect on the way that people listen to music. The sound from the headphones of a portable player is more intimate and immediate than the sound coming from the loudspeaker of the home stereo; the listener can hear a wider range of frequencies and more of the lower amplitudes of music, while the reverberation caused by sound bouncing off walls is reduced. The listening public have become accustomed to the Walkman sound and expect it to be duplicated on commercial recordings. Recording studios that once mixed the balance of their master recordings to suit the reproduction characteristics of car or transistor radios now mix them for Walkman headphones.

See also **Audio Systems**

ANDRE MILLARD

Further Reading

Abbate, J. The electrical century: getting small—a short history of the personal comp. *Proc. IEEE*, 87, 1695–1698, 1999.

Klein, L. Happy Anniversary, Sony Walkman! *Radio Electron.*, 60, 72–73, 1989.

Lyons, N. *The Sony Vision*. Crown, New York, 1976.

Millard, A. *America on Record: A History of Recorded Sound*. Cambridge University Press, New York, 1995.

Morita, A., Reingold, E. and Shimomura, M. *Made in Japan: Akio Morita and Sony*. Hutton, New York, 1986.

Read, O. and Welch, W.L. *From Tin Foil to Stereo: Evolution of the Phonograph*. Sams, Indianapolis, 1976.

Schiffer, M. *The Portable Radio in American Life*. University of Arizona Press, Tucson, 1991.

Pest Control, Biological

Insect outbreaks have plagued crop production throughout human history, but the growth of commercial agriculture since the middle of the nineteenth century has increased their acuteness and brought forth the need to devise efficient methods of insect control. Methods such as the spraying of insecticides, the application of cultural methods, the breeding of insect-resistant plants, and the use of biological control have increasingly been used in the twentieth century. Traditionally limited to checking the populations of insect pests through the release of predatory or parasitic insects, biological control now refers to the regulation of agricultural or forest pests (especially insects, weeds and mammals) using living organisms. It also includes other methods such as the spraying of microbial insecticides, the release of pathogenic microorganisms (fungi, bacteria or viruses), the release of male insects sterilized by radiation, the combination of control methods in integrated pest management programs, and the insertion of toxic genes into plants through genetic engineering techniques. Biological control is also directed against invasive foreign species that threaten ecological biodiversity and landscape esthetics in nonagricultural environments.

The Chinese are known to have used natural enemies to control insect pests as far back as the ninth century and European naturalists and agriculturists were proposing the use of entomophagous insects in the eighteenth century. Biological control only seriously took off at the end of nineteenth century, however, following a successful insect control campaign by the U.S. government. In 1888, Charles Valentine Riley, chief entomologist of the U.S. Department of Agriculture, sent one of his field entomologists, Albert Koebele, to Australia to search for natural enemies of the cottony cushion scale, an insect pest accidentally introduced in the citrus groves of California in 1869. Koebele found a lady beetle of the genus *Vedalia* that fed on the scale. Specimens of the beetle were sent to California, then bred and distributed throughout the state. Suppression of the scale followed within two years of the beetle's introduction.

This achievement incited other countries to initiate their own campaigns of biological control. Australia, Canada and the U.S. became especially proficient, as these countries were the most affected by the accidental introduction of foreign insects; insect pests rapidly reached outbreak levels in environments that contained none of the adverse conditions that would have normally limited their multiplication. Under the leadership of Riley's successor, Leland O. Howard, the U.S. Bureau of Entomology (USBE) set up laboratories for breeding parasites of destructive insects such as the gypsy moth, the Japanese beetle and the European corn borer, and it posted entomologists abroad to collect natural enemies. At the University of California, Harry S. Smith, the entomologist who coined the expression biological control in 1919, organized a Division of Beneficial Insect Investigations where he trained generations of practitioners. In England, the Imperial Bureau of Entomology established the Farnham House Laboratory (renamed the Commonwealth Institute of Biological Control (CIBC) in 1948) to coordinate and conduct biological control campaigns in the British colonies and dominions.

Figure 1. Sevenspotted Lady Beetle, Coleoptera:Coccinellidae, a predator of aphids. [*Photography : courtesy of Daniel Coderre, Laboratoire de lutte biologique, Université du Québec à Montréal.*]

However, the instantaneous and spectacular success of *Vedalia* was rarely repeated, and support for biological control dwindled in the interwar period. The introduction of synthetic organic insecticides during World War II significantly transformed applied entomology and further challenged the relevance of biological control. Their extreme toxicity proved to be an important drawback, however, and it soon became evident that they were not the panacea that many thought they would be.

Figure 2. Trichogramma, Hymenoptera: Trichogrammatidae, a parasitoid wasp. Larvae develop inside the host egg, eventually destroying it. [*Photography : courtesy of Daniel Coderre, Laboratoire de lutte biologique, Université du Québec à Montréal.*]

Microbial control—the use of pathogenic organisms—offered an attractive alternative. During the 1930s, the fortuitous spread of two insect diseases demonstrated the rapidity and efficiency of microorganisms in controlling outbreaks: a bacterium drastically reduced the Japanese beetle population in the U.S. and a polyhedral virus ended the European spruce sawfly outbreak in Canada. Advances in the microbiological sciences and research on diseases of beneficial insects like silkworms and bees provided the knowledge to propagate pure cultures of entomopathogens and to maintain their virulence in the field. In the 1950s, the commercialization of a microbial insecticide using a toxin derived from the *Bacillus thurigiensis* (Berliner) bacteria paved the way for the aerial spraying of large forest areas in Canada and of agricultural fields in the U.S. At the time of writing, *Bacillus thurigiensis* toxin genes are inserted directly into crop plants. Although selective to certain insect pests, microbial control is not self-sustaining, and certain entomologists do not consider it a method of biological control.

Integrated control is another technique that entomologists popularized after World War II. As in biological control, integrated control relies on natural mortality factors, but it also involves the spraying of selective chemical insecticides that, applied in a timely manner, avoid inhibiting the activities of natural enemies. The timing depends on an economic threshold: microeconomic and population dynamic models are used to determine when crop damage is expected to surpass a tolerable level of loss. During the 1940s, Alison D. Pickett in eastern Canada and Harry H. Smith in California had already devised spraying programs that encouraged the survival and multiplication of natural enemies. After chemical insecticides came under public scrutiny during the 1960s, integrated control and its successor, integrated pest management, became widely accepted in policy and agricultural circles.

A characteristic of the growth of biological control in the twentieth century was the international cooperation of national and individual actors. From the outset, the USBE and the CIBC encouraged entomologists from different countries to collect and exchange specimens of natural enemies and information on the ecology of insects. With comparatively few cases of agricultural crops damaged by the introduction of foreign insects, biological control in Europe revolved mainly around the utilization of native arthropods and vertebrates. However, the need to introduce natural enemies from foreign environments eventually led European entomologists to create the International Commission for Biological Control in 1955. Founded under the aegis of the International Union of Biological Sciences, the Commission (renamed the Organization in 1965) served Western Europe, the Mediterranean basin and the former colonies of France and Belgium. In 1971, it enlarged its geographical scope and became a worldwide organization for the identification, collection and distribution of insects.

See also **Crop Protection, Spraying; Pesticides**

STÉPHANE CASTONGUAY

Further Reading

Evenden, M.D. The laborers of nature. Economic ornithology and the role of birds as agents of biological pest control in North American agriculture, ca. 1880–1930. *Forest Conserv. Hist.*, 39, 172–183, 1995.

Garcia, R., Caltagirone, L.E. and Gutierrez, A.P. Comments on a redefinition of biological control. *Bioscience*, 38, 10, 692–694, 1988.

Hagen, K. A history of biological control, in *History of Entomology*, Smith, R.F., Mittler T.E., and Smith, C.N., Eds. Annual Reviews, Palo Alto, 1973.

Huffaker, C.B., Eds. *New Technology of Pest Control.* Wiley, New York, 1980.

Kingsland, S.E. *Modeling Nature: Episodes in the History of Population Ecology*. University of Chicago Press, Chicago, 1985.

Marchal, P. *Les Sciences Biologiques Appliquées à l'Agriculture et la Lutte Contre les Ennemis des Plantes aux États-Unis [Biological Sciences Applied to Agriculture]*. Librairie l'Homme, Paris, 1916.

Palladino, P.S. *Entomology, Ecology and Agriculture: The Making of Scientific Careers in North America, 1885–1985*. Harwood, Amsterdam, 1996.

Perkins, J.H. *Insects, Experts, and the Insecticide Crisis: The Quest for New Pest Management Strategies*. Plenum Press, New York, 1982.

Sawyer, R. *To Make a Spotless Orange: Biological Control in California*. Iowa State University Press, Ames, 1996.

Smith, H.S. On some phases of insect control by the biological method. *J. Econ. Entomol.*, 12, 288–292, 1919.

Thompson, W.R. *The Biological Control of Insect and Plant Pests: A Report on the Organisation and Progress of the Work of Farnham House Laboratory*. King's Printer, London, 1930.

Pesticides

A pesticide is any chemical designed to kill pests and includes the categories of herbicide, insecticide, fungicide, avicide, and rodenticide. Individuals, governments, and private organizations used pesticides in the twentieth century, but chemical control has been especially widespread in agriculture as farmers around the world attempted to reduce crop and livestock losses due to pest infestations, thereby maximizing returns on their investment in seed, fuel, labor, machinery expenses, animals, and land.

Until the twentieth century, cultural pest control practices were more popular than chemicals. Cultural methods meant that farmers killed pests by destroying infested plant material in the fields, trapping, practicing crop rotation, cultivating, drying harvested crops, planting different crop varieties, and numerous other techniques. In the twentieth century, new chemical formulations and application equipment were the products of the growth in large-scale agriculture that simultaneously enabled that growth.

Large scale and specialized farming provided ideal feeding grounds for harmful insects. Notable early efforts in insect control began in the orchards and vineyards of California. Without annual crop rotations, growers needed additional insect control techniques to prevent build-ups of pest populations. As the scale of fruit and nut production increased in the early decades of the century, so too did the insect problem.

In the early 1900s, chemical control became one of two lines of research for entomologists. The first, biological control, was the search for natural enemies of the insect pests. Although American entomologists scored a noted success in controlling the cottony cushiony scale (an insect that devastated California's citrus crops) with Australian ladybugs in the late 1800s, most biological control efforts were not so successful. Chemical control was also a mix of failure and success. Farmers applied solutions containing arsenic, sulfur, and mercury to rid their crops and livestock of boll weevils, coddling moths, and ticks that carried Texas fever. Some farmers applied an inappropriate chemical or the wrong amount. Many, however, urged on by manufacturers, scientists, and their own need to maximize yields, still had confidence in chemical solutions.

The 1930s and 1940s were crucial decades for the development of pesticides. In 1939, Paul Mueller of the Geigy Company (Switzerland) developed an effective insect killer, dichloro-diphenyl trichloroethane (DDT). American armed forces used it to reduce the number of noncombat related casualties to record lows for soldiers and civilians, most notably controlling an outbreak of typhus in Naples in 1943 by killing body lice, the disease vector. U.S. War Research Service scientists experimented with 2, 4 dichlrophenoxyacetic acid (2,4-D), a synthetic hormone used in the 1930s to promote uniform ripening of fruit. This growth regulator, when administered in large doses, actually killed plants by stimulating them to grow themselves to death. As World War II ended, the widespread use of DDT and experiments with 2,4-D were signal accomplishments of American industry and government.

After the war, the U.S. Department of Agriculture (USDA), the agricultural extension service, and manufacturers promoted the use of pesticides among farmers and other groups including local governments, businessmen, and household decision makers. Armed with powerful new killers, Americans attempted to reduce or even eradicate pest species on golf courses, lawns, farms, and in homes, lawns, offices, restaurants, warehouses, and grain elevators. Agricultural Research Service leaders promoted fire ant eradication in the southern states, while cooperative extension experts in the Midwest promoted the elimination of various fly species. Similarly, farmers and governments around the world turned to pesticides, notably on plantations dedicated to single crops and to kill mosquitoes that carried malaria. Pesticides became popular because people liked quick, easy, and inexpensive pest control.

During the 1960s and 1970s the widespread optimism about the value of pesticides waned, especially in developed nations. In 1946, scientists cautioned farmers not to use DDT and other chlorinated hydrocarbon insecticides directly on dairy animals or beef cattle immediately before they were sold, since these chemicals were stored in fat and ended up in food products. By the late 1950s the U.S. Food and Drug Administration inspected food products and compelled farmers who sold products with residual DDT to dump their milk. In the 1960s, municipal governments aggressively used DDT to destroy the insect that carried Dutch elm disease across the U.S., but the DDT also killed songbirds, provoking public concern about the wisdom of using pesticides. Farmers and ranchers found that some plant and insect species became resistant or were already resistant to certain chemicals. Once farmers suppressed a targeted species, resistant species filled the ecological vacuum. In 1962 Rachel Carson's critique of pesticides, *Silent Spring*, attracted the attention of entomologists and the public. Carson's clear explanations and moving prose showed the harmful unanticipated consequences of chemical use by focusing on ecosystems. Concerns about the health of ecosystems and wildlife, not human health, prompted the public discussion that led to the U.S. ban of DDT in 1972. In the 1970s and 1980s, American Vietnam veterans and scientists claimed that exposure to trichlorophenoxyacetic acid (2,4,5-T), a herbicide and defoliant known as Agent Orange that included dioxin, caused illness.

Critics of the high economic and potential health costs of pesticides practiced organic farming. Although the definition of organic was a moving target, depending on USDA policy and pressure from retailers and producers, the idea of pesticide-free products was attractive to a minority of consumers and producers. Organic producers substituted labor and fuel costs for pesticides, hoping to capitalize on high-return niche markets of affluent or environmentally concerned citizens.

By the end of the twentieth century, pesticides were essential in agriculture and industry, despite some skepticism about their value. In many places in the world, chlorinated hydrocarbon insecticides continued to play a valuable role in controlling disease vectors, saving millions of lives. Herbicides played a role in reducing child labor in developing nations, allowing more children to attend school. Many industrial and agricultural chemical users of the late 1900s practiced a balance of chemical, cultural, and environmental controls called integrated pest management (IPM) to reduce reliance on chemicals. Genetically modified organisms (GMOs), such as roundup-ready soybeans, became a significant portion of crops grown, although some nations refused to import GMO crops because of fears over unknown consequences and protectionist trade policies. Fields planted with roundup-ready plants could be sprayed with herbicide, killing weeds but preserving crops. Corn with the *Bacillus thuringensis* gene spliced into its DNA infects the European corn borers that eat the corn. These practices actually reduced the need for chemical insecticide. Consumers, producers, distributors, and retailers all found uses for pesticide in the twentieth century, although they climbed steep learning curves in adopting and adapting it.

See also **Agriculture and Food; Chemicals; Farming, Agricultural Methods; Pest Control, Biological**

J. L. ANDERSON

Further Reading

Dunlap, T.R. *DDT: Scientists, Citizens, and Public Policy*. Princeton University Press, Princeton, NJ, 1983.

Russell, E. *War and Nature: Fighting Humans and Insects with Chemicals from World War I to Silent Spring*. Cambridge University Press, Cambridge, 2001.

Schlebecker, J.T. *Whereby We Thrive: A History of American Farming, 1607–1972*. Iowa State University Press, Ames, 1975.

Sharrer, G.T. *A Kind of Fate: Agricultural Change in Virginia, 1861–1920*. Iowa State University Press, Ames, 2000.

Stoll, S. *The Fruits of Natural Advantage: Making the Industrial Countryside in California*. University of California Press, Berkeley, 1998.

Photocopiers

The photocopier, copier, or copying machine, as it is variously known, is a staple of modern life. Copies by the billions are produced not only in the office but also on machines available to the public in libraries, copy shops, stationery stores, supermarkets, and a wide variety of other commercial facilities.

Over the years, various processes have been employed. By far the most common type of photocopier today is the electrostatic, or xerographic. It is the type most people are familiar with, and arguably that with which most people associate the term photocopier. It was the electrostatic process that revolutionized copying as a part of everyday life. The modern photocopying era began in 1960 with the introduction of the Xerox 914, the first push-button, plain-paper copier. Within one year, sales doubled, and *Fortune* magazine called the 914 "the most successful product ever marketed in America."

While "photocopying" to most people means copying with the degree of the quality and speed that has existed since the 1960s, copying technology in a rudimentary form can be dated to the pantograph of the seventeenth century. The standard of the later eighteenth century was a copying machine patented by James Watt in 1780. It was a device more accurately known as a copying press, and consisted of a mechanism for exerting pressure on a dampened sheet of tissue placed over the document to be copied. Written with ink based on gum arabic or sugar, the image on the original was transferred to the tissue copy, albeit in reverse, when the two sheets were interfaced under pressure. The reason for the tissue was so that the copy could be viewed correctly looking through it from the back.

A copying press that Benjamin Franklin brought back to the U.S. from Europe was apparently one of this type. George Washington was known to have had two copying presses. Thomas Jefferson, a prolific letter writer, made copies of his correspondence using a copying machine he considered not only "a most precious possession" but "the finest invention of the present age." Jefferson first used a copy press and then a pantograph, or "polygraph" as his was known.

The polygraph was a mechanical apparatus that used wires and movable wooden arms holding a pen or pens to duplicate, on a separate page or pages, the motion of the human arm writing out the original. One of the earliest U.S. patents was for a device of this kind—a "machine for writing with two pens" patented in 1799 by Marc Isambard Brunel.

An advantage to the polygraph was that it made exact (not image-reversed) copies on plain paper, the same paper as the original if desired. But the polygraph was a delicate, fragile mechanism that was difficult and clumsy to use.

It was versions of the copying press, rather than the polygraph, that generally became the standard of nineteenth century. Besides being simpler to use, the copying press could be made small and rugged enough to be easily transportable, and it was sometimes used by travelers.

A device known as the electric pen, patented by Thomas A. Edison in 1876, led to the most common form of copying machine of the early twentieth century. Developed as part of Edison's automatic telegraph, the electric pen, working at a rate of roughly 8000 pulses per minute, could make minute perforations in the form of letters or drawings on a stencil. A plain sheet of paper was then placed under the stencil, and ink was pressed through with a roller, making exact copies, albeit in small quantity. This led to the mimeograph machine, the mainstay of copying over approximately the first half of the twentieth century. At first, copies were made by hand, one by one. The rotary mimeograph, which was introduced by A.B. Dick in 1904, made copies automatically using a revolving cylinder, at a great increase in speed.

The photostat process, using a special camera to produce an image directly on photosensitized paper without going through a negative, was developed in the early twentieth century. Photostats remained a common way of copying documents until the coming of the modern photocopier. A shortcoming of the photostat, in comparison to the photocopier, is that it was not ordinarily consumer-operated technology.

Do-it-yourself copiers suitable for the small office began to appear in significant numbers by the 1950s. Ordinarily, photosensitized paper was required. Some processes worked with a liquid and others with fumes, and still others with infrared rays to produce heat by which the image was transferred to photosensitized paper (thermography). In general, copies were clearly identifiable as such, the paper usually having a sheen and glossy feel unlike, and inferior to, modern plain-paper copies.

Meanwhile xerography, the principal modern form of photocopying, was also under development. Generally speaking, xerography (from the Greek *xeros* meaning "dry") uses a dry powder as opposed to ink or liquid chemicals. Static electricity attracts and bonds the powder to form an image on paper, and heat then makes the bond permanent.

When xerography came into use, it was generally only for large office applications. The first commercial XeroX Copier (then spelled with a capital X at the end) was introduced in 1949 by the Haloid Company. It was based on a 1938 invention of Chester Carlson. Haloid later became Haloid Xerox and subsequently Xerox Corporation.

The first XeroX was messy and difficult to use. Most of the process was carried out manually, and it often misprinted. By modern standards, it was also a notoriously slow process. At an early demonstration, according to one newspaper account, observers timed the operation at 45 seconds per copy.

In 1960 a vastly improved model, the Xerox 914, was introduced, and with it the modern era of photocopying. The push-button 914 worked automatically and printed on plain paper as opposed to the less-desirable photosensitized paper. However, it weighed 290 kilograms, limiting its use to large-office applications. Its bulky size notwithstanding, the 914 caught on quickly and revolutionized photocopying.

Modern xerographic copiers, produced by a number of manufacturers, are available as desktop models suitable for the home as well as the small office. Many modern copiers reproduce in color as well as black and white, and office models can rival printing presses in speed of operation.

See also **Electronics; Printers**

MERRITT IERLEY

Further Reading

A.B. Dick Company. *The Edison Mimeograph*. Pamphlet, A.B. Dick Company, Chicago, 1889.

A.B. Dick Company. *The Story of Stencil Duplication*. Pamphlet, A.B. Dick Company, Chicago, 1935.

Bedini, S.A. *Thomas Jefferson and His Copying Machines*. University Press of Virginia, Charlottesville, VA, 1984.

Carlson, C. The original duplicator. *Technology Review*, September/October 1998.

Ierley, M. *Wondrous Contrivances: Technology at the Threshold*. Clarkson Potter, New York, 2002.

Rhodes, B. and Streeter, W.W., *Before Photocopying: The Art and History of Mechanical Copying*, 1780–1938. Oak Knoll Press, New Castle, DE, 1999.

Verry, H.R. *Document Copying and Reproduction Processes*. Fountain Press, 1958.

Williams, E.M. *The Physics and Technology of Xerographic Processes*. Wiley, New York, 1984.

Photosensitive Detectors

Sensing radiation from ultraviolet to optical wavelengths and beyond is an important part of many devices. Whether analyzing the emission of radiation, chemical solutions, detecting lidar signals, fiber-optic communication systems, or imaging of medical ionizing radiation, detectors are the final link in any optoelectronic experiment or process.

Detectors fall into two groups: thermal detectors (where radiation is absorbed and the resulting temperature change is used to generate an electrical output) and photon (quantum) detectors. The operation of photon detectors is based on the photoelectric effect, in which the radiation is absorbed within a metal or semiconductor by direct interaction with electrons, which are excited to a higher energy level. Under the effect of an electric field these carriers move and produce a measurable electric current. The photon detectors show a selective wavelength-dependent response per unit incident radiation power.

The photoeffect takes two forms: external and internal. The former involves photoelectric emission where the photogenerated electrons escape from the material as free electrons. In the latter process, the excited carriers remain within the material, usually in semiconductors, thereby increasing its conductivity (photconductive detectors or photoresistors).

The principle of photoemission was first demonstrated in 1887 by Heinrich Hertz who noticed that the breakdown voltage of spark gaps decreased when ultraviolet light was shone on the metal cathode. Two years later, Julius Elster and Hans Geitel revealed that if an alkali metal electrode was used, this effect could be produced with visible radiation. Although these effects could be demonstrated reproducibly, no satisfactory explanation was offered until Einstein proposed his theory of photoemission in 1905.

The first internal photon effect, the photoconductive effect, was discovered by Willoughby Smith in 1873 when he experimented with selenium as an insulator for submarine cables. The material Tl_2S (thallium sulfide) was the first infrared photoconductor of high responsivity and was developed by Theodore W. Case in 1917 as an infrared sensor for signaling. The years during World War II saw the origin of modern detector technology because of the need for signaling and aircraft detection.

Depending on the nature of the interaction, photon detectors can be further subdivided into different types as shown in Table 1. The most important are intrinsic detectors (in which an electron or hole moves from the valence band to the conduction band), extrinsic detectors (in which transitions are between the doped level in energy gap and conduction or valence bands), photoemissive detectors (like PtSi Schottky barrier detectors), and quantum well detectors (with intersub-band transitions inside bands).

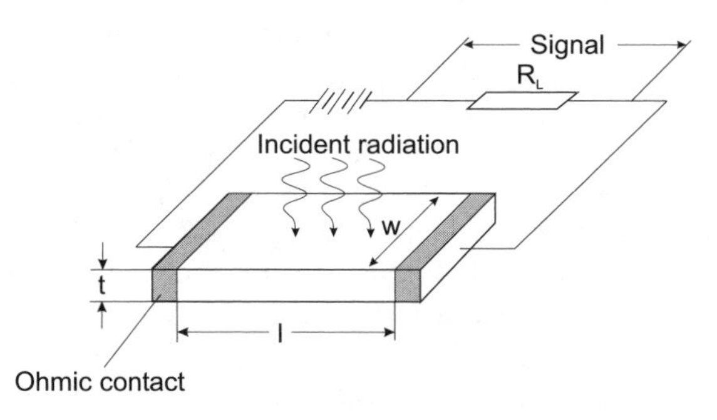

Figure 3. Geometry and bias of a photoconductive detector.

The light-generated carriers can produce an electric signal in two ways. In one case they can increase the sample conductivity. Such devices are called photoconductive detectors or photoresistors (Figure 3). The other way to produce a signal is to use a semiconductor device that has an internal electric field. It may be a *p–n* junction diode (Figure 4), or a Schottky barrier (metal semiconductor contact) diode (Figure 5). In any of these cases, electrons and holes can be separated by the built-in electric field or, more typically, by the electric field that is a combination of the external and internal electric fields. Such a photovoltaic detector is called a photodiode.

In comparison with photoconductive detectors, photodiodes exhibit four important advantages:

1. Low or zero bias voltage
2. High impedance, which aids coupling to read-out circuits in imaging arrays
3. Capability for high-frequency operation

Table 1 Photon detectors with internal photoeffect.

Type	Transition	Electrical output	Example
Intrinsic	Interband	Photoconductive Photovoltaic Capacitance (MIS)	Si, GaAs, GaN, PbSe, InSb, HgCdTe Si, GaN, InGaAs, InSb, HgCdTe Si, GaAs, InSb, HgCdTe
Extrinsic	Impurity to band	Photoconductive	Si:As, Si:Ga, Ge:Cu, Ge:Hg
Free carriers	Intraband	Photoemissive	PtSi, Pt_2Si, IrSi Schottky barriers GaAs/CsO
Quantum wells	To or from spatially quantised levels	Photoconductive Photovoltaic	GaAs/GaAlAs, InAs/InGaSb

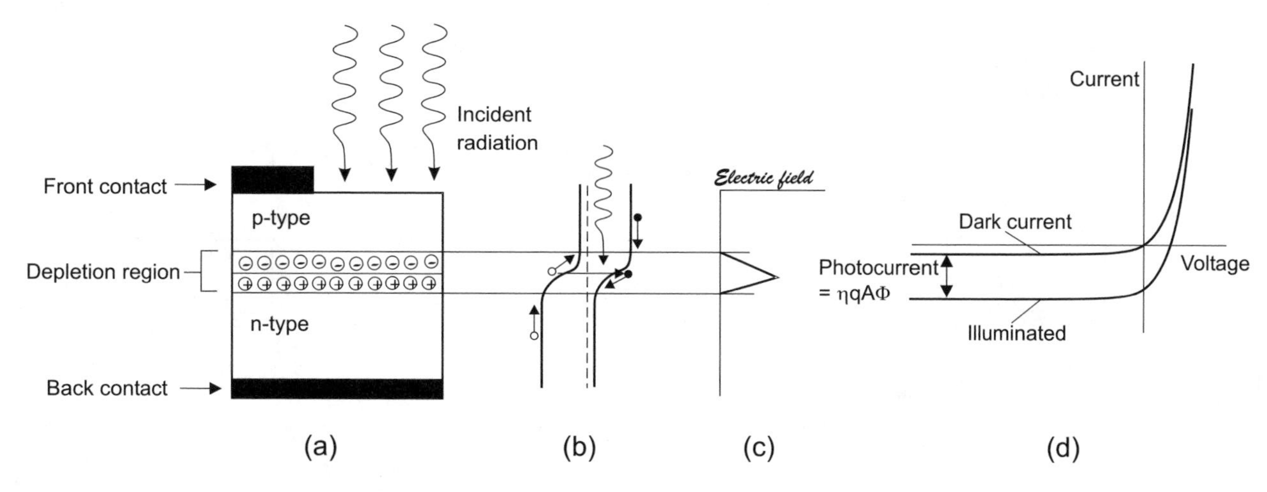

Figure 4. *p–n* junction photodiode: (a) structure of abrupt junction, (b) energy band diagram, (c) electric field, and (d) current–voltage characteristics.

4. The compatibility of the fabrication technology with planar-processing techniques.

Photoemissive detectors are generally the detectors of choice for ultraviolet (UV), visible, and near-infrared applications where high efficiency is available. They usually take the form of vacuum tubes called phototubes. Electrons are emitted from the surface of a cathode and travel to an electrode (anode), which is maintained at a higher electric potential (Figure 6(a)). As a result of the electron transport between the cathode and anode, an electric current proportional to the photon flux incident on the photocathode is created in the circuit. The photoemitted electrons may also impact other specially placed metal or semiconductor surfaces in the tube, called dynodes, from which a cascade of electrons is emitted by the process of secondary emission. The result is an amplification of the generated electric current by a factor as high as 10^7. This device is known as a photomultiplier tube (Figure 6(b)).

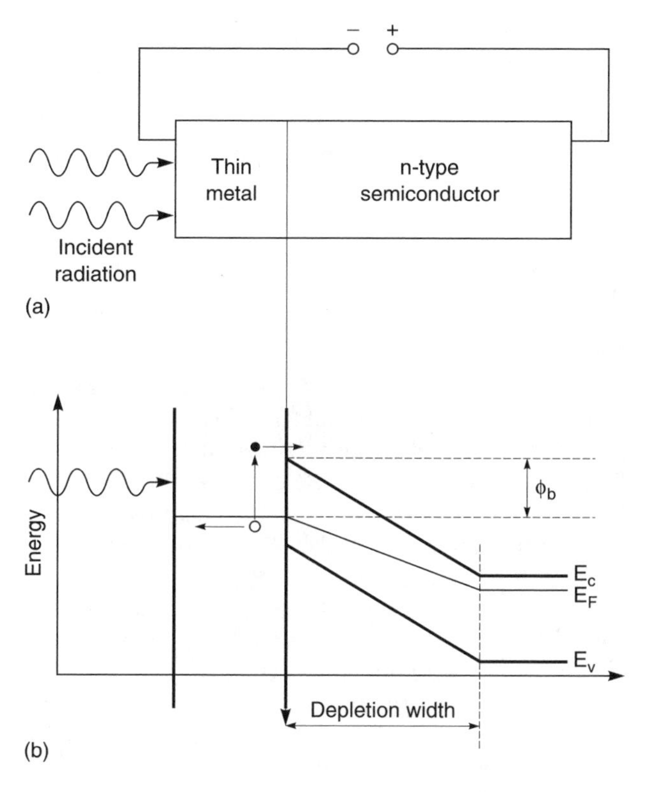

Figure 5. Schottky barrier photodiode: (a) structure and (b) energy band diagram.

The secondary emission is also used in a modern imaging device called the microchannel plate, widely used to detect UV radiation, and soft x-ray fluxes. It consists of a honeycomb array of millions of glass capillaries with an internal diameter of around 10 micrometers drawn out by fiber optic techniques, in a glass plate of a thickness of around 1 millimeter. Both faces of the plate are coated with thin metal films that act as electrodes and a voltage is then applied across them (Figure 6(c)). The interior walls of each capillary are coated with a secondary-electron-emissive material and behave as a continuous dynode, multiplying the photoelectron current emitted at that position (Figure 6(d)). In such a way, the local photon flux can be converted into a substantial electron flux that can be measured directly. Furthermore, the electron flux can be reconverted into an optical image by using a phosphor coating as the rear electrode to provide electroluminescence; this combination provides an image intensifier.

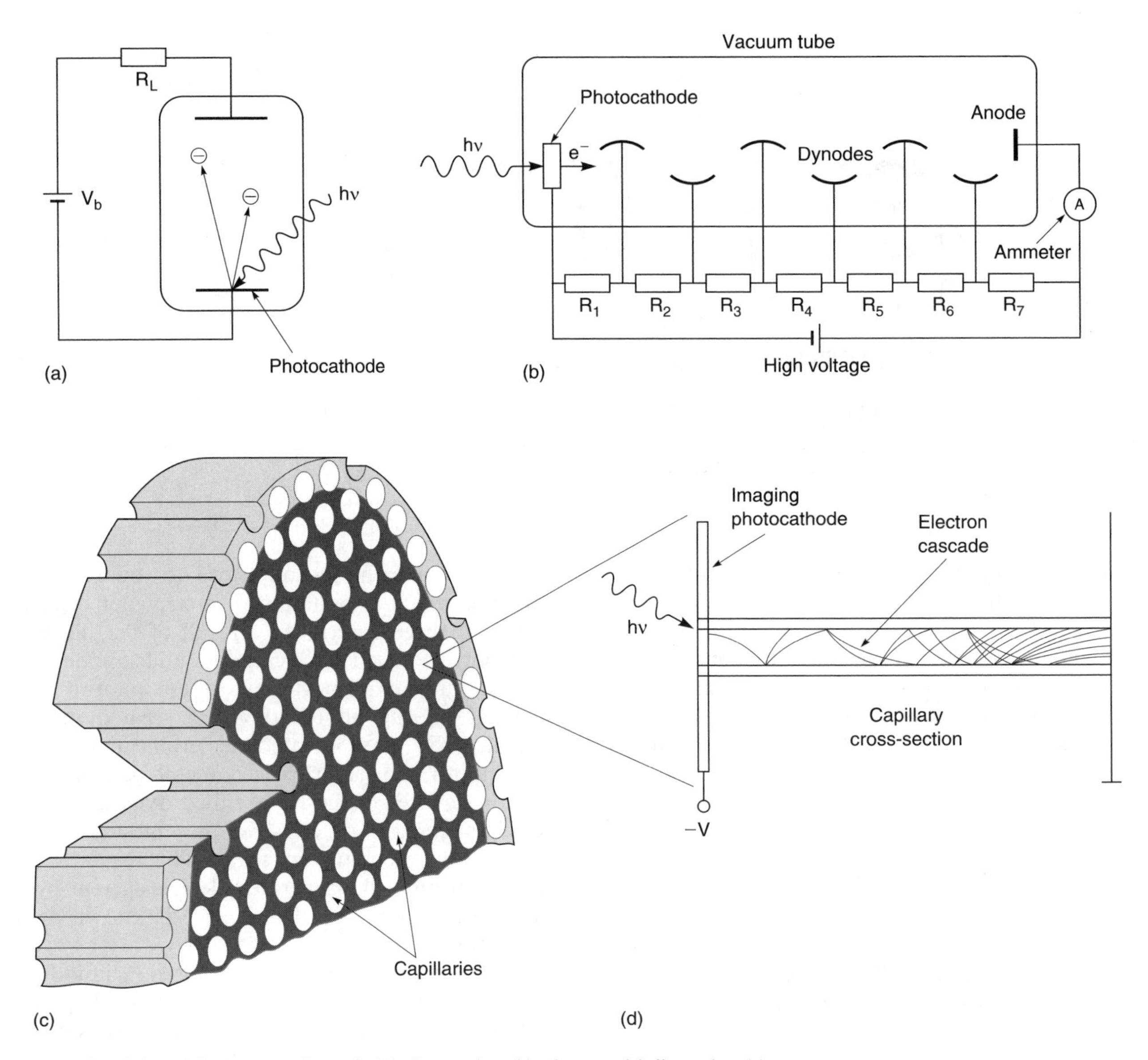

Figure 6. Schematic presentation of: (a) phototube, (b) photomultiplier tube, (c) cutaway view of microchannel plate, and (d) single capillary in a microchannel plate.

Photoconductive detectors and especially photodiodes (with their very low power dissipation) are combined with electronic readouts to make detector arrays used in imaging devices. They are often called as focal plane arrays (FPAs) because they are located in the focal plane of imager optics.

There are a number of architectures used in the development of FPAs. In general, they may be classified as monolithic or hybrids. In the monolithic approach, both detection of light and signal readout (multiplexing) is done in the detector material. The integration of detector and readout onto a single monolithic piece reduces the number of processing steps, increases yields, and reduces costs. For visible and near-infrared detection, monolithic structures can be built in silicon, forming arrays with more than 10 million high-performance pixels. The most highly developed of these visible detectors is the charge-coupled device (CCD). This approach to image acquisition was first proposed in 1970 by Bell Lab researchers Willard S. Boyle and George E. Smith.

In effect, a CCD is a light-sensitive device that stores electrical signals. The CCD technique relies on the optoelectronic properties of a well-established semiconductor architecture: the metal oxide semiconductor (MOS) silicon capacitor (Figure 7). An MOS capacitor typically consists of an extrinsic silicon substrate (in this case, *p*-type) on which is grown an insulating layer of silicon dioxide (SiO_2).

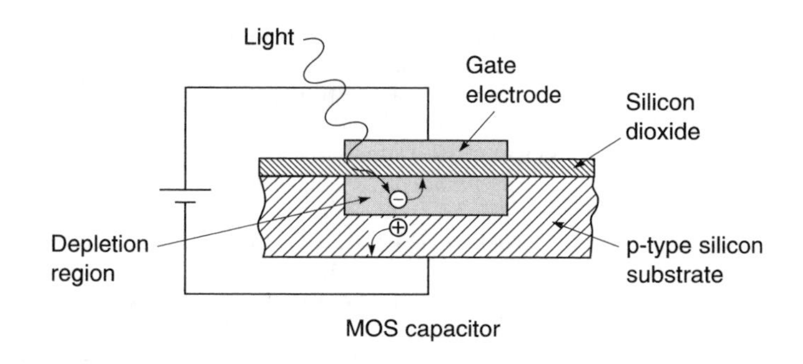

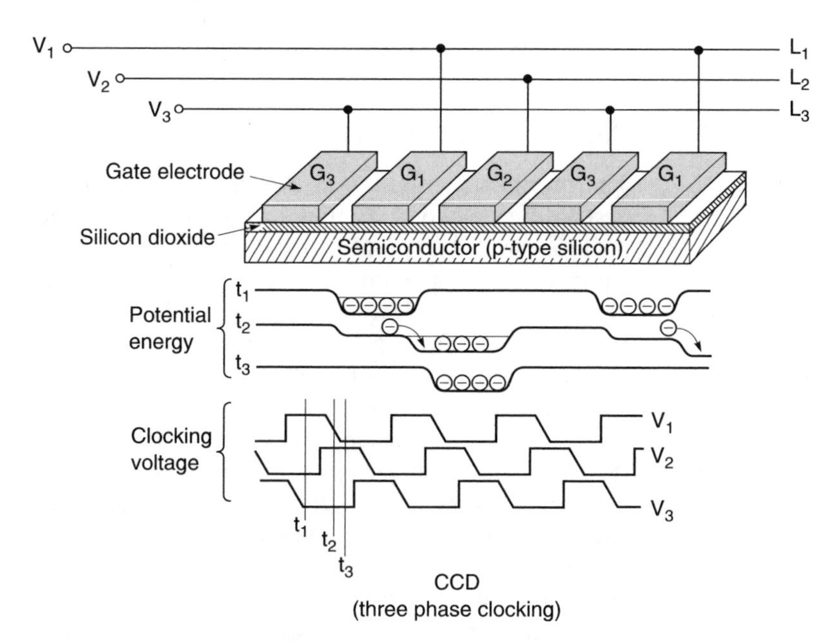

Figure 7. Charge-coupled device. Light absorbed by a *p*-type metal oxide semiconductor (MOS) silicon capacitor (top) creates mobile electrons that become trapped in the depletion region under the gate. Three-phase CCDs (bottom) use a timed sequence of three gate voltages to move accumulated electrons across an array of MOS capacitors. Note the potential energies at times t_1, t_2, and t_3. Gates G_1, G_2, and G_3 define a single pixel.

When a bias voltage is applied across MOS structure, majority charge carriers (holes) are pushed away from the Si–SiO_2 interface directly below the gate, leaving a region depleted of positive charge and available as a potential energy well for any mobile minority charge carriers (electrons). Electrons generated in the silicon through absorption will collect in the potential energy well under the gate. Linear or two-dimensional arrays of these MOS capacitors can therefore store images in the form of trapped charge carriers beneath the gates. In the CCD solution the accumulated charges are transferred from potential well to the next well by using sequentially shifted voltage on each gate. One of the most successful voltage-shifting schemes is called three-phase clocking (Figure 7, bottom). Column gates are connected to the separate voltage lines (L_1, L_2, L_3) in contiguous groups of three (G_1, G_2, G_3). The setup enables each gate voltage to be separately controlled.

The first CCD imager sensors were developed in the 1970s primarily for television analog image acquisition, transmission, and display, and CCD cameras were first used in telescopes in 1979. CCD cameras have subsequently been developed for scanners, video cameras, and digital cameras. Modern CCD detectors have excellent x-ray response, and have been used in orbiting x-ray telescopes. With increasing demand for digital image data, the traditional analog raster scan output of image sensors is of limited use, and there is a strong motivation to fully integrate the control, digital interface, and image sensor on a single chip. In particular, silicon fabrication advances for computer processors and memory now permit the implementation of complementary MOS (CMOS) transistor structures that are considerably smaller than the wavelength of visible light and have enabled the practical integration of multiple transistors within a single image sensor. At each CMOS imager sensor pixel, there is a photosensor with some integrated MOS transistors that are used to multiplex and possibly manipulate the analog signal from the photosite (Figure 8). Analog output signals from any CMOS imager

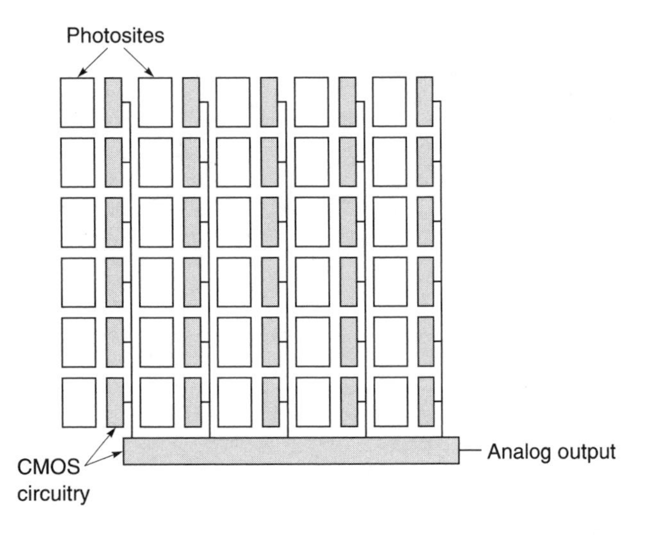

Figure 8. CMOS imager architecture.

pixel can be randomly accessed through wires and multiplexing circuitry.

Developing rapidly, CMOS imaging technology has several potential advantages over CCD technology: low-voltage operation and low power consumption, compatibility with integrated CMOS circuitry, random access to image data, and a lower cost for both digital video and still camera applications. The processing technology for CMOS image sensors is typically two to three times less costly than standard CCD technology.

Both CCD and CMOS imagers described above are monolithic devices. In infrared spectral range, hybrid architectures are usually used. Hybrid FPA detectors and multiplexers are fabricated on different substrates and mated with each other by the flip-chip bonding or loophole interconnection. In this case we can optimize the detector material and multiplexer independently. Other advantages of the hybrid FPAs are near 100 percent fill factor and increased signal-processing area on the multiplexer chip. In the flip-chip bonding, the detector array is typically connected by the contacts via indium bumps to the silicon multiplex pads.

See also **Iconoscope; Infrared Detectors; Semiconductors, Elemental; Semiconductors, Pre-Band Theory; Solar Power Generation**

Antoni Rogalski

Further Reading

Bhattacharya, P. *Semiconductor Optoelectronic Devices*. New Jersey, Prentice Hall, 1997.

Csorba, I.P. *Image Tubes*. Sams, Indianapolis, 1985.

Donati S. Photodetectors: Devices, Circuits and Applications. Prentice Hall, New Jersey, 2000.

Goldstein, A. Charge coupled device, in *Instruments of Science*, Bud, R., Ed. Garland, London, 1998.

Levine, B.F. Quantum-well infrared photodetectors. *J. Appl. Phys*. 74, R1–R81, 1993.

Rieke, G.H. *Detection of Light: From the Ultraviolet to the Submillimeter*. Cambridge University Press, Cambridge, 1994.

Rogalski, A. *Infrared Detectors*. Gordon & Breach, Amsterdam, 2000.

Sequin, C. and Tompsett, M. *Charge Transfer Devices*. Academic Press, New York, 1975.

Smith, G. and Boyle, W. The invention and early history of the CCD, in *Optical Detectors for Astronomy*, Amico, P. and Beletic, J.W. Kluwer, Dordrecht, 1998.

Plastics, Thermoplastics

Thermoplastic polymers are perhaps some of the best-known plastics, as they are used for a variety of household items and packaging materials. They comprised 10 percent of the global chemical industry in 1995 (around 90 million tons). Initially made from a process based on coal, they are now usually made from oil-based products.

Thermoplastics include plastics developed as early as 1877 when polymethylmethacrylate (PMMA, or acrylic) was first formed by Rudolph Fittig, developed by Otto Rohm, a German chemist, but not commercially developed by Rohm & Haas until 1928 and later by Rowland Hill and John Crawford at ICI in 1934. In the 1930s other thermoplastics were developed, such as polyethylene by ICI in the U.K., and nylon 6 and nylon 6,6 by Wallace Hume Carothers at DuPont in the U.S. Polyethylene is now commonly used as a packaging material and synonymous with the plastic bag. Nylon was first used to make toothbrush bristles in 1938, then ladies' stockings in 1939. Thermoplastic polypropylene, polyvinyl chloride (PVC), and polystyrene are common packaging materials today. Carothers had worked on polyesters and his work was progressed by John R Whinfield and James T Dickson at the Calico Printers Association in the U.K., who produced the polyester fiber Terylene in 1941. This material is widely used to make clothing and soda bottles. Thermoplastics also include certain resins such as polyether ether ketones (PEEK) which are used to make composite material with superior mechanical properties although at a high cost, therefore they are normally only used in high-technology fields such as aerospace (see Composite Materials).

With thermoplastics, increased strength and thermal resistance comes at an increased price, with polystyrene as the cheapest, least thermal resistant and weakest thermoplastic, followed by low-density polyethylene, high-density polyethylene, polypropylene, polymethyl methacryalate (PMMA, or acrylic), high-impact polystyrene, acrylonitrile butadiene styrene (ABS*) (used for car bumpers), polyesters, polycarbonates and polysulfones to nylon (a polyamide) which has the highest strength and thermal resistance but is also the most expensive. Polycarbonates have high impact strength, hardness, toughness, and resistance to temperatures between about −40°C and 145°C. Polysulfones are heat-resistant at temperatures of up to 150°C

Thermoplastics can be molded to set into a certain shape, but they can then be reshaped after reheating. This is due to their molecular structure, which consists of long chains of molecules held together by weak intermolecular forces. They have a structure almost resembling spaghetti, bound together by the weak Van der Waals forces. Some cross-linking can occur with side groups such as the vinyl group in PVC. The orientation of these side groups will have a great influence on how the polymer will behave and its properties such as strength.

Thermoplastics have to be processed at a higher temperature than thermosetting plastics; this may be more than 400°C for "high temperature" thermoplastics.

The weak intermolecular forces in thermoplastics make them relatively easy to process using a variety of methods ranging from extrusion, vacuum and blow molding to perhaps the most common process, injection molding. They are pliable, and easily shaped and molded. The molding process becomes easier as thermoplastics become hotter, but at some point they will melt.

Thermoplastics have a glass transition temperature (Tg) above room temperature, whereas that of elastomers is below room temperature. The glass transition temperature is where a polymer changes from a rigid solid to a rubber. Below this temperature the polymer becomes hard and brittle, like glass. Some polymers are used above their glass transition temperatures while others such as polystyrene and PMMA are used below.

Thermoplastics can also be blended with other polymers, such as elastomers, to enhance certain properties such as toughness. These are known as copolymers and they can be engineered for specific purposes.

There is also a class of thermoplastic elastomers with physical rather than chemical cross-links. An example of such a material is a thermoplastic elastomer, polyurethane.

Thermoplastics are easy to mold and can be shaped. They are resistant to deformation but will eventually permanently deform or break. They are hard and brittle below their Tg, but pliable and soft above it. Crystalline polymers melt below their glass transition temperature, whereas amorphous polymers are hard and brittle below their Tg, but become flexible and rubbery above it. Most thermoplastics are a mix of a crystalline and an amorphous structure. Hard plastics such as polystyrene and PVC are used at temperatures below their Tg whereas flexible plastics such as polypropylene are used above it. Some thermoplastics such as polypropylene, nylon, polyketones and syndiotactic polystyrene are highly crystalline, whereas others such as PMMA, polycarbonates and atactic polystyrene are highly amorphous due to their polymer structure and the intermolecular forces.

Thermal or mechanical methods can be used to alter the crystalline structure of a thermoplastic. If a polymer is cooled slowly from its melting point, a higher degree of crystallinity is more probable. However when polymers are cooled quickly from the melt, the amorphous chains may be frozen into the solid. This is because the chains will not have had enough time to untangle from their melted form to separate and form crystals.

Thermoplastic composites are frequently tougher than thermoset composites, but although they do not possess enhanced fatigue or static properties and they may also have worse compression strength, they are more resistant to moisture and a range of industrial solvents than thermosets.

Thermoplastics are made from molecular chains with various types of stereochemical arrangements possible that can confer different properties on the polymer. These are called atactic, isotactic and syndiotactic. In an isotactic polymer, all the substituent groups lie on one side of the molecular chain. In isotactic polypropylene, for example, the methyl (CH_3) groups are on the same side of the molecular chain. This arrangement allows the molecular chains to pack together more easily giving a crystalline structure that is strong, stiff and brittle (see Figure 9).

Atactic thermoplastics have substituent groups that are randomly placed on both sides of the chain for example the methyl groups in atactic polypropylene (see Figure 10). This arrangement means that the chains are unable to pack together, resulting in an amorphous structure that makes the polymer tough and rubbery. This results in a form of toughened polypropylene.

Syndiotactic polymers have repeating units on either side of the molecular chain backbone,

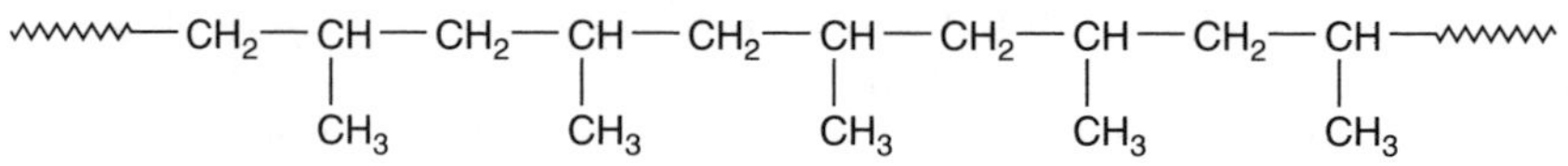

$\sim\sim CH_2-CH(CH_3)-CH_2-CH(CH_3)-CH_2-CH(CH_3)-CH_2-CH(CH_3)-CH_2-CH(CH_3)\sim\sim$

Isotactic polypropylene

Figure 9. Chemical formula. [*after http://www.psrc.usm.edu/macrog/pp.htm*]

alternating with each other (see Figure 11). In syndiotactic polystyrene, the chains are able to pack tightly together giving a crystalline structure whereas the atactic form of polystyrene is amorphous as the chains cannot pack together so tightly.

Thermoplastics can be recycled, but require careful sorting to separate the various different polymers. If different grades are mixed the resultant recycled polymer will have variable properties. Normally, recycling a polymer will involve some deterioration in its properties.

Thermoplastics have flexibility and so are useful where this is needed, for example in squeezable washing-up bottles. The first of these made was the 1958 "Squeezy" bottle, which was made of polyethylene (although with metal ends at this point). Thermoplastics can be blow-molded into a variety of shapes and polyethylene terepthalate (PET) is a popular thermoplastic used to make "pop" bottles.

Thermoplastics are excellent materials for coextrusion—a process widely used in packaging to make multilayered sheet with different properties in the different layers—tough on the outside and impermeable on the inner layers. This technique is popular for producing packaging materials.

Although polyethylene is self-toughening, different formulations of polyethylene have been developed such as low-density polyethylene, the earliest type developed first in 1933, high-density polyethylene, and now ultrahigh molecular weight polyethylene.

Polypropylene is used for packaging and in cars and polypropylene fibers are also used for clothing, carpets and nonwoven fabrics. Impact-resistant copolymers of polypropylene are used to make car bumpers as well as for medical use.

Teflon (polytetrafluoroethylene or PTFE) is used to make products that operate at high temperatures because of its heat-resistant properties. It has excellent chemical, electrical, mechanical and thermal properties and is also chemically inert. Almost nothing sticks to Teflon hence its use for non-stick frying pans. As it is very heat resistant, it is often used in space applications, for example as a material to protect the deployment rods of the solar arrays which replaced the original arrays during the first servicing mission of the Hubble space telescope in 1993.

The future of thermoplastics demands improved recyclability, toughness, repairability and "smart applications" such as self-healing abilities and perhaps inbuilt obsolescence and, ideally, biodegradability. Bioengineered thermoplastics—the class of so-called biopolymers—are an exciting new development. Polyesters are already engineered to make specialist fibers required for sporting applications. More improved engineered fibers are continuously being developed. In the future they will also have to possess an environmentally friendly life-cycle.

See also **Biopolymers; Fibers, Synthetic and Semi-Synthetic; Plastics, Thermosetting; Synthetic Resins**

SUSAN MOSSMAN

Further Reading

Arlie, J.-P. *Commodity Thermoplastics*. Editions Technip, Paris, 1990.
MacDermott, C.P. and Shenoy, A.V. *Selecting Thermoplastics for Engineering Applications*, 2nd edn, vol. 42. 1997.
Olabisi, O. *Handbook of Thermoplastics*, vol. 41. 1997.
Scheirs, J. *Polymer Recycling*. Wiley, Chichester, 1998.

Plastics, Thermosetting

Thermosetting plastics, or thermosets are a type of polymer, usually made by mixing two or more suitable liquids together. Thermosets are usually supplied in the form of partly polymerized pre-

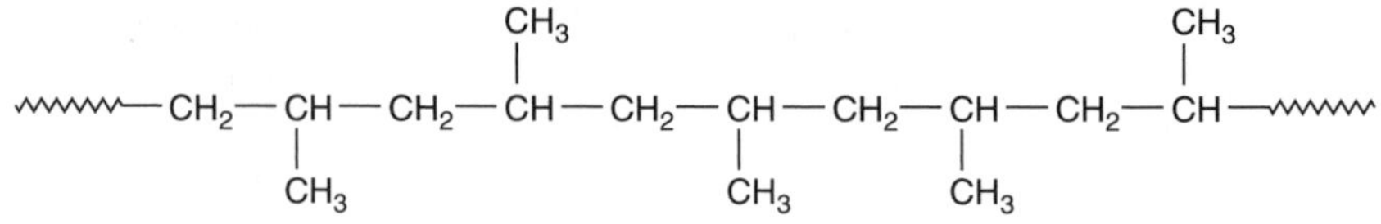

$\sim\sim CH_2-CH(CH_3)-CH_2-CH(CH_3)-CH_2-CH(CH_3)-CH_2-CH(CH_3)-CH_2-CH(CH_3)\sim\sim$ (methyl groups on the 2nd and 5th CH above the chain; on the 1st, 3rd and 4th below)

Atactic polypropylene

Figure 10. Chemical formula. [*after http://www.psrc.usm.edu/macrog/pp.htm*]

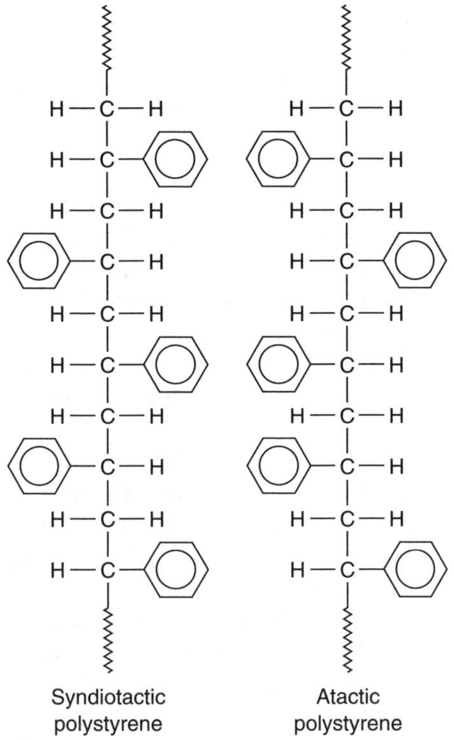

Syndiotactic polystyrene

Atactic polystyrene

Figure 11. Chemical formula.
[*after http://www.psrc.usm.edu/ macrog/styrene.htm*]

cursors or as mixtures of monomer and polymer. When mixed, the liquids undergo a chemical reaction and form a hard solid. Polymers are made up of long molecules that resemble chains. In thermosetting polymers, the long molecular polymer chains are linked by covalent bonds located in three dimensions by cross-linking. When the liquid precursors are mixed together, bonds start to form between the polymer chains during the curing process. This process continues until all the chains have joined together forming a single giant molecule. The molecule's gigantic size makes it solid. Chemicals, applied heat or radiation may be used during manufacture to bring about the polymerization process (curing).

Thermosets have a critical temperature and once heated above this temperature and molded into shape, they will stay in that form. Usually, thermosets cannot be altered by further heating. However, although the two-step curing process forms a three-dimensional structure with cross-linked bonds that do not break down on heating, at very high temperatures thermosets will break down permanently.

The molecular structure of thermosetting polymers determines their properties. The cross-links that exist in their molecular structure stop the molecular chains from sliding past one another, and give thermosets a higher modulus and better creep resistance. Once it has begun, the cross-linking process cannot be reversed. This results in materials that cannot be recycled by remelting. Thermosets are usually more brittle, less flexible and impact-resistant than thermoplastics although they possess better abrasion and dimensional properties.

Thermosets are similar to elastomers—at room temperature, the polymer chains in thermosets are below their glass transition temperature (Tg), making them hard and brittle. However, the polymer chains in elastomers are above their Tg at room temperature, and this factor makes them rubbery. The Tg is the point at which a polymer changes from a rigid solid to a rubber. Some polymers are used above their Tg while others are used below it.

Thermosets in the form of phenolic resins were first developed in the U.S. by the Belgian émigré chemist, Leo Baekeland in 1907 and first patented in 1909. They were developed contemporaneously in Britain by Sir James Swinburne. Baekeland named his invention Bakelite. He reacted phenol and formaldehyde under controlled conditions, producing an amber-colored resin to which he added a filler such as wood flour or cotton flock, producing dark-colored moldings: normally dark brown, dark red or dark green. Products made from phenolic resins possess good electrical resistance and mechanical properties, and Bakelite was immediately utilized as an insulating material in electrical applications such as plugs and insulators. As phenolics are very difficult to ignite and are thermosets, they were also used for purposes that require materials that need to be able to resist high temperatures, such as in thermos flasks.

Other thermosetting polymers developed since the advent of phenolic resins in the early twentieth century include amino resins (including thiourea-urea-, urea- and melamine-formaldehydes), unsaturated polyesters, polyurethanes, and epoxy

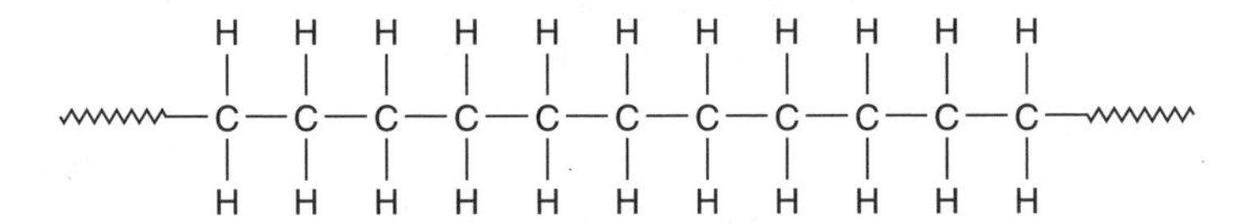

Figure 12. The polyethylene molecule.

resins. Thiourea-urea formaldehydes were discovered in 1924 by Edmund Rossiter, while working for the British Cyanides Company and became popular for a range of decorative tableware called Bandalasta. By 1929, urea-formaldehyde had been developed with better properties. Baron Justus von Liebig, a German chemist, discovered melamine resin in 1834, but melamine-formaldehyde polymer was not patented until 1935. The American Cyanamid Company produced it commercially in 1939. As melamine-formaldehyde was water resistant and tougher than urea-formaldehyde and transparent, it became possible to impregnate patterned papers for surfacing decorative laminates such as Formica and Warerite. These led the way to easy-to-care for surfaces, particularly in the kitchen. Polyimides were introduced in the 1960s and cyanate esters are now under development (see Figure 13).

As thermosets are cross-linked polymers that do not melt, they have to be molded under pressure; for example, by compression molding, transfer molding and various injection molding techniques. Thermoset composite materials can be made in a number of ways, for example by using processes such as resin transfer molding, centrifugal molding and autoclave molding. The manufacture of thermoset composite materials may be rather labor intensive in cases where some components have to be hand laid. For example when making a carbon fiber–thermosetting resin composite, the carbon fiber might be laid by hand, usually in the form of a prewoven fiber. The thermosetting resin, such as epoxy, will then have to be applied and the system allowed to cure. Another type of thermoset composite material such as glass-reinforced fiber (GRP) might use short fibers that are embedded in a polyester or epoxy resin matrix.

The chemistry of processing of thermosets is more complicated than that needed to process thermoplastics, which need only to be melted and cooled. As thermosets are transformed from materials that can be fused and are soluble to highly inflexible cross-linked resins that are impossible to mold, they have to be manufactured during the cross-linking process.

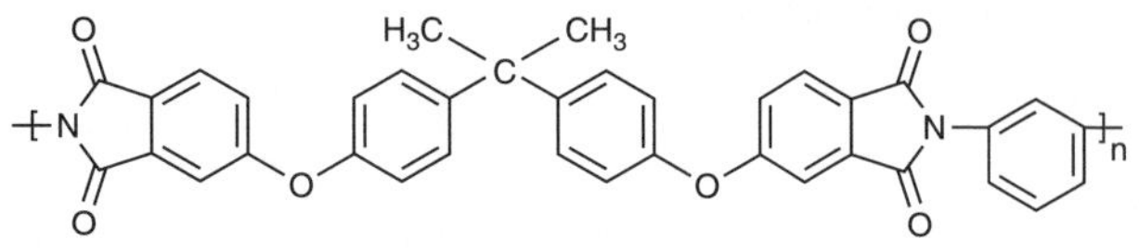

Figure 13. A polyimide molecule
[*after http://www.psrc.usm.edu/macrog/imide.htm*]

Unsaturated polyester resins have a maximum service temperature of around 100°C and vinyl esters and epoxies a temperature of about 150°C. For this reason, high-performance systems such pyromelltimide-type polyimides, cyanate esters and *bis*-maleimides have been developed, as resin matrices that can be used at higher temperatures are needed in certain applications.

These high-performance systems form thermosets that can be used at temperatures of up to 250°C and molded at around 250–300°C. In contrast, high-performance thermoplastics have to be processed at higher temperatures of up to 450°C.

Jon Jakob Berzelius produced the first polyester resin (polyglycerol tartrate) in 1847. A range of polyester resins is now available, including the Marco and Crystic unsaturated polyester resins, which were developed by Scott Bader in 1946. The first commercial use of low-pressure resins to make reinforced polymer composites occurred in 1942 in the form of glass-cloth-reinforced resin radomes made for aircraft in the U.S. By the late 1940s GRP, more commonly known as fiber glass, was used commercially, with many of the earliest developments derived from making hulls. The first car with a fiber-glass body, the Corvette, was made in 1953. Fiber glass was also used for corrugated roofing, decorative moldings and (unsuccessfully) for window frames and baths. Certain items of furniture, such as stackable chairs, are often made using fiber glass and first appeared in the 1950s.

Epoxy resins are also used with glass fiber to make a composite low-pressure reinforced molding material. Epoxy resins were first developed in the 1930s by Pierre Castan and became commercially viable in 1939 with IG Farben's patent concerning liquid polyepoxides. The initially high cost of production of epoxy resins as compared with polyesters limited their use until later improvements in production methods. Today they are particularly advantageous for space applications, due to their light weight and excellent electrical properties.

Thermosets are particularly suitable as materials of choice for making big components such as boat hulls, as the liquid precursors from which they are made are not very viscous and so can flow easily, filling up large moulds without needing great injection pressures. Once they have filled up the mould, the precursors then react to form a solid product.

Thermosets such as phenolics are very difficult to recycle because of their high temperature resistance, inability to soften on melting, and their insolubility. However Bakelite (phenol formaldehyde) will biodegrade if broken, provided it has been filled with organic fillers such as woodchip, which can become the site for biodegradation to occur. Thermoset composites are used in a variety of products from GRP furniture, boat hulls, and sheet-molded compounds (SMC) to carbon fiber-reinforced plastic (CFRP) tennis rackets, Formula One car bodies and aircraft radomes. These are currently difficult to recycle. Furniture and sports equipment may be recycled in the form of repair or reuse, but eventually, once beyond repair, they end up on the scrapheap for the foreseeable future. Thermoset composite materials can be burned as fuel or to give energy (pyrolysis). However, because many of these fiber-based products such as CFRPs are so expensive to produce, ways have been found to recycle them by grinding and reusing them as fillers, and by selective chemical degradation.

See also **Plastics, Thermoplastics; Synthetic Resins**

SUSAN MOSSMAN

Further Reading

CES. *Cambridge Engineering Selector*. Granta Design, Cambridge, 2000.

Goodman, S.H., Ed. *Handbook of Thermoset Plastics*. Noyes Publications, Westwood NJ, 1998.

Mayer, R.M. *Design with Reinforced Plastics*. Design Council, London, 1993.

Morton-Jones, D.H. *Polymer Processing*. Chapman & Hall, London, 1989.

Scheirs, J. *Polymer Recycling*. Wiley, Chichester, 1998.

Positron Emission Tomography

Medical imaging is the term for a number of techniques for viewing anatomical structures and, by the end of the twentieth century, physiological functioning of the body. Medical techniques such as x-ray, computed tomography, and magnetic resonance imaging yield exquisitely detailed images. Such images can be acquired by viewing the decay of radioisotope bound to molecules with known biological properties. This class of imaging techniques is known as nuclear medicine imaging or emission computed tomography (ECT). In ECT multiple cross-sectional images of tissue function can be produced, thus removing the effect of overlying and underlying radioactivity. Experimentation with x-ray tomography began as early as the 1930s, but the technology was not commonly used in clinical medicine until the 1980s.

The techniques of ECT are usually considered as two separate modalities.

1. Single photon emission computed tomography (SPECT), invented in the 1970s, involves the use of a radioisotope such as technitium-99 ($^{99}Tc^{m}$), where a single γ-ray is emitted per nuclear disintegration.
2. Positron emission tomography (PET), the development of which began in the 1950s. PET is a radiotracer imaging technique, in which tracer compounds labeled with positron-emitting radionuclides are injected into the subject of the study. These tracer compounds can then be used to track biochemical and physiological processes *in vivo*.

One of the prime reasons for the importance of PET in medical research and practice is the existence of positron-emitting isotopes of elements such as carbon, nitrogen, oxygen and fluorine, which may be processed to create a range of tracer compounds that are similar to naturally occurring substances in the body. Among many applications, these radiotracers, such as carbon-11 (^{11}C), nitrogen 13 (^{13}N), oxygen 15 (^{15}O), and fluorine-18 (^{18}F), are used clinically to identify and diagnose cancers, epilepsy and other movement disorders, and heart and cardiovascular problems.

After the injection of a tracer compound labeled with a positron-emitting radionuclide, the subject of a PET study is placed within the field of view (FOV) of a number of detectors in a scanner that are capable of registering incident gamma rays. The radionuclide in the radiotracer decays, and the resulting positrons subsequently annihilate on contact with electrons after traveling a short distance (around 1 millimeter) within the body. Each annihilation produces two 511-kiloelectron-volt photons traveling in opposite directions, and these photons maybe detected by the detectors surrounding the subject. The detector electronics are linked so that two detection events occurring within a certain time window may be called coincident and thus be determined to have come from the same annihilation (Figure 14). These coincident events can be stored in arrays corresponding to projections through the patient and reconstructed using standard tomographic techniques. The resulting images show the tracer distribution throughout the body of the subject. Photon isotopes may decay via positron emission, in which a proton in the nucleus decays to a

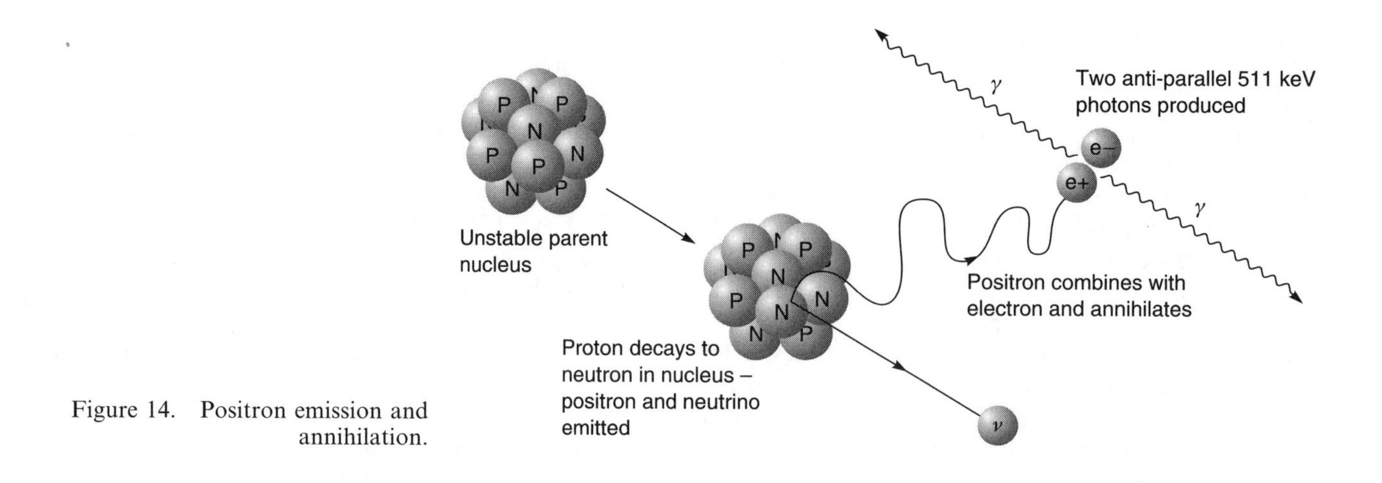

Figure 14. Positron emission and annihilation.

neutron, a positron, and a neutrino. The daughter isotope has an atomic number one less than the parent.

In the PET medical technique, the subject is surrounded by a cylindrical ring of detectors with a diameter of 80 to 100 centimeters and an axial extent of 10 to 20 centimeters. The detectors are shielded from radiation from outside the field of view by relatively thick lead end-shields. Most scanners can be operated in either a slice collimated mode, where axial collimator is provided by thin annual rings of tungsten called septa or in a fully three-dimensional mode where the septa are retracted and coincidences can be collected between all possible detector pairs. The usual configuration of detectors in PET is a rectangular bundle of crystals, termed a block, optimally coupled to several photomultiplier tubes (PMTs). When a photon interacts in the crystal, electrons are moved from the valence band to the conduction band. These electrons return to the valence band as impurities in the crystal, emitting light in the process. Since the impurities usually have metastable excited states, the light output decays exponentially at a rate characteristic of the crystal. The ideal crystal will have a high density so that a large fraction of incident photons scintillate, high light output for positioning accuracy, fast rise time for accurate timing, and a short decay time so that high counting rate can be handled. Most current scanners use bismuth-germanate (BGO), which generates approximately 2500 light photons per 511-kiloelectron-volt photon and has a decay time of 300 nanoseconds. One such block couples a seven by eight array of BGO crystals to four PMT where each crystal is 3.3 millimeters wide in the transverse plane, 6.25 millimeters wide in the axial dimension, and 30 millimeters deep. The block is fabricated in such a way that the amount of light collected by each PMT varies uniquely depending on the crystal in which the scintillation occurred. Hence integrals of the PMT outputs can be decoded to yield the position of each scintillation. The sum of the integrated PMT outputs is proportional to the energy deposited in the crystal.

Coincident events in PET fall into four categories: true, scattered, random, and multiple (Figure 15):

- True coincidence: occurs when both photons from an annihilation event are detected by detectors in coincidence, neither photon undergoes any form of interaction prior to detection, and no other event is detected within the coincidence time window.
- Scatter coincidence: occurs when at least one of the detected photons undergoes at least one scattering event prior to detection. Scatter coincidences add a background to the true coincidence distribution that changes slowly with position, decreasing contrast, and causing the isotope concentrations to be overestimated. The number of scattered events detected depends on the volume and attenuation characteristics of the object being imaged and on the geometry of the camera.
- Random coincidence: occurs when two photons not arising from the same annihilation event are incident on the detectors within the coincidence time window of the system. Also, the number of random coincidence depends on the volume and attenuation characteristics of the object being imaged and on the geometry of the camera.
- Multiple coincidence: occurs when more than two photons are detected in different detectors within the coincidence resolving time.

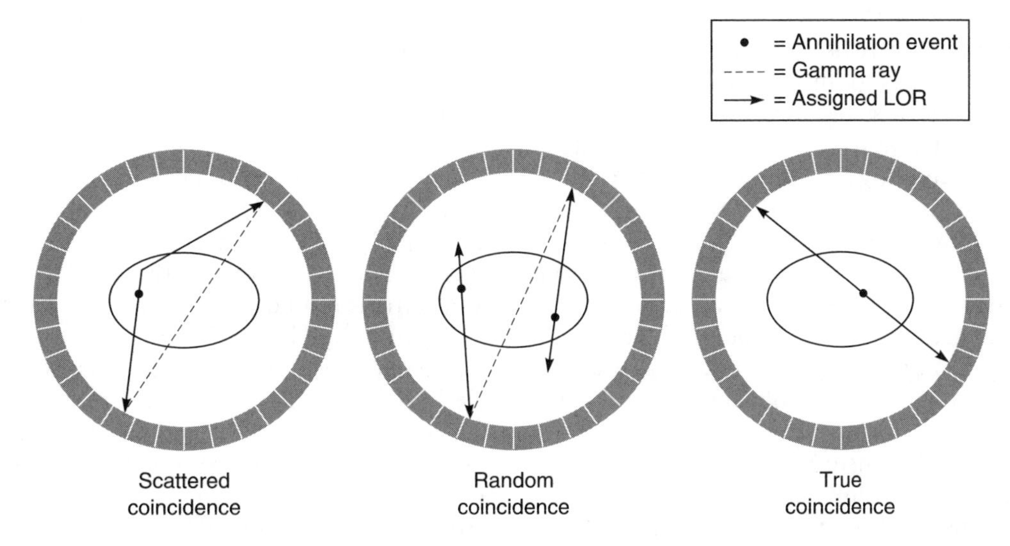

Figure 15. Types of coincidences in positron emission tomography.

See also **Neurology; Nuclear Magnetic Resonance (NMR, MRI); Tomography in Medicine**

Stelios Zimeras

Further Reading

Bailey, D.R. and Meikle, S.R. A convolution–subtraction scatter correction method for 3D PET. *Phys. Med. Biol.*, 39, 411–424, 1994.

Raylman, R.A., Hammer, B.E. and Christensen, N.L. Combined MRI-PET scanner: a Monte Carlo evaluation of improvements in PET resolution due to the effects of a static homogeneous magnetic field. *IEEE Trans. Nuc. Sci.*, 43, 4, 2406–2412, 1992.

Webb, S. *The Physics of Medical Imaging*. Institute of Physics Publishing, Bristol, 1996.

Webb, S. *From the Watching of Shadows: The Origins of Radiological Tomography*. Institute of Physics Publishing, Bristol, 1990.

Power Generation, Recycling

Recovering energy from wastes from municipal or industrial sources can turn the problem of waste disposal into an opportunity for generating income from heat and power sales. The safe and cost-effective disposal of these wastes is becoming increasingly important worldwide, especially with the demand for higher environmental standards of waste disposal and the pressure on municipalities to minimize the quantities of waste generated that must be disposed.

Energy from Waste (EfW) Technologies

There are a number of ways of recovering energy from waste:

- Landfill gas: collecting and using some of the gas that is produced as waste decay in landfill sites. The landfill gas can be used to run engines, fire boilers or kilns, and generate electricity.
- Anaerobic digestion: capturing the gas produced when organic wastes such as food preparation wastes and garden prunings break down in an air-free (i.e. anaerobic) environment. Anaerobic digestion improves on gas collection from landfill sites by using a sealed system where all of the gas can be collected for use as a fuel.
- Refuse-derived fuel: turning the combustible portion of waste, such as paper and plastics, into a fuel that can be stored and transported as pellets or briquettes or directly used on site to produce heat and power.
- Incineration with energy recovery: burning mixed waste in sophisticated plants where the heat given off is harnessed for hot water and electricity generation. Combined heat and power (CHP), which not only produces electricity from the generators, but also captures waste heat by using the steam to heat local buildings by installing a network of pipes, maximizing the recovered value.

The most commonly used incineration technology is so-called mass burn, where mixed waste is burned in large furnaces on moving, inclined grates. This incineration technology is well established worldwide and proven over many years. The first incineration plants in the UK that burnt mixed fuel producing steam to generate electricity were built in Nottingham in 1874. Ongoing research and development continues to improve the technology, for example in terms of increasing efficiency, reducing emissions, and lowering capital costs. Fluidized bed technology is thermally more efficient than incineration, but incinerators have not yet been fully developed in the U.K. or U.S. In many municipal waste sites, a plasma arc is used to heat solid waste to very high temperatures (3,000–10,000°C). The product is a hydrogen-rich gas and an inert glassy residue from the nonorganic waste. EfW incineration is not the whole answer, but—as with recycling and composting—when sensibly applied, it can help ensure that the maximum value is recovered from materials that would otherwise be consigned to landfill and lost.

There is a very wide range of municipal or industrial wastes that may be used as fuel. The nature of the waste, and the waste disposal method, will determine the way that energy can be recovered. Dry household, commercial, or industrial wastes can either be burned (combusted) as raw waste, or they may first undergo some sorting or processing to remove some of the waste components that may be recycled separately. Waste incineration is an established way to dispose of wastes. Incineration consumes the biodegradable fraction of the waste, decreases the volume of the waste, and allows for recovery of metals and other potentially recyclable fractions, so that the remaining residue, which must be sent to landfill, is more stable and does not generate potentially harmful emissions such as methane. The heat recovered from waste incineration can be used to generate electricity or used for industrial heat applications. The size of an EfW plant is designed to meet the waste-disposal needs of the community that it serves. Such a plant can typically process between 100,000 and 600,000 tons per year, and from this it can generate up to 40 megawatts of electricity. Power is produced from these wastes by using the steam raised in the combustion process to drive a steam turbine to generate electricity. This technique uses the same proven technologies as those in power plant fueled with fossil fuels such as coal.

Gasification or pyrolysis is one of the later twentieth century technologies that is increasingly being used for waste disposal. It is a thermochemical process in which biomass is heated, in the absence of air, to produce a low-energy gas containing hydrogen, carbon dioxide, and methane, together with tar or ash. The gas can then be used as a fuel in a turbine or combustion engine to generate electricity. Gasification was first developed in the late nineteenth century to produce gas from heated coal or coke. A process developed by Ludwig Mond of the Brunner, Mond & Company (which later became Imperial Chemical Industries, or ICI) was used commercially from 1901 to generate "producer gas" for large industrial furnaces. Gasifiers are now increasingly being developed to accept more mixed fuels, including wastes. The first plants in the U.S. to use gasification to treat municipal solid waste were built in the 1970s, although by 2002 there were no commercial-scale solid waste gasification systems operating in the U.S. Modern gas treatment technology ensures that the resulting gas is suitable to be burned in a gas engine. Gasifiers operate at a smaller scale than a mass-burn EfW plant, and they can also be provided in modular form to suit a range of different scales of operation.

Strict environmental standards now apply in all European countries governing the atmospheric emissions from EfW plants, particularly of heavy metals and dioxins. All energy from waste plants must now meet these standards, which can be achieved through the installation of extensive state-of-the-art gas cleaning systems. An EfW plant managing a guaranteed proportion of residual waste helps to underpin recycling activities by providing a treatment option when markets for recyclables are poor. Moreover, ferrous and nonferrous metals recovery at EfW plants enables many thousands of tons of metal to be recycled. The ash produced from such plants can also be used as an inert aggregate in the construction industry, avoiding the need for disposal of the ash to landfill. Experience in a number of European countries shows that EfW and high recycling rates coexist happily. Indeed, many of the countries achieving high recycling rates also tend to have high-energy recovery rates.

EfW also has an important role in reducing the pollution of carbon dioxide (CO_2), sulfur dioxide (SO_2), and oxides of nitrogen (NO_x) caused by electricity generation from coal, oil, and gas. All power generation processes produce emissions; EfW processes, however, are very well regulated and controlled and, as a result, are far less polluting than fossil sources used for electricity generation. European Directives governing the operation of EfW plants, for example, are far more rigorous and detailed in scope than those applicable to conventional power stations.

More than 60 percent of the heat value of waste is derived from carbon-based renewable biomass sources; that is, wood, paper, and vegetable matter (material that has absorbed CO_2 when it was growing). When these materials are burned in an EfW facility, the carbon that was trapped in them is rereleased as CO_2. Because there is no net increase in CO_2 however, such a process is CO_2-neutral.

See also **Biomass Power Generation; Electricity Generation and the Environment; Energy and Power**

IAN BURDON

Further Reading

Department of the Environment, Transport and the Regions. *Waste Strategy 2000 for England and Wales*. Her Majesty's Stationary Office, London, 2000.

MacNeill, J.R., Ed. *Something New Under the Sun: An Environmental History of the Twentieth Century*. Penguin, London, 2001.

Petts J. Incineration as a waste management option, in *Waste Incineration and the Environment*, Hester, R.E. and Harrison, R.M., Eds. Royal Society of Chemistry, Cambridge, 1994.

Williams, P.T. *Waste Treatment and Disposal*. Wiley, New York, 1998.

Useful Websites

http://www.wasteguide.org.uk/

National Energy Technology Laboratories, Gasification Technologies: http://www.netl.doe.gov/coalpower/gasification/

UNEP United Nations Environment Program, Division of Technology, Industry, and Economics, Newsletter and Technical Publications: Municipal Solid Waste Management, Regional Overviews and Information Sources: North America: http://www.unep.or.jp/ietc/ESTdir/Pub/MSW/RO/contents_North_A.asp

Power Tools and Hand-Held Tools

While the basic hand tools—hammers, saws, planes, and wrenches—used in construction during the twentieth century changed little from those available for generations, there was a revolution in power tools. Developments in power technology led to the mechanization of tools of all types. Coupled with efforts to use new materials that made tools both lighter and more manageable, construction work became more efficient and cost effective.

Few power tools were available at the beginning of the twentieth century. Those driven by compressed air had their roots in the pneumatic rock drills of the nineteenth century. That technology was easily modified into chipping hammers, paving breakers, and jack hammers. Rivet hammers, developed before 1900, were never more effective than when used to assemble the steel frames of skyscrapers during the 1920s and 1930s.

The evolution of hand-held power tools was dramatic. The first portable electric drill weighed more than 20 kilograms and was in use shortly after 1900. It not only took two men to manage the drill, but a third was needed to control the power supply. A breakthrough in the technology came in 1917 when the Black & Decker Company in the U.S. introduced its first small portable electric drill. Assembled in an aluminum housing, the drill could be held and operated by one worker. Expanding electrification and the miniaturization of electric motors made it possible as well as practical to motorize all sorts of tools. In 1924, the Michel Electric Handsaw Company (later Skilsaw) produced the first portable electric handsaw. This rotary blade device changed the construction industry by dramatically reducing the time needed to cut and fit lumber. Although many of the early developments came from the U.S., Germany was also active in tool design. During the mid-1920s, Gottlieb Stoll produced electric drills and saws. In 1932, the Robert Bosch Company produced an electropneumatic hammer, which was the first in a long line of power hand tools.

The widespread manufacture of small air-cooled internal combustion engines following World War II had a remarkable impact on the construction industry. They made it possible to motorize an even greater variety of tools than ever before. Tools as varied as shovels and wheelbarrows were motorized or self-propelled. One of the first examples of a powered wheelbarrow was introduced in the U.S. by the Bell Aircraft Company in 1948. The operator walked behind and steered the "Bell Prime Mover" as it transported loads of up to 450 kilograms. The tool's usefulness could be broadened by replacing its bucket with either a platform for carrying stacked loads or dozer blade. Similar devices with a wide range of accessories were built in Europe.

Concrete work was improved with the introduction and use of several power tools. A vibrator was devised to remove voids in large pours. The number of workers needed to finish concrete was reduced when the power trowel was marketed in the late 1940s. Traditionally, large expanses of freshly poured concrete were smoothed by workers who pulled wooden or metal floats across its surface. A single laborer using a power trowel could accomplish the same work in less time. Powered by a gasoline engine, the spinning blades of the propeller-like trowel skimmed across and trued the surface of partially set concrete.

The process of setting fasteners in concrete slabs or block walls was simplified with the development of powder-actuated tools (PATs). Ramset in the U.S. and Hilti in Liechtenstein developed these tools in 1947 and 1948, respectively. Expanding gases from the explosion of a precise powder charge drove a piston against the fastener with such force that it would penetrate concrete.

In the late 1950s and 1960s, nailing was made easier with hand-held nail guns. The gun contained a magazine of nails that were set with the pull of a trigger. These time saving and easy to use devices, powered by compressed air or electricity, were a fast way of joining wooden framing and plywood.

One of the more significant late twentieth century developments was the introduction of cordless battery powered hand tools. This took place in 1962 when the Porter Cable Company in the U.S. marketed its line of "Big 10" cordless drills. Each was powered by 10-volt (hence the name) battery packs worn by its operator. However, the batteries held a charge that lasted no more than 5 or 6 minutes of use. Nonetheless, this new type of tool gave workers previously unknown freedom of movement.

The power for these tools came from nickel–cadmium (NiCad) batteries which were first developed in Germany during World War II. Not only could the cells be readily recharged, they also produced a constant voltage as they discharged. As batteries were improved, recharge time was reduced and the duration of a useful charge increased. Long life and quick recovery made it practical to adapt an increasing variety of tools to battery operation. A significant advancement occurred in the late 1960s when the nickel metal hydride battery was developed. It maintained its charge 50 percent longer than the NiCad batteries it replaced. By the end of the twentieth century rechargeable batteries had improved to the point where they contained energy sufficient to power high-speed circular saws and hand-held planers. The Japanese firm Makita became a leading international manufacturer and innovator of these cordless power tools.

The construction industry benefited from the development of the laser during the 1960s when, in 1968, the rotary laser was used for the first time. That battery-operated tool was used to determine rectilinearity and it soon displaced the optical levels that builders had used for decades. The perfectly straight line indicated by the laser beam had interior and exterior applications. A laser incorporated into the end of a traditional spirit level provided the means for projecting a straight line beyond the reach of the level itself.

During the 1990s the versatility of PAT-type fastener drivers was improved upon with the introduction of a new propulsion system. Unlike PATs that had no more than 10 shots per load, the new gun powered by a cartridge of highly flammable compressed gas had as many as 1200 shots. The interior of the revised fastener gun was similar to an internal combustion engine in which a piston was moved by the explosion of a charge of gas and air ignited by an electric spark.

How tools were used and their impact on the user, led to changes in the design of many handles, grips, and triggers. Concern for the overall weight of tools led to a greater use of plastics and alloys. The distribution of weight within tools led to some overall redesigns in which centers of gravity were repositioned for better balance. The 1990s was a period during which the ergonomics of hand and power tools were scrutinized.

WILLIAM E. WORTHINGTON, JR.

Further Reading

Allhands, J.L. *Tools of the Earthmover*. Sam Houston College Press, Huntsville, 1951.

Cacha, C.A., Ed. *Ergonomics and Safety in Hand Tool Design*. Lewis, Boca Raton, FL, 1999.

Mcdonnell, L.P. *Portable Power Tools*. Delmar Publishing, Albany, 1962.

Nagyszalanczy, S. *Power Tools*. Taunton Press, Newtown, 2001.

Presentation of Technology

Representations of technology in various media—including expositions, theme parks, fiction, and film—have played central roles in the way human beings have imagined technology and its relationship to progress. Beginning with the scientific revolution in the seventeenth century, when the idea of progress became increasingly secularized and associated with individual and social betterment here on earth, the modern definition of technology as the application of scientific knowledge for human betterment began to emerge. But, as technology became increasingly associated with industrialization in the eighteenth and nineteenth centuries and its negative social consequences, the assumption that better technology would inevitably lead to human progress, came into question. As doubts increased (think of William Blake's "The Songs of Experience") and fed political movements like Luddism, positive representations of technology became increasingly important to the builders of industrializing nation-states who sought to counter growing political opposition to their authority and power with exhibits that made human progress seem synonymous with industrialization, consumerism, and the globalization of capitalism. Though typically less overtly political in aim and message, presentations of technology in fiction and, later, film frequently echoed and reinforced the optimistic views of the international exhibits. However, novels and movies also offered a means for expressing doubts about the dominant view as they tapped into the persistent current of fear and anxiety about the industrial or postindustrial future that always ran just beneath the surface.

Expositions and World's Fairs

Anxieties about the future became especially acute in mid-nineteenth century England in the aftermath of the 1848 political revolutions in continental Europe that deepened concerns between England's industrialists and their political supporters about the upsurge of Chartist reforms that threatened their prerogatives. To counter political opposition at home and to establish a cultural breakwater against the seas of revolution abroad, England's political authorities, led by Prince Albert, announced plans to organize a "Great Exhibition of the Industry and Works of All Nations." Christened the Crystal Palace Exhibition, the 1851 fair proved tremendously successful in building popular support for the English nation-state and its colonial expansion. The Crystal Palace Exhibition launched the world's fair movement that, over the next century and a half, swept the globe and helped give meaning to technology in the modern and postmodern world. Subsequent exhibitions in France promoted a view of the world that drew sharp distinctions between an industrialized "civilization" and a technologically backward "savagery," thus giving the meaning of progress a definition that was at once technologically and racially inflected. In the U.S., where the devastating industrial depression that began in 1873 inspired waves of political and social upheaval, the 1876 Philadelphia Centennial Exposition helped to build confidence that an industrializing nation could promote progress through economic growth.

However important the equation of "technology equals economic growth equals progress" was to the Victorian era, it was never easy to maintain. In the first half of the twentieth century, two world wars bracketing the Great Depression provided new challenges to sustaining this belief system. Even before World War I ended, political and economic leaders in Europe and the U.S. began laying plans for a new generation of world's fairs that would rebuild Europe as well as the public's faith in scientifically and technologically based progress. Their efforts met with little public enthusiasm until the world's capitalist economies collapsed during the Great Depression. Threatened by the rise of Soviet communism, western capitalists dedicated millions of dollars to revitalizing the world's fair medium as a means toward securing popular faith in capitalism. As the American fairs of the 1930s made clear, representations of technology were central to this effort.

At the 1933–1934 Chicago Century of Progress Exposition, the theme of the fair: "Science Finds, Industry Applies, Mankind Conforms" inspired an artist to sculpt a heroic trinity depicting a robot hunched over and nudging forward figures representing mankind and womankind. At all of the U.S. fairs of the 1930s, representations of labor-saving devices like dishwashers and air conditioners along with new entertainment technologies such as television, held out the promise of a science-based, technologically driven consumerist utopia in America's future if only the existing political and economic order remained essentially intact. The 1939–1940 New York World's Fair brought the era to close with its "Dawn-of-a-New Day" theme and its Futurama and Democracity exhibits that depicted America's arrival at perfection within a lifetime.

The devastation of World War II, especially the Nazi extermination camps and U.S. use of nuclear weapons, made glib assertions about the future very difficult to sustain, but promoters of the 1958 Brussels Universal Exposition pulled out all the stops in their efforts to promote nuclear power as a safe, energy-efficient fuel; as did the promoters of the 1962 Seattle Century 21 Exposition who settled on space exploration as the theme for their fair. Not until Montreal's Expo '67 did a major world's fair suggest that basic assumptions about the meaning of technology and progress needed reconsideration. With its "Man in Control?" subtheme, Expo '67 led to a generation of postmodern international expositions that gave increasing attention to the nuclear and ecological crisis confronting humanity. However, these fairs, dominated by multinational corporations, remain showcases for cultural technologies dedicated less to finding long-term solutions to global problems than to propping up confidence in corporation-run nation-states. Whether the upcoming world's fairs planned for Japan and Shanghai break the mold remains to be seen at the time of writing.

Fiction

Fictional representations of technology became increasingly common in the twentieth century as industrialization and the machine emerged as central and, in the eyes of some, defining characteristics of modern societies. Immediately after their introduction, innovations such as automobiles, radios, and airplanes most frequently figured in novels and stories as creative plot devices or convenient symbols of modernity, though often with little substantial commentary on the broader social or political significance of the technology. Once technologies had lost their novelty, often with astonishing rapidity, they continued to appear without comment in the background of fiction of all types. This ultimate transformation of the new into the unremarkable was perhaps the most telling sign of the centrality of technology to the twentieth century, suggesting that technological change was increasingly viewed as simply an inevitable aspect of life in the modern age.

Among authors whose work commented expressly on the significance of technology in society, views ranged from unalloyed technophilia to dire predictions of machine-driven Armageddon. Many works of fiction elaborated on basic themes that emerged during the previous century as authors first began to ponder the potentially revolutionary effects of industrialization. The British author Mary Shelley's 1818 novel, *Frankenstein*, in which a hubristic scientist creates and loses control of a technological monster was subsequently echoed in countless books and movies that explored the theme of what Langdon Winner has called "autonomous technology": Are humans in control *of* or controlled *by* their machines? In the U.S., the tumultuous process of industrialization inspired both unreasonable hopes and ominous fears. A decade after the Philadelphia Centennial Exposition of 1876, Edward Bellamy's *Looking Backward* (1888) reinforced the idea of industrial utopianism, suggesting that the economic and social dislocations of the Gilded Age could be solved not by turning away from industrialization and technology but by more fully embracing the rationality of the machine. Set in Boston in the year 2000, Bellamy's novel prophesizes that a utopian age of leisure, peace, and unbounded consumer affluence might be achieved if only the productive power of the factory system is properly used for the greater good of society. Though considerably less influential than Bellamy's book (which inspired thousands of Americans to join clubs dedicated to realizing his prophesy), Mark Twain's novel of the subsequent year, *A Connecticut Yankee in King Arthur's Court* (1889), reflected the darker fears of many Americans that Bellamy's work and the industrial expositions sought to allay. In Twain's pessimistic fantasy, a time-traveling Yankee engineer's introduction of modern machinery into Arthurian England brings not leisure and peace but mass destruction and carnage.

Twain's cautionary tale notwithstanding, technological optimism and exuberance remained the dominant fictional theme well into the twentieth century, perhaps suggesting the efforts of industrial and political elites to equate technology, economic growth, and progress were succeeding. In Great Britain, France, and the U.S., the inventor, engineer, and industrialist became popular public heroes, and their quasifictional counterparts frequently played starring roles in popular adventure and romance novels. Tellingly, other writers demoted actual human beings to secondary roles, establishing an enduring genre of fiction (one thread in the complex weave of twentieth-century science fiction) in which science and machines were the real stars. The French novelist Jules Verne pioneered this approach in the previous century with his series of novels envisioning a not-too-distant future of undersea ships, rapid circumglobal air travel, and the human exploration of space. Unapologetically technophilic, Verne's novels gloried in the sheer adventure and excitement of powerful machines while also providing reasonably accurate technical and scientific explanations of their principles. Verne's twentieth century literary progeny found a new home in the pulp fiction science magazines inaugurated in 1926 with Hugo Gernsback's *Amazing Stories*. Graced with vividly colored and imagined illustrations of interplanetary rockets, flying buzzsaws, and mysterious alien cities, the pulp fiction magazines deliberately sought to spark the sense of wonder and awe that historian David Nye calls the "technological sublime." Out of dramatic necessity the stories often explored the multitude of ways in which new technologies could go awry, yet the overall message of the magazines was resolutely optimistic: technology can and will create a better and infinitely more exciting world of tomorrow.

The events of the twentieth century, however, made the pulp fiction promise of progress appear increasingly naïve. The economic misery of worldwide depression, the rise of fascist and totalitarian states, and the appalling mechanical carnage of two world wars forced even the most faithful to question whether technology truly equated with progress. The British author H.G. Wells' novel,

The Shape of Things to Come (1933), envisions a future war of such massive technological destruction that civilization reverts to the conditions of the Middle Ages. Two of the most influential dystopian novels of the century—George Orwell's *1984* (1949) and Aldous Huxley's *Brave New World* (1932)—warned that the exciting new consumer products of the electronic, chemical, and medical industries, might also be used to extend the power of totalitarian states to control every facet of human existence. In Orwell's society of 1984, televisions and surveillance cameras allow "Big Brother" to constantly watch over and propagandize an almost perfectly controlled citizenry, while Huxley's new world government uses cloning and mechanical wombs to rationally manage even the most intimate realms of the human body.

After World War II, atomic weapons and power, digital computers, space rockets, and other new products of the military-academic-industrial complex each inspired a share of adoring fictional treatments. Yet the sheer destructive and transformative power of these new devices also suggested that earlier predictions of future technological disasters might now be coming true. By the time of Walter Miller's 1959 novel, *A Canticle for Liebowitz*, nuclear weapons provided an all-too-plausible means for realizing H.G. Wells' earlier preatomic idea of a massive war that sends civilization reeling back into the Middle Ages. Likewise, Kurt Vonnegut's *Player Piano* (1952) returns to the oft-explored theme of technological unemployment with the difference that the skilled motions of a soon-to-be-unemployed mechanic are now copied into the memory of EPICAC (an obvious reference to ENIAC, the first electronic digital computer). Yet if technology could now realize disaster, perhaps it could also create paradise. Growing public concerns about pollution and overpopulation inspired a new generation of authors to critique technology from a distinctly environmental perspective. Ernest Callenbach's 1975 novel, *Ecotopia*, echoed Edward Bellamy's attempt nearly a century earlier to envision a reasonably realistic alternative to industrial society. Drawing on new ideas from the appropriate technology movement, Callenbach imagines a future nation that deliberately rejects certain advanced technologies as environmentally unsustainable or socially corrosive.

In the final decades of the century, new technologies continued to inspire new fictional treatments. With his 1984 novel *Neuromancer* William Gibson sparked the science fiction subgenre of cyberpunk, dedicated to exploring emerging technologies like virtual reality and global information networks. Perhaps one of the most telling fictional portraits of technology and society in the postmodern age was Don DeLillo's 1985 novel, *White Noise*. Set in a small college town in the middle America of the 1980s, *White Noise* suggests modern technology has brought neither the utopian joys nor the dystopian destruction envisioned by earlier writers. Instead, it has become the largely unnoticed fabric of everyday human existence, a constant "white noise" of babbling TVs, airwaves filled with electromagnetic radiations, and mysterious chemicals that may or may not be slowly killing us. In the postmodern age of technological abundance and unknown risks, DeLillo seems to suggest, age-old human fears and hopes remain unchanged as the latest new thing is embraced and then forgotten, quickly assimilated into the fabric of a society that ultimately remains stubbornly and splendidly merely human.

Film

A technological revolution in their own right, motion pictures proved to be one of the most influential means of exploring the meaning of twentieth-century technology. Early film makers eager to demonstrate the medium's unique ability to portray movement were drawn to the dynamism of trains, autos, and trolleys. Thomas Edison's 1903 western, *The Great Train Robbery*, included the first of what would be a long line of dramatic film fights staged on top of a careening train. The pioneering actor and director Buster Keaton brilliantly developed the comic potential of steamships, automobiles, and trains in *The Navigator* (1924) and *The General* (1926). But other early films offered considerably darker views. The German film *Metropolis* (1926) depicts a future society where the elite few live in luxurious ease while the mass of workers toil ceaselessly below in a sunless subterranean world of dangerous machines. Inspired in part by director Fritz Lang's visit to New York City, *Metropolis* reflects growing concerns about the power of technocratic elites and the fear that mass production was an enslaving rather than liberating force. The American film *Modern Times* (1936) explores similar themes, though director–actor Charlie Chaplin softens his message with a humor completely absent from the darkly serious *Metropolis*. A pointed if ultimately ambiguous critique of Taylorism, Fordism, and the deskilling and subordination of labor, *Modern Times* vividly conveys fears of industrial domination when

Chaplin's character is dragged into a mass of gigantic gears, almost literally becoming a mere "cog in the machine."

In the post-World War II period, the atomic bomb and Cold War tensions with the Soviet Union made unalloyed celebration of technology increasingly difficult to sustain. In Japan the atomic devastation of Hiroshima and Nagasaki provided a subtext to Inoshiro Honda's 1954 film *Godzilla* in which radioactivity from American bomb tests in the Pacific awakes a gigantic reptile that proceeds to crush much of downtown Tokyo. In *Them!*, an American film of the same year, nuclear radiation causes ants to mutate into marauding truck-sized monsters who threaten Los Angeles. Other films of the 1950s dealt more bluntly with the genuine peril of the Cold War nuclear arms race, as in *The Day the Earth Stood Still* (1951) where an alien messenger warns humans not to extend their war-like ways into outer space or risk utter destruction. Abandoning the fantastic all together, *On the Beach* (1959) uses stark realism to offer a sobering tale of the final days of the last survivors of a global nuclear war.

As if the dangers of nuclear Armageddon were not enough, directors explored dozens of less obvious ways in which the countless new postwar technologies might threaten humanity. Stanley Kubrick's homicidal computer HAL in the epic *2001* (1968) suggests the dangers of over-reliance on machines, though the film ultimately portrays advanced technology as the stepping stone to human transcendence. Fears of intelligent machines rebelling against humanity combined with uneasiness about the radical potential of genetic engineering to produce some of the most disturbing representations of technology in the final decades of twentieth century. The cybernetic assassins and guardians of the three hugely popular *Terminator* films (1984, 1991, 2003) and the corporate-created cyber-policeman of *Robocop* (1987) became the latest incarnation of Dr. Frankenstein's monster-machine that rebels against its creator. However, in all these films humanity is ultimately triumphant, either destroying or taming the machine and thus perhaps reaffirming the ultimate rightness of technological progress. An exception was the 1982 film *Blade Runner*, director Ridley Scott's disturbing story of a race of genetically engineered "replicants." Created by a bioengineering corporation as human slaves, the replicants eventually come to be seen as a threat that must be destroyed, even though they prove to be more human—and humane—than their creators in many ways. Given the rapid progress in genetic engineering that resulted in the first map of the human genome in 2000, *Blade Runner* might well be seen as emblematic of a time when the boundaries between the biological and the mechanical or between the human and the machine became increasingly difficult to define.

If films like *Blade Runner* raised troubling questions about technological progress in the last quarter of the twentieth century, technological exuberance and optimism by no means disappeared. The short-lived but influential television drama *Star Trek* (1966–1969) and the subsequent legions of small- and big-screen sequels offered a generally optimistic future in which advanced technology has eliminated human hunger, poverty, and a host of other social ills. Likewise, the immensely popular *Star Wars* saga (begun in 1977) deliberately revived the uncomplicated technological enthusiasm of earlier film and television serials like Flash Gordon. Thus the worldwide popularity of the *Star Trek* and *Star Wars* films may suggest that, wisely or not, many citizens of the late twentieth-century postindustrial and postmodern world ultimately continued to be confident that technology remained an essentially beneficial and beneficent force for human progress.

ROBERT W. RYDELL AND TIM LECAIN

Further Reading

Basalla, G. Keaton and Chaplin: the silent film's response to technology, in *Technology in America: A History of Individuals and Ideas*, Pursell, C.W. Jr., Ed. MIT Press, Boston, 1986.

Findling, J.E. and Pelle, J.D. *Historical Dictionary of World's Fairs and Expositions, 1851–1988*. Greenwood Press, New York, 1990.

Hendershot, C. *Paranoia, the Bomb, and 1950s Science Fiction Films*. Bowling Green State University Popular Press, Bowling Green, Ohio, 1999.

Nye, D.E. *Narratives and Spaces: Technology and the Construction of American Culture*. Columbia University Press, New York, 1997.

Rydell, R.W. *World of Fairs: The Century of Progress Expositions*. University of Chicago Press, Chicago, 1993.

Rydell, R.W., Findling, J.E. and Pelle, K.D. *Fair America: World's Fairs in theUnited States*. Smithsonian Institution Press, Washington D.C., 2000.

Telotte, J.P. *A Distant Technology: Science Fiction Film and the Machine Age*. Wesleyan University Press, London, 1999.

Telotte, J.P. *Science Fiction Film*. Cambridge University Press, New York, 2001.

Tichi, C. *Shifting Gears: Technology, Literature, and Culture in Modernist America*. University of North Carolina Press, Chapel Hill, 1987.

Trachtenberg, A. *The Incorporation of America: Culture and Society in the Gilded Age*. Hill and Wang, New York, 1982.

Printers

Although numerous types of computer printers were developed during the last half of the twentieth century, printing technology emerged during antiquity. The earliest methods of printing included human scribes in Egypt and Greece, but other forms of printing technology existed around the globe including that developed by the Indians and the Mayans. It is China, however, that is notable for its contributions to automatic printing technology. Block printing emerged in China during the ninth century and baked-clay moveable type was introduced in the early eleventh century. In Europe, the Gutenberg press introduced moveable type in the fifteenth century. Moveable type was a revolutionary advancement in printing technology, but it also was the introduction of the telegraph in 1844 that significantly contributed to the development of automatic printing. In fact, the American Wheatstone Automatic Telegraph (1858) and the French Baudot Telegraph (1874) made it possible for telegraph sites to automatically receive messages in Morse code without the need for skilled operators.

In 1910, American engineers Charles and Howard Krum oversaw the installation between New York City and Boston of the first teletypewriter system. Unlike telegraph systems, teletype did not utilize Morse code. Instead the teletypewriter was an input device in which the user could transmit alphanumeric characters typed in one at a time. The system also was equipped with an early teleprinter that was composed of a roll of paper that automatically printed incoming messages. Teletype technology steadily improved, and by 1914, the Associated Press used teletype systems to communicate information to U.S. newspapers. After World War I, these systems were adopted for use throughout the world and remained active for much of the twentieth century. For example, AT&T had 60,000 teletype centers in the U.S. by 1962. The decline of teletype use can be attributed to the introduction of other automated communications systems, including fax machines and email. However, teletype features remain in use today in TTY communication technologies.

Automated printing technology undertook its most significant leap with the advent of the electronic computer. Modern digital computers emerged during the World War II era but reading the output of these early computers was a laborious process. For example, results from the first incarnation of the Manchester Mark I, a computer developed in the U.K., were presented in binary on the face of a display tube—a cathode-ray tube, with bright spots representing 1 and dim spots representing 0. Punched paper tape output was also available. By 1949, however, another British computer, the EDSAC, began displaying its output via an attached teleprinter, to rolls of ordinary paper. Within three years, the American company Remington Rand introduced the first printer designed for use with a computer. These early printers were essentially typewriters attached to a computer; but as computer hardware technology evolved, so did computer printers. In 1971, Gary Starkweather, a researcher at Xerox Palo Alto Research Center, adapted xerography technology in his development of the first laser printer. In the late 1970s, Hewlett Packard began developing inkjet printer technology, which culminated in the introduction of relatively inexpensive color inkjet printers two decades later. During the last 30 years of the twentieth century, the major developments in printing technology emerged primarily from research groups working within the U.S. computer industry.

Printers generally can be categorized as either impact or nonimpact. Like typewriters, impact printers generate output by striking the page with a solid substance. Impact printers include daisy wheel and dot matrix printers. The daisy wheel printer, which was introduced in 1972 by Diablo Systems, operates by spinning the daisy wheel to the correct character whereupon a hammer strikes it, forcing the character through an inked ribbon and onto the paper. Dot matrix printers operate by using a series of small pins to strike a matrix or grid ribbon coated with ink. The strike of the pin forces the ink to transfer to the paper at the point of impact. Unlike daisy wheel printers, dot matrix printers can generate italic and other character types through producing different pin patterns. Nonimpact printers generate images by spraying or fusing ink to paper or other output media. This category includes inkjet printers, laser printers, and thermal printers. Whether they are inkjet or laser, impact or nonimpact, all modern printers incorporate features of dot matrix technology in their design: they operate by generating dots onto paper or other physical media.

Printers are equipped with a print head, which is the device that moves back and forth across the media producing upwards of thousands of dots. There are four types of print heads which can be used: impact, thermal, inkjet, and electrostatic. In addition to the print head, basic printing technology utilizes printer language, bitmaps, and outline fonts to determine the placement of the dots.

Computers use printer language to tell the printer the dot formatting specifications. These commands compress data and manage color, font size, and graphics. Two influential printer languages are page description languages and printer command languages.

Page description language (PDL) is a device-independent language that describes the appearance of a printed page to the printer. Postscript is perhaps the most well-known PDL. Postscript was introduced by Adobe in 1985 for use with laser printers, and it introduced features such as vector graphics. Although commonly used with printers, Postscript can be used with any image-creating device including image setters, screen displays, and slide recorders.

Unlike PDLs, printer command languages (PCLs) are usually proprietary and thus are designed for work with specific models of printers. Hewlett Packard's PCL is a well-known example of such a language. Their PCL originally was for use with dot matrix and inkjet printers, but has since expanded for use with laser printers.

In addition to printer languages, printers can utilize bitmaps and outline fonts to determine dot placement on the paper. Bitmapped fonts are analogous to Gutenberg's moveable type. These fonts are typefaces of a specific size with attributes such as boldface or underline. Most printers come with a limited choice of bitmapped fonts installed into their permanent memory (ROM). However printers also are designed with random access memory (RAM), which permits a user's computer to send a larger selection of bitmapped fonts to the printer's font library. Printer language enables the computer to send the printer the directions for what bitmap table to use, and the printer uses that bitmap to generate the desired typeface. A limitation of bitmapped fonts stems from the fact that they are not scaleable since they are digital representations of typeface. Since bitmaps are composed of a set size or a limited set of sizes, these printers do not have the ability to completely control the resultant generated image. Printers that use outline fonts are much more functional. Outline fonts are scaleable and are used with Postscript and PCL printer languages. The printer language directs the printer to treat all output as if it were a graphic instead of a typeface. This functionality generates output that is more attractive and versatile than that produced by bitmapped fonts.

Early computer printers were developed for use by government agencies, large corporations, and universities. However, since Centronics Data Corporation introduced the first dot matrix printer in 1970, printers have become increasingly smaller and affordable. In addition, their operation has become increasingly user-friendly. Much of the crossover from institutional to individual use can be illustrated by personal printing developments in the 1980s. In 1984, Hewlett Packard began marketing both personal inkjet and laser printers for use in homes and smaller offices. Inkjet printers are equipped with an ink-filled print cartridge that is attached to the print head. The cartridge contains nozzles through which droplets of ink are sprayed onto paper to generate an image. Hewlett Packard had been working with inkjet technology since the late 1970s, and their innovations included miniaturizing inkjet printers and reducing the level of printer noise. The quality of inkjet output was superior to dot matrix which explains why dot matrix began to lose its appeal to consumers. A variation of the inkjet printer is bubble jet technology. Where inkjet printers use non-heating crystals to generate its output, bubble jet printers use heating elements to shoot ink from nozzles onto the paper.

The 1991 introduction of color inkjet printers by Hewlett Packard is a significant contributing factor as to why inkjet technology retained its popularity at the close of the twentieth century. Color inkjet printers are equipped with four ink cartridges—magenta, cyan, yellow, and black—attached to its print head. These printers use thermal technology to generate images on a wide variety of physical media. Each ink cartridge possesses a thin resistor through which an electrical current flows; this heats a thin layer of ink to more than 480°C for several millionths of a second. The ink is then rapidly sprayed onto the printing medium.

Laser printers are high-resolution printers that use a version of electrostatic reproduction or xerography technology found in copying machines to fuse images onto paper. Laser printers essentially operate by reflecting a laser beam from a spinning mirror to attract ink or toner from a rolling drum onto targeted areas of the paper. As laser printers generate over 300 dots per inch, consumers were able to afford typeset-quality output. Laser printers, and personal computer printers in general, illustrate one of the more important trends in printing technology: the role played by users. Printing technology during the last half of the twentieth century offered users increased opportunities to undertake their own print jobs. Before the development of personal computer printers, clerical staff and other individuals were forced to send their printing jobs elsewhere. Computer printers now offer the means to quickly generate high-quality, independently produced newsletters, business cards, brochures, and so on. With the advent of digital cameras at the end of the twentieth century, consumers could begin developing their own digital photographs. Although automatic printing was originally directed for use by governments and corporations, consumers were able to adopt this technology. Computer printers have revolutionized printing in much the same way as that of the moveable type.

See also **Computers, Personal; Electronic Communications; Fax Machine**

JULIA CHENOT GOODFOX

Further Reading

Campbell-Kelly, M. and Aspray, W. *Computer: A History of the Information Machine*. Basic Books, New York, 1996.

Hiltzik, M. Dealers of Lightning: Xerox PARC and the Dawn of the Computer Age. Harper Business, New York, 2000.

Lang, H. *A Phone of Our Own: The Deaf Insurrection Against Ma Bell*. Gallaudet University Press, Washington D.C., 2002.

Lubar, S. *InfoCulture: The Smithsonian Book of Information Age Inventions*. Houghton Mifflin, Boston, 1993.

Packard, D. *The HP Way: How Bill Hewlett and I Built Our Company*. Harper Business, New York, 1995.

Standage, T. *The Victorian Internet: The Remarkable Story of the Telegraph of the Nineteenth Century's On-Line Pioneers*. Berkley Publishing Group, New York, 1999.

White, R. *How Computers Work*, 6th edn. Que, Indianapolis, 2002.

Useful Websites

The Associated Press History: http://www.ap.org/pages/history/history.html

American Telephone &Telegraph History: http://www.att.com/history

Processors for Computers

A processor is the part of the computer system that manipulates the data. The first computer processors of the late 1940s and early 1950s performed three main functions and had three main components. They worked in a cycle to gather, decode, and execute instructions. They were made up of the arithmetic and logic unit, the control unit, and some extra storage components or registers. Today, most processors contain these components and perform these same functions, but since the 1960s they have developed different forms, capabilities, and organization. As with computers in general, increasing speed and decreasing size has marked their development.

The early computing machines of the twentieth century relied on various and mechanically complex ways to handle data. The first digital computers, such as the ENIAC built for the U.S. Army and completed in 1945, relied on processing units constructed from thousands of thermionic valves (or vacuum tubes) plugged into connectors (see Computers, Early Digital). Valves were connected in circuits and transferred electronic signals to enable mathematical and logical operations. The thermionic valve increased the speed of a computer's calculations. However, the valves required large amounts of power and were very expensive; they were approximately the size of small light bulbs and made computers very large and difficult to maintain and operate.

The data processing of the ENIAC was hampered not only by the fragility, size, and cost of the valves but also by its inability to store a program and data. In 1945, mathematician John von Neumann synthesized the research he and his colleagues conducted on the ENIAC and outlined the construction for a new computer in his seminal paper *A Draft Report on the EDVAC*, in which he proposed the stored program concept as an answer to the ENIAC's computing problems. The paper also described the basic components and functioning of the processor. While this paper articulated the operating concepts of the modern processor, its physical manifestation would take several more years of work and the use of valves would dominate computer construction into the late 1950s.

The EDVAC processor required a circuitry based on binary logic. Precedents for computing machines based on binary logic existed earlier but were not widely known. Machines built in the late 1930s and early 1940s by pioneers such as Konrad Zuse, Alan Turing, John Atanasoff and Clifford Berry employed a binary system. Claude Shannon, a researcher at the Massachusetts Institute of Technology, also noted the applicability of a computer's system of ones and zeros to the Boolean logic values of TRUE and FALSE. His 1938 paper, *A Symbolic Analysis of Relay and Switching Currents*, was an analysis of this relationship between computer logic and Boolean algebra, a mathematical system devised in the mid-nineteenth century by British mathematician, George Boole.

Binary logic offered a more efficient construction for computer processors. In the first computers, "logic gates," which were groups of valves in early computers and transistors in later ones, were placed to form circuits and operate together under a binary system in which information was transmitted in ones and zeros. The information traveled via electric current and the voltage of the current determined whether the signal was a one or a zero. The gates received the signals and according to their configuration would output a new signal. The complex circuitry formed by these gates allowed computers to process instructions given by the user. Most early processors contained sets of circuits known as accumulators, registers and control units. Accumulators performed simple arithmetic and stored sums; registers provided temporary storage for data and instructions; and control units directed the processor's operations.

Basing their work on these early processing units, many designers in the 1950s and 1960s developed processors for the large computer companies such as IBM, Honeywell, General Electric, Burroughs, Univac, and Digital Equipment Corporation. However, the physical form and capabilities of the computer processor were most dramatically changed with the invention of the transistor and the integrated circuit. In 1948, at Bell Laboratories in New Jersey, the work of three physicists, John Bardeen, Walter H. Brattain, and William Shockley resulted in the production of the first transistor. It was, however, nearly a decade before transistorized computers would appear for commercial use. The U.S. military supported the construction of the first fully transistorized computer in 1952. Built by Bell Laboratories, the computer was named TRIDAC (transistorized digital computer). Due to the early transistor's cost and unreliability, it was not until the late 1950s that companies devised more sturdy devices and built commercial computer processors using transistors.

Following the construction of commercial transistorized computers, research continued on making electronic components smaller and faster. In 1959, Jack Kilby at Texas Instruments and Robert Noyce at Fairchild Semiconductor, working separately to improve microelectronics and circuit design, invented and filed patents on a device called the integrated circuit or "chip." Integrated circuits were made up of transistors etched onto the surface of a semiconductor, most commonly a thin wafer of silicon. Processors were now made up of groups of integrated circuits placed onto printed circuit boards. Integrated circuits improved the speed and decreased the size of computer processors. After the introduction of an improved semiconductor called MOS (metal oxide semiconductor), several engineers and researchers worked to fit processor components onto a single integrated circuit. This process resulted in the single-chip central processing unit (CPU). The single-chip CPU heralded a new generation of computer processors and the development of the personal computer.

Intel, a company founded by former integrated circuit makers Robert Noyce and Gordon Moore, commercially produced the first single-chip CPU or microprocessor in 1971. This microprocessor, named the 4004 was designed and constructed by Intel employees Ted Hoff, Stan Mazor, and Federico Faggin. The 4004 included 2300 transistors and contained the registers and arithmetic and control units of a basic general-purpose computer processor. It performed 600,000 instructions per second.

Increases in speed were constant after the introduction of the Intel 4004 and several companies competed with Intel to produce smaller and faster microprocessors. In accordance with the popularly termed "Moore's law," in which Gordon Moore observed that approximately every eighteen months after the invention of the integrated circuit the number of transistors on a circuit doubled, the number of transistors in a microprocessor increased significantly with each new model. For example, the Pentium 4, introduced by Intel in 2000, contained 42million transistors and performed 2 billion instructions per second. These increases in processor speed, in conjunction with improvements in computer memory, allowed more and larger software applications to run on a computer. The decreasing costs of manufacturing processors and their increasing compactness has made possible very powerful personal computers.

See also **Computer Memory; Computers, Early Digital; Computers, Mainframe; Computers, Personal; Computers, Uses and Consequences; Integrated Circuits; Transistors; Valves/Vacuum Tubes**

Jessica R. Schaap

Further Reading

Aspray, W. *John von Neumann and the Origins of Modern Computing*. MIT Press, Cambridge, MA, 1990.

Ceruzzi, P. *History of Modern Computing*. MIT Press, Cambridge, MA, 1998.

Kilby, J.S. Invention of the integrated circuit. *IEEE Trans. Electron. Devices*, 23, 648–654, 1976.

Noyce, R. and Hoff, M. A history of microprocessor design at Intel. *IEEE Micro*, 1, 8–22, 1981.

Rojas, R. and Hashagen, U., Eds. *The First Computers: History and Architectures*. MIT Press, Cambridge, MA, 2000.
Smith, R.E. A historical overview of computer architecture. *Ann. Hist. Comp.*, 10, 4, 277–303, 1989.

Prospecting, Minerals

Twentieth century mineral prospecting draws upon the accumulated knowledge of previous exploration and mining activities, advancing technology, expanding knowledge of geologic processes and deposit models, and mining and processing capabilities to determine where and how to look for minerals of interest. Geologic models have been developed for a wide variety of deposit types; the prospector compares geologic characteristics of potential exploration areas with those of deposit models to determine which areas have similar characteristics and are suitable prospecting locations. Mineral prospecting programs are often team efforts, integrating general and site-specific knowledge of geochemistry, geology, geophysics, and remote sensing to "discover" hidden mineral deposits and "measure" their economic potential with increasing accuracy and reduced environmental disturbance. Once a likely target zone has been identified, multiple exploration tools are used in a coordinated program to characterize the deposit and its economic potential.

The field of geophysics uses physical property measurements of the earth's surface and subsurface to locate mineral deposits. Taken as a whole, electrical and electromagnetic methods represent a large class of geophysical methods, measuring current flow, electrical potential (voltages), and electromagnetic fields. Electrical potential measurement can be traced back to the mid-1800s when Robert Fox of Cornwall, England, first used the self-potential (SP) electrical conductivity method to detect extensions of known copper deposits in the 1830s. By the beginning of the twentieth century, electrical conductivity in rock and soils was widely recognized, sonic wave transmission was understood, and magnetic susceptibility and radioactivity were recognized. Conrad Schlumberger (France) discovered the principle of induced polarization (IP) in 1912 by noticing that soil resistivity measurements varied with frequency for some subsurface materials (e.g., clays and pyrite), but not others. By the 1940s, Canadian and later American and European geophysicists were experimenting with active (systems with both source and receiver) and passive (receiver only) electromagnetic (EM) surveying techniques, using both ground-based and aircraft-based platforms, to differentiate highly conductive massive sulfide orebodies from their less conductive host rocks. EM surveying began to flourish after World War II with greater availability of pilots and aircraft, increasing global demand for minerals, and the rise of the integrated mining and exploration company, which could afford to fund such expensive methods. The U.S. Geological Survey and the U.S. Navy conducted the world's first full-scale airborne magnetic survey in 1945.

Over the next 25 years, EM surveys proved useful in the discovery of massive volcanogenic sulfide base-metal deposits and diamondiferous kimberlite pipes, to the point that the frequency of discovery of such deposits dramatically increased. Canadian prospectors made extensive use of EM techniques in their exploration of the Canadian Shield, with the result that Canada became a world-class producer of base metals by the end of the century. Aircraft-borne geophysical systems had other long-term impacts; observation of magnetic banding in the Atlantic seafloor was a key element leading to the development of the concept of plate tectonics during the late 1960s and the early 1970s.

Many techniques routinely used in mineral exploration were originally developed for use in hydrocarbon or mineral-fuels exploration. During the 1930s, geophysicists (nicknamed "doodlebuggers" for the devices used to locate underground water, gas, or ores) developed the first primitive dynamite-source and inductive-coil-response seismic methods in their search for oil and gas. By the 1940s, airborne radiometric surveys had been conducted by the uranium industry to remotely map "radioactive" rocks. Initial experiments with measuring the strength of the gravity field of hydrocarbon-rich salt domes required development of mechanical–optical systems that could detect changes in the earth's gravitational field as small as one part in 10 billion. By the 1950s, methods to measure gravitational field differences reflecting the density contrast between mineralized areas and surrounding material had been expanded to many mineral deposit types. Chromite, iron, and massive sulfide deposits produce positive density contrasts. Mineral deposits that produce negative contrasts include gypsum, potash, and salt. By the end of the century, telemetry and laser ranging techniques were sufficiently advanced that airborne gravimetric systems were being used in regional land and ocean surveys. Although seismic three-dimensional signal processing and modeling is extensively used in oil and gas exploration, application to nonfuel mineral exploration is currently limited.

Remote sensing, broadly defined as obtaining information about an object without coming into physical contact with it, has been around since Galileo first used a telescope to observe celestial objects in 1609. The photographic camera has been a prime remote sensor for more than 150 years, since its development in France by Jacques Daguerre and Nicephore Nepce in 1839. Aerial photography, used initially from tethered balloons during World War I, began to play a part in mineral exploration in the 1920s, but did not come into prominence until after World War II. Improvements in helicopters and development of spacecraft-mounted cameras during the latter half of the twentieth century permitted exploration programs to operate in remote regions. With the advent of earth satellites with reflectance imaging systems (such as the Landsat satellite series first launched by the U.S. in 1972), remote sensing began to use instrumentation that indirectly detected minerals by their contrasting brightness levels (air photos) and spectral reflectance measurements. By the 1990s, hyperspectral scanners such as the airborne visible/infrared imaging spectrometer (AVIRIS) operated by the National Aeronautics and Space Administration (NASA) could record up to 200-plus spectral channels simultaneously, improving the instrument's ability to identify individual minerals or rocks (Figure 16). Remote sensing analysis is currently used to characterize geomorphology, lithology, mineralogy, and structure. When processed results are compared with mineral deposit models, prospecting targets can be inferred and field checked. Confidence in target selection can be increased by integrating remotely sensed results with geophysical data, geologic maps, geochemical sampling, topography, and other data.

In addition to the acquisition of various physical properties data, interpretation of geophysical and remotely sensed data made great strides during the second half of the twentieth century. Increases in computational power and data storage capability led to new ways of modeling data and the development of new analysis methods to provide information not apparent in initial imaging of raw data. Powerful imaging algorithms and fast, high-resolution displays allow prospectors to generate

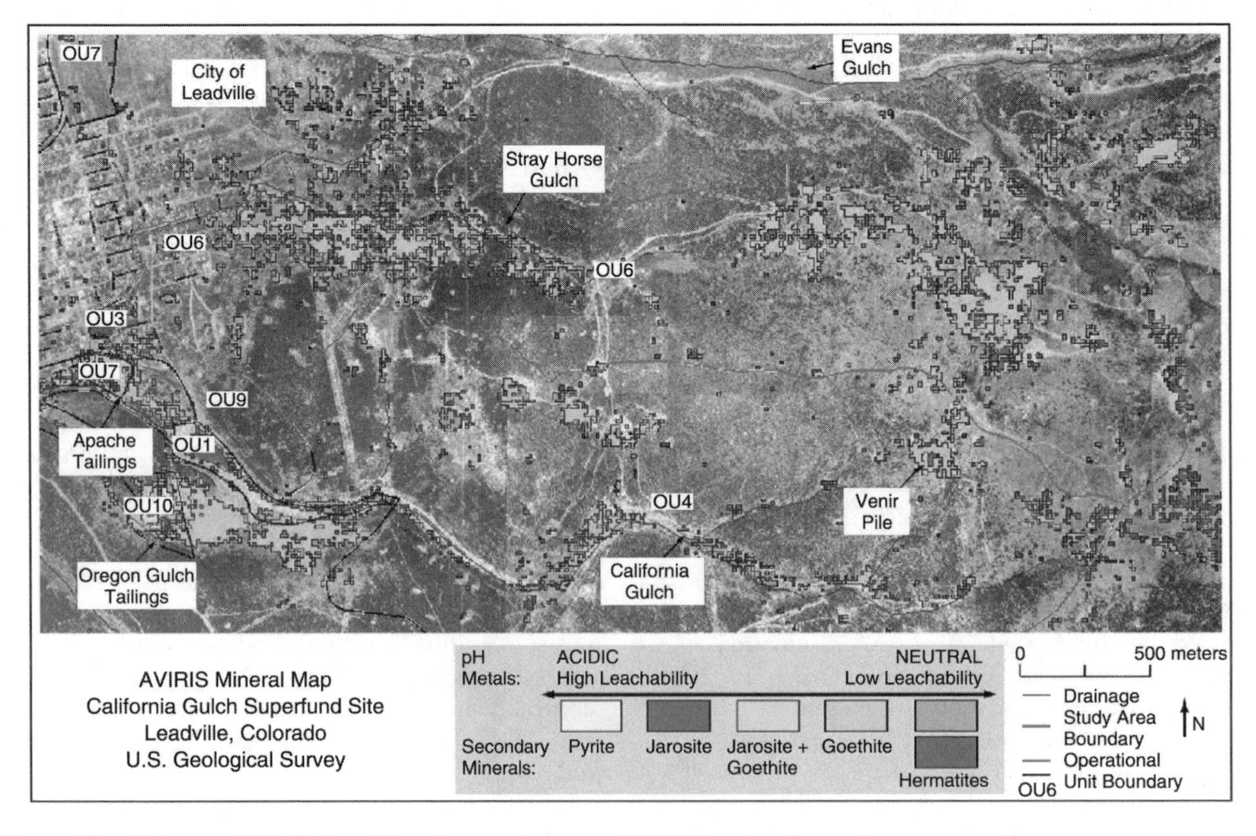

Figure 16. Airborne visible/infrared imaging spectrometer (AVIRIS) mineral map from a hyperspectral scanner operated by the National Aeronautics and Space Administration (NASA).

three-dimensional images of subsurface rocks and structures. Computers allow field teams to transmit data instantaneously, and speed up acquisition and analysis of large volumes of data.

Exploration geochemistry applies chemical principles, properties, and measurements to the discovery, delineation, and development of mineral resources. The evolution and increased contribution of exploration geochemistry to mineral discovery during the twentieth century has been primarily dependent on the increase in demand for and diversity of materials used by society, the decrease in the number of mineral deposits exposed at the surface, and adaptation of technological advances to mineral exploration and extraction.

Trace element analysis as a supplement to visual observation was first used in the mid-1930s when Victor Goldschmidt and J.H.L. Vogt (Scandinavia) and Vladimir I. Vernadsky and Alexander E. Fersman (USSR) independently carried out experiments using spectrographic analysis of soils and plants as a prospecting method. The development of dithizone (a colorimetric chemical reagent) for rapid chemical analyses in the mid to late 1940s has been attributed as one of the most important factors for the subsequent development of geochemical prospecting methods in North America. H.E. Hawkes, T.S. Lovering, and Bloom (U.S.) initially developed colorimetric field analytical techniques in 1947; during the 1950s, J.S. Webb (Great Britain) and other international researchers enhanced these techniques. Alan Walsh (Australia) introduced atomic absorption spectrometry in 1955 and subsequent research by Ward (U.S.) led to the rapid, accurate, sensitive and relatively interference-free analysis of many materials of interest in minerals exploration.

Another key advancement in exploration geochemistry was the development of inductively coupled plasma-mass spectrometry (ICP-MS) during the mid-1980s. This technique allows for cost-effective detection of many solid materials at parts per billion (ppb) or water at parts per trillion levels. The method formed the impetus not only for renewed interest in hydrogeochemistry in exploration, but also the renaissance in selective and partial extraction techniques used to assess processes that control metal migration in fluids. Since less than 1 percent of the metal may be retrieved by such extractions, accurate determination of ppb metal concentrations by ICP-MS technology is essential. As mineral exploration increasingly focuses on subsurface resources, measurement of gases (e.g., carbon dioxide, helium, radon, sulfur, and volatile mercury) in soils and the atmosphere became important. The dramatic increase in diamond exploration during the last decade was aided by the geochemical identification of specific indicator minerals (such as the G10 garnet) that are directly associated with diamondiferous deposits.

Although geochemical and geophysical prospecting can suggest the presence of anomalous mineral occurrences, delineation of the geometry, quality, and size of mineral prospects may best be accomplished by exploration drilling. Major strides in exploration drilling techniques came about with the need to delineate deeper and more geologically complex mineral deposits. The diamond drill, developed by Georges Auguste Leschot (France) in the 1860s, is considered one of the most important drilling tools. This early device used steel bits set with natural diamonds, capable of penetrating hard rock. Drilling was advanced with improved, more efficient drill design, and engine technology during the early part of the twentieth century. With the development of tungsten carbide bits in the 1950s, diamond core drilling advanced. By the 1970s, drill bits using more advanced alloys and impregnated with manufactured diamonds became the industry standard.

See also **Satellites; Environmental Monitoring**

David R. Wilburn, Donald I. Bleiwas, Karen D. Kelley, K. Eric Livo, and Jeffrey C. Wynn

Further Reading

Bloom, L.B. and Hall, G.E.M. Advances in exploration geochemistry; successful application of selective extraction techniques. *Explor. Min. Geol.*, 6, 3, 281, 1997.

Boyle, R.W., Ed. *Geochemical Exploration*. Canadian Institute of Mining and Metallurgy (CIM) Special Volume 11, Montreal, 1971.

Cox, D.P. and Singer, D.A., Eds. *Mineral Deposit Models, U. S. Geological Survey Bulletin 1693*. U.S. Government Printing Office, Washington D.C., 1992.

Govett, G.J.S. Geochemical remote sensing of the subsurface, in *Handbook of Exploration Geochemistry*, vol. 7. Elsevier, Amsterdam, 2001.

Hanna, W.F., Ed. Some historical notes on early magnetic surveying in the U.S. Geological Survey, in *Geologic Applications of Modern Aeromagnetic Surveys, U.S. Geological Survey Bulletin 1924*. U.S. Government Printing Office, Washington D.C., 1987.

Knepper, D.H., Jr. Mapping hydrothermal alteration with landsat thematic mapper data, in *Remote Sensing in Exploration Geology, 28th International Geological Congress, Field Trip Guidebook T182*, Keenan Lee, Ed. American Geophysical Union, Washington D.C., 1989.

Levinson, A.A. *Introduction to Exploration Geochemistry*. Applied Publishing, Wilmette, IL, 1974.

Lowe, C., Thomas, M.D., and Morris, W.A., Eds. *Geophysics in Mineral Exploration: Fundamentals and Recent Developments*, short course notes vol. 14.

Geological Association of Canada, St. John's, Newfoundland, 2000.
Rencz, A.N. and Ryerson, R.A., Eds. *Manual of Remote Sensing, Volume 3: Remote Sensing for the Earth Sciences*. Wiley, London, 1999.
Rose, A.W., Hawkes, H.E., and Webb, J.S. *Geochemistry in Mineral Exploration*. Academic Press, New York, 1979.
Rowan, L., Abrams, M., Clark, R., Goetz, A., Kruse, F., Lee, K., Lyon, R., Peters, D., Podwysocki, M., and Raines, G. Mineral resources, in *Airborne Remote Sensing for Geology and the Environment—Present and Future: U. S. Geological Survey Bulletin 1926*, Watson, K. and Knepper, D., Eds. U.S. Government Printing Office, Washington D.C., 1994.
Sharma, P.V. *Geophysical Methods in Geology*. Elsevier, Amsterdam, 1986.
Swayze, G.A., Smith, K.S., Clark, R.N., Sutley, S.J., Pearson, R.M., Vance, J.S., Hageman, P.L., Briggs, P.H., Meier, A.L., Singleton, M.J. and Roth, S. Using imaging spectroscopy to map acid mine waste. *Environ. Sci. Technol.*, 34, 47–54, 2000.
Telford, W.M., Geldart, L.P. and Sheriff, R.E. *Applied Geophysics*. Cambridge University Press, Cambridge, 1990.
Ward, S.H., Ed. *Geotechnical and Environmental Geophysics*. Society of Exploration Geophysicists, Tulsa, Oklahoma, 1990.

Useful Websites

Boyd, T.M. *Introduction to Geophysical Exploration*: http://www.mines.edu/fs_home/tboyd/GP311/introgp.shtml
Pees, S.T., *Oil History*: http://www.oilhistory.com/pages/diamond/inventor.html
Short, N.M., Sr. *The Remote Sensing Tutorial*, National Aeronautics and Space Administration: http://rst.gsfc.nasa.gov
Wilburn, D.R., Goonan, T.G. and Bleiwas, D.I. Technological Advancement: A Factor in Increasing Resource Use, U.S. Geological Survey Open File Report 01-197: http://pubs.usgs.gov/openfile/of01-197

Psychiatry, Diagnosis and Non-Drug Treatment

Psychiatry is the medical specialty for diagnosing and treating mental, emotional, and behavioral disorders. During the twentieth century, psychiatry adapted and abandoned several technologies in the treatment of mental illness with the notable exception of electroconvulsive therapy, which from its invention in 1938 has remained an important treatment for mental illness.

At the beginning of the twentieth century, there were few nondrug treatments available in psychiatry. Psychotherapy was available for the well-to-do, but others went untreated or were consigned to asylums that were little better than prisons.

One method used for the first two decades of the century was electrotherapy. Electrotherapeutic devices used static electricity, induction coils, or direct current electricity drawn from simple batteries ranging in size from small wooden boxes to the room-sized static electric machines. By passing an electrode, or wand, along the spine and nape of the patient's neck, physicians claimed to cure intractable neuroses such as neurasthenia or hysteria. In other cases patients were seated in front of giant static electrical machines and received a static wind. Electrotherapy fell out of favor after World War I when psychiatrists redefined mental illness and the symptoms clustered under neurasthenia and hysteria were reclassified as psychodynamic rather than physical ailments. Throughout the twentieth century variations of electrotherapy survived as part of physical therapy.

Hyperthermal cabinets, used from the late 1930s through to the early 1950s, were also once part of the psychiatric treatments. The idea for fever therapy, as it was known, came from Julius Wagner von Jauregg who injected patients with blood-borne malaria to induce high fevers and successfully treated general paresis. Rather than risk the side effects of malaria, hyperthermal cabinets were manufactured and sold to hospitals as a safer, more controlled means of raising the body's temperature. The devices, shaped much like the iron lung, encased the patient below the neck and raised the body to temperatures above 106°F (41.1°C). Despite clinical success on both schizophrenia and depression, this treatment was abandoned with the advent of electroconvulsive therapy.

Electroconvulsive therapy, or ECT, was invented in 1938 by Ugo Cerletti who sought to replicate the convulsive therapy pioneered by Ladislas von Meduna. Meduna believed that epilepsy was antagonistic to mental disease and so contrived a convulsive medication called metrazol or cardazol which sent patients into convulsive fits that seemed to cure schizophrenia. Cerletti's research team set about to find a technologically cleaner and simpler method to create convulsions and succeeded in 1937 when Cerletti's assistant Lucio Bini invented a simple electroconvulsive machine. The machine ran on a standard household alternating current of 125 volts and sent electricity through the brain using a pair of calipers attached to the temples. In April 1938 Cerletti gave ECT to a mental patient and his published findings confirmed his success.

In the U.S., machines were first built by psychiatrists themselves and had several common features. Machines tended to be small devices encased in wood with a glass-covered meter that gave the voltage of the electricity, while another dial allowed the physician to increase the voltage, and a button that when pushed began or ended a treatment. "Home made" devices of wood gave way to metal and a number of manufacturers competed for the practitioners by including a variety of designs and features between 1940–1960. Lektra, Medcraft, Offner, and Reiter were important manufacturers of ECT machines throughout this period and each added technical refinements. The size could vary from 13 by 8 centimeters to that of a large suitcase. The standard models incorporated instruments that could read the patient's electrical resistance and provide a slow rise in current (or glissando); some included so-called reverse glissando because they started with a higher current that stepped down gradually. There were also machines that used the so-called brief stimulus method that delivered electricity in short bursts or waves which presumably lessened the prevalent side effects such as marked memory loss. It was not uncommon in the 1950s to find machines that would also provide what was called electrosleep or electronarcosis. This was a subconvulsive dose of continuous electricity that led to unconsciousness without convulsion. Some physicians used this a treatment for mental or nervous disorders, while others used it to put patients to sleep prior to administering ECT.

ECT fell from favor after the advent of psychopharmacology, and beginning in the 1960s, manufacturers abandoned the field one by one. In the 1980s Medcraft, now called Hittman-Medcraft, was the only original manufacturer remaining in the U.S. However, the psychiatrist Paul Blachy modified a machine in 1973 to incorporate an electroencephalogram (EEG) and electrocardiogram (EKG) to monitor patient's vital signs before and after the treatment. He called his design the monitored electroconcovulsive therapy apparatus (MECTA) and began manufacturing his machines under the MECTA name in the mid-1970s. In the mid-1980s, Richard Abrams and Conrad Swartz combined their efforts to create Somatics, which is the leading manufacturer and seller of ECT devices (particularly its Thymatron line). The new devices incorporated computer chips to provide a number of read-outs and deliver precise amounts of electricity at precise time intervals. Although using higher voltage, some estimates suggested 180 or more volts, these devices cycled at faster rates and had shorter pulses of electricity over a longer duration. Devices constructed in the 1980s also allowed physicians to set the machine according to the age of the patient, an important factor in determining the body's electrical resistance. With the advent of so many new features, the devices have become larger and approximately the size of a large stereo receiver. ECT, which in the 1960s seemed destined for the same end as electrotherapy, hydrotherapy, and fever therapy, made a dramatic resurgence in the 1980s and is particularly useful for medically resistant forms of depression. Estimates suggest as many as a 100,000 treatments are given annually.

The popularity of ECT technology has inspired the use of the experimental treatment known as transcranial magnetic stimulation (TMS), which employs some of the ideas of electrotherapy. In TMS an electromagentic coil is placed on the scalp and an electric current of high intensity is turned on and off through the coil. This creates a magnetic field lasting from 100–200 microseconds, which is repeated in a process known as rTMS. Clinical reports show efficacy among treatment-resistant depressions. Unlike ECT, which employs a general electrical charge through the head, rTMS can be situated to deliver electricity to a more specific region of the brain. Side effects such as headache and memory loss are lessened, patients remain awake throughout the treatment and, though still not approved by the Food and Drug Administration (FDA) as a treatment for depression, this technology remains promising.

The last major medical breakthrough in twentieth-century psychiatry may well be the introduction of vagus nerve stimulation (VNS) to treat depression and anxiety. In the late 1990s psychiatrists began employing this technique, which involves implanting a pacemaker-sized device into the chest wall. The device delivers a short pulse of electricity to the vagus nerve located in the neck, which in turn stimulates a series of nerves that seem to be involved in brain chemistry. Since 1998 clinical results have demonstrated the promise of this new technique.

See also **Electroencephalogram (EEG); Psychiatry, Pharmaceutical Treatment.**

TIMOTHY W. KNEELAND

Further Reading

American Psychiatric Association. *The Practice of ECT: Recommendations for Treatment, Training, and Privileging*. American Psychiatric Press, Washington D.C. 1990.

Braslow, J. *Mental Ills and Bodily Cures: Psychiatric Treatment in the First Half of the Twentieth Century*. University of California Press.

George, M., Harold Sackein, *et al.* vagus nerve stimulation: a potential therapy for resistant depression. *Psychiatr. Clin. N. Am.*, 23, 4, 757–782, 2000.
George, M., Lisanby, S.H. and Sackeim, H. Transcranial magnetic stimulation: applications in neuropsychiatry. *Arch. Gen. Psychiatr.*, 56, 300–311, 1999.
Kneeland, T. and Warren, C. *Pushbutton Psychiatry: A History of Electroshock in America*. Praeger, West Port, CT, 2002.
Valenstein, E. *Great and Desperate Cures: The Rise and Decline of Psychosurgery and Other Treatments for Mental Illness*. Basic Books, New York, 1987.

Psychiatry, Pharmaceutical Treatment

Psychiatry, the medical specialty that deals with emotional and behavioral disorders, underwent a profound transformation in the mid-twentieth century with the development of psychopharmacology, the use of drug therapy to treat specific mental illnesses ranging from schizophrenia to depression and anxiety. Following the serendipitous discovery of drugs such as chlorpromazine in 1952, researchers examined the specific effects of these drugs, in turn spawning a revolution in the scientific understanding of the brain and the role of neurotransmitters in the expression of mood and behavior.

Prior to the twentieth century, private practice and asylum physicians used a combination of drugs such as opiates, barbiturates, bromides, chloral hydrate and hyoscine to treat the mentally ill. These nonspecific drugs were used to sedate agitated patients or elevate the mood of the melancholic. Although these attacked the symptoms of the illness, they were adjunct to other treatments that were seen as curative. During the first half of the twentieth century, psychiatric treatments incorporated drug regimens to alleviate the symptoms of ailments classified as either psychosis or neurosis. One such regimen was sleep therapy, or prolonged sleep, popularized by Jacob Klaesi in 1920. Using various bromides or barbiturates, Klaesi put patients to sleep for days at a time and found that prolonged sleep decreased the symptoms of the illness but did not cure patients. In their search for a cure some psychiatrists turned to shock or convulsive therapies. Based on the notion that epileptic patients were immune to schizophrenia, psychiatrists tried a host of convulsive agents to cure that schizophrenia. Beginning in the 1930s Ladislas Meduna pioneered the use of the chemical agent pentylenetetrazol, also known as Metrazol or Cardiazol, to cause patients to convulse. Similarly, Manfred Sakel developed insulin therapy, another shock or convulsive therapy. In this treatment, popular from the 1930s to the mid-1950s, synthetically produced insulin was employed over the course of hours to induce comas and convulsions that seemed to alleviate the symptoms of schizophrenia. The development of electroconvulsive therapy (ECT) and psychopharmacology made Metrazol and insulin therapy obsolete after 1950.

Drug therapies appeared as a byproduct of research in the chemical and pharmaceutical industry. Chlorpromazine (CPZ), better known by the trade name Thorazine, had been explored for its properties as a dye in the late nineteenth century and for its antihistamine properties in the first half of the twentieth century. Its sedative properties were noted and several clinicians including French surgeon Henry Laborit, as well as Jean Delay and Pierre Denker, were to independently confirm its ability to calm agitated mental patients and alleviate the symptoms of schizophrenia. Thus was born a new class of drugs known as neuroleptics in Europe and antipsychotics in North America. Chlorpromazine set the pattern for the psychopharmacological revolution and followed the life-cycle taken by almost every other pharmaceutical innovation: drugs were introduced with high expectations and enthusiasm, but after the emergence of side effects they were derided and in some cases abandoned.

After its introduction in North America by SmithKline & French in 1953 and its approval for use in the U.S. by the Food and Drug Administration (FDA) in 1954, CPZ became a standard therapy in the state mental institutions leading to the dramatic decline of inmates from 500,000 to 150,000 within a short period of time. However, within a decade severe side effects that created a Parkinson-like disorder known as tardive dsykinesia were discovered and alternative drugs were sought.

The initial success of CPZ led to further research on histamine blockers and in 1957, Roland Kuhn would slightly alter the chemistry of CPZ and create imipramine hydrochloride, (Tofranil), a drug found to be useful in the treatment of depression. Imipramine was the first of the tricyclic antidepressants, so called because of their three-ring chemical bond, but others followed including amitriptyline (Elavil) in 1961. On this medication, depressed patients became more vivacious, had restored interest in their favorite activities, and experienced restful sleep and healthier appetites.

A third class of drugs known as the monoamine oxidase inhibitors (MAO-I) was developed in pharmaceutical laboratories. Nathan Kline pioneered phenelzine sulfate, marketed by such names as Nardil and Marplan. Its effectiveness was similar to other antidepressants but severe side effects were noted when patients took the medication in conjunction with foods containing tyamine, such as cheese, wine, and pickled foods.

The drug lithium, used during World War II as a substitute for table salt, was discovered to have antipsychotic properties by John Cade who worked in Australia. Although not approved by the FDA in the U.S. until the 1970s, it quickly became a standard treatment for bipolar disorder, or manic depression.

In the 1950s the first antianxiety drugs or minor tranquilizers were developed, the diphenylemythanes such as meprobamate (marketed under the trade name Miltown) was introduced in 1954. Later the benzodiazepine tranquilizers such as Librium or Valium were produced. These seemed to eliminate anxiety with few side effects, however the addictive properties of these drugs would receive a great deal of scrutiny in Congressional hearings and the popular press in the 1970s. In response, pharmaceutical companies sought less addictive versions of these drugs and produced Xanax in 1981, which was found to be effective for the newly named disorder panic attacks.

All of these drugs inspired interest in the neurotransmitters. First identified in 1920 by Otto Loewi, neurotransmitters are found in the synaptic gaps between the neurons of the brain. Although there are hundreds of them in the brain, research focused on two that seemed to be particularly important in both mood and behavior: dopamine and serotonin. Depressed patients were found to have less serotonin available than those who were not depressed. This led to the now prevalent belief that depression was caused not by environment but by changes in the brain chemistry such as a lack of free serotonin. Depression was therefore assumed to be as treatable as a biological disorder rather than to a psychological one. Researchers at Eli Lilly focused on creating a drug that would block the reuptake of serotonin into the neuron, thus was created an entirely new class of drugs—the selective serotonin reuptake inhibators (SSRIs). Lilly developed fluoxetine hydrochloride in 1974, but the drug was not approved by the FDA until 1987 when it was marketed under the name Prozac and became the established industry standard, perhaps the most well-known psychiatric drug of all time.

The psychopharmacological revolution that began with the accidental discovery of the clinical possibilities of a variety of drugs generated an interest in the specific effects of those drugs on brain chemistry, leading to the discovery of more specific treatment agents such as the SSRIs. For some this confirmed their belief that mental illness is biological rather than psychological.

See also **Medicine; Psychiatry, Diagnosis and Non-Drug Treatments**

TIMOTHY W. KNEELAND

Further Reading

American Psychiatric Association. *Essentials of Clinical Psychopharmacology*, Schatzberg, A.F. and Nemeroff, C.B. American Psychiatric Press, Washington, 2001.

Grob, G. *From Asylum to Community: Mental Health Policy in Modern America.* Princeton University Press, Princeton, 1991.

Healy, D. *The Creation of Psychopharmacology*. Harvard University Press, Cambridge, MA, 2002.

Shorter, E. *A History of Psychiatry from the Era of the Asylum to the Age of Prozac.* Wiley, New York, 1997.

Speaker, S. From "happiness pills" to "national nightmare": changing cultural assessment of minor tranquilizers in America, 1955–1980. *J. Hist. Med. Allied Sci.*, 52, 338–376, 1997.

Valenstein, E. Blaming the Brain: The Truth about Drugs and Mental Health. Free Press, New York, 1998.

Q

Quantum Electronic Devices

Quantum theory, developed during the 1920s to explain the behavior of atoms and the absorption and emission of light, is thought to apply to every kind of physical system, from individual elementary particles to macroscopic systems such as lasers. In lasers, stimulated transitions between discrete or quantized energy levels is a quantum electronic phenomena (discussed in the entry Lasers, Theory and Operation). Stimulated transitions are also the central phenomena in atomic clocks. Semiconductor devices such as the transistor also rely on the arrangement of quantum energy levels into a valence band and a conduction band separated by an energy gap, but advanced quantum semiconductor devices were not possible until advances in fabrication techniques such as molecular beam epitaxy (MBE) developed in the 1960s made it possible to grow extremely pure single crystal semiconductor structures one atomic layer at a time.

In most electronic devices and integrated circuits, quantum phenomena such as quantum tunneling and electron diffraction—where electrons behave not as particles but as waves—are of no significance, since the device is much larger than the wavelength of the electron (around 100 nanometers, where one nanometer is 10^{-9} meters or about 4 atoms wide). Since the early 1980s however, researchers have been aware that as the overall device size of field effect transistors decreased, small-scale quantum mechanical effects between components, plus the limitations of materials and fabrication techniques, would sooner or later inhibit further reduction in the size of conventional semiconductor transistors. Thus to produce devices on ever-smaller integrated circuits (down to 25 nanometers in length), conventional microelectronic devices would have to be replaced with new device concepts that take advantage of the quantum mechanical effects that dominate on the nanometer scale, rather than function in despite of them. Such solid state "nanoelectronics" offers the potential for increased speed and density of information processing, but mass fabrication on this small scale presented formidable challenges at the end of the twentieth century.

Discovery of Quantum Wave Effects in Electronics

The most important quantum effect used for electronic devices is the tunneling effect, the ability of electrons whose energy is not otherwise high enough to overcome or "tunnel" through an energy gap (arising for example from a junction with some other material). This energy gap creates a barrier that contains the electron's movement. Electrons (and holes) actually go through (rather than over) a potential energy barrier. The tunneling effect is associated with the wave characteristics of electrons and the Heisenberg Uncertainty Principle. This effect was first noticed in a theoretical calculation of tunneling between bound states in an atom by Friedrich Hund in 1927, who termed it "barrier penetration."

In 1928, quantum tunneling was shown by Ralph Howard Fowler and Lothar W. Nordheim to explain electron emission from a metallic surface under an intense applied electric field. The field lowered the potential barrier that normally prevented electrons from escaping the metal, allowing tunneling across this lowered barrier. In 1934 Clarence Zener modeled the electrical breakdown of insulators (i.e. they began to conduct) with the idea of tunneling. The Zener effect in semiconductors occurs with application of a reverse voltage to a junction between *n*-type and *p*-type semiconductor material that is heavily doped. Beyond a certain breakdown voltage, large numbers of new carriers (electrons) "tunnel" across the junction to form the avalanche current that occurs at breakdown. In the mid-1950s after the development of techniques for growing heavily doped monocrystal germanium with precise geometries and controlled doping, the Zener effect was used to build Zener diodes with a predetermined "breakdown" voltage. The Zener diode could be used to stabilize and limit direct current (DC) voltages in circuits, and they became the earliest commercial quantum electronic devices.

After World War II, various striking new tunneling phenomena were discovered, and new devices based on tunneling were invented. In 1954 Pierre Aigrain made theoretical calculations of the tunneling effect in high-impurity germanium semiconductors. Building on these preparatory works, Leo Esaki invented the tunnel diode (also known as the Esaki diode) in 1957 while working for Sony Corporation. In a tunnel diode—a highly doped germanium *p–n* junction with a small depletion region—the *p*-material valence band and *n*-material conduction band nearly overlap. At low bias the tunneling of electrons and holes can occur. The diode has an unusual current–voltage characteristic curve as compared with an ordinary junction diode: current initially increases as the bias voltage is increased from zero, but when the voltage reaches a value comparable to the band gap of germanium, the tunneling current decreases. Therefore an anomalous negative resistance region appears, which allows oscillator action and the construction of high-frequency electronic oscillators and logic circuits based on tunnel diodes. The quantum-confinement effects of tunnel diodes, while having much increased circuit switching speeds and reduced power consumption, did not find widespread use due to difficulties in processing devices with heavy doping.

In 1962 Brian D. Josephson predicted the existence of a tunneling supercurrent that traversed a gap separating two superconductors. Ivar Giaever and others confirmed this superconducting tunnel effect experimentally using Josephson junctions consisting of superconducting thin films separated by a thin oxide barrier. In 1973, Josephson, Esaki, and Giaever were jointly awarded the Nobel Prize for Physics for the discovery of tunneling phenomena. Josephson junctions can be used for the construction of rapid single-flux quantum (RSFQ) electronics (proposed by Konstantin Likharev, Oleg Mukhanov, and Vasily Semenov in 1985). RSFQ electronics allows ultrafast circuitry with switching rates over 100 gigahertz (i.e., at least 300 times higher than the fastest similar semiconductor circuits) but are hampered by the necessity of their cooling to 4 to 5° Kelvin.

The concept of quantum confinement of electrons is used to trap electrons in specific locations in nanometer-scale structures. Quantum "islands" or "wells" between two closely spaced barriers (usually layers of two different III/V semiconductor alloys) are used in resonant tunneling diodes and transistors (invented by Esaki, R. Tsu, and L.L. Chang in 1974). Controlling the composition, shape, and size of the island permits control of the energy levels of the confined electrons such that electrons may pass through the device only by quantum tunneling. Resonant tunneling diodes can achieve very high switching speeds and have the speed performance of Josephson devices without the necessity of cryogenics. As MBE growth techniques improve in the early twenty-first century, chips incorporating resonant tunneling diodes and transistors will provide increased density and speed of information processing, consuming far less power than conventional integrated circuits.

The single-electron tunneling transistor, first proposed by Dmitri Averin and Konstantin Likharev in 1985, was developed by Theodore Fulton and Gerald Dolan at Bell Laboratories. Single-electron transistors, just a few nanometers wide, control single electrons by localizing them on nanoscale circuit elements linked by tunnel junctions. Such devices could represent bits of information with a single electron and be used in ultrahigh-density memory devices. However single-electron transistors have low voltage gain, are sensitive to random background charges, and until the late 1990s could not operate at room temperature.

The next generation of electronics will see advances in nanofabrication technology such as the ability of MBE to artificially engineer the band structure of a device and reliably produce nanometer-scale islands, barriers, and "heterojunctions" between islands and barriers. This will make it possible to constrain electrons so that they travel in a one-dimensional quantum wire only, or to create quantum dots ("zero-dimensional"), further increasing the performance and reducing the size of quantum electronic devices for integrated circuits.

See also **Clocks, Atomic; Josephson Junction Devices; Nanotechnology, Materials and Applications**

GILLIAN LINDSEY

Further Reading

Brown, L.M., Pais, A., Pippard, A.B. *Twentieth Century Physics*. American Institute of Physics Press, New York, 1995.

Hund, F. *The History of Quantum Theory*. Harrap, London, 1974.

Lundqvist, S., Ed. *Physics 1971–1980*. World Scientific, Singapore, 1992. Nobel lectures, including presentation speeches and laureates' biographies.

Nobel Foundation. *Physics 1963–1970*. Elsevier, Amsterdam, 1972. Nobel lectures, including presentation speeches and laureates' biographies

Turton, R. The Quantum Dot: A Journey into the Future of Microelectronics. Oxford University Press, New York, 1995.

R

Radar aboard Aircraft

As with much else in radar, airborne radar sets were first developed during World War II, and most of the modern uses for such sets were explored during that war. While airborne radar shares much in common with surface and naval sets, there are many factors involved that make airborne installations very different from either of the latter.

Generally speaking, the radar set itself needs to be made smaller to fit in the limited space available within an airframe. Antennas also need to be smaller; too large, and the drag on the aircraft will make it dangerous to fly. During World War II, a Luftwaffe directive hindered German airborne radar development by prohibiting external radar antennas on aircraft; it was rescinded in 1941. The concern was well founded as at least one test-bed aircraft crashed due to the performance problems caused by the external antennas. There are also only so many places on an airframe where a larger antenna can be placed, possibly affecting the usefulness of the radar. Another major consideration is power; an aircraft is limited in how much power it can produce, and it needs much of that power to fly safely. Additional power supplies can be added to the airframe, but these add to the weight of the aircraft. The size and weight of the set, the size and placement of the antenna(s), and the power consumed must all be balanced against the purpose of the radar set.

In the early days of World War II, the British used their ground-based radar network ("Chain Home") to direct fighters into position to attack German bombing raids. This was easy enough during the day and in good weather as the Chain Home sets had a margin of error of around 5 kilometers, which was close enough for the bombers to be seen by the fighter. At night, the visual range was reduced to around 300 meters. Something needed to bridge that gap if the fighters were to defend against night raids, so British scientists (under Sir Robert Watson-Watt) began work on an airborne intercept (AI) radar. AI needed to be at least accurate enough to get the fighters within that 300 meters. In order to increase the accuracy, AI radar needed to use centimetric radio waves (with a wavelength close to 10 centimeters) as opposed to the "long wave" (1- to 1.5-meter wavelengths) used by Chain Home.

Using a shorter wavelength helps to increase the accuracy of radar but causes other problems: all else being equal, a shorter wavelength fades more over a distance than a longer one (therefore needing more power in the transmission). The AI only needed a maximum range slightly exceeding the Chain Home margin of error, reducing power requirements, but existing wave generators still could not provide the necessary power at the shorter wavelength. Existing wave generators (klystron valves and magnetrons) either could not produce the centimetric waves needed or produced only very low-powered waves.

The necessary power at short wavelengths came in 1940 with the discovery of the resonant cavity magnetron by John Randall and Henry Boot at the University of Birmingham, U.K. Their invention produced a centimetric wave nearly 1000 times stronger than any previously produced. Like the magnetron, the cavity magnetron uses a magnetic field to produce oscillations in a stream of electrons moving through a vacuum. These oscillations are the source of the radio waves used by the radar.

However, the magnetron cannot efficiently produce wavelengths much shorter than 1 meter. The cavity magnetron adds a cylindrical metal shell with "resonant cavities" around its circumference. Under the influence of the magnetic field, several electron streams "pinwheel" around the central cathode. As the tips of the pinwheel pass over the resonant cavities, they produce oscillation in each cavity—the process has been compared to blowing into a whistle. This construction allows the cavity magnetron to oscillate much faster than the ordinary magnetron, resulting in shorter wavelengths.

In addition to greater accuracy, the shorter wavelengths also allowed antennas to be more compact, since antennas work best at a whole multiple of the wavelength being used. In aircraft, the antenna is most commonly a dipole one-half of the wavelength used. Early long-wave sets used antennas attached longitudinally to the sides of the fuselage, but these limited the radar to "seeing" only to the side. Forward-looking radar depended on long, cable-like antennas affixed along the leading edges of the wings.

Yagi antennas (named for a Japanese physicist who published papers about the antenna type in English) were also used in many designs. They resemble an old rooftop television antenna: a series of parallel rods in a plane, often tapering in length toward one end. German night fighters used four of these extending from the nose of the airplane, which was a crude method of electronically "steering" the radar beam. With centimetric radar, the Allies used a mechanically directed, parabolic "dish" antenna. Because they were using such short wavelengths, these antennas could be placed inside a fairing or within the nose of the aircraft.

Fast-acting duplexers were also important to AI development, as these allowed the radar set to transmit and receive using the same antenna. The duplexer is like a check valve that prevents the powerful transmitter energy from going directly to the sensitive receiver, which would overload it. Without a duplexer, a radar set would require two very accurately aligned antennas: one to transmit and one to receive.

Collaboration with American researchers at the Massachusetts Institute of Technology Radiation Laboratory (Rad Lab) resulted in AI sets operating at 10 centimeters with fine resolution and long range. Initially, these were installed in small bombers and large twin-engine fighters, but as the war progressed, the AI sets steadily got smaller and simpler to use, until they could be installed in single-seat fighters without greatly affecting their performance. German AI benefited from examining Allied radar sets in crashed airplanes and then producing some very good airborne radars. In general, the German effort lagged behind that of the Americans and British.

The centimetric AI sets were soon adapted for air-to-surface vessel (ASV) use. The Germans used long wave (1- to 1.5-meter) ASV radars that helped them locate the Allied convoys to Russia, but the primary Allied use of ASV was detecting submarines. The British had used long wave ASV since 1940, but those sets could only get the bombers within range of the submarine, which then had to be located visually, often giving alert U-boat crews a chance to crash-dive and get away. Also, once the Germans realized that the Allies were tracking their subs with radar, they began equipping the U-boats with "Naxos," a radar detector. Centimetric sets not only defeated the Naxos (it was not designed to detect radio emissions much shorter than 1 meter), but also increased resolution to the point where the periscopes and snorkels of submerged submarines could be picked out of the background clutter and the bomber could get close enough to drop depth bombs using ASV alone.

Each of the aforementioned radars used oscilloscopes to display information: range, height, and direction were often shown as spikes in a flat line. The next development was the British H_2S navigational radar, which used a plan position indicator (PPI). The PPI is the most familiar radar display: a circular screen with a line sweeping around it like the hand of a clock. The PPI shows graphically the horizontal relationship between the radar (at the center of the screen) and anything it detects. By connecting a PPI to the ASV radar, the ground clutter that made AI useless at too low an altitude instead made a picture in which water and roads showed up as dark areas and buildings and steep cliffs showed up brightly. By comparing the radar picture to a map, a radar operator could navigate by noting the location of prominent radar landmarks.

From 1946 to 1950, an American version of H_2S, designated AN/APS-10 was used in an experiment by American Airlines, whose pilots found it useful in avoiding dangerous weather, leading to the weather avoidance radar common to all modern commercial carrier aircraft. Commercial and general aviation aircraft also carry a radar altimeter that gives an accurate measurement of height above terrain that is very useful when landing using instruments alone.

One descendant of AI is the airborne early warning (AEW) aircraft. The most recognizable is the U.S. E-3 Sentry, often called AWACS (for airborne warning and control system), which entered service in 1977. But the AEW idea goes back to the postwar period and is not limited to the U.S. Several other countries also build AEW airplanes and helicopters, and many more countries buy them. In the late 1980s the American military began developing the E-8 JSTARS (joint surveillance target attack radar system) aircraft. The E-8 is equipped with air-to-ground synthetic aperture radar that looks to the side and uses the motion of the airplane and lots of computing power to simulate a much larger antenna (the "synthetic aperture"). This produces radar images at a very high resolution, allowing the system to detect ground vehicles at a great distance.

At the close of the twentieth century, combat aircraft continued to use radar for the same purposes as in World War II: navigation, air and surface search, and targeting. Using computers and guided munitions, they could also automatically release bombs or launch missiles at the appropriate time. The big difference between 1945 and 2000 is that most, if not all, of these functions can be done by a single aircraft carrying a single radar with a range and resolution much greater than any airborne set used during the war.

See also **Radar, Defensive Systems in World War II; Radar, Displays; Radar, High Frequency and High Power**

LAURENCE BURKE

Further Reading

Brown, L. *A Radar History of World War II: Technical and Military Imperatives*. Institute of Physics Publishing, Bristol, 1999. A good history of radar developments in all countries during World War II.

Buderi, R. *The Invention that Changed the World: How a Small Group of Radar Pioneers Won the Second World War and Launched a Technological Revolution*. Simon & Schuster, New York, 1996. Concentrates on the MIT's Rad Lab and its people, but also includes good information on the British and Bell Labs efforts. Also follows many postwar "spinoffs" of wartime radar research.

Frieden, D.R., Ed. *Principles of Naval Weapons Systems*. Naval Institute Press, Annapolis, MD, 1985. Intended as a textbook for students at the U.S. Naval Academy, the descriptions of different systems (including radar) begin with basic information and quickly move toward equations that describe performance.

Jane's Avionics. Jane's Publishing, London, 2003. Published annually, this volume attempts to list every commercial avionic system (including radar) made in every country, whether civil or military.

Jane's Weapon Systems. Jane's Publishing, London, 1988. This volume includes military airborne radar systems, with a separate listing of air- and surface-launched missiles using radar guidance. Some time after 1988, *Jane's* began publishing separate volumes for air-launched weapons, missiles and rockets, naval weapons systems, electronic mission aircraft, and radar and electronic warfare systems.

Pritchard, D. *The Radar War: Germany's Pioneering Achievement 1904–1945*. Patrick Stephens, Wellingborough, 1989. Pritchard presents a comprehensive picture of German radar development, though he also gives a lot of technical details and model designations that might daunt the casual reader.

Rhodes, M. RADAR: wartime development—postwar application, AN/APS-10. *Aerospace Historian*, December, 231–240, 1981. Presents a brief history of the airborne search radar along with a description of its early postwar applications (including the American Airlines test).

Scott, P. Self-control in the skies: an onboard device for flying closer together more safely. *Scientific American*, 282, 1, 36, January, 2000. Using radar technology to prevent aircraft collisions.

Radar, Defensive Systems in World War II

With the onset of war in September 1939, Britain, Germany, and the U.S. had advanced radar designs while France, Russia, The Netherlands, Italy, and Japan had little of value in comparison although they had made research efforts along those lines. Of these endeavors, only Britain had proceeded past the prototype stage into a state of war readiness in the form of the Chain Home air defense. Germany had technically the best radar designs, but the Wehrmacht intended to wage a war of aggression and initially gave little support to a technology whose strength lay overwhelmingly in defense. In the U.S., because of the contentious battleship–bomber disputes of the 1920s, the Navy had pressed for any new technical method to defend ships against air attack, and the Army had sought to perfect its anti-aircraft artillery with methods of combating bombers at night.

Radar was not a factor during the opening months of the war except to provide Britain her narrow margin of victory in defeating the Luftwaffe's attempt to destroy the Royal Air Force. The Germans then resorted to night attacks on cities, intending to damage industrial production and intimidate the population. The Chain Home radars were ineffective for tracking planes over land, and the attacking formations met with impotent anti-aircraft artillery. This led to the introduction of new radar techniques in early 1941. A 1.5-meter radar formed a "searchlight" beam that could follow aircraft overland, allowing the bomber and the pursuing fighter to be tracked close enough for the fighter pilot to observe it, visually at first and later with airborne radar. By May 1941 this method began to cause serious losses to the bombing fleets.

When the air war began to flow toward Europe, Germany built a system of radar-based fighter and anti-aircraft gun defenses that caused serious losses to RAF Bomber Command at night and the U.S. Army Air Forces during the day. In this they were greatly helped by the 50-centimeter equipment provided by the Telefunken Company for anti-aircraft guns and night fighters. Meterwave equipment with large arrays of dipoles were able to locate the attackers at ranges of 250 kilometers or more. From 1943 to 1945 the contending forces fought a highly technical electronic war in which new techniques were quickly met with countermeasures.

Britain developed airborne radar for use by the Fleet Air Arm that operated at 1.5-meter wavelength. This set, ASV Mk II, proved to be one of the most powerful air weapons in the war at sea. Mounted on the sturdy carrier biplane Swordfish, it located the battleship *Bismarck*, leading to her destruction. More importantly it located the Italian convoys carrying supplies to North Africa at night, resulting in their interdiction and in Rommel's defeat. The ASV Mk II, redesignated ASE for American service, became the U. S. Navy's eyes in the Pacific War when mounted on the long-range Catalina.

In September 1940 Britain sent a secret scientific technical mission to the U.S., called the Tizard Mission after its leader Henry Tizard. It opened Britain's technical war secrets, specifically radar, to the Americans on the assumption that the hoped-for future ally was lagging in this new field. It also opened up the vast American electronic industries for production, which was impossible if the technological details remained secret. American radar engineers found only two designs of interest: ASV and the resonant magnetron, a revolutionary new generator for waves of a few centimeters (microwaves). To exploit the latter a completely new research group was established at the Massachusetts Institute of Technology—the Radiation Laboratory.

A wavelength one-tenth of that of the 1-meter-wave sets previously in use allowed targets to be located on the surface of the earth, a difficult task for longer wavelength sets because of the confusion caused by surface reflections. This led to airborne sets that provided a map-like presentation of the ground, a development that was hoped might permit bombing German cities at night with some semblance of accuracy. Despite the technical sophistication of the equipment, this hope was not fulfilled, other than that the city was hit rather than the surrounding countryside. Much more useful was the installation of microwave sets for sea search aboard ships. The strong initial motivation for this was in the struggle with submarines, but it was soon found that the device was most useful in navigation. Ships could enter harbors in thick fog, hold their position in blacked-out convoys, and approach unmapped Pacific islands with greater safety.

Decimeter waves had been applied from the beginning of the war for directing fire against ships and aircraft, but microwaves improved the accuracy of such sets to a high degree. The most remarkable radar of this kind was the American SCR-584, a 10-centimeter, automatic tracking set intended primarily for directing anti-aircraft guns. This weapon was deployed in early summer 1944 in time to be used against the German V-1 flying bombs. When combined with an artillery fuse that sensed the presence of the target by its effect on a radio signal emitted by the fuse, anti-aircraft artillery became astonishingly effective. In the final days of the V-1 attacks, 95 percent of the bombs were brought down, mostly by anti-aircraft fire.

The entire nature of naval warfare in the Pacific was changed by radar. Most importantly, radar provided aircraft carriers with warning adequate to clear decks of planes, move bombs below, and flush petrol lines with carbon dioxide. When carriers were caught unaware with their thin decks over such explosive cargo, they simply blew up from a bomb that would not ordinarily have proved fatal. The Japanese fleet at Midway lost four carriers in this way. It was this extreme vulnerability that had given carriers a subservient role to heavily armored battleships in naval doctrine before radar's function became clear.

Japan did not deploy radar until mid-1942, generally in sets distinctly inferior to those of the Allies. However, Japan did invent the resonant magnetron independently, before the British in fact, and used it on occasion to the consternation of the Americans. Although the Soviet Union had also discovered the resonant magnetron in the late 1930s (dismissing it as of no importance and publishing details in open literature in 1940), it had only a few third-rate radars at the beginning of their struggle with Germany. Bombing of industrial targets by both sides did not figure in those gigantic land actions, so radar had little effect on the outcome. Germany used its superior radar to help maintain air superiority.

See also **Computers, Analog; Radar Aboard Aircraft; Radar, Long-Range Early Warning Systems; Radar, Origins to 1939**

LOUIS BROWN

Further Reading

Brown, L. *A Radar History of World War II*. Institute of Physics Publishing, Bristol, 1999.

Buderi, R. *The Invention that Changed the World*. Simon & Schuster, New York, 1996. A history of microwave radar.

Burns, R, Ed. *Radar Development to 1945*. Peter Peregrinus, London, 1988. A collection of 40 competently written articles.

Guerlac, H.E. *Radar in World War II*. Tomash, American Institute of Physics Publishers, New York, 1987. A history of the MIT Radiation Laboratory.

Howse, D. *Radar at Sea: The Royal Navy in World War 2*. Naval Institute Press, Annapolis, 1993.

Jones, R.V. *The Wizard War: British Scientific Intelligence, 1939–1945*. Coward, McCann & Geoghegan, New York, 1978. American edition of *Most Secret War*. Treats key elements of radar history.

von Kroge, H. *GEMA: Birthplace of German Radar and Sonar*. Institute of Physics Publishing, Bristol, 2000.

Lobanov, M.M. *Nachalo Sovetskoy Radiolokatsii* [The Beginnings of Soviet Radar]. Sovetskoye Radio, Moscow, 1975.

Nakagawa, Y. *Japanese Radar and Related Weapons of World War II*. Aegean Park Press, Laguna Hills, CA, 1997.

Price, A. *The History of US Electronic Warfare: The Years of Innovation—Beginnings to 1946*. Association of Old Crows, Washington D.C., 1984.

Reuter, F. *Funkmess: Die Entwicklung und Einsatz des RADAR-Verfahrens in Deutschland bis zum Ende des Zweiten Weltkrieges* [The development and deployment of radar in Germany to the end of World War II]. Westdeutscher Verlag GmbH, Opladen, 1971. Exhausts the material of the German archives.

Rowe, A.P. *One Story of Radar*. Cambridge University Press, Cambridge, 1948. Memoirs of the director of Air Ministry radar research.

Watson-Watt, R. *The Pulse of Radar: The Autobiography of Robert Watson-Watt*. Dial Press, New York, 1959. American edition of *Three Steps to Victory* by the inventor of Chain Home.

Radar, Displays

Those who first conceived of radar early in the century often envisioned systems that would simply indicate, perhaps by sounding a buzzer or lighting a lamp, that a target had been detected and where it was located. Those who first reduced radar to practice in the 1930s, however, were radio scientists who knew that the returning radar signals would somehow have to be distinguished against a background of radio-frequency interference and noise. They were accustomed to displaying signals visually on cathode ray tube (CRT) oscilloscopes, and they naturally turned to such means for radar. This made the operator an essential part of the radar system, responsible for the final stages of the detection process and extraction of target data.

The CRT was an invention of the late nineteenth century (Karl Ferdinand Braun, 1897) that would become familiar to people in the late twentieth century in the form of television display tubes and computer monitor screens (see Computer Displays). The CRT was typically housed in an evacuated glass envelope roughly conical, or funnel-shaped, in form. Near the apex or neck is the cathode, a source of electrons. By means of magnetic or electrostatic fields (either may be used, depending on application) the electrons are formed into a very narrow, tightly focused beam aimed toward the front or screen end of the tube, to which they are drawn by the anode. Additional sets of magnetic or electrostatic elements are provided to deflect the beam. Since the mass of the beam is very small and the forces relatively large, it can be moved from spot to spot anywhere on the screen at a rate fast enough to seem instantaneous to a human observer. The point at which the beam intersects the screen is made visible by coating the inside of the CRT face with a phosphor that emits photons on absorbing electrons.

The basic radar display was what came to be called an A-scope, which provided a plot of signal amplitude against range. On the basis of signal strength (resulting in a peak rising above the grass) and width, operators learned to distinguish radar returns from the "grass" (so called from its appearance on the A-scope display, see Figure 1)

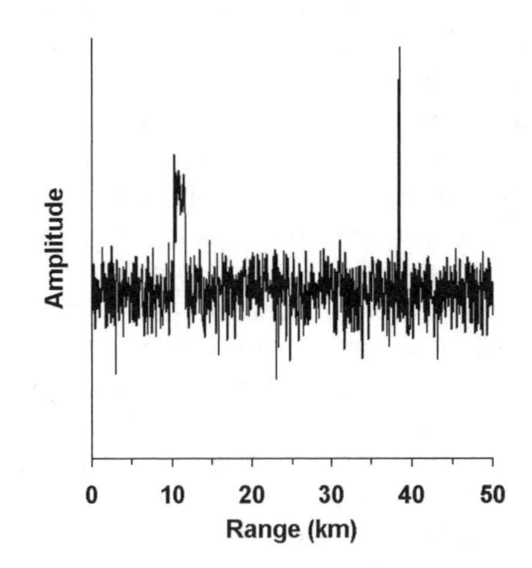

Figure 1. Typical A-scope trace showing noise or "grass" at all ranges, clutter return between ranges of 10.200 meters and 11,600 meters, and strong aircraft target return at range of 38,300 meters.

resulting from noise and from the "clutter" of interference and unwanted or spurious targets. The A-scope provided an immediate measurement of range, and the azimuth or elevation could be determined by noting the position of the antenna at the time of detection.

Although such a system worked, it was of course cumbersome and slow. The 1940 British invention of the plan position indicator (PPI) provided a far more practical display for search and general orientation purposes. On a PPI display, range was proportional to distance from the center point, azimuth was directly translated to azimuth from an index on the CRT face, and signal strength was displayed by modulation of the intensity of the beam. By reducing sensitivity or gain to a level that caused most of the noise background to disappear, targets could be seen as bright spots. Range and azimuth could be read directly from the display; and by marking the point on the screen at which the target appeared during successive sweeps of the radar antenna and connecting the dots using grease pencil, the target's track relative to the radar position could be displayed directly on the face of the PPI. This greatly improved the efficiency of radar as an aid to directing fighters to intercept bombers, as well as for navigating ships in order to approach targets and avoid collisions. The A-scope and a variety of other specialized displays remained valuable for particular uses, however. A-scopes gave more precise range determination and discrimination between targets and clutter, for instance.

Radar displays improved as a result of development of better focusing and deflection systems for CRTs as well as superior phosphors. More fundamental, however, were improvements in the electronic processing of the signals that were input to the display. Physical considerations as well as careful measurements provided information about electronic characteristics that could be used to distinguish wanted signals from unwanted noise and clutter. A simple example is the case in which the target of interest is expected to move toward or away from the radar at a velocity that differs from that of clutter and interfering sources. If the radar is able to measure the Doppler frequency shift—proportional to velocity—then processing can suppress returns with the "wrong" shift values, resulting in great improvements in the ratio of signal to interference and noise ($S/(I+N)$) on the display. Improvements in knowledge of signal, interference, and noise characteristics as well as advances in electronics brought about progress in displays.

From 1975, digital technology had an increasing role in radar displays. As digital circuit speeds increased, approaching and even exceeding the radio frequencies of radars, it became possible to digitize more and more of the radar's processing chain, allowing for far more complex computations. In many cases, the radar's computer calculated the most likely estimated target positions and characteristics and presented a "synthetic" picture of this on a display that was really a computer monitor. When noise and interference levels were especially high, the computer might be instructed to try to form coherent tracks of successive observations before displaying anything at all, so that the first thing the operator saw was a track already extending seconds or even minutes into the past.

Owing to the increasing dominance of computers, displays largely disappeared as a separate topic in texts on radar after about 1980. Thus over the course of the century, radar displays in a sense came full circle, returning to the original concept of a system that simply reported the desired target data and left all else aside. Where maximum discrimination is needed, there often remained a role for operators in examining and evaluating data, but only after the data had first been subjected to intensive computer processing. As in many other fields of technology, by the end of the twentieth century, computers usually stood between humans and their radars.

See also **Computer Displays; Radar, Origins to 1939**

WILLIAM D. O'NEIL

Further reading

Berg, A.A. Radar indicators and displays, in *Radar Handbook*, 1st edn., Skolnik, M.I., Ed. McGraw-Hill, New York, 1970.

Ridenour, L.N., *et al.* The gathering and presentation of radar data, in *Radar System Engineering*, Louis N. Ridenour, Ed. McGraw-Hill, New York, 1947.

Skolnik, M.I. *Introduction to Radar Systems*, 3rd edn. McGraw-Hill, New York, 2000.

Radar, High Frequency and High Power

High-Power Microwave Radar

While early radar designers were driven to frequencies of more than 1000 megahertz by considerations of the availability of high-power components, it was appreciated very early on that higher frequencies and thus shorter wavelengths would allow better precision. Frequency and wavelength are inversely related according to the equation

$$\text{Wavelength} = c/\text{frequency}$$

where c = velocity of light = 3×10^8 meters per second.

The need for higher frequencies became acute with sets carried by aircraft, where practical antenna dimensions gave very inadequate accuracies at the only available (i.e., low) frequencies that were possible until 1940.

This need was met through one of the most important of radar technology breakthroughs, the 1939 U.K. development of a high-power version of the earlier magnetron. This "cavity magnetron" was a thermionic valve (vacuum tube) in which an arrangement of tuned cavities in the anode structure modulated electron motion through their effect on magnetic field geometry. It was quickly applied to radars operating at frequencies of around 3 gigahertz, rising to around 9 gigahertz later in World War II. Because the wavelengths of these radars were so much shorter than those of earlier sets (approximately 3 to 10 centimeters versus 1 meter), they were termed "microwave" radars. The compactness and accuracy of powerful microwave radars, an achievement not duplicated by the German and Japanese enemies, played a significant role in the Allied victory. By the end of the war, peak powers of more than 1 megawatt were available at a frequency of 3 gigahertz.

Microwave radar became the dominant type for the majority of applications after 1945, employing a variety of forms of power tubes. Microwave frequencies generally were not used, however, in applications requiring very long range where the higher power and lesser atmospheric attenuation available at lower frequencies made them preferable. Antennas were often space-fed parabolic reflectors, producing a flat wave front from a single point source of radiation.

An important post-World War II development was the "monopulse" technique for precise tracking. This normally involved the use of multiple feeds with a common reflector to form several separate beams looking in slightly different directions. By comparing the amplitudes (or occasionally the phases) of the signals on different beams, the radar was able to calculate the precise azimuth and elevation of the target on the basis of a single pulse detection.

From the 1950s, phase-coherent microwave radars were progressively developed to permit detection and tracking of small moving targets against strong backgrounds of stationary clutter on the basis of the Doppler shift of the frequency of the received signal as compared with that transmitted. The first coherent radars were generally at lower frequencies and used continuous-wave rather than pulsed signals. As technology developed, the technique was extended to microwave and pulsed radars. Depending on the pulse repetition frequency employed, either the range or Doppler velocity information would be ambiguous; the choice would thus depend on the application's needs. By the 1960s, pulse Doppler microwave radars were beginning to be used in airborne service.

The 1970s brought the first phased-array microwave radars, allowing a single antenna to look in multiple directions simultaneously or in very quick succession. This involved hundreds or thousands of separate antenna elements, each separately fed by a signal whose phase could be independently varied in a systematic manner. Phased arrays were particularly attractive in applications where a computer steered the beams to track many targets at once. By the 1980s, the power outputs possible with microwave solid-state devices began to make it feasible to consider them as transmitter elements for phased arrays in place of elements fed from a central power tube through intervening phase-shifters. Even at the end of the twentieth century, however, microwave solid-state radars were in limited use owing to high costs of early generation components.

The 1990s saw the introduction of microwave active electronically scanned array (AESA) radars. In these, each transmitter element was coupled on the same electronic module or even the same chip, with a complete receiving element that detected and digitized its portion of the incoming signal, thus allowing great flexibility and precision of operation. Again, component costs slowed its acceptance.

High-Frequency Over-the-Horizon Radar

Radars operating in the high-frequency (HF) band (3 to 30 megahertz) may detect targets well beyond the nominal horizon through two mechanisms: "sky wave" and "surface wave." Early in the century, it was discovered that high-frequency radio waves were strongly refracted by the ionosphere. A HF beam aimed near the horizon would, under suitable conditions, be effectively reflected, returning to sea level some hundreds to thousands of kilometers from its transmission site. From the 1940s, interest developed in using this sky-wave transmission phenomenon to provide surveillance at great ranges. Early HF over-the-horizon radars (OTHRs) were bistatic "forward scatter" systems in which a widely separated transmitter and receiver detected and tracked targets lying between them. Ballistic missile tracking was a major application.

By the 1980s technology had advanced sufficiently to support development of "backscatter" OTHRs having transmitter and receiver relatively close together to provide surveillance at distance. Typical transmitter arrays covered a line hundreds of meters long and broadcast at an average power of hundreds of megawatts. Receive arrays generally were more than 1 kilometer in length. Using Doppler processing to distinguish moving targets from stationary clutter, OTHRs searched annular sectors typically around 60 degrees in width, with a depth of several hundred kilometers, at ranges which varied (with frequency and ionospheric conditions) from about 500 to 3500 kilometers. Spatial accuracy was coarse, but Doppler velocities could be distinguished finely. Propagation of signals was affected by ionospheric conditions and could be disrupted completely at some periods. Reliability of operation was a function of location and orientation, being worst for systems in high latitudes looking toward the pole, where auroral phenomena affect the ionosphere (see Figure 2).

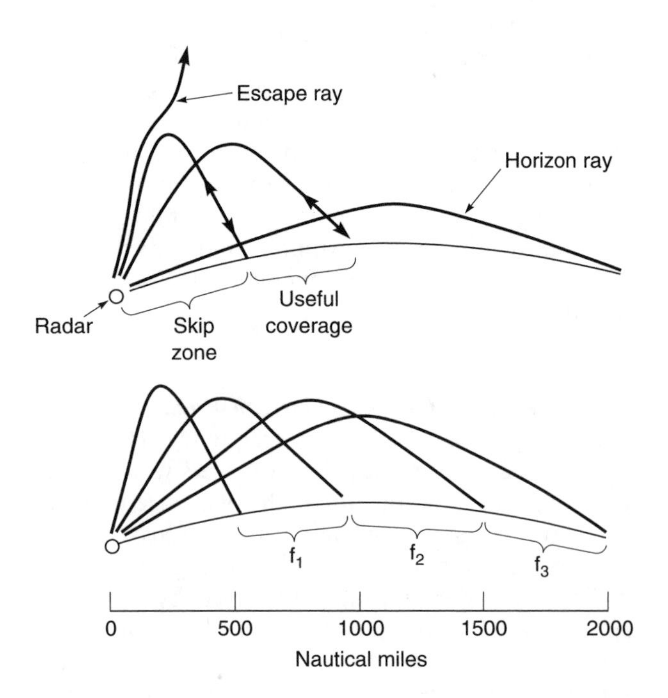

Figure 2. Vertical paths for sky-wave over-the-horizon radar (OTHR). Different frequencies, f_1, f_2, and f_3, cover different range intervals. (Ranges shown in nautical miles = 1852 meters; vertical scale exaggerated for clarity.)

Surface-wave high-frequency OTHRs saw less use. They depended on the diffraction of vertically polarized HF waves around the curve of the earth, especially over the smooth conductive surface of the sea. With high powers and Doppler processing, such radars could detect sea-level targets at ranges up to several hundred kilometers. OTHR applications included surveillance of aircraft, missiles, and ships for military purposes and control of smuggling as well as measurement of phenomena associated with ocean waves.

See also **Radar Aboard Aircraft; Radar, Origins to 1939; Radar, Defensive Systems in World War II**

WILLIAM D. O'NEIL

Further Reading

Boyle, D. OTH radar clinging to the surface. *Int. Defense Rev.*, 4, 501–503, 1989.

Brookner, E., Ed. *Radar Technology*. Artech, Dedham, MA, 1977.

Ferraro, E. and Ganter, D. Cold War to counter drug. *Microwave J.*, March, 82–92, 1998.

Foxwell, D. Beyond the horizon, but not out of sight. *Jane's Int. Defense Rev.*, 8, 48–51, 1997.

Guerlac, H.E. Radar in World War II, in *History of Modern Physics*, vol. 8. Tomash, 1987.

Headrick, J.M. HF over-the-horizon radar, in *Radar Handbook*, 2nd edn., Skolnik, M.I., Ed. McGraw-Hill, New York, 1990.

Headrick, J.M. Looking over the horizon. *IEEE Spectrum*, July, 36–39, 1990.

Headrick, J.M. and Thomason, J.F. Applications of high frequency radar. *Radio Science*, 33, 1045–1054, 1998.

Klass, P.J. HF radar detects Soviet ICBMs. *Aviation Week and Space Technology*, December 6, 38–40, 1971.

Kolosov, A.A., *et al. Over-the-Horizon Radar*. Artech House, Boston, 1987.

Ridenour, L.N., Ed. *Radar System Engineering*. McGraw-Hill, New York, 1947.

Skolnik, M.I. *Introduction to Radar Systems*, 3rd edn. McGraw-Hill, New York, 2000.

Radar, Long-Range Early Warning Systems

During the 1930s, Great Britain was one of several countries, including most notably Germany and the U.S. that experimented with radar for early warning of air attacks. The British "Chain Home" system, designed by Sir Robert Watson-Watt and established by 1939, included a string of stations along the east and south coasts. By mid-1940, most of the stations featured two 73-meter wooden towers, one holding fixed transmitter aerials and the other receivers. When it was discovered that low-flying aircraft could slip undetected beneath the original fence, Britain created a second string of "Chain Home Low" stations, beginning with

Truleigh Hill. The latter sites consisted of two separate aerials, one to transmit and the other to receive, mounted on 6-meter-high gantries and short enough to allow an operator inside the equipment hut beneath the gantry to manually rotate the arrays. Together, Chain Home and Chain Home Low provided a detection range of 40 to 190 kilometers depending on an incoming aircraft's altitude. This early warning capability contributed immeasurably to the RAF victory over the Luftwaffe in the Battle of Britain.

American forces failed to make equally beneficial use of radar to warn of the Japanese attack on Pearl Harbor on December 7, 1941. During late 1941, U.S. Army personnel in Hawaii had been field-testing SCR-270-B mobile units capable of detecting aircraft more than 160 kilometers away. Unfortunately, the system was undermanned by relatively inexperienced people. An hour prior to the December 7 attack, three of the five operational radar sites—Kaaawa, Opana, and Kawailoa—saw large reflections approximately 220 kilometers north of Oahu. All locations reported these sightings to the Information Center, where plotters put the data on a master board. Nobody at the Information Center interpreted the plots as unusual. The Opana unit continued to track the incoming target for more than 30 minutes, until ground clutter interfered with the signal just 32 kilometers from the coast. By that time, no one was on duty at the Information Center to plot the data. Not until several days later did the Americans realize that their radar stations had accurately detected the approach of the Japanese aerial armada, which might have enabled the Army, Air Force, and Navy aircraft to respond earlier and thereby limit the disaster.

Germany, Japan, and the Soviet Union also employed early warning radar during World War II with varying degrees of success. A network of Freya radars known as the "Kammhuber Line" enabled the Germans to detect approaching British and American bombers at a range of 95 kilometers and, when augmented by Würzburg radars, proved especially effective in determining exact range and height as the aircraft came closer. Japan relied on Mark-1 and Tachi-6 units—with ranges of 120 and 200 kilometers, respectively—to alert the home islands of impending B-29 raids late in the war. In early 1942, the Soviet Union installed RUS-2 radars, with a range of 95 to 145 kilometers, to aid locally in the defense of Moscow and Leningrad. While these various radar applications for early warning probably had little influence on the course of World War II, they were seedlings from which greater capabilities would spring.

The use of long-range, early-warning radar equipment for air defense grew from these roots to blossom into complex, integrated systems during the Cold War era. Since the most direct route by which Soviet aircraft could attack North America was across the polar cap, the U.S. and Canada cooperated during the 1950s to orient a multi-layered early warning capability in that direction. In 1951, they began construction of the CADIN/Pinetree line, a series of more than 30 stations situated approximately along the U.S.–Canada border. To this was added the distant early warning (DEW) line in 1956—a string of as many as 70 sites stretching along the Arctic Circle from Alaska to Greenland. In 1958 the mid-Canada line—a Doppler electronic fence containing as many as eight sector stations and 90 unmanned sites along the fifty-fifth parallel—completed the original system (Figure 3). Upgraded over the years, the entire system was eventually replaced in 1985 by the north warning system, which was comprised of 13 long-range sites and 39 shorter-range stations. In 2001, the U.S. Air Force contracted with Lockheed Martin Corporation to upgrade operating software for long-range, atmospheric early warning radar systems at 33 locations by August 2006.

Concerned that the U.S. needed even earlier detection of incoming Soviet bombers or low-flying cruise missiles, the U.S. Air Force had contracted with General Electric Aerospace to begin building a prototype over-the-horizon-backscatter (OTH-B) radar in Maine during the mid-1970s. That system, designated AN/FPS-118, achieved limited operational capability in 1988. By bouncing signals off the ionosphere, the powerful OTH-B radar could locate and track targets thousands of miles from U.S. air space. The U.S. Air Force accepted both the East Coast and West Coast OTH-B systems from the contractor in 1990, but the end of the Cold War and a desire on the part of Congress to cut operating costs soon led to shutdown of the western portion. In 1994, Congress directed that the eastern system be used for counter-narcotics detection as well as for monitoring weather and the ocean environment. Three years later, that system was also mothballed.

To counter the threat of surprise intercontinental ballistic missile (ICBM) attacks by the Soviet Union, the U.S.—as a matter of the highest national priority—began construction of the ballistic missile early warning system (BMEWS) in 1959. Otherwise known as the 474L system, it

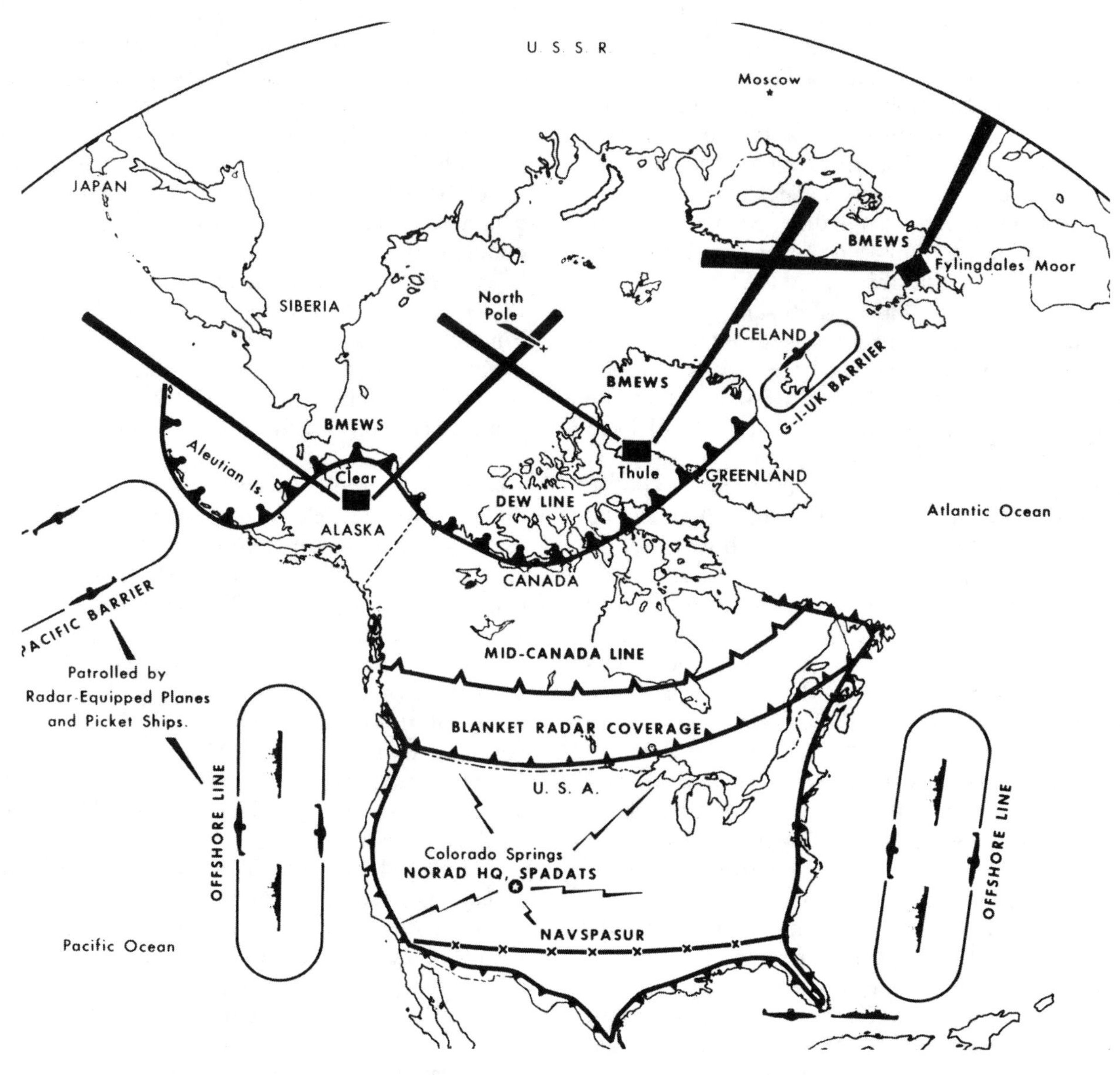

Figure 3. Location of U.S. early warning sensors, circa 1962. [*Source: Courtesy of U.S. Air Force.*]

included three sites: Thule Air Base, Greenland (1960); Clear Air Force Station (AFS), Alaska (1961); and RAF Fylingdales Moor, England (1963). The original BMEWS equipment consisted of two principal items: the AN/FPS-50 detection radar, which included three fixed antennas, each 50 meters high and 120 meters long; and the AN/FPS-92 tracking radar, a mechanical antenna 25 meters in diameter. During the 1970s, the increased threat of a massive sea-launched ballistic missile (SLBM) attack caused the U.S. to supplement BMEWS coverage with a long-range, solid-state phased-array radar (SSPAR) system called PAVE PAWS. Two of the latter, designated AN/FPS-115 and possessing a detection range of 3000 nautical miles, became operational in 1980 at Otis Air Force Base (AFB) in Massachusetts and Beale AFB in California. Others were built later at Robins AFB in Georgia and Eldorado AFS in Texas, but they were shut down in the 1990s as a cost-saving measure. Between 1985 and 2001, the U.S. also replaced the original BMEWS equipment with SSPAR technology.

During the Cold War, the USSR maintained an even more extensive early warning system for both ballistic missile and air defense. By 1985, the Soviets had deployed over 7000 air surveillance radars of various types, including over-the-horizon

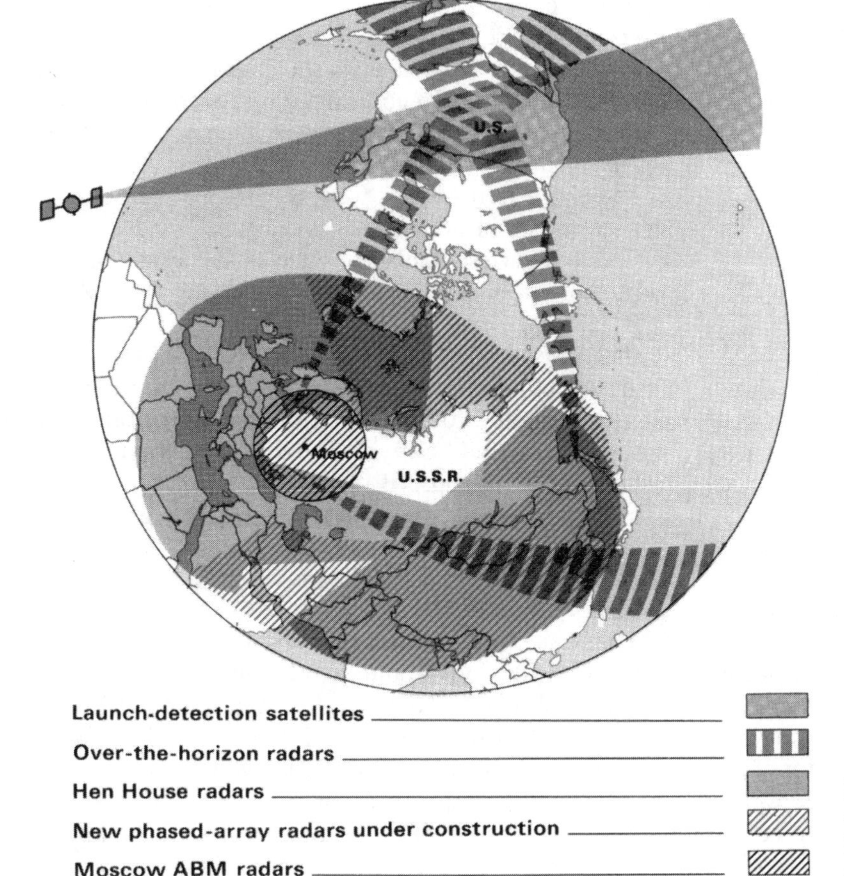

Figure 4. Location of USSR ballistic missile and tracking systems in 1985. [*Source: Courtesy of the U.S. Department of Defense.*]

models, at approximately 1200 locations, which gave them coverage at medium to high altitudes over all their territory and in some areas for hundreds of kilometers beyond their borders (Figure 4). They were also deploying new systems for early warning of cruise missile and bomber attacks. For ballistic missile warning, the USSR operated eleven large "Hen House" detection and tracking radars at six peripheral locations, as well as new phased array equipment. With the USSR's collapse in the early 1990s, however, the missile warning radar system became too fiscally burdensome. In 2001, Russia announced its intention to scrap many of the fixed installations and rely on mobile stations.

See also **Radar, Defensive Systems in World War II; Radar, High Frequency and High Power; Radar, Origins to 1939**

RICK W. STURDEVANT

Further Reading

Brown, L. *A Radar History of World War II: Technical and Military Imperatives.* Institute of Physics Publishing, Bristol, 1999.

Clark, D.L. Early advances in radar technology for aircraft detection. *Lincoln Lab. J.*, 12, 167–179, 2000.

Guerlac, H.E., Ed. *Radar in World War II*, 2 vols. Tomash, American Institute of Physics Publishers, New York, 1987.

Jockel, J.T. *No Boundaries Upstairs: Canada, the United States, and the Origins of North American Air Defence*, 1945–1958. University of British Columbia Press, Vancouver, 1987.

Nakagawa, Y. *Japanese Radar and Related Weapons of World War II.* Aegean Park Press, Laguna Hills, CA, 1998.

Schaffel, K. *The Emerging Shield: The Air Force and the Evolution of Continental Air Defense, 1945–1960.* Office of Air Force History, Washington D.C., 1991.

Stone, M.L. and Banner, G.P. Radars for the detection and tracking of ballistic missiles, satellites, and planets. *Lincoln Lab. J.*, 12 217–243, 2000.

U.S Department of Defense. *Soviet Military Power 1985.* Government Printing Office, Washington D.C., 1985.

Winkler, D.F. *Searching the Skies: The Legacy of the United States Cold War Defense Radar Program.* U.S. Air Force Air Combat Command, Langley AFB, June 1997.

Useful Websites

Belmont, L. *Radar in WWII*: http://www.strandlab.com/radar/

Canadian Minister of Public Works and Government Services. *Post World War II Radar*: http://www.dnd.ca/commelec/brhistory/anxf_e.htm
Federation of American Scientists, AN/FPS-115 PAVE PAWS Radar: http://www.fas.org/spp/military/program/track/pavepaws.htm
McManus, G.P. *BMEWS—510 Full Days*: http://www.bwcinet.com/thule/index.html
Toon, T.H. *World War Two—The Chain Home Low Radar*: http://schoolsite.edex.net.uk/468/radar.htm

Radar, Origins to 1939

Reflection was an important part of Heinrich Hertz's 1887 demonstration of the existence of electromagnetic waves, and the idea of using that property to "see" in darkness or fog was developed shortly afterwards.

Christian Hülsmeyer constructed a device in 1902 that he hoped might prevent collisions at sea. It used two cylindrical paraboloid reflectors to transmit and receive waves of decimeter length generated from spark oscillations with the reflected signals detected by coherer. The equipment was demonstrated successfully at a conference of ship owners in Rotterdam, but it was only capable of showing the direction of an object and not its range, as this required accurate timing of signals at the microsecond level, a technique that lay years in the future. Thus a liner eight kilometers distant was indistinguishable from a tug at 500 meters, and the remarkable device aroused no enthusiasm among seamen. Hülsmeyer's set was soon forgotten, although it was adequately patented and demonstrated to numerous witnesses. The idea recurred to no effect in World War I but seems to have been discussed informally among engineers. Guglielmo Marconi proposed using reflected radio waves for the location of objects in a paper delivered at a meeting of the Institute of Radio Engineers in New York in 1922.

By 1920 the vacuum triode had revolutionized the generation and detection of radio waves, and broadcast radio was transforming many aspects of everyday life. However the circuit elements needed to measure the time between the emission and reception of a pulsed train of waves—the key to radar ranging—were still missing. Triodes were able to work at high frequencies, and in the early 1930s various experimenters built equipment capable of measuring the speed of an automobile, if not its range. If the target were moving, the reflected wave would have its frequency shifted by the Doppler effect and through interference with the transmitted signal within the receiver would produce an easily recognizable signal, the beat frequency of which was proportional to the speed. Police radar was therefore possible before air-warning radar. A revival of the Hülsmeyer idea returned in 1935 in equipment designed by Camille Gutton and mounted in the new transatlantic liner *Normandie*. The equipment did not do well at sea, and the watch officers were not impressed because there was still no range information.

Triodes were unable to follow the rapid changes in signal amplitude that characterize a short wave train. Furthermore, operators required a device that presented the time elapsed between emission and reception. It was recognized that a cathode-ray oscillograph would perform this function, but those available before 1930 were inadequate for a variety of reasons. Both of these functions were also vital for television of sufficiently high definition to rival the cinema, and both were the subject of research in the electronic industry, which demonstrated all-electronic television in 1930. The necessary elements were multigrid amplifier valves and high-vacuum, low-voltage cathode-ray tubes. When these two circuit elements became available for the video amplifier and for the picture tube, respectively, serious radar work could begin.

The radar sets first envisioned were to use wavelengths of a few centimeters, which allowed the beam to be shaped into a form of "radio searchlight" with a reflecting dish of practical size. This approach faced a serious obstacle: the absence of any generator working at these short wavelengths with sufficient power or frequency stability. There were, however, many observations (through the Doppler effect) of aircraft and ships seen at wavelengths of a few meters, an effect that was particularly pronounced in experiments studying the propagation of waves intended for the transmission of television. At these frequencies, antennas of manageable size could be fashioned from arrays of dipole radiators so that the individual radiations would constructively interfere to form the desired radio searchlight.

By the early 1930s, serious efforts were underway in the U.S., Germany, and Britain to construct radio-location devices using relatively long wavelengths. (Russian efforts were ahead in the early 1930s, but they yielded little as a result of serious organizational problems and purges that sent key engineers to the gulag.) The German company GEMA built the first device that can be called a functioning radar set in 1935 with Britain and America following only months behind. Two groups in the U.S.—the Signal Corps and the Naval Research Laboratories—proceeded independently but on lines very similar to those of the

Germans in using dipole arrays. They had air-warning and searchlight-pointing prototype sets ready for production in 1939.

The British physicists Robert Watson Watt and Arnold Wilkins proceeded along a different line using wavelengths of tens of meters with broadcast rather than "searchlight" transmission. This equipment, although inferior to that working on shorter wavelengths, was seen by Air Vice-Marshal Hugh Dowding as the key to the air defense of Britain from expected German attack. As commander of the newly created Fighter Command, he created a system of radar stations and ground observers linked by secure telephone lines to the fighter units. He drilled Fighter Command to use the new technique, and when the Luftwaffe came in the summer of 1940, the attacking squadrons were ambushed by defending fighters positioned by radar.

Triodes capable of generating significant amounts of power at decimeter wavelengths had been developed by the late 1930s, and by 1940,the Bell Laboratories in the U.S., the Royal Navy in the U.K., and the Telefunken Company in Germany had sets that worked at 50 centimeters.

All these designs went into production with the onset of World War II and furnished, with various modifications, the radar used by the combatants until 1943, when centimetric-wave equipment was developed, which used electronics of a completely different nature.

See also **Radar, Defensive Systems in World War II; Radar, Long-Range Early Warning Systems**

LOUIS BROWN

Further Reading

Allison, D.K. *New Eye for the Navy: The Origin of Radar at the Naval Research Laboratory*. NRL Report 8466, U.S. Government Printing Office, Washington D.C., 1981.

Brown, L. *A Radar History of World War II: Technical and Military Imperatives*. Institute of Physics Publishing, Bristol, 1999.

Burns, R.W. *Television: An International History of the Formative Years*. Institution of Electrical Engineers, London, 1998.

Guerlac, H. The radio background of radar. *J. Franklin Inst.*, 250, 285–308, 1950.

Jones, L.F. A study of the propagation of wavelengths between three and eight meters. *Proc. Inst. Radio Eng.*, 21, 349–386, 1933. The remarkable behavior of waves later to be used for radar.

von Kroge, H. *GEMA: Birthplace of German Radar and Sonar*. Institute of Physics Publishing, Bristol, 2000.

Nakagawa, Y. *Japanese Radar and Related Weapons of World War II*. Aegean Park Press, Laguna Hills, CA, 1997.

Swords, S.S. *Technical History of the Beginnings of Radar*. Peter Peregrinus, London, 1986.

Taylor, A.H. *Radio Reminiscences: A Half Century*. U.S. Naval Research Laboratory, Washington D.C., 1948, 1960.

Watson-Watt, R. *The Pulse of Radar: The Autobiography of Robert Watson-Watt*. New York: Dial Press, 1959. American edition of Three Steps to Victory.

Radio: AM, FM, Analog, Digital

The term "radio" includes many different modes of wireless transmission. In the late nineteenth and early twentieth century, wireless telegraphy used spark-gap technology and intermittent waves to transmit Morse signals. Only with the development of continuous wave transmission in the early twentieth century did wireless telephony become possible, allowing effective transmission of the human voice and music. All wireless or radio signals consist of a carrier wave (a continuous wave, altered by amplitude or frequency, to which is attached the intended voice or music intelligence being transmitted) and one or more sidebands (the band of frequencies produced by modulation). The intended information or content of a given signal is carried on one of the modulated sidebands. While only analog amplitude modulation (AM) was used until about 1940, frequency modulation (FM) became increasingly important after that date. Digital transmission developed beginning in the 1980s.

AM

Amplitude modulation indicates that the strength of the sideband signal is modulated (several thousand times per second) in accordance with the amplitude or strength of the carrier wave. Broadcast radio, which was developed in the 1920s, was assigned to medium-wave frequencies in most nations. In the U.S., AM radio stations are assigned to 10-kilohertz channels; in much of the rest of the world, AM or medium-wave radio uses 9-kilohertz channels. While AM stations use narrower channels (and thus less frequency space) than FM stations, AM is prone to natural and most man-made electrical interference, or noise, which cannot be separated from the desired signal. Attempts to overcome static using more transmitter power failed. In part due to their narrow bandwidth and crowding on the AM band, signal response or "sound quality" is much poorer for AM (up to about 5 cycles per second) than FM, which regularly provides a signal response of up to 15 cycles per second due to its greater bandwidth.

Because they are located on medium-wave frequencies in most nations (540 to 1705 kilohertz in the U.S.), AM broadcast stations utilize ground wave propagation during daylight hours and sky wave propagation at night. Sky wave propagation, in which signals are bounced back to earth from the ionosphere, can carry a signal hundreds of kilometers, especially on cold, clear nights, though not in a predictable fashion. Station coverage or "reach" therefore varies by time of day and season.

FM

Frequency modulation signals vary by a swing of frequency rather than power output within the assigned channel. Edwin Howard Armstrong found that using a channel 20 times wider than an AM channel (200 kilohertz) would allow an analog signal with excellent frequency response (up to 15,000 cycles per second) that could avoid atmospheric interference (e.g., static from electrical storms) and much man-made interference as well. An FM signal needs to be only twice as strong as a more distant competing transmitter to suppress the interfering signal. Utilizing very high-frequency (VHF, or VHF radio outside of the U.S.), FM signals are propagated by direct line-of-sight means day or night, limiting transmitter coverage to a radius of no more than 95 to 110 kilometers depending on local terrain, but eliminating multipath AM interference.

FM radio was developed from 1928 to 1933 by Edwin Armstrong, a professor at Columbia University in New York and prolific radio inventor. Armstrong fought many patent battles, as did most early American radio inventors. Two were very important and lasted for years: the fight with inventor Lee de Forest from 1914 to a Supreme Court decision two decades later over the rights to the regenerative circuit, which he lost; and the battle over his basic FM patents, fought against RCA and only settled after his death in 1954. After considerable experimentation by Armstrong and others, FM was introduced as a broadcasting service in the U.S. in 1941, on 42–50 megahertz. About 50 stations were on the air before a wartime freeze was imposed on new construction in 1942. Television standards approved in 1941 required FM for the sound portion of the signal. After extensive research and hearings, the Federal Communications Commission (FCC) in early 1945 reallocated FM to its present 88–108 megahertz range, thus providing more channels but at a cost to the medium of "starting over" in the face of television and revived AM competition.

In 1955 the FCC allowed FM stations to multiplex an additional nonbroadcast signal as a means of generating revenue. The service usually provided background music to offices and stores. Six years later, stereo multiplex FM transmission standards were approved for commercial operation in the U.S. The National Stereophonic Radio Committee has been formed by the radio manufacturing industry in 1959 to test various proposals. The system developed by General Electric was recommended to and approved by the FCC in early 1961. Stereo transmission is downward compatible, meaning it allows monaural reception in radios not equipped with stereo reception.

Digital

AM and FM transmissions are analog signals, subject to an inherent background electrical noise (or "hiss"), although FM suffers less than AM. Most consumers were introduced to digital sound and its superior signal-to-noise ratio with the success of the digital compact disc in the 1980s, and began to seek similar quality over the air.

By the late 1990s, most American and many international radio stations used the Internet to stream their signals, making worldwide reception possible. Audio streaming quality of sound was most often limited by the computer speakers used.

Based on research that began in the early 1980s, digital audio broadcasting (DAB) became operational in Europe and Canada, which in the 1990s agreed to use the "Eureka 147" technical standard, allowing transmission of digital-quality sound as well as information and data. Depending on the country, stations operated on Band III (around 221 megahertz) or L-band (1452 to 1492 megahertz) frequencies, well above current FM/VHF frequencies. L-band was allocated for digital radio at the World Administrative Radio Conference in 1992. The first commercial DAB receivers became available in 1998; and terrestrial DAB service was available by late 2001 to nearly 300 million people from more than 400 digital radio stations, with plans for satellite delivery in the future.

Declining to adopt Eureka 147, U.S. manufacturing companies formed the National Radio Systems Committee (NRSC) to determine which of a half-dozen schemes to recommend for approval by the FCC. To ease the eventual transition to digital radio from existing analog stations, the industry and FCC sought an in-band, on-channel (IBOC) scheme whereby the new service would operate during a transition period side-by-side the analog stations that it would eventually replace. However, agreement on a workable technical standard was continually delayed. In the meantime, digital audio radio service (DARS), transmitted from domestic satellites, was authorized by the FCC in the late 1990s. XM Satellite Radio began to offer its subscription-based 100-channel service in late 2001 while competitor Sirius planned to begin advertising-free transmissions by early 2002. Both required the purchase of special digital receivers.

See also **Radio, Early Transmissions; Radio Receivers, Valve and Transistor Circuits; Radio Transmitters, Continuous Wave**

CHRISTOPHER H. STERLING

Further Reading

NAB Engineering Handbook, 9th edn. National Association of Broadcasters, Washington D.C., 1999.

Inglis, A.F. *Behind the Tube: A History of Broadcasting Technology and Business*. Focal Press, Stoneham, MA, 1990.

Lessing, L. *Man of High Fidelity: Edwin Howard Armstrong*. Lippincott, Philadelphia, 1956.

McNicol, D. *Radio's Conquest of Space*. Murray Hill Books, New York, 1946. Reprinted by Arno Press, 1974.

Pawley, E. *BBC Engineering 1922–1972*. BBC, London, 1972.

Sterling, C.H. and Kittross, J.M. *Stay Tuned: A History of American Broadcasting*, 3rd edn. Lawrence Erlbaum Associates, Mahwah, NJ, 2002.

Useful Websites

Digital Radio: http://infoweb.magi.com/~moted/dr/

World Forum for Digital Audio Broadcasting: http://www.worlddab.org/

Radio, Early Transmissions

In the transition from experiments to regular broadcasting, important pioneering efforts in radio transmission took place before the beginning of World War II. Records conflict, and there is some controversy over the primacy of some "firsts." Some pioneers have likely been forgotten for lack of the promotion available to others. Most of the experimental precursors transmitted telegraphy code; only slowly did the ability to send speech and music through the air become possible.

Wireless Telegraphy

Any consideration of pioneering transmissions must begin with German physicist Heinrich Hertz who, in a series of experiments in late 1887 and early 1888, was the first to demonstrate that prior theorizing about wireless was indeed correct. He succeeded in transmitting telegraph code across a room (using what soon became known as "Hertzian waves") without wire connections. Though he was not interested in the commercial potential of his findings, their publication triggered research by many others.

Guglielmo Marconi learned of his work while reading Hertz's obituary in early 1894. Within a year, using Hertz's spark-gap technology with the important addition of an antenna wire, Marconi was transmitting telegraphy signals for several hundred meters on his father's estate near Bologna, Italy. By 1896 he was in England, demonstrating wireless transmissions over several kilometers to government officials. Transmissions across the English Channel followed on March 27, 1899. On December 12, 1901, Marconi and several assistants (see Figure 5) succeeded in transmitting the Morse code signal for the letter S (three dots) across the North Atlantic from Poldhu in Cornwall to Signal Hill in St. Johns, Newfoundland. This was the first public demonstration of the feasibility of long-range radio communications, though regular commercial transatlantic wireless telegraph services did not commence until October 17, 1907.

That wireless would play an important role at sea was acknowledged early on. When the East Goodwin lightship in the English Channel was struck during heavy fog in 1898, a wireless transmission from the vessel helped to save lives—the first of many such radio feats. Regular radio transmissions to and from naval vessels began with Royal Navy operations in 1900, the same year Marconi first equipped several German liners with wireless transmitters. After several years of experimentation, regular Marconi commercial wireless telegraph transmission to and from merchant shipping began in 1904. Other maritime rescues using wireless became common, including the January 23 1909 saving of more than 1500 people from the *S.S. Republic*. However, the April 14–15, 1912, *Titanic* disaster and the role of wireless in saving some 700 survivors captured the public imagination as to the benefits and potential of wireless transmissions and hastened developments. It also ensured adoption of regulations that required passenger vessels to have a radio operator on duty at all times.

In October 1899 the British Army first used wireless transmission during the Boer War in South Africa. Both the Russians and the Japanese made extensive use of wireless transmis-

Figure 5. Marconi and his assistants launching the kite-supported aerial at Signal Hill, St. John's, Newfoundland, December 1901. [*Courtesy of the Marconi Corporation.*]

sions, at sea and on land, during the brief but fierce Russo–Japanese War of 1904–1905. By World War I, wireless transmission was widely used by all combatants, though wireless was still supplementing wired telegraph and telephone connections.

Broadcasting: 1906 to 1941

By connecting a telephone to a high-frequency alternator of his design, Reginald Fessenden became the first to transmit speech and music on Christmas Day and again on New Year's Eve of 1906, from Brant Rock, Massachusetts south of Boston. His success (which followed wireless telephony experiments dating back to 1902) with an "audience" of shipboard radio operators was widely publicized at the time and was among the first indications that wireless might move beyond telegraph signals. In the next few years, there were many other one-time broadcast demonstrations. Audion inventor Lee de Forest transmitted speech and music in 1908 from the Eiffel Tower in Paris and the voice of Enrico Caruso from New York's Metropolitan Opera on January 13, 1910. The first transmissions to and from an airplane in flight took place in both Britain and the U.S. in 1910; the first airliner equipped with radio flew the London to Paris route a decade later. On the West Coast of the U.S., Charles D. Herrold began occasional, and soon regularly scheduled, voice and music broadcasts from a transmitter at his radio school in San Jose, California. The outbreak of World War I in 1914 put an end to all of this experimentation.

Radio broadcasting resumed only after hostilities concluded in late 1918. During 1919 and 1920, initial and occasional broadcast transmissions of speech and music emanated from stations in The Netherlands, Canada, Britain, and the U.S., most of them initiated by amateur radio operators. Marconi began daily broadcasts from Chelmsford, England, on February 23, 1920. While the election night broadcast of November 2, 1920, from station KDKA in Pittsburgh, Pennsylvania, is often credited with being the first regular broadcast transmission, several other stations can make the same claim and cite earlier dates. The first coast-to-coast broadcast in the U.S. on October 24, 1924, combined telephone wires to transmit a presidential speech from New York to 22 stations across the country, previewing the rise of regular commercial network transmissions just two years later.

The first short-wave radio transmissions were undertaken by Marconi researchers in the early 1920s. Amateur operators sent the first such signal across the Atlantic on December 11, 1921. Regular two-way amateur traffic began on December 8, 1923, and has continued since. Westinghouse used short-wave transmissions to interconnect U.S. broadcasting stations in 1923–1924. By the end of the decade, the first regular cross-border, or international, short-wave propaganda broadcasts were being transmitted by the Soviet Union. Soon other nations began their own such services, including the BBC's Empire Service on December 19, 1932. On January 7, 1927 regular commercial transatlantic radio-telephone transmissions began between London and New York using AT&T and British Post Office long-wave circuits. These were supplemented with short wave in early 1927, and full two-way short-wave links opened June 1, 1929, soon replacing the long-wave transmissions. These were the first regular voice telecommunication links across the Atlantic to operate commercially and, updated from time to time, were the only such link until the first submarine telephone cable was opened for service in 1956.

The first public transmission of frequency modulation (FM) radio (prior developments were in AM, or amplitude modulation, radio) came on November 6, 1935, when U.S. inventor Edwin Howard Armstrong demonstrated his new system to a meeting of engineers at the Institute for Radio Engineers in New York. While commercial FM broadcasts began in the U.S. on January 1, 1941 (and on their present VHF spectrum band in 1945), FM stations in Europe appeared only slowly after the war, operating first in Germany. FM took decades to become a commercial success, achieving substantial audiences only in the 1970s in both Europe and the U.S.

See also **Radio: AM, FM, Analog, Digital; Radio Transmitters, Continuous Wave; Radio Receivers, Early; Radio Transmitters, Early; Radio Receivers, Valve and Transistor Circuits**

CHRISTOPHER H. STERLING

Further Reading

Aitken, H.G.J. *The Continuous Wave: Technology and American Radio, 1900–1932*. Princeton University Press, Princeton, 1985.

Dalton, W.M. *The Story of Radio, Part I: How Radio Began*. Adam Hilger, Bristol, 1975.

Douglas, S.J. *Inventing American Broadcasting, 1899–1922*. Johns Hopkins University Press, Baltimore, 1987.

Fleming, J.A. *The Principles of Electric Wave Telegraphy and Telephony*, 2nd edn. Longmans, London, 1910.

Hong, S. *Wireless: From Marconi's Black-Box to the Audion*. MIT Press, Cambridge, MA, 2001.

Maclaurin, W. R. *Invention and Innovation in the Radio Industry*. Macmillan, New York, 1949. Reprinted by Arno Press, 1971.

McNicol, D. *Radio's Conquest of Space*. Murray Hill Books, New York, 1946. Reprinted by Arno Press, 1974.

Useful Websites

A site sponsored by Marconi Corporation that explores the life, science and achievements of Guglielmo Marconi: www.marconicalling.com/

Radio Receivers, Coherers and Magnetic Methods

From 1900 to about 1914, when Morse code was the normal language of radio communication, a variety of methods for the detection of electromagnetic radiation were employed. Two of the more successful utilized discoveries made during the nineteenth century, namely (1) the property possessed by some types of loose contacts to cohere when activated by radio-frequency currents and go from a non-conducting to a conducting state and (2) the phenomenon of hysteresis in the magnetization and demagnetization of magnetic materials. Detectors based on these properties were in common use until their gradual replacement by crystal diodes and by vacuum tubes (or valves), which were becoming common components of electromagnetic wave detector circuits just before World War I. Coherers and magnet materials were not in themselves detectors, but they acted in conjunction with other circuit elements to render radio waves sensible to human beings by producing visible or audible signals. While coherers could be made to operate galvanometers, printers, and earphones, the output of magnetic detectors was sufficient only to operate earphones.

In 1890 the French experimenter Edouard Branley was the first to note that metallic powders, iron filings for example, exhibited the property of coherence. Subsequently, several workers investigated the phenomenon and produced their own devices, examples of which are shown in Figure 6. Some of the more significant contributions to the development of coherers were made by Oliver Lodge in Britain and by Alexander Popov in Russia, and both men devised their own circuits to be used in conjunction with their detectors. There was no standard form of coherer, nor did there appear to be particular suppliers of the devices. Many types were developed, in fact almost as many as there were experimenters, and they ranged in structure from solid materials in contact to powders. They were often precisely engineered, although it is not clear that precision was really required. Experimenters had to take into account the fact that once the coherer was put into the conducting state it did not return to its starting condition of nonconduction. In many devices the original condition could be obtained only by mechanical interference, shaking or striking the coherer for example. Some elaborate arrangements were used to achieve correct and reliable operation and to overcome the disadvantage of slow operating speeds that were contingent upon a mechani-

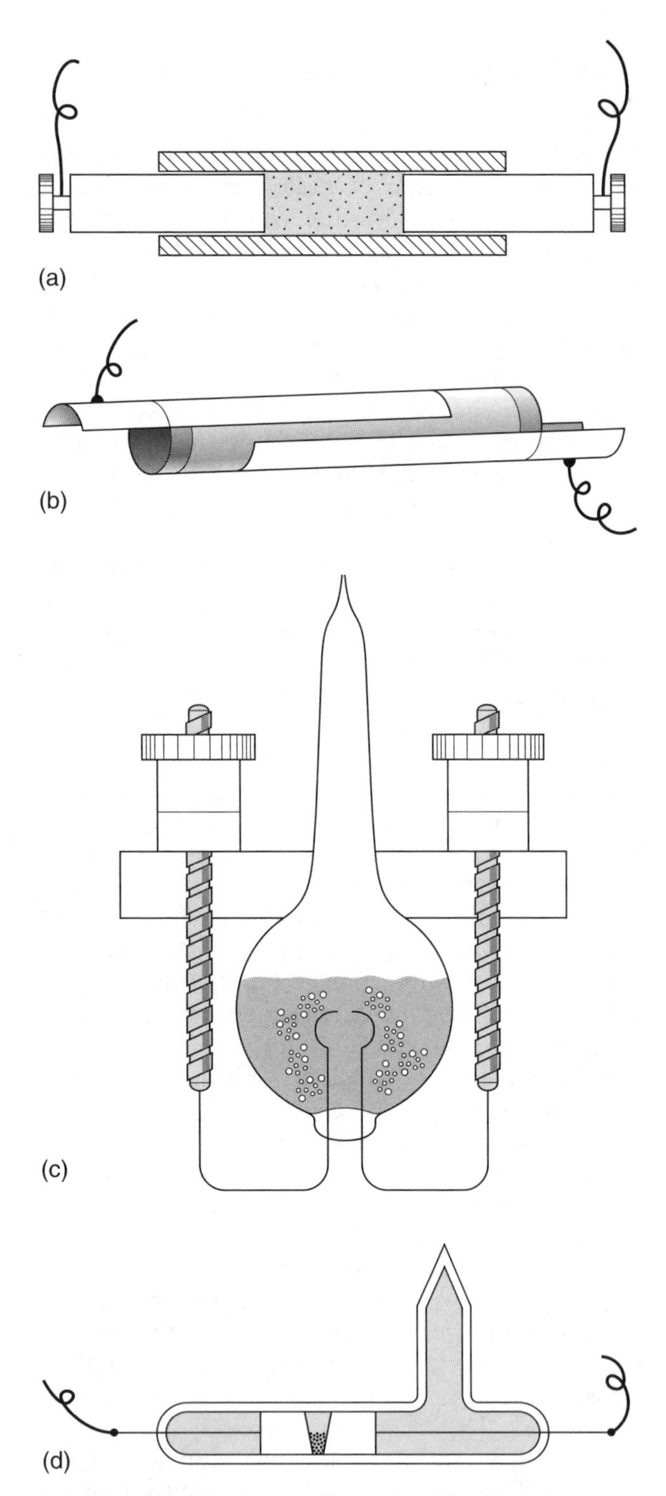

Figure 6. Types of filings coherers: (a) Branley, 1890; (b) Popov, 1895; (c) Lodge, 1895; and (d) Marconi, 1895. [*Source: Phillips, V.J. Early Radio Wave Detectors, IEE, London, 1980, p. 31.*]

cally operated device. If wireless telegraphy were to be a serious competitor to the existing terrestrial and submarine telegraph cables, then high signaling speeds would be essential. Terrestrial telegraphy could operate at hundreds of words per minute, and even very long submarine cables could carry messages at scores of words per minute. Mechanically operated coherers found it difficult to achieve speeds greater than 10 to 20 words per minute, which did not present serious competition to existing means of telegraphy.

There were some materials that displayed the properties of coherence and returned without external action to the nonconducting condition. Substances used in such self-restoring or autocoherers were most commonly steel, carbon, and mercury in various combinations (Figure 7). Other such devices were constructed by instrument makers such as E. Ducretet of Paris. In spite of great interest in coherers and their common use as radio detectors, no satisfactory theory of their operation has been given.

There were several methods of using coherers, and a typical circuit is shown in Figure 8. The battery served two purposes: to provide current through the galvanometer and earphone and to keep the coherer in a high-resistance, or sensitive, condition. In this state very little current flowed around the circuit abcd. However, when a radio-frequency signal was present in the aerial circuit, the resistance of the coherer dropped significantly, resulting in a sudden increase in the current, which caused the galvanometer to deflect or to be heard as a click in the earphone.

Magnetic detectors, a second type of early detectors, fell into two broad groups: (1) electrodynamic detectors, in which the presence of a radio signal was shown by the movement of a needle or a mirror, or (2) magnetic hysteresis detectors, in which the oscillatory field changed the state of magnetization of a piece of ferromagnetic material due to an aerial current. A common form of the magnetic hysteresis detector was that developed by Guglielmo Marconi (Figure 9). A loop of iron wire driven by a clockwork motor passed under two magnets and was cycled around a hysteresis loop (see Figure 10). Any oscillating decreasing currents in the aerial coil would tend to move the iron to a

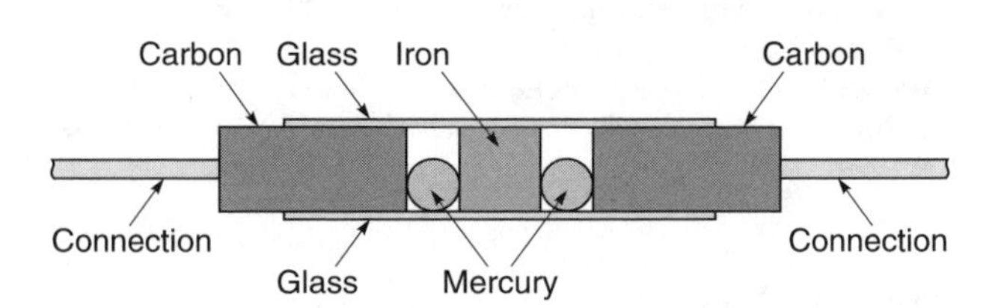

Figure 7. Auto-coherer.
[*Source: Adapted from Sewall, C. H. Wireless Telegraphy, Crosby Lockwood, London, 1904, p. 161.*]

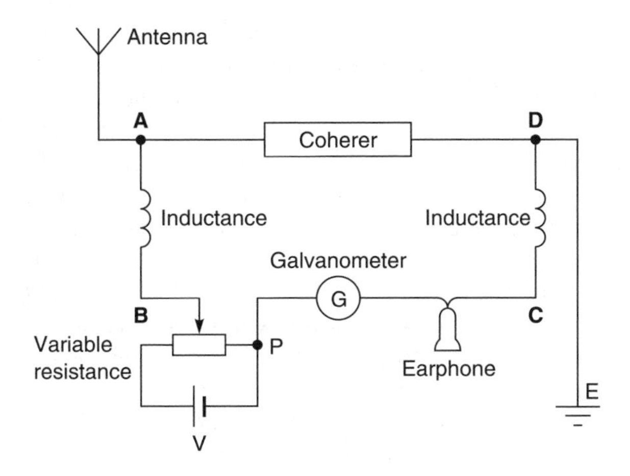

Figure 8. Basic coherer circuit.
[*Source: Adapted from Phillips, V. J. Early Radio Wave Detectors. IEE, London, 1980, p. 20.*]

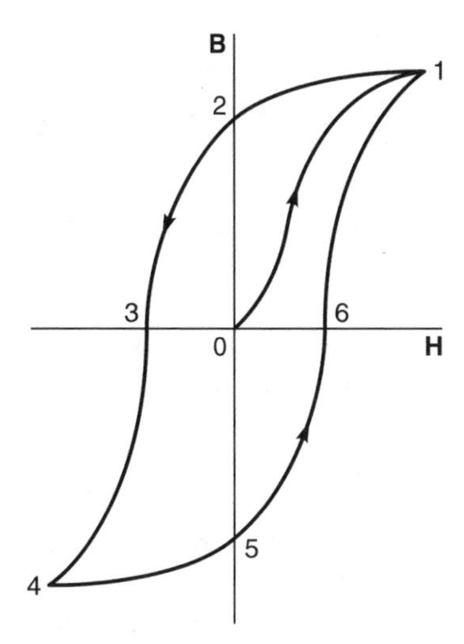

Figure 10. Hysteresis loop.

magnetization state corresponding to the path 1 to 0, and the change would be heard as a signal in the earphone via the shorter secondary coil surrounding the wire.

Before World War I, radio communication had largely been restricted to ship-to-shore communication and to military uses, and coherers and magnetic detectors were adequate for these applications. Although widely used for the reception of Morse signals, these devices were unsuitable for the detection of continuous waves. In the first decade of the twentieth century, continuous waves had been generated, typically by high-speed alternators. Although speech had been broadcast, these experiments had been desultory. During World War I, much attention was devoted to developing wireless telephony, and vacuum tubes were found to be the most effective way of generating continuous radio

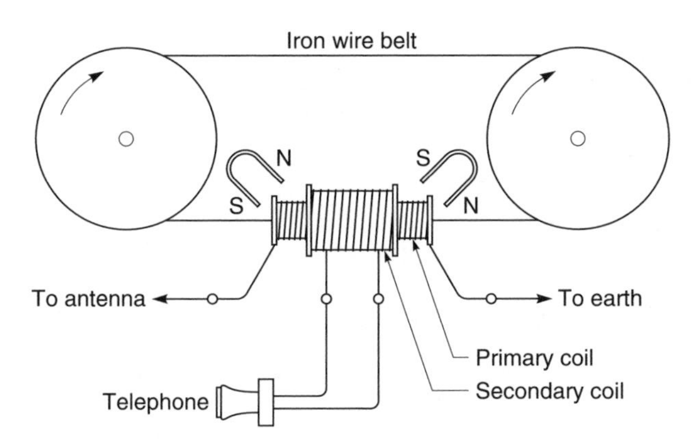

Figure 9. Schematic drawing of Marconi's hysteresis detector.
[*Source: Adapted from Dalton, W. M., The Story of Radio, Part 1: How Radio Began. Adam Hilger, Bristol, 1975, p. 95.*]

waves that could carry speech. After the war, particularly when the market for radio widened in the 1920s to become a fully commercial consumer-oriented industry, coherers and magnetic detectors vanished to be replaced by vacuum tubes and the crystal detector.

See also **Radio Receivers, Crystal Detectors and Receivers; Radio Receivers, Valve and Transistor Circuits; Radio Transmitters, Early; Rectifiers; Vacuum Tubes/Valves**

COLIN HEMPSTEAD

Further Reading

Beauchamp, K. *History of Telegraphy*. Institution of Electrical Engineers, London, 2001.
Burns, R.W. Discovery and invention: lodge and the birth of radio communications. *IEE Review*, 40, 131–33, 1994.
Dalton W.M. *The Story of Radio, Part I: How Radio Began*. Adam Hilger, Bristol, 1975.
Fleming J.A. *Fifty Years of Electricity: the Memories of an Electrical Engineer*. Wireless Press, London, 1921.
Phillips V.J. *Early Radio Wave Detectors*. Peter Peregrinus, Stevenage, in association with the Science Museum, 1980.
Phillips, V.J. The "Italian Navy Coherer" affair: a turn of the century scandal. *Proc. IEEE*, 86b, 248, 1998.
Pickworth, G. Coherer-based radio. *Electronics World and Wireless World*, 100, 563–567, 1994.

Radio Receivers, Crystal Detectors and Receivers

Crystal detectors based on the phenomenon of rectification were commonly employed in commercial radio receivers in the years following World War I. They were very effective in the detection of modulated radio waves that carried speech and musical signals; and, because they could be employed directly to operate earphones, they were particularly useful in locations where a supply of electricity was not easily available. A circuit consisting of a crystal rectifier, a variable capacitor in parallel with an inductance, and an earphone, was sufficient to detect signals from commercial transmitters (Figure 11). However, even with a modicum of tuning they were lacking in selectivity; that is, the ability to separate transmissions from stations that were close in frequency (Figure 12).

In the 1870s several researchers noted the phenomenon of "unilateral conductivity," now termed "rectification." In 1874 Ferdinand Braun discovered that the electrical resistance of metal sulfides was direction dependent, and Arthur Schuster found a similar effect in circuits of copper and brass (where copper wires, for example, were connected with brass terminals). Desultory work continued through the last years of the nineteenth century, and more examples of rectification were found. Between 1900 and 1910 several systematic studies were made of crystal rectifiers with Henry Dunwoody of the U.S. Army patenting a device using carborundum. Between 1908 and 1910 Wichi Torikata and E. Yokoyama in Japan, and George W. Pierce and L.W. Austin in the U.S. added a large number of elemental and compound semiconductors that could be used as crystal detectors. It is clear that all crystal rectifiers utilized surface contacts either between two semiconductors or between fine metal points and semiconductors. The structure of a metal point and a crystal became known as a "cat's whisker." Simple crystal rectifiers were used commercially until well into the 1950s, although the cat's whisker had been replaced by then with encapsulated, fixed crystals, which were developed during World War II. The typical form of a crystal detector was a relatively large crystal against which could be placed either a thin wire (the cat's whisker) or a smaller crystal (see Figure 13).

Crystal sets were widely available in the early days of radio broadcasting when the technology became available to the home market. Nevertheless, despite the unreliability of devices that depended on the delicate physical contact between different materials, it was these that brought radio to a relatively large audience in the industrialized world. A large number of manufacturers produced crystal sets, some elaborate in their construction. Prices varied depending on the quality of the cases, for there was very little difference in the performance of the receivers. The quality of the cases was cosmetic rather than electronic in purpose. Nevertheless, many amateurs constructed their own crystal sets from the 1920s. Although crystal rectifiers were capable of possessing excellent characteristics, reliable results often depended on the user to apply considerable skill and patience in finding the best point of contact between the cat's whisker and crystal, or between two crystals. The sound quality was poor, and it was not easy for more than one listener to use a receiver, although acoustic amplifiers were sometimes employed. One homely method of amplification was a set of earphones placed in a bowl so that faint sounds could be heard by several listeners.

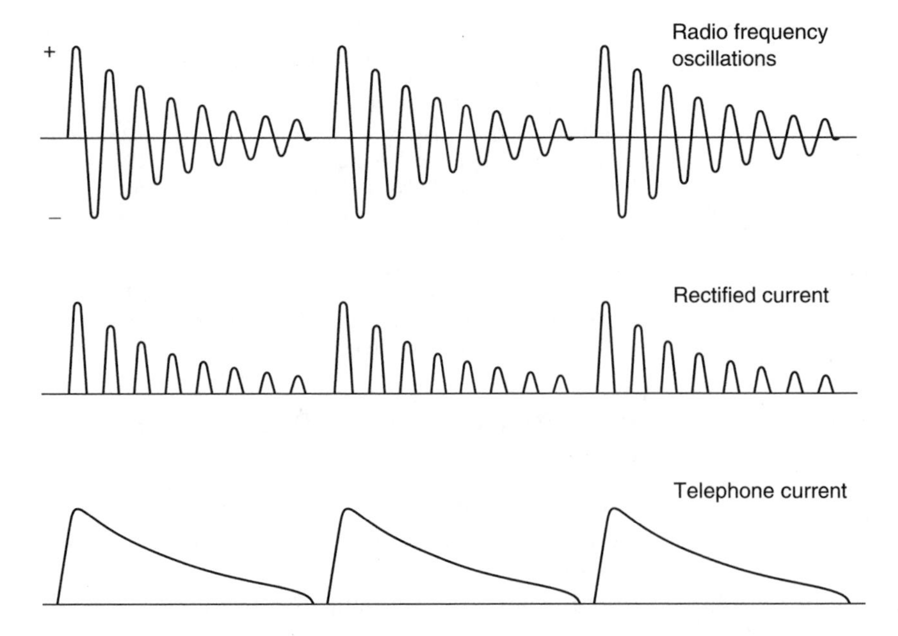

Figure 11. Rectification in action. [*Source: Adapted from Bucher, E. E. Practical Wireless Telegraphy. Wireless Press, New York, 1917, p. 132.*]

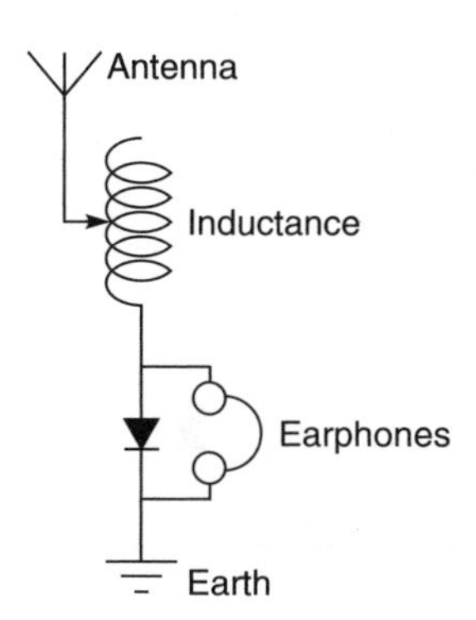

Figure 12. Simple crystal receiver.
[*Source: Adapted from Bucher, E. E. Practical Wireless Telegraphy. Wireless Press, New York, 1917, p. 132.*]

By 1917 a wide range of detectors was available, most of which were of limited applicability and many were unsuitable for domestic purposes. Table 1 lists many of the types available and indicates their general utility.

Table 1 Types of crystal detectors.

	Detector Type
No local power	Galena; Zincite-Bornite; Carborundum; Fleming Diode
Local power	Carborundum, Zincite-Bornite; Fleming Diode; Triode; Silicon
Rectification	Galena; Silicon; Carborundum; Cerusite; Zincite-Bornite; Fleming Diode; Triode; Electrolytic
Detector: damped oscillations	Galena; Silicon; Zincite-Bornite; Carborundum; Fleming Diode; Triode; Marconi Magnetic
Detector: undamped oscillations	Tikker; Tone Wheel; Heterodyne Receiver; Vacuum Tube Oscillator

Source: Adapted from Bucher, E. E. Practical Wireless Telegraphy. Wireless Press, New York, p. 141.

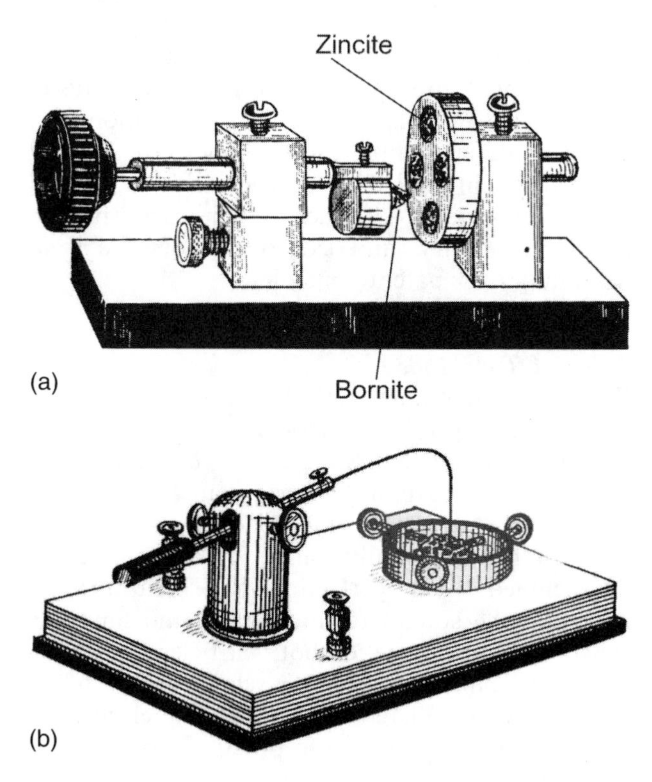

Figure 13. Two forms of crystal detectors. (a) a pointed crystal of Bornite touches one of four flat crystal of Zincite. (b) a fine wire touches flat crystals of silicon or galena. Each detector allows the most sensitive region to be found; on the left, different flat crystal can be brought in contact with the Bornite, and the pressures between the two crystals can be changed; on the right, a cats whisker contact can be adjusted while in contact with crystals of silicon or galena.
[*Source: Adapted from Bucher, E. E. Practical Wireless Telegraphy. Wireless Press, New York, 1917, p. 140.*]

The simple circuit was modified because while it was an effective detector, it was not particularly sensitive or discriminating. Therefore, a variety of inductively and conductively coupled crystal receivers were developed. Basic examples of these are shown in Figure 14. Some crystal detectors required the application of biasing voltages—

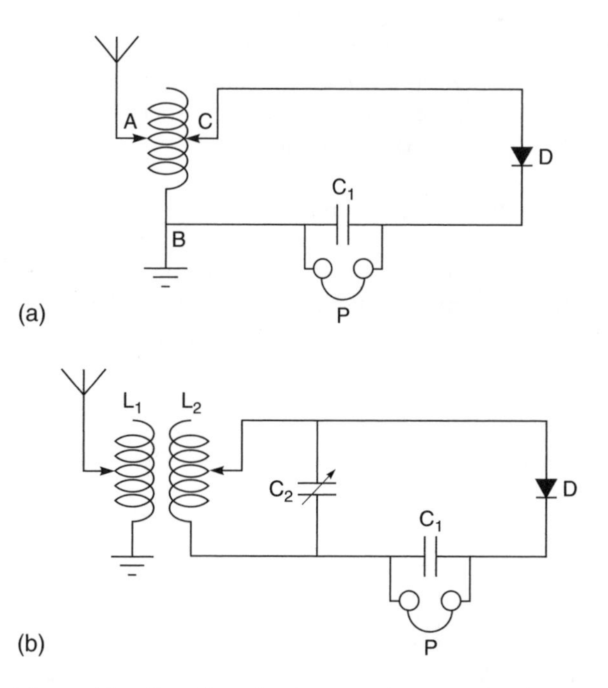

Figure 14. Coupling circuits: capacitance and inductance.
[*Source: Adapted from Bucher, E. E. Practical Wireless Telegraphy. Wireless Press, New York, 1917, p. 133.*]

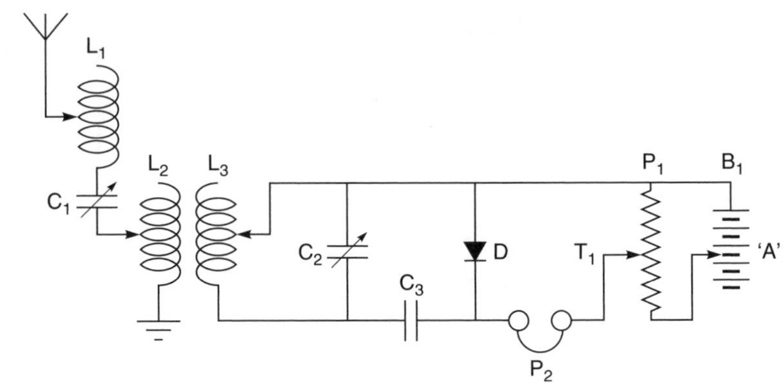

Figure 15. Biasing circuit used with carborundum.
[*Source: Adapted from Bucher, E. E. Practical Wireless Telegraphy, New York, Wireless Press, 1917, p. 134.*]

carborundum in particular operated well with a current flowing through the junction—and a number of circuits were developed. Figure 15 illustrates one such circuit.

Crystal receiving sets were an important element in the development of public broadcasting. They were simple in construction, robust (although tricky to use), and above all, relatively cheap to buy. However, in spite of much work, effective theories of the phenomena of rectification were elusive. It was made evident very early in the history of crystal rectifiers that surface properties determined their actions, or at least the rectifying effects lay close to the edges of contacts. However, only when the modern theory of semiconductors was put forward in the 1930s did it become clear that chemical and crystalline purity were essential if reliable solid-state devices were to be made. A group of new technologies arose—zone-refining, solid-state diffusion, and so on—which allowed crystal rectifiers to be fully understood and to be subsumed into the modern industry of semiconductors.

See also **Radio Receivers, Coherers, and Magnetic Methods; Radio Transmitters, Early; Radio Receivers, Valve and Transistor Circuits; Rectifiers; Transistors; Valves/Vacuum Tubes**

COLIN HEMPSTEAD

Further Reading

Beauchamp, K. *History of Telegraphy*. Institution of Electrical Engineers, London, 2001.

Blake, G.G. *History of Radio Telegraphy and Telephony*. Radio Press, London, 1926.

Burns, R.W. Discovery and invention: lodge and the birth of radio communications. *IEE Rev.*, 40, 131–133, 1994.

Bucher, E.E. *Practical Wireless Telegraphy*. Wireless Press, New York, 1917.

Dalton, W.M. *The Story of Radio, Part 1, How Radio Began*. Adam Hilger, Bristol, 1975.

Erskine-Murray, J. *A Handbook of Wireless Telegraphy*. Crosby, Lockwood, London, 1913.

Fleming J.A. *Fifty Years of Electricity: the Memories of an Electrical Engineer*. Wireless Press, London, 1921.

Phillips V.J. *Early Radio Wave Detectors*. Peter Peregrinus, Stevenage, in association with the Science Museum, 1980.

Radio Receivers, Early

While early magnetic detectors such as coherers provided an effective means of receiving wireless telegraph (code) signals, the continuous waves used to transmit wireless telephony (voice, music) required both improved means of reception and effective signal amplification. The crucial and related inventions that allowed this appeared within two years of each other early in the twentieth century, and led to considerable legal wrangling over patent control.

Edison Effect (1883)

While experimenting with his new incandescent electric light, prolific American inventor Thomas A. Edison noted that the glowing (because it was heated) lamp carbon filament somehow created conditions that slowly blackened the insides of light bulbs, eventually rendering them useless. Preoccupied with perfecting his breakthrough invention, he sought to understand and limit the causes (he presumed carbon particles had been thrown off), filed a patent on a related device in late 1883 (now considered to be the first electronics patent), but did not pursue practical applications of his finding.

What became known as the "Edison effect" soon fascinated researchers on both sides of the Atlantic, one of whom was John Ambrose Fleming, a London-based electrician for the British branch of Edison's company. Edison demonstrated the effect to Fleming while the latter visited the U.S. Fleming and others tried to determine what caused the effect and whether it had any potential use, and published brief reports on their limited findings. All noted that a narrow strip of glass was not blackened, where one leg of the light filament shielded the other.

By the late 1890s, building on a growing understanding of both conduction through gasses and emissions from hot elements within vacuums, researchers finally understood that streams of electrons, not carbon particles, were responsible for the Edison effect.

Fleming Valve (1904)

Within a few years, devices to improve wireless technology were often the focus of experimental electrical work in many countries. Ambrose Fleming was one of many researchers who began to seek whether newly discovered electrons could be effectively harnessed to detect wireless signals. He had a particular reason for seeking an improved detector that could display its results on an electrical meter—he was partially deaf and often missed weak signals.

Fleming and others determined that electrons were negatively charged, and thus a positively charged plate could be used to attract them. Put another way, electrical current would "flow" only when a plate was positively charged. Inserting such a plate into a small glass vacuum tube along with a filament conductor created what looked very much like an incandescent light, but was in reality a one-way gate or "valve" which could change (rectify) alternating current (such as wireless or radio waves) to pulsating direct current. Using such a two-element (filament and plate) valve or "diode," an operator could detect the presence of wireless telegraphy signals.

Fleming (by then a scientific advisor for the Marconi Company in London) was granted British, American and German patents for his "oscillating valve" device in 1904–1905. Though his valve worked as well as the earlier magnetic detectors, and was somewhat more stable in operation, it demonstrated no greater sensitivity to weak signals. It was simply an alternative approach to wireless signal detecting. The Marconi Company made wide use of the device in many of its radio stations up to World War I. Fleming was knighted in 1929. Over time, his device (indeed all that followed it) became better known as "thermionic" valves as they involved the use of incandescent (heated) electrons.

De Forest Audion (1906)

In America, experimenter Lee de Forest (1873–1961) was also seeking improved means of wireless signal detection, especially those that might avoid further patent infringement problems that already plagued his operations. He had laboriously developed a series of detectors (one even used small gas flames) and soon turned to variations on the Fleming valve as a promising approach. However, he lacked a scientific understanding of its principals, as over the space of several months he sought to improve its operation by experimenting with added elements, soon focusing on placing an additional electrode inside the tube. This took the form of a second plate or "wing," and was patented in 1906. The resulting device was little better than Fleming's valve. Indeed, Fleming became bitter over de Forest's ignoring of his prior work (de Forest claimed not to know of the Fleming valve patent before 1906) as well as his own lack of any royalties on the diode (as the Marconi Company owned the patent rights).

De Forest's added tube element soon changed, taking the form of a tiny inserted wire in the form of a grid, and the true three-element tube or "triode" was born, dubbed the "Audion" by one of his assistants, and patented in 1907 (confusingly, the term Audion was applied by de Forest to many diode and triode devices). Even this latest version was initially perceived as just another detector among many available options.

The new Audion tube would, however, turn out to be a far more capable device—indeed it can be argued that it lay the foundation for electronics development until the era of solid-state devices; but realization of the full potential of the Audion took time, and much of the effort to make that so came from others. The improved tube's ability to amplify weak signals came to the attention of the American Telephone & Telegraph Company (AT&T), which for years had sought an amplifier for long-distance wired telephone circuits. AT&T purchased partial rights to the Audion in 1909 and using a modified Audion device, opened transcontinental telephone service in 1914.

Meanwhile the man who would become de Forest's primary radio rival, Edwin Howard Armstrong (1890–1954), had discovered the "feedback" potential of the Audion as a student at Columbia University. When an Audion's output was fed back into its own input, the circuit could become a generator of electrical signals—effectively a radio transmitter. While virtually all engineers credited Armstrong's genius, the U.S. Supreme Court terminated two decades of patent battles in 1934 when it held that de Forest deserved credit.

See also **Radio Receivers, Coherers and Magnetic Methods; Radio Receivers, Crystal Detectors and Receivers; Radio Receivers, Valve and Transistor Circuits; Radio Transmitters, Continuous Wave; Vacuum Tubes/Valves**

CHRISTOPHER H. STERLING

Further Reading

Aitken, H.G.J. De Forest and the Audion, in *The Continuous Wave: Technology and American Radio, 1900–1932*. Princeton University Press, Princeton, 1985.

Blake, G.G. The thermionic valve and some of the fundamental valve circuits, in *History of Radio Telegraphy and Telephony*. Chapman & Hall, London, 1928. Reprinted by Arno Press, 1974.

Chipman, R.A. De Forest and the triode detector. *Scientific American*, 212, 92–100, 1965.

Fleming, J.A. *The Thermionic Valve and its Developments in Radiotelegraphy and Telephony*. *Wireless Press*. London, 1919.

MacGregor-Morris, J.T. *The Inventor of the Valve: A Biography of Sir Ambrose Fleming*.Television Society, London, 1954.

McNicol, D. The magic bulb, in *Radio's Conquest of Space*. Murray Hill Books, New York, 1946. Reprinted by Arno Press, 1974.

Shiers, G. The first electron tube. *Scientific American*, 220, 104–112, 1969.

Thermionic Valves 1904–1954: The First Fifty Years. Institution of Electrical Engineers, London, 1954.

Tyne, G.F.J. *Saga of the Vacuum Tube*, 2nd edn. Sams, Indianapolis, 1977.

Radio Receivers, Valve and Transistor Circuits

The direction of radio receiver circuit development between the 1920s and the early twenty-first century has been closely linked to two interrelated factors. The first is the inseparability of the development of radio components and radio hardware.

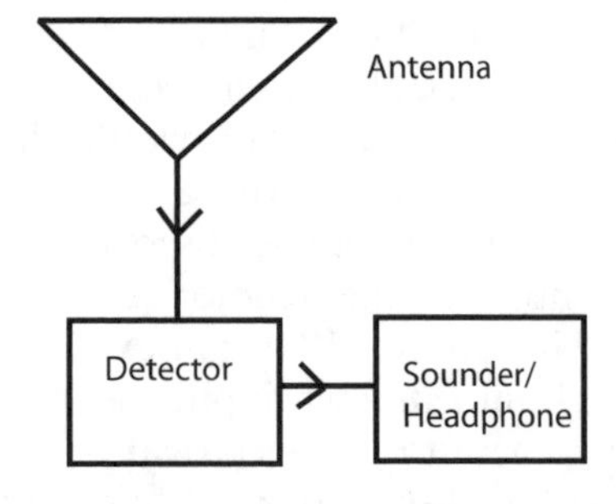

Figure 16. A simple radio receiver, consisting of an antenna, detector (such as a crystal diode, coherer, or other device), and a means of listening to the detected signals, often an electromechanical sounder or ink printer for wireless telegraphy.

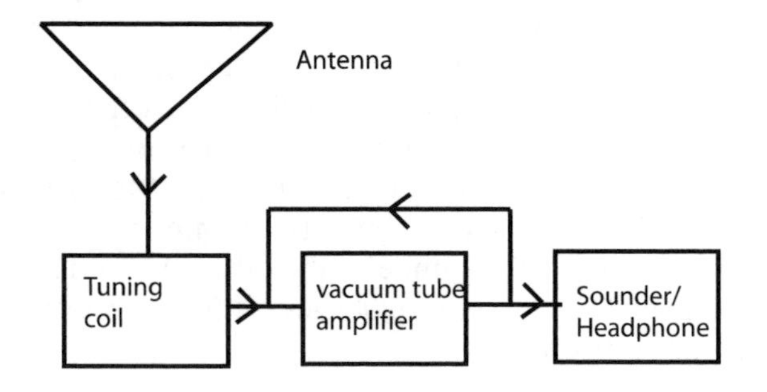

Figure 17. A regenerative receiver, utilizing an electronic vacuum tube amplifier with part of the output fed back to the input, thereby increasing the volume. By the time of the regenerative circuit, it was also necessary to include a tuner because of the large number of stations on air.

Improvements in components such as vacuum tubes and transistors have alternately stimulated or been stimulated by developments in circuit design. The second factor is the influence of social factors, sometimes seemingly unrelated to purely technical issues. Regulatory changes, for example, have determined how radio can be used and have been reflected in circuit designs. In some cases, such social factors have spelled success or failure for particular circuits, but more common is the phenomenon of national or international regulations extending the life of certain types of circuits by reigning in technological changes that would otherwise have made particular designs obsolete.

The earliest radio transmissions consisted simply of an electric spark generated through the use of a battery connected to a high-voltage electrical transformer (or coil), a capacitor, a means of switching current on and off, and miscellaneous other components such as resistors and interconnecting wires. The reception at increasingly greater distances of the resulting electromagnetic waves was of great concern to early radio engineers, stimulating many innovations in receiving circuits. Until the invention of the Fleming valve in 1904, most receiving circuits relied on electromechanical detectors of some kind, which had very limited sensitivity.

Basic spark transmitters in use around the turn of the twentieth century each broadcast a wide bandwidth signal, but these gave way to tunable transmitters because of the desire to conserve the electromagnetic spectrum. In fact, tunable transmitters were made necessary in order to comply with international treaties that attempted to reduce station interference. This change demanded circuits capable of restricting the frequency of the generated radio signal at the transmitter, as well as

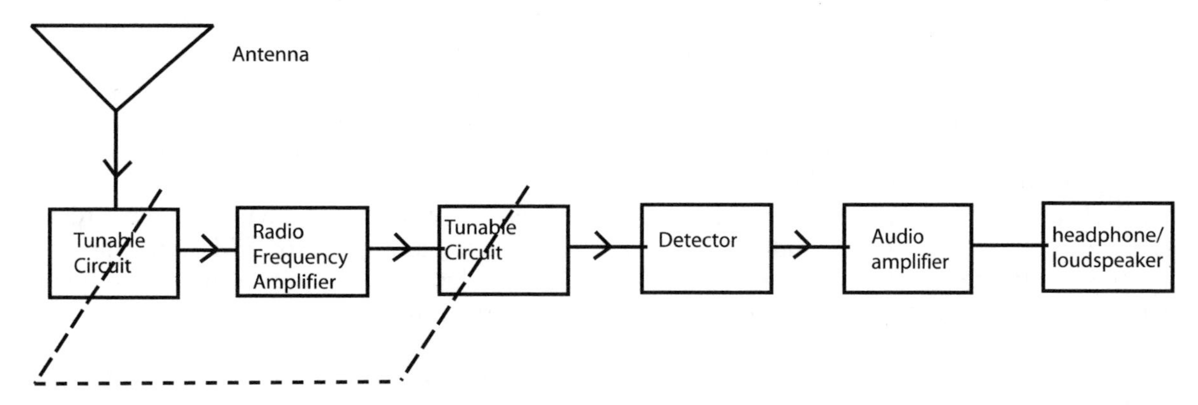

Figure 18. The neutrodyne circuit was the most successful variation of an earlier receiver design called the tuned radio-frequency (TRF) circuit. In a TRF receiver, the incoming signal passed through two or more stages of amplification before being detected. The tuning devices in each stage had to be mechanically linked together to operate in unison (shown by the dotted line). The neutrodyne overcame a major feedback problem by adding a "neutralizing" capacitor (not shown) in each radio frequency amplification stage.

circuits that would allow a receiver to select from a range of signals being broadcast at different frequencies. The latter task was almost universally accomplished by using a variable, air-gap capacitor in the receiving circuit so that the user could change the frequency at which the circuit received waves most efficiently.

Such receivers, using what was generically called a "tuned radio frequency" circuit, were common through the late 1920s. However, the advent of the amplifying vacuum tube in 1906–1907 radically changed the way radio circuits were designed: because the incoming signal could be electronically amplified, engineers had greater options. Receiver designers changed their emphasis from preserving the strength of the incoming signal to other issues, such as the ability to receive both strong and weak signals equally well. Key innovations of the vacuum tube era included the 1912 "regenerative" receiver of Edwin H. Armstrong of the U.S., which in effect tapped off a portion of the incoming signal and fed this back into the amplifying circuit, thereby creating a powerful feedback loop capable of great signal amplification (and hence capable of receiving weak, distant signals easily). Regenerative receivers were inexpensive, making them popular among amateur radio enthusiasts, but they were also unstable and could easily be overloaded, leading to unpleasant howling sounds. The major competitor of the regenerative receiver

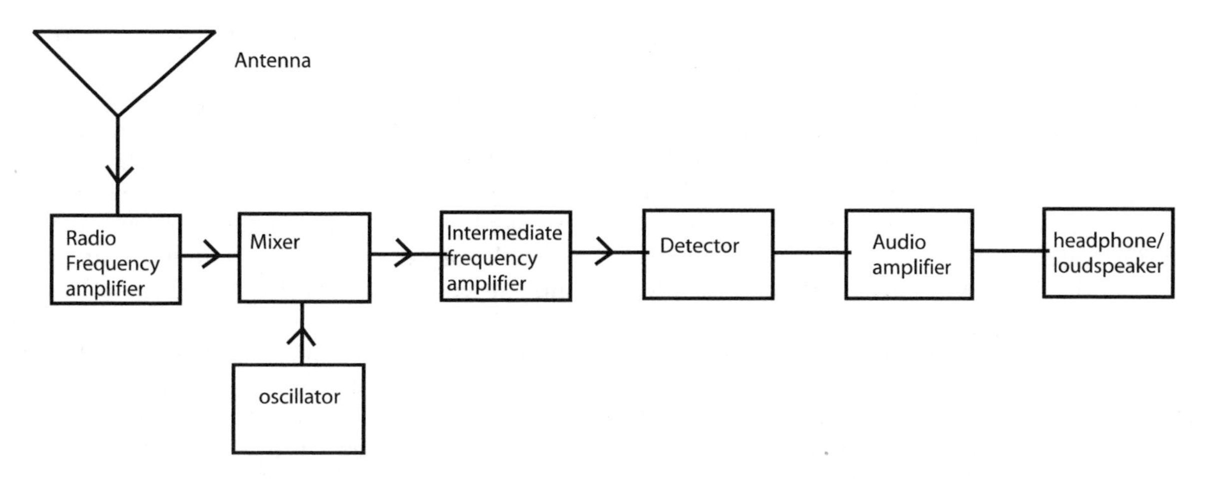

Figure 19. Superheterodyne receiver. The electronic oscillator produces a high-frequency signal, which is mixed or "heterodyned" with the incoming signal from the antenna. The product is kept within a narrow band of frequencies, allowing the following amplifier and detector stages to be optimized to a narrow bandwidth.

was the tuned radio frequency (TRF) receiver, which used several stages of electronic amplification to boost the incoming radio-frequency signal before feeding the signal to a detector. TRF receivers also tended to break into uncontrolled oscillation, but the "neutrodyne" circuit invented in 1922 by Harold Wheeler and Alan Hazeltine of the U.S.) used carefully chosen induction coils at key points in the circuit to counteract this tendency. The neutrodyne circuit, licensed by several radio manufacturers, did not have the problem of sudden loud howls caused by uncontrolled feed back loops as in the regenerative circuit.

Edwin Armstrong, however, had been busy at work while engaged in the U.S. Army Signal Corps during World War I. In 1918, he invented the superheterodyne circuit, which took a new approach to the issue of receiver sensitivity. While radio signals span a wide range of frequencies, radio tubes and circuits amplify some frequencies better than others. The superheterodyne first changed the frequency of the incoming signal, whatever it was, to a single "intermediate" frequency. The remaining receiver circuits, optimized to process signals in a band near the intermediate frequency, thus always had maximal sensitivity and performance. The superheterodyne circuit was widely licensed or copied in the 1930s and remains a standard feature of nearly every radio device.

By the 1930s, radio broadcasting authorities worldwide had settled on amplitude modulation (AM) as the standard form of radio transmission. Most countries worked to ensure that all their citizens were within the range of the radio networks. As part of this effort, a regulatory structure emerged that had the effect of petrifying AM technology. Other forms of modulating a radio wave had been proposed, but they were rejected for technical or economic reasons, or simply because it was desirable to support AM. The now-familiar frequency modulation (FM) form of broadcasting was proposed in the 1920s, but it was not until the 1930s that vacuum tube technology was sufficiently advanced to permit FM sound quality to exceed that of AM. Furthermore, implementing FM required a considerable political struggle because it competed with AM and the new technology of television.

Once again, Armstrong was at the forefront of the movement to reintroduce FM as a broadcast technology, demonstrating an improved system in 1934. Many countries around the world introduced FM broadcasting after World War II. Regulatory agencies gave FM broadcasters such wide swaths of bandwidth compared to AM stations that all sorts of additional services could be interleaved with ordinary audio transmissions. For example, many FM stations in the 1950s simultaneously broadcast radio-facsimile or background music services, which were not detectable on consumer receivers. These were abandoned in order to use the extra bandwidth for FM stereophonic broadcasts, which were approved by regulators as early as 1961 in the U.S.

The use of transistors had less of an effect on radio circuit design than might be expected. AM superheterodyne receivers using a minimal complement of five tubes were nearly universal by the 1950s, and in some cases transistors were substituted for tubes in virtually identical circuits. However, transistors are smaller and much more energy efficient than tubes, and their use resulted in home and portable receivers that were smaller and cooler in operation. The pocket-sized transistor radios that helped bring semiconductor technology to the attention of the public also used minimally modified versions of vacuum tube circuits. A memorable feature of the early transistor receiver designs was the installation of the transistors into plug-in sockets, mimicking vacuum tube designs. In the 1950s, the extremely high reliability of transistors was not yet known, so it was assumed that transistor replacement would be common and that repair would be made easier through the use of these sockets.

FM receivers required many more vacuum tubes than AM receivers, and hence were quite expensive before the introduction of the transistor. The price reductions possible with the use of transistors helped spur sales of FM receivers (which were not selling well before that) and hence popularize FM broadcasting. Many of the circuit innovations of the transistor period were related to the low cost of transistors, which encouraged manufacturers to add features to existing designs. In the 1970s, for example, signal processing tasks such as "active," amplified tone control circuits replaced the simpler passive filter networks used earlier. Such devices as the elaborate graphic equalizer tone controls introduced as early as the 1960s would have been prohibitively expensive for the consumer market if rendered in vacuum tube circuitry. Since the 1960s, the integrated circuit has extended this trend, reducing the number of components in receiver, amplifier, and tone control circuits. It is possible today to put all or most of the circuit elements necessary to make a functional radio receiver on a single integrated circuit requiring only an external battery, antenna, and earphone or loudspeaker.

See also **Radio Receivers, Crystal Detectors and Receivers; Radio Receivers, Early; Radio Transmitters, Early**

DAVID MORTON

Further Reading

Aitken, H.G.J. *The Continuous Wave: Technology and American Radio, 1900–1932*. Princeton University Press, Princeton, 1985.

Morton, D. *A History of Electronic Entertainment Since 1945*. IEEE History Center, New Brunswick, NJ, 1999.

Radio Club of America. *The Legacies of Edwin Howard Armstrong*. N.p.: Radio Club of America, Inc.

Radio Transmitters, Continuous Wave

In the 1900s, most contemporary wireless telegraphy transmitters were spark-gap generators. Because of their high radiation resistance, the waveforms produced by a spark gap died out, or damped, quickly. While damped waveforms were not a problem for telegraphy, they presented a serious obstacle for telephony because continuous voice signals needed to ride on continuous waves. Continuous wave transmitters had to be developed for wireless telephony.

The arc generator was an early step in the development of continuous wave transmitters. The arc phenomenon appeared when two conductors previously joined were separated; the current kept flowing between the separated conductors, and a flame-like arc was established. In the 1890s in Britain, Hertha and William Ayrton discovered that an electric arc had a negative resistance. Connecting an arc to an ordinary oscillation circuit, the arc's negative resistance compensated for the positive radiation resistance and thus reduced the damping effect. The electric arc could therefore be used to construct a continuous wave oscillator. William Duddle, Ayrton's student, implemented this idea with his "singing arc" circuit, in which an arc lamp was connected with a capacitor and an inductor in parallel (Figure 20). This circuit could generate continuous waves up to 15 kilohertz. In 1903, the Danish physicist Valdemar Poulsen transformed the Duddle arc into a megahertz oscillator by replacing air with hydrogen gas, applying a magnetic field, using a water-cooled positive copper electrode, and rotating the carbon electrode. In the 1910s, the Poulsen arc became an important technology for wireless telegraphy, but it was not a genuine continuous wave transmitter. British engineer John Ambrose Fleming demonstrated that an arc generator produced a series of short pulses instead of continuous waveforms.

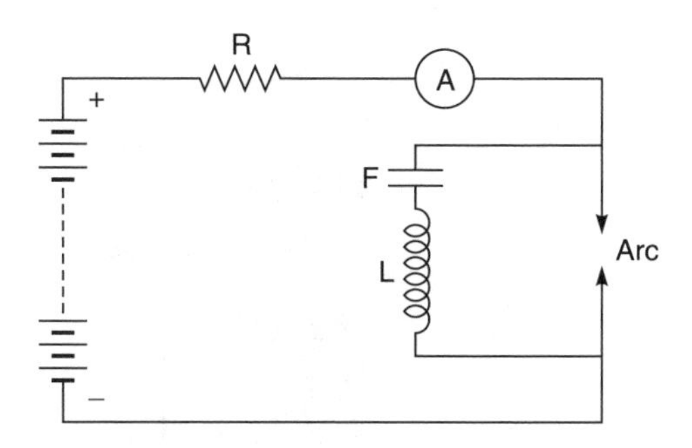

Figure 20. Duddle's singing arc circuit.
[*Source: William Duddle, J.Inst. Elec. Eng., 30, 232–283, 1900. Reproduced with the permission of the Institution of Electrical Engineers.*]

A promising candidate for a true continuous wave radio transmitter was the high-frequency alternator. The alternator, a technology for producing alternate electricity, used the principle of electromagnetic induction. By rotating an armature wound with coils in a fixed magnetic field, an electromotive force was induced in the coils. An alternator was typically used to generate high-power electricity below 150 hertz. In 1901, Reginald Fessenden proposed to generate radio-frequency waveforms with a high-speed alternator and to couple these waveforms directly to the antenna circuit. Unlike the spark-gap transmitter that produced damped waves, the alternator-based transmitter would provide fully undamped waves as long as its armature's motion was maintained. To design this kind of alternator, Fessenden contracted with the General Electric Company. The most critical technical challenge for its development was speed. A radio alternator required the armature to rotate at 50,000 cycles per second or more so that the signal frequency could exceed 50 kilohertz. At the time, no known mechanical armature could withstand the centrifugal force produced by such a high rotational velocity, but General Electric engineer Ernst Alexanderson was eventually able to overcome this problem. In his design, the rotary armature was replaced by two stationary armatures and a rotary steel disk, the circumference of which was embedded with iron teeth, or poles of an electromagnet. An electromotive force could be induced in the coils, and one could increase the frequency of the electromotive force by adding more poles on the disks' circumferences. Rapid disk rotation did not pose the mechanical problem of a rotating armature.

A schematic drawing of Alexanderson's alternator is shown in Figure 21. A magnetic flux, *M*, that is produced by a field coil, *S*, passes through

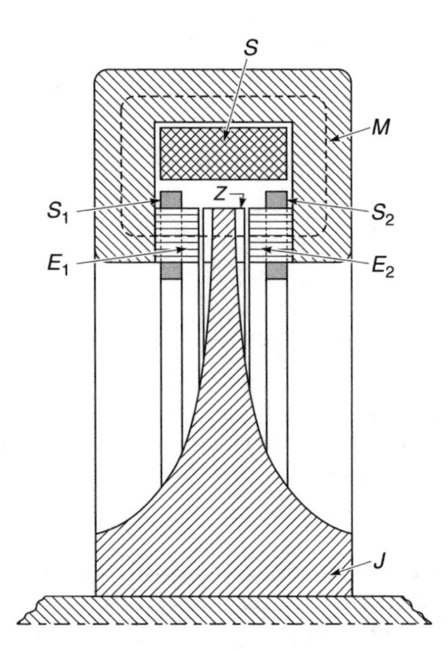

Figure 21. The Alexanderson alternator.
[*Source: Zenneck, J. and Seeling, A. E. Wireless Telegraphy. McGraw-Hill, New York, 1915, p. 214. Reproduced with the permission of McGraw-Hill.*]

the iron cores *E1* and *E2* of the armature coils *S1* and *S2*. The rotary iron disk, *J*, has teeth, *Z*, on its periphery that can extend to the space between the two iron cores *E1* and *E2*. When a tooth of the iron disk is in the gap between the iron cores, the overall reluctance of the magnetic circuit is small, so the magnetic flux, *M*, is large and the induced voltages on *S1* and *S2* are large. When there is no iron tooth between the cores, the air gap contributes a large reluctance to the magnetic circuit, so the magnetic flux, *M*, is small and the induced voltages on *S1* and *S2* are small. Therefore, as the disk rotates, the induced voltages vary periodically between a maximum value and a minimum, and the frequency of this variation is the number of teeth multiplied by the angular velocity of the disk.

Based on Alexanderson's design, General Electric successfully made a 2-kilowatt 100-kilohertz alternator in 1909 and delivered it to Fessenden. Following the 1909 machine, alternators with higher power and frequency were further developed. Although gradually replaced by tube transmitters, alternators had greater survivability in long-distance communications. The U.S. Navy's alternator in Hawaii remained operational until 1957. Another alternator in Massachusetts was used by the U.S. Air Force until 1961.

Another continuous-wave transmitter technology used vacuum tubes. Invented by Fleming and American Lee de Forest in the mid-1900s, vacuum tubes were first used in radio detection. Their potential in radio transmission was not explored until the early 1910s. By about 1912, de Forest, Fritz Lowenstein, and Edwin Armstrong in the U.S, H.J. Round in Britain, and Alexander Meissner in Germany had all conceived the idea of vacuum-tube oscillators. They discovered that when connecting the output of a tube amplifier with its input via a resistive–inductive–capacitive network an oscillatory electric current would appear at the circuit. The reason, de Forest argued, was that the amplifier and the feedback network made the entire circuit a regenerative system. Signals from the tube's output were directed to its input and were further enhanced by amplification. Because the feedback network directed only a narrow frequency band to the input, the self-regenerative wave was nearly sinusoidal. The explanation implied that the square of the resonating frequency of a tube oscillator was inversely proportional to the inductance and capacitance of the overall circuit. Therefore, by adjusting the inductors and capacitors in the feedback circuit, one could vary the resonating frequency of a vacuum-tube oscillator.

A diagram of a primitive vacuum-tube oscillator, proposed by Meissner in 1913, is shown in Figure 22. When a tiny resonance is produced at the circuit of *L1*, *L2*, and *C*, it is coupled to the vacuum tube's grid, *G*, through the inductor *L3*. The resonating signal is then amplified at the tube's plate, *P*. The amplified plate current is coupled to *L1* through the inductor, *L4*, and hence enhances the original resonance. Such a feedback cycle goes on and on until the resonating signal gains its maximum.

The vacuum-tube oscillator became the dominant radio transmitter after 1912. But when engineers began to explore waves beyond 10

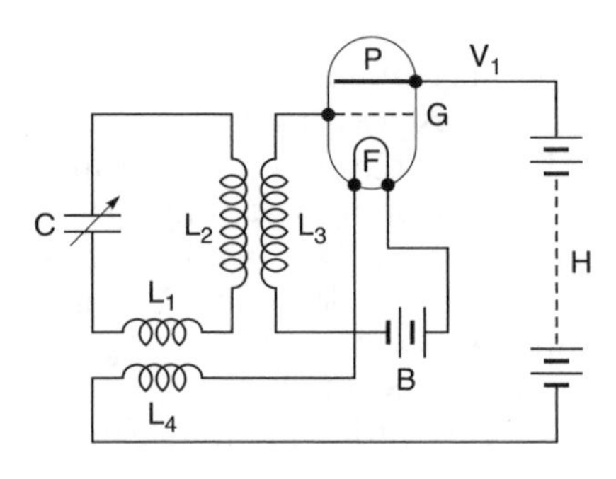

Figure 22. Meissner's oscillating circuit.
[*Source: Scott-Taggart, J. Thermionic Tubes in Radio Telegraphy and Telephony. Iliffe & Sons, London, 1924, p. 289. Reproduced with the permission of Iliffe & Sons.*]

megahertz during World War I, the de Forest-type vacuum tube oscillator was found to be inadequate because it was necessary to either shrink the tube's size or change the circuit structure in order to obtain high frequencies. In 1920, H. Barkhausen and K. Kurtz in Germany proposed raising the oscillator's frequency to 1500 megahertz by giving a strong positive bias to the tube's grid. The Barkhausen–Kurtz oscillator was the most popular ultrashortwave transmitter until the invention of microwave devices in the 1930s.

See also **Radio Receivers, Early; Radio Transmitters, Early; Radio Receivers, Valve and Transistor Circuits; Telephony, Long Distance; Valves/Vacuum Tubes**

CHEN-PANG YEANG

Further Reading

Aitken, H.G.J. *The Continuous Wave: Technology and American Radio, 1900–1932.* Princeton University Press, Princeton, 1985.
Brittain, J. *Alexanderson: Pioneer in American Electrical Engineering.* Johns Hopkins University Press, Baltimore, 1992.
Douglas, S.J. *Inventing American Broadcasting: 1899–1922.* Johns Hopkins University Press, Baltimore, 1987.
Fleming, J.A. *The Principles of Electric Wave Telegraphy and Telephony.* Longman, Green & Company, New York, 1916.
Hong, S. *Wireless: From Marconi's Black-Box to the Audion.* MIT Press, Cambridge, MA, 2001.
Scott-Taggart, J. *Thermionic Tubes in Radio Telegraphy and Telephony.* Iliffe, London, 1924.
Zenneck, J. and Seeling, A.E. *Wireless Telegraphy.* McGraw-Hill, New York, 1915.

Radio Transmitters, Early

According to James Clerk Maxwell's theory, the very existence of electromagnetic waves entailed the principle that matter can act on other matter at a distance without any intervening matter. Before Heinrich Hertz's experimental work with electromagnetic waves, no one was confident about this implication. Hertz proved it so by detecting the wave generated by a spark from an electrical discharge across the gap formed by two electrodes. Hertz was driven to this experiment by the challenge posed by Maxwell's theory, but as a theoretician himself, he had no interest in (and therefore could never have predicted) the application of his discovery—the spark transmitter and the early form of radio transmission.

Hertz was not the first to produce electromagnetic waves, but history regards him as the first to detect the waves he produced. In 1887 he created what he called "electric waves" by generating a sufficiently high voltage to create a spark. He detected the resulting electromagnetic wave by means of a loop of wire, which he called a "resonator," the ends of which terminated in a small gap. When the wave passed through the loop, an electric current was induced in the wire causing a spark to leap across the gap.

Hertz's discovery provoked much speculation about the use of electromagnetic waves for signaling, especially between land and ships in danger of running aground in fog. His apparatus, however, did not produce waves that traveled far enough or were stable enough for multiple simultaneous transmissions in the same locale. In fact, because Hertz analogized the propagation of electromagnetic waves to the propagation of light, he never would have overcome the limitations of his initial invention.

Early spark-gap transmitters consisted of a pair of electrodes and a combination of condenser and coil that were tuned to a certain frequency. The spark generated across the gap provided a short burst of electrical energy that shocked the tuned circuit into oscillation producing an electromagnetic wave that would eventually die out. Applying continuous shocks to the tuned circuit, however, would sustain the wave, much like a flywheel is kept in rotation by the continuous but intermittent force of a piston.

These transmitters produced waves that were inefficient, variable in frequency, and broad-banded. When a spark occurred across the gap, it caused some of the metal in the electrodes to vaporize. Being conductive, this vapor allowed an electrical arc to form across the gap, which lowered the efficiency of the transmitter and caused it to generate waves of multiple frequencies. Some means of dissipating the vaporized metal around the electrodes was needed—a process later called "quenching."

In 1889 Oliver Lodge demonstrated electrical resonance, the property that made possible the generation of a wave of one stable frequency. From that point the search was on to find the optimum combination of configuration and composition for the design of spark transmitter electrodes. Marconi's designs included iron electrodes shaped like mushrooms. A variety of other metals went into the electrodes: silver, aluminum, graphite, tungsten, and others. In an effort to quench the arc, some electrode designers tried forcing the spark through running liquids or encasing the electrodes in water-cooled jackets. Nikola Tesla and other researchers developed rotary gaps to produce sparks with higher repetition rates and thus more stable waves. Such continuous waves could produce musical notes; in fact, some early wireless stations in Great Britain were known to have transmitted electromagnetic renditions of "God Save the King." As these designers produced purer sinusoidal waves, the prospects for transmitting voice seemed within their reach. No designer, however, could overcome the spark transmitter's inherent noisiness in order to make radio telephony practical.

Spark transmitter designs did eventually make radio telegraphy practical. In 1897, Marconi secured a patent in Great Britain for a wireless telegraphy system that could transmit messages up to 3 kilometers. Most of the components of his system were adaptations of existing devices—Righi's spark transmitter, a Branly coherer, and a Morse inker, among others. Marconi's contribution that broke the limitation on transmission distance was his decision to liken his wireless system to the existing wired telegraphy systems. Transmitters and receivers in a wired telegraphy system used an earth ground as one of their interconnections; Marconi grounded his spark-gap transmitter and its receiver, and thereby created the first practical wireless telegraphy system. Marconi went on to develop his spark transmitter system to produce a 5-centimeter spark from 100,000 volts that transmitted a series of Morse dots from Cornwall to Newfoundland—the first transatlantic radio transmission.

On the American side of the Atlantic, Lee de Forest and Reginald A. Fessenden were developing their own spark transmitter systems. De Forest's design consisted of tungsten electrodes with an adjustable gap; Fessenden's was a rotary spark-gap design. Both of these inventors attracted the attention of the U.S. Navy, whose patronage was a major force in the development of wireless telegraphy. At the same time they were also testing the Poulsen arc transmitter, which covered greater distances and was freer from noise.

Spark transmitters were pressed into service in the early years of the twentieth century in a variety of settings. The armed forces of many nations used them, and spark transmitters first played an important role in war in the conflict between Japan and Russia in 1904. Cruise and merchant ships at this time also used spark transmitters; the *Titanic* broadcast being one of the first distress messages in the form of SOS in 1912.

Ultimately, however, de Forest's invention of the triode vacuum tube, or thermionic valve, in 1912 made possible the efficient generation of pure sinusoidal electromagnetic waves and rendered the spark transmitter with its noisy broadbanded output obsolete. In its brief lifetime the spark transmitter was essential for jump-starting the theory and practice of radio communication. Today, however, it is an historical curiosity, having been banned from the ether in 1927 by U.S law because of concerns about fire safety.

See also **Radio, Early Transmissions; Radio Receivers, Coherers and Magnetic Methods; Radio Receivers, Crystal Detectors and Receivers; Radio Receivers, Early**

LAWRENCE SOUDER

Further Reading

Archer, G.L. *History of Radio*. New York, 1938.
Beauchamp, K. *History of Telegraphy*. London, 2001.
Blake, G.G. *History of Radio Telegraphy and Telephony*. London, 1926.
Hertz, H. *Electric Waves*. London, 1900.
Hong S. *Wireless: From Marconi's Black-Box to the Audion*. Cambridge, 2001.
Howe, G.W.O. *Handbook of Wireless Telegraphy and Telephony*. London, 1921.

Useful Websites

Belrose, J.S. The Sounds of a Spark Transmitter, Telegraphy and Telephony: http://www.hammondmuseumofradio.org/spark.html

Radioactive Dating

There are natural radioactive isotopes that have half-lives comparable to the age of the earth, the most familiar being uranium-235 and -238(^{235}U, ^{238}U) and thorium-232 (^{232}Th). The decay of these isotopes proceeds sequentially through intermediate products having much shorter half-lives to the stable isotopes of lead, ^{207}Pb, ^{206}Pb, and ^{208}Pb, respectively. There are 14 other isotopes scattered through the periodic table that have half-lives spanning times from 10^{15} to 10^{9} years. All can be used for dating geologic materials. There are other isotopes, termed cosmogenic isotopes, that are produced by cosmic rays and that have half-lives that are useful for measuring periods of historical interest. The best known of these is carbon-14 (^{14}C), which has important uses in archaeology because of its half-life of 5760 years.

The age of the earth has been the subject of study since the beginning of human thought, and until the eighteenth century age of Enlightenment, theological and philosophical values were generally accepted, which for Western civilization gave an age of about 10,000 years. Geologists reasoned from observations of sedimentary rocks and from the time required for the oceans to reach their present salinity. Based on the rate at which salt is delivered to the oceans by rivers, an age of many millions of years was determined. Toward the end of the nineteenth century, ages based on rather inappropriate models of the cooling of a hot primordial earth by Lord Kelvin disputed the ages preferred by geologists, but the discovery of radioactivity at the end of that century quickly altered the situation. Ernest Rutherford was the first to suggest the principle of using uranium to helium (U/He) ratios to compute the age of rocks. In 1906 he estimated from the amount of helium found in uranium-bearing rocks that they must be at least 500 million years old. Bertram Boltwood ascertained in 1907 that lead was the other final product of uranium decay and made an age estimate based on U/Pb ratios, but neither the accumulation of helium nor of lead yielded satisfactory ages for rocks when the method of determination was chemical (rather than isotopic) analysis. The existence of isotopes was not confirmed until 1913.

It was the development of mass spectrometry immediately after World War II that opened the way to accurate age determinations, as it then became possible to measure the concentration of a decay product relative to its parent isotope. Crucial to the technique is the separation of rubidium (^{87}Rb) and strontium (^{87}Sr) from one another with ion-exchange resin, as the mass spectrometer cannot distinguish the two mass-87 isotopes from one another. Consider the example of rubidium 87 decaying to strontium 87, perhaps the simplest in concept and one of the most important in application. Otto Hahn and colleagues proposed using the decay of rubidium for age determinations, but they were unable to do so because the mass spectrometer was not a tool in their experimental kit.

Age measurements followed a 1956 paper by L. Thomas Aldrich *et al.* in which he and his colleagues determined the laboratory techniques and decay constant. Although the rubidium is generally present in amounts too small for accurate chemical analysis, it can be determined by adding a known quantity of isotopically enriched rubidium, called the "spike," to the dissolved sample. The mass spectrometer yields rubidium isotope ratios from the mixture that allow one to determine simply and accurately how much natural rubidium is present. The measurements of strontium-87 are made relative to one of the other strontium isotopes, usually strontium-86, that are not of radiogenic origin. Comparison of the strontium-87 with the rubidium-87, allows the calculation of the age from the known rubidium-87 decay rate. An igneous rock to be dated is generally broken mechanically into a number of its mineral phases, each having a different concentration of the parent rubidium-87 and for which there is a measurement of the excess strontium-87. Presenting the data for each mineral graphically gives a plot, called an isochron, that must be a straight line if the data are consistent. Deviations from the line indicate that the rock has been subjected to heat, pressure, or liquid flow that has allowed the two chemically different atoms to migrate, invalidating the age determination.

Other isotope pairs with different geochemistry have application to dating of particular rocks. Using the decay of potassium (^{40}K) into argon (^{40}Ar) came at about the same time as Rb–Sr. Somewhat later, the decay of samarium (^{147}Sm) into neodymium (^{143}Nd) became as important as that of Rb–Sr. Because their refractory nature provided insight into certain geochemical systems, measurement of the decay of rhenium (^{187}Re) into osmium (^{187}Os) was long sought, but being refractory they could not be run on thermal ion source machines as atomic ions. This limitation came to a welcome end in 1989 when it was learned that these two elements form negative molecular oxygen ions that run very stably and accurately by the thermal method.

The thermal method utilizes the low ionization potentials of periodic table Groups 2a, 3b, and the lanthanides to effect ionization by temperature alone. The chemical element to be examined is extracted from the bulk sample and deposited on a ribbon-shaped filament, usually of tantalum, rhenium, or tungsten, which is placed in the mass spectrometer ion source. On heating to the temperatures attainable with such filaments, the deposited atoms are evaporated with many having been ionized. This technique, which has proved to be of great value in radioactive dating, has two advantages: (1) the ionization is selective, so that impurities that may have been deposited with the sample will not generally be ionized, and (2) the emitted ions have very little kinetic energy, specifically only that resulting from the temperature of the filament, which is of the order of 1 electron volt. Other methods of ionization yield distributions of energies of tens and hundreds of electron volts, requiring energy filters preceding the magnet.

Geologic ages rest on knowledge of the decaying parent isotopes, and the important isotopes of uranium, thorium, rubidium, and potassium have half-lives in the range of 10^{10} and 10^{11} years. Except for uranium and thorium, which proceed to lead through the emission of a number of energetic and easily identifiable alpha particles, such long half-lives are difficult to measure in the laboratory. For this reason, very accurate measurements of uranium have been accepted as the standard. The other decay constants are determined by comparing age determinations of the same rock, meteorites being the best. Radiometric ages are generally believed accurate to better than 0.1 percent.

The use of cosmogenic isotopes differs from this simple procedure because there are no parent isotopes with which to compare them. The production of carbon-14 takes place in the upper atmosphere where it quickly bonds with oxygen to become carbon dioxide (CO_2) and is incorporated into photosynthesizing systems. After death there is no further carbon-14 replenishment, and it decays. To a first approximation the production rate is constant, so the ratio of carbon-14 to the stable carbon-12 can be used directly for the calculation of the age of the material containing the radioisotope. The rate of carbon-14 generation has varied over the past, so a correlation has to be made by studying a wide variety of archaeological materials of known age, especially tree rings of overlapping time spans. Radioactive counting of beta decays was initially the method for determining the amount of carbon-14 in a sample, but a special kind of mass spectrometry using nuclear particle accelerators has improved accuracy and extended the limits of maximum age determination.

The method is a variation of the magnetic sector type that uses a tandem Van de Graaff accelerator to measure isotope ratios of the order of 1×10^{-12}, ratios that are common in measuring cosmogenic isotopes. Such ratios are too small to be measured with conventional machines because the signal beam is "swamped" by ions scattered from the electrodes, chamber walls, and residual chamber gas. The accelerator furnishes ion energies a thousand times greater than in conventional spectrometers. Such high energies allow nuclear particle detector methods to identify individual ions of the cosmogenic atoms in the presence of huge background beams. The technique requires negative ions for the tandem Van de Graaff and by chance, five elements that have cosmogenic isotopes form such ions without interfering masses, although each requires a special experimental arrangement.

For uranium and thorium to be useful for dating, minerals must retain not only these elements but also their intermediate decay products and their lead. A mineral that best satisfies these requirements is zircon ($ZrSiO_4$), which has a strong affinity for uranium but excludes lead, and these characteristics have made it a workhorse for dating. In minerals in which the retention of uranium and lead differs sufficiently so that parent–daughter relationships cannot be used, researchers make use of uranium's unique property of having two isotopes with significantly different half-lives that give rise to two different isotopes of lead: uranium-235 to lead-207, and uranium-238 to lead-206. In the lead–lead method, the mass spectrometrist measures the lead isotopes relative to one another. Rocks of the same age that have lead-rich minerals that exclude uranium, such as galena (PbS), draw on sources with different uranium–lead compositions at the time of their hardening. For such a rock the lead isotopes taken from these minerals form an isochron that marks the time of the mineral's deposit. Claire Patterson measured the age of the earth in 1956 to be 4.55×10^9 years using the lead–lead method; this required determining the primordial isotopic composition of lead, the composition at the time of the earth's formation.

Procedures for measuring isotope composition have the disadvantage of destroying the extracted portion of the sample used. The phenomena that allow nondestructive analysis rely on exciting optical radiation or x-radiation from the atoms of the sample, most commonly induced by electron bombardment. In these methods, isotopic effects are too small to be accurately observed.

See also **Isotopic Analysis; Mass Spectrometry**

LOUIS BROWN

Further Reading

Aldrich, L. T., Wetherill, G.W., Tilton, G.R. and Davis, G.L. The half-life of ^{87}Rb. *Phys. Rev.*, 103, 1045–1047, 1956. Established the techniques of Rb/Sr dating.

Boltwood, B.B. On the ultimate disintegration products of the radioactive elements. part II: the disintegration products of uranium. *Am. J. Sci.*, 23, 77–88, 1907. Of historical importance.

Dalrymple, G.B. *The Age of the Earth*. Stanford University Press, Stanford, CA, 1991. A thorough history of the subject.

Harper, C.T., Ed. *Geochronology: Radiometric Dating of Rocks and Minerals*. Dowden, Hutchinson & Ross, Stroudsburg, PA, 1973. Reprints with comments on the basic papers.

Nier, A.O., Thompson, R.W. and Murphey, B.F. The isotopic constitution of lead and the measurement of geological time: III. *Phys. Rev.*, 60, 112–116, 1941.

Noller, J.S., Sowers, J.M. and Lettis, W.R., Eds. *Quaternary Geochronology: Methods and Applications*.

American Geophysical Union, Washington D.C., 2000. Among other things, a survey of radiocarbon dating but also many nonradioactive methods of dating.
Patterson, C. Age of meteorites and the earth. *Geochim. Cosmochim. Acta*, 10, 230–237, 1956. The age of the earth finally determined.
Wetherill, G.W. Discordant uranium–lead ages. *Trans. Am. Geophys. Union*, 37, 320–326, 1956. Mastering uranium–lead dating.

Radio-Frequency Electronics

Radio was originally conceived as a means for interpersonal communications, either person-to-person, or person-to-people, using analog waveforms containing either Morse code or actual sound. The use of radio frequencies (RF) designed to carry digital data in the form of binary code rather than voice and to replace physical wired connections between devices began in the 1970s, but the technology was not commercialized until the 1990s through digital cellular phone networks known as personal communications services (PCS) and an emerging group of wireless data network technologies just reaching commercial viability. The first of these is a so-called wireless personal area network (WPAN) technology known as Bluetooth. There are also two wireless local area networks (WLANs), generally grouped under the name Wi-Fi (wireless fidelity): (1) Wi-Fi, also known by its Institute of Electrical and Electronic Engineers (IEEE) designation 802.11b, and (2) Wi-Fi5 (802.11a).

Bluetooth replaces the physical connection between devices, such as between a personal desk assistant (PDA), digital camera, or printer to a personal computer (PC), or a cell phone headset with a cell phone. Bluetooth allows these devices to automatically interact; a PDA would automatically synchronize its database with its host PC whenever the devices come within range of each other. A Bluetooth WPAN operates on the 2.4-gigahertz-frequency band and allows devices to interact within a 10-meter area at 1 megabit per second (Mbps).

Wi-Fi (802.11b) is an Ethernet WLAN for the interconnection of PCs and mobile devices equipped with Wi-Fi transceivers, such as cable modems or other PCs, at 11 Mbps within 100 meters. The newer Wi-Fi5 (802.11a) operates at 54 Mbps within 100 meters. The propagation of individual Wi-Fi networks has resulted in larger wireless "hot spots" or wireless Internet service providers (ISPs). These anarchic hot spots theoretically allow an urban user to move from Wi-Fi node to Wi-Fi node for uninterrupted Internet access, in much the same way that a cell phone call is handed off from cell to cell.

Although the concept of wireless data networks is fairly new, the basic technology was created during World War II by two unlikely "scientists": Austrian-born actress Hedy Lamarr and orchestra leader George Antheil. Their U.S. patent described frequency-hopping, radio-controlled torpedoes that could not be jammed by the Nazis.

In the late 1950s, Sylvania's Electronic Systems Division devised an electronic version of frequency hopping and applied it to securing the military's satellite communications used during the 1962 Cuban Missile Crisis. Frequency hopping, now called spread spectrum, remained highly classified until the mid-1980s, by which time it had been developed for digital signal transmission, particularly for the code division multiple access (CDMA) protocol used by cell phones in North America and Asia.

In the late 1980s, the U.S. Federal Communications Commission (FCC) adopted rules for the commercial use of spread spectrum for low-power, unlicensed usage in the 902 to 928 megahertz, 2400 to 2483.5 megahertz, and 5752.5 to 5850 megahertz bands of the FM spectrum at a maximum of 1 watt peak output—the ISM (industrial, scientific and manufacturing) bands. Several European and Asian companies successfully lobbied their governments to adopt similar rules. Consumer telephone makers used these rules for longer-range cordless phones.

Equatorial Communications first commercialized spread spectrum data communications using a variation called direct sequence, which enabled access to multiple signals from synchronous satellites. Direct-sequence spread spectrum is used mostly in broadband applications such as Wi-Fi, while frequency hopping is used primarily for narrowband applications such as Bluetooth.

In 1997, the IEEE approved the first Wi-Fi specification, 802.11, which offered transmission speeds of 2 Mbps within 30 meters. Further improvements led in September 1999 to the adoption of Wi-Fi (802.11b) and Wi-Fi5 (802.11a). Using the wider 5-gigahertz bands instead of the 2.4-gigahertz bands, 802.11a accommodates more users, and its higher 54-mpbs speed enables multiple channels of video streaming. The 802.11g, expected to be adopted in early 2003, also offers 54 Mbps but uses the 2.4-gigahertz spectrum, making it compatible with 802.11b and enabling mixed Wi-Fi networks.

In 1994, Ericsson began research on Bluetooth, a narrowband WPAN frequency-hopping spread spectrum system named after a tenth century Danish king, Harald Blaatand (Bluetooth). In May 1998, Ericsson, IBM, Motorola, Intel, and Toshiba formed the Bluetooth Special Interest Group (SIG). In 1999, 3Com, Lucent, Microsoft, and Motorola joined the group, and the first Bluetooth specifications were published. These included specifications for a broad array of applications, or "profiles," which include wireless interconnection between stereo audio components and speakers.

In the summer of 1999, the IEEE began efforts to standardize Bluetooth specifications. On 21 March 2002, the IEEE 802.15 Working Group for Wireless Personal Area Networks approved IEEE 802.15.1, which is based on Bluetooth SIG specification version 1.1.

While Bluetooth and Wi-Fi are designed to provide clean and secure connections within the 2.4-gigahertzband, the two technologies can interfere with one another. The IEEE 802.15.2 specification, which was scheduled for adoption sometime in 2003, is designed to facilitate coexistence between the two. The Bluetooth SIG also is considering Wi-Fi coexistence.

In February 2002, the FCC approved limited deployment of a faster and broader wireless pipeline called ultrawideband (UWB), which could operate anywhere between 3.1 to 6 gigahertz at speeds of 100 to 500 Mbps within 10 meters. UWB was initially approved for applications such as collision avoidance, ground-penetrating radars, and wall-penetrating imaging systems, but questions persist about potential interference with other wireless technologies, such as global positioning satellite (GPS) systems. The IEEE is exploring the integration of UWB into 802.15.3, also called WiMedia, offering 110 Mbps in the 2.4-gigahertz band for use in imaging and multimedia applications.

There are a number of other wireless data transmission standards on the drawing boards, including IEEE 802.16, a metropolitan area network (MAN) that features 10 gigabits per second (Gbps) transmission speeds. Several worldwide spectrum regulatory bodies also have set aside spectrum in the 5-gigahertz band designed for high-speed Internet access in industrial applications. At the time of writing, only a handful of devices designed to take advantage of these bands has been announced.

See also **Electronic Communications; Mobile (Cell) Telephones; Satellites, Communications**

STEWART WOLPIN

Further Reading

Garretson, C. FCC leaning toward deregulated broadband. *InfoWorld*, 14 February, 2002.
Harmon, A. Good (or unwitting) neighbors make for good Internet access. *The New York Times*, 4 March, 2002.
Lansford, J. Can Bluetooth, Wi-Fi coexist in the future? *EE Times*, 1 February, 2002.
Leopold, G. FCC gives cautious nod to ultra-wideband. *EE Times*, 14 February, 2002.
Mannion, P. IEEE approved Bluetooth-based wireless PAN standard. *EE Times*, 22 March, 2002.
Markoff, J. The corner Internet network vs. the cellular giants. *The New York Times*, 3 March, 2002.
Reuters. FCC okays limited use of ultrawideband technology. *InfoWorld*, 14 February, 2002.
Vaughan-Nichols, S. Too many wireless LAN standards. *Enterprise*, 11 December, 2001.
Yoshida, J. UWB MACs: build or borrow? *EE Times*, 1 February, 2002.
Yoshida, J. and Leopold, G. Ultrawideband companies gear up for FCC ruling. *EE Times*, 1 February, 2002.
Zyren, J. 802.11g spec: covering the basics. *EE Times*, 1 February, 2002.

Useful Websites

Bluetooth, The Official Bluetooth Web site: http://www.bluetooth.com/
Hytha, D. Bourk, T., Linsky, J. Bluetooth adaptive hopping: tech challenges and regulatory headaches. *Comms. Design*, 13 February, 2002: http://www.commsdesign.com/story/OEG20020213S0015
IEEE 802.11 Wireless Local Area Networks, Working Group: http://ieee802.org/11/
IEEE 802.15 WPAN Task Group 1 (TG1): http://www.ieee802.org/15/pub/TG1.html
Siep, T.M., Gifford, I.C., Braley, R.C., and Heile, R.F. Paving the way for personal area network standards: an overview of the IEEE P802.15 Working Group for Personal Area Networks. *IEEE Personal Communications*, February 2000: http://grouper.ieee.org/groups/802/15/pub/2000/Mar00/99012r6P802-15_WG -PCM-Article.pdf
TI brings Wi-Fi (802.11b) to the masses. Texas Instruments press backgrounder: http://www.ti.com/corp/docs/press/backgrounder/80211.shtml
Wireless Ethernet Compatibility Alliance (WECA) website: www.wi-fi.org

Radionavigation

Astronomical and dead-reckoning techniques furnished the methods of navigating ships until the twentieth century, when exploitation of radio waves, coupled with electronics, met the needs of aircraft with their fast speeds, but also transformed all navigational techniques. The application of radio to dead reckoning has allowed vessels to determine their positions in all weather by direction finding (known as radio direction finding, or RDF) or by hyperbolic systems, discussed below.

Another use of radio, radar (radio direction and rangefinding), enables vessels to determine their distance to, or their bearing from, objects of known position. Radionavigation complements traditional navigational methods by employing three frames of reference. First, radio enables a vessel to navigate by lines of bearing to shore transmitters (the most common use of radio). This is directly analogous to the use of lighthouses for bearings. Second, shore stations may take radio bearings of craft and relay to them computed positions. Third, radio beacons provide aircraft or ships with signals that function as true compasses.

Radio developments accelerated before and during World War I. The *Mauretania* was the first ship to carry a radio receiver for taking bearings (1911). At the Battle of Jutland, Royal Navy ships tracked German fleet movements by radio. However, it was not until after 1921, when the first land-based transmitters were installed for continuous navigational use, that RDF enabled aircraft and ships to take bearings in any weather without dependence on actual sightings. In fact, bearings can be taken from virtually any detectable transmission source. The earliest receiver found on aircraft was the loop antenna, developed by the Marconi Company, also called a "radio compass." The loop rotated electronically to detect a transmission and enable a bearing to be taken (see Figure 23). Figure 24 represents a common pre-World War II short-wave RDF receiver employing antennas at right angles. Similar systems are still used, particularly common in automatic direction-finding equipment. Radionavigation, despite its advantages, is affected by atmospheric refraction, attenuation, polarization, quadrantal error (reflection from the craft itself), and differences between a visual line of bearing and one obtained by radio (see hyperbolic systems below). Furthermore, frequencies also vary in range, ionospheric distortion, and accuracy. By the 1930s, mid- to high-frequency ranges were employed for most radio applications.

World War II created an industry for improved methods of radionavigation for night bombing. The Royal Air Force deployed very high frequency (VHF) RDF stations for this purpose, while the Germans installed radiobeacons that broadcasted distinctive pulses along compass points to direct bombers to British targets.

Hyperbolic systems developed to meet tactical needs. In a hyperbolic system, a master transmitter broadcasts a distinctive series of pulses that is repeated at intervals by two slave stations. A ship, receiving signals from all transmitters, measures the time differences in their arrival at the receiver

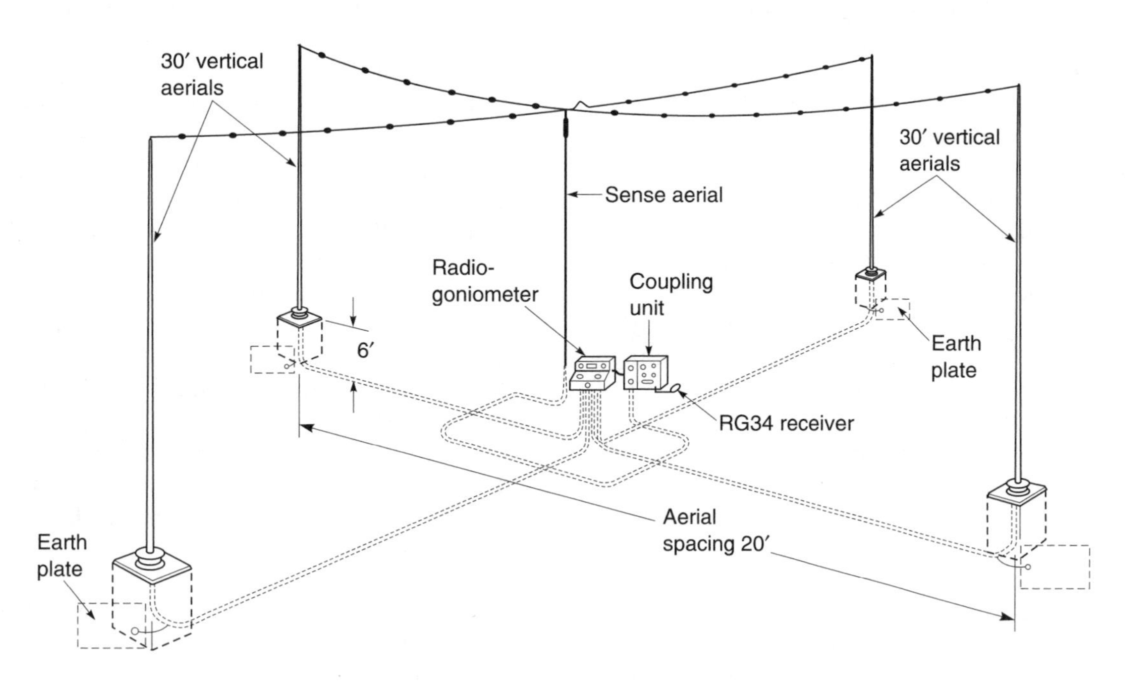

Figure 23. A loop antenna commonly found on aircraft before and during World War II. The loop couples to an autosyn that gives a relative bearing to the transmitting source. The loop aligns to a signal either manually or electrically.
[*Source: U.S. Navy. Air Navigation Part Four: Navigation Instruments, Flight Preparation Training Series. McGraw-Hill, New York, 1944, p. 103.*]

Figure 24. Marconi–Adcock RDF equipment from the 1930s. The goniometer determines the direction of signals received.
[*Source: Weems, P.V.H. Air Navigation, 2nd edn. McGraw-Hill, New York, 1938, p. 225.*]

to determine distance from each station. On a chart or plot, points at which signals show the same time difference define a hyperbola, a line of position. The points of intersection of multiple hyperbolas define a fix.

By 1942, the U.K. had established the first hyperbolic system, Gee, operating between 30 and 80 megahertz. Before the war, the Germans developed their hyperbolic system, Sonne, later adapted as Consol by the Royal Air Force. The U.S. followed closely with Loran (long-range navigation) in 1943. This method, later called Loran-A, developed after the war into Loran-C, which covers 1300 to 2300 kilometers through broadcast at 90 to 110 kilohertz, and although considered obsolete by world navies by the end of the century, remains the most successful and widely used marine radionavigation system. A British hyperbolic system, Decca, operating between 70 and 130 kilohertz (with a range under 500 kilometers), relies not on the timing of pulsed signals, but instead compares phase shifts between signals from multiple transmitters against a reference signal to determine location. First used under the name QM to support the D-Day invasion in 1944, Decca remains in use for coastal navigation, especially commercial fishing. The last hyperbolic application of the century was Omega, (still in global use), which operates between 10 and 14 kilohertz (with a range of up to a few thousand kilometers) and, like Decca, measures phase shifts.

Various researchers and inventors theorized that the Doppler shift of radio waves along the spectrum owing to motion toward or away from the measuring device had a navigational application. Following the launch of Sputnik-1, researchers at the Applied Physics Laboratory, Johns Hopkins University, developed Doppler techniques (already in aircraft navigational use) to track satellite orbits and determined that if a satellite's orbit was known, then the measurement of the satellite signal's Doppler shift with respect to an earthbound point could determine location. The U.S. Navy immediately applied the principle to the development of a satellite-based navigational system known as NAVSAT or TRANSIT, which became operational in 1964. Operating at 400 megahertz, TRANSIT satellites occupy polar geosynchronous orbits 1000 kilometers from the earth and deliver an accuracy of 0.1 nautical mile (around 185 meters). TRANSIT, however, applies mainly to slow-moving ships, and began to be phased out by the late 1990s, when it was superseded by the global positioning system, or GPS. GPS became operational in 1995 with 24 satellites in six orbital planes. In both military and commercial applications, GPS can provide time, location, and speed data to any user. A user's receiver selects signals from four satellites and synchronizes them to obtain locational data based on a comparison of time signals emanating from clocks internal to the satellites and receiver. With the launch of additional satellites, GPS promises reliability and accuracy, and will doubtless supersede other navigational methods.

Radar followed a concurrent development with RDF and hyperbolic navigation. During the 1920s, the detection of ships by reflected radio waves had been demonstrated, and by the outset of World War II, radar detection was established in the U.K. Radar equipment measured a time interval between the transmission of a signal and its reception as an echo as reflected from an object. Radar had been installed on naval ships during the 1930s, but its accuracy was poor. British scientists at Birmingham University pioneered radar's wartime breakthrough with the invention of the multicavity magnetron, which permitted the Royal Navy to deploy imaging at centimeter wavelengths, a huge improvement in accuracy. British authorities shared the invention with the U.S., where scientists developed microwave radar during the war. Following the war, radar saw wide navigational applications, and by the end of the century, most vessels and aircraft were required to carry radar.

Aircraft and ships today receive continuous positional data through radio fixing. Ships rely on radionavigation in conjunction with other methods, while aircraft have been more radio-dependent. Submarines and missiles rely on inertial methods, although the advent of GPS may eventually replace all other methods of radionavigation.

See also **Global Positioning System (GPS); Gyrocompass and Inertial Navigation.**

ROBERT D. HICKS

Further Reading

Defense Mapping Agency Hydrographic/Topographic Center. *The American Practical Navigator*, Pub. No. 9, Bethesda, MD, 1995.

May, W.E., *A History of Marine Navigation*. W.W. Norton, New York, 1973.

Page, R.M. *The Origin of Radar*. Anchor, New York, 1962.

Sonnenberg, G.J. *Radar and Electronic Navigation*, 6th edn. Butterworth-Heinemann, London, 1988.

Tetley, L. and Calcutt,D. *Electronic Navigation Systems*, 3rd edn. Butterworth-Heinemann, London, 2001.

Williams, J.E.D. *From Sails to Satellites: The Origin and Development of Navigational Science*. Oxford University Press, Oxford, 1994.

Wylie, F.J. *The Use of Radar at Sea*. Hollis & Carter, London, 1952.

Rail, Diesel, and Diesel-Electric Locomotives

Gasoline and oil-engine trolleycars appeared in the late nineteenth century. Internal combustion locomotives were first tried in dockyards, on military and contractor's railways, and where steam traction was undesirable, but these were of low power and limited use. Diesel traction owed much to the thermal electric principle demonstrated by J.J. Heilmann in the 1890s using locomotives that carried their own electricity-generating steam plant. Between 1900 and 1920 the oil engine was put on rail: passenger vehicles and baggage cars were equipped with gasoline or oil motors for branch line work; road vehicles were fitted with rail wheels; and a few new intercity trains were constructed. Oil-engine locomotives were, however, relatively rare.

Pioneering work in Scandinavia around 1912, Germany in 1914, and the U.S. about 1917 was crucial. Direct mechanical drive to the wheels via shafts and gear boxes, sometimes with hydraulic drive, was suitable for low or moderate power, and electric transmission was favored for greater power, especially in the U.S. General use of the diesel engine with electric transmission, was retarded by the lack of a system to control engine, generator, and traction motors working together. This was eventually provided by Lemp in the U.S.

Electric and diesel-electric locomotives used common components, and multimode units were constructed in the 1920s with a diesel engine, batteries, current collection shoes, and pantograph. However, a considerable increase in power-to-weight ratio was needed for the diesel locomotive to become a general traction unit. The Winton Engine Company, engine suppliers to the General Motors-owned Electro-Motive Company, produced the motor needed. Harold Hamilton and Richard Dilworth founded the Electro-Motive Company and built over 500 rail cars between 1924 and 1933. Electro-Motive was taken over by General Motors (GM) in 1930 but remained an autonomous concern. During the interwar period, General Electric (GE) led diesel locomotive work and produced several noteworthy demonstration units but lost the lead to General Motors in the 1930s.

The U.S. led the dieselization of railways after 1930. Working relationships were established between steam locomotive builders (Alco, Baldwin, and Lima), the big electrical combines such as Westinghouse and GE, and the makers of internal combustion prime movers such as Ingersoll and GM. Development projects were controlled by large companies that had the range of engineering skills and funds. In the U.S., electric transmission was standard from the beginning, because low-voltage direct current (DC) systems dominated railway traction and street trolleys. The DC motors were ideal for railway service, and were therefore used for the diesel locomotive in the U.S.

In Europe the suitability of electric transmission was questioned. Many forms of transmission were tried, including steam, air, exhaust gas, and mechanical systems. These "compound diesels" demanded radical departures from the orthodox form of oil engine. Hybrid engines, like the Still, combined the diesel cycle with steam using an exhaust-heated boiler, but the system failed to become established. In Russia, Lenin funded investigations by G.W. Lomonossoff into the main types of transmission, with considerable German help. He found that electric transmission was the best method of transmitting work from the engine to the wheels over a wide speed range. In Russia, there was considerable theoretical analysis of locomotive thermodynamics and design, and one of the first university chairs devoted to the diesel locomotive was set up in Moscow. In

Germany there was close cooperation between the railway company's research departments, technical university specialist schools, and the manufacturers' research teams. However, no European establishment matched the organized program mounted by GM to dieselize railways. Several lightweight, high-speed passenger trains powered by diesel engines were demonstrated in the 1930s, but these were not powerful enough to displace the best steam locomotives of the time.

GM used automobile industry lessons of manufacture, marketing, publicity, sales technique, and training and applied them in launching diesel traction. Production designs were standard, hire-purchase terms encouraged railways to accept diesels, demonstration units were offered on loan to work prestige train routes. In 1935, two 1800-horsepower Electro-Motive units coupled together worked the Sante Fe Super Chief straight through from Chicago to Los Angeles. Free training of engine crew was available, and freight diesels toured North America demonstrating their superiority to steam traction. Although, dependent at first on GE for manufacture of electrical parts, GM set up its own electrical factory to meet its locomotive needs. Independent after 1938, GM built up a lead that other manufacturers could not overtake. GE failed to displace the GM diesel with a steam-electric locomotive called the Steamotive, but using its considerable resources was able to capture a fair share of the market. Other builders in the U.S. tried to introduce diesels of their own, but went out of business in the 1950s.

Following World War II, postwar recovery led to dieselizing of nonelectrified railways in Europe. Experiments with gas-turbine-electric locomotives and advanced diesel-mechanical and hydraulic locomotives failed to displace the diesel-electric as the best complement to electric traction. In the U.S., diesel-electric traction was regarded as "cheap electrification" that avoided fixed works.

In Europe, the diesel engine became a power unit in all forms of train. The pioneer work of Franz Kruckenberg in Germany in the 1930s, led to fast diesel trains in the postwar Benelux countries, continuing development that had started with a train called the Flying Hamburger. By the 1960s, British high-speed trains showed that diesels could work long-distance express services faster than steam and could rival contemporary electric operations. Present day electric *Train á Grande Vitesse* (TGV) operations far exceed the speed of any diesel service, but the modern diesel train can regularly work at 250 kilometers per hour. In the 1930s diesel began to displace steam and electricity in freight hauling. In the U.S. and Russia, diesels run in multiple have moved test trains of over 100,000 tons in weight controlled from one cab.

Electronics has transformed the diesel locomotive. Component size has been reduced. Solid-state rectifiers enabled the DC generator to be replaced by the smaller alternating current (AC) alternator. The solid-state inverter and control apparatus enabled the asynchronous motor to replace the traditional DC traction motor. Globalization of industries in the 1980s led to designs for a universal diesel locomotive able to work anywhere in the world. This philosophy gave way to the modular policy of building diesel and electric units, from heavy goods engines to TGV, using common components. At the close of the twentieth century, European companies cooperated with North American combines to produce global designs, and GM diesel-electric locomotives made in North America were at work in Europe. The diesel locomotive continued to develop: driverless working was demonstrated; and experiments to operate diesel locomotives with hydrogen, coal-based fuels, methane and other gases, and coal-slurry have been carried out. Though it cannot match the very high performance of the electric locomotive, the diesel is ideal for all operations outside electrified zones.

See also **Rail, High Speed**

MICHAEL C. DUFFY

Further Reading

Brown, H.F. Economic results of diesel-electric motive power on the railways of the United States of America. *Proc. Inst. Mech. Eng.*, 175, 5, pp 257–317, 1961.

Cummins, L. *Diesel's Engine, Vol.1: From Conception to 1918*. Carnot Press, Oregon, 1993.

Duffy, M.C. The Still engine and railway traction," *Trans. Newcomen Soc.*, 59, pp 31–59, 1987–1988.

Dunlop, J. Internal combustion locomotives. *Trans. Inst. Eng. Ship. Scotland*, 68, 234–278, 1925.

Fell, L.F.R. The compression ignition engine and its applicability to British railways. *Proc. Inst. Mech. Eng.*, 124, 1–66, 1933.

Hinde, D.W. *Electric and Diesel Locomotives*. Macmillan, London, 1948.

Kirkland, J.F. *Dawn of the Diesel Age*. Interurban Press, Glendale, CA, 1983.

Morgan, D.P. Diesel traction in North America, in *Concise Encyclopedia of World Railway Locomotives*. Hutchinson, 1955, pp 107–142.

Tufnell, R.M. *The Diesel Impact on British Rail*. Mechanical Engineering Press, 1979.

Vickers, R.L. The beginnings of diesel-electric traction," *Trans. Newcomen Soc.*, 71, 1, 115–127, 1999–2000.

Rail, Electric Locomotives

Electric locomotion grew from nineteenth century electrification of quarries, mines, light railways, and cable car systems, and basic forms were established in the street trolleys of the 1880s in the U.S. Heavy-duty electric traction evolved in the 1890s on rapid transit railways in large cities like Chicago and New York. Low-voltage direct current (LVDC) technology, using rotary converters and conductor rails, was needed for terminal and tunnel lines, where locomotives were required. Locomotives equal in power to any steam locomotive were working in the period from 1900–1910. Overall energy efficiency was low if power came from coal stations and electric traction was employed to meet legal requirements or operational needs. Early locomotives took many forms: some used body-mounted motors driving through rods and shafts; others used geared axle-mounted nose-suspended motors. Other than heavy-duty rapid-transit railways, electric traction could not replace steam traction on general-purpose railways. The Pennsylvania Railroad and New York Central terminal lines and the Detroit and Baltimore tunnel lines were early locomotive worked railways.

From 1900 to 1920, other electric transport systems appeared, including cableways (Germany, Switzerland), monorails (Langen, Kearney), and funiculars (Italy). Electric rail "mules" were used to haul ships through the Panama Canal locks.

The rail-less trolleybus, locomotive, and train, which drew power from an overhead line and ran on road wheels, appeared before World War I (1914–1918). Road locomotives powered from batteries found use. A hybrid vehicle, with engine, generator, and accumulators, was invented, and found use throughout the twentieth century in passenger transport over routes too lightly used to justify electric tramways. Heavy dumper trucks that draw power from an overhead supply line continue to be of use in quarries. The use of electric rail-less road vehicles declined with the rise of motor transport. Many electric trolley systems were dismantled, but there has been a general revival of electric railways since 1970.

The early heavy-duty electric railways, with third-rail direct current (DC) distribution were not suitable for long-distance mainline work. The three-phase railway dated from the Lugarno tramway (1896), and was extensively used in Italy, on the Simplon Tunnel route (Switzerland), and through the Cascade Tunnel (U.S.). The need for two overhead wires and speed-control limitations prevented widespread use. By the early 1930s this system had been replaced by monophase systems in Switzerland and the U.S., although it survived in Italy until after World War II. The single-phase alternating current (AC) railway dates from 1902 with the introduction of a reliable, powerful single-phase commutator motor. The 1905 tests of 11-kilovolt single-phase AC locomotives on the Pennsylvania Railroad in 1905 established the system, and in 1907 it was used on the New Haven Railroad lines into New York. In 1914 it was installed on the Philadelphia Paoli Division of the Pennsylvania Railroad.

In Europe, the 1908 Loetschberg line in Switzerland used 15 kilovolts and 16.66 hertz, which became standard in Austria, Switzerland, Germany, Sweden, and Norway until replaced by the 25 kilovolt, 50 hertz European standard in the 1950s. Tests on the Seebach–Wettingen line between 1901 and 1905 with phase converter locomotives investigated various motor and control systems drawing power from a single contact wire. The first locomotive to use industrial frequency (50 hertz), at 15 kilovolts, was tested there.

By 1913, locomotives of 2500 horsepower were in use. AC motors of great power were large and often body-mounted, with rod and jackshaft drive and many engineers favored the DC motor. Converter locomotives used single-phase AC supply with DC motors by having an onboard rotary converter. These were also heavy, large machines. The first in the U.S. was constructed in 1925 by Henry Ford at the Rouge River Plant for the Ford-owned Detroit, Toledo and Ironton railroads. Its supply was 11 kilovolts and 25 hertz; its traction power was 3100 kilowatts; and its weight was 393 tons.

Converter locomotives were never numerous. The final American examples were supplied in 1948 to the Great Northern Railway (320 tons, 5000 horsepower) and to the Virginian Railway (454 tons, 7800 horsepower). It was rendered obsolete by the locomotive mercury-arc rectifier, and the solid-state rectifier.

The Hungarian Kando experimented between 1927 and 1941 to combine single-phase AC supply with DC motors, but size, weight and complexity prevented general use. By the mid-1930s rectifier locomotives using mercury vapor were under trial. Phase-splitting locomotives, which converted single-phase AC to three-phase supply to induction-type traction motors, were used on the Norfolk and Western Railway in 1915 and on the Virginian Railway. Its supply was single-phase 11 kilovolts and 25 hertz.

The high-voltage direct current (HVDC) railway generally used a supply voltage of 1500 or 3000 volts. This was proven in 1911 on the Butte, Anaconda, and Pacific Railway in Montana, and used on the Montana Division electrification scheme (700 kilometers) of the Chicago, Milwaukee, and St. Paul Railway, completed in 1915. This showed that a general-purpose steam railway could be better worked by electric traction. All sections used 3000 volts DC. These installations influenced strategy worldwide, and HVDC became a world standard.

The spread of railway electrification was encouraged by the construction of high-efficiency thermal power stations, the use of the mercury-arc rectifier as a static rectifier and phase converter, and the standardization of grid supply after 1930. In Europe, electrification was aided by national ownership of railways The wars, economic difficulties, and political crises of Europe and Asia between 1914 and 1950 delayed many electrification projects. In the U.S., the diesel-electric locomotive met the needs of general-purpose railways, and electric locomotive development ceased until the 1980s. By the late 1930s, electric locomotives were outperforming steam traction in every field. In the 1950s there were increasing demands for much higher speeds, well in excess of 160 kilometers per hour, and the superiority of electric traction became evident. The ability of electric locomotives to integrate closely with electric signaling and control technologies was advantageous.

After World War II, the 25-kilovolt 50-hertz single-phase system became the world standard. High-speed trials in France, under the 1500-volt DC system, set a world speed record of 331 kilometers per hour with locomotive BB9004 in March 1955. This record stood until 1981 when a SNCF *Train á Grande Vitesse* (TGV) train reached 371 kilometers per hour. The locomotive-mounted mercury-arc rectifier combined the single-phase AC supply with DC traction motors. Hibbert conducted pioneer experiments in 1913, but regular use did not begin until 1950. Locomotive mercury rectifiers proved unreliable and were replaced by solid-state devices from the mid-1960s. The success of solid-state rectifiers, inverters, and controllers enabled any type of supply system to be used with any kind of traction motor. In the 1970s Brown–Boveri developed variable speed control for modern three-phase traction motors. Solid-state devices provided multistage transformations of energy between supply and motors with minimal loss. They introduced reliable, compact technologies for suppressing interference between traction equipment and signaling and control networks. Onboard information processing systems shifted trackside signaling into the locomotive and made possible the automated, driverless mainline railway. Driverless railways were previously simple systems like the London Post Office Railway of the 1920s. Rapid transit lines in San Francisco, London, and other large cities were automated after the 1960s. Mine railways were worked without drivers in the U.S., and driverless goods trains tested on main lines in Germany. By the close of the twentieth century it was possible to monitor all locomotive operations from a control satellite in geostationary orbit. Automatic control of all operations is likely to be introduced in the early twenty-first century.

In the 1970s the "universal" electric locomotive was designed for most duties. The DB Class 120, introduced in 1979, is one example. The "modular" philosophy became dominant, and a range of locomotives, from shunting engines to very high-speed trains use components in common. Design has been "globalized" by the formation of worldwide combines. European-based industries such as Adtranz and Alstom are moving toward partnership with American companies like General Electric and General Motors. In Great Britain, privatization of the nationalized railways has slowed new electrification because private operators find diesel traction cheaper. In North America, the diesel remains the norm, despite occasional proposals to use dual-mode locomotives over partly electrified routes. On the high-speed railways found in most industrialized countries, both multiple-unit and locomotive-hauled trains generally run at speeds of 350 kilometers per hour, and speeds of 515 kilometers per hour have been attained. The limits of wheel-on-rail transport have not yet been reached, and conventional TGV trains have outperformed maglev (magnetic levitation), jet-trains, tracked hovercraft, and monorails. Apart from short run passenger conveyers, long-distance maglev routes are unlikely, despite the engineering success of the German and Japanese test installations. In terms of passenger-carrying capacity, loads moved, and speed, the electric railway worked with conventional locomotives and multiple units has much unrealized potential. The effects on railways of applying electronic control and guidance to road vehicles and modern road-trains (the "automated highways" concept) have yet to be determined.

See also **Rail, Diesel and Diesel Electric Locomotives; Rail, High Speed; Rail, Steam Locomotives; Railway Mechanics**

MICHAEL C. DUFFY

Further Reading

Agnew, W.A. *Electric Trains: Their Equipment and Operation*, 2 vols. Virtue, London, 1937.

Bailey, M.R. The tracked hovercraft. *Trans. Newcomen Soc.*, 65, 129–145, 1993–1994.

Bezilla, M. *Electric Traction on the Pennsylvania Railroad, 1895–1968*. Pennsylvania State University Press, 1980.

Binney, E.A. *Electric Traction Engineering*. Cleaver Hume, 1965.

Carter, F.W. *Railway Electric Traction*. Arnold, 1922.

Dover, A.T. *Electric Traction*. Pitman, London, 1954, 1929, 1918, and 1917. History of electric traction theory and practice traced through successive editions of standard text.

Duffy, M.C., *Electric Railways 1880–1990*. Institution of Electrical Engineers, London, 2003. History of progressive integration of traction, power supply, signaling, and control elements of electric railway system.

Haut, F.J.G. *The History of the Electric Locomotive*. George Allen & Unwin, 1969.

Hinde, D.W. *Electric and Diesel Locomotives*. Macmillan, 1948.

Johnson, J. and Long, R.A. *British Railways Engineering 1948–1980*. Mechanical Engineering Publications, 1981. Contains review of British electrification policy and electric traction equipment.

Kaller, R. and Allenbach, J.-M. *Traction Electrique*, 2 vols. Presses Polytechniques et Universitaires Romandes, Lausanne, 1995. Review of current railway and tramway practice.

Machefert-Tassin, Y.M., Nouvion, F and Woimant, J. *Histoire de la Traction Electrique, Tome 1: Des Origines a 1940*. La Vie du Rail, Paris, 1980; *Tome 2: De 1940 a Nos Jours*, La Vie du Rail, Paris, 1986. The most thorough and authoritative history of electric traction yet available.

Manson, A.J. *Railroad Electrification and the Electric Locomotive*. Simmons Boardman, New York, 1925.

Middleton, W.D. *The Interurban Era*. Kalmbach, 1961.

Middleton, W.D. *The Time of the Trolley*. Kalmbach, 1967.

Middleton, W.D. When the Steam Railroads Electrified. Kalmbach, 1974.

Raven, V. Railway electrification. *Trans. North-Eastern Coast. Eng. Shipbuild.*, 38, 173–240, 1921–1922.

Raven, V. Electric locomotives. *Proc. Inst. Mech. Eng.* Paris Meeting issue, 1922, pp. 735–781, 1057–1082.

Semmens, P and Machefert-Tassin, Y. *Channel Tunnel Trains: Channel Tunnel Rolling Stock and the Eurotunnel System*. Eurotunnel, 1994. Review of equipment and operations of modern electric railway.

Rail, High Speed

High speed railway operations in the twentieth century resulted from deliberate attempts to work a system at speeds much higher than the maximum speeds attained on conventional railways, with competition for the shortest time on routes between large cities; for example, passenger train rivalry on routes between New York and Chicago in the 1930s. High-speed railways required construction of new lines; segregation of operations; new forms of rolling stock, and unorthodox methods of propulsion relying on techniques developed outside the railway industry. However, the great success of the segregated high-speed railway is a post-1960s phenomenon.

In the 1890s the "crack" express trains might have reached 145 kilometers per hour (km/h) for short periods, but sustained working at 95 km/h (or 60 miles per hour) was considered adequate. A first-class railway could be proud of its "mile-a-minute" express timetables, start-to-stop. It was a wiser objective to increase the number of express trains averaging 95 km/h, and briefly touching a maximum of 145 km/h, than to seek ways of running at speeds far in excess of 150 km/h. In the years from 1900 to 1914, a few very fast trains were run by companies in Europe and the U.S. using conventional rolling stock, which may have exceeded 160 km/h during unscheduled attempts at a record. These reports were not authenticated and remain doubtful. Scheduled running in excess of 160 km/h tested contemporary technology and made little economic sense.

In the closing years of the nineteenth century, some experiments were made to see how fast trains could run. There were various schemes for constructing completely new railways, powered by electricity, and intended to run streamlined expresses at speeds well in excess of 200 km/h. Many of these schemes were fraudulent with a main aim of robbing unwise investors of their money, and few made engineering sense. One early project that displayed engineering skill if not economic acumen was due to David G. Weems, an American dentist who built a narrow-gauge electric railway in 1889 to attract funds for a full-sized system. A model car of some 2 meters in length reached a speed of 192 km/h over a 2-mile (3.2-kilometer) circuit. It was driven by electricity picked up from overhead wires with return through the rails.

In some quarters the steam-electric locomotives of J.J. Heilmann, built in the 1890s, were seen as potential motive power units for high-speed railways, despite their intended role in the electrification of ordinary railways. Radically different systems intended to replace steam railways were proposed by competent engineers driven by a utopian vision of rapid communication fostering universal brotherhood. Some concepts failed technically; others were of limited use. None of the proposed alternatives was particularly fast.

Serious, scientific engineering research into very high-speed railways began with investigations in Germany by Siemens & Halske and by AEG between 1899 and 1903. These trials demonstrated the potential of electric traction in sustained running at speeds that no steam locomotive could match. The trials had state backing and took place on a military railway between Marienfeld and Zossen, near Berlin. Three vehicles were tried: one locomotive not intended for high-speed working, and two motorized carriages which were to set long-standing records. The vehicles were fitted with three-phase motors, with a 10 kilovolt supply picked up by a triple collector from three overhead contact wires. Speed was controlled by varying the speed of the steam engine driving the alternator according to signals telegraphed from the cars to the power house, which varied frequency: either 25 or 50 hertz. Speeds of the order of 210 km/h were reached by both cars. An attempt to match this performance in 1904 by a specially constructed 4-4-4 steam locomotive failed, though the design, by Kuhn, was to influence later German attempts to work steam railways at very high speeds. There was no compelling demand between 1900 and 1950 for very fast express passenger trains able to run above the maximum set by existing engineering limits and the need to work many kinds of trains, at very different speeds, over a common system. The years of war, economic difficulty, political uncertainty, postwar reconstruction of basic industries, and reorganization of many national railways, were not favorable to high-speed railways. The emphasis was on improving, or making do, with what existed. Some extremely valuable research into fast running and vehicle design was undertaken by Franz Kruckenberg in Germany in the 1930s. In Europe and the U.S. improved regular services between major cities were provided by short, lightweight trains powered by internal-combustion engines. The Zephyrs in the U.S., and the Flying Hamburger and similar trains in Germany and the Netherlands were examples. For high-speed operating of heavy trains over several hundred kilometers, only steam traction was available in the 1930s. The Chicago North Western's Hiawatha trains, the British LMS Coronation Scot, and LNER Coronation and Silver Jubilee demonstrated steam traction at a peak that would be difficult to surpass. These services severely taxed existing signaling and made the routing of slower trains difficult. Record-breaking runs with steam locomotives, like the LNER Mallard, which reached 203 km/h, questioned the wisdom of trying to work trains by steam at sustained speeds of over 160 km/h.

During World War II, a detailed study of high-speed railways was undertaken by the German state in the years when it was confident of victory. Plans were made for a network of 250 km/h services throughout the Greater German Reich and its conquered territories, some using improved standard-gauge routes and others using a proposed broad-gauge system promoted by Hitler. Very thorough studies were completed, and the broad-gauge project continued until halted by the defeat of Germany in 1945. Many modes of traction were considered, but electric traction was not favored. The standard-gauge designs included development of the orthodox steam locomotive; steam-electric traction; gas-turbine-electric traction; high-speed steam motors, and various internal-combustion engines with electric, hydraulic, and mechanical transmission. The most favored design, from Lubeck Technical University, was made up of two 4-8-4 five-cylinder compound-expansion locomotives, placed back-to-back with a condensing tender between. The general disposition owed something to Kuhn's earlier design and the design of each locomotive was derived from Chapelon's Sorbonne proposal. None of the designs was realized, though one of them was based on the General Electric "Steamotive" which was tested in the U.S. in the late 1930s. Even more gigantic locomotives were designed for the broad-gauge project.

In 1939, the railway speed record was 230.2 km/h (143 mph) set between Ludwigslust and Wittenberge, Germany, in June 1931 by the propeller-driven Kruckenberg Rail Zeppelin. The record stood until February 1954. The Kruckenberg Car was an experimental vehicle to test vehicle behavior at high speeds and had a steerable axle controlled from the cab. In the 1930s the speed record for steam traction was 161 km/h set in November 1934 by 4-6-2 engine 4472 on the LNER between Grantham and Peterborough, England. This was raised in March 1935 to 173.8 km/h over the same route. A speed of 181 km/h was equaled by Milwaukee Road 4-4-2 and LNER 4-6-2 in 1935, and LMS 4-6-2 in 1937. In May 1936 German 4-6-4 O5.002 reached 200.4 km/h at Neustadt an der Dosse, and in July 1938 the record for steam traction was set at 202.8 km/h on the LNER between Grantham and Peterborough by an A4 4-6-2 engine. These records were less than the speed gained by electric traction in 1903 (Marienfelde-Zossen). They were surpassed by the diesel records of 181 km/h set by the Pioneer Zephyr in the U.S. in May 1934; and 205 km/h, reached by the Leipzig railcar between Ludwigslust and Wittenberge in February 1936. In June 1939, a Kruckenberg diesel set reached 215 km/h between

Hamburg and Berlin. Very-high-speed running then ceased during the war and in the long recovery period afterward. When research into high-speed railway operations was resumed in 1954, the initiative was largely with electric traction.

By the mid-1950s, the industrial economies had recovered sufficiently from the war to make use of airlines and motor transport on an increased scale. Railroads needed to increase speed of services to win back passengers from airlines and automobiles for journeys between 400 and 600 kilometers. It was necessary to determine how fast trains could be operated over existing routes within the standard loading gauge. A separate issue concerned the forms guided land transport might take. There were three general strategies: to run trains over existing railways at much higher speeds; to build new high-speed railways of conventional form, reserved for trains to run at very high speeds unhindered by slower operations; and to seek a completely new form of guided high-speed transport. The former strategy was pursued by British Rail via its Advanced Passenger Train project. This was a new, tilting electric train designed to take curves at higher than normal speeds, and to run at a maximum well above conventional train speeds; that is, faster than the high-speed diesel train introduced in the 1970s. The same concept has been used in the Spanish "Talgo" trains since the 1950s. It enabled the existing infrastructure to be used, but created difficulties in routing trains at lower speeds. Gas-turbine power was used during the experimental stage, but electric traction was chosen for production units. Unfortunately the project failed for nontechnical reasons. The strategy of building segregated high-speed routes (*Lignes a Grande Vitesse*, or LGV) was pioneered by the French and Japanese and taken up by other nations. Standard gauge was the norm, so these lines could be linked into the existing network in most instances, and the *Trains á Grande Vitesse* (TGV) could run over old and new lines and use city stations. On segregated routes, trains ran through curves at a set speed, so the rails could be superelevated to eliminate passenger discomfort. Tilting coaches were unnecessary and design was simplified. Segregated lines were the best means of getting high-speed rail services, but were extremely expensive and required state aid. Availability of funds sets the limits to the extent of the network. The third strategy, to build completely new kinds of guided land transport systems, has not been successful despite experiments with many different forms, none of which (with the possible exception of maglev, or magnetic levitation) approached the performance of the conventional TGV. Their proposers failed to anticipate the services offered on the electric LGV. The British guided or tracked hovercraft, the French jetrain, and the present-day German and French "maglev" projects are the best-known examples. At the close of the twentieth century, only maglev was being pursued with serious commitment and full-scale test facilities in Japan.

Research into very high-speed running was resumed in 1954. In February the SNCF ran locomotive hauled carriages on the 1500-volt DC Paris–Lyons main line. Locomotive CC 7121 reached 243 km/h with a three-coach train between Dijon and Beaune, thereby breaking all previous speed records. The rolling stock was relatively unmodified. In 1955, further SNCF speed trials were held on the electrified Bordeaux–Hendaye main line, with modified rolling stock and supply system. Mobile substations increased supply voltage from 1500 to 1900 volt DC. In March 1955, locomotive CC7107 reached 326 km/h, and BB9004 set a new record of 331 km/h, which lasted until 1981 when the SNCF TGV reached 371 km/h. The tests showed that a great deal had to be done to improve pantograph–catenary and vehicle–rail interactions, and to reduce aerial disturbance, which raised dust, disturbed ballast, and generated noise. Research was begun in several countries to achieve these goals and operate wheel-on-rail systems at increased speeds. The Japanese *Shinkansen* system originated in plans for a new electric railway between Tokyo and Osaka to provide greatly increased passenger-carrying capacity when the existing narrow-gauge link could no longer cope. A high-speed segregated line was opened in 1964 to provide a start-to-stop average of 161 km/h. Plans to work freight trains by night were abandoned: short nighttime periods were used for repairs; the rest was needed for passenger working. Success of the first line led to construction of the *Shinkansen* network, which stimulated similar high-speed railways elsewhere. The Japanese used lightweight sets with motors distributed throughout the train. Though successful, these networks were opposed on economic and environmental grounds. They required expensive noise suppression screens and embankments. State subsidies were essential. French trials in 1954–1955 resulted in the first regular passenger services in Europe to run at 200 km/h begun in 1967 by the *Capitole* express between Paris and Toulouse. These first trains used locomotive haulage and ran over ordinary railways with improved signaling, control, and communications. In the 1970s these SNCF services were faster than any outside Japan. The need to reduce axle loading called into question the use of locomotives on LGV, but both locomotives and lightweight sets have found use. To reduce axle loading, and to provide very fast services on nonelectrified routes, the SNCF tested gas turbine-powered sets with hydromechanical transmission. In 1969, a five-car set was built with three trailers between two outer gas turbine-powered cars, containing traction alternators. The train was articulated, with traction motors on each axle. Eventually power-to-weight ratio was raised to 23.3 kilowatts per ton. Tests began in April 1972 and in December a speed of 318 km/h was reached. During four years running, this unit reached 250 km/h over a thousand times. Features now common on high-speed rolling stock were developed during these trials. The French plans to use such trains over existing railways met with routing problems, and eventually a network of segregated LGV was constructed. Changes in relative fuel costs prompted a switch to electric traction, using the new international standard of 25 kilovolts and 50 hertz. Some TGVs could also run under the 1500-volt DC of older lines. The French LGV, like the Japanese, employed cab signaling, and there were no lineside signals apart from marker boards at the start of each block. Signal blocks were indicated for five blocks ahead.

In Great Britain a different philosophy was pursued because new, segregated lines were ruled out on grounds of excessive costs and disruption to built-up areas. The High Speed Train (HST) was developed in the late 1960s to provide services at 200 km/h over improved existing lines using two diesel electric power units at the end of a rake of coaches. This proved most successful. To provide much higher speeds, of 250 km/h and above, the Advanced Passenger Train was developed, drawing on research into vehicle–rail interaction begun at the Railway Technical Centre, Derby, in 1964 to resolve problems related to high-speed goods operations. Such research showed that wheel-on-rail systems could operate at at least 320 or even 400 km/h. The present record is 515 km/h held by a French TGV. Because of this, British Railways Board rejected projects for unusual schemes and launched the Advanced Passenger Train (APT) program in 1968. The project received very little funding compared to that allocated by the French and Japanese states to the LGV and *Shinkansen*.

The engineers' story is related by Williams (1985). Special research and development facilities were set up in Derby and a disused line converted into a test route. Members were recruited to the design team from outside the railway industry. The APT experimental unit was completed in 1972 and used to test the tilt mechanism, bogies, suspensions, and train–track interaction. Authority for three prototypes was given in 1974. An electric and gas turbine-powered version were considered. However, the final test run was made in April 1976 and the project was terminated. Succeeding operations were left to the HST, and a new generation of conventional electric locomotives able to run at 225 km/h. Tilting trains have been used in Scandinavia, Italy, Switzerland, Norway, and Japan to reduce journey time over routes with heavy curves, but most TGVs at work in many lands are nontilting. The success of the Japanese and French TGV led to similar trains, working over routes partially or largely segregated, in Germany, Italy, Belgium, Netherlands, Spain, and South Korea. The cross-Channel Eurostar trains are an example. The TGVs associated with France and Germany have been used to advance the international business interests of the railway engineering industries of these countries, though globalization is reducing such national identification.

In the U.S., the primacy given to air transport for moderate to long distances and the reliance on automobiles for much else discourages construction of LGV on the European scale. In the closing years of the twentieth century, the U.S. introduced TGV, based on European practice, for service in the Northeast Corridor (Boston, New York, and Washington D.C.). These were tilting trains and could operate at 240 km/h. The future of LGV is uncertain, granted their great expense and the success of airlines in meeting competition over distances between 200 and 600 kilometers. The TGV work best between large centers of population where total journey times are about 3 or 4 hours maximum, and time is important.

See also **Railway Mechanics**

MICHAEL C. DUFFY

Further Reading

Anonymous. Les tres grandes vitesses ferroviaires en France. *Revue d'Histoire des Chemins de Fer*, 12–13, Paris, 1995. Special issue devoted to French TGV.
Hughes, M. *Rail 300*. Ian Allan. 1988.
Rahn, T., Ed. *ICE: High-Tech on Rails*. Hestra-Verlag, Darmstadt, 1986.
Tassin, Y., Nouvion, F., and Woimant, J. *"Histoire de la traction electrique, tome 2: De 1940 a nos jours*. La Vie du Rail, Paris, 1986.
Williams, H. *APT: A Promise Unfulfilled*. Ian Allan, 1985.

Rail, Steam Locomotives

In 1900 British influence on the design of steam locomotives was strong but existed alongside distinct French, German, and American traditions. In the twentieth century, American designs and organization of steam traction became global standards due to research by universities, railroads, and the engineering industry. After 1900, North American designs departed radically from their European counterparts and became heavier in size, weight, tractive effort, and power. The basic "Stephenson" locomotive remained the norm throughout the twentieth century and was improved by superheating, piston valves, and the rational proportions pioneered by W.F.M. Goss and William Woodard in the U.S. Radical departures from this norm failed and were always rare. For general-purpose railways outside rapid transit systems, there was no alternative to steam traction before the 1940s. Throughout the century, express passenger locomotives worked heavier trains at sustained speeds higher than in the period from 1890 to 1910, but top speeds did not increase much apart from a few record runs of over 160 kilometers per hour (km/h), and steam traction coped until the 1950s. (The world speed record for steam was set at 203 km/h in 1935 by the British 4-6-2 engine Mallard.) From 1900 to 1930 freight trains of 3000 to 6000 tons were moved at 8–16 km/h mph and required engines with high tractive effort, low speed, and moderate power. The compound-expansion Mallet type, weighing up to 350 tons, was employed in the U.S. for this duty. After 1930, operations favored lighter, faster trains, and the high-powered, high-speed, rigid-framed 2-8-4 or 2-10-4 wheel arrangement type was widely used.

Woodard, Goss, and others increased performance using the basic Stephenson type, demonstrated by engine No. 50,000 of 1910. In Britain, the locomotives designed by H.N. Gresley after 1922 were derivatives of engine 50,000. Woodard's philosophy was applied to 4-6-2, 2-8-4, 2-10-4 and the later Mallet types, and by 1930 the American steam locomotive had virtually reached its final form. Locomotives exported to Russia, Japan, and China established the American exemplar in those countries in the 1930s. High-powered designs reached finality in the S1 6-4-4-6 Duplex locomotive of 1939, built to sustain 6500 horsepower and speeds of 160 km/h mph with 1000 tons on the level. Operated on the Pennsylvania Railroad, its great size (43 meters long), weight (482 U.S. tons), and limited route availability suggested that conventional steam traction was approaching its limit and that another mode of traction would be needed to raise power and speed standards after the 1940s.

Steam traction reached practical perfection by integrating the locomotive with the track through better balancing technique; rationalizing locomotive proportions, and applying scientific management to all parts of the traction system. This system worked best with simple, standardized locomotives with two outside cylinders and all components distributed for ease of maintenance. Mixed traffic locomotives were favored; that is, those able to work both passenger and freight trains. This was essential in the 1930s Depression years, during the 1939–1945 World War II years, and in the postwar period when labor was costly and funding scarce.

Locomotive science advanced after 1930 and led to new forms of steam locomotive using innovations in power station technology and marine practice. These combined high-pressure boilers with turbines, vacuum condensers, and electric transmission. Extremely complex machines were tried without success. Many projects were delayed by the war and revived afterward, when diesel and electric traction were proven and affordable. In France, Chapelon developed a high-performance locomotive, based closely on the orthodox form, which used compound expansion. Remarkable performances were demonstrated by Chapelon's 4-8-0, 4-6-2, 4-8-4, and 2-12-0 types between 1930 and the late 1940s. These were more promising designs than the contemporary high-pressure units developed by Muhlfeld in the U.S., but they had no significance once postwar economies permitted use of diesel and electric traction.

In the 1940s and early 1950s final efforts were made to develop a new form of steam locomotive free from the defects intrinsic to the orthodox type. The steam-electric locomotives of the Chesapeake and Ohio, and Northern and Western railways, tried between the late 1940s and early 1950s, were the last efforts made in the U.S. The Northern and Western railway retained steam traction using well-planned servicing stations, but the system became anachronistic, and belated dieselization proved essential in the 1960s. In the British Isles the unsuccessful experiments of Bulleid on locomotives with steam motors mounted in bogies concluded the search for a new form by the mid-1950s. Steam traction ended with the acceptance of the North American model, worked on the common-user system according to principles of scientific management. Monthly "best mileages" increased from about 6,000 miles (9656 kilometers) in 1910 to 12,000 miles (19,300 kilometers) per month in 1939 for 4-6-4 locomotives on the high-speed Chicago and Northwestern's Hiawatha services, and 24,000 miles (38,600 kilometers) per month for 4-8-4 locomotives on New York Central express trains in the late 1940s. The need for increased speeds and productivity in an age of motor transport, cheap air services, and increasing labor costs accelerated the shift to diesel and electric traction. Steam traction was intrinsically obsolete, and all attempts to save it, including one-man operations, automatic boiler regulation, multiple-unit working, and electric transmission, failed. The problem remained that the old forms were destructive of the track at high speeds, as the Chicago and Northwestern rail working tests had shown as early as the 1930s.

In its final form the American exemplar set the world standard. Typical weights for a North American 4-8-4 in running order were:

Engine 220,000–230,000 kilograms
Tender: 160,000 to 210,000 kilograms
Length: 33 to 37 meters

Sustained cylinder power was 5500 to 6000 horsepower, with a maximum of around 7000 horsepower. The largest articulated locomotives (e.g., Union Pacific 4-8-8-4 class) had an engine weight of 327,600 kilograms; tender of 198,000 kg; and a length of 40 meters. A modern steam locomotive would have most of the following features: a cast-steel bed; a welded boiler and steel firebox; rocking grate; self-cleaning smokebox; regulator valves in superheater header; compensated springing; mechanical lubrication; antidistortion disk wheels; roller bearings on all axles; roller bearings in rods and motion; and modern front end. Some engineers would add a combustion chamber, syphons and circulators, and Franklin wedges in the driving axle boxes. Most American engines would have a stoker (unless they were oil-burning), power reverse gear, and two outside cylinders. The final types built later in Europe, Russia, and China were derived from this American exemplar of the 1940s, but were much smaller.

Improvements to the steam locomotive were applied in backward economies with cheap labor until the 1990s. The work of David Wardale in South Africa and Livio Dante Porta in South America provide examples. Other improvements, like the Giesel ejector invented in the late 1940s in Austria, enabled older engines to run more economically in the years left before scrapping. Recent improvements incorporated into new steam locomotives built in the 1990s for tourist railways do not change the status of steam traction. The fuel crises of the 1970s and later led to plans for coal-fired locomotives, which included triple-unit steam-electric machines with condensing tenders, but none was constructed, and the coal-fired diesel electric and the coal-fired gas-turbine are favored instead. The schemes for atomic-powered steam-electric locomotives, proposed in the late 1940s and 1950s were never taken seriously by railwaymen.

See also **Rail, Diesel and Diesel Electric Locomotives; Rail, Electric Locomotives; Rail, High-Speed; Railway Mechanics**

MICHAEL C. DUFFY

Further Reading

Atkins, P. *Dropping the Fire: The Decline and Fall of the Steam Locomotive*. Irwell Press, 1999.

Ahrons, E.L. *The British Steam Railway Locomotive 1825–1925*. Locomotive Publishing, London, 1927. Reprinted by Ian Allan, 1966.

Carling, D.R. Locomotive testing stations. *Trans. Newcomen Soc.*, 42, 105–182, 1972.

Dalby, W.A. *The Balancing of Locomotives*. Edward Arnold, London, 1930.

Duffy, M.C. Rail stresses, impact loading & steam locomotive design, in *History of Technology*, 9th ann. vol. Mansell, London, pp. 43–101.

Durrant, A.E. *The Mallet Locomotive*. David & Charles, Newton Abbot, 1974.

Gresley, H.N. High pressure steam locomotives. *Proc. Inst. Mech. Eng.*, 121, pp. 101–206, 1931.

Hodgson, J.T. and Lake, C.S. *Locomotive Management: Cleaning, Driving, Maintenance*. Revised by Oaten, W.R., Tothill Press, London, 1954. Detailed, illustrated description of modern British locomotives. A standard railwayman's text.

Johnson, R.P. The Steam Locomotive: Its Theory, Operation & Economics including comparisons with Diesel-Electric Locomotives. Simmons-Boardman, New York, 1944, reprinted 1981.

Lucas, W.A., Ed. *100 years of Steam Locomotives*. Simmons-Boardman, New York, 1957. Evolution of American locomotive shown through engineering drawings.

Nock, O.S. *The British Steam Railway Locomotive 1925–1965*. Ian Allan, 1969.

Ransome-Wallis, P. *Concise Encyclopedia of World Railway Locomotives*. Hutchinson, 1959. Contains survey of final generation of steam locomotives.

Sauvage, E and Chapelon, A. *La Machine Locomotive*. Beranger, 1947. Reprinted by Les Editions du Layet, 1979.

Wiener, L. *Articulated Locomotives*. Constable, London, 1930.

Railway Mechanics

Railway mechanics grew from the nineteenth-century analysis of the union of locomotive and track. Improving the "fit" between engine and track was essential for safe and progressive operation. Analysis of the new machine stimulated general engineering mechanics. Investigators studied the balancing of locomotive mechanisms in the nineteenth century, and the phenomenon of fatigue failure was first identified in a locomotive context. In the early twentieth century, the study of shatter cracks in rail heads resulted in new regulations governing the manufacture of rails. Sylvester and Tschebyshev considered the kinematics of mechanisms, including valve gears. O. Reynolds studied balancing of reciprocating parts, and coupling rod failure. Railway studies of materials are reflected in the work of Mohr, whose analysis is still used in strength of materials.

In the 1890s, mechanics was supplemented by thermodynamics, and quantified analysis of energy flows were carried out in locomotives. The electric power industry used energy analysis, which passed into railway engineering after 1900. In the 1880s A. M. Wellington published his general theory of railway location, linking engineering and operation to climate, topography, commerce and economics. The influence of this work is still felt. In 1900, the machine-ensemble (traction system) was integrated closely with signaling via interlocking. Systematic investigations into locomotive proportions were conducted by W.F.M. Goss at Purdue, following earlier studies by Borodin in Russia, and Le Chatelier in France. Le Chatelier was a pioneer of rational management, efficient energy use, and scientific organization, as were Taylor, Gilbreth, Gantt, and Ford. Their theories transformed railway engineering management and design, and American exemplars transformed every department of the machine ensemble: traction, permanent way, signaling, and control. New devices were in part responsible: track design, rail shape, track circuits, power signaling, automatic signals, mechanized equipment in depots, and electrical safety mechanisms; but revolutionary concepts in management and organization drove much of the technical changes. Electric railways witnessed early attempts to model the technical and economic features of large industrial systems.

The work of Wellington in the U.S. modeled the railway in its general environment, and the work of G. W. Lomonossoff in Russia resulted in a comprehensive theory of railway mechanics in the early twentieth century. He advocated research departments dedicated to railway mechanics to frame rational strategies for engineering development. Lomonossoff's theories and experiments showed the difficulty of assigning accurate, measured values to parameters in theoretical expressions, without which comprehensive theories were of little value. Modern testing techniques were developed to overcome this problem. Experiment and experience were insufficient to solve major problems, and theoretical analysis was needed. The Russian contribution was considerable. Grinevetsky, Syromyatnikov, Nikolayev, and Shelest developed academic courses on locomotive design between 1900 and the 1960s. Similar academic work was done in Germany and the US. Steam locomotive science was developed in France by Chapelon and by William Woodard and R.P. Johnson in the U.S.

The effects on track of out-of-balance forces in steam locomotives was made in the 1940s tests carried out on the Chicago and Northwestern Railroad. The behavior of rails, trackbed, and locomotive components were measured using electrical recording apparatus and high-speed cameras. Wheel lift was observed during slipping. A mechanical model of wheel–rail interaction was constructed using rollers and springs, which could represent different rail dimensions, ballast characteristics, axle loads and train speeds. This helped compare theory with observations. Guidelines were derived governing springing, wheel design, balancing and number of cylinders. The tests were repeated under British conditions by the LMS railway in 1941.The LMS set up a research center in the 1930s after the American model, which investigated train resistance, aerodynamics, and vehicle–rail interaction.

F.W. Carter pioneered general analysis of rail–vehicle interaction in 1916. Hunting or sustained oscillations due to dynamic instability lacked adequate theoretical interpretation, as did creep and the dynamics of curving; these affected all vehicles on all kinds of track. The growth of oscillations at high speeds, even on straight track, limited very high-speed operations. Graphical analysis was often needed to solve equations. Carter's work, though of great importance, was limited in value because the configurations of vehicle body, wheels and trucks that he considered were inherently unstable. The analysis was subsequently developed by Rocard in France. Many railway engineers tolerated hunting as inevitable, but the high-speed tests of the 1930s in Germany, Great Britain, and the U.S. drew attention to the lack of understanding. In Great Britain the LMS railway contracted Prof. Inglis at Cambridge University. The Cambridge work, published in 1939, included equations of motion for a two-axle bogie, but the twelfth-order expressions could not be easily solved by the techniques then available. Advances were made after World War

II by Matsudaira at the Railway Technical Institute, Tokyo, who developed the concept of stiffness to stabilize spring-restrained wheelsets at any critical speed. By the 1960s this work enabled the Japanese railways to operate very high-speed *Shinkansen* trains. The problems investigated concerned all forms of train, and in the 1960s British Railways contracted Imperial College to develop adequate analysis. In both the Japanese and British work, dynamicists with a background in aeroelasticity played vital roles (e.g., Matsudaira and Wickens). The theoretical analysis was tested by experiment using roller test stands and instrumented trains and track, greatly aided by the availability of computers to solve expressions that were difficult or impossible to solve by earlier methods. Developing adequate theories of vehicle behavior on curves was difficult and had been investigated from 1883 to 1903 in Britain, France, and Germany. A fundamental contribution was made in Britain by Porter whose 1934 studies on mechanics of a locomotive on curved track became classics. By the late 1960s, fairly complete theories of curving were in existence, which included vehicle form, creep, and stability. In 1977, a general theory of curving was formulated by Elkins and Gostling.

At the end of the twentieth century, research and development involved cooperation among states, manufacturers, universities, and railway companies. The trend was toward international cooperation, encouraged by globalization of leading manufacturers. The German program in advanced railway technology, begun in 1972, is one example. Government funding varied between 50 and 100 percent depending on the project. Federal funding in the U.S. also varied with the project. The German program covered vehicle and rail interaction, vehicle design (including aerodynamics and control), track and trackbed design and behavior, control technology, power technology including pantograph-contact wire interaction, and environmental matters. The ME-DYNA program was structured for nonlinear simulation of vehicle track interaction. The Munchen–Friedmann test facility was constructed from 1977 to 1980 to investigate four-axle vehicles. Wheelsets have been tested up to 503 km/h; techniques for predictive analysis of bodywork stresses, modes of vibration, and distortion during all modes of motion were all improved; and aerodynamic behavior of vehicles and pantographs was optimized. Without this analysis, dependent on electronics for its execution, the modern railway, with its enhanced performance and close integration of traction system, signaling, and automatic control would not be possible.

See also **Rail, Diesel and Diesel Electric Locomotives; Rail, High Speed; Rail, Steam Locomotives**

MICHAEL C. DUFFY

Further Reading

Carter, F.W. *Railway Electric Traction*. Arnold, 1922.
Carter, F.W. On the stability of running of locomotives. *Proc. Royal Soc. A.*, 121, 585–611, 1928.
Duffy, M.C. Rail stresses, impact loading, and steam locomotive design, in *History of Technology*, vol. 9. Mansell, London, 1984, pp. 43–101.
Gilchrist, A.O. The long road to solution of the railway hunting and curving problems. *Proc. Inst. Mech. Eng.* 212, Part F, 219–226, 1998.
Lomonossoff, G.V. *Introduction to Railway Mechanics*. Milford/Oxford University Press, Oxford, 1933.
Majumdar, J. Economics of Railway Traction. Gower, 1985.
Rahn, T., Hochbruck, H., Moller, F.W. and Lubke, D., Eds. *ICE: high-tech on rails*. Hestra-Verlag, Darmstadt, 1986.
Wellington, A.M. *The Economic Theory of Railway Location*. Wiley/Chapman & Hall, 1887, 1889.
Wickens, A.H. The dynamics of railway vehicles—from Stephenson to Carter. *Proc. Inst. Mech. Eng.*, 212, Part F, 209–217, 1998.

Rectifiers

Rectifiers are electronic devices that are used to control the flow of current. They do this by having conducting and nonconducting states that depend on the polarity of the applied voltage. A major function in electronics is the conversion from alternating current (AC) to direct current (DC) where the output is only one-half (either positive or negative) of the input. Rectifiers that are currently, or have been, in use include: point-contact diodes, plate rectifiers, thermionic diodes, and semiconductor diodes. There are various ways in which rectifiers may be classified in terms of the signals they encounter; this contribution will consider two extremes—high frequency and heavy current—that make significantly different demands on device design.

Diodes assumed a special importance after the invention of wireless. The conventional method of transmitting audio signals was to use amplitude modulation (developed by Reginald A. Fessenden in 1906). The wireless signal consisted of a single high-frequency signal (carrier wave), whose amplitude was changed in direct proportion to the audio signal that was to be transmitted. The carrier frequency is well outside the range of the human ear. However, once the signal has been passed through a detector, such as the silicon point contact diode patented by Greenleaf Whittier Pickard in 1906, the much slower changes of the positive "envelope" can be heard with the aid of headphones.

In the days before transistors, the amplification of the acoustic signal was achieved using thermionic valves such as the triode invented by Lee de Forest in 1906. These devices required a high-voltage DC supply. This was normally derived from the 50- or 60-hertz AC mains via one or more thermionic diodes (invented by John Ambrose Fleming in 1904). A metal plate and an electric filament are separated within a vacuum enclosure. When a current is passed through the filament (normally tungsten), electrons are emitted from its surface. If the voltage on the metal plate (called the anode) is positive, then current will flow. However, if the voltage on the plate is negative, then electrons are repelled and current does not flow.

Researchers L.O. Grondahl and P.H. Geiger discovered the copper oxide rectifier in 1927 and noted that the flow of current through a layer of copper oxide on metallic copper is not symmetrical with respect to bias; more current flows if a battery is connected one way than if it is connected with the reverse polarity. Similar observations were made about nickel oxide in contact with nickel, but the market for low-power, low-frequency applications was dominated by selenium rectifiers, which were invented by Charles E. Fitts in 1933. Selenium has a reverse breakdown voltage in the region of 20 to 30 volts, so higher voltages such as the AC mains supply could be converted to DC using a series of selenium diodes separated by metal plates. A stack of such diodes could be arranged as a bridge for full-wave rectification and as such were used in radio and TV receivers into the1960s.

The advent of silicon *p–n* junction diodes and transistors marked a revolution and consigned thermionic diodes and other valves to the history books. However, variants of some of the old devices are still in use today. The cat's whisker was a rudimentary type of metal-semiconductor diode. Structures of this type were extensively studied by the German physicist, Walter Schottky. Unlike ordinary semiconductor diodes the electrical conduction in this diode involved one carrier type only and this allowed them to switch very quickly between conducting and nonconducting states. It was therefore possible to use them at much higher frequencies, well outside the range of conventional junction diodes. Suitably designed point contact diodes are still used in radar and microwave telephone links. A particularly fast device called the "snap-off" diode is an essential component in frequency multiplier circuits and at one time this was the only way to generate signals in the range 50 to 90 gigahertz.

The conversion of mains supply to DC for heavy current applications presented a particular problem for rectification. The arc lamps in cinema projectors were one such example and the mercury-arc rectifier valves that were used were enormous in terms of volume occupied per amp supplied. Silicon junction power diodes were a relatively late development within the semiconductor revolution. Large currents required large-area silicon dice. In the early stages, the level imperfections, measured as defects per unit area in the starting crystal was sufficiently large that the yield of good-quality marketable devices was unacceptably low. Technological developments changed this situation and it is now possible to obtain diodes that can block in excess of 2000 volts when reverse biased. Low- and medium-power devices with slow or fast turn-off are now readily available at reasonable prices.

A silicon *p–n* junction diode has a property called the forward voltage drop. This is generally between 0.6 and 0.7 volts and would mean that if a diode were handling 1000 amperes, one would need to dispose of heat equivalent to approximately 700 watts, and this continues to present a challenge for design engineers. However, a variant of the original metal-semiconductor junction, called the Schottky diode, has two very useful properties. The forward voltage drop is significantly less than for *p–n* junction diodes, so that the equivalent size of diode package can carry much larger currents. Second, because they have very fast turn-off times, they can rectify high-frequency signals. Accordingly, they are a key component in the switched-mode power supply unit (SMPSU), a revolution in rectification. The incoming main is rectified using a conventional bridge circuit. This is then converted back to AC, but at very high frequency. High-frequency transformers are much lighter and more efficient than conventional 50- or 60-hertz transformers. The output from such a transformer is rectified using a Schottky diode. The power supply in a desktop computer will have several transformer or Schottky pairs, and a typical 250-watt PSU may have to deliver up to 32 amperes at 5 volts, 0.5 amperes at −5 volts, 10 amperes at 12 volts, and 0.5 amperes at −12 volts.

The thyristor is the ultimate in rectification. It has a gate contact that can be operated by currents between 10 microamperes and 200 milliamperes. The device remains "off" if the gate has not been triggered. The application of a current pulse to the gate during a positive half-cycle of the applied AC voltage "fires" the thyristor. Current then flows between the anode and cathode terminals from that instant until the end of the half-cycle. This provides a variable level DC output. The thyristor is used in domestic lamp dimmers. Stacks of thyristors, each rated in excess of 1000 amperes, with a 2000-volt reverse breakdown are used in the high-voltage DC link between Britain and France.

See also **Control Technology, Electronic Signals; Radio Receivers, Crystal Detectors and Receivers; Semiconductors, Compound; Transistors**

DONARD DE COGAN

Further Reading

Braun, E. and Macdonald, S. *Revolution in Miniature: The History and Impact of Semiconductor Electronics.* Cambridge University Press, Cambridge, 1982.

Grondahl, L. O., and Geiger, P. H. A new electronic rectifier. *Proceedings of the American Institution of Electrical Engineers, Winter Convention, New York, 1927*, p. 357.

Refrigeration, Absorption

William Cullen's seminal experiments on the evaporation of liquids in the 1750s served as a precursor to the development of absorption refrigeration. Cullen, who accepted a chair in chemistry at the University of Edinburgh in 1756, demonstrated the possibility of refrigeration when he produced a small amount of ice through vaporization by reducing the pressure under a bell jar. One of Cullen's students, Joseph Black, made an important contribution to both physics and refrigeration engineering with his articulation in 1761 of the theory of latent heat, which accounts for the release and absorption of heat as a substance changes state. Another of Cullen's students at Edinburgh, Edward Nairne, improved upon the pioneer's laboratory experiments by placing a cup of sulfuric acid along with a cup of water under a partially evacuated bell jar. The sulfuric acid absorbed the vaporized water and thereby hastened the evaporation process. Nairne's successor at Edinburgh, John Leslie, and later a British brewer named John Vallance made further improvements to Nairne's sulfuric acid–water absorption device in the early part of the eighteenth century, but neither developed their devices beyond the experimental stage.

Edmund Carré, a French inventor, developed a sulfuric acid absorption refrigeration device in the 1850s. Although it was installed in a number of bars and restaurants in France, the device malfunctioned and underwent disabling corrosion. Edmund's brother, Ferdinand Philippe Eduard Carré, concluded that the problem was not the device itself, but the choice of refrigerant. By replacing the sulfuric acid and water mixture with an ammonia and water mixture, Ferdinand Carré developed the first successful absorption refrigeration system. The initial Carré device had just two components. One component combined the functions of a generator and an absorber; the other combined the functions of a condenser and an evaporator. Because of the dual functions of each component, the machine operated intermittently rather than continuously. Soon thereafter, however, Carré developed a continuous machine.

The Carré absorption unit had all of the components of a modern day absorption machine, including: an evaporator, an absorber, a generator, an expansion valve, and a pump. In ammonia–water absorption systems, the ammonia refrigerant flows at low pressure through the evaporator where it absorbs heat from the refrigerator cabinet. It then moves to the absorber where it mixes with water. The water and ammonia mixture is then pumped into the generator, and the heat of the generator separates the mixture into water and high-pressure ammonia gas. The water returns to the absorber, and the ammonia gas then flows into the condenser, where cool water passing over the outside of the coils removes heat from the ammonia and allows it to return to a liquid form. From there, the ammonia gas moves through an expansion valve and is ready to flow back into the evaporator where, once again, it absorbs heat from the refrigerator cabinet (see Figure 25).

The absorption machine gained quickly in popularity. It was adopted and produced in Germany, Great Britain, and the U.S. Carré refrigeration units were shipped through Union blockades during the U.S. Civil War to a military hospital in Augusta, Georgia, and King Ranch near Brownsville, Texas. Manufacturers readily improved upon the Carré design. The use of wood chips to heat the generator was the most glaring problem with the machine. To gain greater control over the heating process, manufacturers of industrial machines replaced the wood-burning system with steam coils. Manufacturers developed more efficient rectifiers and new types of absorbers throughout the late nineteenth and early twentieth centuries. The refrigeration units were developed almost exclusively for large-scale commercial establishments, such as ice-making plants and breweries. The production process, therefore, resembled a construction operation. Without the

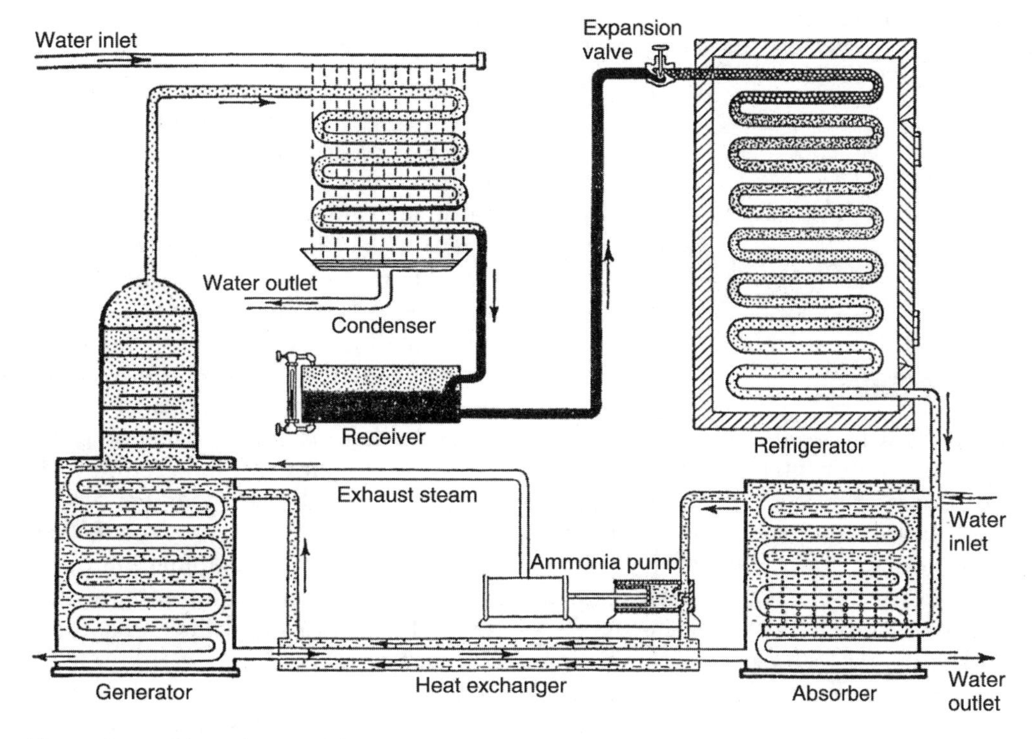

Figure 25. Absorption refrigeration machine.
[*Source: From Anderson, O.E. Refrigeration in America: A History of a New Technology and Its Impact. Princeton University Press, Princeton, 1953.*]

possibility of stable designs, refrigeration companies depended on specialty manufacture, craft skill, and on-site engineering improvements to build their systems.

A number of characteristics distinguish absorption refrigerators from mechanical-compression systems. Both force the evaporation and condensation of a refrigerant to transfer heat across physical boundaries, but the absorption system uses heat to generate the pressure necessary to enable the evaporation process, while the mechanical-compression system relies on an electric-powered compressor. Rather than mechanical processes and electrical energy, the absorption system depends on heat and physiochemical processes for its operations. The absorption system, furthermore, has more components than its competitor, has no moving parts, and has a refrigerant consisting of two substances, not one.

Absorption refrigeration remained competitive with compression systems up to World War I in the U.S. During the interwar years, however, absorption refrigeration for both industrial and domestic settings plummeted in popularity. The higher initial cost of absorption machines and the inability of manufacturers to match the innovations of producers of compression systems are only part of the story. Household absorption refrigerators depended on gas heat, while the compression systems depended on electrical power. The emerging giants of the electrical industry in the U.S.—General Electric, Westinghouse, and the Frigidare Division of General Motors—and their well-financed partners in the electrical utility industry were able to muster far greater financial and technical resources than producers of absorption refrigeration systems. The single major producer of absorption refrigerators, Servel, was at too great a financial disadvantage to compete with the electrical industry in research and development, and marketing. Absorption refrigeration, however, did not vanish from the marketplace. Despite its early eclipse, it occupied an important niche in the global refrigeration market throughout the twentieth century.

See also **Air Conditioning; Refrigeration, Mechanical**

GLEN ASNER

Further Reading

Anderson, O.E. *Refrigeration in America: A History of a New Technology and Its Impact*. Princeton University Press, Princeton, 1953.

Cowan, R.S. *More Work for Mother: The Ironies of Household Technology from the Open Hearth to the Microwave*. Basic Books, New York, 1983.
Sauer, H.J. Jr. Refrigeration, in *Encyclopedia of Energy, Technology, and the Environment*, vol. 4, Bisio, A. and Boots, S., Eds. Wiley, New York, 1995.
Thévenot, R. *A History of Refrigeration Throughout the World*. Translated by Fidler, J.C. International Institute of Refrigeration, Paris, 1979.
Woolrich, W.R. *The Men Who Created Cool: A History of Refrigeration*. Exposition Press, New York, 1967.

Refrigeration, Mechanical

Over 170 years elapsed between William Cullen's experiments on refrigeration at the University of Glasgow and the mass production of mechanical-compression refrigerators for the consumer market in the 1920s. Beginning with Cullen in 1748, scientists and inventors across the Western world manipulated the basic properties of the elements to produce ice and cold air. The most prominent methods of refrigeration included air-cycle compression, vapor compression (or mechanical compression), thermoelectric, and absorption. While no single method proved superior in all circumstances, mechanical-compression refrigeration emerged by the 1940s as the dominant method for both domestic and industrial refrigeration.

Inventors throughout the Western world contributed to the development of mechanical-compression refrigeration throughout the nineteenth century. Oliver Evans, the American inventor of the automated flour mill, laid the foundation for the continuous-cycle vapor-compression refrigerator (see Figure 26) when he conceived a method in 1805 for recycling vaporized refrigerant. After removing heat from the surrounding environment, vaporized refrigerant would move through a compressor, and then a condenser, where it would revert back into a liquid form and begin the process again. Although Evans failed to transform his idea into a practical device, Jacob Perkins, an American-born resident of London who befriended Evans during a stay in Philadelphia, built on Evans' ideas to construct a cyclic vapor-compression machine in 1834. The Perkins machine, which used ether as a refrigerant, was the first full-scale machine to contain a compressor, a condenser, an expansion valve, and an evaporator—the basic parts of the modern mechanical-compression refrigeration system. Perkins, however, never developed his machine for commercial use.

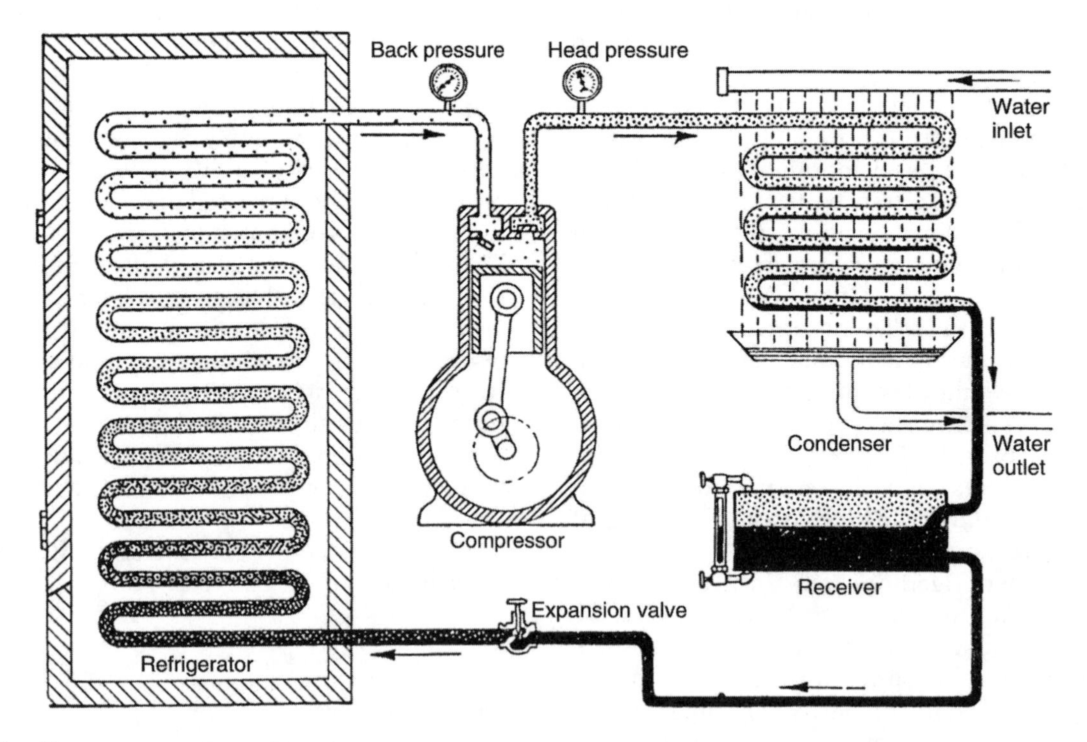

Figure 26. Vapor-compression refrigerator.
[*Source: From Anderson, O.E. Refrigeration in America: A History of a New Technology and Its Impact. Princeton University Press, Princeton, 1953.*]

Dr. John Gorrie, the director of the U.S. Marine Hospital in Apalachicola, Florida, received a patent in 1851 for the first refrigeration system operated for practical use. Gorrie, however, installed his machine only at his own hospital, and it was an air-cycle compression machine rather than a vapor-compression machine. Whereas the refrigerant in a vapor-compression machine alternates between a gaseous and a liquid state, the refrigerant in an air-cycle compression machine—air—remains gaseous throughout the compression and expansion cycles. Alexander Twining, a college professor and practicing civil engineer, received a patent in 1853 and built the first commercially viable continuous-cycle vapor-compression refrigeration machine based on Jacob Perkins' original design. Several inventors subsequently developed large-scale compression machines for industrial settings that varied from Twining's machine only in component design and choice of refrigerant. Many of the early machines used ether, a highly volatile and toxic substance, as a refrigerant. In the 1870s, David Boyle in the U.S and Carl von Linde in Germany introduced ammonia, a less toxic refrigerant that changed states more readily than ether. However, other more toxic refrigerants remained in use alongside ammonia. Reciprocating compressors were first developed in 1880 for industrial applications, such as ice making, brewing, and meat packing.

Demand for domestic refrigeration emerged by the 1880s, as indicated by the widespread use of iceboxes in households. Rudimentary mechanical refrigeration units for the home first appeared in 1910, and the Kelvinator Corporation began quantity production of the first domestic refrigerator unit with automatic controls in 1918. These early machines, however, were unreliable, expensive, plagued with technical problems, and even dangerous. Most systems used sulfur dioxide or methyl chloride as refrigerants, both of which are highly toxic. Even less-toxic ammonia could prove lethal if it leaked, and leakage was a constant problem with the early machines. They also required regular servicing due to chronic problems with thermostats, motors, and compressors. The separation of the refrigerating machinery from the refrigeration compartment exacerbated these problems by forcing the compressor to work harder.

The redesign of the refrigerating unit and the development of a new class of refrigerants helped to increase the reliability and safety of mechanical-compression refrigerators. General Electric developed the first hermetically sealed motor-compressor for domestic refrigeration in the mid-1920s. The GE Monitor Top, which was based on a 1905 design by Audiffren in France, contained all its mechanical parts in a single unit, ingeniously placed on top of the refrigerator box. The machine was air cooled and made of steel rather than wood, which was commonly used for most machines at the time. Other manufacturers soon replicated GE's design. By 1940, almost all domestic refrigerators were self-contained units and made of steel with no external parts.

The discovery of chlorofluorocarbon (CFC) refrigerants by Thomas Midgely and a team of researchers at General Motors in the late 1920s eliminated the immediate health and safety hazards of refrigeration. The company announced its discovery in 1930. Soon thereafter, the Kinetic Chemical Company, a joint venture of GM and DuPont, began manufacturing the CFC refrigerant known commercially as Freon. Although General Motors initially intended to restrict use of Freon to machines produced by its Frigidaire Division, the benefits and the potential profits from widely distributing the nontoxic and nonflammable refrigerant were too great to resist. Within a few years Freon became the refrigerant of choice for mechanical-compression refrigerators.

Other important landmarks in the history of refrigeration include the development of the humidity drawer for fruits and vegetable in 1930, General Electric's 1939 introduction of the dual temperature refrigerator, which had separate compartments for chilled and frozen foods, and Frigidaire's replacement of reciprocating compressors with quieter and smaller rotary compressors in 1933. Automatic defrosting was introduced in the early 1950s.

Whereas immediate health and safety concerns stimulated the development of chlorofluorocarbon refrigerants, long-term environmental concerns led to the abandonment of CFCs. In 1974 two chemists at the University of California at Irvine, F. Sherwood Roland and Mario Molina, discovered that chlorine atoms catalytically break down ozone when exposed to high-frequency ultraviolet light. Roland and Moline concluded that CFCs were stable enough to pass through the troposphere but would decompose and release chlorine in the stratosphere, where they would deplete the earth's ozone shield. Not until the British Antarctic Survey discovered a hole in the ozone layer in 1985, however, did concern over CFCs become widespread. In keeping with their commitments under the 1987 United Nation's Montreal Protocol, industrialized nations abandoned the use of CFCs by 1996. The search for refrigerants to replace CFCs yielded new chemical compounds, such as hydroflourocarbons (HFCs), which were less damaging to the ozone layer. However, no single refrigerant emerged as a global standard. Research on refrigerants and refrigeration methods at the end of the twentieth century was as vibrant as it was at anytime during the century.

See also **Air Conditioning; Cryogenics; Refrigeration, Absorption; Refrigeration, Thermoelectricity**

GLEN ASNER

Further Reading

Anderson, O.E. *Refrigeration in America: A History of a New Technology and Its Impact*. Princeton University Press, Princeton, 1953.

Cooper, G. *Air-Conditioning America: Engineers and the Controlled Environment, 1900–1960*. Johns Hopkins University Press, Baltimore, 1998.

Cowan, R.S. *More Work for Mother: The Ironies of Household Technology From the Open Hearth to the Microwave*. Basic Books, New York, 1983.

Hård, M. *Machines are Frozen Spirit: The Scientification of Refrigeration and Brewing in the 19th Century—A Weberian Interpretation*. Campus Verlag, Frankfurt am Main; and Westview, Boulder, CO, 1994.

Thévenot, R. *A History of Refrigeration Throughout the World*. Translated by Fidler, J.C., International Institute of Refrigeration, Paris, 1979.

Sauer, H.J., Jr. Refrigeration, in *Encyclopedia of Energy, Technology, and the Environment*, vol. 4, Bisio, A. and Boots, S., Eds. Wiley, New York, 1995.

Woolrich, W.R. *The Men Who Created Cool: A History of Refrigeration*. Exposition Press, New York, 1967.

Woolrich, W.R. The history of refrigeration: 220 years of mechanical and chemical cold, 1748–1968. *ASHRAE Journal*, July, 31–39, 1969.

Refrigeration, Thermoelectricity

The scientific principles underlying thermoelectric refrigeration were understood by the mid-nineteenth century. In 1822, German scientist Thomas Johann Seebeck discovered that a needle would move when held near the junction of two dissimilar metals maintained at different temperatures. Seebeck misidentified the effect as magnetic, but Hans Christian Oersted, the father of electromagnetism, and James Cumming, a Cambridge chemist, correctly categorized Seebeck's discovery, known as the Seebeck effect, as an electrical phenomenon. A Parisian clockmaker, Jean Charles Athanese Peltier, made the second important discovery in the field of thermoelectricity in 1834 while performing an experiment to measure the conductivity of bismuth and antimony. As he had predicted, the temperature at the junction of the two conductors changed with the application of an electrical current. He also discovered that the temperature of the metals differed at their ends and that the current absorbed heat at one end and released it at the other. Like Seebeck, however, Peltier misinterpreted his results. With an ingeniously simple experiment—placing a drop of water at the junction of the two conductors and watching it freeze and melt depending on the direction of the current—Emil Lenz first demonstrated and correctly interpreted the Peltier effect in 1838.

From the time Lord Kelvin clarified the relationship between the Peltier and Seebeck effects in 1854 until the 1950s, research on thermoelectricity moved along at a languid pace. Bold efforts to develop practical devices based on thermoelectric principles met with little success. Known materials allowed for efficiencies of just 1 percent, far too low to justify any serious development efforts. The most important contribution to the study of thermoelectricity in the early twentieth century came from E. Altenkirch, a German scientist, who determined that progress in thermoelectricity depended on finding materials that exhibited three characteristics—(1) high electrical conductivity, (2) high voltage capacity, and (3) low thermal conductivity. Since he knew of no such materials, Altenkirch abandoned his search.

Thermoelectric researchers developed a greater understanding of the possibilities of semiconductors in the 1930s, and the positive–negative (*p–n*) junction, a crucial component of thermoelectric devices, was developed in 1942. However, not until the 1950s, after researchers at the Soviet Institute of Semiconductors declared the inevitability of a thermoelectric breakthrough and H.J. Goldsmid of General Electric's London laboratory provided a rationale for studying the heaviest semiconductor compounds such as bismuth telluride and lead telluride, was there a concerted worldwide effort to overcome the technical barriers to the development of thermoelectric devices.

At the core of technologies based on thermoelectric principles is a simple solid-state device that facilitates the exchange of thermal and electrical energy through the movement of electrons and holes (Figure 27). An electrical current passing through the device will draw heat from one side of the *p–n* junction and release it on the other side. If the current is reversed, so too is the heat transfer process. Employed in this way, as a solid-state heat pump, the thermoelectric device can be used for refrigeration, air conditioning, and heating. The

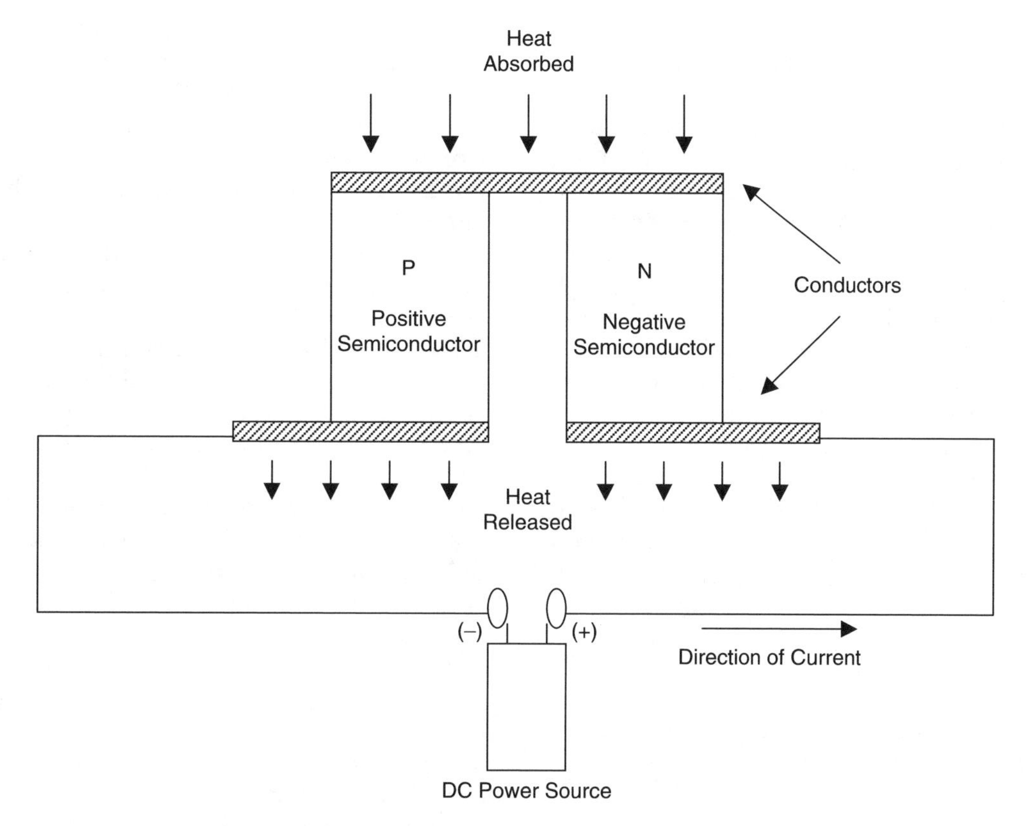

Figure 27. Schematic drawing of a simple thermoelectric device.

same device also can be used to exploit temperature differentials of opposite sides of the *p–n* junction to produce electricity. In thermoelectric generators, thermal energy is transformed into electrical energy as it passes from one side of the device to the other.

The 1950s witnessed an explosion of interest in thermoelectric research, especially for refrigeration. Optimistic about the potential of thermoelectricity, researchers in corporate laboratories convinced electrical industry executives and government research agencies to devote substantial resources to thermoelectric research and development for military and commercial applications. Researchers worldwide published twice as many papers on thermoelectricity in the four years between 1956 and 1960 as they had during the previous 130 years. In the U.S. alone, over 75 organizations, including research universities, government laboratories, private research institutes, and major corporations, maintained substantial thermoelectric research programs.

Major electrical appliance manufacturers, such as General Electric, RCA, Westinghouse, and Whirlpool, pinned their hopes on the large, growing, and highly profitable market for standardized household refrigeration units in the late 1950s. Thermoelectric refrigerators, however, could not compete in price or efficiency with existing mechanical refrigerators. Furthermore, manufacturers were unwilling to adjust their expectations and to shift their focus to smaller markets for specialty devices in which thermoelectric units might have been competitive. As the promise of big profits in the household refrigeration market evaporated by the mid-1960s, all the major appliance manufacturers and all but a handful of specialty firms abandoned thermoelectric research and development for the consumer market. Research on military applications of thermoelectricity continued nonetheless and proved productive over the long term. Thermoelectric devices have been used to power electronic equipment in spacecraft, to cool submarines, and to provide a number of other military applications in which cost and efficiency are of limited concern.

Interest in commercial applications of thermoelectricity refrigeration reemerged in the 1990s. Niche markets developed for portable coolers, and at least one automobile manufacturer embedded thermoelectric cooling units in its car seats.

Manufacturers in the U.S., Europe, and Japan began to experiment with using thermoelectric chips to cool computers and medical devices such as blood analyzers. Most of these devices used bismuth telluride alloys in simple *p–n* junctions, which were still too inefficient to employ in larger units such as household refrigerators. The technical breakthrough that researchers had been seeking for 40 years appeared to have arrived in 2002, through U.S. government-funded research begun in 1993 at the Research Triangle Institute (RTI) in North Carolina. RTI researchers used a thin film deposition process to lay down thermoelectric alloys in alternating layers. The microscopic structure, known as a superlattice, enhances electron flow while hindering heat transfer, resulting in efficiencies two-and-a-half times greater than allowed with simple *p–n* junctions. Whether devices based on superlattice structures will finally bring about the long-promised thermoelectric revolution in refrigeration and cooling remains to be seen.

See also **Refrigeration, Mechanical; Semiconductors, Compound; Semiconductors, Elemental**

GLEN ASNER

Further Reading

Finn, B.S. Thermoelectricity, in *Advances in Electronics and Electron Physics*, vol. 50. Academic Press, New York, 1980.

Goldsmid, H.J *Applications of Thermoelectricity*. Methuen & Company, London; and Wiley, New York, 1960.

Goldsmid, H.J. *Electronic Refrigeration*. Pion, London, 1986.

Joffe, A.F. The revival of thermoelectricity. *Scientific American*, 199, 5, 31–37, 1958.

Lynch, C.J. Thermoelectricity: the breakthrough that never came. *Innovation*, 28, 48–57, 1972.

Thévenot, R. *A History of Refrigeration Throughout the World*. Translated by Fidler, J.C. International Institute of Refrigeration, Paris, 1979.

Skrabek, E.A. Thermoelectric energy conversion, in *Encyclopedia of Energy, Technology, and the Environment*, vol. 4, Bisio, A. and Boots, S., Eds. Wiley, New York, 1995.

Renewable Power Generation, *see* **Biomass Power Generation; Hydroelectric Power Generation; Power Generation, Recycling; Solar Power Generation; Wind Power Generation**

Reppe Chemistry

Reppe chemistry refers to a group of high-pressure reactions used in industry to make various organic chemicals. Most of these reactions were based on acetylene, and as acetylene declined in popularity (because of the cheapness of ethylene) in the 1960s, most of the Reppe reactions have also lost their significance. However, the formation of butanediol from acetylene has survived and is now one of the few acetylene-based processes used in the chemical industry.

Walter Reppe who was working for the German firm of IG Farben at Ludwigshafen on the Rhine, discovered in 1932 that alcohols could be added to acetylene under considerable pressure to form the corresponding vinyl ether. The dangers of using acetylene were well known, and the success of this reaction was unexpected. The vinyl ethers were used as the starting point for the production of polyvinyl ethers, which were considered to be possible alternatives to polyvinyl chloride (PVC). Subsequently they turned out to have only limited applications, for instance, as a synthetic substitute for chewing gum.

Six years later, Reppe discovered a different type of reaction that preserved the triple bond of acetylene. Two molecules of formaldehyde were added to acetylene under pressure using copper acetylide, a detonator, as the catalyst. The initial product, butyne-1,4-diol was converted into butane-1,4-diol which was then used to make synthetic rubber during World War II. Since 1945, butanediol has been used to make the well-known polyurethanes and the important solvent tetrahydrofuran (THF). Reppe then attempted to make aldehydes by adding carbon monoxide to acetylene, but the reaction produced acrylic acid, a chemical that is used extensively in paint but had previously been very expensive. Learning of similar research at Ruhrchemie (the OXO process) in 1940, Reppe extended the reaction to ethylene (yielding propionic acid) and methanol (yielding acetic acid). A semi-industrial plant for the manufacture of propionic acid, used to make fine chemicals such as mold inhibitors, started up at Ludwigshafen in 1951, and a full-scale acetic acid plant followed in 1957.

In 1940, Reppe also achieved a remarkable cyclization of acetylene to the exotic compound cyclooctatetraene (COT), using a nickel cyanide catalyst. Much hope was held out for the industrial development of COT, which was not fulfilled. BASF, the postwar successor company to IG Farben at Ludwigshafen, maintained a small pilot plant for many years and provided free COT to academic researchers.

See also **Chemicals**

PETER MORRIS

Further Reading

Achilladelis, B.G. Process Innovation in the Chemical Industry. University of Sussex D.Phil. thesis, 1973, pp. 75–90.

Copenhaver, J.W. and Bigelow, M.H. *Acetylene and Carbon Monoxide Chemistry*. Reinhold, New York, 1949.

Hummel, H.G. Walter Reppe, 1892–1969. *Chemische Berichte*, 117, 1–21, 1984.

Miller, S.A. *Acetylene: Its Properties, Manufacture and Uses*, vol. 1. Ernest Benn, London, 1965.

Morris, P.J.T. The Development of Acetylene Chemistry and Synthetic Rubber by I.G. Farbenindustrie Aktiengesellschaft, 1926–1945. Oxford University D.Phil. thesis, 1982, pp. 111–117 and 123–133.

Morris, P.J.T. *Ambrose, Reppe and the Emergence of Heavy Organic Chemicals in Germany, in Determinants in the Evolution of the European Chemical Industry, 1900–1939: New Technologies, Political Frameworks, Markets and Companies*, Anthony S. Travis, et al., Eds. Kluwer, Dordrecht, 1998, pp. 89–122.

Morris, P.J.T. An industrial pioneer: Walter Reppe. *Chemi. Br.*, 29, 38–40, 1993.

Morris, P.J.T. The technology–science interaction: Walter Reppe and cyclooctatetraene chemistry. *Br. J. Hist. Sci.*, 25, 145–167, 1992.

von Nagel, A. *Äthylen, Acetylen*. Firmenarchiv der BASF, Ludwigshafen, 1971.

Reppe, W. *Neue Entwicklungen auf dem Gebiete der Chemie des Acetylen und Kohlenoxyds*. Springer, Berlin, 1949.

Reppe, W. *Chemie und Technik der Acetylen-Druck-Reaktionen*. Verlag Chemie, Weinheim, 1951; expanded edition, 1952.

Stokes, R.G. *Opting for Oil: The Political Economy of Technological Change in the West German Chemical Industry, 1945–1961*. Cambridge University Press, Cambridge, 1994, pp. 35–39.

Research and Development in the Twentieth Century

By the end of the nineteenth century, the emergence of large multinational industrial companies had begun. Along with this growth, companies established different departments for each specific aspect of the industrial process. This is what Alfred Chandler called the "visible hand" in industrial management in that period, and he showed that this phenomenon took place in countries such as the U.S., Britain, and Germany. One of the tasks that became the main focus for a separate department was the research and development (R&D) function. In various countries, large companies set up laboratories to conduct R&D that would create new products or improve existing ones. The first sector in which such laboratories emerged was the chemical industry, soon followed by the electrotechnical sector. In the U.S., for example, General Electric (GE) started a central laboratory in 1900, DuPont in 1902, and Eastman Kodak in 1910. In Germany, the Siemens company set up its main laboratory in 1905; in the U.K., the General Electric Company (GEC) started a central laboratory in 1919; and in The Netherlands the Philips Company initiated its *Natuurkundig Laboratorium*, or "physics laboratory," in 1914.

One of the driving factors for R&D was the redefinition of patent laws. In the U.S. this was the 1890 Sherman Antitrust Law. In order to assure that industries were free to act and to maintain a sound position in the market, companies had to develop their own knowledge bases. This was one of the tasks of the new laboratories.

Research and development in the twentieth century will be discussed mainly for the chemical and electrotechnical industries because they are knowledge-intensive and have therefore traditionally received most attention in terms of their R&D. Among other areas that have also become knowledge intensive is the pharmaceutical industry. Considerable developments have taken place through R&D; for example, vitamins in the 1930s, steroids, antibiotics, and cardiovascular products in the 1940s and 1950s. The area of materials can also be mentioned as one in which much R&D was undertaken. However, much less has been written about the history of R&D in these sectors than for the chemical and electrotechnical sectors.

In many companies with a new central laboratory, the first effect of the lab activities was the diversification of the company's product portfolio. This can be seen clearly with the examples of GE and Philips. Both companies were started at the end of the nineteenth century to produce light bulbs only. However, as their laboratories started undertaking research to study phenomena related to the functioning and production of light bulbs, they soon realized that the same phenomena also accounted for the functioning of other devices, such as x-ray and radio vacuum tubes. As a result, increased insight into the light bulb phenomena also contributed to the development of other products. Usually there was close contact with the company's management so that the decision to diversify the product portfolio was often a matter of cooperation between the company managers and the lab managers. When both GE and Philips had moved into the market of radio vacuum tubes, they took the next step in the diversification process by extending R&D activities to other elements of the radio: electric circuits, loudspeakers, and later sound recording. In the beginning all these components were treated independently. This can be called "device research." Later, the integration of these parts into a whole (e.g., a complete radio set, or a complete radio broadcast and receiving system) became a separate concern for research activities. This was the beginning of what later would be called "systems research" as a third type of research alongside materials and devices research. Most of these developments took place in the first half of the twentieth century.

Although World War II had a large impact on many aspects of social life, it did not seem to have much influence on the research and development activities of most large companies. Even in The Netherlands, which was occupied by the Germans from 1940 to 1945, the research program of the Philips Company to a large extent was continued. Some activities were terminated because the outcomes might be useful for military use by the Germans, and others were continued secretly. In countries that the Germans did not occupy, such as the U.K. and the U.S., World War II stimulated research for military purposes. The main impact of the war, however, was the successful development of the atomic bomb, which created an awareness that research into fundamental phenomena could result in dramatic breakthroughs. This caused the American president's scientific advisor, Vannevar Bush, to produce a report titled *Science, the Endless Frontier*. The main message of this report was that basic research was to be stimulated by the government and be taken up in industrial companies, because in the end it would always lead to industrial applications, often of a breakthrough character. This claim was soon supported by another important example, namely the invention of the transistor at the Bell Labs in 1947. This invention took place in a research group that fitted well with the ideas that Vannevar Bush had formulated. Earlier efforts to make a transistor by creating a solid-state analog of the triode had failed, and it was not until solid-state physics was applied to the problem that the transistor was invented. Not long afterward, the laser was invented; this too was a result of the application of solid-state physics, thus confirming the Vannevar Bush philosophy of basic research resulting in industrial application.

The ideal of basic research led to a new role for the research laboratories of the industrial companies. A substantial part of R&D was now dedicated to research into phenomena without any concrete relationship to possible applications. Sometimes, as in the case of the Philips Company, this change was accompanied by establishing development laboratories in the product divisions. Thus basic research was conducted in the central laboratory, while the division labs undertook applied research. The basic research approach did yield some great successes; two discoveries at Philips illustrate this point. First, the Plumbicon, a television pickup tube, was invented in 1958, and it would replace earlier tube types in all professional television cameras. Second, Philips researchers invented local oxidation of silicon (LOCOS), a technique for producing very compact integrated-circuit (IC) structures in silicon. Both were the outcome of research that was oriented to fundamental phenomena, and the patents on these inventions created a large income for the company as competing companies realized that they were almost forced to use these inventions to maintain their position on the market. The field of ICs is related to another trend, the internationalization of R&D activities from 1950 to 1970. The Philips Company, for example, established research laboratories in the U.K., France, Germany, and the U.S. in the first decade after World War II. There were regular contacts between the research laboratories of the various companies. This sometimes led to important exchanges of knowledge; for example, the Bell Labs transistor knowledge was made available to Philips based on the fact that Philips previously had made available to Bell certain knowledge of ferrites (nonmetallic magnetic materials). In the last decades of the twentieth century, globalization of R&D increased rapidly. For example, R&D expenditures by foreign-owned companies in the U.S. more than doubled, from $6.5 billion in 1987 to $14.6 billion in 1993. By 1999 this amount had further increased to $17 billion.

To be successful in basic research, the research laboratories claimed they needed a large degree of independence and freedom. In a number of companies the research laboratories received their budgets directly from the company's management, which made them financially independent of the product divisions. The result was that product divisions had almost no say in the research program. At the same time, the product divisions felt no obligation to adopt research output for which they had not paid. This mutual lack of commitment in companies with separate research facilities often frustrated cooperation between research and development in this period. A striking example of this was the research on the hot-air engine that for many years took place in the Philips *Natuurkundig Laboratorium*. The origin for this research had been the need for a small, silent energy source for radio transmitters and receivers in developing countries. This need, however, had fallen away after the transistor had replaced vacuum tubes in radios, and a simple battery would do for the energy supply. However, the Philips researchers decided to continue the hot-air engine research with car engines as a new application. There was not a single product division in the company for which that would have been a useful development. The financial independence of the research lab allowed it to continue the work on its own. It never yielded any profit, and it was not terminated until an entirely new situation for the research lab was established in the late 1970s.

In some companies, DuPont for example, the labs in the various company divisions also undertook a certain amount of long-term basic research. In this company, such research in earlier years had led to the development of nylon. The hope was that basic research in the division labs would result in finding similar new products. When this expectation was not fulfilled, the company management began to suspect the validity of basic research promises. The economic decline of the late 1960s forced companies to critically review the basic research of central laboratories. In this review it became clear that the number of major breakthroughs due to basic research in the 1950s and 1960s had been fairly small compared with the effort that had been spent on this type of research. This result was found not only at DuPont but also in electronics companies such as Philips, GE, and RCA.

In addition to economic factors, increasing social criticism of technological developments caused companies to reconsider their research and development activities. The first report from the Club of Rome in 1972 had placed concerns about environmental damage by technological developments on the political agenda. Together with increased doubts about the value of basic research in industrial companies, these concerns would lead to a new position for R&D within companies. Society, not science, was to be the new "Endless Frontier" (a term from the 1988 European Commission report, titled *Society, the Endless Frontier*).

The transition to the new situation can be suitably illustrated by the work that was undertaken on optical recording in various companies. At both Philips and RCA, the research labs led this technological development. The concrete outcome of this effort, the videodisc, was a commercial failure for all companies that were involved in the field. The same knowledge was later used at Philips and Sony to develop the compact disk (CD). In that development, however, the product divisions had the leading role and the research labs were called in "only" for delivering specific knowledge on optical recording. This would establish the new relationship between research and development in such companies: the division in which the technological development work took place would decide on the desirability and feasibility of new products, and the research laboratories were commissioned to deliver the specific knowledge to be used by the divisions to develop the products. The idea of basic research leading to development work on new products was abandoned. This new approach also led to an almost complete "divisionalization" of research programs; only a very small amount of basic research was kept.

Market requirements became increasingly important for both research and development activities. In the 1950s and 1960s, economic growth enabled companies to market all sorts of new products, and rarely was the introduction of a new product a commercial failure. However, the economic decline in the early 1970s made customers aware that they could only spend their money once, and they became more critical of new products. Not only were they more hesitant to buy every new gadget that appeared, they were also more critical in comparing products offered by competing companies. Taking into account the customers' desires therefore became a crucial issue for product development. This led to a new way of considering the concept of quality. In the past, quality had mainly been a concern for production departments, and it generally meant that the output of production lines was checked for obvious defects. With increased attention to customers' wishes, however, came the awareness that the customer had to be taken into account not only in the production phase but also in the development phase, and perhaps even in the research phase. Quality was to be "built in" to the design of the product. For that purpose, new methods for supporting product development were created. The term total quality management (TQM) was often used to describe the overall effort of assuring quality throughout the development and production activities. It became evident in time that the application of such methods presented other problems. In general, the concept of quality management had previously been used in the context of production activities that had a repetitive and predictable nature. Often statistics played a role in such methods. R&D activities were much less repetitive and predictable, and this created a tension between quality methods and the nature of such activities. Yet quality management clearly found its way into the development work of industrial companies, and even in research laboratories, quality managers were appointed to stimulate and monitor quality-oriented efforts. For researchers, particularly those who had been used to a large degree of freedom for their work, this was a significant change in culture.

Quality methods helped product developers take into account the whole life-cycle in the development of new products and try to create designs such that all phases would go smoothly. This resulted in a collection of methods called "design for *x*," where *x* could be any phase in the life-cycle of the product: "design for manufacturing," "design for assembly," "design for logistics," "design for disassembly," "design for recycling," and so on. A method was developed in Japan that specifically focused on the customer's requirements and was called quality function deployment (QFD). In general, Japanese companies were the first to take the customer's requirements into account, while Western companies still focused on technical perfection. One example of the success of the customer-oriented approach in the marketplace is seen in the competition between the digital compact cassette (DCC) that Philips released in the late 1980s and the minidisc that Sony brought out almost simultaneously. Philips had directed its efforts toward the technical operation of the product, while Sony had carefully looked at possibilities for attracting the customer's attention by adding certain functions. For example, the minidisc was almost immediately produced in a portable version, and unlike the DCC, the user could program the title of a music track into the disc so that it would be displayed while playing. Striving for technical perfection had also caused Philips to abandon the plan to release the DCC before Sony could bring out its minidisc. The result was that both came on the market almost simultaneously, allowing customers to compare them and recognize the advantages of the minidisc. Such stories made it clear that three issues were crucial for success in R&D in the 1980s and 1990s: cost, time-to-market, and quality (in the sense of customer satisfaction).

A second trend that related to the need to consider the product's life-cycle in industrial R&D was the concern for environmental effects of technological developments. This concern became a political issue in the early 1970s, and by the 1980s and 1990s it was "operationalized" into guidelines for industrial activities, not only in production but also in R&D. Here, too, the idea was that environmental concerns should be built into the design for new products. Rather than rejecting polyvinyl chloride (PVC) because it caused polluting emissions in the production phase, researchers would reconsider its merits and take into account the fact that it could be recycled. To make a balanced, overall evaluation of the environmental impact of a product, the "life-cycle analysis" (LCA) was developed as a new method in product development. In an LCA, the sum of a number of environmental effects for all phases in the life-cycle is presented. Such effects may include the contribution to ozone layer depletion, contribution to the greenhouse effect, and contribution to solid waste. LCAs allow designers to compare different materials and processes with respect to their impact on the environment. Even though there are several methodological problems, as in the field of quality methods, LCAs have found their way into industrial practice.

The changes in approach to R&D in the twentieth century have sometimes been described in terms of generations. All R&D before the 1970s is called first generation R&D, and technology is seen as the main asset in this generation. Second generation R&D is related to the rest of the business functions, as before World War II, and attention is shifted toward individual projects as the main asset. By the 1980s R&D sought to invade the whole enterprise more systematically. Fourth generation R&D means that learning from the customer becomes crucial in the R&D function. The newest trend at the end of the twentieth century was called fifth generation R&D, where knowledge in general is regarded as the main asset. Of course, different companies moved through these generations in different ways, but they do give an overall indication of the dynamics of R&D in the late twentieth century.

MARC J. DE VRIES

Further Reading

Bralla, J.G. *Design for Excellence*. McGraw-Hill, New York, 1996.

Chandler, A.D., Jr. *Scale and Scope. The Dynamics of Industrial Capitalism*. Belknap Press of Harvard University Press, Cambridge, MA, 1990.

Clayton, R. and Algar, J. *The GEC Research Labs, 1919–1984*. Peter Peregrinus, London, 1989.

Graham, B.W. *The Business of Research. RCA and the Video Disc*. Cambridge University Press, Cambridge, 1986.

Hounshell, D.A. and Smith, J.K., Jr. *Science and Corporate Strategy: DuPont, 1902–1980*. Cambridge University Press, Cambridge, 1988.

Reich, L.S. *The Making of American Industrial Research. Science and Business at GE and Bell, 1876–1926*. Cambridge University Press, Cambridge, 1985.

Rosenbloom, R.S. and Spencer, W.J., Eds. *Engines of Innovation: US Industrial Research at the End of an Era*. Harvard Business School Press, Boston, 1996.

Stokes, D.E. *Pasteur's Quadrant: Basic Science and Technological Innovation*. Brookings Institution Press, Washington D.C., 1997.

De Vries, M.J. *80 Years of Research at the Philips Natuurkundig Laboratorium (1914–1994). The Role of the Nat. Lab. at Philips*. Stichting Historie der Techniek, Eindhoven, 2001.

Rocket Planes

Once the Wright brothers had proved that controlled and sustained powered flight was possible, two related avenues of development were immediately apparent: increasing the distance and the speed of flight. From that first flight in 1903 until the late 1930s, aircraft control and propulsion technology was developed to improve all aspects of flight, culminating in the production of the subsonic turbojet. Although further development of the jet engine would revolutionize both military and commercial aviation and boost speeds into the supersonic region, what became known as the "sound barrier" (Mach 1) was first broken by a rocket-powered aircraft, the XS-1.

Subsonic World War II aircraft had approached Mach 1 in steep dives, but the results were often disastrous since the materials and control systems of the time were unable to cope with the aerodynamic forces encountered. Part of the solution was to ensure that an aircraft passed through the transonic zone as quickly as possible, which meant providing thrust levels higher than the jet engines of the time could deliver. The obvious solution was the rocket engine.

A number of experiments in rocket-assisted flight were conducted in Germany from 1928, when experimenter Friedrich Stamer flew 1.2 kilometers in a small glider propelled by an elastic rope and two small rockets. In 1929, a Junkers 33 seaplane made a rocket-assisted take-off at Dessau, Germany, and by the late 1930s the Heinkel 176, the first rocket-powered aircraft designed for sustained flight, had been tested successfully. This led to the development of the Messerschmitt Me-163 research plane, which first flew in 1940 and set a new world speed record of 917 kilometers per hour (km/h) in 1941. This attracted the attention of the military authorities, who proceeded to develop the Me-163 as a fighter, renaming it Komet. More than 360 were built and some were used effectively in World War II, but by then rocket research had switched to the development of ballistic missiles, the most notable result of which was the V-2 "vengeance" weapon, which became the technical basis for both Soviet and American rocket programs after the war.

Meanwhile in the U.S., the need for similar research and development was being debated. In 1943, the National Advisory Committee for Aeronautics (NACA, the forerunner of the National Aeronautics and Space Administration, NASA) conceived a rocket-propelled research aircraft designed to exceed Mach 1 in horizontal flight. Thus on October 14, 1947, pilot Charles E. (Chuck) Yeager exceeded the speed of sound in the XS-1 (Experimental Supersonic 1). Between them, the XS-1 and two later planes designated X-1, made 156 flights, setting a speed record of 1540 km/h (about Mach 1.45) and an altitude record of 21,916 meters. The aircraft, which were dropped from the modified bomb-bay of a B-29 bomber for their test-flights, were designed and built by Bell Aircraft Corporation and were sometimes known as the Bell X-1. They were powered by a Reaction Motors rocket engine, burning ethyl alcohol and liquid oxygen, which produced about 2700 kilograms of thrust.

The aircraft themselves were designed to have unprecedented structural strength to survive the aerodynamic forces of the transonic region (up to 18 g (the acceleration due to gravity) and –10 g), and carefully shaped to reduce drag. By reducing the cross-sectional area of the fuselage toward the rear, the air compressed by the plane's motion and forced back along the fuselage in a shock wave would have a chance to expand slightly, thus easing the aircraft's progress through the sound barrier. The resulting improvement in aerodynamic efficiency meant that engines generating much lower thrust levels could achieve supersonic speeds (Figure 28).

Later X-planes were designed to go higher and faster than the X-1 and to test new materials that would survive the frictional heating effects of ever-higher velocities. This culminated in the development of the X-15, a rocket-propelled research aircraft designed and built by North American Aviation to investigate the effects of high speed at high altitude on future space planes (Figure 29). The main engine burned the propellants, anhydrous ammonia and liquid oxygen, while the altitude control thrusters used for steering consumed hydrogen peroxide. NASA's fleet of three X-15s was dropped from beneath the wing of a converted B-52 bomber in a series of 199 flights (the first and last powered flights occurred in September 1959 and October 1968, respectively). Designed to reach speeds of up to Mach 6 in

Figure 28. The Bell Aircraft Corporation X-1-1 in flight, showing a shock wave pattern in its exhaust plume. This aircraft was nicknamed "Glamorous Glennis" by Chuck Yeager in honor of his wife, and is now on permanent display in the Smithsonian Institution's National Air and Space Museum in Washington, D.C.
[*Photo courtesy of NASA.*]

Figure 29. An X-15 rocket plane after landing. As ground support personnel begin post-flight activities, the NASA B-52 launch aircraft and two F-104 chase aircraft provide an aerial salute.
(*Photo courtesy of NASA.*)

horizontal flight, the X-15 set a record of 7274 km/h (about Mach 6.7) and, separately, an altitude record of 107,960 meters. One of its more famous pilots was Neil A. Armstrong, the first man to walk on the moon.

Despite their success as research and technology demonstrators, rocket planes turned out to be a technological cul-de-sac. Interest in rocket-powered aircraft waned as the performance of jet engines improved to the point where, less than ten years after Yeager's historic flight, jet-powered fighters capable of Mach 2 were production-line items. Initial plans to fly the X-15 into orbit were cancelled in favor of vertically launched rockets, which were to become the preferred delivery system for NASA's Mercury program and those that followed. Nevertheless, X-15 research did engender an extensive program of winged lifting-body research, which led eventually to the Space Shuttle. Moreover, it is widely believed that expendable rockets will one day be superseded by reusable space-planes.

See also **Aircraft Design; Rocket Propulsion, Liquid Propellant; Space Launch Vehicles**

MARK WILLIAMSON

Further Reading

Caidin, M. *Wings into Space*. Holt, Rinehart & Winston, New York, 1964.

Godwin, R. *X-15: The NASA Mission Reports*. Apogee Books, Ontario, 2000.

Johnston, A.M. and Barton, C. *Tex Johnston: Jet-Age Test Pilot*. Smithsonian Institution Press, Washington D.C., 2000.

Rotundo, L. *Into the Unknown: The X-1 Story*. Smithsonian Institution Press, Washington D.C., 1994.

Yeager, C. and Janos, L. *Yeager: An Autobiography*. Century, London, 1985.

Rocket Propulsion, Liquid Propellant

The world's first rocket, probably originating in China in the eleventh century AD, was solid-propelled with ordinary gunpowder as the propellant. Eight hundred years later, from the late nineteenth to the early twentieth centuries, Konstantin Tsiolkovsky of Russia, Robert H. Goddard of the U.S., and Hermann Oberth of Transylvania independently conceived the idea of the liquid-propellant rocket. These pioneers made the important theoretical discovery that the rocket operates according to Isaac Newton's third law of motion: "For every action there is an opposite and equal reaction." This means that reaction propulsion operates both on the ground and in the vacuum of space. This was in total contradiction to the widely held but erroneous belief that the rocket works because the gases push against the air. The pioneers went further and realized that liquid propellants were potentially a far more efficient way to achieve space flight for several reasons.

Liquid propellants consist of a fuel (the substance burned) and the oxidizer (the liquid containing oxygen atoms needed for combustion of the fuel). The liquids are kept in separate tanks and pumped into a combustion chamber at regulated pressures by pumps, in a process of controlled combustion. The two ingredients burn in the chamber and expel their gases out of the rear opening of the rocket, driving the rocket forward by reaction propulsion. Using valves, liquid propellants can be stopped and started at will. An important extra advantage with liquids is that they are more powerful than solids. All three of the pioneers found that the combination of liquid oxygen (oxidizer) and liquid hydrogen (fuel) was the ideal propellant, although these liquefied gases are extremely cold (below 0°C) and were then very difficult to manufacture. For this reason, Goddard, the only one of the pioneers to undertake extensive experiments, chose to use readily obtainable but less powerful liquid oxygen ("LOX") and gasoline. From 1921 until his death in 1945, Goddard engaged in liquid-propellant rocketry experiments.

On March 16, 1926, Goddard launched the world's first liquid-fuel rocket at Auburn, Massachusetts. It went up to 12.5 meters in 2.5 seconds, but this effort was not publicized. Earlier, Tsiolkovsky first published his theories of the liquid-propellant (LOX/hydrogen) space rocket in 1903 in a Russian popular science magazine. Although he continued to publish his theories, his work was almost unknown in the West. In 1923, Oberth's seminal book, *Die Rakete zu den Planetenräumen [The Rocket into Planetary Space]* appeared. It was enormously influential and virtually created a space flight and rocket fad. Oberth described several models of LOX/alcohol rockets in detail, including pumps and methods of cooling, and the possibility of manned space missions.

As a result, several amateur rocket groups were formed around the world. The most prominent was the *Verein für Raumschiffahrt*, or VfR (Society for Space Travel). They began crude experiments with liquid-propellant rockets from 1930. The German Army began its own secret rocket program and in 1932 hired Wernher von Braun, a young and gifted VfR member to be the technical director. By 1940, the German Army had developed the A-4 (Aggregate 4) rocket, later known during World War II as the V-2. Using LOX and alcohol as propellants pumped into an hourglass-shaped combustion chamber with 18 separate propellant injectors on its dome, the V-2 produced 27,000 kg of thrust with a range of 320 kilometers. Cooling was by a combination of regenerative cooling, in which the fuel flowed around a cooling jacket surrounding the chamber before it was injected from the top, and "film cooling," in which the fuel was sprayed onto the inside of the nozzle wall, resulting in a thin film of cool fuel. The V-2 was the world's first large-scale liquid-fuel rocket. However, despite a popular misconception, V-2 development was more influenced by Oberth than Goddard. Under self-imposed secrecy, Goddard worked in the New Mexico desert and achieved remarkable results, notably gyroscopically stabilized, pump-driven rockets and thrusts of almost 450 kilograms. The highest altitude in his flight tests was 2740 meters on March 26, 1937. Goddard was unaware that the Germans had attained three-dimensional gyroscopic control and thrusts of 1485 kilograms pounds with their A-3 test rocket, which was not meant for altitude flights. The Germans also developed other missiles.

In 1941 Reaction Motors, Inc. (RMI) was formed by four American Rocket Society (ARS) members and was based on the success of James H. Wyld's regeneratively cooled rocket motor design. Similar to the V-2 design, the fuel in the Wyld motor circulated around a cooling jacket before injection. RMI adapted this technique to JATOs (jet-assisted-take-off) rockets for heavily loaded seaplanes and to small, experimental missiles. Following the war, both the U.S. and the Soviet Union captured German V-2 rockets, acquired V-2 technicians, and began developing their own long-range missiles. In 1946 the North American Aviation Company initiated development of its 800-kilometer-range Navaho. Navaho was eventually upgraded to an intercontinental missile, and it established the most important milestones in the history of U.S. rocket technology.

At first, the Navaho was designed as a modified V-2 with wings. When the Air Force requested a change to a 1600-mile-range weapon, the Navaho was designed as an air-breathing ramjet-powered cruise missile with large liquid-propellant boosters. To learn how to build and handle large rocket engines, North American made three copies of the V-2 engines. By 1950 they had created an entirely new and more compact cylindrical, single flat-plate injector engine, the XLR-43-NA-1. Still using LOX/alcohol, this engine produced 34,000 kilograms of thrust. This was the first large-scale liquid fuel engine developed by the U.S.

Also in 1950, the U.S. Army, working with the emigrant German scientist Wernher von Braun, began development of its 800-kilometer-range Redstone missile, essentially a super-V2. In the following year, when Navaho was further upgraded and required more powerful boosters, the Army adopted the XLR-43 for the Redstone. The Redstone subsequently used a modified XLR-43 and became the U.S.'s first operational surface-to-surface ballistic missile. Later, with a more powerful hydrazine-based propellant and producing 37,700 pounds of thrust plus upper stages, the modified Redstone (also known as the Jupiter C) launched the U.S.'s first artificial satellite Explorer 1 on January 31, 1958. The Redstone also launched America's first astronaut, Allan B. Shephard, into space on May 5, 1961, on a suborbital flight.

Meanwhile, the North American Aviation Company produced the more powerful XLR-43-NA-3 for the Navaho on November 19, 1952. It was the first U.S. liquid-fuel rocket engine to reach a thrust of more than 100,000 pounds (45,360 kilograms), with a maximum thrust of 120,000 pounds (54,430 kilograms). Among the engine's major technological breakthroughs was the "spaghetti" or tubular form of combustion chamber, in which the cooling tubes formed the walls of the chamber itself. This reduced the weight of the chamber by 50 percent. The spaghetti configuration became adopted throughout the U.S. rocket industry for large-scale engines such as the Space Shuttle main engine (SSME). Edward A. Neu, Jr., of RMI conceived the idea about 1947, but other companies also developed the concept.

In 1952, the U.S. Air Force Rocket Engine Advancement Program (REAP) saw another important development applied to the Navaho and later engines, the use of LOX/hydrocarbon propellants like JP-4, or kerosene. Hydrocarbon fuels replaced V-2 era alcohol and increased the specific impulse, a unit of measurement for the efficiency of rocket engines. North American used the regeneratively cooled, tubular chamber with flat-head injector and LOX/JP-4 as a building-type engine. In the upgrade of the Navaho, two of the engines were coupled together to produce a 240,000-pound (108,862-kilogram) thrust engine, tested at full thrust in 1954.

The rapid success of North American's large-scale engine development led to contracts to develop similar engines for the Atlas intercontinental ballistic missile (ICBM) and the 2400-kilometer Jupiter and Thor intermediate range ballistic missiles (IRBMs). In the Jupiter program, another breakthrough was made. G.V.R. Rao, a mathematician at Rocketdyne (a division of American Aviation, later Rockwell), found that a bell-shaped nozzle theoretically provided optimum thrust over standard conical nozzles. Rocketdyne successfully tested this design in January 1956 with a Jupiter engine. The bell-shaped chamber became another industry-wide feature in large-scale U.S. rocket engines.

Navaho first underwent flight testing in November 1956, but several of the vehicles failed. On July 11, 1957, the program was cancelled, mainly due to its enormous cost of almost a billion dollars. These medium-range ballistic missiles nonetheless left an enormous technological legacy in rocket propulsion. From its building-block engine derived directly from Navaho, beginning in 1959 Rocketdyne developed the F-1 engine, which powered the first stage of the three-stage Saturn-V launch vehicle used in ten Apollo missions. This single-chamber liquid-fuel engine with 680,000-kilogram thrust was the largest and most powerful single liquid-fuel engine ever utilized in rockets. The Space Shuttle main engine was part of the same lineage. (The use of LOX/hydrogen in rocket engines was pioneered by the Pratt & Whitney Company, which developed the Centaur high-energy upper-stage engine. Rocketdyne adapted this technology to their H-1 and J-2 engines for the Saturn-1B and Saturn-V, respectively.)

In other countries, notably the former USSR, the development of the liquid-fuel rocket evolved along quite different paths. Following World War II, the Soviets also acquired captured V-2 hardware and German technicians. However, the Soviets, principally under the technical leadership of Sergei P.Korolev, used V-2 variant designs in rocket engines and vehicles for a far longer period than the Americans. The Soviets did not make the same breakthroughs in lightweight engines but used such techniques as the "cluster of clusters" approach; that is, their large rockets utilized many engines firing together to achieve large thrusts. These rockets, such as the Vostok series, were thus extremely large and inefficient for their size, but they did function. A Vostok vehicle enabled the Soviets to place the first artificial satellite, Sputnik-1, into orbit on October 4, 1957, thereby opening the space age.

Liquid-propellant rocket engines have been used for several applications including sounding, or upper atmospheric research, rockets, rocket-sled propulsion, rocket research aircraft, and small attitude control and course correction thrusters for spacecraft. The latter are the simplest types of motors, often using single or monopropellants, pressure-fed systems rather than pumps, and high heat-resistant walls for cooling.

See also **Missiles, Long Range Ballistic; Missiles, Surface-to-Air and Anti-Ballistic Missiles; Space Launch Vehicles**

FRANK H. WINTER

Further Reading

Baker, D. *The Rocket*. Crown Publishers, New York, 1978.
Bilstein, R.E. *Stages to Saturn*. National Aeronautics and Space Administration, Washington D.C., 1980 and 1996 editions.
Dornberger, W. *V-2*. Viking Press, New York, 1954.
Goddard, E.C. and Pendray, G.E. Eds. *The Papers of Robert H. Goddard*. Macmillan, New York, 1970.
Smith, M. *Space Shuttle*. Haynes, Sparkford, 1985.
Uhl, M. *Stalins V-2*. Bernard & Graefe-Verlag, Bonn, 2001.
Winter, F.H. *Rockets into Space*. Harvard University Press, Harvard, MA, 1990.

Rocket Propulsion, Solid Propellant

Solid propellants typically consist of a mixture of fuel and oxidizer, whereas liquid-propellant components are generally stored in separate tanks and combined in a combustion chamber. Although solid and liquid rocket propellants are usually employed separately, they are used together in the hybrid rocket, the most common combination being solid fuel and liquid oxidizer. The propellants used in the cold-gas thrusters employed to position some spacecraft in orbit are unusual in that they are generally stored as a gas.

The rocket in its simplest form is believed to have been invented by the Chinese in the eleventh century AD using a solid propellant similar to what was later called gunpowder. The Englishman William Congreve initiated modern rocketry research and development in the early nineteenth century when he built the first gunpowder rocket weapon systems to challenge Napoleon's invasion plans for England. Solid rockets continued to be developed throughout the nineteenth century and were used during World War I, but their power and efficiency (now measured by the quantity "specific impulse") were limited by the available propellants. Thus it was that Russian rocket pioneer Konstantin Tsiolkovsky suggested the use of liquid propellants in 1903, and American researcher Robert Goddard began their development in the early 1920s. Goddard had conducted tests with solid fuels to refine his technique, and in 1915 he found that a tapered nozzle increased the expulsion efficiency of the exhaust gases by some 64 percent. He subsequently used the nozzle in all his rocket work, and it was adopted by others. Since the 1920s, the majority of rocket systems required to carry significant payloads, especially into space, have been based on liquid propellants. Despite this, solids have their uses.

By the end of the twentieth century, solid-propellant rockets were available in all shapes and sizes and were often chosen over liquid systems because they are simpler and can be stored long-term without maintenance. They are widely used, for example, by the military services to power missiles, torpedoes, and aircraft ejection seats. On a larger scale, their inherent storability has made them ideal for intercontinental ballistic missiles (ICBMs), which may be required at very short notice. Solid-propellant rockets are also used in the space industry as launch vehicle stages, strap-on boosters and stage separation devices, and sometimes as "upper stages," which boost spacecraft to their final orbit or toward the planets.

There are two main types of solid propellant: homogeneous or double-base propellants, which are limited in power (e.g., nitrocellulose plasticized with nitroglycerin plus stabilizing products); and the heterogeneous or composite type. Composite propellants consist of a mixture of fuel and oxidizer, the latter providing the oxygen for combustion. An oxidizer in common use is crystalline ammonium perchlorate (NH_4ClO_4 or AP). It is mixed with an organic fuel, such as polyurethane or polybutadiene, which also binds the two components together. The inclusion of a plastic binder produces a rubbery material, making the propellant relatively easy to handle. Performance is enhanced by adding finely ground (10 micrometers) metal particles of aluminum, for example, which increase the heat of the reaction due to the formation of metal oxides. As an example, the propellant used in the Space Shuttle solid rocket boosters is comprised of 14 percent polybutadiene acrylic acid acrylonitrile (binder and fuel), 16 percent aluminum powder (fuel), 69.93 percent ammonium perchlorate (oxidizer), and 0.07 percent iron oxide powder (catalyst).

A propulsive device that uses solid propellants to move a vehicle is known as a rocket motor or solid rocket motor (SRM). Although the terms motor and engine are interchangeable in colloquial English, it is customary to call propulsive devices using a solid propellant motors, and those using a liquid propellant engines.

A typical SRM comprises only a few major components: a motor case that contains the propellant grain (or charge), a surrounding insulating blanket or propellant liner, an ignition system, and an exhaust nozzle (Figure 30). The motor case is typically made from a carbon composite or Kevlar composite by a process known as filament winding, which produces a strong yet lightweight structure. The propellant liner acts as both a thermal insulation blanket and a flame inhibitor and supports the propellant grain during manufacture, as it is poured into the open-ended motor case.

A typical ignition system operates as follows: a power supply sends a current pulse to a pyrotechnic cartridge, which ignites a small sample of the propellant in a steel or glass-fiber housing. This produces a controlled amount of hot gas that ignites the main motor, much like a detonator in an explosive device. The typical exhaust nozzle, or

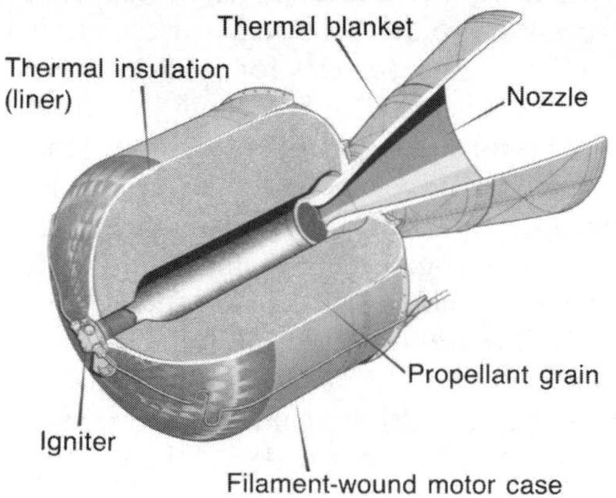

Figure 30. A solid rocket motor used as a satellite apogee kick motor; the diagram shows the major components. [*Reproduced with permission from Snecma.*]

expansion nozzle, consists of two sections: one convergent, the other divergent. The narrowest part of the nozzle, called the throat, is designed to maintain the required pressure within the motor case and regulate the outflow of combustion gases; it is also the region of transition from subsonic to supersonic flow. The exhaust gases expand and accelerate rapidly as they leave the motor, and it is the exit cone that controls the expansion of the exhaust plume.

In operation, the SRM has no need of the complex piping, pumps, and pressurization systems of the liquid rocket engine. The SRM must, however, be designed to give precisely the required amount of thrust and total impulse since it cannot be throttled, easily stopped, or restarted. Where two SRMs are used together, as on the solid rocket boosters of the Space Shuttle, they must also be precisely matched to provide the same thrust profile and to ensure that they cease firing together.

Since the thrust available from a block of propellant is proportional to the area of the combustion surface, the only way to control the magnitude of the thrust is to offer varying surface areas for combustion throughout the burn. The active area can be increased by making a cylindrical hole through the center of the block so that the hole enlarges radially and the thrust increases as the propellant burns. Constant thrust, which is more often required, can be provided by a star-shaped bore (Figure 31). A variation in thrust throughout the burn can be arranged by varying the cross-section along the length of the propellant grain.

Although the performance of solid propellants, in terms of thrust derived per unit mass, is generally inferior to that of liquid propellants, solids continue to be used in many applications. This is mainly because they are easier to handle and store and are simpler, therefore more reliable, in their mode of operation.

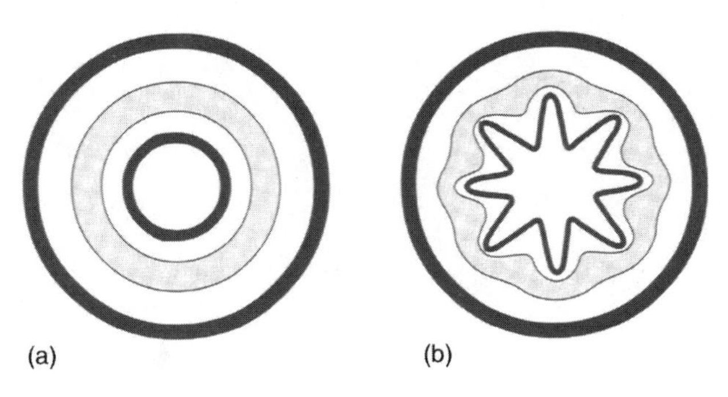

Figure 31. Solid-propellant grain cross-sections: (a) thrust increases throughout the burn; (b) constant thrust. [*Reproduced with permission from the Dictionary of Space Technology. Adam Hilger, 1900.*]

See also **Rocket Propulsion, Liquid Propellant; Space Launch Vehicles; Space Shuttle**

MARK WILLIAMSON

Further Reading

Brown, C.D. *Spacecraft Propulsion*. American Institute of Aeronautics and Astronautics, Washington D.C., 1996.

Davenas, A., Ed. *Solid Rocket Propulsion Technology*. Pergamon Press, Oxford, 1993.

Sutton, G.P. *Rocket Propulsion Elements: An Introduction to the Engineering of Rockets*. Wiley, New York, 1956.

Stoiko, M. *Pioneers of Rocketry*. Hawthorn Books, New York, 1974.

Turchi, P.J., Ed. *Propulsion Techniques: Action and Reaction*. American Institute of Aeronautics and Astronautics, Reston, 1998.

Williams, B. and Epstein, S. *The Rocket Pioneers*. Lutterworth Press, London, 1957.

Williamson, M. Spacecraft propulsion systems. *Phys. Technol.*, 15, 6, 284–290, 1984.

Winter, F.H. *The Golden Age of Rocketry: Congreve and Hale Rockets of the Nineteenth Century*. Smithsonian Institution Press, Washington D.C., 1990.

S

Satellites, Communications

Arthur C. Clarke, a science-fiction writer and an early member of the British Interplanetary Society, is credited as first stating the theoretical possibility of satellite communications. The February 1945 issue of the British technical journal *Wireless World* included a letter from Clarke under the title "V-2 for Ionospheric Research" in which he explained that three artificial satellites positioned 120 degrees apart in geosynchronous orbit could provide television and microwave coverage to the entire planet. Three months later, he privately circulated six typewritten copies of a paper titled "The Space Station: Its Radio Applications." A refined version, where Clarke gave a detailed, technical analysis of the orbital geometry and communications links, appeared under the title "Extra-Terrestrial Relays" in the October 1945 issue of *Wireless World.* Clark, however, acknowledged that the forerunner of the concept came from Hermann Potocnik, whose 1929 book *Das Problem der Befahrung des Weltraums* [*The Problem of Space Travel*], published under the pseudonym Noordung, described "stationary circling."

During the next decade, others promoted the idea of satellite communications. In their May 1946 report titled "Preliminary Design of an Experimental World-Circling Spaceship," Project RAND engineers at the Douglas Aircraft Company's plant in Santa Monica, California, told the U.S. Air Force that satellites could significantly improve the reliability of long-range communications and might spawn a multibillion-dollar commercial market. Subsequent RAND studies by James E. Lipp in February 1947 and Richard S. Wehner in July 1949 further developed the concept of equatorial-orbiting communications satellites. Writing under a pseudonym in *Amazing Science Fiction*, John R. Pierce of AT&T's Bell Telephone Laboratories suggested a communications satellite system in 1952. He became one of the first people outside defense-related circles to evaluate systematically both technical options and financial prospects. In a 1954 speech and a 1955 article, Pierce assessed the utility of passive "reflector" and active "repeater" satellites at various orbital altitudes and estimated a single satellite's capacity for handling simultaneous telephone calls as nearly 30 times greater than the original transatlantic telephone cable.

The USSR's launch in 1957 of Sputnik-1 (Figure 1), which transmitted an electronic signal back to earth simply for tracking purposes, sparked serious efforts by the U.S. to develop satellite communications for both military and commercial use. Score, developed by the Advanced Research Projects Agency and launched in December 1958, became the world's first active communications satellite. It received messages transmitted from a ground station and stored them on a tape recorder for retransmission back to earth. Launched in August 1960, the National Aeronautics American and Space Administration's (NASA's) Echo-1 first tested the merits of using a passive "reflector" satellite—a 30-meter-diameter, aluminized mylar balloon—for the transmission of voice, data, and photographs between ground terminals. The U.S. Army's Courier satellite, launched in October 1960, operated on much the same principles as Score but carried solar cells and rechargeable batteries to extend its potential lifetime. Stemming from its interest in transoceanic

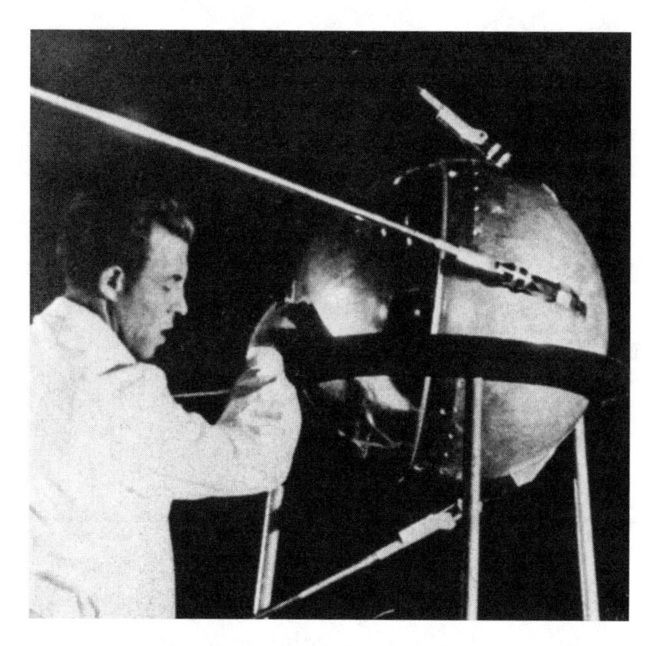

Figure 1. The Sputnik-1 satellite on a rigging truck in the assembly shop in the fall of 1957 as a technician puts finishing touches on it.
[*Credit: Courtesy of NASA.*]

communication, AT&T launched Telstar-1 in June 1962 to experiment with telegraph, facsimile, television, and multichannel telephone transmissions between the U.S. and Europe, as well as Japan. With the launch of NASA's Syncom-2 in July 1963, the world finally had its first geosynchronous communications satellite. Not until 1975 would the USSR achieve a similar feat with its Raduga spacecraft for military and governmental communications, followed by its Ekran series in 1976 for direct television broadcasting and its Gorizont series in 1979 for domestic and international telecommunications.

Throughout the remainder of the twentieth century, most communications satellites were placed in geosynchronous orbit, because that afforded the greatest coverage with the smallest constellation. Established in 1964, the International Telecommunications Satellite Consortium (Intelsat)—with Communications Satellite Corporation (Comsat) as its U.S. component—opted for a geosynchronous satellite produced by Hughes Aircraft Company instead of a medium-orbit model proposed jointly by AT&T and RCA. Early Bird (Intelsat-1), launched on 6 April 1965, marked the beginning of global satellite communications networks open to all nations. Government-controlled, regional satellites to service a group of geographically proximate or culturally similar nations soon appeared—one of the first being Indonesia's Palapa-A in 1976.

Privately owned satellites represented yet another type of international system—the first being Japan Communications Satellite's JCSat-1 in 1989. The first domestic communications satellite was Telsat Canada's Anik-A in 1972, followed two years later by Western Union's nearly identical Westar-1, the first U.S. domestic communications satellite. During the 1980s, the International Mobile Satellite Organization (INMARSAT) operated a mobile telecommunications network primarily for maritime users but began expanding voice and facsimile services to aircraft on international routes during the 1990s.

While the U.S. military has relied heavily on commercial satellite communications over the years, it has also developed and launched dedicated systems to satisfy unique national security requirements. On 16 June 1966, the U.S. Air Force launched the first seven of 26 Initial Defense Communications Satellite Program (IDCSP) satellites, which provided high-speed digital data links from Vietnam to Washington D.C. That success led to the geosynchronous Defense Satellite Communications System (DSCS) II constellation during the 1970s (Figure 2) and the even more improved, jam-resistant, secure DSCS III in the 1980s (Figure 3). In December 1990, DSCS III links became the primary means for transmitting missile-warning data from key sensors worldwide to processing and command centers in the central U.S. To ensure survival of communications links, even during nuclear conflicts, the U.S. Air Force launched the first satellite in the Milstar program on 7 February 1994. A Fleet Satellite Communications (FLTSATCOM) constellation, provided links for the U.S. Navy from the late 1970s until replaced by ultrahigh frequency follow-on (UFO) satellites in the 1990s. The US also provided assistance in the development of Britain's Skynet military satellite communications system as well as a dedicated North Atlantic Treaty Organization (NATO) satellite communications system.

A competitive rush during the 1990s to provide global, commercial satellite telecommunications services to individual and corporate mobile phone users led to constellations of numerous, small satellites in low or intermediate (medium), rather than geosynchronous, orbit. Testing for Orbcomm's 26-satellite, data-transfer network began with the launch of an experimental package in 1991. Teledesic envisioned an 840-satellite design in 1994, but it was scaled back to 288 in 1998 and further reduced to only 30 in medium-earth orbit (MEO) in February 2002. Motorola's Iridium

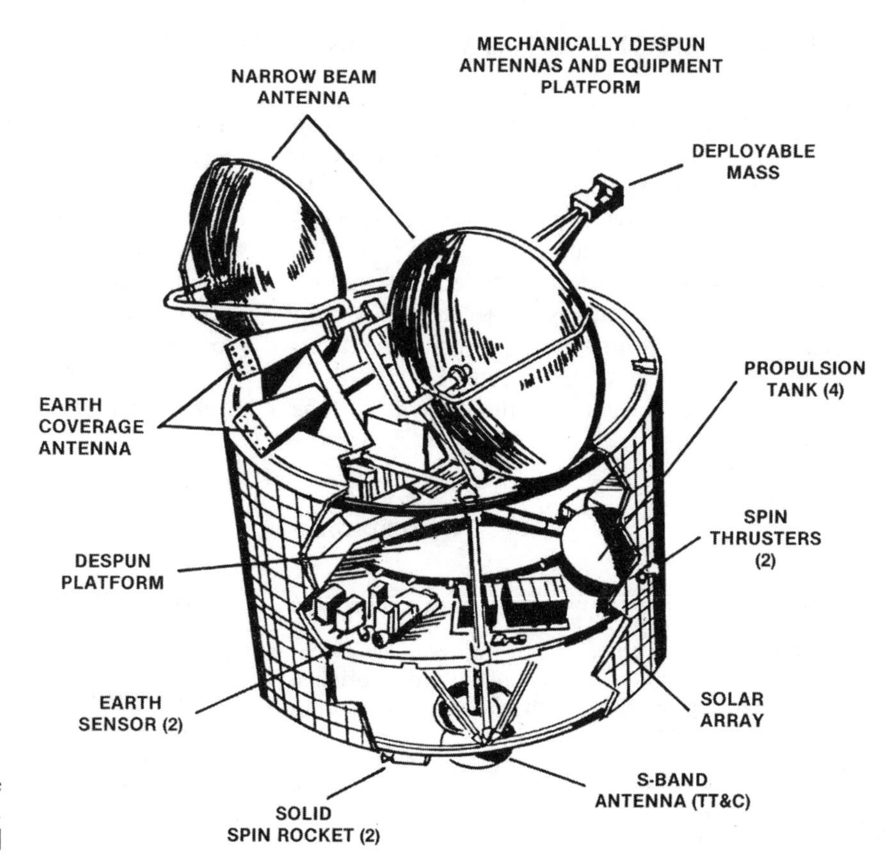

Figure 2. DSCS-II satellite diagram. [*Courtesy U.S. Air Force.*]

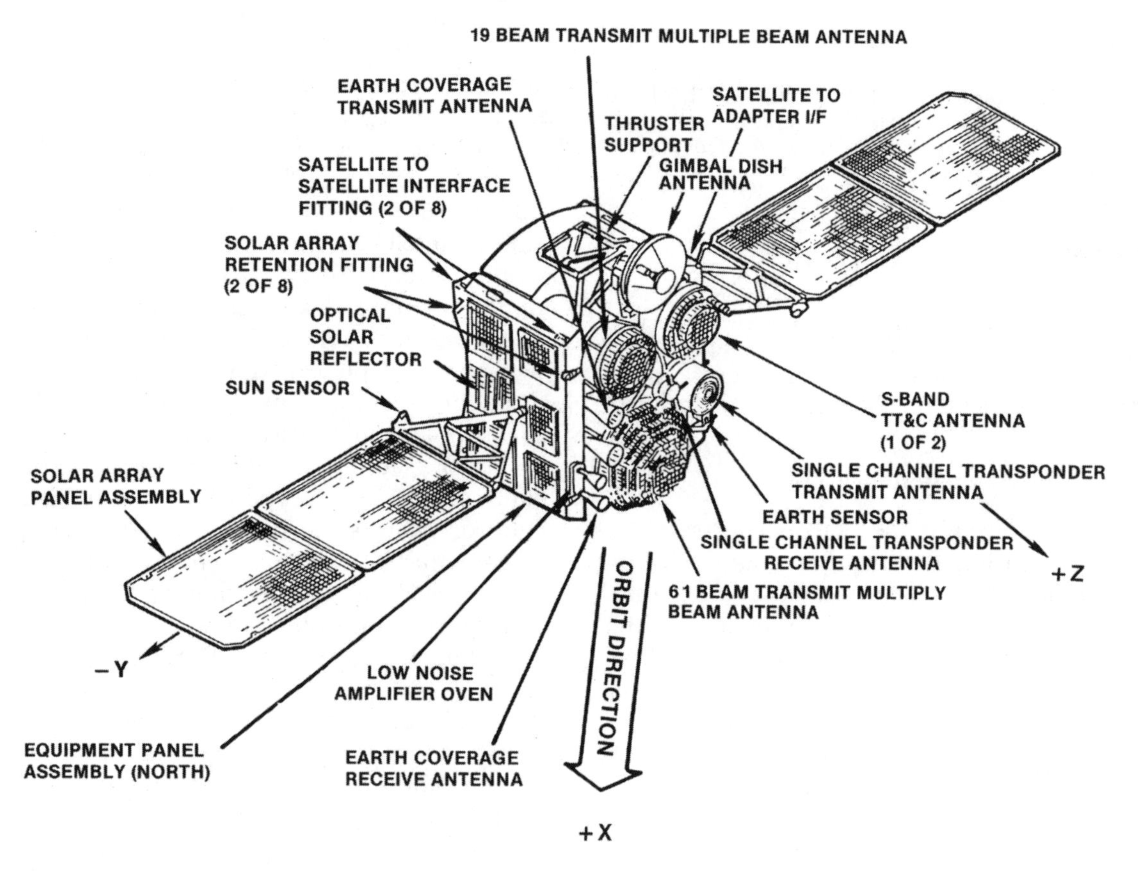

Figure 3. DSCS-III satellite diagram. [*Courtesy U.S. Air Force.*]

system, which began with the launch of five satellites in May 1997, relied on 66 identical satellites in circular, low-earth orbit (LEO). Globalstar, which envisioned a 64-satellite constellation, entered the arena by launching four in February 1998. When customers failed to purchase services at the pace originally projected, a series of bankruptcies, buyouts, and reorganizations occurred. Nevertheless, prospects for ultimate success remained bright at the dawn of the twenty-first century.

See also **Sputniks**

RICK W. STURDEVANT

Further Reading

Clarke, A.C. *Ascent to Orbit: A Scientific Autobiography*. Wiley, New York, 1984. Contains all of Clarke's original technical writings, together with historical annotations in Clarke's own words.

Dorsey, G. *Silicon Sky: How One Small Start-Up Went Over the Top to Beat the Big Boys Into Satellite Heaven*. Perseus, Cambridge, 1999.

Hudson, H.E. *Communication Satellites Their Development and Impact*. Free Press, New York, 1990.

Johnson, N.L. and Rodvold, D.M. *Europe & Asia in Space, 1993–1994*. Kaman Sciences Corporation, Colorado Springs, ND.

Martin, D.H. *Communication Satellites*, 4th edn. Aerospace Corporation, El Segundo, 2000.

Spires, D.N. and Sturdevant, R.W. From ADVENT to Milstar: the United States Air Force and the challenge of military satellite communications. *J. Br. Interplanet. Soc.*, 50, 207–214, 1997.

Useful websites

National Aeronautics and Space Administration (NASA). *Communications Satellite History*: http://www.hq.nasa.gov/office/pao/History/commsat.html

Wood, L. *Lloyd's Satellite Constellations*: http://www.ee.surrey.ac.uk/Personal/L.Wood/constellations/index.html

Satellites, Environmental Sensing

Aerial photography was the precursor of satellite imagery for recording earth-surface characteristics using instrumentation that is located at a distance from (i.e., remotely), rather than in contact with, the subject. Established in the late 1800s, the first aerial photographs were obtained from cameras aboard balloons and kites until the advent of aircraft in the early 1900s. These early aircraft made a substantial impact on the collection of land-use and land-cover information, resource inventory, and archaeological data as well as being an important tool in wartime. In the late 1940s the first attempts at space exploration provided new opportunities for remote sensing of the earth. The first satellite for remote sensing was TIROS-1 launched in 1960; it was designed to collect meteorological data such as cloud cover. As improvements in sensors occurred, the range and quality of the data collected by meteorological satellites also improved. In particular the physics of the sensors became sufficiently sophisticated to compensate for distortions due to turbulence and refraction of light in regions of varying densities and temperatures in the atmosphere.

> "*The prospect of looking through,* not just *at,* the earth's atmosphere had begun." [*Lillesand and Keifer, 2000, p. 48*]

Further developments occurred as space flights increased. In the 1960s astronauts used hand-held cameras to capture images of the earth. Experiments involving color-reversal film to determine geological features were conducted successfully as part of the Gemini program. Subsequently, other aspects of the earth's surface became the focus of photographic record; for example, the oceans and the world's vegetation cover. New frontiers were breached with the Apollo space program, especially the use of mechanically operated Hasselblad cameras in conjunction with multispectral photography. The latter involves different types of film with various filters; for example, panchromatic (multicolored) film, infrared monochrome (black and white) film, and infrared color film. Piloted and robotic spacecraft with radar on board began as part of Russia's military intelligence program, but by the late 1970s weather radar (looking for radar echoes of precipitation) and radar altimeters (measuring heights of the earth's surface from the time delay of a signal) were deployed from satellites.

Radar sensor systems are "active" in that they emit the same microwave radiation that is used for the remote sensing. "Passive" sensors, on the other hand, are dependent upon receiving reflected sunlight or thermal infrared emission. All passive sensors record electromagnetic radiation (i.e., radiant energy) which is emitted from the earth's surface or atmosphere and which is classified according to wavelength. These types of radiant energy, listed in order of increasing wavelength, are gamma radiation, x-rays, ultraviolet radiation, visible radiation, microwaves, and radiowaves. As instrumentation have become increasingly sophisticated, sensors that can record a variety of radiant-energy wavelengths, (i.e., multispectral) have been developed. Skylab, launched in 1973 and carrying the earth resources experiment package (EREP) was an early example of space-based multispectral remote sensing involving photographic and electronic measurements. From this the National Aeronautics and Space Administration (NASA) began to develop a program of recording earth resources from satellites. The first satellite in what later became the Landsat programme was launched in 1972. Landsat has become one of the most enduring earth observation missions, the latest satellite being Landsat 7 launched in 1999. The instruments on the Landsat satellites record data in the electromagnetic range between visible light and thermal radiation and with resolutions between 80 and 15 meters. The detail revealed by these missions confirmed the value of satellite remote sensing for monitoring the status of earth-surface features and resources. The Landsat program was the start of true satellite remote sensing. It shed light on a range of features, contributed to the explanation of their structure and formation and highlighted the magnitude of environmental change in many regions, including the rapidly declining extent of tropical forests.

Subsequent satellites for remote sensing include "Seasat," launched in 1978. Carrying radar sensors, it focused on the oceans, their circulation and sea-ice cover. The French *Centre National d'Etudes* (CNES) in combination with Belgium and Sweden launched the first non-U.S. satellite in 1986. This was one of a series known as *Systeme Pour l'Observation de la Terre* (i.e., SPOT), which subsequently developed into an international project involving five satellites launched between 1986 and 2002. Soviet satellite monitoring facilities included the Cosmos-1870 and ALMAZ-1 in 1987 and 1991 respectively. Both collected radar images and ALMAZ-1 was the first satellite to operate commercially. The European Space Agency (ESA) also launched a satellite, ERS-1, in 1991, and ERS-2 in 1995. The primary sensors record microwave, radar and radiometric data. Other nations that have developed satellite monitoring systems include Japan whose space agency launched JERS-1 in 1992, Canada launched its first Radarsat in 1995 and China launched the satellites YZ-1 and YZ-2 in 1999 and 2000. There were at least 31 satellites in orbit in 2000.

These satellites transmit images to earth and sophisticated computer programs translate data from various wavelengths into information on geology, water resources, vegetation, soils, minerals, tectonics, agriculture, forestry and urban environments. The end products are images from which maps of particular characteristics can be constructed. The continuous recording of earth-surface features provides detailed information on global environmental change at various scales. Such data can facilitate management, the prediction of future change and the assessment of the quality and quantity of biological and mineral resources.

The varied applications of satellite imagery have turned what began as government-funded endeavor into a commercial activity. However, satellites have also been developed for other purposes, two of which include weather and climate, and military applications. The recording of weather and climatic data involves monitoring of the atmosphere; today's efforts have advanced considerably on those of TIROS-1 (see above), and now weather and climate prediction, including advance warning of extreme conditions, is a primary objective. Satellites so designed are known as metsats; examples include those of the U.S. National Oceanic and Atmospheric Administration (NOAA), the Geostationary Operational Environmental Satellite (GOES), and the U.S. Air Force Defense Meteorological Satellite Program (DMSP). Military satellites differ from nonmilitary satellites insofar as they are always government funded, provide encrypted data that requires a special and secure receiver, and operate at a higher resolution.

Recent developments include a new family of agile satellites, such as IKONOS (1-meter resolution and Quickbird (0.61-meter resolution), which exploit emerging optical technologies to compete in the market for high-resolution images normally provided by aerial photography. The rapid and continuous collection of varied environmental data, the increasing resolution possible by sensors and increasing sophistication of analytical techniques mean that satellite monitoring has a bright future.

See also **Environmental Monitoring**

A. M. MANNION

Further Reading

Campbell, J.B. *Introduction to Remote Sensing*, 3rd edn. Taylor & Francis, London.

Dury, S.A. *Images of the Earth: A Guide to Remote Sensing*, 2nd edn. Oxford Science Publications, Oxford, 1998.

Konecny, G. *Geoinformation: Remote Sensing, Photogrammetry and Geographic Information Systems*. Taylor & Francis, London, 2003.

Lillesand, T.M. and Keifer, R.W. *Remote Sensing and Image Interpretation*, 4th edn. Wiley, New York, 2000.

Monmonier, M.S. *Spying with Maps: Surveillance Technologies and the Future of Privacy*. University of Chicago Press, Chicago, 2002.

Useful Websites

Canada Center for Remote Sensing. *Learning Resources*: http://www.ccrs.nrcan.gc.ca/ccrs/learn/learn_e.html

Short, N.M. *The Remote Sensing Tutorial*: http://rst.gsfc.nasa.gov/start.html

Semiconductors, Compound

Compound semiconductors are semiconductors made up of two or more elements. Since the 1960s, group III and group V compound semiconductors (sometimes called intermetallic semiconductors) such as gallium arsenide (GaAs), gallium nitride (GaN), and indium phosphate (InP) have been important in semiconductor and optoelectronic devices such as lasers (fiber-optic transmission, CD players), light emitting diodes, and microwave integrated circuits (cellular phones), but early compound semiconductors featured in turn of the century devices are considered here first.

Experimental work on compound semiconductors began as early as 1874 when Karl Ferdinand Braun (Germany) described the rectifying properties of lead sulfide and iron sulfide crystals. The invention of radio communication acted as a stimulus to the development of the crystal detector, but due to the amplifying properties of the thermionic valve, interest in crystal detectors diminished during most of the interwar period. During World War I, lead sulfide and selenium photodetectors were developed in Germany capable of operating in the infrared spectrum, although conversion efficiencies were low (less than 1 per cent). Furthermore, selenium suffered from a slow response to changes in light, rendering it unsuitable for many applications. Work on lead sulfide detectors was resumed in Germany in 1932, where lead selenide and lead telluride were also investigated. In order to increase conversion efficiency, solid carbon dioxide and liquid nitrogen were used as coolants. Shortly afterwards, work on lead sulfide detectors began in Poland. An important commercial development was the introduction of the copper oxide rectifier (1927) and selenium rectifier (1931), both having advantages over thermionic diodes as low-frequency power rectifiers. Although manufactured in large quantities, no adequate theory existed at the time to explain their action. An important handicap affecting the manufacture of these devices prior to the early 1950s was the lack of suitable techniques of material purification. During the 1940s and 1950s, attention shifted towards germanium and silicon, initially in response to wartime demands for efficient high-frequency detectors for use in radar equipment.

The semiconductor industry has been largely dominated by the element silicon (Si) since the development of the planar process of device manufacture by Jean Hoerni of the Fairchild Corporation in 1959. However, group III–V compound semiconductors in particular have been increasingly employed from the 1960s onward in a growing number of applications. Some group III–V compound semiconductors have important advantages over silicon where high-frequency response and high-speed switching are important; they are also able to withstand higher temperatures and radiation damage. Silicon, because of its limited frequency response, is also restricted in its use as a detector of electromagnetic radiation in both the visible and infrared spectrum, and compound semiconductor materials are much more widely used in this field. Given these advantages, it is not surprising that military requirements in particular have stimulated research into the properties of compound semiconductors and their applications.

Interest in gallium arsenide (GaAs) began in the early 1950s as part of a search for higher band-gap materials that could operate at higher frequencies than silicon, in order to manufacture devices to meet military requirements for microwave communications and radars. Gallium arsenide was the first group III–V semiconductor compound to find widespread commercial application. It possesses the advantages mentioned above but, unlike silicon and germanium, suffers the disadvantage that (in common with other III–V semiconductor compounds) it exhibits a very low vapor pressure at its melting temperature. This causes additional problems in material preparation. Also, most compound semiconductor materials evaporate at different rates, altering their proportions during crystal growth. To prevent this, the required material is grown under pressure or prevented from evaporating by means of a nonreactive covering. Further disadvantages are that GaAs is difficult to dope with the required impurities, does not grow an electrically stable oxide, and crystal defects tend to be higher than those for silicon. Consequently, GaAs, like other III–V semiconductor compounds, is more difficult to manufacture than silicon, has taken more time to develop, and is therefore more expensive; this factor in particular so far preventing its wider use.

One characteristic of GaAs is that electron carrier mobility (and therefore electron transit time) is higher than in silicon. This fact can be put to advantage in the manufacture of *n*-channel field-effect transistors, since channel conduction is entirely by electrons. Such devices may be three to five times faster than their Si equivalents. However, since hole mobility has a lower value in GaAs than in Si, *p*-channel GaAs field-effect transistors do not have this advantage.

GaAs has a band-gap of 1.43 electron volts. Devices have recently been developed made from wider band-gap compounds such as silicon carbide (SiC) and gallium nitride (GaN) (3.26 and 3.39 electron volts, respectively). They can therefore withstand a much wider range of temperatures ranging from the heat of a jet engine to conditions met with in outer space. Other advantages are high-voltage breakdown and low noise levels. A major application for such materials will be in the construction of high-power, high-frequency devices, although at present fabrication problems still remain.

Electrical conductivity may also be achieved by means of light. This phenomenon is known as photoconductivity, and was discovered in selenium by Willoughby Smith in 1873. The absorption of photons arriving on the surface of a suitable material may possess sufficient energy to raise electrons across the forbidden gap to the conduction band. Since different materials have different band gaps, they vary in response to photons of different wavelengths, and therefore vary from each other both in efficiency and spectral response. For example, Si and GaAs have peak spectral responses in the infrared (1100 and 870 nanometers, respectively), well outside the visible range (approximately 390 to 770 nanometers) Widely used compounds peaking near or within the visible spectrum are GaP (550 nanometers), GaN (370 nanometers), and SiC (470 nanometers).

Another important use of III–V compounds has been in the fabrication of light emitting diodes (LEDs), since silicon is very poor at emitting light. It was known by about 1960 that semiconductor materials could emit light. GaAs and GaAsP (with phosphorus) pulse-operated single *p–n* junction devices were made by Nick Holonyak in 1962, emitting red light. These materials were succeeded during the subsequent decade by indium arsenide (InAs), indium phosphate (InP), and indium antimonide (InSb). More complex structures then followed; for example, InGaAs, InAsP and InGaP. A significant advance (giving a two-fold increase in spectral efficiency) was made in 1990 when AlGaInP LEDs were constructed. By the end of the twentieth century they remained the most efficient LEDs within the range 570 to 650 nanometers (red–orange–yellow), achieving quantum efficiencies of up to 24 percent at 635.6 nanometers. These devices are claimed to be more efficient than unfiltered incandescent lamps, with projected lifetimes over a magnitude greater than conventional light sources.

By the mid-1990s, LEDs were being used over a color range from red to green. However, an unfilled gap existed in the blue region of the spectrum. In 1996, a breakthrough came with the production of high quantum-efficiency GaN devices and the blue laser diode by Shuji Nakamura at Nichia Chemical Industries in Japan. Previously, the best that could be achieved was by using SiC diodes, giving a peak quantum efficiency of 0.02 percent. Using GaN, quantum efficiencies of up to 10 percent in the blue region and 5 percent in the green was now possible, the latter figure also being a considerable improvement over other types.

In order to improve their characteristics, more complex multijunction devices have evolved from the simple *p–n* junction. Increased quantum efficiency and a wider range of spectral coverage has led LEDs to being used in an increasing number of applications in, for example, advertising displays, traffic signals, automobile stop lights, and indicators.

A further important use for compound semiconductor materials is as a laser light source, since stimulated emission at a definite wavelength may be produced by an applied voltage. Coherent radiation from a forward biased GaAs *p–n* junction was first observed in the U.S. (Gunter Fenner, Robert Hall, and Jack Kingsley at General Electric, 1962). The first diode laser optical recording system was produced in Holland by the Philips Company in 1978. It used an AlGaAs laser capable of delivering a pulsed light output equivalent to a large gas laser, despite its small size. It has since (as the CD player) become the major domestic application for lasers.

Another widely growing use for lasers is as a light source in the field of optical fiber transmission. In this application, the device must be optically matched to the transmission line in order to achieve maximum efficiency. To achieve this, advantage is taken of the property of group III–V compounds to undergo variation in peak spectral response when doped with minute amounts of further components whilst in the molten state. For example, phosphorous is used in this manner to dope GaAs. The ternary mixture thus formed is GaAsP. A further example is the widely used material AlGaMP.

Following work in the U.K., it was demonstrated in the U.S. (by John Gunn, 1963) that when an electrical field is applied to certain group III–V semiconductors, microwave oscillations take place when the field exceeds a certain value. Such a device is known as the Gunn diode oscillator. III–IV compounds have also found applications in the field of infrared technology. In 1960, E.H. Putley, of the U.K. Royal Radar Establishment, observed photoconductivity at low temperatures in *n*-type InSb at wavelengths between 0.1 and 4.0 millimeters. Subsequent work in this field has been largely in connection with military applications.

See also **Lasers, Applications; Lasers in Optoelectronics; Light Emitting Diodes; Semiconductors, Elemental; Semiconductors, Postband Theory; Semiconductors, Preband Theory**

ROBIN MORRIS

Further Reading

Bar-Chaim, X., Ury, I. *et al.* Integrated optoelectronics. *IEEE Spectrum*, 5, 38, 1982.
Bell, T.E. Japan reaches beyond silicon. *IEEE Spectrum*, 10, 46–52, 1985.
Braunstein, R. Radiative transitions in semiconductors. *Phys. Rev.*, 99, 195, 1892.
Denbaars, S.P. Gallium-nitride-based materials for blue to ultraviolet optoelectronics devices. *IEEE Proc.*, 85, 11, 1740–1749, 1997.
Dettmar, R. GaAs on silicon—the ultimate wafer? *ME Review*, 4, 136–37, 1989.
Kenney, G.C., Lou, D.Y.K. *et al.* An optical disc replaces 25 mag tapes. *IEEE Spectrum*, 2, 33, 1979.
Loebner, E.E. Subhistories of the light emitting diode. *IEEE Trans.*, ED-23, 675–698, 1976.
Martin, R. In search of the blue light. *Chem. Ind.*, 5, 173–177, 1999.
Nakamura, S. and Fasol, G. *GaN Based Light Emitters and Lasers*. Springer, New York, 1997.
Vanderwater, D.A., Tan, I.H. *et al.* High brightness AlGaInP light emitting diodes. *IEEE Proc.*, 85, 11, 1752–1764, 1997.
Voelcker, L. The Gunn Effect. *IEEE Spectrum*, 7, 24, 1989.

Semiconductors, Crystal Growing and Purification

Crystals of high purity are essential to the manufacture of semiconductor devices, including integrated circuits, and much effort has gone into their refinement. The first single semiconductor crystal was drawn in 1948 and, from 1949 or 1950 onward, monocrystalline material has been used exclusively. The conditions imposed are extremely stringent and have become progressively more so as manufacturing technology has advanced. From the time of the invention of the transistor in 1947 until the mid-1960s most were made using germanium, although since then silicon has been mainly employed. By the 1980s silicon accounted for 98 percent of all semiconductor devices sold worldwide.

The three basic stages in the production process are extraction, purification, and crystal growing.

Extraction

The extraction of metallurgical grade silicon takes place within an electrode-arc furnace. Quartzite, coal, coke, and wood chips are fed into the furnace and a series of chemical reactions take place, finally resulting in the extraction of silicon of about 98 percent purity. The major contaminant is usually boron, although carbon and other impurities are also present. Only a small fraction (about 1 percent) of silicon produced by these means is used in the semiconductor industry.

Purification

This stage is necessary in order to produce electronic grade silicon, a material polycrystalline in structure with impurity levels of only a few parts per billion. Such low impurity levels are necessary to ensure that contamination does not occur during the following crystal-growing stage. The purification process currently used is basically as follows:

1. Pulverised metallurgical grade silicon is heated with anhydrous hydrogen chloride at 300°C, yielding trichlorosilane (a gas), together with hydrogen.
2. The trichlorosilane is cooled and becomes liquid (at 32°C).
3. Fractional distillation is employed to remove impurities present (especially phosphorous and boron).
4. The reaction is reversed by passing a mixture of trichlorosilane and hydrogen over a resistance-heated silicon rod at a temperature of between 1000 and 1200°C.
5. Small crystals of electronic grade silicon are deposited on the rod during the reaction. Pure polycrystalline rods so formed are a few meters long and several inches in diameter.

Due to the complexity and cost of the technology, it had become limited to the U.S., Japan and West Germany well before the end of the twentieth century.

An earlier method of material purification, zone refining, was first described by the Russian physicist Petr Kapitza in 1928 and perfected at Bell Laboratories by William G. Pfann in 1951. The impure silicon or germanium polycrystalline rod is supported in a quartz tube in an inert ambient, and a small zone melted by radio-frequency heating. The impurities then collect at the zone edge where the temperature of the melt is lowest. When the zone is gradually swept from one end of the rod to the other the impurities are swept along the rod and collected at the end. This effect occurs because most impurities have a lower melting point than silicon or germanium, which therefore crystallize out of the melt before the impurities. Although a great advance upon earlier methods, zone refining has the disadvantage that only relatively small amounts of material can be purified and the removal of boron in particular is time consuming.

Crystal Growing

First, it is necessary to ensure that the material is as dislocation-free as possible, since crystal defects affect the rate at which impurities diffuse. Dopant atoms (introduced in highly accurate amounts during subsequent production processes) travel much more quickly at the surface of grain boundaries due to crystal dislocations. Dopant atoms introduced into material with an unacceptable level of dislocations would result in a highly uneven diffusion profile, and consequent device failure. It is also important to control the crystal orientation. This is because: (1) crystals cleave easily in certain directions; (2) during device manufacture, it is invariably necessary to put down uniform layers of dopants and oxides; and (3) chemical etching is frequently used to cut windows in oxides grown on the surface of wafers. Etching will only proceed uniformly if the crystal surface presents the correct orientation.

The favored crystal-growing technique is the Czochralski process, developed by Jan Czochralski in 1918, but perfected for growing germanium and silicon single crystals by Gordon K. Teal and John E. Little at Bell Laboratories in 1950. By the end of the twentieth century it accounted for 70 to 80 percent of production. Three other methods of crystal growing are also used, float-zone, Bridgman, and epitaxy. Each has its advantages and disadvantages.

In the case of the Czochralski method, the crystal is "pulled" from molten material (melt), which is contained within a crucible. There is therefore a possibility of material contamination. With germanium (melting point 960°C) it is possible to use a graphite crucible. However, this cannot be done with silicon (with a melting point of 1420°C) and therefore quartz is used. The process takes place within an enclosed inert ambient (helium or argon). The temperature of the melt is controlled externally by radio-frequency heating. The seed crystal, mounted in the desired orientation, is held in a chuck fixed at the end of a shaft, which is driven by a motor mounted externally above the enclosure. The crystal is then lowered to make contact with the melt, which is set at about 15°C above its melting point (about 1435°C). The crystal itself will not melt because of its higher resistivity. The seed crystal is then slowly raised and rotated, causing the molten silicon to freeze onto the seed crystal with the same orientation as the seed itself. Finally, the fully grown crystal is cooled to about 300°C before being exposed to the external atmosphere.

The float-zone method developed by P.H. Keck in 1953 overcomes the difficulty of crucible contamination by vertically suspending a polycrystalline silicon rod in an inert ambient. A radio-frequency coil melts a narrow cross-section of the rod above the seed crystal and is slowly raised. The silicon below the molten zone resolidifies in single-crystal form with the same orientation as the seed crystal. The molten zone does not flow out, due to the geometry and surface tension. This process is more expensive and difficult to control than the Czochralski method.

The Bridgman method (named after Percy W. Bridgman) involves a boat containing seed and melt being slowly pulled through a horizontal tube furnace, the melt freezing in the required orientation as the crystal enters the cooler zone. Towards the latter years of the twentieth century this was still the preferred technique for growing gallium arsenide (GaAs) crystals. Single crystals of 75 millimeters diameter and low dislocation density were by then being produced on a production basis.

Conclusion

All the above processes involve crystal growth from the melt. Epitaxy is the technique of depositing single-crystal material in successive layers upon an atomically flat single crystal substrate. The substrate most commonly used is silicon, which enables many compound semiconductors to be deposited; for example, crystalline GaAs and cadmium telluride (CdTe). During the epitaxial process the deposited material takes up the same crystalline orientation as the substrate. Fabricating transistors by impurity doping of epitaxial single crystals grown from the gas phase was first achieved at Bell Laboratories in 1960. In the following year, N. Nielson of RCA described the process of epitaxially growing GaAs and Ge crystals from the liquid phase. Molecular beam epitaxy, although originating in elementary form in the 1960s, only came into production use by the early 1980s. It is however capable of greater precision.

See also **Integrated Circuits; Semiconductors, Compound; Semiconductors, Elemental; Semiconductors, Postband Theory; Thin Film Materials and Technology**

ROBIN MORRIS

Further Reading

Gilman, J.J. *The Art and Science of Growing Crystals*. Wiley, New York, 1963.

Harper, C.A. Handbook of Materials and Processes for Electronics. McGraw-Hill, New York, 1970.

Keck, P.R. and Golay, M.I.E. Crystallisation of silicon from a floating liquid zone. *Phys. Rev.*, 89, 1297, 1953.

Pearce, C.W. Crystal growth and wafer preparation, in *VLSI Technology*, Sze S.M., Ed. McGraw-Hill, Singapore, 1985, pp. 9–50.

Pearce, C.W. Epitaxy, in *VLSI Technology*, Sze S.M., Ed. McGraw-Hill, Singapore, 1985, 51–92.

Petritz, R.L. Contributions of materials technology to semiconductor devices. *IRE Proc.*, 1025–1037, 1962.

Pfann, W.G. Techniques of zone melting and crystal growing, in *Solid State Physics*, Seitz, F. and Turnbull, D., Eds. Academic Press, New York, 1957, pp. 423–521.

Teal, G.K. and Buehler, E. Growth of single crystal silicon crystals and of single crystal silicon *p–n* junctions. *Phys. Rev.*, 87, 190, 1952.

Teal, G.K. and Little, I.E. Growth of germanium single crystals. *Phys. Rev.*, 78, 647, 1950.

Thomas, R.N., *et al*. Status of device-qualified GaAs substrate for GaAs integrated circuits. *IEEE Proc.* 76, 7, 778–791, 1988.

Semiconductors, Elemental

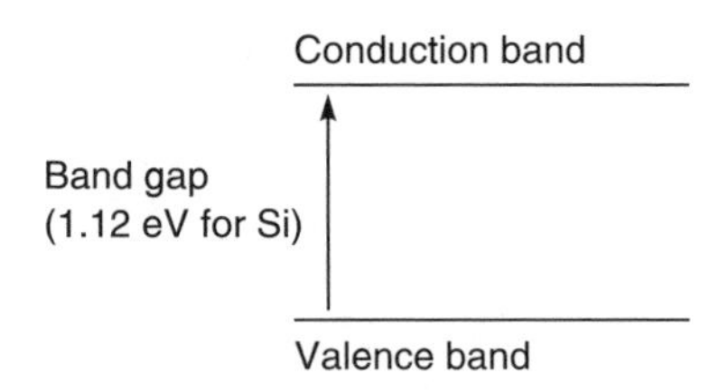

Figure 4. Electrons are raised from the valence band to the conduction band provided they have enough energy to jump the gap. Charge conduction is therefore directly related to band gap width.

Elemental semiconductors may be classified either as intrinsic or extrinsic. An intrinsic semiconductor is one that exists in pure crystalline form, its conductivity depending entirely upon its intrinsic properties. An extrinsic semiconductor is one containing impurity atoms within its crystalline structure. The addition of minute quantities of impurities (dopants) greatly affects the electrical properties of semiconductors, and they are therefore described as being structure sensitive. Because semiconductor materials only naturally occur in an extrinsic state, their value has largely depended upon developing techniques of refinement in order to make efficient use of their electrical properties.

An important factor in the selection of a semiconductor material is that its characteristics strongly depend upon the structure of the energy bands occupied by electrons. The band gap for a semiconductor is the energy gap between the valence (ground state) band and the conduction band (Figure 4). Materials with wide band gaps such as carbon in the form of diamond (band-gap energy 7 electron volts) may offer in theory the prospect of very high temperature operation since its wide band-gap means much higher energies are required to change the conducting state (other semiconductors change state at high voltage or high power). However diamond has so far remained virtually unused in electronic devices, due to practical difficulties in fabrication. Those semiconductors with narrow band gaps such as grey tin (0.1 electron volts) have also so far been found to be unsuitable. The principal elemental semiconductors used in device manufacture to date have been the group IV elements germanium (Gi) (0.72 electron volts) and silicon (Si) (1.12 electron volts), the material being initially purified to an intrinsic state and controlled amounts of dopants then added in order to meet the desired specifications.

The first successful application of elemental semiconductor materials was as point contact rectifiers for the detection of electromagnetic waves in the early twentieth century (see Radio Receivers, Crystal Detectors). From 1900 to 1910 a wide range of elements and compounds were investigated, lead sulfide (PbS) and silicon proving the most efficient. Although the characteristics of these devices were substantially improved over the following years, this was almost entirely as a result of empirical affect. Nevertheless, a variety of theories were put forward in an effort to explain the phenomenon of rectification. Early explanations included, for example, the thermoelectric

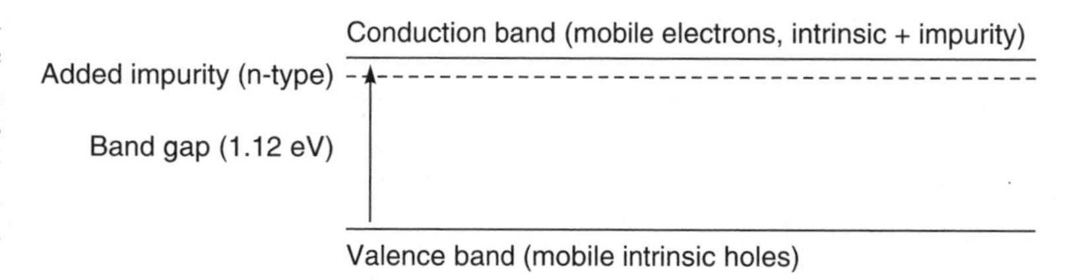

Figure 5. By adding minute controlled amounts of certain impurities to, for example, intrinsic silicon (group IV in the periodic table), the conductivity may be varied over a wide range. With the addition of *n*-type impurities (group V in the periodic table), electrons now become the majority carriers as the *n*-type impurity level donates electrons to the conduction band. Little energy is required due to the proximity of the impurity and conduction bands (0.05 electron-volt gap).

effect (already rejected by Karl Ferdinand Braun), the electrostatic barrier theory, propounded by M. Huizinga (Germany) in 1920, and the cold emission theory of G. Hoffman (Germany) in 1921. About this time, Walter Schottky (Germany) also put forward a semiconductor theory with blocking layers and potential thresholds, but did not yet realize that holes would act as charge carriers. However, none proved satisfactory prior to the band theory of electronic conduction in solids, developed by Alan H Wilson (U.K.) in 1931. By the beginning of World War II a theoretical picture was beginning to emerge that explained the movement of charge carriers in semiconductor junctions. B. Davydov (USSR) produced the first model of a *p–n* junction in 1938, which included the concept of minority carriers (holders) in the semiconductor conduction. This was followed by an explanation of matter and to semiconductor junctions by Shottky in 1939.

A major problem facing the early investigators was their lack of appreciation of the extremely high degree of purity required in order to control the equilibrium concentration of mobile charge carriers within semiconducting material, and hence keep its resistivity correct. Consequently, experimental results were unpredictable and difficult to evaluate. In any case, no suitable process of refining semiconductor material to the level of purity required was then available. A further important factor delaying progress was that, due to the overwhelming success of the thermionic valve, little stimulus existed until almost the end of the 1930s to carry out investigations into the properties of semiconductor materials. Gordon Teal (U.S.), recalling his doctoral thesis on germanium during that decade remarked, "Its complete uselessness fascinated and challenged me."

The urgent need for efficient ultrahigh frequency detectors for radar applications at the outbreak of World War II stimulated renewed interest in crystal detectors within the major advanced industrial countries. Isolated effort by individual researchers was now supplanted by the power of large research establishments and commercial corporations, with major emphasis being placed on the production of silicon and germanium point-contact diodes. In Britain, by 1940, the first commercially produced silicon diodes for use in radar frequencies were being made at Thomas Houston Limited, using polycrystalline silicon of 98 percent purity. Later, crystals were grown at the General Electric Company from highly purified silicon powder, to which had been added controlled fractions of aluminum and beryllium. Although great efforts were now being made in material preparation, the work largely proceeded on an ad hoc basis.

After 1941, the center of attention of semiconductor materials research shifted to America and

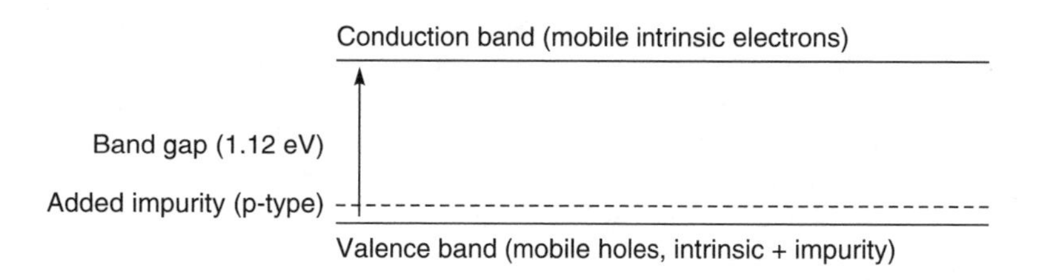

Figure 6. With the addition of *p*-type impurities (group III in the periodic table), holes now become the majority carriers as electrons are raised from the valence band to the *p*-type impurity band. Little energy is required due to the proximity of the impurity and valence bands (0.05 electron-volt gap).

to Bell Laboratories in particular. One early wartime method of material preparation was to produce pure germanium and silicon films by pyrolitic deposition on to tantalum filaments. It was then used to manufacture infrared detectors and photoconductors. The technique of material purification by the segregation of impurities through the repeated freezing was also employed at this time. From February 1942 onwards, germanium microwave diodes were being manufactured in quantity at various establishments and a basic research program on germanium was instituted. Pennsylvania University concentrated on silicon, and cooperating closely with industry, was soon producing material with a spectroscopic purity better than 99.9 percent using a process involving the reduction of silicon tetrachloride with zinc. Other firms, including Sperry, were also manufacturing microwave diodes.

Throughout the period of the war, polycrystalline material was used exclusively in the manufacture of semiconductor devices. However, in October 1948, Gordon Teal and John Little (Bell Laboratories) grew the first single germanium crystals by a crystal-pulling process developed initially by Jan Czochalski in 1918. Within months, crystals were being doped with impurities to produce germanium "grown junction" diodes. The period 1952–1953 was particularly important in terms of material purification and the consequent evaluation of the physical properties of semiconducting materials. The process of zone refining was introduced by Willian Pfann in 1952, and float-zone refining by Henry C Theurer in 1953. From then on, it was possible to produce material of a consistently high quality. In 1952 Teal and Ernest Buehler also produced large high-quality silicon crystals by pulling from the melt, which was impurity doped to form single crystal diodes. With such high purity material available, much more accurate measurements of electron and hole carrier mobility could now be carried out, consequently achieving a deeper understanding of the properties of intrinsic semiconductors.

In 1954, silicon grown-junction transistors, with a high degree of lattice orientation, were first fabricated by Gordon Teal at Texas Instruments, using a crystal-pulling process. The ability of silicon (due to its wider band gap) to operate at higher temperatures than germanium was of particular importance. Although methods of refining semiconductor materials have since been greatly improved, progress has been of a steady, incremental nature. The technique of growing ever-larger silicon crystals on a mass production basis, free from dislocations and other defects, has also constantly advanced. In 1975, crystal diameters were typically around 50 to 75 millimeters. By the end of the century, diameters had risen to 300 millimeters. From the mid-1960s onwards, silicon has almost entirely replaced germanium in device manufacture, a major factor having been the ability to grow an electrically stable oxide on its surface, thus rendering it suitable for planar fabrication.

The group VI element selenium (Se) has played an important role in the fabrication of electrical and electronic devices. From 1931 onwards, selenium rectifiers were used commercially on an increasing scale. The first practical solar cells, using selenium, were constructed as early as 1883. Their conversion efficiency was very low (less than 1 percent) and further development was neglected until the early 1930s, when they were introduced commercially. Cheap to produce, selenium photovoltaic cells were then employed in an increasingly wide range of applications, including photographic work. This use was aided by a peak spectral response (556 nanometers) approximating to that of the human eye. However, when used as photo detectors, conversion efficiencies still remained at about 1 percent and response time was limited. A significant breakthrough came with the development of the silicon photovoltaic cell by Russell Ohl (U.S.) in 1941. By 1944, silicon photovoltaic cells with conversion efficiencies of 6 percent were being manufactured, this figure rising to about 12 percent by 1960 and to about 16 percent by 1996. Silicon photocells have therefore replaced selenium, being more efficient, mechanically robust, and cheaper to manufacture.

The group IV element carbon (C) has also been widely used within the radio and electronics industry. Apart from possessing a negative temperature coefficient, it has the property of decreasing its resistance under applied pressure. This effect was utilized in the invention of the carbon microphone by Thomas A. Edison (U.S.) in 1877. A transverse-current carbon microphone, developed in Germany by G. Neumann in 1924, was used by the BBC between 1926 and 1935.

Carbon-film resistors were first made by T.E. Gambrell and A.F. Harris (U.K.) in 1897. The high-stability cracked-carbon type was invented at Siemens and Halske (1925) and the sprayed metal-film type a year later by S. Lowe (Germany). Other uses include automatic voltage regulators, electrodes, and brushes in the electrical machinery.

See also **Semiconductors, Compound; Semiconductors, Preband Theory, Semiconductors, Postband Theory**

ROBIN MORRIS

Further Reading

Brice, J.C. and Joyce, B.A. The technology of semiconductor materials preparation, in *Radio Electron. Eng.*, 43, January February, 21–28, 1973.

Fraser, D.A. *The Physics of Semiconductor Devices.* Clarendon Press, Oxford, 1977.

Fuller, C.S. and Ditzenburger, L.A. Diffusion of donor and acceptance elements in silicon. *J. Appl. Phys.*, 27, 544, 1956.

Pfann, W. Zone refining. *Scientific American*, 12 63–72, 1967.

Ryder, J.D. *Electronic Fundamentals and Applications.* Pitman, London, 1977.

Shaw, D.F. *An Introduction to Electronics.* Longman, London, 1962.

Shockley, W. *Electrons and Holes in Semiconductors.* Van Nostrand, NJ, 1963.

Teale, G.K. Single crystals of germanium and silicon—basic to the transistor and Integrated Circuit. *IEE Trans. Electron. Devices*, 23, 7, 621–639, 1976.

Semiconductors, Postband Theory

The theoretical explanation of transistor action rests upon the concepts of postclassical physics. Important advances took place from the early 1930s onward, when Arnold Sommerfeld, Felix Bloch and co-workers applied quantum theory concepts to the theory of metals. A study of the structure of semiconductor materials applying quantum mechanics was made in Britain by Alan Wilson, 1931–1932. However, only after World War II did clearer pictures begin to emerge which could explain the movement of charge carriers in semiconductor *p–n* junctions and metal-to-semiconductor contacts.

Semiconductor research was stimulated during the 1930s by the possibility of military applications in radar detectors and signaling. The major impetus came from the inability of thermionic valves, or vacuum tubes, to operate as fast switches, due to high interelectrode capacitance. This limited their use as signal detectors in radar equipment. Since the point contact diode proved superior in this respect, great efforts were made under wartime conditions to improve their characteristics. Consequently, large well-funded research programs were rapidly instituted by Britain and Germany, and also by the U.S. after 1941.

Although it was known that the addition of minute amounts of certain impurities to bulk semiconductor material altered the characteristics of semiconductor diodes, the effect was difficult to predict. This was because techniques of material purification were not sufficiently developed and therefore inconsistency in batch production was unavoidable. Consequently, work proceeded largely on an ad hoc basis. A method of obtaining a high degree of material purification (zone refining) was described by Petr Kapitza in 1928 but its significance was not realized until much later in the U.S. by William G. Pfann in 1952. The effects of various doping elements was still not clearly understood in the immediate postwar period. Petritz (U.S.) mentions that even as late as 1948 rectifiers were made with tin-doped germanium in the belief that tin was the doping element, although in fact doping levels were due to impurities within the tin, the tin itself having no electrical effect on the germanium.

Perhaps the most important semiconductor research during World War II was carried out at Purdue University and Bell Laboratories. Purdue concentrated on the study of germanium, Bell on improving silicon point-contact diodes. These devices had already been substantially improved in Britain and elsewhere since the early "cat's whisker" and now consisted of purified *p*- or *n*-doped crystalline material, the metal to semiconductor contact being made by a pointed wire (usually of tungsten or molybdenum) electrically attached by a "forming" process, resulting in a low capacitance rectifying *p–n* junction. The assembly was then sealed within an inert ambient.

The first *p–n* junction silicon rectifier was produced by Russell Ohl in 1941 and the presence of group III and V impurities in germanium and silicon was also discovered at about this time (Jack Scaff and William Pfann at Bell Labs). Consequently, it was now possible to make rectifying devices with much higher reverse breakdown voltages and handling much higher powers, as well as multijunction devices such as thyristors. Instead, efforts were concentrated on satisfying wartime requirements and by 1945 Bell (Western Electric) were producing over 50,000 rectifiers monthly.

The point-contact transistor (see Figure 7) was invented at Bell laboratories by Walter Brattain and John Bardeen, (1947), following an investigation of the surface properties of semiconductors and an understanding of the role of minority current carriers in electrical conduction. It consisted of two closely spaced wires in electrical contact with a substrate of *n*-type single-crystal germanium. This device was the product of a two-year goal-oriented research program. Realizing its significance, the U.S. government immediately

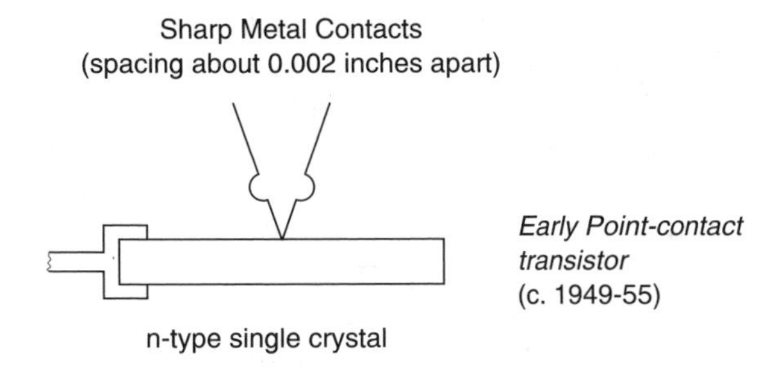

Figure 7. Early point-contact transistor, c. 1949–1955.

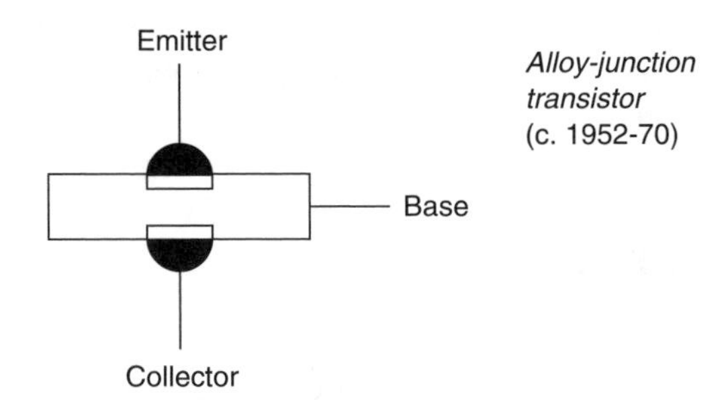

Figure 8. Alloy-junction transistor, c. 1952–1970. The germanium-alloy junction transistor has indium emitter and collector beads alloyed in a germanium base. Parameters are controlled by doping levels, geometry, and surface conditions.

allocated funds for further development. Germanium was chosen because it had been investigated in detail during the war years; it was simpler in structure than compound semiconductors and its melting point (96°C) was lower than silicon (143°C).

Limited production of germanium point contact transistors began in 1951, but the device suffered from considerable defects. Its characteristics varied widely from one device to another, it was electrically unstable, and noise levels were high. Also, device action was extremely complex due to surface effects, making theoretical analysis very difficult. Efforts were therefore concentrated on the more electrically stable junction transistor.

The Czochralski crystal-growing process developed by Jan Czochralski in 1918 was a major advance allowing single crystals to be drawn from the molten state in the form of a *p–n–p* or *n–p–n* structure. This method enabled germanium junction transistors to be manufactured in quantity with uniform characteristics. An alternative approach, the germanium alloy-junction transistor (see Figure 8), had the advantage of superior frequency response. The U.S. military realized the significance of these improvements and immediately instituted a "million a month" manufacturing program with Western Electric.

The next major step was the successful manufacture of silicon grown-junction transistors by Gordon Teal in 1954. These were single crystal *n–p–n* junctions grown by the Czochralski process.

Again, this device was of great interest to the military, because it could operate at higher temperatures than germanium and from this time onwards large-scale military involvement took place within the industry, concentrating on silicon technology.

Two significant advances now followed: (1) precise control of junction depth by vapor diffusion; and (2) oxide masking during the diffusion process. Precise control of junction depth resulted in greatly extending high-frequency performance, while oxide masking electrically stabilized the device surface, greatly lowering leakage currents and improving voltage breakdown levels. These developments led to major advances in device construction, most significant by far being the planar transistor (see Figure 9). This device was so described because the insulating oxide covering the semiconductor surface formed a flat, planar layer. The planar approach offered increased reliability at decreased cost and also permitted large-scale integrated circuit manufacture. Apart from applications such as high-voltage rectifiers and thyristors, it has rendered previous methods of construction obsolete.

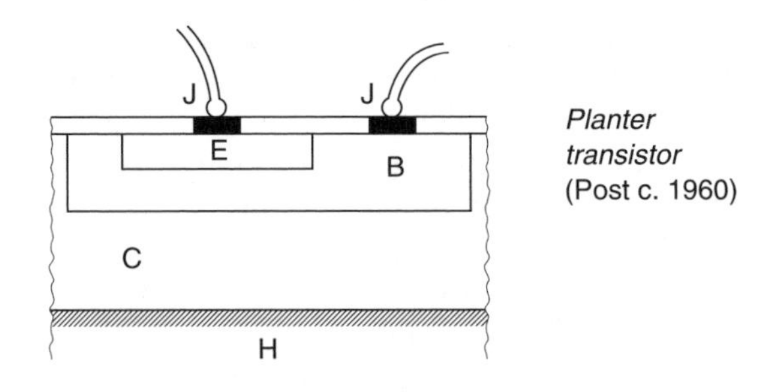

E = Emitter (heavily doped n-type Si.)
B = Base (p-type Si.)
C = Collector (lightly doped n-type Si.)
D = Silicon Dioxide surface passivation
H = Header
J = Gold wire thermocompression bonds

Figure 9. Planar transistor, post-1960. Note that the chip is welded to a gold-plated metallic substrate (header). Gold wires are bonded by thermocompression onto the aluminum bonding pads attached to emitter and base. The wires are then connected to electrically isolated posts on the header. The whole assembly is encapsulated in an inert ambient. Device parameters are controlled by geometry and doping levels.

In 1960, Bell Laboratories succeeded in growing very thin layers of doped semiconductor crystalline materials upon a silicon semiconductor substrate by vapor deposition. This process is termed epitaxial and it enabled high-resistivity layers to be deposited upon low-resistivity substrates, which cannot be done employing conventional diffusion techniques. It was now possible to construct devices within the epitaxial layer using advantageous doping profiles unattainable by previous means, this greatly assisting the subsequent development of integrated circuitry.

Planar technology is now in almost universal use and enables transistors, capacitors and resistors to be fabricated on a single slice with great precision. Windows are opened in the oxide by selective etching, enabling successive *p–n* junctions to be formed by impurity diffusion and also allowing metallic interconnections to be deposited. Integrated circuits use these metallic interconnections to link components electrically and achieve the desired circuit configuration. Since all the devices or integrated circuits on the slice are fabricated simultaneously, variation in their characteristics is minimal and only at the end of the production process are they separated to form "chips." Electrical connections from the chip to its external packaging are usually made using gold wires, which are attached to the bonding pads by thermocompression bonding. Development of integrated circuitry has largely rested upon silicon, since a stable germanium oxide cannot be grown successfully. However, a number of compound semiconductor materials were increasingly used.

Planar fabrication is particularly suited to the manufacture of metal-oxide field-effect transistors. This led to a whole new development in integrated circuitry from the late 1960s onward. Subsequently, component density per chip has approximately doubled every 18 months (Moore's law), resulting in great improvements in reliability, speed of operation, reduction in power dissipation and attainment of greater circuit complexity. Cost per bit has been vastly reduced. For example, a 1-kilobit dynamic random access memory (DRAM) manufactured in 1974 cost one cent per bit. By 1985, a 1-megabit DRAM was being produced as a cost of one thousandth of a cent per bit.

See also **Semiconductors, Compound; Semiconductors, Elemental; Semiconductors, Preband Theory, Integrated Circuits, Fabrication; Semiconductors: Crystal Growing, Purification; Transistors**

ROBIN MORRIS

Further Reading

Aschner, L.E., Bittman, C.A. *et al.* A double diffused silicon high frequency transistor produced by oxide masking techniques. *J. Electrochem. Soc.*, 106, 1145–1147, 1959.

Braun, E. and Macdonald, S. *Revolution in Miniature: The History and Impact of Semiconductor Electronics.* Cambridge University Press, Cambridge, 1978.

Frosch, C.J. and Derrick, L. Surface protection and selective masking during diffusion in silicon. *J. Electrochem. Soc.*, 104, 547, 1957.

Goldstein, A. and Aspray, W. Eds. *Facets: New Perspectives On the History of Semiconductors.* IEEE Center for the History of Electrical Engineering, New Brunswick, NJ, 1997.

Gorton, W.S. The genesis of the transistor. *IEEE Proc.*, January, 50–52, 1998.

Morris, P.R. *A History of the World Semiconductor Industry*. Peter Peregrinus, London, 1990.

Petritz, R.L. Contributions of materials technology to semiconductor devices. *IRE Proc.*, May 1025–1038, 1962.

Riordan, M. and Hoddeson, L. *Crystal Fire: the Birth of the Information Age.* W.W. Norton, London, 1997.

Roberts, D.H. Silicon integrated circuits. *Electronics and Power*, April, 282, 1984.

Ross, I.M. The invention of the transistor. *IEEE Proc.*, January, 1998.

Shockley, W. Twenty-five years of transistors. *Bell Labs. Record*, December 340–341, 1972.

Sparkes, J.J. The first decade of transistor development. *Radio Electron. Eng.*, 43, 8, 1973.

Semiconductors, Preband Theory

Investigations into the properties of what we now define as semiconductors began in the early nineteenth century. No satisfactory theoretical explanations were possible, and little effort was made to put discoveries to practical use. However, by the beginning of the twentieth century the main distinguishing characteristics of semiconductors (negative temperature coefficient of resistance, asymmetric conduction of electricity in solids, photoelectric effect, photovoltaic effect, large Hall effect, and high thermoelectric power) had been discovered. The one-way conductance, or rectification, property of semiconductors is the characteristic that makes semiconductor devices useful today.

Charles E. Fitts (U.S.) constructed the first practical photocell around 1883–1886 by spreading a semitransparent sheet of gold leaf onto the surface of a selenium layer on a copper backing plate. However, this early work was not pursued for some decades. The main stimulus leading to the development of semiconductor devices came through the need to develop an efficient detector of electromagnetic waves. Jagadis C. Bose (India) patented the first solid-state point-contact detector in 1904, mentioning a variety of substances including galena (lead sulfide). In 1906 Henry Dunwoody and Greenleaf Whittier Pickard (U.S.) developed detectors using silicon carbide and silicon. In the first decade of the twentieth century many workers were engaged in efforts that led to improvements in detection efficiency for radio receivers (see Radio Receivers, Crystal Detectors and Receivers). This resulted in production of the so-called "cat's whisker," consisting of a rectifying metal to semiconductor contact, the polycrystalline surface of the material being probed by a pointed wire until a rectifying metal-to-semiconductor contact was obtained. The principal substances then being used were silicon and lead sulfide (in the form of galena). By 1909, Karl Baedeker was using the Hall effect to study systematically semiconductor behavior. J. Konigsberger published papers in 1907 and 1914, classifying silicon, selenium, and tellurium as semiconductors (germanium was not added until 1926).

The invention of the thermionic valve and its success as an amplifier soon overshadowed the crystal diode, finally displacing it in radio receivers by about 1926. Consequently, interest waned, although rectification characteristics were still being steadily improved by empirical means. However, a theoretical explanation for the phenomenon of rectification was still lacking. A major factor limiting progress in this respect was that the electrical properties of semiconductor materials are extremely sensitive to the introduction of minute amounts of impurities, and this was not realized at the time. Since varying amounts of impurities are present within unrefined semiconductor material, device characteristics varied considerably as a consequence. Even if the problem had been appreciated, techniques of crystal purification were not then sufficiently advanced to permit manufacture of material of a suitable quality in order to construct solid-state amplifying devices.

The introduction of the copper or copper oxide rectifier by L.O. Grondahl and P.H. Geiger (Germany) in 1927 was soon followed by the development of efficient selenium rectifiers, which were smaller and lighter for the same operating conditions, and could be operated at higher temperatures. Connected in series and provided with means of cooling, high reverse breakdown voltages were now possible. Their invention stimulated further interest in semiconductor rectifiers, since they extended existing commercial applications to include rectification in battery chargers and radios. It also raised interest in the possibility of controlling their current by means of a third electrode. Further uses for semiconducting devices followed the introduction of the selenium photovoltaic cell in 1931 by Bruno Lange and colleagues in Germany. This device operates by generating a voltage across a semiconductor junction when exposed to light, and was soon widely used in photographic exposure meters and applications such as control of artificial lighting, burglar alarms, and the opening and closing of lifts and doors. Photoconductive detectors, whose widespread use also dates from about this time, are devices that vary their conductivity when their surface is exposed to light. They have the advantage of low cost and quite high sensitivity, although their range of applications is restricted by their relative slowness in operation. A photosensitive material such as cadmium sulfide or cadmium selenide is usually deposited as a polycrystalline film onto a suitable substrate with a honeycomb of electrodes arranged to ensure maximum contact. These cells found an application as photographic exposure meters.

By 1930, little progress had been made in achieving a satisfactory theoretical explanation for the behavior of either semiconductor junctions or metal-to-semiconductor junctions, although a number of rival explanations were now being put forward, including an electrolytic barrier theory, the existence of a Peltier voltage generated by local Joulean heating, and also cold emission across a gap. Theoretical progress continued to be limited by the lack of reproducibility of experimental results, due to such factors as imperfections in crystal structure, bulk impurities, contact potentials, surface states, photoeffects and heating. Main techniques used in the investigation of semiconductor material at this time included Hall effect, conductivity, and thermoelectric power measurements. A further factor delaying development was that work in the field of solid-state devices was largely restricted to isolated individuals who lacked the resources necessary to mount a sufficiently large program of investigation.

Attempts were made in Germany, Britain, and elsewhere during the interwar years to construct various devices including field-effect transistors (FETs). However, these efforts met with little success. (In field-effect devices a current flows through a semiconductor channel, its width being controlled by means of a gate voltage). Probably the most well-known attempt was that made by Julius Edgar Lilienfield, then professor of physics at Leipzig, who took out patents in 1926 and again in 1928 for a voltage-controlled multilayer structure; one using a thin magnesium layer between

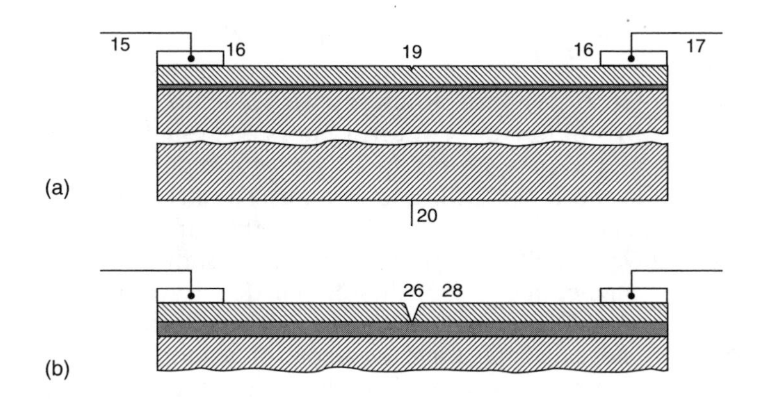

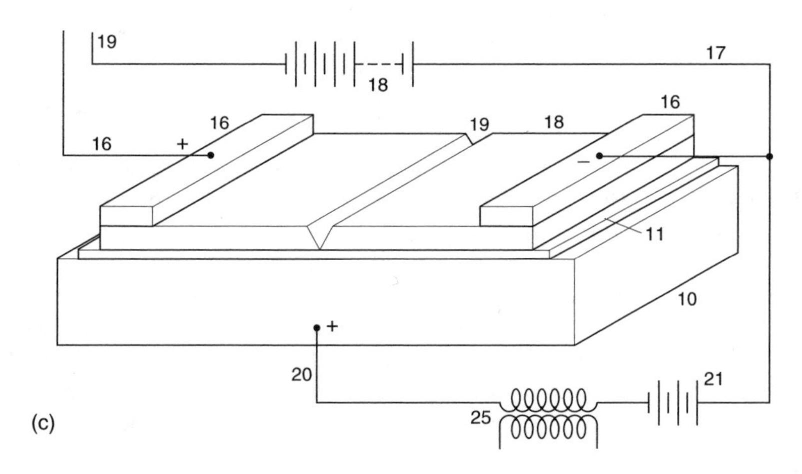

Figure 10. (a) and (b): The construction of a field effect transistor patented by J.E. Lilienfeld in the U.S. in 1933. The construction uses a thin layer of magnesium sandwiched between semiconducting copper sulfide layers. The notch shown in the center of (b) is intended to achieve a narrow cross-section, increasing the electric field at this point and hence the control effect. (c): This device shown with associated circuitry. [*Reproduced with permission from the Patent Office, U.K.*]

semiconducting copper sulfide layers (see Figure 10). There is no evidence that his devices actually worked. Nor did those of Oskar Heil (Germany) who in 1934 described a field-effect transistor with an insulated gate (see Figure 11). These patents were nevertheless important because their possession conferred priority. Rudolf Hilsch and Robert W. Pohl (Germany), working with alkali halide crystals in 1938, inserted a platinum control grid into the junction space-charge layer of a potassium

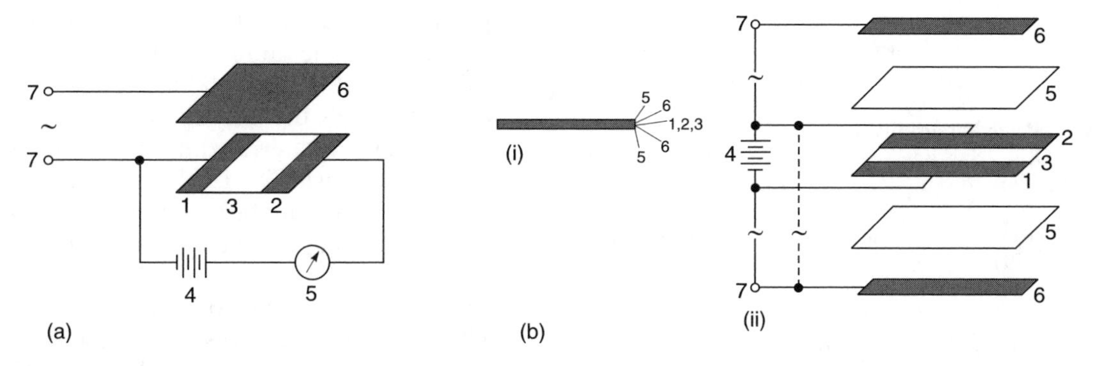

Figure 11. (A) and (B): The operation of a field effect transistor patented by O. Heil in Germany in 1934 and Britain in 1935. Heil states therein that "one or more thin layers of semiconductor traversed by current is or are varied in accordance with control voltage applied to one or more control electrodes arranged to and insulated from said semiconductor layer or layers." (A) illustrates the principle of modulation, the current flowing within the circuit in series with the semiconductor layers being controlled by an alternating voltage applied to a control electrode situated above it. (B) illustrates the circuitry in more detail The unshaded plates shown in the illustration act as insulators.
[*Reproduced with permission from the Patent Office, UK.*]

bromide crystal (the width of which could be calculated), and modulating a signal by applying a grid voltage in similar fashion to that used in thermionic triodes. Alkali halides were chosen because large crystals of high purity could be obtained. However, the physical limitations of this approach prevented a useful frequency response being achieved, although their work demonstrated that it was possible to construct a solid-state amplifier. Further efforts to do so were made by William Shockley and Alan Holden (U.S.), working at Bell Laboratories in 1939, but little agreement was found between existing theory and experimental results. This was due to the lack of appreciation of the effect of minute amounts of unwanted impurities and other crystal imperfections, as well as the importance of the behavior of charge carriers at semiconductor surfaces.

See also **Photosensitive Detectors; Radio Receivers; Semiconductors, Compound; Semiconductors, Elemental; Semiconductors, Postband Theory; Transistors**

ROBIN MORRIS

Further Reading

Braun, E. and Macdonald, S. *Revolution in Miniature*. Cambridge University Press, Cambridge, 1982.

Braun, E. Selected topics from the history of semiconductor physics and its applications, in *Out of the Crystal Maze*, Hoddeson L.H., Ed. Oxford University Press, New York, 1992, 443–488.

Gibbons, M. and Johnson, C. Science, technology and the development of the transistor, in *Science in Context: Readings in the Sociology of Science*. Oxford University Press, Oxford, 177–185, 1982.

Gosling, W. The pre-history of the transistor. *Radio Electron. Eng*., 43, 1/2, 10, 1973.

Loebner, E. Subhistories of the light emitting diode. *IEEE Trans. Electron. Devices*, ED23, 7, 675–697, 1976.

Morris, P.R. *A History of the World Semiconductor Industry*. Peter Peregrinus, London, 1990.

Sengupta, D.L., Sarkar, T.K *et al.* Centennial of the semiconductor diode detector. *IEEE Proc*., 86, 23, 5–243, 1998.

Weiner, C. How the transistor emerged. *IEEE Spectrum*, 10, 24–35, 1973.

Wilson, A.H. Theory of electronic semi-conductors. *Proc. R. Soc. A*, 133, 458–491, 1931.

Wilson, A.H. Theory of electronic semi-conductors. *Proc. R. Soc. A*, 134, 277–287, 1931.

Ships, Bulk Carriers and Tankers

The era of the steam-propelled ship began in 1807 when Robert Fulton's *Clermont* paddle-wheeled up the Hudson River to Albany, New York. Steam application to ship technology progressed so rapidly that in only 30 years the *Great Western*, a 72-meter wooden paddle-wheeler owned by the Great Western Railway, was able to sail from Bristol to New York in 15 days and establish itself as the first steamship to perform regular service between the U.S. and Britain. The next seagoing milestone came in 1843 when the cargo ship *SS Great Britain*, with an iron hull, a state-of-the-art engine, and a propeller-driven thrust system, was launched in Bristol. Although it still carried masts and sails, the three elements of a modern cargo ship were present: iron hull, steam engine, and propeller drive. By 1914 the triple expansion reciprocating steam engine was developed and remained in use through the century. The first steam turbine for marine propulsion, built for demonstration purposes only, was Charles Parsons' *Turbinia* in 1894. Turbines were adopted for Navy ships such as *HMS Dreadnought* and in 1905 the first turbine-powered large ocean liner was built. The *Titanic* also carried a Parsons steam turbine engine. In 1912 the first cargo ship to be powered by an internal combustion engine was built. At the conclusion of World War I, some 300 "motor ships" were in operation. In addition to engine technology, improvements were made in hull design, cargo handling, fuel usage, and ease of repair.

The days of the wind-powered ship were over. Gradually the schooner and the sleek "clippers," represented by the *Cutty Sark* and *Flying Cloud*, were driven from the sea. The completion of the American transcontinental railroad ended the necessity of a 100-day ride around Cape Horn from New York to San Francisco, especially following the discovery of gold in California. The opening of the Suez Canal cut as much as 2500 kilometers off the sailing trip from China and India to Europe and the British Isles. In 1800, 2026 ships passed through the 162-kilometer ditch that connected the Mediterranean and Red Seas. The necessities of war in 1914 brought some of the schooners and smaller clippers out of "mothballs.," but they provided easy targets for submarine gun crews.

By 1914 the basic layout for dry-bulk carriers as well as oil tankers was established. Both were about 100 to 114 meters in length with a beam between 12 and 15 meters; metal hulled; equipped with a triple-expansion steam engine or possibly a steam turbine; were propeller driven and boasted no auxiliary masts and sails. The dry-bulk carrier had its housing set in the midst of five holds, three holds forward of the engine room and two holds to the rear. The two aft holds were penetrated by a protective drive shaft housing which connected the engine via the drive shaft to an exterior propeller. About 11,000 cubic meters of cargo could be carried in these vessels. Within 30 years of Edwin Drake drilling the first oil well at Titusville, Pennsylvania in 1859, the first oil tanker was floated (1886). Dimensionally it differed little from the dry-cargo vessel. Safety dictated that the engine be set to the rear of the vessel so that no piercing of tanks and bulkheads was necessary for the propeller shaft housing. The ship's superstructure was set fore and aft with the hull tanks in the middle. Both bulk carriers could travel about 400 kilometers per day maximum; not much by twenty-first century standards.

The 1920s and 1930s were not decades of great prosperity. Some new ships were built and some old ships were refurbished, but no large construction orders were received until Henry J. Kaiser agreed to construct 2700 Liberty ships shortly after the outbreak of World War II. About 2500 Liberties survived the conflict. Eventually these were sold to a variety of purchasers, and only two were still afloat at the turn of the twenty-first century.

If a new ship is desired, the purchaser should know what he wants his ship to accomplish. A reputable naval architect should be contacted and his services engaged. The architect is responsible for determining the ship's design: size, shape, cargo capacity, and propulsion. The hull design likely would be determined by a 32-point dimensioning system utilizing computer-aided design (CAD) software. Ensuring the ship's buoyancy in water is essential. Displacement weight and volume has to be calculated so the vessel will float properly. Mathematical and computer models are used to work out the shape and displacement requirements and establish the ship's draft; that is, how deep the vessel sits in the water. Ship stability in the rolling sea, which can generate sideway and endway tilts, must be provided in the structural scheme. Resistant forces present when the ship is in motion must be countered by proper hull design and propulsion technologies. The ship will probably be furnished with a two-stroke diesel engine directly connected to the propeller shaft. The propeller will be either a fixed-pitch propeller or a newer controllable-pitch propeller. The fixed-pitch propeller has passed the test of time and is a solid design. It will be energized to the hull by a thrust block thus limiting its impact on the engine machinery.

The architect, with the help of a marine engineer, will combine the hull; superstructure; propulsion system machinery; ventilation, air, and heating apparatus; cargo handling equipment; and any other necessary gear into a profit-generating commercial vessel. A fine example of cost-ineffectiveness came after World War II with construction of the chic, sleek *NS Savannah*, a bulk and break-bulk carrier with passenger accommodation. It had 17,800 cubic meters of cargo space and could carry 9,400 tons of cargo at 20 knots. It required a large crew of more than 100 technically trained sailors, as it was powered not by a diesel engine, but by a nuclear reactor. It was launched in 1959 and retired 10 years later after sailing 730,000 kilometers and burning 74 kilograms of uranium ore. The cost of nuclear fuel and the salary requirements of a specially trained crew of 100 sailors made it very cost-ineffective when compared with a conventionally powered motor ship. It is now a deactivated museum piece in South Carolina. Nuclear powered vessels still have their adherents but there were no more Savannahs built.

One of the concerns of bulk carrier owners was speed and proficiency in the loading and unloading of cargo. New York shipbroker, Ole Skaarud, and a few collaborators dedicated themselves to improving the process. They increased the size of holds for better cargo handling and raised the hatches, especially for grain loading, several feet above the decks. Ship's housing and machinery were placed aft so that unobstructed holds could be installed in the central portion of the ship. Some vessels had cranes, booms, conveyors, and even pneumatic tubes attached to the deck while others relied upon off-ship loading facilities. A newly developed conveyor boom attached to a ship could swing over the holds and convey cargo off the ship. Skaarud also realized that overhead loading of bulk holds usually created a pointed load with the cargo sloping down to the hull sides and bulkheads. To improve safety and control cargo movements in rough seas, ballast tanks were installed in the upper reaches of each hold. Improved hull construction had lessened hatch failures, which in turn lessened breeched and flooded holds. Obviously, larger holds led to larger ships. There are four classifications for bulk carriers based on their dead weight tons (DWT) carrying capacity: Handysize, 10,000 to 35,000 DWT; Handyman, 35,000 to 50,000 DWT; Panamax, 50,000 to 80,000 DWT; and Capesize, 80,000 to 200,000 DWT. Panamax and Capesize cannot pass through the Panama Canal. Some bulk carriers move unusual cargo that requires special safety and ease of handling devices. These may be "reefers," with refrigerated cargos, multidecked and well-ventilated livestock carriers, and roll on/roll off (RO/

RO) ships which are designed to allow motor cars to drive on and off ships through doors in the hull. Many sailors consider the RO/RO carriers break-bulk vessels. The "banana boats" of fiction actually exist and service customers. The silhouette of the old cargo ship on the horizon has changed dramatically in the last half of the twentieth century.

No new development in sea traffic since the end of the World War II was more significant than the creation of containers and container ships. These were a product of intermodal transport invented by Malcolm McLean, a struggling truck, or lorry, operator in the Depression of the 1930s. He placed a motor truck trailer on a railway flat car destined for a seaport (or a railway terminal) where the trailer was removed to a ship (or was driven away). Soon the trailer was joined by a steel container 8 feet high by 8 feet wide (2.5 by 2.5 meters) and either 20 feet or 40 feet (6 or 12 meters) in length. The containers were placed on the ship, above and below deck, using standardized loading devices; a considerable difference from the old winch and crane loading system. Ships were soon designed and constructed specifically to carry containers. The first such ship, with 226 containers, set sail in 1957. Measurements in the ocean container trade are based on twenty-foot equivalency units (TEU). One TEU equals an 8 by 8 by 20 container. There are now many container ships that boast 6,000 TEUs; but it is prophesied that 12,500 TEU ships are not far in the future. Container ships are quite fast, ranging from 20 to 25 knots—much faster than tankers. Containers are offered for sale or lease in many varieties including dry bulk, break-bulk, liquid (tanktainer), insulated, and refrigerated. It is said that 10 percent of all sea traffic today is containerized, but this may be an understatement.

Large oil tankers have been described as floating behemoths. A few current tankers can carry 350,000 DWT of liquid cargo, usually crude or product (refined) oil. Gas tankers are essential where pipelines do not or cannot exist. Carrying natural, butane, or propane gas, these ships come in many sizes and are often easily recognizable by domed tanks rising above the ship's deckline.

Like container ships, tankers are classified by size. The largest are called ultra large crude carriers (ULCC) and run 350,000 DWT or higher. The inability of many oil ports to handle ships of this size and their high cost of operation has led, in recent years, to a substantial decline in their construction. Often they must be unloaded, via a floating pipeline, 2 or 3 kilometers offshore. The very large crude carriers (VLCC) are next in capacity with 200,000 to 350,000 DWT. These tankers are more port-acceptable, and thus more popular than the ULCC. They cannot, however, traverse the Suez Canal if fully loaded. It requires a Suezmax at 130,000 to 160,000 DWT to do this. New tankers, for safety, are now required by law to be double-hulled. Some tankers, classified FPSO (floating, production, storage, and offloading), anchor above an oil source and draw up the crude oil by pipeline hose from tanker to ocean floor. The 1967 closure of the Suez Canal did not cause great inconvenience because most ULCC and VLCC tankers are too large for this facility.

Europe and the Russian Federation offer good examples of efficient control of bulk traffic movement on rivers and canals. The use of barges and motor ships on inland waterways is more common in Europe than in the Americas. The Danube, Volgo, Rhine-complex, and Elbe Rivers are constantly dredged and kept in good repair. Interconnecting canals such as the Main Danube give the waterways system considerable reach and ease of movement. Barges and motor ships transport bulk cargos of coal, oil, ores, construction materials, fertilizers, and foodstuffs in, out, and through the system.

Barges are of various sizes, dependent upon the width and depth of the waterways in which they operate. Some are self-propelled, but most are pushed or pulled by towboats or tugs. Barges may be open, closed, tanked, or hoppered. Lighters can be seen—these are covered barges unloaded from an oceanic LASH (lighter aboard ship) vessel at a river, lake, or canal port and pushed or pulled by a tug or towboat to their final destination on the river, lake, or canal. After the cargo is unloaded, the procedure may be reversed.

The use of diesel or turbine-electric motor ships in Europe is also common, but take second place to barges in total tonnage moved. Motor ships range in capacity from 100 to 5,000 DWT, the latter with capabilities for movement in the Baltic, Mediterranean, and Black Seas. Most of these vessels resemble their oceanic colleagues, only smaller. They have as many as four holds, with machinery and housing in the rear and offer a speedier, but costlier, alternative to the slower barges.

The Great Lakes in the U.S. offer a variety of clumsy-looking ships, up to 300 meters in length, with many holds, and topped off with superstructures fore and aft. Some possess very modern self-unloading equipment. They mingle with pleasure craft, sailing vessels, ocean-going tankers, bulk, and break-bulk ships. The latter, called "salties," enter the Lakes through the St. Lawrence Seaway. The Great Lakes freighters, or "longships" are 150 to 300 meters in length and move around the Lakes at 10 to 15 knots carrying as much as 60,000 tons of iron ore from the Mesabi Range in Minnesota, coal from Appalachia, or grain from the Great Plains. Any cargo ship that moves in and about the Lakes must fit into the locks at Sault St. Marie, Michigan or the locks in the Welland Canal system in Ontario. "Salties" cannot ascend Niagara Falls without utilizing the Welland Canal. As a result of size some "salties" get only this far. The "longships" and "salties" sometimes travel in very rough weather.

Lloyd's Register of Shipping lists over 85,000 vessels of more than 100 gross tons. No small group of nautical corporations can dominate the shipping trade, for they lack sufficient vessels for control. The Nippon Yusen Kaisha (NYK Line) may be the world's largest seagoing cargo carriers. It has three container groups, two bulker groups, and three petroleum and gas groups. Nevertheless, NYK can float no more than 560 ships. The Maersk SeaLand and SafMarine companies in the A.P. Moller Group (Denmark) total 290 container vessels, the world's largest inventory of these ships. It is interesting to note that Moller acquired this unit—Maersk SeaLand—in 1999 from CSX Corporation, an American railway conglomerate. P&O Ned Lloyd Container Line Ltd. (U.K.) boasts the second largest container fleet with about 146 vessels. Many large seagoing carriers are, in fact, conglomerates with divisions and subsidiaries outside the shipping industry.

The greatest growth in the shipping industry is in container vessels. In the past two years dry-bulk and oil carriers have remained quite stable; that is, no growth and no loss. Container ships appear to have the most promising future.

ERNEST M. TEAGARDEN

Further Reading

Armstrong, R. *The Merchantmen*. Praeger, New York, 1969. Fine work on steamship development.

Batio, C. *Super Cargo Ships*. MBI Publishing, Osceola, WI, 2001.

Hornsby, D. *Ocean Ships*. Ian Allen, Shepperton, Surrey, 2000.

Know Your Ships. Marine Publishing, Sault Ste, Marie, MI, 2002.

Labaree, B.W. *et al. America and the Sea: A Maritime History*. Mystic Seaport Museum, Mystic, CT. 1998. A 620-page history of seafaring Americans from colonial times to the late 1990s.

Nersesion, R.L. *Ships and Shipping: A Comprehensive Guide*. Penn Well, Tulsa, OK, 1981.

Thompson, M. *Queen of the Lakes*. Wayne State University Press, Detroit, 1994.

Valkar, I. Investment in inland waterways? Infrastructure needs. *Seminar on the Inland Waterways of Tomorrow on the European Continent*, Paris, 30 January, 2002. Document No. 5.

Useful Websites

Nuclear Merchant Ships: http://www.atomicengines.com/ships.html.

Ship Design and Construction: http://www.mi.mon.ca/mi-net/shiptech/shipbldg.htm.

Ship's Propulsion Layout: http://members.shaw.ca/diesel-duck/machinery_page/propulsion_layout/propulsion_layout.htm

Skyscrapers

Skyscrapers are the world's tallest buildings. One-hundred-eighty- and 200-meter-high buildings that were considered to be exceptionally tall in 1910 were overshadowed by skyscrapers of more than 300 meters in a matter of 20 years. Advances in construction techniques enabled engineers to build ever-taller structures throughout the twentieth century. However, the principle reasons for erecting exceptionally tall buildings changed little over time. Densely populated cities with escalating land values called for maximum utilization of available space, and tall buildings are one of the most economical means of assembling large numbers of workers in one place. While the majority of skyscrapers were built for the profits they could generate, other reasons included self-aggrandizement, prestige, image, and recognition.

The skyscraper is a late nineteenth century American innovation that was more aptly defined by its height than any characteristic of construction or particular style. While the genre first appeared in Chicago in the 1880s, New York City became the focal point of tall building construction shortly after 1900 and remained so for much of the twentieth century.

The increase in the height of New York's tall buildings was rapid. One of the first exceptionally tall structures was the 187-m, 47-story Singer Building erected in 1906. In 1912, the 241-meter-tall Woolworth Building boasted 60 stories. The height of the tallest had vaulted to over 300 meters in 1930 when the Chrysler Building was completed. Only a year later its height was eclipsed by the Empire State Building, standing 381 meters high.

At that point it was as if an upper limit had been reached and the Empire State Building remained the world's tallest building for over 40 years. There were no extraordinary increases in height in the world record buildings that followed and they tended to be only fractionally taller than the previous record holder. In 1972, the 417-meter twin towers of the World Trade Center were completed but their supremacy was brief as Chicago's Sears Tower, standing 19 meters taller, opened in 1974.

Despite the burgeoning technology of tall building construction in America, Europe did not follow suit. Five- or six-story buildings seemed to be the limit there until after World War II. It was not until 1990 that the first large-scale skyscraper, the 257-meter-tall high-strength concrete Messeturm, was constructed in Frankfurt, Germany. Although the situation differed little in Britain, efforts were made in London to impose height limits on new buildings out of concern that historic buildings not be overshadowed. However, at the end of the twentieth century there was active planning for a number of skyscraper-size buildings.

Structurally, most skyscrapers were basic steel frame buildings assembled of columns and beams. Columns passing through the interior of the structure from bottom to top carried the structural load down to the building's foundation. Beams supported both concrete floors and lightweight curtain walls of terracotta, brick, glass, or metal. The first major exception to this configuration was the hollow tube design of the towers of the World Trade Center in which the outer walls were load bearing. The design featured a perimeter of closely spaced columns and the elimination of all but a main core of heavy internal columns created large open floors. This same form of tube of construction was used to build the Sears Tower, which until 1998 was the world's tallest building.

Throughout much of the century, hot rivets were used to assemble the structural steel members of skyscrapers. During the 1940s, the first of several significant technical developments occurred beginning with the increased use of welded joints. Yet another important advance came about during the 1950s when high-strength bolts were introduced for use in connections. The ductility of joints consisting of a combination of both welding and bolting made them so superior to existing methods that by the early 1960s riveting had been all but abandoned for building construction. Welded and bolted frames could be further enhanced with eccentric bracing that better resisted the lateral forces of wind pressure and earthquakes. Earthquake-resistant building design continued to be one of the most complex and challenging problems faced by engineers. Possible solutions included the use of shear walls designed to resist sideways forces and the isolation of a structure from its foundation by separating the two with layers of rubber.

From the 1950s, new high-strength concrete provided engineers with a reasonable economic alternative to traditional steel frame construction for tall buildings. Without compromising strength, columns could be smaller and in turn they freed up rentable floor space. Nonetheless, concrete structures had higher mass and a greater damping effect in wind loads and movement caused by earthquakes, and they outperformed their structural steel counterparts. Concrete had the added advantage of being both fire and blast resistant.

Skyscrapers require the same systems for heating, air conditioning, lighting, and power that are used in their shorter counterparts; however, while the installation of most systems pose no particular problems, supplying water to tall buildings does require special equipment. As typical city mains pressure would raise water only a few stories, pumps are needed to move water to integral storage tanks which in turn supply the structure. However, of all the systems that go into a skyscraper, none is more critical than the elevator. With five or six stories being the maximum that can be reached readily by foot, the elevator is essential to the utilization of a major part of a skyscraper. Bulky power sources of early elevators typically took up valuable floor space. The system was revolutionized in 1903 with the introduction of electric gearless traction and an elevator's motive power could be located at the top of the shaft.

In the early 1990s, the center of tall-building construction moved from North America to Asia where there was great activity in building skyscrapers. The century of the skyscraper was brought to a close in 1998 with the completion of the world's tallest buildings, the two 452-meter-high concrete Petronas Towers in Kuala Lumpur, Malaysia.

The destruction of both towers of New York City's World Trade Center by terrorist attacks in 2001 raised anew questions regarding the desirability and safety of skyscrapers as a building form. Technologically, the upper limit to the height to which buildings may rise has yet to be reached. Although economics remains key among the factors deciding how high buildings rise, it is politics that may play a pivotal role in determining if towering skyscrapers are built in the future.

See also **Fire Engineering; Vertical Transportation (Elevators)**

WILLIAM E. WORTHINGTON, JR.

Further Reading

Tauranac, J. *The Empire State Building: The Making of a Landmark*. Scribner, New York, 1995.
Landau, S.B. and Condit, C.W. *Rise of the New York Skyscraper, 1865–1913*. Yale University Press, New Haven, 1996.
Goldberger, P. *The Skyscraper*. Alfred A. Knopf, New York, 1981.
Beedle, L.S., Ed. *Second Century of the Skyscraper, Council on Tall Buildings and Habitat*. Van Nostrand Reinhold, New York, 1988.
Douglas, G.H. *Skyscraper: A Social History of the Very Tall Building in America*. McFarland, Jefferson, 1996.
Elliott, C. *Technics and Architecture: The Development of Materials and Systems for Buildings*. MIT Press, Cambridge, MA, 1992.

Smart and Biomimetic Materials

Smart materials react to stimuli such as heat, light, electricity, or some other factor from their external environment. They include shape-memory alloys and polymers, piezoelectric polymers and ceramics, biomimetic polymers and gels, conductive polymers and controllable fluids, magnetostrictive materials, and chromogenic materials. Smart materials can be used in a smart materials system, such as micromachined electromechanical systems and fiber-optic sensor systems. Smart materials systems are a late twentieth century design methodology that integrates the actions of sensor, actuator, and control circuit elements in systems that can respond and adapt to changes in their environment or condition in a helpful way. They bestow smartness, or a functionality that enhances the value of materials, technologies, or the final products. Smartness can improve the performance of a system in a way that cannot be achieved using the more established non-smart approaches.

Many other smart materials systems are continually researched and developed, notably by the defense and aerospace industries. The National Aeronautics and Space Administration (NASA), for example, is developing the ultimate smart structure: a shape-changing airplane wing that will reorder and optimize its shape during flight to match the atmospheric conditions or the mission it must carry out.

Shape-Memory Materials

Shape-memory materials behave in a predetermined way when exposed to a particular stimulus. They can exist in different shapes at two or more different temperatures once their transition temperature has been reached. Shape-memory alloys have very different shape-changing characteristics compared with shape-memory polymers. Devices made from shape-memory alloys can provide force when they are exposed to their transition temperature, as is the case with actuators. However shape-memory polymer devices undergo mechanical property loss while exposed to their transition temperature, as when used to make releasable fasteners.

Shape-memory alloys are much more costly than polymers. It takes time to program a metal alloy that also needs heat treatment at temperatures of hundreds of degrees Celsius, resulting in a maximum deformation of only about 8 percent. Shape-memory polymer systems are being developed that are easier to shape, with more applications than the shape-memory materials already in use, such as the nickel–titanium alloy Nitinol, used to make items such as flexible spectacle frames. The polymers can be programmed into shape in seconds at about 70°C and can undergo major deformations of several hundred percent.

Fiber-Optic Sensors

Fiber-optic sensors are based on optical fibers attached to sensing devices, forming smart systems that may be built into structures such as airplane wings and bridges. Fiber-optic sensor technology is used to measure mechanical properties such as strain, pressure, and temperature. Structures incorporating fiber-optic sensors react to or warn of imminent failure and demonstrate the state of the structure following damage.

Piezoelectric Polymers and Ceramics

Piezoelectric polymers and ceramics are materials that exhibit new properties when they are exposed to an electric current (piezo-electricity). Many polymers, ceramics, and molecules such as water are continuously polarized, with some parts of the molecule being positively charged and other parts negatively charged. By applying an electric field to these materials, the polarized molecules will align themselves with the electric field, producing induced dipoles within the molecular or crystal structure of the material.

Piezoelectric materials are used in acoustic transducers that change acoustic (sound) waves into electric fields and electric fields into acoustic waves. These find an application in devices such as speakers and drums.

Semicrystalline polyvinylidene fluoride (PVDF) has been the only piezoeletric polymer commercially available until recently. Lightweight, flexible and easily produced as sheets or in complicated shapes, its low mechanical and acoustic impedance makes it highly suitable for use in underwater and medical applications. However PVDF has limited temperature use and poor chemical stability in extreme environments. Polyimides may be an alternative as they have excellent thermal, mechanical and dielectric properties combined with high chemical resistance and stability.

Magnetostrictive Materials

Magnetostrictive materials can convert magnetic energy into mechanical energy and also transform mechanical energy into magnetic energy. Magnetostriction is a property of the material that does not lessen over time. These materials expand when exposed to a magnetic field, an effect known as the Joule effect or magnetostriction after James Prescott Joule. In the early 1840s Joule identified the phenomenon when he observed a change in length of an iron sample as its magnetization altered. This effect is due to the lining up of the magnetic domains in the material with the magnetic field. A change in the size of the width occurs together with the change in length produced by the Joule effect. When the material is stretched or compressed, it undergoes strain and its magnetic energy changes. This is known as a magnetomechanical or Villari effect, commonly used in magnetostrictive sensors. Magnetorestrictive materials include iron, cobalt, nickel, ferrite, metglass, and terbium alloys (Terfenol-D). They are used in a range of modern devices such as sensors, sonar and ultrasonics, speakers, vibration and noise control, and drills and reaction mass actuators.

Chromogenic Materials and Systems

Chromogenic materials and systems can change their optical properties in response to an electrical, photo, or thermal stimuli. Electrically activated chromogenic systems are used for smart windows and mirrors in the automotive and architectural industry and for low-information content displays. Electrically activated chromogenics can be controlled by the user, unlike photochromic and thermochromic devices which are self-regulating and passive.

Electrochromic materials are chromogenics that change color on electrical stimulation. They are actuator elements that need sensor and control circuits to be added to make the system smart. Reflective hydrides may also be regarded as electrochromics, but they differ in a number of ways from the more commonplace oxide electrochromics. Originally deposited as a metal, they can be converted to a partially transparent hydride by injection of hydrogen from the gas or solid phase when they switch to a reflective state, which has several potential advantages in terms of energy performance and durability. Transition-metal hydrides have now also been developed. Liquid crystal windows switch quickly from a transparent state to a diffuse white state; however, they have little control over solar heat gain. Suspended particle displays are also under development. Photochromics darken in sunlight and are therefore mainly used to make sunglasses that darken automatically. Thermotropic materials respond primarily to heat.

Heat-sensitive polymers (thermochromic) are used for children's toys, tee-shirts, and toothbrushes that change color when touched. These materials have additional functions that are visible in real time and can provide a variety of intelligent responses.

Biomimetic Materials

Biomimetic materials are based on nature's best designs and attempt to mimic them. Nature solves problems by looking for a solution that works using the minimum amount of energy. Engineers solve problems by searching for an effective solution with the lowest cost. Plants and animals require a great deal of energy to produce the basic materials they need for survival, but they can use almost any shape. Engineers are able to produce a wide variety of materials cheaply, but shapes are often expensive to make. By copying nature's designs and shapes, researchers can make more efficient structures that can be used to solve tomorrow's engineering problems as well as to develop innovative new materials. Biomimetic polymers are being developed based on natural materials; for example, spider's silk, which is in fact a biopolymer. Biomimetic gels; for example, those based on the sea cucumber, are being researched and may result in a number of biomedical applications.

Microelectromechanical Systems (MEMS)

Microelectromechanical systems (MEMS) are miniaturized devices. They may be as small as a silicon semiconductor chip, which is able to integrate sensors, information or signal processing, and control circuits in a single device. A range of MEMS -based devices have been developed including pressure sensors, transducers, transmitters, microrelays, optical attenuators and photonic switch components, and smart security and tagging systems. They are under development for further biomedical applications.

Controllable Fluids

Controllable fluids have properties that depend on an electric or magnetic field. They are a smart technology that may be used instead of piezoelectric transducer-controlled semiactive suspension systems.

Conclusion

Among the more interesting biomedical developments in smart systems in the 1990s were synthetic muscle actuators, which included shape memory alloys, piezoelectrics, and electroactive polymers. A particular example of a smart material used for such an application is IPMC (ion-exchange polymer membrane metallic composites). The synthetic muscles contract or bend when exposed to an electric current and can be made into wires that are as thin as a human hair.

See also **Ceramic Materials; Photosensitive Detectors; Plastics, Thermoplastics**

SUSAN MOSSMAN

Further Reading

Friend, C. *From Sensuous Aeroplane to Cuddly Toaster*. Institute of Materials, London, 1997.

Culshaw, B. *Smart Structures and Materials*, Artech House, London, 1996.

Holnicki-Szulc and Rodellar, J., Eds. *Smart Structures: Requirements and Potential Applications in Mechanical and Civil Engineering*. Kluwer Academic, London, 1999.

Srinivasen, A.V. and McFarland, D.M. *Smart Structures: Analysis and Design*. Cambridge University Press, Cambridge, 2000.

Vincent, J.F. Stealing ideas from nature. *RSA Journal*, August/September, 36–43, 1997.

Vincent, J.F. Naturally new materials. *Mater. Today*, 1, 3, 3–6, 1998.

Social and Political Determinants of Technological Change

The question of whether and to what extent social and political factors determine or influence technological change is a major issue in the historical and social understanding of technology, one that took special form during the twentieth century. Precisely because technological change has been so persistent and so closely associated with obvious social and political transformations, questions have regularly been asked about both the drivers of the inventive creativity and the parameters of its social and political interactions.

Focusing on the U.S., which has often been a site for the leading edge of such interactions over the course of the twentieth century, one may distinguish three major periods of technological change.

First, the century opened in the presence of emerging industrial conglomerates controlled by powerful families such as Carnegie, Rockefeller, and Vanderbilt and the rise of corporations increasingly committed to technological research and development such as Bell Telephone, Eastman Kodak, Edison Electric, Ford, and General Motors. These new social constellations appeared in conjunction with new technological systems of unprecedented scale and complexity. Coal and iron ore mines were linked by railroads with smelters and mills to make steel for use in the assembly-line production of automobiles and other goods to be consumed by an increasing number of people in urban centers, the populations of which for the first time surpassed those of rural areas.

During the broad middle third of the century, post-Depression and especially in conjunction with World War II, technology became increasingly science dependent and rationalized. Examples ranged from physics for the development of radar, the atomic bomb, and computers; to chemistry for the creation of dyes, synthetic, and agricultural chemicals; biology for hybrid seeds and antibiotics; and aerodynamics for the design and manufacture of airplanes and rockets. At the same time as the government established a system of national laboratories, research and development activities were increasingly promoted on university campuses. The need for more efficient industrialization and effective management of large human–machine complexes such as battle groups incorporating naval, army, and air force components stimulated the development of operations research and analytic management techniques well beyond traditional accounting and marketing procedures. After the war, the reconstruction of Europe and Japan, the rise of international capitalism (Bretton Woods, the World Bank, and the International Monetary Fund) and politics (the United Nations, the Cold War, and Third World development programs) was further associated with extensions of technological change as consumer goods multiplied in availabilities and technical sophistication.

A final short third of the century witnessed the emergence of sometimes competing criticisms of various aspects of these massive technological and social developments. The environmental movement and technological disasters (Three Mile Island, Challenger, Chernobyl, Bhopal) raised questions with respect to any simple assumptions about the unqualified beneficence of technological change. The antinuclear, consumer, anti-Vietnam War, and AIDS activist movements called for greater public participation in technical decision making and reorientations in social priorities. Economic and political assessments of bureaucracies challenged the efficacy of large-scale hierarchical organizations and regulatory agencies. Together with the energy shortages during the 1970s and the end of the Cold War in 1989, environmental and economic analyses were associated with transformations toward flexible systems of production distributed across many countries, designer consumer goods with niche marketing, and the promotion of new electronic communications, from computers and cable or satellite television to the Internet and cell phones. The century closed with the rise of interactive digital technologies that created virtual realities, and of biotechnologies and nanotechnologies that challenged traditional boundaries between the natural and the artificial.

Theories of Science, Technology, and Society Interaction

Across the twentieth century technological change appeared, paradoxically, to be both inevitable and the result of the free choice of human beings. The common belief has certainly remained that sociopolitical factors do in fact influence both the direction and rate of technological change. Precisely for this reason, early twentieth century historians sought to explain the singular rise of modern technology in Europe since the 1500s by appealing to the influence of social and political factors such as agricultural productivity, resource availability, population, economic structures, religion, or democracy; while sociologists and economists sought to specify the circumstances under which technology could effectively be transferred from one context to another, either from one country to another or from laboratory to market. At the same time, people in the twentieth century sometimes experienced technology as functioning in a semiautonomous manner, determining or influencing more than being influenced by society, politics, or economics. In response, governments have often looked for ways to control and regulate technological change, but have not always found themselves successful in doing so.

Ironically, the notion of technology as a semiautonomous force, itself determining social and political phenomena, has been at once hailed as good and criticized as bad. The motto of the Chicago Centennial of 1933 celebrating "A Century of Progress" was "Science Finds, Industry Applies, Man Conforms." During this same period, however, University of Chicago sociologist William Fielding Ogburn coined the term "cultural lag" to describe disharmonies that resulted when technological change outstrips social and political development. One of his examples was how technological increases in productivity and medically promoted decreases in infant mortality combined to undermine the rationale for large families, although it normally takes one to two generations for family size to drop appropriately. Attempts to clarify the precise senses in which technological change is and is not an independent variable in relation to society and politics was thus a recurring theme in twentieth century reflections on technology.

The bottom line is that there exists no comprehensive theory of the ways technology determines or influences society and politics or of the ways society and politics influence technology. Indeed, there is not even any consensus about how to define technological change, which because of its complexity is often termed "technosocial" or "sociotechnical change"—thus finessing rather than answering the theoretical question. This makes it impossible to review the phenomenon of technology–society relations themselves; the best that can be done is to survey theoretical approaches to such relationships. From this perspective, scholars during the first half of the century tended to concentrate on searches for macro theories about technological influence, while those of the second half stressed micro theories about how social and political factors determine or influence technology. Near the close of the century scholars began advancing theories that attempted to bridge the two positions, but without fully satisfying the challenging complexities of the multiple questions at issue.

Theories of Technological Determinism

Theories emphasizing the influence of technology on society and politics, often termed theories of technological determinism, inherited from the nineteenth century three basic stories about the history of technology. These three competing and collaborating stories are

- The idea of the progressive human conquest of nature (insofar as technologies are well-defined extensions of human action).
- The massive industrial proliferation of goods and services (through increased division of labor and standardized production routines).
- A tendency of technology to itself escape human control (as artifacts progressively tap non-human energy resources, formalize human behavior routines, and overwhelm the social world with the products of human labor that produce manifolds of unintended consequences).

The work of American cultural critic Lewis Mumford and French sociologist Jacques Ellul explored, refined, and criticized these three inherited stories, and in the process presented theories of technology as an underappreciated and often unwelcome determinant of many aspects of society and politics.

Mumford's *Technics and Civilization* (1934) took the anthropologist's periodization of human prehistory as defined by materials (Stone Age, Bronze Age, Iron Age) and extended it into history, proposing to complement the more traditional cultural and political distinctions (Greek and Roman Ages, Middle Ages, modern period of nation states) with one defined by changes in technologies. Mumford's proposal was to distinguish the "eotechnic" use of animal power and increasingly specialized tools during the Middle Ages, the "paleotechnics" of steam power and the machines of the Industrial Revolution, and the "neotechnics" of electric power in the twentieth century, each sponsoring special social and political orders. Although his terminology never took hold, the book itself virtually created the notion of a social history of technology, all previous attempts being more internalist technological histories. In one of his vivid illustrations of the power of technologies to influence social order, Mumford argued that the invention of the clock transformed the experience of time; before the clock, for instance, working days had varied in length with the seasons of the year. The clock made possible the standardized working day and thus the mass production factory in which all workers for a particular shift are required to show up at the same time. A later example was nuclear energy, which virtually required centralized, hierarchal, anti-democratic management.

For Mumford the twentieth century was in a situation comparable to that of the Industrial Revolution, having created powers that it now needed to learn to guide and manage. Making alliances with socialists and liberal progressives, he called for a more conscious appreciation of this circumstance and an effort to bring a broad spectrum of human values to bear on what he saw as a monomaniacal technological world focused too exclusively on the pursuit of power. In his last work, however, *The Myth of the Machine* (2 vols, 1967 and 1970), Mumford appeared to despair of the possibilities for this exercise in human responsibility, although in reality his arguments almost certainly had an influence on the environmental and economic criticisms of technology.

Whereas Mumford's intellectual background was literary-based cultural criticism, Ellul sought to extend perspectives inherited from Karl Marx. Even though he was not a strong technological determinist, Marx credited technology with exercising a powerful social and political influence. In a famous phrase from *The Poverty of Philosophy* (1847), Marx remarked, "The hand-mill gives you society with the feudal lord, the steam-mill society with the industrial capitalist." Of course, one then has to ask, but what gives you the technologies of hand-mill or steam engine. To claim, as Marx did, that such technological change is a more or less unconscious process of human interaction with the natural world and existing technologies, which then have unanticipated social consequences, constitutes a kind of soft determinism.

This is the view developed at length by Ellul in *La Technique* (1954), which begins by distinguishing between "technical operations" and the "technical phenomenon." Polytechnical operations are present throughout history, always subordinate to the contexts in which they occur. The technical phenomenon, by contrast, is a unified method for the guidance of making and using that arises more or less by accident. However, once it comes on the historical scene, this phenomenon begins to transcend all particular contexts with appeals to decontextualized notions of efficiency. The public faith that human beings then place in "technological efficiency" furthers its dominance. To appreciate the determining power of this new phenomenon, which may also be described as technology turned into a social institution, requires a special act of consciousness. It is Ellul's aim to contribute to the emergence of this consciousness with what he calls a "characterology" of technology.

According to Ellul's characterology, technology in the mid-twentieth century exhibits the distinguishing features of artificiality, self-augmentation, universality, and autonomy. It replaces the natural milieu with one increasingly fabricated by human beings, thus occluding natural orders and influences. It repeatedly extends itself ("The only solution to the problems of technology is more technology"). It is more and more the same everywhere ("If you've seen one Holiday Inn, you've seen them all"). It becomes increasingly independent of external values ("Why do something in a way that is inefficient?"). By means of these characteristic features, technology transforms economics, politics, medicine, education, sports, and entertainment, driving all such traditional human activities to incorporate technical means and seek efficiency in their respective areas of activity.

These two versions of technological determinism emphasized, in turn, the primacy of technology as physical artifact (Mumford) and human process (Ellul). In a more abstract interpretation, the German philosopher Martin Heidegger (1889–1976) argued that modern technology was constituted by a stance at once willful and epistemological; technology takes up an attitude toward the world that forces it to reveal itself as resources available to be controlled and manipulated. From each perspective—artifact, activity, will, and knowledge—technology is argued to exercise deep influences on politics and society. Langdon Winner's *Autonomous Technology* (1977) and *The Whale and the Reactor* (1986), along with a Merrit Roe Smith and Leo Marx edited volume titled *Does Technology Drive History?* (1994), provide fitting reviews of the determinist tradition in the midst of the emergence of a major alternative research program called social constructivism.

Theories of the Social Construction of Technology

During the latter half of the twentieth century, a new generation of scholars turned away from both grand theories and the idea of technological determinism in order to explore with case study detail the myriad ways in which social and political factors can, and indeed do, influence technology. In part, this change resulted from the conceptual maturity of academic programs in science and technology studies (STS), programs that first arose more or less simultaneously and independently during the 1970s in the U.S., Europe, and other parts of the world. Within this context, two leading proponents of the new contextualized approach that promoted theories of the social construction of technology were the Dutch historian Wiebe Bijker and the French social scientist Bruno Latour.

In an influential study of technological change in the bicycle, Bijker (1987) argues that this artifact never played the dominant role in all the social changes with which it was associated that technological determinism might have implied. The early bicycle, with its large front wheel and small back wheel, was quite unstable and known as a "bone breaker." Young men rode it in the park to display their masculine daring. The development of the "safety bicycle" occurred in conjunction with women asserting their right to appear in public places doing many of the things that men do, but maybe not just like men. The fact that boys still refuse to ride a drop frame "girl's bike," despite its structural superiority, clearly shows how social interpretation dominates technological change.

For Bijker the social meaning of an artifact is always underdetermined by the artifact itself. Technologies are thus subject to interpretative flexibility, and function as sites of competing adaptations. The history of technologies is a history not just of hardware, but of social contests about the meanings attached to the hardware and how it is going to be deployed in society. In Bijker's hands, social constructivism shows how different social groups (e.g., designers, regulators, users) negotiate the meaning and function of a technology. According to Bijker:

> "The sociocultural and political situation of a social group shapes its norms and values, which in turn influence the meaning given to an artifact" [*Bijker 1987, p. 46*].

However, Bijker also recognizes that one interpretative stratagem has been to describe technology as determining of social and political institutions. In an effort to account for this experience of consumer alienation in which technology is viewed as independent and out of human control when analysis reveals extensive dimensions of human construction, Bijker (in 1997) simply proposed that people with low inclusion in the construction process are often faced a "take-it-or-leave-it" choice when new technologies come on the scene. Technologies appear to them predetermined and autonomous because they had no part in the determination; but those with high inclusion in the design and construction process—engineers and trend-setting consumers—are quite conscious of their constitutive powers and thus do not feel so alienated.

In a complementary theoretical stance derived from the social sciences, Latour set forth an actor-network theory of technological change. Since the middle part of the twentieth century, social scientists had been complementing functionalist and structuralist analyses of society with descriptions of actor networks. The character of a social institution such as a church or school was fully explained neither by its social function nor its structural features. The network of people involved in the institution was often of complementary importance. In a similar manner, Latour argued in *Science in Action: How to Follow Scientists and Engineers through Society* (1987) that what is central to understanding technologies is to see the relations between the actors involved. For actor-network theory, the settlement of controversies between actors over the final design of a technological artifact is the result of an effective deployment of allies and resources behind the winning design.

The often appealed to search for efficiency never resolves any design issue by itself, because efficiency can mean multiple things in multiple contexts. A car that uses less gasoline but fails to attract consumers is not an efficient product for the corporation that manufactures it. Through the positioning and deployment of other actor and networks, engineers constantly (re)create society ("society in the making") as they struggle to legitimate their designs. In *The Golem at Large: What You Should Know about Technology* (1998), two colleagues of Latour apply this perspective to a series of case studies—from debates about Patriot missile defense and the Challenger Space Shuttle disaster, and questions concerning nuclear fuel flasks and the origins of oil to economics, Chernobyl, and AIDS treatments—noting in each situation how engineers worked in diverse ways to enroll various interests behind different interpretations of technologies. For social constructivists the message is always that things could have been otherwise, and the need to recognize differences in situations where determinists and others tend to see only similarities and uniformity.

One way to summarize the upshot of social constructivism is to describe it as an attempt to deconstruct the inherited stories of technology as a progressive conquest of nature, increased production of goods and services, or a semiautonomous force, in order to write a new story with multiple authors seeking coherent narratives. Sometimes a narrative strategy succeeds in enrolling others, sometimes not; but whether successful or not, the contributors keep writing. What exists in the history of technology for this complex interactionist model is not so much progress or regress as sideways slipping and sliding; yet one cannot help but wonder why the notion of determinism persists to be argued against, or why the authors of technologies continue to write.

Beyond Determinism and Constructivism

Three other theories about relationships between technology, society, and politics that call for mention are those involved with feminism, evolution, and economics. Feminist theories, as introduced in Judy Wacjman's *Feminism Confronts Technology* (1991), can be related to both determinist and social constructivist perspectives. Insofar as women are excluded from the technological design process, some feminists argue this creates for women a historically contingent determinism. But what is desirable are situations in which women and their perspectives are allowed to contribute to the social construction of technology.

The idea that technological change can be understood as an evolutionary process has exercised persistent appeal. The most extensive development of this perspective is found in John Ziman's edited volume, *Technological Innovation as an Evolutionary Process* (2000). Ziman and colleagues explore the possibilities for analyzing technological change in terms of the selective retention of invented variations in products and processes, although they note that with technology neither variation nor selection is as blind as in the organic world.

The Ziman research program is dependent on a distinction initially developed by economist Joseph Schumpeter between invention and innovation, technical creation, and its economic exploitation. Yet even more important in Schumpter than this trope is his critique of equilibrium economics in the name of disequilibrium economics, the latter of which is characterized by an internal commitment to technological change. An economy so intimately involved with technological change becomes characterized by repeated phases of "creative destruction": water power, textiles, iron (early 1800s), steam, railroads, steel (late 1800s), electricity, chemicals, internal combustion engine (early 1900s), petrochemicals, electronics, aviation (mid- to late 1900s), digital networks, software, new media (late 1900s and early 2000s). As Marx and Engels noted in *The Communist Manifesto* (1948), in such a world "everything that is solid melts into thin air." For later economists, however, the thin air had become itself a phenomenon to be analyzed by rational choice theory, as Jon Elster's influential *Explaining Technical Change* (1983).

Conclusion

The twentieth century witnessed major changes in technology as well as in social and political affairs. Examples range from the creation of technological power in both economic and military forms associated with the U.S. becoming the dominant world power, through the technocratic industrialization of countries such as Spain under Franco that may have set the stage for later democratic developments, to the information technology revolution in the Soviet Union that (according to interpreters such as President Ronald Reagan) contributed to its complete collapse between—not to mention the industrial and information technology transformations in Japan, China, and India that can be correlated with wide ranging sociopolitical transformations. By contrast, the failures of technological development in much of Africa and significant parts of South America appear to be aligned with differing degrees of social and political failure.

Looking back at attempts to comprehend the creative destructions of such scientific, technological, and social interactions, scholars during the first half of century appear more concerned with how societies became technological in the first place, how technology burst into history and was then passed from one society to another. Scholars during the last half of the century, living in the triumphal wake of a new order of artifice, aspire to a more internalist comprehension of existing processes. Their case studies focus on situations that are themselves already highly technological and undergoing continuous technological transformations; and whereas macro theorists were often animated by grand issues of social and political justice, micro constructivists localized such questions and then turned them over to applied ethicists. The hope was that by acting locally one might have global impact, in ethics and politics as much as in technology.

CARL MITCHAM AND JUAN LUCENA

Further Reading

Bijker, W.E. *Of Bicycles, Bakelites, and Bulbs: Toward a Theory of Sociotechnical Change*. MIT Press, Cambridge, MA, 1997.

Bijker, W. and Law, J., Eds. *Shaping Technology/Building Society: Studies in Sociotechnical Change*. MIT Press, Cambridge, MA, 1994.

Bijker, W., Hughes, T and Pinch, T. Eds. *The Social Construction of Technological Systems: New Directions in the Sociology and History of Technology*. MIT Press, Cambridge, MA, 1989.

Collins, H.M. and Pinch, T. *The Golem at Large: What You Should Know about Technology*. Cambridge University Press, New York, 1998.

Ellul, J. *La Technique ou l'Enjeu du Siecle*. A. Colin, Paris, 1954. Translated by Wilkinson, J. for the English version: *The Technological Society*. Knopf, New York, 1964.

Elster, J. *Explaining Technical Change: A Case Study in the Philosophy of Science*. Cambridge University Press, New York, 1983.

MacKenzie, D.A. and Wajcman, J., Eds. *The Social Shaping of Technology*, 2nd edn. Open University Press, Buckingham, 1999.

Mumford, L. *Technics and Civilization*. Harcourt Brace, New York, 1934.

Mumford, L. *The Myth of the Machine, vol. 1: Technics and Human Development; vol. 2: The Pentagon of Power*. Harcourt Brace Jovanovich, New York, 1967 and 1970.

Schumpter, J.A. *Capitalism, Socialism, and Democracy*, 6th edn. Unwin, London, 1987.

Smith, M.R and Marx, L., Eds. *Does Technology Drive History? The Dilemma of Technological Determinism*. MIT Press, Cambridge, MA, 1994. Twelve essays from a conference on the theme. Bruce Bimber's "Three Faces of Technological Determinism" provides an insightful analysis.

Wajcman, J. *Feminism Confronts Technology*. Pennsylvania State University Press, University Park, PA, 1991.

Winner, L. *Autonomous Technology: Technics-out-of-Control as a Theme in Political Theory*. MIT Press, Cambridge, MA, 1977.

Winner, L. *The Whale and the Reactor*. University of Chicago Press, Chicago, 1984.

Ziman, J., Ed. *Technological Innovation as an Evolutionary Process*. Cambridge University Press, New York, 2000.

Software Application Programs

At the beginning of the computer age around the late 1940s, inventors of the intelligent machine were not thinking about applications software, or any software other than that needed to run the bare machine to do mathematical calculating. It was only when Maurice Wilkes' young protégé David Williams crafted a tidy set of initial orders for the EDSAC, an early programmable digital computer, that users could string together standard subroutines to a program and have the execution jump between them. This was the beginning of software as we know it—something that runs on a machine other than an operating system to make it do anything desired. "Applications" are software other than system programs that run the actual hardware. Manufacturers always had this software, and as the 1950s progressed they would "bundle" applications with hardware to make expensive computers more attractive. Some programming departments were even placed in the marketing departments.

The manufacturers could not create an application for every need. Programming a computer was still an expensive and largely mysterious art. A group of International Business Machine (IBM) users, many of whom were competitors, gathered in a Los Angeles hotel room to form SHARE in 1955. This group, as indicated by its name, was a way of sharing expensive applications programs among its mostly aerospace members.

At first, most applications were created for the government, which could afford them. Business would make do with the applications included with the machines and they often had to change their business practices to match the programs. In the U.S., a large custom application was the semi-automatic ground environment, or SAGE, an air defense system. Making this application gobbled up 700 of the roughly 2000 programmers in the U.S. After it was completed, a company was formed from the remnants. This System Development Corporation kept making applications for the government, but it established a separate software industry. Computer manufacturers still threw in the software for free, but the principle of independent construction was made and the number of independent contractors and user groups therefore increased.

Also around the late 1950s, higher-level programming languages appeared. FORTRAN (FORmula TRANslator) and COBOL (Common Business Oriented Language) came into use and reduced the cost and variety of applications for government and civilians alike. Both were quickly adopted, and are alive today.

In the 1960s, there were a number of big projects using all levels of languages, such as air traffic control, the airline reservation system SABRE, and the space program. There was also a continuation of bundling. Toward the end of the decade, the government pressured computer manufacturers to unbundle software. This also made economic sense, as software was not getting cheap, but hardware was. IBM, the largest manufacturer, led the way by offering its CICS successful business product. Others soon followed suit.

After a few years, the various suppliers of application software realized that certain domains asked for the same collections of programs. They hit on the idea of supplying "packages" of relevant programs. This mirrored the sorts of routines supplied as part of bundling. This became a big industry, but custom programs were bigger. In 1970, $70 million of packages were sold, while $650 million of consulting was done.

All of these applications were primarily on mainframe or minicomputers (in other words, large ones), whereas today we are surrounded by microcomputers. The main reason for this is the development of a series of "killer apps" or highly successful and useful programs, that ran on personal computers. The first killer app was Visicalc, developed by Daniel Bricklin. This was a spreadsheet program that allowed managers to do calculations and reports on their desktop that formerly required an expensive machine and a programmer to obtain.

Along with the development of microcomputer hardware, an operating system needed to be developed to run the machine. The Microsoft Cooperation supplied the DOS (disk operating system) for the IBM PCs (personal computers), later the industry standard. Later, Microsoft developed a family of individual applications that ran both on DOS and on the Apple Computer Corporation's MacOS for the Macintosh series. Applications companies that had developed software for other operating systems ported (translated) their applications to DOS, or died. These included a word processor, spreadsheet, presentation software, and a simple database application. Some of these were combined to yield Microsoft Office, a package for microcomputers.

At the end of the twentieth century the application software with the highest sales was Microsoft Office, the database from Oracle, and the enterprise package SAP. Application software is simply what makes computers useful, and the computer revolution would not have existed without it.

See also **Software Engineering; Systems Programs**

JAMES E. TOMAYKO

Further Reading

Aspray, W. and Campbell-Kelly, M. *Computer: A History of the Information Machine*. Basic Books, New York, 1996.

Ceruzzi, P. *A History of Modern Computing*. MIT Press, Cambridge, MA, 1998.

Software Engineering

Software engineering aims to develop the programs that allow digital computers to do useful work in a systematic, disciplined manner that produces high-quality software on time and on budget. As computers have spread throughout industrialized societies, software has become a multibillion dollar industry. Both the users and developers of software depend a great deal on the effectiveness of the development process.

Software is a concept that didn't even pertain to the first electronic digital computers. They were "programmed" through switches and patch cables that physically altered the electrical pathways of the machine. It was not until the Manchester Mark I, the first operational stored-program electronic digital computer, was developed in 1948 at the University of Manchester in England that configuring the machine to solve a specific problem became a matter of software rather than hardware. Subsequently, instructions were stored in memory along with data.

In the early days of software development in the 1950s and 1960s, that memory was expensive and precious. This inspired programming tricks that could reduce memory usage as well as increase efficiency of program execution, although there was frequently a tradeoff. Programming was labor intensive, especially when done in the binary "machine language" native to the hardware or in "assembly language" that substituted alphabetic symbols for the binary machine codes. Systematic development was not a priority as programmers focused on the essentials of making programs small enough and fast enough. English-like high-level "procedural" programming languages such as FORTRAN (FORmula Translator, designed for scientific problems) and COBOL (Common Business Oriented Language) designed for business tasks) helped make software more understandable and programmers more productive, but even then the imperatives of size and speed often remained.

Those imperatives began to change in the 1960s. Computer hardware costs began steadily decreasing while software costs (and hardware capabilities) steadily increased. Software soon became the dominant cost component of an information system. Worse yet, software was also increasingly plagued by schedule slippages and quality issues including functional defects and poor usability. These problems were epitomized by IBM's OS/360 operating system, which was released in 1966 considerably late, seriously over budget, and full of flaws. The mounting problems prompted both a label—"the software crisis"—and a response, a conference on what was provocatively termed "software engineering," sponsored by the North Atlantic Treaty Organization (NATO) Scientific Committee at Garmisch, Germany in 1968.

Participants from academia, industry, and government, traced the software crisis to several key characteristics of software as well as a general lack of discipline among programmers. First, software of any significant size is highly complex, thereby straining human cognitive capacities. Second, software is easily changed, owing to the fact that it is notation rather than a physical artifact. Poorly conceived and implemented changes can degrade quality over time. Finally, software is used to solve problems and perform work in an incredibly wide variety of areas, which makes it difficult to generalize techniques and tools as well as the software itself so as to reuse it.

Several approaches were soon proposed to address these various problems. To impose more discipline on programmers and counterbalance software's malleability, development would be guided by a life-cycle model defining each stage of the process—requirements specification, design, coding and implementation, verification (of correct implementation of the specification, usually through testing) and validation (that the software meets the user's needs), and maintenance (correction, adaptation, and enhancement). Hierarchical decomposition of programs into functionally independent modules (stepwise refinement) that hid their implementation details from other modules (information hiding) represented systematic methods for coping with complexity. Modularity and information hiding ultimately found their fullest expression in object-oriented programming, in which data and operations are bundled into "objects" that model the problem elements and whose interactions are strictly controlled. Structured programming aimed to render programs more intellectually manageable and amenable to rigorous analysis through the exclusive use of well-understood constructs of sequence, iteration, and selection. Structured programming (coding) was soon accompanied by structured (requirements) analysis and structured design. Various programming languages such as Pascal in the early 1970s, and Ada, C, and C++ in the mid-1980s emerged to better support these techniques, as did various types of computer-aided software engineering (CASE) tools.

Many of the proposed techniques, though important and useful, were perceived as lacking a certain depth. It often seemed difficult to get more specific than principles of general problem solving. Things like hierarchical decomposition (essentially divide and conquer) and abstraction (the hiding of unnecessary detail), though vital to the development of software, contrasted with the underpinnings of established scientific and technical disciplines. This concern reflected the fact that attempts to establish a discipline for software development (just like other fields) were as much about social and professional status as about practical necessity. Furthermore, individuals with a mathematical or scientific background often had visions of a discipline very different from the ideas of those with an engineering or other type of background.

Nowhere was this tension more apparent than in arguments over formal verification of software, which sought to prove mathematically that a program satisfied its specified requirements. Formal verification was the most contentious manifestation of a broader formal methods movement that sought to apply mathematical notations and techniques, the ultimate in rigor, to virtually all aspects of software development. Those who took a more scientific or mathematically oriented view of software development argued that not only would formal verification solve software quality problems, it would make software development superior to other technical endeavors. Others strongly disputed both the feasibility and the usefulness of formal verification, given its laborious nature and dependence on correctly specified requirements.

The argument over formal verification (and formal methods generally) was emblematic of what Fred Brooks, Jr., manager of the OS/360 project, dubbed the "silver bullet syndrome." Brooks and others perceived a tendency among software technologists to seize upon a single technique or approach as the solution to the software crisis; but many of the same characteristics that engendered the software crisis in the first place also made a single, comprehensive solution unlikely.

At the end of the twentieth century, increasingly powerful software continued to be developed and used, despite the persistent problems that were once deemed a crisis. A greater appreciation for variety of technique can be seen in every phase of every software life-cycle model; and the professional status of software engineers was still debated. Software engineering's unique flavor reflected the mixtures of knowledge and practitioners that converged behind the creation of operational technical artifacts consisting only of notation.

See also **Computer Science; Software Application Programs; Systems Programs**

STUART S. SHAPIRO

Further Reading

Bergin, T. and Gibson, R., Eds. *History of Programming Languages, vol. 2.* Addison-Wesley/Longman, Boston, 1996. The product of a conference on the topic, with chapters covering significant modern programming languages.

Brooks, Jr., Frederick. No silver bullet: essence and accidents of software engineering. *Computer*, 20,4, 10–19, 1987. Insights on the nature and future of software engineering.

Brooks, Jr., Frederick. *The Mythical Man-Month: Essays on Software Engineering*, 2nd edn. Addison-Wesley Longman, Boston, 1995. Lessons from OS/360 and other reflections on software engineering.

Campbell-Kelly, M. Programming the EDSAC: early programming activity at the University of Cambridge. *IEEE Annals. Hist. Comp.*, 20, 4, 46–67, 1998. An analysis of the development of important early programming techniques.

Ceruzzi, P. Crossing the divide: architectural issues and the emergence of the stored program computer, 1935–1955. *IEEE Ann. Hist. Comp.*, 19, 1, 5–12, 1997. An analysis of the transition from programming in hardware to programming in software.

Ensmenger, N. The "question of professionalism" in the computer fields. *IEEE Ann. Hist. Comp.*, 23, 4, 56–74, 2001. An analysis of professionalization efforts in computing in the US in the 1950s and 1960s.

Pour, G., Griss, M. and Lutz, M. The push to make software engineering respectable. *Computer*, 33, 5, 35–43, 2000. An assessment of the professional state of software engineering at the turn of the century.

Sammet, J. *Programming Languages: History and Fundamentals*. Prentice Hall, Englewood Cliffs, NJ, 1969. A discussion of the origins and characteristics of early programming languages.

Shapiro, S. Boundaries and quandaries: establishing a professional context for IT. *Inf. Technol. People*, 7,1, 48–68, 1994. An analysis of professionalization efforts in computing in the U.S. and the U.K. focusing primarily on the 1970s and 1980s.

Shapiro, S. Splitting the difference: the historical necessity of synthesis in software engineering," *IEEE Ann. Hist. Comp.*, 19, 1, 20–54, 1997. An overview and analysis of the technical development of software engineering.

Solar Power Generation

The emergence of solar power generation is part of the overall movement toward renewable energy production. Interest in this type of energy production grew in the early 1970s with an increased public awareness of the negative impact of technological developments on the environment. The use of solar power, of course, was not new. Heat produced by the sun was used for all sorts of purposes from the early history of humankind. In the search for renewable energy sources, the direct use of the sun's heat has continued in the use of solar panels. In these panels, heat from the sun is absorbed by water flowing in pipes, and the hot water can then be used for heating purposes. In the twentieth century, two types of thermal solar energy systems developed: (1) active systems that used pumps or fans to transport the heat; and (2) passive systems that use natural heat transfer processes. In 1948 a school in Tucson, Arizona, with a passive solar energy system was built by Arthur Brown. In 1976 the Aspen-Pitkin County airport was opened as the first large commercial building in the U.S. that used a passive solar energy system for heating. However, the original idea of using passive solar energy goes back to ancient times. Archeologists have found houses with passive solar energy systems dating back to the fifth century AD (see Buildings, Designs for Energy Conservation).

What was new in the renewable energy trend of the twentieth century was the conversion of solar energy into electricity in order to replace the energy that was produced from fossil, or nonrenewable sources. The device that was developed to realize this conversion is the photovoltaic (PV) cell. This cell is based on the PV effect of a number of semiconducting materials, first discovered in selenium by Willoughby Smith in 1873. In 1877 William G. Adams and Richard E. Day discovered this effect in a selenium–platinum junction, and went on to build the first selenium solar cell. The effect was subsequently seen in a variety of other semiconducting materials such as germanium and silicon. In 1954 Bell Laboratories researchers demonstrated their first solar cells, primarily for space applications. The following year Western Electric sold the first licenses for producing silicon PV cells. Commercial production of PV cells started in the same year by Hoffman Electronics Semiconductor Division. In 1958 the satellite Vanguard I was the first to be powered by PV solar cells.

Silicon is still often used for producing PV cells. The most important types of silicon cells are monocrystalline (based on single crystals), polycrystalline (based on numerous grains of monocrystals), and amorphous silicon (no crystals but thin homogeneous layers). The monocrystalline cells have the highest efficiency, but the polycrystalline cells are cheaper. Amorphous cells are the cheapest and also the thinnest type, which has advantages when the cell is to be integrated into a device. Apart from silicon, cadmium telluride and copper indium diselenide are used for making thin film cells. For thick films, gallium arsenide is an alternative material for silicon. Production processes are different for different types. Crystalline cells are produced in wafers, and amorphous cells are made by depositing the silicon on a substrate (a steel or a glass sheet covered with a layer of tin oxide).

The earliest PV cells had efficiencies of just a few percent. For example, Hoffman Electronics first cells in commercial production had an efficiency of only 2 percent, and by 1957 this had increased to 8 percent. In the course of the second half of the twentieth century, considerable research and development were done to improve the efficiency of the PV cells and to reduce the price of PV electricity. This was not without success. The average efficiency of the monocrystalline and polycrystalline silicon cells increased from 11 percent in 1985 to 16 percent in 1995. The efficiency of the amorphous silicon cells increased from 5 to 10 percent in the same period. The price of all types of silicon cells dropped to less than half the 1985 price in these years. As a result, PV electricity can now be produced for $0.25 to 0.40 per kilowatt-hour (kWh), but this is still five times as much as electricity produced by burning coal and gas.

The structure of a PV cell is shown in Figure 12. There are several layers in the cell, and in the middle there are two semiconducting layers, one *n*-type (negative) and one *p*-type (positive). (See the entry on Semiconductors for further information on the functioning of *n*- and *p*-type semiconductors). When sunlight hits the cell, electrons in the semiconducting layers receive energy that makes them free to move. An electric field in the semiconducting layers forces them to move; and when a load is connected to the cell, an electric current can flow. This current is the PV electricity that is produced by the cell. There are two conducting layers on both sides of the pair of semiconducting layers for connecting the load. To protect the cell, there is a covering glass layer. An

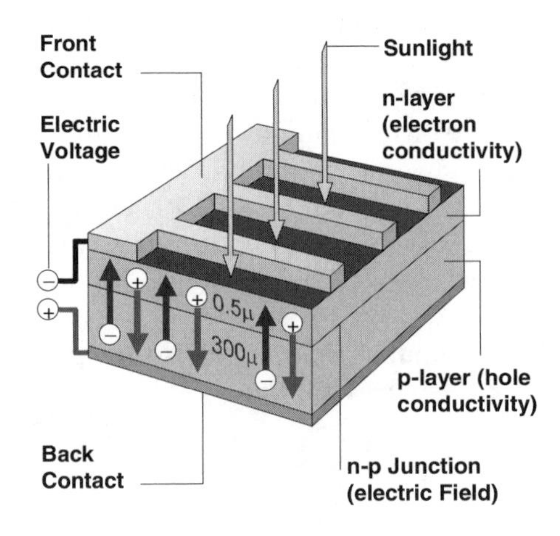

Figure 12. Diagram of a photovoltaic cell.
[*Source: From the EU Report 20015: Photovoltaics, 2001. Reprinted with permission from the Joint Research Center, Ispra, Italy.*]

antireflection layer prevents incoming sunlight from being reflected away from the semiconducting layers. The size of such cells varies from 1 to 10 centimeters in diameter.

Individual PV cells can only serve as an energy source for low-power applications. Several cells, however, can be connected and used together to form a module, also referred to as a "solar panel," not to be confused with the water-based solar panels mentioned above. The PV cells produce direct current (DC). As most electric appliances are based on the use of the electricity network with its alternating current (AC), PV systems usually have a converter that transforms DC into AC. The output of an average module is 12 volts, which is converted into the 110 or 230 volts that most electric appliances need. The power that is generated by an average module is around 50 to 80 watts. The average electricity demand of a household is 1.5 to 2 kilowatts, so it is common to have around 20 to 30 modules in a PV system for supplying electricity in a house. This requires an area of around 15 by 15 square meters.

The efficiency of a complete PV energy system not only depends on the quality of the cells in the panels but also on the extent to which the changing position of the sun can be taken into account. In a number of applications the position of the panels is fixed; for example, when they are integrated into the roof of a house; in which case the house can be oriented for the optimal use of sunlight. In other cases, when a tracking system is used, the position of the panels can change, as when solar panels are used in a satellite. Tracking systems can have one or more axes in order to capture the optimum amount of sunlight. Concentrators (Fresnel lenses with concentration ratios of 10 to 500 times, or mirrors) for focusing the sunlight allow the panel to use the available sunlight more effectively. Such concentrating systems started to be used in the late 1970s.

Two types of PV cell applications can be distinguished: stand-alone and grid-connected. In stand-alone applications, the PV system functions independently of the electricity network. This type of application is usually found where connection with the network is problematic because of large distances. As there is no backup energy source in this case, a battery has to be part of the PV system. Some examples of practical applications are: energy supply for villages in developing countries; energy supply for water pumping systems; lighting of beacons in the sea; and energy supply in satellites and electrically driven boats and cars. In the case of a grid-connected system, there is an exchange of energy between the PV system and the grid. In case there is a lack of energy in the PV system (at night or dark sky days), energy can be retrieved from the electricity network. When there is a surplus of energy in the PV system, however, this surplus can be sold to the energy company by feeding it into the grid. Figure 13 is a schematic drawing of a grid-connected PV panel application in a house.

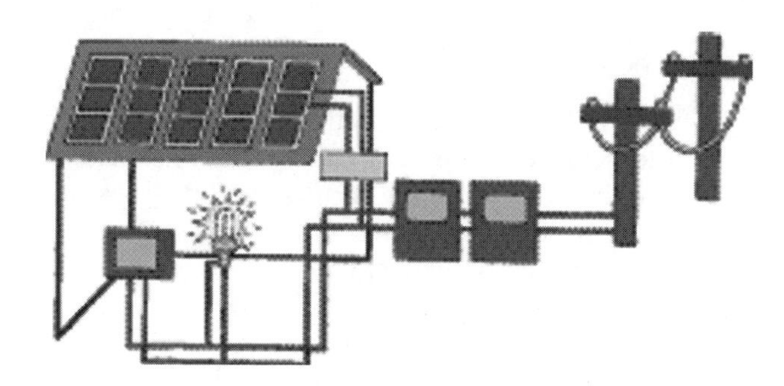

Figure 13. Grid-connected photovoltaic panel application in a house. Between the grid and the house are two meters for incoming and for delivered electricity. The converter is shown under the roof panels (right), and two electricity-using devices are shown in the house (left).

The total amount of solar energy as a contribution to the total energy production in most countries at the end of the twentieth century is still relatively small, even though it has been calculated that the sales of PV systems have increased from $2 million in 1975 to more than $750 million in 1993. For the year 1996 it was estimated that of the off-grid residential PV systems operational worldwide, about 10,000 were in remote vacation homes in Scandinavia. Among the reasons for the relatively low numbers is that the price of solar cells and the often-needed batteries is still too high to make a PV system economically competitive with nonrenewable (fossil fuel) energy production. A second, but less important, reason is that it is not yet clear if some types of solar cells really have better environmental properties over the whole life-cycle of the systems.

See also **Buildings: Designs for Energy Conservation; Electricity Generation and the Environment; Photosensitive Detectors; Semiconductors; Technology, Society, and the Environment**

MARC J. DE VRIES

Further Reading

Markvart, T., Ed. *Solar Electricity*. Wiley, Chichester, 2000.

Partain, L.D., Ed. *Solar Cells and their Applications*. Wiley-Interscience, Chichester, 1995.

Zweibel, K. *Harnessing Solar Power: The Photovoltaics Challenge*. Plenum Press, New York, 1990.

Solvents

Solvents are the hidden element in a broad range of technological activities, including chemical processes, paint, dry cleaning, and metal degreasing. They are used to dissolve organic compounds (water is the usual solvent for inorganic compounds) to enable reactions or polymerization, spreading or ease of use, or to extract compounds from a matrix such as plant material. Dry cleaning is a specialized form of the last group, as it removes fatty substances that adhere dirt to clothing. Many organic compounds either react with or do not dissolve in water, hence the need to find suitable organic chemicals to act as the solvent. It is hard to find compounds that dissolve a wide range of substances, but are also relatively cheap, nonflammable and nontoxic. In practice, the solvents used represent a compromise, often an unsatisfactory one.

The first solvent to be readily available was ethanol (ethyl alcohol) made by fermentation and distillation since the late Middle Ages. Amyl alcohol (fusel oil), a byproduct of ethanol manufacture, also became a popular solvent. Oil of turpentine, made from pine resin, became an important solvent in the eighteenth century and was used in paint and varnishes, and to dissolve rubber. It was also the basis of the earliest form of dry cleaning, which started around 1825. Wood spirit, made by dry distilling wood, first appeared in the early nineteenth century, but its purification into methanol (wood alcohol, methyl alcohol), acetone, and methyl ethyl ketone only took place in the middle of that century. As late as 1914, acetone and methanol were the only significant solvents in the U.S. (by volume), which had access to extensive virgin woodland. Synthetic methanol, made by treating carbon monoxide with hydrogen under pressure, was first made by the German chemical firm BASF in 1923 and eventually replaced the natural product.

Carbon disulfide, made by heating sulfur with charcoal, was the first synthetic solvent. From the 1840s onwards, it was used to dissolve oils, fats, waxes, and rubber. At the end of the nineteenth century, it became important as a major component of viscose rayon manufacture. Despite its value as a solvent, commercial carbon disulfide smells of rotten eggs and is also toxic.

Around 1820, the Scottish chemist Charles Macintosh replaced oil of turpentine by coal-tar naphtha (obtained from the growing coal-gas industry) as a rubber solvent and successfully used this solution to make the waterproof overcoats that bear his name. The real breakthrough, however, was the separation of benzene from coal-tar, pioneered by Charles B. Mansfield in the late 1840s. Benzene is an excellent solvent and available in large quantities from coal-tar. Although Mansfield died from burns caused by his distillation apparatus in 1855, benzene's flammability was tolerated and its toxicity was not considered to be a problem until the 1960s. Somewhat later in the nineteenth century, petroleum naphtha became available and competed with benzene in the dry-cleaning market. The less flammable "white spirits" fraction of petroleum became popular in the paint industry as a substitute for oil of turpentine (hence its popular if misleading name "turps") after 1900. In the mid-1920s, a variant of white spirits, Stoddard solvent (named after W.J. Stoddard, president of the U.S. National Institute of Dry Cleaning in 1928), was introduced as a less flammable version of petroleum naphtha, particularly in the dry-cleaning industry. It is still used in dry cleaning in the U.S. and Australia.

Stoddard solvent competed with a wholly different class of solvents—the chlorinated hydrocarbons. The first widely used member of the group, chloroform, was discovered in 1831. It was made cheaply from ethanol and chlorine, but its role as a solvent was soon overshadowed by the discovery of its anesthetic properties in 1847. The German chemist Hermann Kolbe made carbon tetrachloride in 1843 by reacting carbon disulfide with chlorine. First produced commercially by Chemische Fabrik Rheinau of Mannheim in 1892, it has much the same solvent properties as carbon disulfide and was also used in dry cleaning, especially for removing stains. As well as being toxic, it is also rather unstable, producing corrosive hydrochloric acid. However, the chlorinated solvents industry really took off in the early 1900s, when cheap chlorine became available as a byproduct of electrolytic caustic soda production. By 1913, the Bavarian firm Alexander Wacker, and its British associate Weston Chemicals (partly owned by the Castner-Kellner Company, which became part of ICI), were already marketing tetrachloroethane, trichloroethylene and perchloroethylene. Tetrachloroethane ("tetra" or Westron) was taken up very quickly by the dry-cleaning industry as it was an excellent solvent for fats. It was soon replaced by trichloroethylene ("tri" or Westrosol), which is more resistant to hydrolysis and less toxic. Trichloroethylene also found an important niche in the 1920s as a degreasing agent for metals. Perchloroethylene slowly became the most important dry-cleaning solvent, having beaten off stiff competition from the chlorofluorocarbons R-11 (trichlorofluoroethane) and R-113 (trichlorotrifluoroethane) in the 1960s and 1970s, but which are now banned due to their ozone-destroying properties.

The rapid growth of car production after 1900 led to a great increase in the use of nitrocellulose-based lacquers. Initially the solvents used were based on amyl alcohol, which was expensive and in short supply. Auguste Fernbach and Chaim Weizmann developed a fermentation process for the production of butanol and acetone with the aim of making synthetic rubber. Initially, the attempt to make synthetic rubber having failed, it was the acetone that was the more valuable component, as it was used to make explosives and canvas lacquer ("dope") used to smooth and waterproof aircraft canvas frames in the World War I. The Commercial Solvents Corporation in the U.S. operated the process in the 1920s to obtain butanol for the car lacquer market. In Germany, BASF developed a route to butanol from coal-based acetylene. Ethyl acetate, produced by reacting acetic acid with ethanol, also grew in importance in the 1920s although it had been available since the mid-nineteenth century.

The development of the petrochemical industry in the 1920s and 1930s, mainly in the U.S., led to the introduction of new solvents and new routes to already established solvents such as acetone. Alcohols such as ethanol, isopropanol and isobutanol could be made by treating refinery gases with sulfuric acid. The key intermediate, ethylene oxide (initially made by treating ethylene with chlorine water and more recently by direct oxidation of ethylene), could be converted into ethylene glycol (which was a solvent for dyes and inks as well as an antifreeze) and diethylene glycol. Several exotic solvents appeared in the 1960s. The oldest member of this group, tetrahydrofuran (THF), is an important solvent for plastics. It is made from butanediol using Reppe chemistry but it can also be derived from maize, thus is capable of being a renewable chemical (see Green Chemistry). Dimethylformamide (DMF) is a powerful solvent that can be used for inorganic salts as well as plastics and pigments. Dimethyl sulfoxide (DMSO) is extremely efficient at extracting substances, and is used as a paint remover and a medium for reactions.

By the 1950s concerns were growing about the safety of several important solvents, notably benzene, carbon tetrachloride and carbon disulfide. The British factory inspector Ethel Browning was a pioneer in this field, publishing *Toxic Solvents* in 1953. Other halogenated solvents were later implicated in destruction of the ozone layer. At first, the harmful solvents were replaced by safer alternatives (benzene by toluene, carbon tetrachloride by trichloroethane and then trichloroethylene), but efforts are now focused on the replacement of organic solvents altogether. This can be achieved by using water-based emulsions (a method used in polymerization processes since the 1920s) and wherever possible, eliminating the solvent altogether ("green" chemistry). The use of ethylene dibromide (which is both carcinogenic and ozone-destroying) to extract caffeine from coffee beans has been largely displaced by a new technique, supercritical fluid extraction, which uses supercritical liquid carbon dioxide under pressure. This technique has also been adapted for dry cleaning, but the equipment is expensive.

See also **Cleaning, Chemicals and Vacuum Cleaners; Coatings, Pigments, and Paints; Green Chemistry**

PETER MORRIS

Further Reading

Cammidge, E.M. Benzene and turpentine: the pre-history of drycleaning. *Ambix*, 31, 79–84, 1991.
Durrans, T.H. *Solvents*, 6th edn. Chapman and Hall, London, 1950.
Wacker Chemie. *Im Wandel Gewaschsen: Der Weg der Wacker-Chemie, 1914–1964*. Verlag für Wirtschaftspublizistik, Wiesbaden, 1964.

Sonar

The word "sonar" originated in the U.S. Navy during World War II as an acronym for "SOund NAvigation and Ranging," which referred to the systematic use of sound waves, transmitted or reflected, to determine water depths as well as to detect and locate submerged objects. Until it adopted that term in 1963, the British Admiralty had used "ASDIC," an abbreviation for the Anti-Submarine Detection Investigation Committee that led the effort among British, French, and American scientists during World War I to locate submarines and icebergs using acoustic echoes. American shipbuilder Lewis Nixon invented the first sonar-type device in 1906. Physicist Karl Alexander Behm in Kiel, Germany, disturbed by the *Titanic* disaster of April 1912, invented an echo depth sounder for iceberg detection in July 1913. Although developed and improved primarily for military purposes in World War I, sonar devices became useful in such fields as oceanography and medical practice (e.g., ultrasound).

The earliest sonar devices used in military operations were called "passive" systems because they did not transmit signals. One such instrument was the hydrophone, essentially a submerged microphone hanging off the side of a ship. Developed in 1915 by French physicist Paul Langévin and Constantin Chilowsky, a Russian living in Switzerland, the hydrophone used the piezoelectric properties of quartz for powerful ultrasonic echosounding to detect submarines. The hydrophone's transducer consisted of a mosaic of thin quartz crystals glued between two steel plates with a resonant frequency of 150 kilohertz. On April 23 1916, the German submarine UC-3 became the first ship confirmed sunk after detection by hydrophone. Hydrophone efficiency improved significantly when engineers realized a pair of highly directional microphones, separated by a few feet on a connecting bar that could be rotated at the center, yielded a more accurate bearing on the sound source. Langévin's invention formed the basis of the development of naval pulse-echo sonar.

By 1918, both Britain and the U.S. had produced "active" systems that transmitted and received sound waves. Those systems owed much to the ingenuity of individuals seeking to improve iceberg detection. English meteorologist Lewis Richardson filed the first patent for an underwater, echoranging design at the British Patent Office shortly after the sinking of the *Titanic*. Reginald Fessenden, a Canadian living in the U.S., developed the workable "active" sonar, the Fathometer, in 1914. His system, which employed an electromagnetic moving-coil oscillator that emitted a low-frequency signal, and then switched to a receiver to listen for echoes, detected an iceberg underwater at a distance of more than 3 kilometers.

Although the development of sonar came too late to have any significant effect on the course of World War I, its improvement and widespread shipboard installation during the interwar years profoundly influenced naval operations during World War II. In 1921, a sonar system installed aboard the *HMS Antrim* could detect a shutdown submarine lying on the ocean bottom at a distance of 1800 meters. The following year, the U.S. Navy employed on the survey ship *USS Stewart* an echosounder designed by Harvey Hayes. Many French ocean liners, beginning with the *Ile de France* in 1928, carried Langévin's echosounding devices. Meanwhile, Rudolf Kühnhold at the German navy's Nachrichtenmittel-Versuchsanstalt (NVA) in Kiel began devising a sum-difference method of sound location for directing gunfire at surface or underwater targets and, by the early 1930s, developed a close working relationship with Tonographie company founders Paul-Günther Erbslöh and Hans-Karl Freiherr von Willisen. They subsequently created a spin-off firm, GEMA, to produce the sophisticated electronic sounding equipment needed by the German military. In 1931, the U.S. Navy Underwater Sound Group produced the "QB" echoranging sonar, which was effective below 6 knots, and the first installation of echoranging equipment on American destroyers occurred in 1934. By the beginning of World War II, over 200 British warships carried sonar equipment that was effective up to 15 knots. Although sonar was responsible for approximately 60 percent of all submarine kills during the first two years of World War II, it proved to be an ineffective tool for sweeping broad areas.

Sonar improvements during World War II enabled ever more precise detection, tracking, and targeting of ever more capable, deeper-diving submarines. In 1942, the "Q" attachment permitted tracking of U-boats at closer ranges than the earlier "searchlight" sonar allowed. Before the end of 1943, introduction of the Type-147 or "Sword" system added the capability to track deep-diving U-boats laterally and thereby to make last-second targeting adjustments. Meanwhile, American submarines employed the bathythermograph to detect thermoclines, layers of water where the temperature gradient is greater than that of the warmer layer above and the colder layer below, beneath which they could escape sonar detection by German or Japanese surface ships. Sonar also became useful in guiding depth charges and torpedoes to their targets, as well as in detecting subsurface mines.

The Cold War that characterized the latter half of the twentieth century brought advances in sonar technology. Cognizant of the success of German U-boats during World War II and concerned about Soviet expansionism, the U.S. Navy relied for many years on the sound surveillance system (SOSUS), an essentially passive, worldwide network of underwater microphones, to detect enemy submarines. Helicopters, specially equipped with both passive and active sonar gear, also became important platforms in antisubmarine warfare.

During the 1980s, the U.S. and other naval powers began experimenting with active, long-range sonar systems that used high-intensity, low-frequency sound waves. Approximately 39 ships participated in the low-frequency active (LFA) sonar testing, which proved an LFA-equipped vessel could detect an enemy ship hundreds of kilometers distant. In March 1995, the U.S. Navy also began testing aboard the *USS Asheville* a high-frequency system that enabled the submarine to navigate more effectively in shallow water. All the high-intensity, active sonar systems drew protests, petitions for action, and lawsuits from conservation and animal welfare organizations, because they caused widespread harm to marine mammals and other ocean life. In October 2003, the U.S. Navy agreed to restrict peacetime use of its surveillance towed array sensor system (SURTASS) LFA system, leaving environmentalists to seek similar limitations on mid- and high-frequency sonar.

See also **Submarines, Military; Submersibles; Ultrasonography in Medicine**

RICK W. STURDEVANT

Further Reading

Gannon, R. *Hellions of the Deep: The Development of American Torpedoes in World War II.* Pennsylvania State University Press, University Park, PA, 1996.

General Accounting Office. Report to the Acting Secretary of the Navy; Undersea Surveillance: Navy Continues to Build Ships Designed for Soviet Threat (GAO/NSIAD-93–53). General Accounting Office, Washington D.C., 1992.

Gerken, L. *ASW Versus Submarine Technology Battle.* American Scientific Corporation, Chula Vista, California, 1987.

Hill, J.R. *Anti-Submarine Warfare.* Naval Institute Press, Annapolis, MD, 1993.

von Kroge, H. *GEMA: Birthplace of German Radar and Sonar.* Institute of Physics Publishing, Bristol, 2000.

Urick, R.J. *Principles of Underwater Sound.* McGraw-Hill, New York, 1983.

Useful Websites

Bellis, M. *The History of Sonar*: http://inventors.about.com/library/inventors/blsonar.htm

Department of the Navy, U.S. *SURTASS LFA*: http://www.surtass-lfa-eis.com/index.htm

Derencin, R. *Counter Sonar Measures—A Bathythermograph*: http://uboat.net/articles/?article=45, http://uboat.net/articles/?article=45

Derencin, R. *Underwater Sound Detector in WW1 and WW2*: http://uboat.net/articles/index.html?article=52

Federation of American Scientists. *Introduction to SONAR*: http://www.fas.org/man/dod-101/navy/docs/es310/uw_acous/uw_acous.htm

IUSS-CAESAR Alumni Association. *The Cable.* http://www.iusscaa.org/ide.htm

McDaniel, J.T. *Hydrophones, Sonar, and Other Listening Gear*: http://www.authorsden.com/externalsite.asp?authorID=2136&destURL=www%2Efleetsubmarine%2Ecom&msg=

Moreavek, L.T., Leonard, J.G. and Brudner, T.J. *USS Asheville Leads the Way in High Frequency Sonar*: http://www.fas.org/man/dod-101/sys/ship/docs/uss_asheville.htm

Proc, J. *Asdic, Radar and IFF Systems Aboard HMCS Haida*: http://webhome.idirect.com/~jproc/sari/sarintro.html

Tanaka, T. *et al. Low Frequency Synthetic Aperture Sonar*: http://www.nec.co.jp/techrep/en/r_and_d/r03/r03-no2/rd04.pdf

USS Francis M. Robinson Association. *Sonar Equipment*: http://www.de220.com/Electronics/Sonar/Sonar.htm

Space

Space technology has a unique place in the history of twentieth century technology, since it concerns the application of technology beyond the confines of the earth. In engineering terms, this has involved communications (since signal travel time may be measured in minutes), ruggedness (since spacecraft must survive a hostile thermal and radiation environment), and reliability (since spacecraft beyond low earth orbit cannot be retrieved or repaired).

Space technology is also unusual because it is both a user of other technologies—electrical, electronic, mechanical, power, computing, and telecommunications—and a developer of these technologies. The challenge of developing systems that operate reliably in the space environment and hardware that is sufficiently light and compact to be launched in the first place is so great that developments in space technology often led to improvements in earthbound systems. One example is the microminiaturization of electronic components required for early spacecraft, which drove the development of electronics and computer systems of the late 1950s and early 1960s.

The degree to which this "spin-off" occurred is difficult to prove, since it is impossible to remove the influence of space exploration and development from the history of the twentieth century. However, the subject has proved itself to be important in at least two main ways. First, space is important as a place to go: as the mountains and the oceans have to past explorers, it provides a goal and a challenge with which to satiate mankind's inherent desire to explore (summarized for many in the phrase "the final frontier"). Second, space is important as a platform from which to provide services: from satellite communications and GPS navigation, to remote sensing of the earth and the cosmos.

Here, we look at the development of space technology in the twentieth century, from its roots in science fiction to its industrialization and later commercialization. We discuss the impact of its various applications on society and its influence on culture in the second half of the century. Although these factors have varied in degree from nation to nation, space technology has had a global impact; indeed it has been a key factor in the development of "globalization" itself.

Fictional Roots

It seems likely that space travel has been part of human imagination since mankind realized that space was "somewhere to go." As a planet, earth is unusual in that it has a moon that appears relatively large in its sky, and even with the naked eye, considerable detail is visible on its surface. This alone must have evoked speculation as to the possibilities of traveling there, even before the invention of the telescope.

Although thirteenth century China is credited with the invention of the rocket, Cyrano de Bergerac is believed to have been the first to propose that rockets should be used as a form of space propulsion. In his novel *The Comic History of the States and Empires of the Moon and the Sun*, published in 1649, he imagined being raised aloft by rockets attached to a flying machine.

Jules Verne, however, is credited with writing the first fictional account of spaceflight based on scientific fact: *From the Earth to the Moon*, which appeared in 1865. Since the bullet was by far the fastest thing in Victorian life, his fictional spacecraft was a bullet-shaped projectile propelled by a cannon. Verne's first astronaut crew would have been pulverized by the acceleration, but at least he had recognized the need for speed to escape the earth's gravity as described by Newton in the seventeenth century.

Once the technology of "moving pictures" had been developed, it was inevitable that it should be used to portray what we now call science fiction. Indeed, a number of science fiction "shorts," of only a few minutes in length, were made prior to 1900. However, the first feature-length (21-minute) science fiction film was *Le Voyage dans la Lune* (A Trip to the Moon), made by George Méliès in 1902; it was inspired by Verne's *From the Earth to the Moon* and H.G. Wells' *The First Men in the Moon*. Thus began a long line of space-related films which, for some, culminated in the 1968 classic *2001: A Space Odyssey*, released the year before Neil Armstrong became the first man to set foot on the moon.

Military Funding

The early pioneers such as Konstantin Tsiolkovsky and Robert Goddard (see entry on Rocket Propulsion, Liquid Propellant) received little public support or government funding for their early work, and in common with many technologies, the development of the rocket was driven by the needs of warfare. In 1932 Wernher von Braun was employed by the German army's rocket artillery unit to develop ballistic missiles. In World War II, the result was the V-2, which later became the basis for both Russian and American rocket programs. Indeed, it was Wernher von Braun who became the leading proponent of, first, the U.S. Army's and later, the National Aeronautics and Space Administration's (NASA's) space exploration programs of the 1950s and 1960s, culminating in his work on the Saturn-V moon rocket (see Figure 14).

Even today, it is difficult to reach a consensus on so-called "dual-use technologies," such as rockets and remote sensing satellites: the former can be used as spacecraft and weapon delivery systems and the latter can provide high-resolution images for both civil and military applications. Although the same dichotomy exists in the use of aero-engines, terrestrial surveillance systems and computer technology, a satellite-targeted, global positioning system- (GPS-) guided, nuclear-tipped intercontinental ballistic missile (ICBM) has far greater destructive potential.

Politics and the Space Age

It is undeniable that the early developments of the space age, which began with the launch of Sputnik-1 on 4 October 1957, were driven as much by politics as they were by science or engineering. Both the USSR and the U.S. were technically capable of launching the first satellite in the late 1950s. The reason the Soviet Union won what became known as the "space race" boils down to two complementary factors: a desire on the part of the Soviet leadership to prove superiority in at least one high-technology field; and an incapacity on the part of the American leadership to believe that the Soviet Union was capable of such a feat (despite the fact that the launch of Sputnik-1 had been trailed several months before it occurred).

Indeed, it was the American public (led no doubt by the media) that was most outraged, and worried, by the Soviet triumph. They believed that if the USSR could place a satellite into orbit, it could just as easily target a nuclear warhead on an American city. These feelings of vulnerability and inferiority, fuelled by further Soviet "space firsts"—including the first moon probes, first spacecraft to Venus, first animal in space, and first man in space—led to President Kennedy's

Figure 14. The Apollo-17 astronauts (Harrison Schmitt, Ron Evans and Gene Cernan) pose in their lunar roving vehicle (LRV) in front of the Saturn-V rocket that took them to the moon. Apollo-17, launched in December 1972, was the final mission in the Apollo program and the third to carry an LRV to the moon.
[*Source: Courtesy of NASA.*]

historic decision to commit his nation to landing a man on the moon before the end of the 1960s.

This political decision, and the realization of its goal in July 1969, provided a spur to the development of space technology that has never been seen since. It is significant not only in the scope of the history of technology, but of history itself, that only eight years after Yuri Gagarin became the first human being to orbit the earth and less than 12 years after Sputnik-1, two men were standing on the surface of the moon (Figure 15).

It has often been suggested since the end of the Apollo program that a similar venture, if attempted today, could not achieve the same result in the same timescale. This is probably true because no one can envisage reproducing the level of political and financial support that existed in the 1960s.

Figure 15. Edwin "Buzz" Aldrin, the second man to set foot on the moon in July 1969. Although often captioned incorrectly as a picture of Neil Armstrong, this photograph was taken by Armstrong, who appears reflected in the visor of Aldrin's helmet.
[*Source: Courtesy of NASA.*]

Industrialization

While space was a mainly scientific endeavor in the late 1950s, typified by investigations of radiation belts and solar particles, in the 1960s it became industrialized. Although government laboratories and rocket test ranges operated by the military could handle the requirements of the early space age, the level of organization required to send manned spacecraft to the moon was something else. Indeed, apart from any technological spin-offs, the NASA-led Apollo program is credited with much of the development of modern management practice. Never before had it been necessary to coordinate so many different industrial contractors, in so many states, to produce so many leading-edge technologies on such a short timescale. When Kennedy made his Apollo commitment in May 1961, America had just 15 minutes and 22 seconds of experience in manned spaceflight, accrued on just one suborbital hop, which reached an altitude of 187 kilometers. The moon, by contrast, was about 375,000 kilometers away and a successful lunar mission would last at least eight days.

The developments required to meet Kennedy's challenge were undertaken largely by the nascent space industry, an outgrowth from the existing aerospace industry. Even after the glory days of Apollo, companies appeared proud of their space divisions. Especially for defense contractors, which were obliged to keep much of their work secret, the civilian space industry offered a "shop window" for their capabilities. Likewise, companies that were known mainly for earthbound products, such as cars and refrigerators, used their involvement in space to indicate their advanced, high-tech status. For example, it is not widely known outside the space industry that Ford, renowned for its motor vehicles, was one of America's top three satellite manufacturers until 1990, when its space division was bought by the Loral Corporation. And Bendix, which among other things manufactured washing machines, was the losing bidder (to Boeing) for the Apollo lunar roving vehicle. In the 1960s and 1970s, corporations of all different shapes and sizes wanted a piece of the action in space.

As enthusiasm for manned spaceflight waned in the U.S. (no American astronauts were launched between the Apollo–Soyuz docking mission in 1975 and the first Space Shuttle launch in 1981), the space industry polarized its efforts in two main directions: military and civilian. Already familiar with the needs of military agencies, contractors continued to develop military communications and reconnaissance satellites, while their civil divisions concentrated on similar satellites for civil government and commercial customers. Still other divisions continued to develop launch vehicles in a type of symbiotic relationship: as satellites grew larger and more complex, they needed more capable launch vehicles; and as launch vehicles grew more capable, it became possible to specify heavier satellites.

America was the undisputed leader in most aspects of space technology in the 1960s, but it was challenged later by other nations as they too developed a space industry. For example, although Russia operated under an entirely different political and industrial system, it took the lead in manned spaceflight in the 1970s and 1980s with its deployment of the Salyut and Mir space stations. Meanwhile, Europe's introduction of the Ariane launch vehicle in 1979 led to its domination of the commercial launch industry in the late 1980s and 1990s.

Several other nations developed their own space industries, predominantly based on government as opposed to commercial programs. Japan, for example, has an industry founded on science applications that has benefited from U.S. technology transfer programs for both satellites and launchers. Many of its satellites have been built jointly with U.S. contractors and its early launch vehicles were effectively American vehicles built under license by domestic contractors. By the end of the century, this had placed Japan in a position to embark on the commercialization of its satellite industry and its H-IIA launch vehicle.

India, too, is a nation that recognized the advantages of space systems. For example, the unique advantage of the communications satellite in geostationary orbit is that it can provide coverage to an entire nation, interconnecting villages that will probably never be linked by terrestrial telecommunications cables. At the same time, satellites can provide television signals to an entire nation, an advantage that has been used by India to provide education services and health advice, as well as entertainment, to millions of people who would otherwise remain deprived. By the same token, remote sensing satellites, which provide wide geographical coverage on a repeatable basis, are used for weather forecasting, flood monitoring, crop-health monitoring and many other applications in support of an ever-growing population. Despite its status as a less developed country with a large underprivileged and undereducated population, India made a political decision to develop both satellites and launch vehicles manufactured by a domestic space industry.

Global and Cultural Impact

The development of space technology has had an undeniable impact on our understanding and appreciation of the earth, largely because it allowed us to take a global view. For example, the first pictures showing earth as a planet in space were taken from spacecraft in lunar orbit, the most famous being the "earthrise" sequence from Apollo-8 (see Figure 16). It is often suggested that these images provided the inspiration for the green movement and kick-started a new phase of global awareness; certainly they have been widely used in the print media ever since.

The communications satellite has, arguably, done as much as any other technology to make the "global village" a reality. Thanks to the satellite, it is now possible to communicate from any position on the earth's surface to any other; uplink news reports from anywhere on earth using equipment the size of a briefcase; and ascertain one's location anywhere on earth using the global position system (GPS). The satellite has also provided a tool for democracy, again because of its global coverage and its insensitivity to political boundaries. It is for this reason that, as the twentieth century closed, individual satellite receiving antennas remained illegal in China.

Apart from its economic, social, and political influences, space technology has had a widespread impact on our increasingly global culture. The evidence that space is part of our culture is there for all to see—in books, newspapers, films, and on television. Incredible though it may seem, after more than 40 years of the space age, space technology is still used in advertising to suggest technical advancement and high reliability, while children's toys still reflect a fascination with space, especially at primary level, where space is reckoned to be second only to dinosaurs in the interest ratings.

On the broadest level, space technology has enhanced our understanding of the "cosmos" and mankind's place in it, an understanding that has both intellectual and religious ramifications. We have absolute proof that the earth is neither flat nor the center of creation, and we know far more about the objects, processes, and scale of the universe than we would without space-based astronomy. From a psychological standpoint, we have proved that mankind is not necessarily limited to living on a single planet, and the development of space technology has shown that "the final frontier" is simply another boundary to cross. Perhaps the ultimate space development is the one that will enable tourists, as opposed to career astronauts and cosmonauts, to cross that boundary. The first step towards this goal was made in 2001, when Dennis Tito paid the Russian authorities a reported US$20 million for a one-week visit to the International Space Station.

As with any technology, space technology can be applied in many ways. It is possible that its future development will allow mankind to conduct wars in space and ultimately destroy the earth. It is equally possible that space technology may one day save mankind from a devastating asteroid impact, a concept that has evoked serious consideration from both film makers and government officials (Figure 17). Certainly, by the time the sun reaches its red giant phase and expands to encompass the earth, only space technology will have offered salvation.

Figure 16. "Earthrise" from lunar orbit, as photographed in December 1968 during the Apollo-8 mission. This was the first time anyone from earth had seen their home planet from the moon. [*Source: Courtesy of NASA.*]

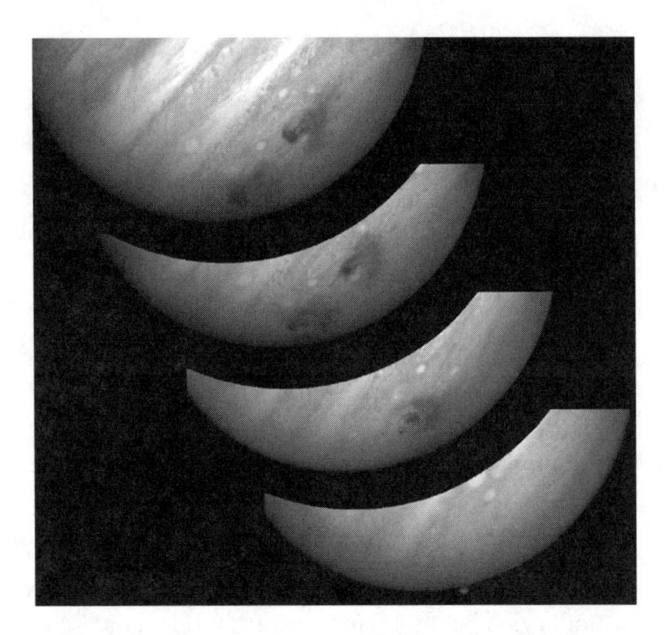

Figure 17. The impact of comet Shoemaker-Levy on Jupiter in 1994, imaged by the Hubble space telescope, showed that earth too is at risk; the circular marking on Jupiter's clouds is approximately the same size as the earth. Space technology is the only technology that offers options to either destroy or deflect such an object.
[*Source: Courtesy of NASA.*]

In the nearer term, and in an historical context, it is relevant to consider what the twentieth century will be remembered for in, say, another thousand years. Unless the planet has suffered another "dark age" and all records have been lost, it will be remembered as the century in which mankind harnessed the techniques of space technology and learned to live and work in space. As Konstantin Tsiolkovsky once said, "earth is the cradle of mankind, but one cannot remain in the cradle forever."

See also **Rocket Propulsion, Liquid Propellant; Satellites, Communications; Space Exploration, Planetary Landers; Space Exploration: Moon, Manned; Space Shuttle; Space Stations, Mir**

MARK WILLIAMSON

Further Reading

Allday, J. *Apollo in Perspective: Spaceflight Then and Now*. Institute of Physics, Bristol, 2000.

Allward, M., Ed. *The Encyclopedia of Space*. Hamlyn Publishing Group, London, 1968–1969.

Bilstein, R.E. *Orders of Magnitude: A History of the NACA and NASA, 1915–1990*. NASA, Washington D.C., 1989.

Canby, C. *A History of Rockets and Space*. Leisure Arts, London, 1964.

Cooper, H.S. F. *Moonwreck—13: The Flight That Failed*. Granada/Panther, St. Albans, 1975.

Editors of Fortune. *The Space Industry: America's Newest Giant*. Prentice Hall, Englewood Cliffs, NJ, 1962.

Furniss, T. *Manned Spaceflight Log*. Jane's, London, 1983.

Gatland, K. *The Illustrated Encyclopedia of Space Technology*. Salamander Books, London, 1981.

Godwin, R. *Rocket and Space Corporation Energia*. Apogee Books, Burlington, 2001.

Hardy, P. *The Encyclopedia of Science Fiction Movies*. Octopus Books, London, 1986.

Harvey, B. *The New Russian Space Programme: From Competition to Collaboration*. Wiley-Praxis, Chichester, 1996.

Harvey, B. *The Chinese Space Programme: From Conception to Future Capabilities*. Wiley-Praxis, Chichester, 1998.

Harvey, B. *The Japanese and Indian Space Programme: Two Roads into Space*. Springer-Praxis, Chichester, 2000.

Krige, J. and Russo, A. *Europe in Space 1960–1973*. European Space Agency, Noordwijk, 1994.

Lewis, R.S. *The Illustrated Encyclopedia of Space Exploration*. Salamander Books, London, 1983.

Lovell, J. and Kluger, J. *Lost Moon: The Perilous Voyage of Apollo-13*. Houghton Mifflin, Boston, 1994.

Miller, R. *The Dream Machines: A Pictorial History of the Spaceship in Art, Science and Literature*. Krieger, Malabar, 1993.

Murray, C. and Cox C.B. *Apollo: The Race to the Moon*. Secker & Warburg, London, 1989.

Ordway, F.I. *Visions of Spaceflight: Images from the Ordway Collection. Four Walls Eight Windows*, New York, 2001.

Ordway, F.I. and Sharpe, M.R. *The Rocket Team*. Heinemann, London, 1979.

Piszkiewicz, D. *Wernher von Braun: The Man Who Sold the Moon*. Praeger, Westport, CT, 1998.

Reeves, R. *The Superpower Space Race*. Plenum Press, New York, 1994.

Schneider, S. *Collecting the Space Race (with price guide)*. Schiffer, Atglen, 1993.

Stoiko, M. *Pioneers of Rocketry*. Hawthorn Books, New York, 1974.

Stuhlinger, E. and Ordway, F.I. *Wernher von Braun: Crusader for Space*. Krieger, Malabar, 1994.

Williamson, M. *The Cambridge Dictionary of Space Technology*. Cambridge University Press, Cambridge, 2001.

Space Exploration, Fly-Past

Flying past a target in space is one of the oldest and cheapest means of space exploration and is still the method engineers choose for first investigations in space. It returns often surprisingly good science with a simple mission profile. At the beginning of space flight, rockets were too weak, navigation too unreliable, and probes too small to do anything but simply fly by. Therefore most of the fly-pasts were in the first decade of the space age, carrying fields and particles experiments, and sometimes, primitive cameras. Later, fly-pasts were continued to reduce the cost and complexity of missions or to gain gravity boosting. Often the unknown nature of the exploration precludes anything but a fly-past. A simple variation on the fly-past mission is the "strike" mission. Such missions were limited to the moon at first because it was thought no life could exist there, and therefore bacteria carried on a spacecraft could not contaminate indigenous life. Also a strike mission would be an early navigational *tour de force*, and the probe would be in line-of-sight of the ground stations here on earth.

Earth's moon was the first of the planets to be investigated by spacecraft. In 1959, the American Pioneer-4, with primitive instrumentation and camera, made the first flight to go near to the moon, if 60,050 kilometers is considered near for a strike mission. No pictures were recorded because the spacecraft did not come close enough to trigger the camera's photoelectric sensor. Following successful fly-past, signals from the spacecraft were received from a record 655,300 kilometers away, amazing for an early mission that was planned to go half as far. Later that year the Soviet Luna-3 swept by, carrying a primitive imaging system that returned a set of photographs of the hidden side of the moon, seen for the first time by inhabitants of the earth. Failed strike missions, Pioneer-3 and Luna-1 and *-2*, preceded both. The Soviets only sent one more intentional flyby mission. The Zond-3 of 1965, which was actually a Mars mission that had missed its launch window and was retargeted, also imaged the far side of the moon, after several failed lander missions. The U.S. sent a barrage of nine Rangers, primarily equipped with cameras, intended for impact on the moon, and several of these became inadvertent flybys when their navigation failed. Both countries quickly graduated to orbiters and landers due to their interest in the size and proximity of the moon.

While Mars and Venus held early fascination for space exploration, both planets are much smaller and farther away. Mariner-II flew past Venus in 1962 with a set of instruments including infrared and microwave sensors, radiometers, magnetometer for exploring space and the planet. The results established the magnitude of the inhospitable surface temperatures, measured at 425°C, and the absence of a magnetic field. No cameras were used because opaque clouds hide the surface of Venus. The Soviets also had several craft fly by Venus, leading to the Venera-3 strike mission (which may have been a failed lander). There were no great discoveries, either because they were too far away for their instrumentation or because of radio failure. Mariner-IV passed Mars in 1964, shooting the first close-up photos of another planet, showing Martian craters, and gathering atmospheric data. The 21 pictures took one week to transmit to earth.

The last of the Mariner missions was Mariner-X, an intentional multiple fly-past of Mercury by way of a Venus near miss to gain gravity-boost, represented the pioneering use of this technique. It carried a high-resolution camera, and the probe imaged Mercury three times, using the planet's gravity to gain the multiple fly-pasts.

During the 1970s, it became apparent that the outer planets would be lined up in such a way that a "grand tour" mission could be planned that would fly past all the gas giants—Jupiter, Saturn, Uranus and Neptune. Financially, this was deemed impossible, so a pair of Voyager spacecraft were prepared for only Jupiter and Saturn fly-pasts. Once launched, and with both probes operating, the grand tour would be deemed viable. Since instrumentation was not intended to view Uranus and Neptune, new software was radioed to the spacecraft, for example, for image compression so that more pictures could be transmitted. Also, because of lower levels of light than expected, new techniques were developed to allow the cameras to track targets during longer exposures. Voyager-2 thus visited Jupiter, Saturn, Uranus and Neptune on one of the most valuable fly-past missions ever, returning spectacular images. The Voyagers had, in addition to the high-resolution cameras, instruments for exploring areas of space such as magnetic fields. They also carried a message from earth on a disk on the spacecraft. The second probe to each of the giants, Galileo to Jupiter, and Cassini to Saturn, were orbiters carrying entry probes. Galileo's entered the Jovan atmosphere, and Cassini's is aimed at the Saturn moon Titan. Both missions depended on data gathered from *the Voyager* fly-pasts for mission planning.

The close approach of Halley's Comet in the mid-1980s spawned a plethora of fly-past missions. The Soviets had two, Vega-I and -II, both of which dropped probes into the Venusian atmosphere on their journeys. The Japanese also had a pair, Sakigake and Suisei. Only Suisei had imaging equipment. They carried instruments to determine the composition of the comet. The European Space Agency sent one probe, Giotto, which went on to another comet later after a period of dormancy. These missions were the first visits to a comet, which has a very weak gravitational field, so fly-pasts were in order for both reasons of novelty and technical limitations.

The only planet remaining to be explored, Pluto, was also targeted for a fly-past, hopefully to arrive before the atmosphere of the planet begins to freeze as it moves farther from the sun. The earliest methods of space exploration will continue to be used, especially when exploring a new part of the universe.

See also **Space Exploration, Unmanned; Space Exploration, Planetary Landers; Space Exploration: Moon, Unmanned; Telescopes, Space**

JAMES E. TOMAYKO

Further Reading

Anderson, Jr, F.W. *Orders of Magnitude: A History of NACA and NASA*, 1915–1980. NASA, Washington D.C., 1981.

Ezell, E. *On Mars: The Exploration of the Red Planet, 1958–1978*. NASA, Washington D. C., 1984.

NASA Historical Data Books, vol. 5. NASA Launch Systems, Space Transportation, Human Spaceflight and Space Science, 1979–1988. NASA, Washington D.C.

Sheldon II, C. *Review of the Soviet Space Program*. McGraw-Hill, New York, 1968.

Siddiqi, A. *Challenge to Apollo*. NASA, Washington D.C., 2000.

Washburn, M. *Distant Encounters: The Exploration of Jupiter and Saturn*. Harcourt Brace Jovanovich, New York, 1984.

Useful Websites

NASA website: http://history.nasa.gov

Space Exploration, Manned Orbiters

Survival in space requires more than just a sturdy ship suitable for the explorers of the past. Manned orbiting vehicles provide the safe, hospitable, and transportable environment found on the surface of the earth that humans need to survive the harsh reality of outer space. From capsules to shuttles to space stations, American, Soviet, and later Chinese space pioneers all took steps toward conquering the unknowns of space in the twentieth century.

Astronauts and cosmonauts alike certainly found the first generation of spacecraft designs cramped and uncomfortable. The designs differed, but their capacities in terms of room to stretch were equally limited. The Soviet Union succeeded in launching the first human into orbit aboard the Vostok spacecraft. The spacecraft measured 7 meters long and included both a spherical re-entry vehicle and an instrument module that was jettisoned as the craft began its descent back into the atmosphere. The re-entry vehicle, which housed the cosmonaut for the duration of the orbital flight, measured only 2.3 meters in diameter. The spherical design of the re-entry vehicle allowed for even distribution of heat across its surface as it entered the atmosphere. Because the Soviets landed their craft on land instead of in the ocean as the Americans did, the cosmonaut ejected from Vostok at a safe altitude and parachuted down to the earth while Vostok pounded into the ground, only slowed by its own parachute. The Americans' first flights into orbit were aboard the Mercury capsule. Largely conical in shape, the bottom surface of Mercury was coated with an ablative heat shield for re-entry. Unlike Vostok, the entire orbiting vehicle returned to earth with the astronaut aboard; but it measured even smaller in size: just 3.3 meters long, including the retro rockets and 1.9 meters across at the heat shield's widest point.

When the two space programs moved toward the moon and began carrying more than one person at a time, the Soviets' concept changed few elements of its design. The idea of ejecting two or more cosmonauts for landing created a more challenging engineering problem than safely ejecting one. The Voskhod was the result. Cosmonauts could land aboard Voskhod once engineers added a retro-rocket system to the vehicle; but given the very tight confines of the vehicle itself, when three cosmonauts were aboard, they could not wear spacesuits during the flight, leaving the crew unprotected from the dangers of depressurization. The Gemini capsule looked much like a larger version of Mercury, at a height of 7.7 meters and weighing 3800 kilograms, two and a half times Mercury's weight. However, its modular component design and the addition of a docking mechanism made Gemini a more flexible and useful design than Mercury.

Part of the Gemini and Voskhod missions included spacewalks in preparation for moon flights. The Voskhod vehicle did not carry enough air to pressurize the entire capsule; so on March 18 1965, Aleksei Leonov used an airlock attached to the side of Voskhod for his spacewalk. With the smaller volume of the Gemini vehicle, Ed White was able to conduct his spacewalk on June 3 1965 without an airlock. James McDivitt stayed inside the Gemini 3 spacecraft wearing his own spacesuit while White floated just outside the hatch.

With the successes of Gemini—they included a spacewalk and a rendezvous with the Agena satellite—the American space program forged ahead with the flight of its newest conical spacecraft, Apollo. Similar to Voskhod, the Apollo spacecraft supported three astronauts. Wider than Voskhod, the Apollo spacecraft, measuring 3.9 meters across at its widest point, allowed the seats to be placed side-by-side. Aboard Voshkod, technicians had to raise the middle seat by several inches to allow the cosmonauts' shoulders to overlap. Apollo also produced a second manned orbiter, the lunar lander.

The lunar landing module, spider-like in appearance, represents the flimsiest of NASA's spacecraft. Designed for the one-sixth gravity and no-atmosphere environment of the moon, the lunar lander stood on spindly legs and was covered by sharp angles. Unlike other manned space vehicles, it had a very limited purpose in terms of crew support. The crew would spend the majority of its flight to the moon and back in the Apollo command module. En route to the moon, the command module pilot turned the command module 180 degrees and docked with the lunar lander. The commander and lunar module pilot only entered the lunar lander prior to their departure for the surface of the moon. More than any other spacecraft, the lunar lander served as a dinghy; the crew used it to get to and from its sailing vessel. No part of the lunar lander returned to earth.

Although the Soviets never achieved a manned lunar landing, they too designed a vehicle for the purpose, which would eventually fly. Sergei Korolëv, the Chief Designer of the Soviet space program, and his team developed Soyuz with docking in mind. Like their previous vehicles, the basic structural design of Soyuz was a sphere. Three modules—the instrument module, the orbital module, and the descent module—connected together to give the crew a ship with both laboratory and living spaces. In January 1969, Soyuz-4 and -5 docked together and transferred crew members for the first time, an achievement that the Americans would have to perform as well for the success of the moon program. The Soviet skills and technical achievements with docking procedures contributed to their next space success, the first space station, Salyut-1, launched on April 19, 1971.

The Soyuz and Apollo spacecrafts both saw service in the 1970s in the cooperative joint mission, the Apollo-Soyuz Test Project. The program brought American and Soviet space enthusiasts together for the first time in friendship instead of competition. However, the shuttle era of the 1980s saw a less intense but seemingly renewed race. In 1983, the U.S. launched the space shuttle Columbia, the first space truck. Designed both as a rocket and a plane, the shuttle launched on the back of solid- and liquid-fuel engines, then landed like an airplane on a flat runway. In November 1988 the Soviets launched their shuttle Buran, almost identical to the American design; but the collapse of the Soviet Union in August 1991 left the Soviet space program without funds and little hope of a future. With the slow destruction of America's own fleet of space shuttles, the shape of the world's future manned space orbiter designs are again flexible. The Chinese entered the boundaries of outer space in the late 1990s. Their vehicle bore a marked resemblance to the Soyuz descent module, an igloo-like capsule. In the years since the Soviets first put a man into orbit, the vehicles going into space have come full circle and the future of manned space orbiters remains to be seen.

See also **Space; Space Exploration, Unmanned; Space Shuttle**

AMY FOSTER

Further Reading

Gatland, K.W. *Manned Spacecraft*. New York, 1976.
Jenkins, D.R. *Space Shuttle: The History of the National Space Transportation System*. Cape Canaveral, FL, 2001.
Neal, V., Lewis, C.S. and Winter, F.H. *Spaceflight: A Smithsonian Guide*. New York, 1995.
Newkirk, D. *Almanac of Soviet Manned Space Flight*. Houston, TX, 1990.

Space Exploration, Moon, Manned

The idea of lunar exploration goes back at least as far as 1638, when the protagonist of Francis Godwin's *Man in the Moon* rode a flock of 25 geese from the earth to the lunar surface. Animals played a rather less fanciful role in man's actual first steps into space, with the dog Laïka riding aloft in the Soviet spacecraft Sputnik-2 in November 1957. Early Soviet space successes prompted the U.S. to accelerate its own space program. The U.S. entered space with the launch of the unmanned probe Explorer-1 on 31 January 1958, and the National Aeronautics and Space Administration (NASA), was founded in October 1958. Manned moon exploration was preceded by a series of unmanned missions by both the USSR and the U.S. The USSR's Luna series of spacecraft successfully reached the moon as early as 1959; the first spacecraft to land on the moon was Luna-2, which touched down on 14 September 1959. The Soviet Luna-9 landed on the moon on 3 February 1966, transmitting pictures of the moon's surface back to earth. The U.S. followed suit with Surveyor-1, which achieved a soft lunar landing on the plain Oceanus Procellarum on 2 June 1966.

Manned space exploration began with the flight of the Russian Yuri Gagarin on 12 April 1961. It continued with the U.S. Mercury program, one of three successive U.S. programs leading up to the first moon landing. The one-manned Mercury capsule was launched into space aboard a Mercury Redstone (MR3) launch vehicle. The first Mercury flight was accomplished by astronaut Alan B. Shepard, Jr. on 5 May 1961. The Gemini program followed. A major goal of the Gemini program was to increase the length of manned space flights. The two-man Gemini craft was lifted into orbit aboard a Titan-2 launch vehicle. Milestones in the Gemini program included the first rendezvous of one spacecraft with another, the first docking maneuver of two spacecraft, and the first "spacewalk" by American astronaut Ed White in June 1965 as part of the Gemini-4 mission (Figure 18). Information gathered on Gemini missions was critical in the Apollo program that would land the first man on the moon in July of 1969. The Apollo missions utilized a three-man space capsule carried into space aboard either a Saturn-4 or Saturn-5 booster rocket. The Apollo spacecraft had three components: the command module (CM) designed to return the astronauts to earth, the service module, housing the electrical and propulsion systems, and the insect-like lunar excursion module (LEM) designed to carry astronauts to the surface of the moon. The 4.6-meter-tall LEM weighed 12 tons, the 7-meter-tall service module, 25 tons. The crew comprised a commander, a LEM pilot, and a CM pilot. The CM pilot remained in orbit around the moon, while the commander and LEM pilot descended to the surface.

There were ten manned flights from 1968 to 1972. The first of the planned Apollo missions ended in tragedy on 27 January 27 1967, when a fire during a launch-pad test killed astronauts Roger Chaffee, Virgil Grissom, and Edward White. The disaster prompted a thorough re-design of the lunar and command modules. The first manned spacecraft to leave the earth's orbit and orbit the moon was Apollo-8, which began its lunar orbit on 24 December 1968. Apollo-10, launched on 18 May 1969, was the "dress rehearsal" for the actual moon landing that followed with Apollo-11. On this mission, which came within 14.5 kilometers of the moon, astronauts investigated the Sea of Tranquility and provided the first live color television broadcast from space. Two months later, Apollo-11 carried the first astronaut to the surface of the moon. Launched on 16 July 1969, the spacecraft landed on Sunday, 20 July at 4:17 p.m. Eastern Daylight Time in the Sea of Tranquility. Aboard the craft were Neil A. Armstrong, the first man to set foot on the moon, Edwin E. "Buzz" Aldrin, Jr., and Michael Collins. At 10:10 p.m., after sleeping for a few

Figure 18. Ed White, first American spacewalker, June 3, 1965.
[*Source: Courtesy of NASA.*]

Figure 19. Astronaut Eugene A. Cernan, Apollo-17 mission commander, makes a short checkout of the lunar roving vehicle during the early part of the first Apollo-17 extravehicular activity. This view of the "stripped down" Rover is prior to load-up. The mountain in the right background is the east end of South Massif.
[*Credit: Courtesy of NASA. Photograph by Geologist-Astronaut Harrison H. Schmitt.*]

hours, Armstrong and Aldrin walked on the moon as Collins remained aboard the command module, Columbia. Armstrong and Aldrin raised the American flag for the television camera and left it behind along with a few instruments and a plaque reading "Here men from planet earth first set foot upon the moon. July 1969 AD. We came in peace for all mankind." It is estimated that roughly 600 million people viewed the televised landing. The lunar module left the moon on 21 July after more than 21 hours on its surface, and the spacecraft returned to earth on 24 July with drilled core samples, photographs and moon rocks.

The Apollo program continued after the July 1969 landing, with the final mission, Apollo-17, launched on 7 December 1972 (see Figure 19). Like the Apollo-11 mission, Apollo-17 astronauts Eugene A. Cernan, Ronald E. Evans, and Harrison H. "Jack" Schmitt left behind a plaque, which read "Here man completed his first exploration of the moon, December 1972 AD. May the spirit of peace in which we came be reflected in the lives of all mankind." In total, the Apollo program executed three earth-orbiting missions, two lunar-orbiting missions, one lunar flyby, and six lunar landings.

See also **Space Exploration, Manned Orbiters; Space Exploration, Unmanned; Space Exploration: Moon, Unmanned Space Exploration, Planetary Landers; Space Launch Vehicles**

TIMOTHY S. BROWN

Further Reading

Chaiken, A. *A Man on the Moon*. Viking Press, New York, 1994.
Collins, M. *Liftoff*. Grove Press, New York, 1988.
Emme, E.M. *A History of Space Flight*. Holt, Rinehart & Wilson, New York, 1965.
Kennan, E.A. and Harvey Jr., E.H. *Mission to the Moon*. William Morrow & Co, New York, 1969.
Murray, C. and Cox, C.B. *Apollo: The Race to the Moon*. Simon & Schuster, New York, 1989.

Space Exploration, Moon, Unmanned

The exploration of the moon by unmanned spacecraft was pursued in phases (flyby, impact, soft landing and orbital injection), which represented increasing accuracy in guidance and control (see Table 1). The first spacecraft to fly near the moon, which is about 400,000 kilometers from earth, was the Soviet Luna-1, launched in January 1959. It passed within 6000 kilometers of the lunar surface and, in doing so, became the first man-made object to escape the earth's gravitational field. In September 1959, Luna-2 became the first spacecraft

Table 1 Lunar spacecraft targeting accuracy: progression from flyby and impact to orbit and landing (selected missions).

Launch	Spacecraft (state)	Achievement
Oct. 1958	Pioneer-1 (U.S.)	Reached distance of 113,854 km (less than one-third of the distance to the moon)
Jan. 1959	Luna-1 (USSR)	Passed within 5,955 km; escaped earth's field
Sept. 1959	Luna-2 (USSR)	First to impact moon
Oct. 1959	Luna-3 (USSR)	First far-side photos/TV images
Apr. 1962	Ranger-4(U.S.)	Far-side impact/spacecraft inoperative
Jul. 1964	Ranger-7 (U.S.)	4,308 photos/impact (first Ranger success)
Jan. 1966	Luna-9 (USSR)	First soft-landing; transmitted photos for 3 days
Apr. 1966	Luna-10 (USSR)	First to enter lunar orbit
May 1966	Surveyor-1 (U.S.)	Soft-landed 2 June 1966, 11,150 photographs
Aug. 1966	Lunar Orbiter-1 (U.S.)	First US orbiter/lunar mapper
Nov. 1970	Luna-17 (USSR)	First automated rover, Lunokhod-1

to hit the moon (note that only successful Soviet probes were given numbers). The mission's main intention—in an intensely competitive period known as the "space race"—was to deliver a 26-kilogram sphere carrying tiny hammer-and-sickle medallions.

The arguably more useful Luna-3, launched in October 1959, was placed in a large elliptical earth orbit which caused it to loop around the moon every 16.2 days. This allowed the Soviet Union another "first" in that it was able to photograph the lunar far side for 40 minutes on its first orbit. The pictures showed that there were many more craters and fewer maria, the large, flat, dark areas on the moon's surface, than on the near side. This was a significant technical achievement since the lunar far side is not visible from earth.

The technology involved in producing those early space photographs makes modern-day electronic systems look simple. Luna-3 was, in effect, a space-based photoprocessing laboratory required to operate under the weightless conditions and wide temperature extremes for which space is renowned. After exposure, the film was automatically developed, fixed and dried, then placed in front of a television scanning system, which converted the photographs into a stream of telemetry data. Once received on earth, this was converted back to a hard-copy photograph.

Following a number of failures, America too succeeded in lunar impact with one of its Ranger spacecraft; it was, however, inoperative at the time. Early Rangers carried a package of science experiments including a cosmic dust detector, magnetometer and x-ray scintillation counters. Later spacecraft (see Figure 20) were redesigned to support the Apollo manned lunar program by returning close-up photographs of the surface. The Ranger program celebrated its first success in July 1964, when Ranger-7 returned 4308 photographs of the lunar surface before impact.

It was the Soviet Union, however, that would perform the first soft landing with Luna-9 in January 1966. Luna-9 was a fairly sophisticated spacecraft comprising two elements, a mother-craft and a lander. The lander, a sphere with four petal-shaped panels which opened after landing to right the spacecraft and point the integral antennas towards earth, carried an instrument package which included a television camera and an ingenious deployable rod with mirrored surfaces which acted as a "rear-view mirror." The spacecraft sent back panoramic views of its surroundings, and measured a temperature of 300°C and a

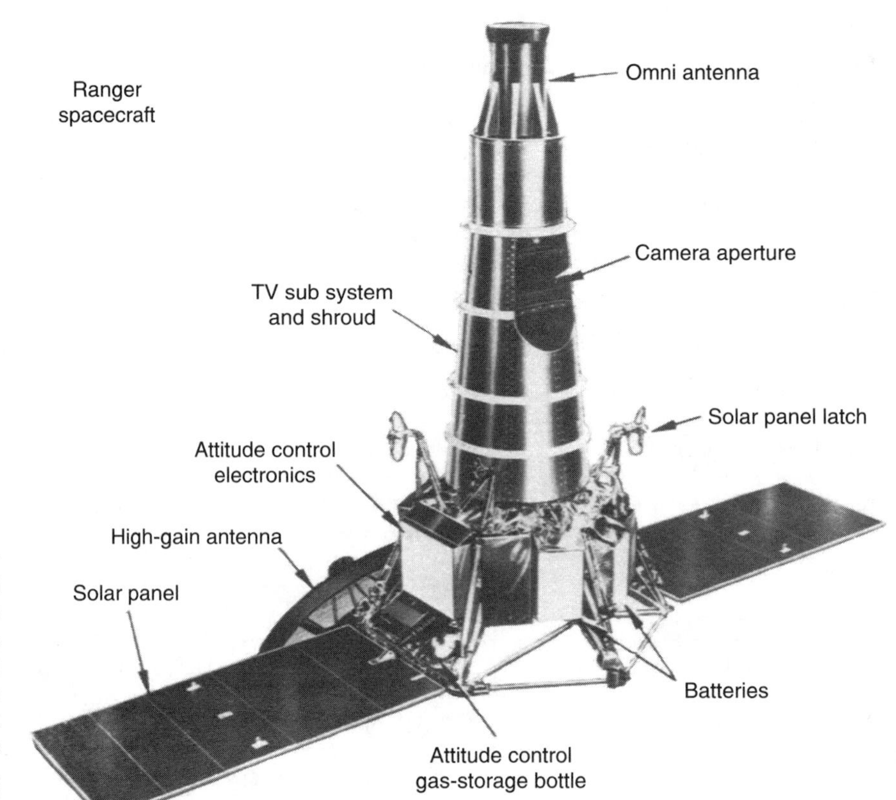

Figure 20. NASA's Ranger spacecraft showing principal equipment. The solar panels were folded against the sides of the spacecraft for launch and deployed once en route to the moon; the circular high-gain antenna (part hidden) was folded up against the base of the spacecraft. [*Source: Courtesy of NASA.*]

radiation level of 30 millirads per day. Importantly, it proved once and for all that a lunar lander would not sink beneath a sea of dust, as some had hypothesized.

After Luna-10 became the first spacecraft to enter lunar orbit in April 1966, the four phases of unmanned lunar exploration were complete. Between August 1966 and August 1967, America succeeded in placing five spacecraft in lunar orbit. The aptly-named Lunar Orbiters returned a wealth of high-quality imagery from orbits between 1600 kilometers and 40 kilometers above the surface. As with earlier spacecraft, the Lunar Orbiters took pictures on film, developed them automatically, then scanned them into a TV system for transmission to earth. The system produced images with some 500,000 times the resolution of the Ranger cameras and showed features measuring about 3 meters across. The fact that the best resolution of earth-based telescopes of the time was about 800 meters proved the advantage of sending spacecraft there. The Lunar Orbiters also discovered mysterious large concentrations of mass—dubbed "mascons"—below the lunar maria. These proved to be extremely important for predicting the orbital paths of the later Apollo spacecraft.

NASA's Surveyor program, operated in parallel with the Orbiter missions, succeeded in soft-landing five out of seven spacecraft, providing a reasonable level of confidence that an Apollo lunar module could be guided to a safe and accurate landing. In the context of the history of technology, it is noteworthy that by mid-1966, when slide-rules and rotary-dial telephones were still the norm, both Russia and America had succeeded in soft-landing spacecraft on the moon.

In November 1970, following the Apollo-11 and -12 manned lunar landings (see Figure 21), the USSR succeeded in delivering the first of two teleoperated rovers, Lunokhod-1, to the moon; Luna-21 deployed Lunokhod-2 in January 1973.

Considering the importance attached to lunar exploration in the 1960s, it is significant that no lunar missions were conducted between Luna-24's landing in August 1976 and the launch of Japan's Hiten/Hagoromo spacecraft in January 1990. This was followed by two U.S. lunar mapping missions—Clementine in 1994 and Lunar Prospector in 1998—the most significant result of which was the suspected detection of water ice at the lunar poles. Although there were plans at the turn of the century to launch further unmanned missions to the moon, it seemed unlikely that any serious lunar exploration would ensue until a decision is made to mount a further manned program of exploration and development.

Figure 21. Apollo-12 astronaut Pete Conrad inspects Surveyor-3 two-and-a-half years after it soft-landed on the moon. Landing demonstrations with the Surveyors helped to refine the techniques that would allow Apollo-12 to land within 200 meters of the spacecraft (the lunar module Intrepid is visible on the horizon). Parts of the Surveyor were returned to earth for analysis.
[*Source: Courtesy of NASA.*]

See also **Space Exploration, Moon, Manned; Space Exploration, Unmanned**

MARK WILLIAMSON

Further Reading

Caidin, M. *Race for the Moon*. William Kimber, London, 1960.

Gatland, K. *Robot Explorers*. Blandford Press, London, 1972.

Kopal, Z. *Exploration of the Moon by Spacecraft*. Oliver & Boyd, Edinburgh, 1968.

Reeves, R. *The Superpower Space Race*. Plenum Press, New York, 1994.

Spudis, P.D. *The Once and Future Moon*. Smithsonian Institution Press, Washington, 1996.

Troebst, C.-C. *Reaching for the Moon*. Hodder & Stoughton, London, 1961.

Wilson, A. *Solar System Log*. Jane's, London, 1987.

Yenne, B. *The Encyclopedia of US Spacecraft*. Hamlyn, Twickenham, 1985.

Space Exploration, Planetary Landers

With the launch of the first earth-orbiting satellites, the use of unmanned planetary landers, probes, and telescopes came under consideration as a means to chart and explore the solar system. The advantages of such machines became clear immediately, as they represented a short cut in the race between the U.S. and the Soviet Union to dominate this new realm of international competition. The design and shape of these instruments varied but displayed similar basic requirements: the maximum use of the area within the superstructure for the instruments and the onboard computer, a means to communicate with earth to receive guiding instructions and send back information collected, camera and sensor pods, and an engine and its fuel capable of functioning for several years.

The first "target" was earth's moon, and both superpowers launched probes beginning in 1958. The first "wave" did not involve landers as such, but machines intended to orbit and perhaps crash on the moon. None of the American Pioneer machines succeeded in this task, but two Soviet Luna spacecrafts came close to earth's satellite. The U.S. then initiated the Ranger project. Begun in 1959, the first four missions failed outright, but Ranger-7 succeeded in sending back photographs of the lunar surface, which paved the way for the Apollo manned moon landings.

The exploration of Venus also began with the Cold War as background. The peculiarity of Venus' cloudy atmosphere as well as its impact on human culture dating back to the earliest civilization made it an important target. The Soviet Union first launched probe Venera-1 on 12 February 1961, but the probe failed on its way to Venus. The U.S. followed suit in summer 1962 with Mariner-1 and Mariner-2. The first failed, but the second one was able to take atmospheric measurements. Several other Venera and Mariner missions followed, all confirming that the planet was completely inhospitable and that the greenhouse effect actually had elevated ambient surface temperature to 430°C. The most recent planetary probe was Magellan, launched in 1989, which mapped most of the planet at high resolution.

Mars drew an equal or even greater interest in its exploration. The Soviet Union reached Mars first, but failed to get any data back from its June 1963 mission. The American Mariner-4 mission flew near the planet in summer 1965, and Mariner-6 and Mariner-7 further mapped the Mars surface in 1969 in preparation for a landing. None of the probes suggested there was any life on Mars. Exploration nevertheless reached a zenith with the Viking-1 and Viking-2 missions. Launched 10 weeks apart in 1975, both machines took about one year to reach Mars. Once there, the landers' missions confirmed the lack of life on Mars due to the high level of ultraviolet radiation that saturates the Martian atmosphere. Although the Viking program ended in November 1976, the probes continued to transmit for another six years. Further planetary probes explored Mars and its satellites. The most successful was the Mars Pathfinder which landed on July 4, 1997, and whose robotic rover, Sojourner, spent 30 days collecting data on the environment and relief of the planet.

The distance between earth and other planets in the solar system has prompted the need for other types of exploration, including solar system probes and space telescopes. The first attempt at surveying outer planets involved the preparation of the Pioneer-10 and Pioneer-11 probes. The first, launched in March 1972, reached the vicinity of Jupiter 21 months later, then pursued a course out of the solar system. Affixed to its side was a plate with a male and female figure and a series of symbols depicting Voyager's origins and travel path. The plaque represented an attempt at considering the possibilities in the search for extraterrestrial intelligence (SETI). Although it was elegantly simple, the design was criticized for depicting the female figure in a passive stance whereas the male raises his hand in greeting, and both figures appear to be Caucasian. As for Pioneer-11, it passed near Saturn twice, and in 1990 left the solar system.

Two other American probes, Voyager-1 and Voyager-2, launched in 1977, conducted a planetary tour of the two outermost planets, Uranus and Neptune, and 48 moons. The last investigation of Jupiter was that of the probe Galileo. Although slower than other probes (it used the gravity of planetary bodies to correct its course and increase its speed, also known as the "slingshot" effect), Galileo reached Jupiter in 1995, six years after its launch. It successfully parachuted a probe into the atmosphere that transmitted for 45 minutes before the atmospheric pressure and the planet's gravity broke it up.

An alternative to expensive probes and the problems of earth-based telescopes (which include visual distortions through the atmosphere) has involved satellite telescopes. Several National Aeronautics and Space Administration (NASA) projects were devoted to this alternative beginning with the Apollo moon missions. The advanced orbiting solar observatory (AOSO), initially scheduled for Apollo moon missions, was turned into the Apollo telescope mount (ATM) and launched aboard the American Skylab space station in 1973. It functioned throughout the station's manned operations (a total of some six months) and recorded solar activity.

Meanwhile, research into a giant space telescope had begun in the 1960s. The NASA project, which later became known as the Hubble space telescope (HST), capitalized on existing technical knowledge used in the manufacture of spy satellites to order a high-resolution mirror. Although the system underwent several redesigns to fit into a space shuttle bay (instead of the top of a Saturn-V moon rocket as had been initially planned), it was ready for launch in the mid-1980s. The shuttle Challenger accident of January 1986 delayed the launch date by four years. Once lofted, however, the HST displayed an aberration in its picture resolution. The error, caused by the improper position of one of the mirrors, required a "corrective lens" that was installed during a 1993 shuttle-servicing mission. The repair was successful and helped HST uncover new information about black holes.

See also **Space Exploration, Manned Orbiters; Space Exploration, Moon, Manned; Space Exploration, Unmanned; Space Stations, Skylab**

GUILLAUME DE SYON

Further Reading

Ezell, E. *On Mars: The Exploration of the Red Planet, 1958–1978*. NASA, Washington D. C., 1984.

Fimmel, R.O., Swindell, W. and Burgess, E. *Pioneer Odyssey*, NASA, Washington D.C., 1977.

Hall, C. Lunar Impact: A History of Project Ranger. NASA, Washington D.C., 1977.

Nicks, O.W. *Far Travelers. The Exploring Machines*. NASA, Washington D.C., 1985.

Smith, R.W. *The Space Telescope: A Study of NASA, Science, Technology, and Politics*. Cambridge University Press, New York, 1994.

Space Exploration, Unmanned

Manned space exploration receives most of the glory and the attention, but unpiloted spacecraft have been used as tools longer and for a greater variety of reasons. Some even believe that humankind should cease piloted space exploration in favor of the much less expensive unpiloted probes when we consider the accomplishments of these robotic craft.

The beginnings of unpiloted space exploration were made by the Soviets. The first artificial earth satellite, Sputnik, or "Fellow Traveler" carried only a radio transmitter and batteries and no other instruments despite its 80-kilogram-plus orbiting weight. Its batteries lasted about two weeks after the 4 October 1957 launch. The Soviets had been working on an earth-orbiting laboratory, Object D, as their first satellite, but Sergey Korolev, the most influential chief designer in their program, was anxious that the U.S. would launch one first. He thus convinced his colleagues to prepare a couple of "simple satellites" for orbit, which eventually became Sputniks-I and -II. Object D was thus pushed down to third in the launch list. When the U.S. orbited its Explorer-I satellite on 31 January 1958, it carried a radiation detector and some other small instruments. The radiation detector picked up evidence of a radiation belt around the earth. Object D, when it became Sputnik-3, confirmed this. The era of robotic exploration had begun.

Several dozen Explorers and hundreds of Soviet Cosmos satellites have explored near-earth space, but bodies such as the moon, Mars, and Venus have attracted the most attention from the public. The moon came under early and frequent exploration. After three American failures in 1958, the Soviet Luna-1 flew past in 1959, discovering the solar wind, and the American Pioneer-4 was launched. The Soviets followed up that year with a hard lander, Luna-2, the first person-built object to land on another celestial object, and Luna-3, which imaged the lunar far side. With Ranger-4, the Americans hit the moon. By then, President John F. Kennedy had pledged America to land a person on the moon and return them to earth by 1970, so the unpiloted programs took on the air of being preparatory to the main event. The Rangers, which flew until 1965, impacted the moon taking pictures all the way down. The Soviets had by then turned their interest to soft landers, suffering several failures that made them into lunar impactors. The Soviets successfully launched a Lander and orbiter to the moon early in 1966. The American Surveyor series began landings soon after. In the summer of 1966 the first American Orbiter arrived at the moon.

During the next two years, America, convinced that it was in a race to make the first piloted lunar landing, sent seven Surveyors, and five Orbiters to the moon, all but a couple of Surveyors successfully. The Soviets continued to fly orbiters, then graduated to lunar fly-arounds with returns to earth on two Zond missions in the second half of 1968, seemingly in further rehearsal of piloted mission. However, the Americans orbited the moon with a piloted mission near Christmas and landed on 20 July 1969. Any race, if there was one, was over.

The Soviets then turned to sample return missions, in which soil samples from the lunar surface would be returned to earth. The first successful one of these was Luna-16, which landed on 20 September 1970. This was followed over the next six years with several more missions, two with Lunokhod rovers. The similarly between the Soviet science accomplishments on their unpiloted missions and the accomplishments of American piloted missions lent support for those who opposed piloted space exploration as too expensive and dangerous for its worth.

After a 14-year hiatus, the Japanese Muses-A orbiters visited the moon the first month of 1990. This was the first non-U.S. or non-Soviet probe to reach the moon. The Japanese are expected to launch penetration missions early in the 21st century. Meanwhile, the U.S. mounted two missions in the 1990s. The first was the Clementine, intended for technology demonstration and launched by the Ballistic Missile Defense Agency. The second was Lunar Prospector, one of the NASA's "faster, cheaper, better" missions searching for water and mineral deposits useful for eventual lunar bases occupied by humans. ("Faster, cheaper, better" is an effort to avoid the excessive costs of space missions.)

Venus was the target of 17 Soviet spacecraft in the 22 years between 1961 and 1983 and five American overall. The early probes revealed Venus to be as inhospitable as it was beautiful. Mariner-2 discovered a 425°C surface temperature and a thick atmosphere. The Soviets decided to land on the planet, regardless. Their Venera-3 spacecraft was intended to explore the atmosphere, but communications failed just after atmospheric entry and it impacted, although this was quite good navigation for 1966. Then followed a series of spacecraft that returned atmospheric data before being crushed by the thick carbon dioxide atmosphere. First Venera-4 on 18 October 1967 returned data on the composition of the atmosphere and the surface temperature, now measured at 500°C. Venera-5 lasted until it reached 26 kilometers of the surface and its sister, Venera-6, went within 11 kilometers. Venera-7 made the first landing on another planet on 15 December 1970 and lasted 23 minutes.

The Soviets then made probes that landed and lasted longer and longer. They carried increasingly complex sensor suites, eventually including imaging equipment. Venera-8 measured wind speeds while descending and lasted 50 minutes after landing. Venera-9 and -10 were orbiters and landers. Venera-9 worked on the surface for 53 minutes, including sending the first picture from another world on 25 November 1975. Venera-10's lander achieved 65 minutes of life on the surface.

The Americans, after a couple of Mariner fly-past missions in the 1960s, sent a pair of Pioneers in 1978. The first, Pioneer-12, was an orbiter with a radar-mapping device that lasted until 1992. The second, Pioneer-13, released a cluster of four atmospheric probes. The Soviets also sent two probes during the 1978 launch window (the period when the planets are best aligned), reaching 95 and 110 minutes of surface life.

The Soviets ended their exploration of Venus with four straight successful probes, Venera-13, -14, -15, and -16. The first pair were fly-pasts with landers, the last two orbiters and landers, all had color imaging equipment, and the orbiter returned a map of Venus's northern hemisphere. Mapping was the chief goal for the 1989–1994 mission by the Magellan American spacecraft that used its synthetic aperture radar on 99 percent of the planet's surface. It was the final probe to reach Venus.

In contrast to the relative success exploring the inhospitable Venus, probing relatively benign Mars proved problematic. Only three of sixteen Soviet probes of Mars have seen full success, while eight of fourteen American probes were satisfactory; less than a third of the combined missions were thus successful. After five Soviet failures and the failure of its partner, Mariner-III, Mariner-IV flew past the Red Planet and returned pictures of craters on Mars. After eleven years and eight failed missions, the Soviet Mars-3 soft-landed on 2 December 1971. Even still, it relayed only 20 seconds of data to its orbiter, which lasted nine months. After a few more fly-pasts, the US Mariner-IX went into Martian orbit at about the same time as Mars-3. A planet-wide dust storm was in progress, and the temperature variations from it inspired the concept of a nuclear winter that scientists theorized would follow any widespread atomic exchange. The Soviet Union succeeded with another orbiter, Mars-5, after the failure of Mars-4, to survey possible Mars-6 and -7 landing sites. However, both probes failed.

The U.S. then had a pair of highly successful orbiters and landers, Viking-1 and -2, which arrived during the 1976 launch window and relayed data until 1980. After over a decade of hiatus, the Soviets tried again with Phobos-1 and -2 in 1988, both including a Phobos lander. Both failed, as did the next American mission, Mars Observer. A replacement spacecraft, Mars Global Surveyor, was launched in November 1996, and entered Martian orbit the next year. The Russians attempted the ambitious Mars-96 mission that same launch window, an orbiter, two landers, and two soil penetrators. The booster failed, leaving the spacecraft to crash into the ocean with 270 grams of plutonium intended to generate electrical power.

The American Mars Pathfinder, one of NASA's "faster, cheaper, better" missions, landed on Mars and released a rover, *Sojourner*. This mission returned a torrent of data on the Martian surface for 83 days after its arrival 4 July 1997. America then suffered a series of embarrassing failures at Mars. The first was Mars Climate Orbiter, which burned up on approaching the planet due to one of its sub-teams calculating in metric units and the other in imperial units. This was followed by the loss of the Mars Polar Lander and its Deep Space-2 Penetrators. All of these Soviet and U.S. failures prompted a panel at a NASA conference, entitled "Why is Mars So Hard?"

Japan sent its Nozomi probe on its way to Mars in 1998, but at the time of writing, it had not arrived. Whether successful or not, it is, however, the first interplanetary probe launched by a country other than Russia or the U.S.

Looking at the deepest end of space exploration is NASA's Great Observatories program, which includes infrared and x-ray observatories, but the most famous is the optical Hubble space telescope. It was placed into orbit by the Space Shuttle on 25 April 1990, and serviced several times, each visit yielding a better spacecraft. It has mapped the entire sky and seen matter at the origins of the universe. It places the exploration of the Moon, Venus, and Mars essentially in our back yard.

See also **Space Exploration, Fly Past; Space Exploration: Moon, Unmanned; Space Stations, International Space Stations; Space Stations, Mir; Space Stations, Skylab; Telescopes, Space**

JAMES E. TOMAYKO

Further Reading

Anderson, Jr., F.W. *Orders of Magnitude: A History of NACA and NASA*, 1915–1980, NASA, Washington, D.C., 1981.

Ezell, E. *On Mars: The Exploration of the Red Planet, 1958–1978*. NASA, Washington D.C., 1984.

Sheldon II, C. *Review of the Soviet Space Program*. McGraw-Hill, New York, 1968.

Washburn, M. *Distant Encounters: The Exploration of Jupiter and Saturn*. Harcourt Brace Jovanovich, New York, 1984.

Useful Websites

Hamilton, C. *History of Space Exploration*: http://www.solarviews.com/eng/history.htm

NASA Headquarters homepage: www.hq.nasa.gov

Space Launch Vehicles

A launch vehicle is a rocket-based vehicle designed to carry payloads into space. Theoretically, there are two main types: expendable launch vehicles (ELVs) comprising a number of propulsive stages which are used and then discarded; and reusable launch vehicles which return from space intact to be refueled for another mission. Although no fully reusable launch vehicles were developed in the twentieth century, the Space Shuttle was designed to be partially reusable in that the orbiter returns to earth and the solid rocket boosters can be refurbished and refueled for subsequent flights.

Following Goddard's 1926 launch of the world's first liquid-fuel rocket, rockets of all types were developed for both military and civilian purposes. As with other technologies, World War II produced a dramatic leap in rocket technology, culminating in the V-2 ballistic missile. After the war, the V-2 and its designers became the basis of both American and Soviet missile research, leading eventually to the intercontinental ballistic missile (ICBM). However, the ultimate spur to the development of space launch vehicles came from the USSR, when on 4 October 1957 it launched Sputnik-1 into orbit. This fuelled the "space race" of the 1960s and the conversion of ICBMs into launch vehicles for manned and unmanned spacecraft.

Apart from improvements in reliability, accuracy and control, the main aim of launch vehicle development since the beginning of the space age has been to increase the payload mass that can be delivered to space. The payload capability of the early vehicles was low, especially the American rockets which were much smaller and less powerful than their Soviet counterparts. This is illustrated by the fact that America's first satellite, Explorer-1, weighed just 8.3 kilograms compared with Sputnik-1's mass of 83.6 kilograms; and whereas America's second satellite, Vanguard, launched by an even smaller rocket, weighed 1.5 kilograms, the USSR's Sputnik-2—carrying the dog Laïka—weighed over 508 kilograms. Little wonder that Soviet Premier Khrushchev nicknamed Vanguard "the grapefruit."

Progress was rapid in the 1960s, however, and by 1968 America had succeeded in sending the Apollo-8 spacecraft weighing more than 30,000 kilograms around the Moon. This was due entirely to the development of the Saturn-V, the first of what came to be known as heavy-lift launch vehicles. Modern launch vehicles can be divided into four groups based on payload capability: small vehicles which can deliver up to about 1000 kilograms to geostationary transfer orbit; medium vehicles (1000 to 2000 kilograms); large vehicles (2000 to 5000 kilograms); and heavy-lift vehicles (more than 5000 kilograms).

The launch vehicle is fundamental to space exploration and development: it is the active element of a space transportation system, in the same way that an aircraft is to an airline. This was made especially clear when the field evolved from government-operated launch vehicles launching scientific and military satellites to commercially operated vehicles launching fleets of commercial communications satellites.

In the U.S., former government-operated ELVs such as the ICBM-based Atlas, Delta, and Titan were commercialized when the Space Shuttle failed to live up to its promise. First launched in April 1981, the Shuttle had been expected to provide cheaper and more reliable access to space, effectively replacing the old-style ELVs. However, following the Challenger accident in January 1986 it was decided not to use it for commercial payloads and the ELVs were given a new lease of life. In fact, the American ELVs in use at the close of the century were essentially upgrades of 1950s ICBM technology.

The European Space Agency (ESA), by contrast, decided to develop an ELV specifically as a commercial satellite launcher; it was named Ariane and conducted its first launch in December 1979. By the end of the century, its fifth variant, Ariane-5 (see Figure 22), was in operation, along with a number of other commercial, or semicommercial, launch vehicles from China, Japan, India and—following the end of the Cold War—from Russia and Ukraine. Several of these vehicles now compete for each satellite launch contract in an international market analogous to the air-cargo business.

Figure 22. A cutaway of the Ariane-5 launch vehicle showing two satellites under the payload fairing and the strap-on solid propellant motors.
[*Reproduced with permission from Arianespace.*]

Most launch vehicles use liquid propellants, because they generally provide better performance than solid propellants. However, solids are often utilized in strap-on boosters, which augment the thrust of a given launch vehicle variant and improve its payload capability. Most launch vehicles employed to deliver payloads to orbit (as opposed to suborbital trajectories) have more than one stage, each with its own engines and propellant tanks. The most common design has three stages, although both two- and four-stage rockets exist.

The main advantage of the multistage rocket (see Figure 23) is that empty tanks and associated structure do not have to be carried all the way to space, with the result that a larger payload can be placed in a given orbit. The disadvantage of the staged design is that most, if not all, of the vehicle is discarded on every mission, which is rather like scrapping an airliner after a single flight.

It has not escaped the notice of the space industry that this is an extremely wasteful and expensive method of accessing space and there have been many proposals, detailed designs and even technology demonstrations aimed at the development of a fully reusable launch vehicle. However, severe technical difficulties associated with the development of appropriate propulsion systems, materials and aerodynamic designs, and most of all the high cost of these developments, have so far obstructed all attempts to find a solution. Vertically, space may only be 100 kilometers away, but it has proved to be one of the

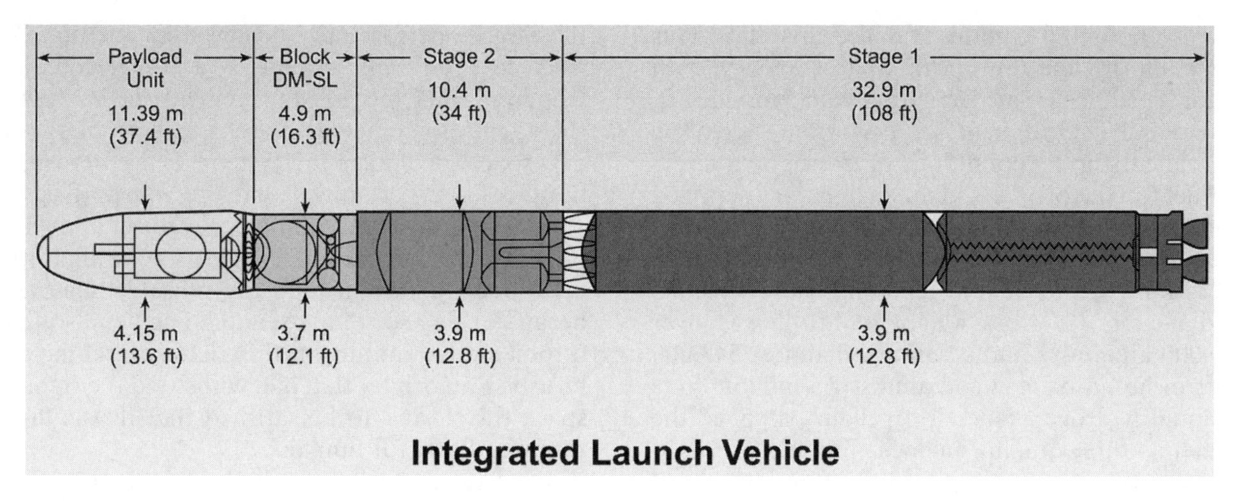

Figure 23. The stages of the Sea Launch expendable launch vehicle showing (from left to right) the payload accommodation, the Block DM-SL upper stage, the second stage and the first stage.
[*Reproduced with permission from Boeing.*]

most challenging journeys technology has ever been required to make.

See also **Rocket Planes; Rocket Propulsion, Liquid Propellant; Rocket Propulsion, Solid Propellant**

MARK WILLIAMSON

Further Reading

Baker, D. *The Rocket: the History and Development of Rocket and Missile Technology*. Crown Publishers, New York, 1978.
Chiulli, R.M., Ed. *International Launch Site Guide*. Aerospace Corporation, El Segundo, 1994.
Hill, C. N. A *Vertical Empire: the History of the UK Rocket and Space Programme, 1950–1971*. Imperial College Press, London, 2001.
Isakowitz, S.J. *International Reference Guide to Space Launch Systems*, American Institute of Aeronautics and Astronautics, Washington D.C., 1995.
Ordway, F.I. and Sharpe, M.R. *The Rocket Team*. Heinemann, London, 1979.
Rycroft, M., Ed. *The Space Transportation Market: Evolution or Revolution*? Kluwer, Dordrecht, 2000.
Stoiko, M. *Soviet Rocketry: The First Decade of Achievement*. David & Charles, Newton Abbot, 1970.

Space Shuttle

The Space Shuttle was developed by the National Aeronautics and Space Administration (NASA) as one component of a program intended to secure the future of the agency after the lunar landings in 1969 and to ensure that the U.S. maintained world leadership and technological superiority in space. Initially conceived as a system to transport humans and materials to a space station, its final configuration embodied a trade-off between NASA's ambitions and the economic and political constraints ensuing on a loss of support for space among the American people and Congress in the early 1970s.

The Shuttle system was intended to be fully reusable, as opposed to the more conventional, expendable launch vehicles (ELVs) or rockets, and was supposed to replace them. In the end it was only partly reusable. NASA had to accept design features demanded by the Department of Defense (DOD) to secure the military's support for the system, and within a few years of its becoming operational the DOD were once again demanding that America's access to space be secured with a mixed fleet of vehicles, including ELVs, striking a further blow to the Shuttle's commercial viability.

Four main components comprised the Shuttle system. First, there was the orbiter itself. Five have been built, Columbia, Challenger (replaced by Endeavor), Discovery and Atlantis (all named after famous exploration sailing ships). The reusable orbiter itself was akin to a delta-winged space plane, which could be brought back to earth and landed on a runway, like a conventional aircraft. It comprised a large crew compartment, a cargo bay about 4.5 meters in diameter and 18 meters long, and three main engines. It could transport up to 29,500 kilograms into near-earth orbit (185 to 1111 kilometers). The orbiter was multipurpose, and in addition to carrying personnel and payloads into space, it could serve as an orbiting service center for other satellites (as in the space walks to repair the defects in the Hubble space telescope) and to return previously orbited spacecraft to earth. Its cargo bay has also housed a laboratory called Spacelab. This was built by the European Space Agency with major funding from Germany, and provided a shirt-sleeve environment for performing scientific experiments under conditions of microgravity.

Fuel for the orbiter's main engines was supplied from an external tank that also served as the structural backbone for the system during launch operations. It was 46.8 meters long and 8.4 meters in diameter. Its gross weight at liftoff was over 750,000 kilograms and it contained almost 543,000 liters of liquid oxygen and almost 1.5 million liters of liquid hydrogen. These propellants supplied the orbiter's three main engines for about eight minutes, whereupon they cut off, at an altitude of about 109 kilometers, just before the spacecraft was injected into low-earth orbit. The orbiter separated from the external tank, which followed a ballistic orbit, and splashed down into the ocean. This tank was not recovered.

Two solid-fuel boosters were strapped on to the sides of the external tank. Each was 45.5 meters long and 3.7 meters in diameter, and contained 500,000 kilograms of propellant (a mixture containing about 70 percent ammonium perchlorate). Together with the orbiters' three liquid-fuelled engines these boosters provided over 3 million kilograms of thrust to lift the Shuttle off the pad. They were designed to burn for just over 2 minutes, and at a height of about 44.5 kilometers they separated from the external tank, falling into the ocean at predetermined points where they were recovered for reuse.

When NASA promoted the Space Shuttle in the early 1970s, the Agency claimed that it would revolutionize space transport in the decade to come. Being reusable, costs would be slashed as compared to ELVs (whose use was compared to operating a railroad by throwing away the locomotive after each trip). Using mission models that predicted that the orbiter would be launched once a week, NASA and its consultants insisted that the Shuttle would make access to space "routine and economical." This proved to be wildly optimistic. Columbia, the first orbiter to be flown, took off from Cape Canaveral on 12 April 1981, landing successfully at the Edwards Air Force Base two days later. By 26 January 1986, when Challenger exploded just over a minute after lift-off, there had been only 24 launches in almost five years, about as many as NASA had originally claimed it could launch in six months. Cost per kilogram into orbit was also far higher than anticipated.

Several reasons account for the huge disparity between the initial estimates made by NASA and the Shuttle performance achieved in practice. The need to "sell" the system to a dollar-conscious Congress and Office of Management and Budget led the Agency to exaggerate the cost-effectiveness of its new space transportation system. There were also a number of uncertainties that are inherent in any new, revolutionary technological project and which only seem evident with hindsight. Above all, it took far longer to turn-around the reusable spacecraft because, being so complex, and being "man-rated," it took several months, rather than several days, to refurbish an orbiter that had withstood the rigors of space travel, and to be satisfied that it was flight-ready and safe for humans.

Even then, risks were always taken. Indeed the detailed analysis of the Challenger accident shows that flight engineers and managers necessarily had to make many micro-decisions about the level of acceptable risk in the process leading up to the launch. Those responsible for the orbiters' solid-fuel boosters knew that the O-rings between their segments could 'freeze' in the cold Florida morning, and not seal efficiently. They deemed this risk to be acceptable, in the light of their previous and ongoing studies of O-ring behavior. Their error cost the lives of seven astronauts, saddened millions of people in the U.S. and around the world, and struck a serious blow to NASA's reputation.

The Shuttle resumed operations in September 1988, and continues to be launched on an approximately bimonthly basis. The low launch frequency and cost, the Challenger accident, and NASA's decision to phase out ELVs created an opportunity for other launch vehicle suppliers to gain an important share of the market in the mid-1980s. Europe's Ariane rocket was a major benefactor, and today Arianespace has 50 percent of the world market. This is one of the ironies of history. The European program was embarked on in the shadow of NASA's insistence that ELVs would soon be obsolete, and it was only thanks to French political determination to break the U.S. monopoly on access to space, notwithstanding the risks, that this alternative to the Shuttle was developed.

See also **Space Exploration, Manned Orbiters; Space Stations, International Space Station; Space Stations, Mir; Space Stations, Skylab**

JOHN KRIGE

Further Reading

Heppenheimer, T.A. *The Space Shuttle Decision. NASA's Search for a Reusable Launch Vehicle.* NASA-SP4221, Washington D.C., 1999.

Krige, J. The decision taken in the early 1970s to develop an expendable European satellite launcher, in *A History of the European Space Agency, 1958–1987. Vol. II: The Story of ESA, 1973–1987*, Krige, J., Russo, A. and Sebesta, L., Eds. European Space Agency, Noordwijk, 2000.

Launius, R. *NASA: A History of the U.S. Civil Space Program*. Krieger, FL, 1994.

Vaughan, D. *The Challenger Launch Decision. Risky Technology, Culture and Deviance at NASA*. University of Chicago Press, Chicago, 1996.

Useful Websites

NASA Historical Data Books. Volume V. NASA Launch Systems, Space Transportation, Human Spaceflight, and Space Science, 1979–1988. NASA SP-4012, Washington D.C., 1999: http://history.nasa.gov/SP-4012/cover.html

Space Stations, International Space Station

The International Space Station (ISS) program is an international collaboration of the U.S., Europe, Japan, Canada, and Russia, each of which is expected to provide modules and equipment designed to support a crew of up to six for at least 15 years. The ISS will be the largest structure ever built in space: at completion its overall dimensions will be approximately 110 by 60 meters, about the size of a soccer pitch. Its mass will be some 420,000 kilograms and its total pressurized volume about 900 cubic meters, equivalent to the passenger cabins of two Boeing 747 aircraft.

U.S. President Reagan directed National Aeronautics Space Administration (NASA) to build a space station in 1984, but redesigns and cost overruns delayed the beginning of orbital assembly until November 1998. Over that period, its original name Freedom was changed first to Alpha and then to ISS, as the team of international partners was amassed. The first module to reach orbit, on 20 November 1998, was Zarya, formally known as the functional cargo block (or, from its Russian name, FGB). It is a self-contained spacecraft—a power, propulsion, and orbital-control module that is pressurized and thus provides habitable accommodation. A U.S.-built connection node called Unity was docked to Zarya on 7 December 1998. Then, following significant delays in its construction, the Russian-built service module Zvezda, a habitable command and control center, was added on 26 July 2000.

Modules expected to join the ISS in later years include the Columbus Orbital Facility (COF), a general-purpose science and technology laboratory supplied by the European Space Agency (ESA), and the Japanese Experiment Module (JEM) supplied by the Japanese space agency NASDA (see Figure 24).

Although the station is being assembled at an orbital altitude of about 380 kilometers, its operational orbit, attained late in the assembly sequence, is expected to have an altitude of 426 kilometers and an inclination of 51.6 degrees. The lower initial orbit allows the Space Shuttle to deliver some 18,000 kilograms of additional payload on each mission, while the inclination is designed to allow the station to over-fly some 85 percent of the earth's landmass and 95 percent of its population.

The six main modules and a number of connecting nodes will be mounted on an "integrated truss structure," the station's structural backbone, along with eight two-panel solar arrays designed to rotate to maximize their exposure to the sun and provide a total of some 110 kilowatts of power. The station's location in low-earth orbit means that it will be eclipsed by the planet on each orbit, obliging its systems to draw power from its nickel–hydrogen (NiH_2) batteries which will be recharged from the arrays during the sunlit portion of the orbit.

The station's orbital track and orientation, or attitude, was determined initially by gyroscopes alone but later used an inertial navigation system incorporating global positioning system (GPS) receivers. Attitude control is engineered using electrically powered momentum wheels, which exchange angular momentum with the station's structure to rotate it in any axis. In common with all unmanned spacecraft, such as communications satellites, adjustments can also be made by bipropellant thrusters, and backup systems are provided for all subsystems in case the primary equipment should fail. The station's altitude is boosted periodically, to overcome atmospheric drag, by firing rocket engines on visiting spacecraft.

In addition to the standard subsystems, there are several novel technological aspects to the station. One is the inclusion of a Canadian-built manipulator arm, which can move along the station's truss on a mobile transporter enabling the crew to perform assembly and maintenance work without leaving the safety of the modules. Although the arm is based on the Space Shuttle's remote manipulator system, it has a "hand" at either end and is not permanently fixed to the station's structure so that it can crawl along the truss. Also new for the ISS is a pair of cupola modules, supplied by ESA, each of which has seven relatively large windows to allow observations of

Figure 24. A view of the station showing the Japanese experiment module (JEM), complete with external payloads and remote manipulator arm. Note the windowed cupola module at center, which will be supplied by Europe.
[*Source: Courtesy of NASA.*]

the earth and stars and monitoring of work on and around the station.

There are many technical challenges inherent in the design and construction of such a large structure in space, not least its structural integrity, which despite numerous computer simulations, remained to be proven in flight. Also, because of its size, the probability of impact from space debris will be statistically greater than for previous spacecraft. As a result, the module shells are designed to withstand impacts from objects less than 1 centimeter in diameter without significant damage, while their life support systems should allow at least three minutes for evacuation following a 10 centimeter puncture.

One of the most important questions posed throughout the station's design and development concerned what it would be used for. The answers usually included microgravity research into growing more perfect semiconductor crystals than available on earth; various aspects of biomedical research that could lead to new drugs and procedures for use on earth; and space science applications such as astronomy.

Over and above such scientific endeavors, the ISS is likely also to be used as an in-orbit model shop for future spacecraft technologies; for example, in the testing of large deployable structures such as solar arrays and communications antennas, or for space-qualifying new materials, thermal control hardware and ion propulsion systems, to name but a few. Interestingly, the ISS has already proved itself as a destination for space tourists, the first of which, Dennis Tito, paid a reported US$20 million for a week on the station in 2001. Moreover, the potential for the ISS to act as a departure point for further exploration of the Moon and Mars should not be ignored.

In addition to the engineering challenge of establishing a small community in low-earth orbit, the ISS program has proved to be a political challenge. Agreement on financing and respective responsibilities among international partners has proved difficult and has resulted in modifications and delays. For example, the inclusion of Russia relatively late in the program—following an end to Cold War hostilities—led to a major redesign to incorporate a significant amount of Russian hardware. Although this led to a station in orbit sooner than the U.S. could have produced alone, it did so at the expense of the Russian station, Mir, which was deorbited in March 2001 because Russia could not afford to operate both. Despite the problems, the ISS seems likely to be a model for future international space programs, such as a manned Mars mission, simply because they are unlikely to prove affordable for a single nation.

See also **Space Shuttle; Space Stations, Mir; Space Stations, Skylab**

MARK WILLIAMSON

Further Reading

Covault, C. International Space Station ready for flight. *Aviation Week and Space Technology*, December 8, 1997.
Messerschmid, E. and Bertrand, R. *Space Stations: Systems and Utilization*. Springer Verlag, Berlin, 1999.
Haskell, G. and Rycroft, M., Ed. *International Space Station: The New Space Marketplace*. Kluwer, Dordrecht, 2000.
Smith, M.S. International Space Station cooperation: a view from the US Congress. *Russ. Space Bull.*, 5, 3, 29–30, 1998.
Williamson, M. New star in orbit. *IEE Rev.*, September, 201-205, 1999.

Space Stations, Mir

On June 1, 1970, the USSR launched the Soyuz-9 spacecraft as part of the initial preparation for a reorientation of the Soviet space program in the direction of establishing a permanent human presence in orbit. The Soyuz-9, manned by Andrian Nikolayev and Vitali Sevastyanov, orbited the earth for 18 days. The purpose of the Soyuz flight was to break the record of days spent in orbit, previously set when the crew of the U.S.'s Gemini-7, Frank Borman and James Lovell orbited the earth for 14 days in 1965. While the USSR, like the U.S., had focused its efforts on lunar landings, it now sought to set up an earth-orbiting space station that would serve as a continuing research center and space science laboratory. It was hoped that the research results would aid the Russian economy as well.

Salyut-1, the first of seven satellite stations, was launched on April 19, 1971. It relied on four solar panels for charging its batteries. Salyut-7 was used from 1982 to 1986, and also employed solar panels. It was the "second generation" of the past Salyut models. In 1975, the U.S. and the USSR established the joint Apollo–Soyuz Project. This relationship established through the Apollo–Soyuz Project was the international precursor to the Mir Space Station. The Mir Space Station was designed as a third generation of the Salyut series, utilizing most of the basic original constructs. The first component of the station complex, the Mir Core, was launched on February 20, 1986. The station complex was built in orbit over a period of ten years, with new modules added slowly over time. The Mir Core contained the living quarters and control center. This first module originally had two solar panels, seven computers running off the Strela system, and six ports for the docking of up to two Soyuz crafts. This meant that the station could house up to six people for short periods of time.

The Mir Core had a mass of 20,100 kilograms, was 13.13 meters long, and had a diameter of 4.15 meters. Its three solar panels supplied up to 11 kilowatts of power. Because of the specifications of the Proton rocket used to launch it, the Mir Core was similar in structure to the earlier Salyut craft, and a Proton SL-13 launch vehicle sent the Mir Core into orbit. It contained sleeping compartments, a toilet with a door (the Salyuts had no door for privacy on the toilet), washing station, table, refrigerator, stove, treadmill, and stationary bicycle. Cosmonauts on the space station had a fairly regimented working day. Everyone was required to bicycle 10 kilometers per day and "walk" (they were harnessed to the treadmill to add friction and stability) 5 kilometers per day to prevent muscle and general health deterioration.

In 1987, the Kvant-1 was launched and added to the space station as the second module. The Kvant-1 was 8 meters long, with a pressurized volume of 40 cubic meters. It contained an astrophysics laboratory, but lacked its own solar panels. In December of 1989, the Kvant-2 module was added. The Kvant-2 had a pressurized volume of 60 cubic meters and two solar panels of 50 square meters total area, supplying 7 kilowatts of power. This module, which had its own propulsion system, did not have a singular purpose like the Kvant-1's astrophysics laboratory. The Kvant-2 added six gyrodynes, two tanks for the Rodnick system, two oxygen generators, a shower, a toilet, and myriad other scientific equipment. Kvant-3, or Kristall, provided even more research capabilities when the module was added in June of 1990. Kristall had no gyrodynes, but it did contain two Rodnick tanks, six Ni–Cd batteries, and newly designed solar panels. Kristall had a pressurized volume of 60 cubic meters and its two retractable solar panels, with a combined area of 72 square meters, provided 9 kilowatts of power. In 1995, the Spektr module was added to expand the complex, although this had originally been scheduled for

1991. It had four solar panels and provided more living quarters, mainly used by American cosmonauts. The Spektr had a pressurized volume of 62 cubic meters and 126 square meters of solar panels, which could generate 16 kilowatts of power. In 1996, the final two modules were added, the Priroda module and the Docking module, providing the station complex with remote sensing instruments and a port for a shuttle, respectively. The Priroda module had a pressurized volume of 66 cubic meters. Its earth remote-sensing instruments examined both environmental and ecological changes, such as the spread of industrial pollutants, the height of waves, temperature changes in the ocean, the structure of clouds, and even plankton concentrations.

The last crew left the Mir complex in June of 2000. Up until that point the station had been almost constantly occupied. In total, 43 cosmonauts lived in Mir and 59 others have visited the station for less than one month. On March 23, 2001, the Mir Space Station was brought back to earth and crashed into the South Pacific Ocean. Mir proved man's ability to live in space for extended periods of time, and provided extensive research in numerous fields of study. The Mir provided information on subjects ranging from new galaxies, patterns of ocean and wind direction, and the influence of gravity on biological processes.

See also **Satellites, Environmental Sensing; Space Launch Vehicles; Space Stations, International Space Station; Space Stations, Skylab**

TIMOTHY S. BROWN

Further Reading

Bonnet, R.M. and Manno, V. *International Cooperation in Space*. Harvard University Press, Cambridge, MA, 1994.

Harland, D.M. *The Mir Space Station: A Precursor to Space Colonization*. Wiley, Chichester, 1997.

Oberg, J. *Red Star in Orbit*. New York, Random House, 1981.

Space Stations, Skylab

Skylab was a U.S. space station designed for earth observation, solar astronomy, and research into how humans would work in, and react to, the space environment. It was launched into low-earth orbit—to an altitude of about 440 kilometers—on 14 May 1973 and visited by three groups of three-man crews for periods of 28, 59, and 84 days, until it was vacated in February 1974. Despite later attempts to raise its altitude, Skylab re-entered the atmosphere on 11 July 1979, having completed 34,981 orbits of the earth.

The Skylab program was, in part, an attempt by the National Aeronautics and Space Administration (NASA) to salvage some of the technological investment made in the Apollo lunar program, which was curtailed when government funding was withdrawn. The station was launched by one of the remaining Saturn-V moon rockets and its crews were launched by the smaller Saturn-1B. The station's main section, the orbital workshop (OWS), was made from a converted Saturn-V third stage, effectively replacing propellant and guidance systems with crew accommodation and life-support systems. From an engineering point of view, this produced a welcome but unusual situation in that most of Skylab's structural design had not only been completed, but also flight-tested, before the station left the drawing board.

Attached to the OWS was an airlock module, which allowed spacewalks (to change film in the external cameras and make external repairs) to be conducted; a multiple docking adapter, at which two Apollo spacecraft could be docked; and the Apollo telescope mount (ATM), designed particularly for solar observations (see Figure 25). With a total mass of some 90,000 kilograms and a total habitable volume of about 360 cubic meters, Skylab was by far the largest spacecraft launched at that time; the OWS alone provided a working volume 15 meters long and 6.6 meters in diameter. The contemporary Soviet Salyut stations were about one-third of the size and mass.

The OWS was split by a gridded floor into two sections—a large laboratory area and a smaller wardroom containing sleeping, cooking, eating, and hygiene facilities. Being weightless, the astronauts could attach their specially adapted shoes to either side of the grid to provide stability for certain tasks or simply to avoid floating away. Running along the axis of the OWS was a removable "fireman's pole" which was used for moving between one section and another. Some care was taken in the interior design of the station, particularly the sleeping and eating areas; there were plenty of storage lockers and hardware was color-coded. One of the innovations of Skylab was an enclosed shower, which is something even the current International Space Station lacks.

Skylab was built with two large, deployable solar panels and four smaller ones, having a combined area of 840 square meters and capable of producing about 12 kilowatts of power. Unfortunately, launch vibrations caused the dis-

Figure 25. The Skylab space station in orbit, showing a single solar array deployed from the side of the Orbital Workshop (OWS) and four smaller arrays deployed from the Apollo telescope mount (ATM). Note also the concertina-shaped sunshade over the main cylinder, which was erected by spacewalking astronauts to replace thermal insulation torn off during launch.
[*Source: Courtesy of NASA.*]

location of a combined meteoroid–thermal shield on the OWS which tore away one of the large solar arrays and restricted the deployment of the other. Telemetry received from the station shortly after launch indicated that no power was being generated from the arrays and that the internal temperature of the station was increasing rapidly. The first Skylab crew was, however, able to restore the station to an operational condition by conducting a series of spacewalks, deploying a temporary sunshade to replace the missing thermal insulation and releasing the remaining solar panel.

The remainder of the 28-day mission, which was twice the duration of any previous U.S. spaceflight, went according to plan. Despite the time spent on repairs, 46 of 55 planned experiments were completed and the crew suffered no significant ill effects.

Together, the three Skylab crews conducted 270 experiments in the fields of life sciences, astrophysics, solar physics, earth observation, and materials processing, as well as demonstrations of engineering and space technology. The solar observations in particular were highly significant at the time. In total, the program added an impressive 513 man-days to the American manned spaceflight record, a figure not exceeded by the Space Shuttle until its fifth year of operation.

The significance of the Skylab program was that only 12 years after Yuri Gagarin became the first man in space, nine astronauts had proved that people were not only capable of living in space for more than a few days, but could also apply themselves to life-threatening situations and make an otherwise uninhabitable station habitable. Although it proved to be another quarter of a century before the U.S.-led International Space Station would be equally habitable, the experience provided by Skylab was invaluable. Indeed, it was not surpassed until Russia began the in-orbit construction of its Mir Space Station in 1986.

Apart from its scientific and technical success, the program provided the first insights into how people might live together in space. It also showed that even highly trained astronauts were not prepared to adopt the same attitudes on long-duration missions as they had on previous week-long missions, when every moment was precious and had to be filled with observations and experiments. Where previously, astronauts had tended to keep their criticisms to themselves, in support of good PR and continued public funding, the third Skylab crew opposed aspects of the flight plan and openly criticized the station's amenities. For example, Gerald Carr complained that the soap was like "dog shampoo;" Bill Pogue likened use of the towels, which were made of a fire-resistant synthetic material, to "drying off with padded steel wool;" and Edward Gibson who, along with the others, suffered cold symptoms and nosebleeds (due to high blood pressure in the upper extremities) disliked the station's paper tissues. In addition, they all agreed that the toilet, as characterized by Pogue, had been designed by someone who had never used one. The Skylab program had unintentionally become a psychological study; it had also shown that the human aspects of space station design and operation are as important as the scientific and technical aspects.

See also **Space Exploration, Moon, Manned; Space Stations, International Space Station; Space Stations, Mir**

MARK WILLIAMSON

Further Reading

Arnold, H.J.P., Ed. *Man in Space: An Illustrated History of Spaceflight*. Smithmark, New York, 1993.
Cooper, Jr. H.S.F. *A House in Space*. Angus & Robertson, London, 1977.
Freeman, M. *Challenges of Human Space Exploration*. Springer-Praxis, Berlin, 2000.
Holder, W.G and Siuru, W.D. *Skylab: Pioneer Space Station*. Rand McNally, Chicago, 1974.
Messerschmid, E. and Bertrand, R. *Space Stations: Systems and Utilization*. Springer Verlag, Berlin, 1999.

Useful Websites

Benson, C. and Compton, W. *Living and Working in Space: A History of Skylab*: http://history.nasa.gov/SP-4208/contents.htm

Spark Transmitters, *see* **Radio Transmitters, Early**

Spectroscopy and Spectrochemistry, Visible and Ultraviolet

Spectrum analysis was launched in 1859–1860 by the physicist Gustav Robert Kirchhoff and the chemist Robert Wilhelm Bunsen. They demonstrated experimentally that each chemical element has its own characteristic set of spectrum lines, which it emits or absorbs when heated to the state of a radiating gas by a Bunsen-burner flame, an electric arc or a spark. Within the context of nineteenth century science, the various patterns in these spectra could only be described, catalogued and mapped—and quite extensively so. An understanding of these patterns of series, bands and the splitting of these lines into components by an electric or magnetic field (Stark and Zeeman effect) had to wait until the twentieth century. Niels Bohr's atomic model of 1913 interpreted spectrum lines as the result of electron jumps between stable orbits around the nucleus. Quantum mechanics abandoned orbits but kept the notion of energy levels henceforth to be calculated on the basis of Schrödinger's equation, leading to an even better agreement between observations and its theoretical predictions concerning the transition terms (selection rules) and energies (line frequencies).

The dramatic success of spectrum analysis after 1860 caused this qualitative analytic technique to make quick inroads into the chemist's or pharmacist's laboratory, the astronomer's observatory, the physician's hospital, and even the judge's court. The most notable use of the new technique involved identifying the presence of the various elements in a given sample. It led to the surprising finding that the metal lithium, for instance, hitherto considered quite rare, was among the more ubiquitous chemical elements. News that it was possible to decipher the chemical composition of the sun and ultimately the stars spread fast: Who could fail to be impressed by the fact that only microscopic amounts of sodium (3×10^{-9} g)—a mere teaspoon of salt in a full swimming pool—were needed for detection of its characteristic yellow D lines in a Bunsen flame? Using a simple pocket spectroscope, a steel caster could now easily identify the exact instant of decarbonization of molten steel, the moment just before it loses its fluidity.

Quantitative emission spectroscopy had a much more difficult start, however. Around 1906, Comte Arnaud de Gramont started to record methodically the detectability of characteristic lines that remain visible as long as the slightest trace of a substance is present. These *raies ultimes*—ultimate or residuary lines—were the most reliable indicators of the respective chemical elements in a sample. For 83 different elements including (besides most metals) a great many rare earths and nonconducting elements, de Gramont listed the low number of 307 ultimate and penultimate lines. As a regular analyst for four French steel mills, de Gramont helped improve their production significantly because he could readily report to them the presence and approximate concentrations of aluminum, boron, cobalt, chromium, copper, manganese, molybdenum, nickel, silicon, titanium, vanadium, and tungsten in their steel. During World War I, de Gramont and a few assistants used the method in a broad array of military applications. Among them were quick and efficient examinations of the structural frames or valves of zeppelins, shrapnel from long-range guns, or ignitors of aircraft.

The spectroscopic laboratory of the National Bureau of Standards in Washington was among the very first to adopt de Gramont's method of "practical spectrographic analysis." After the war, William F. Meggers applied it to the chemical analysis and quality control of noble metals, such as gold and platinum, for the U.S. Mint in San Francisco. Promising analytical and metallurgical applications were explored by the American Brass Company, in Waterbury, Connecticut, as well as by a few other U.S. industrial laboratories. Despite the obvious importance of de Gramont's work for the French war machine, strangely enough, Germany did not implement anything even remotely similar. Carl Friedrich (called Fritz) Löwe was one of the earliest active promoters of its industrial applications among chemists and physicians. Löwe's frankness about the missed opportunities during the recent—lost—World War was effective in arousing renewed interest in the method. His touting of instruments by the Zeiss Company in Jena reached German-speaking audiences. Frank Twyman fulfilled a similar promotional function in the Anglo-Saxon world for his company, Adam Hilger Ltd. in London, and the Stockholm professor of experimental biology Henrik Gunnar Lundegårdh pointed out many applications in mineralogy, biochemistry, plant physiology, and agricultural chemistry. By 1930, a typical spectrochemical procedure took no longer than 20 minutes (including development of the photographic plate). A decade later, the industrial pressure for ever-higher production rates had "super-speed analysts"—as Meggers called them—reduce this time to a minute or two. The laboratories of Ford Motor Company, for instance, carried out large numbers of analyses at high speed: samples were sent by pneumatic tube from the foundry to the spectrographic laboratory and just a few minutes after receipt of the sample, the results were available back on the factory floor. This method left the sample virtually unscathed and allowed close examination of local differences of parts of the surface or various layers of it. By contrast, wet chemical analysis inevitably yielded average results because the sample had to be analyzed in solution. Other applications of spectrochemical analysis after 1930 include:

- Absorption spectrophotometry of organic solutions for identification of hormones, vitamins, and other complicated substances.
- Testing for silver or boron content in the mining industry.
- Routine quality control in the metallurgical and chemical industries, including monitoring of isolation or separation processes.
- Soil analysis for agriculture and plant physiology.
- Applications in the food packing industry (e.g., checking the dissolving rate of inner coating of a tin can by measuring two or three parts of aluminum or lead per ten million, or testing chocolate or chewing gum wrappers, and whiskey distilling vats).
- Forensic analyses or autopsies for detection of trace amounts of toxins (e.g., thallium from rat poison, which is ascertainable in hair samples).
- Analysis of fusible alloys of tin for safety valves or fire sprinklers (to trace impurities such as lead and zinc, which may raise the melting point by undesirable amounts if present in proportions of as little as one part in ten thousand).
- Archaeometric comparisons of the precise composition of metals and alloys from various locations (sometimes enabling archeologists to infer where a certain piece had been manufactured, or conclusions about the geographic and temporal spread of certain technologies or skills).
- Plentiful applications in mineralogical analysis (which, as we have seen in the case of de Gramont, had initiated some of the earliest efforts in quantitative spectrochemical analysis).

The plethora of possibilities turned spectrochemistry into a vibrant and popular field. The industrial world embraced it in the following decades, setting up thousands of spectrochemical laboratories. The boom in this field of research can be gauged somewhat by publication statistics in spectrochemistry: 1467 books and papers, and half a dozen treatises were indexed in the first part of Meggers' and Scribner's bibliographic survey, covering 1920–1939. A total of 1044 contributions were made in the short period of

World War II, 1940–1945, another 1264 in the next five postwar years, and 1866 in the period 1951–1955. A true explosion in the literature followed, with an exponential growth in many scientific fields leading to an estimated total of 10,000 spectrochemical publications by 1963.

Visual resources like atlases were an integrated part of the effective marketing strategy of the major spectrograph manufacturers: Zeiss, Hilger, or Fuess. The most ambitious inventorization effort was the famous Massachusetts Institute of Technology (MIT) table of 100,000 wavelengths. It was compiled with specially developed spectrophotometers capable of automatically measuring, computing, and recording the wavelengths of spectrum lines, thus speeding up these operations some 200-fold. As one of the major teaching and research centers for spectroscopy, the MIT started to host annual summer conferences on spectroscopy in 1933. An initial attendance of 69 persons in the first year increased to 233 in 1938, 250 in 1939, and 302 in 1942. The series was interrupted for the remaining war years but resumed thereafter. The rapidly expanding market for spectrographs and spectrometers led to the initiation of specialized events such as the National Instrument Conference and Exhibit. An overlapping interest in spectrochemical instrumentation and techniques motivated the Society for Analytical Chemistry of Pittsburgh (SACP, founded in 1943) and the Spectroscopy Society of Pittsburgh (SSP, founded in 1946) to combine their annual meetings in 1949. The joint meetings of these hitherto moderately sized societies, held every March since 1950 under the acronym Pittcon (Pittsburgh Conference and Exposition on Analytical Chemistry and Applied Spectroscopy), transformed spectrochemistry to the point that the convention eventually outgrew this steel-producing city and its organizers were forced to find other locations. Whereas the first Pittsburgh Conference offered 56 presentations and 14 exhibits by commercial instrument makers, the 1990 conference (held at the Jacob Javits Center in New York) coordinated more than 1200 talks and 25 symposia, over 3000 instrument exhibits by over 800 commercial instrument makers, and 12,500 hotel bookings.

Both the high demand for spectrochemical techniques during World War II and the ubiquitous pressure for ever-faster results led inevitably to increased substitution of quasi-instantaneous photoelectric detection in photographic recording. This elimination of photographic development and densitometry in favor of photomultipliers and electronic automation was pushed particularly in the U.S. in companies like Dow Chemical Company in Midland, Michigan, Perkin Elmer in Boston, Baird Associates (BA) in Cambridge, Massachusetts, Applied Research Laboratories (ARL) in Glendale, California, and National Technical Laboratories, renamed Beckman Instruments in 1950, whose direct-reading spectrometers flooded the international market in the 1950s. Advertisers claimed these "analysis automats" made "all routine spectrochemical analyses with dispatch and precision," and in the 1950s and 1960s they eventually did. With these improvements also came a rapid expansion of potential applications, especially in infrared spectroscopy.

The near-ultraviolet (UV) had already been explored photographically by Eleuthère Mascart and Alfred Cornu in the nineteenth century. However, glass optics absorb radiation past 3440 Å (1 angstrom = 10^{-10} meters), which could be circumvented by using quartz or Iceland spar prisms; ozone absorbs wavelengths past 2900 Å; and the gelatin emulsions of photographic plates those past 1850 Å. Therefore further progress had to await the development of high-vacuum spectrographs and gelatin-free emulsions. The latter two fields were pioneered by Victor Schumann in Leipzig, who reached wavelengths down to approximately 1000 Å, and Theodore Lyman at Harvard University who discovered the ultraviolet series of hydrogen in 1914. After World War II, grating spectrographs were mounted on rockets and propelled out of the terrestrial atmosphere to record high-resolution solar UV-spectra.

See also **Iron and Steel Manufacture; Materials and Industrial Processes; Spectroscopy, Raman; Spectroscopy, X-Ray Fluorescence**

KLAUS HENTSCHEL

Further Reading

Brand, J.D. *Lines of Light. The Sources of Dispersive Spectroscopy, 1800–1930*. Gordon & Breach, Amsterdam, 1995. Mostly covers the development of theoretical interpretations including quantum mechanics.

Hentschel, K. *Mapping the Spectrum. Techniques of Visual Representation in Research and Teaching*. Oxford University Press, 2002. Mostly covers experiments and applications and further references to primary literature given therein.

Spectroscopy, Infrared

The investigation of invisible optical radiation gained increasing attention during the second half of the nineteenth century, notably with the identification by William Abney of infrared spectra characteristic of different chemical compounds in 1882. Infrared spectroscopy nevertheless grew to become a popular analytical technique only during the mid-twentieth century.

Infrared spectroscopy maps the absorption, emission, or reflection of radiation as a function of wavelength. The infrared spectrum of a chemical absorbing compound constitutes a "fingerprint" which identifies functional groups in its molecules, and can reveal a wide range of physical and chemical properties of matter. Despite such potential utility, its study at the turn of the twentieth century appeared highly unpromising: development was constrained by the very weak signal produced by infrared radiation. Infrared energy passing through a spectrometer causes temperature changes that are typically less than a few thousandths of a degree. The corresponding electrical signal using available detectors was swamped by other influences.

State-of-the-art detection during the 1920s combined an infrared detector with a mirror galvanometer. However, this direct-current signal varied not only with the weak infrared contribution, but with the temperature of the surroundings. To obviate the need for constant zero-adjustment of the galvanometer, in 1929 A. Pfund devised a "resonance radiometer" which detected "chopped" (mechanically interrupted) radiation using a mechanically tuned mirror galvanometer. This modulation technique was a key feature of subsequent instruments. Electronic components developed originally for radio communication began to be applied to instrumentation amplifiers by the late 1920s, and Pfund's method was extended by a capacitively coupled amplifier. Its successors relied solely on electronic components to yield a high-gain tuned amplifier, notably the stable alternating current (AC) amplifier for thermopiles (essentially a compact combination of blackened thermocouples in series) developed by L. C. Roess. As Pfund and Firestone had done, Roess modulated the light beam by interrupting it periodically with a chopper disk, and tuned the amplifier to the modulation frequency to reinforce the very weak infrared signal preferentially. Combined with an automatic pen recorder, the Roess amplifier was quickly adopted by spectroscopic researchers for use with the new thermistor bolometers (blackened semiconductor devices having temperature-dependent resistance and wide spectral response) invented in 1946.

Like spectrometers for the visible portion of the spectrum, the first infrared spectrometers employed either prisms or diffraction gratings as the dispersive elements. Such prisms, fabricated from alkali salts transparent to infrared radiation, were combined with a thermopile detector to extend observations from the visible portion of the spectrum (about 0.7 micrometers) to 40 micrometers. Through the 1930s and early 1940s, designs proliferated as spectrometers were custom-built by laboratory-based researchers. During the World War II, infrared spectroscopy proved valuable for analyzing and quantifying chemical constituents and for monitoring the production of essential materials such as petroleum products and synthetic rubber.

Automatic spectrophotometers for the visible spectrum had been commercialized in the mid-1930s, and infrared versions began to appear increasingly after the war. With the refinement of servo-mechanisms during this period, more spectrometers began to incorporate automatic recording. Increased automation considerably eased the operator's burden of setting amplifier offsets and gains, adjusting the width of slits, and altering the rate of scanning the wavelength to compensate for the dramatic variations in energy across the spectrum.

By the early 1950s there were over a dozen manufacturers of infrared spectrometers, ranging from desk-size research-grade instruments to simpler bench-top spectrometers. Prominent manufacturers during this period included Perkin-Elmer, Grubb-Parsons, Beckman, and Leitz. The market for such instruments centered on chemical research laboratories. The spectrometers were relatively costly and demanded a stable room temperature and knowledgeable operators.

Using such instruments, infrared spectroscopy became increasingly routine at ever-longer wavelengths. Given the low sensitivity of available detectors and the relatively weak infrared sources of radiation, however, there was a limit to such extension. By the late 1960s the best spectrometers, using highly automated control mechanisms and multiple diffraction gratings for different spectral regions, could extend measurement to a wavelength of about 300 micrometers.

To investigate such energy-starved regions of the spectrum and naturally weak emitters such as astronomical sources, a new instrument principle was developed. So-called "Fourier spectroscopy" proved considerably more efficient than conventional methods. Fourier spectroscopy does not disperse the radiation from the light source into separate spectral components, which are then individually measured; instead, the radiation passes through an interferometer, which modulates the radiation and passes it to the infrared detector.

An interferometer consists, in its simplest form, merely of a beamsplitter and two mirrors (see Figure 26). The beamsplitter divides the incoming radiation into two parts. One part passes directly to a fixed mirror, which reflects it back to the beamsplitter. The second part of the radiation passes to a similar mirror, which can be moved backwards to lengthen the optical path. It, too, reflects its beam of light back to the beamsplitter, which recombines the two components to yield an output beam that passes to the optical detector. By moving the adjustable mirror, the light traveling along that arm of the interferometer is successively delayed with respect to that of the other arm. Different wavelengths comprising the radiation are brought into and out of step, interfering and thus changing the intensity of the combined beam. The intensity of the combined beam, modulated as a function of mirror position, is the Fourier transformation of the optical spectrum.

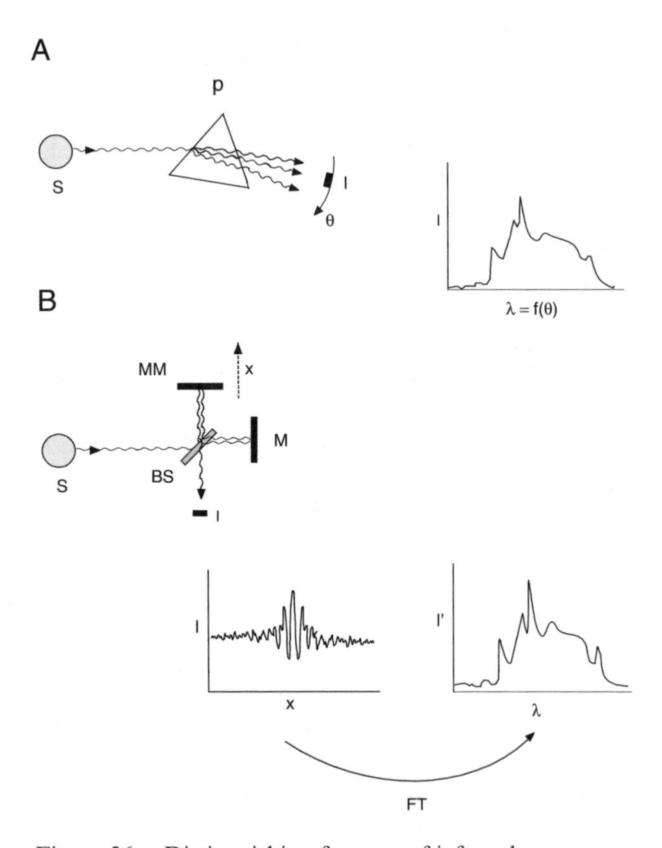

Figure 26. Distinguishing features of infrared spectroscopies. (A): Dispersive infrared spectroscopy: *S*, infrared source; *P*, prism or diffraction grating; *I*, intensity measured by detector swept through angle θ. In the resulting spectrum I(λ), the wavelength λ is a function of θ. (B): Fourier spectroscopy: *S*, infrared source; *BS*, beam-splitting mirror; *M*, fixed mirror; *MM*, moving mirror, translated through distance *x*; *I*, intensity measured by detector. The resulting record of *I* versus *x*, or "interferogram," is related to the spectrum $I'(\lambda)$ (or, more accurately, $I'(1/\lambda)$) by the Fourier transformation *FT*. Both techniques involve specific complications to calibrate the wavelength scale and instrumental response.

Physicists found three distinct advantages of Fourier spectrometers compared to their dispersive predecessors: (1) the replacement of narrow slits with a full optical aperture provides higher optical throughput, or *etendué* (Jacquinot, 1954); (2) the measurement of all wavelengths simultaneously rather than sequentially improves signal quality, or alternatively can reduce measurement time (Fellgett, 1958); and (3) the precise determination of mirror position via a laser reference wavelength provides spectra with much better wavelength precision (Connes, 1966). In combination, these advantages make the technique considerably more sensitive optically and amenable to more extensive data analysis than dispersive techniques.

However, the instrumentation was attractive only to narrow audiences during its first two decades. The operating principles of Fourier spectrometers proved considerably less intuitive than their dispersive counterparts, especially for chemists. Moreover, the Fourier transformation, which converted the modulated signal into an infrared spectrum, demanded expensive and all-too-slow computers. Despite the dramatic speed increase provided by the fast Fourier transform (FFT) from 1966, Fourier spectroscopy made inroads only with isolated groups of physicists. By the mid-1970s, computer power and cost had improved sufficiently to allow commercially packaged and highly automated Fourier spectrometers, rechristened "FTIR" (Fourier transform infrared) instruments, to overtake dispersive instruments in the chemistry market. This process was essentially complete by the late 1980s.

Modern infrared spectroscopy is based principally on Fourier spectrometers coupled to increasingly powerful computers for identification and quantification. Infrared analysis is one of the most commonly used techniques in analytical chemistry (e.g., environmental monitoring), chemical process control, and remote sensing.

See also **Infrared Detectors; Spectroscopy and Spectrochemistry, Visible and Ultraviolet; Spectroscopy, Raman; Spectroscopy, X-Ray Fluorescence**

SEAN F. JOHNSTON

Further Reading

Abney, W.W. and Festing, E.R. On the influence of the molecular grouping in organic bodies on their absorp-

tion in the infra-red region of the spectrum. *Proc. R. Soc. London*, 31, 416, 1882.
Chantry, G.W. *Submillimetre Spectroscopy*. Academic Press, London, 1971.
Connes, J. and Pierre. Near-infrared planetary spectra by Fourier spectroscopy. *J Optic. Soc. Am.*, 56, 896, 1966.
Fellgett, P. A contribution to the theory of the multiplex spectrometer. *J Phys. Radium*, 19, 187, 1958.
Firestone, F.A. A periodic radiometer for eliminating drifts. *Rev. Sci. Instrum*, 3, 163, 1932.
Griffiths, P.R. and de Haseth, J.A. *Fourier Transform Infrared Spectrometry*. Wiley, New York, 1986.
Hardy, A.C. History of the design of the recording spectrophotometer. *J. Optic. Soc. Am.*, 28, 360, 1938.
Jacquinot, P. The luminosity of spectrometers with prisms, gratings or Fabry-Pérot etalons. *J. Optic. Soc. Am.*, 44, 761, 1954.
Johnston, S.F. *Fourier Transform Infrared: A Constantly Evolving Technology*. Ellis Horwood, Chichester, 1991.
Johnston, S.F. In search of space: Fourier spectroscopy 1950–1970, in *Instrumentation: Between Science, State and Industry*, Joerges, B. and Shinn, T., Eds. Kluwer, Dordrecht, 2000.
Kendall, D.N., Ed. *Applied Infrared Spectroscopy*. Reinhold, New York, 1966.
Michelson, A.A. On the application of interference-methods to spectroscopic measurements—I. *Philosoph. Mag.*, 31, 338, 1891.
Roess, L. C. Vacuum tube amplifier for measuring very small alternating voltages. *Rev. Sci. Instrum.*, 16, 172, 1945.
Strong, J. Apparatus for spectroscopic studies in the intermediate infrared region—20 to 40 μm. *Rev. Sci. Instrum.*, 3, 810, 1932.

Spectroscopy, Raman

When Sir C. V. Raman discovered the spectroscopic technique that bears his name in 1928, it was the result of experiments conducted with inexpensive equipment at the Indian Association for the Cultivation of Science (IACS) in Calcutta, India, far removed from the acclaimed laboratories of the Western world. Ironically, the idea for the experiments came during the return trip from his first visit to London in 1921.

Intrigued with the deep blue color of the Mediterranean Sea, Raman was skeptical of Lord Rayleigh's explanation that the color of the sea was simply the reflection of the color of the sky. Formulating his thoughts while still on board the *SS Narkunda*, Raman dashed off a letter to the editors of *Nature* as soon as he reached Bombay. Within a short time, he was able to prove that the ocean's color was not a mirror of the sky above, but the result of sunlight being scattered by the water molecules.

Analyzing the light scattered by a liquid is difficult, and Raman's early work used visual observations of color changes rather than determining the exact wavelength of the light associated with the color, as shown in Figure 27.

Raman's experiment of discovery is outlined in Figure 28. Using a violet filter to isolate the violet portion of sunlight, Raman looked at the scattered light that emerged at right angles to the original beam after it had passed through a liquid sample. Most of the light emerged unchanged as the original violet color. However, when Raman and K.S. Krishnan used a green filter to intercept the scattered light, they found that the spectrum of the scattered light had a weak green component as well as violet, meaning that the wavelength had been shifted. Raman quickly found some 60 samples that exhibited this effect, which soon became known as the "Raman effect."

In spite of this initial success, Raman's problem was that the effect was very weak, as only one in a million of the scattered photons actually changed wavelength. Although sunlight was plentiful in Calcutta, its intensity was barely adequate. In 1927 the IACS obtained a refracting telescope that Raman used to focus the sunlight into a more powerful source. By 1928, the newly available mercury-arc lamp replaced sunlight, and a quartz spectroscope replaced the visual measurement of color. Raman was now able to make precise measurements of the wavelengths of the scattered light. His refined quantitative results had an immediate impact on the scientific community, and resulted in the award of the Nobel Prize in Physics just two years later, in 1930.

At first, the Raman effect found its niche in physics. Seeking to explain the origin of the Raman effect, physicists were unavoidably drawn to using it

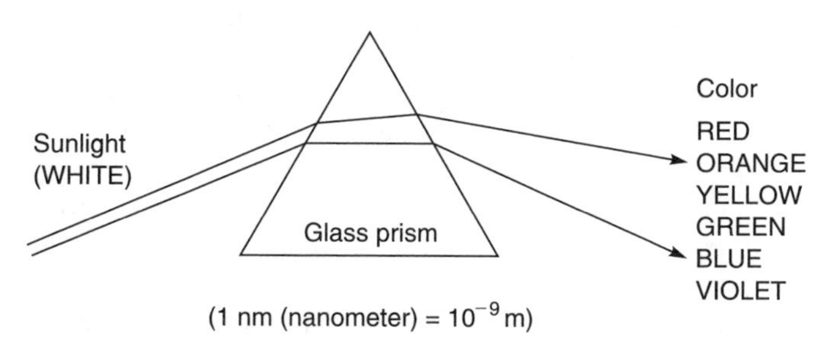

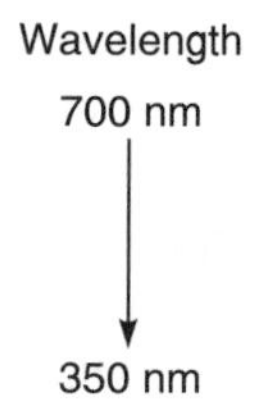

Figure 27. Analyzing light scattered by a liquid using visual observations of color changes.
[*Courtesy of the American Chemical Society.*]

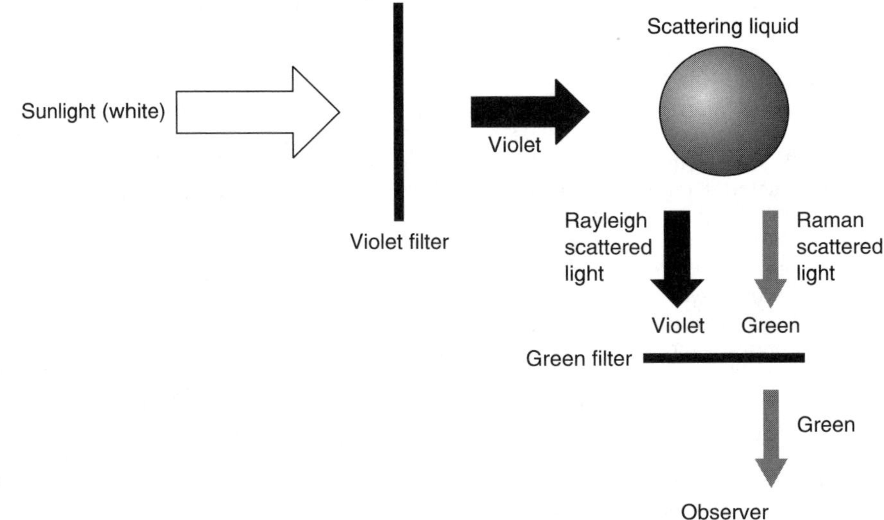

Figure 28. C. V. Raman's experiment of discovery. [*Courtesy of the American Chemical Society*.]

in their study of the vibration and rotation of molecules and ultimately the structure of those molecules. According to R.W. Wood of Johns Hopkins University, Raman's study of light scattering was "one of the most convincing proofs of the quantum theory," a revolutionary new concept that at the time was drastically changing the way that physicists thought about the structure of matter.

After seven years, the intense interest in the Raman effect by physicists waned, but chemists then found that it could be used as an analytical tool because of several characteristics. The Raman effect used by physicists thus became Raman spectroscopy when used by chemists. Since the wavelength of spectral lines of the scattered light emitted by a particular substance was independent of the incident light but characteristic to the sample (related to the internal vibrational motion of the molecule), the lines served as a fingerprint for identification. Moreover, the intensity of those lines was an indication of the amount of the substance in a sample, providing both a qualitative and quantitative determination in a single analysis. Unlike most other analytical techniques then in use, Raman spectroscopy could be applied to solids, liquids or gases, and even aqueous solutions, which had always been difficult to analyze.

Raman spectroscopy fell into disuse after World War II, unable to compete with infrared spectroscopy, a new analytical tool that had been developed during the war. Because infrared spectrometers used sensitive electronic detectors that also came out of wartime research, their operation required little training and became quite routine. By comparison, Raman spectroscopy still needed darkroom facilities and highly trained operators, and attention naturally turned to the newer and easier method.

In the 1960s, however, chemists again became interested in Raman spectroscopy because of a new development that came from physics—the laser. Many years earlier, Raman had constantly sought new and more intense light sources to enhance the scattered light and its subsequent measurement. The laser provided an intensity that far surpassed any previous light source; it also produced quasi-monochromatic light (narrow band of wavelengths), enabling the shifted wavelengths to be distinguished from the source wavelength. When the laser was coupled with Fourier transform techniques and new computers that could quickly process data, a new period of Raman spectroscopy opened in the 1980s when chemists began using the new models that were now commercially available. Today, a Raman spectroscopy system typically includes a laser source (filtered for monochromaticity), focused on a sample; the scattered light is usually detected by a multichannel charge-coupled device array (CCD), which can read the whole scattered spectrum (different wavelengths) simultaneously. Prior to this, point detectors were consecutively positioned at each point in the spectrum.

Raman spectroscopy is used in a myriad of applications, and new ones continue to be found. Micro-Raman spectroscopy is used as an analytical tool for the investigation of objects of art, antiquities, archaeological remains, and other valuables, and has been used to study the embalming techniques of ancient Egyptians. Raman spectroscopy has the potential for rapidly determining the presence of pathogenic bacteria, and will be used to analyze the mineral composition of the Martian surface. Other uses of Raman spectroscopy range from monitoring industrial manufacturing processes in the petroleum and pharmaceutical industries to analyzing illegal drug samples without destroying the evidence seal on the sample bag taken at the crime scene. Nuclear waste materials can be analyzed at a safe distance and paint can be studied while it is drying to improve its adherence properties. Fabric dyeing, foods, polymers, semiconductors, minerals, and medicine are just a few of the other areas currently using Raman spectroscopy. According to Kathy Kincade, Raman spectroscopy even "has the ability to provide specific biochemical information that may foreshadow the onset of cancer and other life-threatening illnesses."

When he described his "new kind of radiation of light emission from atoms and molecules" in 1928, Raman concluded, "We are obviously only on the fringes of a fascinating new region of experimental research which promises to throw new light on diverse problems... It all remains to be worked out." What has been "worked out" since then was recognized by the American Chemical Society in 1998, when it designated the Raman effect as an International Historic Chemical Landmark.

See also **Spectroscopy and Spectrochemistry, Visible and Ultraviolet; Spectroscopy. Infrared; Spectroscopy, X-Ray Fluorescence**

JAMES J. BOHNING

Further Reading

McCreery, R.L. *Raman Spectroscopy for Chemical Analysis*. Wiley, New York, 2000.

Bohning, J.J. *The Raman Effect*. American Chemical Society, Washington D.C., 1998.

Hibben, J.H. *The Raman Effect and Its Chemical Applications*. Reinhold, New York, 1939.

Tu, A.T. *Raman Spectroscopy in Biology*. Wiley, New York, 1982.

Venkataraman, G. *Journey Into Light*. Penguin, New Delhi, 1994.

Kincade, K. Raman spectroscopy enhances *in vivo* diagnosis. *Laser Focus World*, July, 83–91, 1998.

Various. *Nobel Lectures (Physics)*. Elsevier, New York, 1965, 263–277.

Raman, C.V. *The New Physics*. Philosophical Library, New York, 1951.

Useful websites

Aspects of Raman Spectroscopy: http://www.spectroscopynow.com/

Ghent University Raman Group and the Analysis of Artworks: http://allserv.rug.ac.be/~lmoens/Raman/Welcomeraman.html

Raman Research Institute: http://www.rri.res.in/

Nobel Prize in Physics 1930: http://www.nobel.se/physics/laureates/1930/

Spectroscopy, X-Ray Fluorescence

X-ray fluorescence (XRF) spectroscopy is a widely used method for identifying the chemical composition of almost any material, regardless of its quantity or form. The analysis is useful for two primary reasons: it is noninvasive and nondestructive to the sample material; it yields an easily identifiable and reliable atomic composition of materials, excluding only those primarily constituted of elements lighter than aluminum.

The history of XRF spectroscopy begins with Wilhelm K. Roentgen's 1895 discovery of x-rays, which earned him the 1901 Nobel Prize. Less than ten years later, Charles G. Barkla discovered a relationship between the x-rays emitted from a sample material and that sample material's atomic weight. Then in 1913, Henry Gwyn Jeffreys Moseley renumbered the periodic table of elements by measuring the x-ray fluorescence of each element. Prior to his reorganization, the periodic table was arranged by atomic weight rather than atomic number.

Moseley is also credited with inventing the first XRF spectrometer, which used electrons as a rather inefficient energy source. Other scientists continued to build variations of x-ray detectors with little progress in efficiency until 1948 when Herbert Friedman and LaVerne S. Birks built an XRF spectrometer with a highly sensitive x-ray detector that utilized a Geiger Counter. Their innovation increased the number of elements that could be identified. Then in the 1960s and 1970s, computers could rapidly analyze data, allowing for readings of multiple elements at a time and larger quantities of sample materials to be tested. Widespread usage of XRF spectrometers increased as they become more and more efficient and even portable. In fact, the National Aeronautics and Space Administration (NASA) sent XRF spectrometers on both the Apollo-15 and -16 missions, as well as to investigate the asteroid Eros, and to Mars onboard the Pathfinder in 1997 and the Rover in 2004.

XRF spectroscopy is a relatively accessible process for data collection. To conduct research, one needs a sample, an energy source, an x-ray tube, an x-ray detector, and a computer for data analysis. As noted above, one of the advantages of XRF spectroscopy is that nearly any kind of unknown material may be identified, whether it is in solid, gaseous, or liquid form. The choice of energy source is largely dependent on the form of the sample material and the particular application. Energy sources vary from x-rays to alpha-particles to beams of electrons. Via an x-ray tube, the energy source bombards the sample material. An x-ray detector or x-ray fluorescence spectrometer then measures the fluorescent light emitted from the sample atoms. Computers then are used to compute the identity of the element and its quantity within the sample. One disadvantage to this process is that most energy sources for XRF spectrometers are radioactive and must be replaced on a regular basis due to the normal decay of radioactive materials.

The basic process of XRF spectroscopy begins when, depending on the particular application, the sample is initially bombarded through an x-ray tube with an energy source such as x-rays, alpha-particles, or beams of high-energy electrons. When electrons in the innermost orbital shells of sample atoms absorb the energy, they are ejected from the atom. When this happens, electrons from the higher-energy outer shells of the atom move inward to fill vacancies in the lower-energy orbitals to stabilize its atomic structure. The movement of the electrons from higher-energy orbitals to lower-energy orbitals causes energy to be emitted in the form of fluorescent light. This fluorescent light, or x-ray, is the characteristic signature of that particular element.

The x-rays are reliable identifiers of the element because of their relationship with other properties of atomic physics. The energy emitted by the transitioning electron will be equal to the difference between the binding energies of the two orbitals occupied by that electron. Because the difference of two specific orbital shells of a given element is constant, the energy emitted by a transitioning electron is also constant and is therefore characteristic of that element. Thus, XRF spectroscopy is used to qualitatively identify the elemental content of a sample material.

XRF spectroscopy can be used to quantitatively measure the amount of a given element within a sample material. Scientists can assess the count rate or peak intensity of the wavelength of the fluorescent x-ray emitted by the transitioning electron. Specifically, the count rate refers to the number of emitted fluorescent photons per unit of time. The count rate can then be used to establish the quantity of a particular element within the sample. Analysis of count rates is easily accomplished with the help of computers.

Applications from the evaluative processes of x-ray fluorescence spectroscopy touch the everyday lives of people everywhere. Portable versions of XRF spectrometers are widely used for field applications, whether the site is an ancient city long buried by volcanic activity or the surface of Mars. The primary industrial use of XRF spectrometry is for quality control of material composition, from raw forms to finished products. Examples include NASA's use of XRF spectroscopy to evaluate the geology of Mars during the 2004 Rover expedition, and museums' use of the process to identify pigments in rare paintings for purposes of restoration. The U.S. Food and Drug Administration evaluates the content of vitamins and other drugs with XRF spectroscopy while archeologists use XRF spectrometers to identify and date artifacts by their mineral contents.

In just over one hundred years since the discovery of x-rays, XRF spectroscopy has become one of the most reliable and widely used methods for chemical composition analysis. Biologists, chemists, museum curators, health inspectors, forensic scientists, medical doctors, ecologists, mineralogists, archeologists, and many university students all use XRF spectroscopy to identify elemental constituents of a vast range of materials. The process is vital for quality control of raw materials and finished products within numerous industrial settings. By traveling into space to classify yet-to-be explored worlds as well as back in time to identify ancient artifacts, scientists use XRF spectrometers to expand the domain of knowledge.

See also **Spectroscopy and Spectrochemistry, Visible and Ultraviolet**

MARY C. INGRAM

Further Reading

Jenkins, R., Ed. *X-ray Fluorescence Spectrometry*, 2nd edn. Wiley, New York, 1999.

Rath, B.B. and DeYoung, D.J. The Naval Research Laboratory: 75 Years of Materials Innovations. *JOM: The Member Journal of The Minerals, Metals, and Materials Society*, 50, 7, 14–19, 1998.

Scott, D.A. The application of scanning x-ray fluorescence microanalysis in the examination of cultural materials. *Archaeometry*, 44, 475–482, 2002.

Van Grieken, R.E. and Markowicz, A.A., Eds. *Handbook of X-ray Spectrometry*. Marcel Dekker, New York, 1992.

Useful Websites

Mars Pathfinder Instrument Descriptions: http://mpfwww.jpl.nasa.gov/mpf/sci_desc.html

NEAR: Near Earth Asteroid Rendezvous Mission: http://near.jhuapl.edu

Sports Science and Technology

The discipline of sports science in medicine and technology is a combination of biomechanics, kinesiology, and anatomy. Within these parts, sports science dates back to Aristotle and Archimedes, when sport was an integral part of Roman and Greek life and inspired a fascination with the mechanics of the human body. By the twentieth century, sports science also began to include varying disciplines such as engineering, polymer science, psychology, and psychiatry, and performance-enhancing pharmaceuticals inspired by high-stakes competition. Today, the business of sports equipment—from running shoes to stair machines—is a multibillion dollar industry based on the continuing evolution of sports science.

As a regular to Gladiator competitions and a physician to the athletes, the second century AD Greek Galen laid out the basic motor functions of the human body. He identified the processes of muscle contraction and the influence of the mind—or "animal spirits" as he called it—on the performance of the body. Isaac Newton's basic theories of scientific reasoning, specifically the study of modern dynamics, laid the groundwork for understanding the relationship between forces and their effects, based on his laws of rest and movement published in *Principia Mathematica Philosophiae Naturalis*. Not until Leonardo da Vinci studied the structure of the full human body were the specific mechanics necessary for simple movements grasped—from walking and sitting to jumping and sprinting. From that time, interested physicians, engineers, and scientists have pursued elements of biomechanics and kinesiology. By 1865, Gullaume Benjamin Amand Duchenne published *Physiologie des Mouvements*, which identified each individual muscle in relation to its range of movement, making it the basis of most sports science work today. Twenty years later, Etienne-Jules Marey studied the motion of humans and animals through photography, publishing his groundbreaking work, *Le Mouvement*, which examined, frame by photographic frame, the intricate interaction of nerves, muscle, and bone.

With the onset of World War I, an interest in biomechanics swelled, though under unfortunate and sometimes inhumane circumstances: Analysis of soldiers on the march and of prosthesis for amputees and the exhaustive study of prisoners of war and concentration camp prisoners all furthered the understanding of biomechanics and kinesiology. In particular, Frenchman Jules Amar, evaluated the human gait and the task performances of World War I veterans. He employed force and motion measurement techniques to help develop prosthetic limbs. Further research in this subcategory of kinesiology led to studying movement in three dimensions, which led to mathematical analysis of the range of joint forces, movements, and the elements of muscle strain and force against bone and ligaments.

The invention of the ergograph by Angelo Mosso in 1884 assisted greatly in further analysis of human movement in the first half of the twentieth century and evolved into several specialized forms to study very specific muscular functions in the human body. In general, the study of ergonomics as part of the Industrial Age push to better human work performance influenced the development of specialized sports equipment. Track shoes, for example, were designed to allow the foot a "natural" range of motion by using soft leather and a snug fit.

The study of aerodynamics also began to contribute to and employ the burgeoning knowledge of sports science. Likewise, biology has played a part as well: the study of marine mammals has helped shape the swimsuits and technique of many competitive swimmers.

More than any other technical innovation, however, the development of new materials used in sporting goods equipment changed the performance of both athlete and athletic gear. For the first half of the twentieth century, sports equipment was made primarily of steel, wood, or leather; but by the 1990s, composite materials such as graphite, fiber-glass, and new kinds of plastics, metals and ceramics were regularly employed. New metal composites, for example, dramatically altered the way bicycles operated. Virtually unchanged since the 1890's, the frame of bicycles has changed from the classic diamond design—the optimal shape for the strength capacity of steel, historically the most common material. With stronger, lighter materials such as titanium, high-tensile steel, chrome-moly steel, and aluminum, frame shapes could be altered to more aerodynamic shapes without losing strength.

Like other athletic equipment, shoe designs were limited mostly to sneakers, or rubber-soled canvas lace-ups, until the late 1960s. Designs changed and became more specific to sports—the high-top sneaker for basketball, for example—but little in the materials changed until the 1990s. Keeping in mind the lessons from biomechanics and kinesiology, shoe manufacturers tried to meet performance demand, protection, fit, and a saleable style. To protect the most vulnerable part of an athlete—the musculoskeleton—the "impact waves" delivered by each foot strike to the ground need to be dampened. Shoe designers began to employ several different components in the outsole, insole, and midsole. A durable, flexible shock-absorbent rubber, such as elastomer styrene-butadeine is commonly used in outsoles. Insoles were typically made from polymer foam, and midsoles constructed of polymer foam called EVA, for ethyl, vinyl and acetate.

Running tracks also employed different forms of rubbers, polymers and plastics. The goal in track construction, as in shoes, is to reduce shock by increasing viscoelasticity. Other playing fields have evolved with new materials. High-strength tempered glass backboards have replaced most fiberglass, metal, and wood backboards in basketball. Protective gear in American football equipment involved several hard layers of plastic, wire, and soft foam. Astroturf, a plastic and rubber artificial turf, removed the threat of slippery, muddy fields but was not without controversy in some sports, particularly baseball, where "natural" grass was prized.

Like equipment technology, sports science today works largely on chemical innovation. Athletes monitor their nutritional input as much as they emphasize strength and endurance training. A low-fat, high-protein diet, for example, has been known to facilitate the repair of muscle tissue, and the intake of carbohydrates fuels the high-energy demands of a competitive athlete. Today the basics of biomechanics recorded in the early twentieth century have been applied to break down an athlete's movements into their component parts, pinpointing each muscle, nerve and tendon involved in given motion, which can then be targeted for improving performance. Strength training programs are designed for the muscle groups needed for a particular sport. Techniques for this vary, and continue to improve with the further understanding of physiology. For example, "plyometrics" is a popular strategy for pushing strength training beyond traditional weight-resistance techniques. Using a basic understanding of muscle contraction and response, plyometrics involves the stretching and contraction of muscles. Following a stretch, muscles contract more easily and responsively. A sit-up executed from an arched position over a ball, for example, will yield a stronger stomach than a standard sit-up. In the same vein, the importance of psychiatry and honing a "focus" preceding competition brought yoga into the regimen of many athletes.

In addition to increasing muscle strength, a further understanding of aerobic and anaerobic endurance has become a regular component of sports training. Even amateur athletes can now purchase products that measure their blood-oxygen level and provide information on the performance of their body. To push aerobic capacity, athletes alter their training to include intense, short episodes surpassing their peak aerobic mark. This technique—called anaerobic training—has been found to increase the body's aerobic capacity, thereby prolonging endurance at high levels of performance.

Today, the billion-dollar sports-equipment industry drives the majority of research and development in sports science. While the presence of performance enhancing drugs, such as steroids and testosterone derivatives, has been a legal bane, it has also shaped the way sports medicine and science perceive the function and capability of the human body in athletic events. It is likely that the use of drugs will continue to grow—illicit or not—alongside more accepted strategies.

See also **Health**

LOLLY MERRELL

Further Reading

Cicciarella, C. *Microelectronics in the Sport Sciences*. Human Kinetics, Champaign, IL, 1986.

Cavanaugh, P.R. *The Mechanics of Distance Running: A Historical Perspective*. Human Kinetics, Champaign, IL, 1990.

Fung, Y.C. Biomechanics: its scope, history, and some problems of continuum mechanics in physiology. *App. Mech. Rev*., 1–20, 1968.

Haag, H. *Sport Science in Germany: an Interdisciplinary Anthology*. Springer Verlag, Berlin, 1992.

Hoberman, J. *Mortal Engines: The Science of Performance and the Dehumanization of Sport*. Free Press, 1992.

Huxley, A.F. *Reflections on Muscle*. Liverpool University Press, Liverpool, 1980.

Pollack, G.H. *Muscles and Molecules: Uncovering the Principals of Biological Motion*. Ebner, Seattle, 1990.

Wilkerson, J.D. Biomechanics, The history of exercise and sport science. *Hum. Kinet*., 321–353, 1997.

Sputniks

The Sputniks were three artificial earth satellites launched by the Soviet Union in 1957 and 1958. They inaugurated the space age, were a tribute to Soviet science and technology, and caused surprise and panic in the U.S. similar to that after the Japanese attack on Pearl Harbor in December 1941. Space was transformed from a domain of fiction and fantasy into the "new frontier," ripe for scientific exploration, commercial exploitation, military penetration, and international rivalry.

The first Sputnik (the word is Russian for satellite) was launched on 4 October 1957 into a low-earth orbit. It was a small aluminum sphere 58 centimeters in diameter and weighed just under 84 kilograms. Sputnik-I circled the earth about every 96 minutes, burning out after 21 days as it plunged back into the atmosphere. Light reflected from its body, and the fact that it came as close as 228 kilometers to the earth, ensured that it was often visible to the naked eye. The frequency on which it emitted its "bip-bip" signal was easily captured by amateur radios. This public display of scientific and technological prowess was infused with political and ideological meaning. It was taken to demonstrate the superiority of Soviet communism over the capitalism of the West, and the U.S. in particular, and was intended to woo countries newly liberated from the yoke of colonialism into the Soviet camp.

Sputnik-II amplified the message. It was launched into an elliptical orbit on 3 November 1957 to commemorate the fortieth anniversary of the Bolshevik revolution. Cone-shaped, and 1.2 meters high, the satellite weighed 500 kilograms, carried scientific instruments and a small dog named Laïka. Laïka was wired for biomedical research, and she perished in her pressurized cabin after a few days in space. The satellite re-entered the earth's atmosphere on 14 April 1958.

The third and last Sputnik was launched on 15 May 1958. It weighed 13000 kilograms and carried a battery of scientific equipment. By now such exploits caused little stir. The U.S. had launched its first satellite on 31 January 1958 and another, with a different rocket, six weeks later. Even though rockets continued to pose enormous technical difficulties (no less than eight out of twelve U.S. attempts to launch satellites in 1958 ended in failure), it was evident that it was only a matter of time before satellite launches would become routine.

The first Soviet and American launches of scientific satellites into space were officially planned as part of their contributions to the International Geophysical Year (IGY). The IGY was a collaborative program in which scientists from 67 countries took advantage of a period of intense solar activity in 1957–1958 to study together a number of oceanographic and atmospheric phenomena. On 28 July 1955, the White House Press Secretary announced that the launch of small earth-circling satellites would be part of America's contribution to the IGY, providing "scientists of all nations [with an] important and unique opportunity for the advancement of science." Three days later the Soviets indicated that they had similar plans. Behind the U.S. offer was the intention to instrumentalize science for security. The Eisenhower administration wanted to begin a reconnaissance satellite program to spy on Soviet installations as part of the verification process of arms controls agreements. A civilian scientific satellite overflying Soviet territory would, it was thought, create a legal precedent for the "freedom of space," so opening the way for subsequent military-related space activities.

Moscow did not launch Sputnik without warning. An indication that the launch was imminent came on 26 August 1957, when President Khrushchev announced that the Soviet Union had successfully fired an intercontinental ballistic missile (ICBM) with a range of 8000 kilometers. Amateur radio magazines published the frequencies on which the satellite would emit its signal well before the launch date.

Sputnik shocked officials and the public in the U.S. because it was generally believed that the Soviet's were supplying misleading information for propaganda purposes. Once disabused of this misconception, the administration did not only have to deal with the blow to American prestige. It also had to face the fact, even more evident after Sputnik-II, that the Soviet rocket program had probably advanced to the point where ICBMs carrying nuclear weapons could strike New York from Moscow. Putting a dog in space also suggested that the Soviets were collecting data in anticipation of human space flight. The Sputniks were not simply scientific instruments and technological artifacts. They were also weapons in the Cold War rivalry between the superpowers, undermining American national pride, national security and presumed international technological supremacy.

The Sputniks were perceived differently in Western Europe, commensurate with local capabilities and the concerns of medium-sized world powers, and the different significance that space held on that side of the Atlantic. Scientists emphasized the scientific and technological achievements, and capitalized, when and where they could, on the opportunity space research offered for additional funding. The general public was fascinated by the sight and sound of Sputnik-I and, especially in Britain, deeply concerned about the fate of Laïka. (The Royal Society for the Protection of Cruelty to Animals' switchboard was jammed by indignant callers.) Governments, for their part, were relatively unmoved. Only a few, notably Britain and France, had small guided missile or rocket programs, and these were restricted to the upper atmosphere; none had space ambitions. Sheltering under the U.S. nuclear umbrella, and sidelined by superpower rivalry, European governments needed other stimulants to propel them into space. It was the commercial possibilities, notably of telecommunications satellites, that eventually convinced most of them that a major space effort was needed. That was only clear a decade after the Sputniks first circled the globe.

See also **Rocket Propulsion, Liquid Propellant; Rocket Propulsion, Solid Propellant; Space Exploration, Unmanned**

JOHN KRIGE

Further Reading

Divine, R. *The Sputnik Challenge: Eisenhower's Response to the Soviet Satellites*. Oxford University Press, New York, 1993.
Killian, J. *Scientists, Sputniks and Eisenhower*. MIT Press, Cambridge, MA, 1977.
Le Grand Atlas de l'Espace. Encyclopedia Universalis, France, 1989.
Launius, R., Logsdon, J.M. and Smith, R.W., Eds. *Reconsidering Sputnik. Forty Years Since the Soviet Satellite*. Harwood Academic, Amsterdam, 2000.
McDougall, W. *The Heavens and the Earth. A Political History of the Space Age*. Johns Hopkins University Press, Baltimore, 1997.
Siddiqi, A. *Challenge to Apollo. The Soviet Union and the Space Age, 1945–1974*. NASA SP-2000–4408, Washington D.C., 2000.

Strobe Flashes

Scarcely a dozen years after photography was announced to the world in 1839, William Henry Fox Talbot produced the first known flash photograph. Talbot, the new art's co-inventor, fastened a printed paper onto a disk, set it spinning as fast as possible, and then discharged a spark to expose a glass plate negative. The words on the paper could be read on the photograph. Talbot believed that the potential for combining electric sparks and photography was unlimited. In 1852, he pronounced, "It is in our power to obtain the pictures of all moving objects, no matter in how rapid motion they may be, provided we have the means of sufficiently illuminating them with a sudden electric flash."

The electronic stroboscope fulfills Talbot's prediction. It is a repeating, short-duration light source used primarily for visual observation and photography of high-speed phenomena. The intensity of the light emitted from strobes also makes them useful as signal lights on communication towers, airport runways, emergency vehicles, and more. Though "stroboscope" actually refers to a repeating flash and "electronic flash" denotes a single burst, both types are commonly called "strobes."

The stroboscope consists of three basic components. First, a power supply—an electrical outlet or battery—sends electricity into a capacitor that collects the energy. When the stored energy is dumped into the third component, a tube containing electrodes and filled with a rare gas, the excited gas molecules produce a blast of brilliant light and heat. By adjusting the flashing rate, a moving object can appear to be:

1. Frozen in place, if the flash rate equals the speed of the object
2. Moving backward, if the flash rate exceeds the object's speed
3. Moving forward, if the flash rate is less than the speed of the object

The strobe's components originated in the work of eighteenth and nineteenth century experimenters who delved into the secrets of creating electricity and storing the resulting charge. Alessandro Volta built the first battery in 1800 from a stack of silver and zinc disks, each separated by a moist cloth. Another common generator, the Wimshurst machine, built up an electrostatic charge between two hand-cranked glass disks that rotated in opposite directions, but did not touch each other. The charge was collected and transferred by metal combs or chains to a Leyden jar, a glass jar lined with metal foil inside and out. Touching the inner and outer foil layers of this eighteenth-century capacitor simultaneously produce a powerful spark.

Michael Faraday pioneered the development of the final component of a stroboscope, the gas discharge tube, in 1839, using a glass jar and two electrodes (dubbed the "electric egg"). However, it was Johann Wilhelm Geissler who turned Faraday's experimental apparatus into a practical research tool in the 1850s by producing sealed, evacuated tubes containing platinum wires that served as electrodes.

By the early twentieth century, work on stroboscopes was reaching a practical level as generators, capacitors, and gas discharge tubes became more powerful, reliable, and easier to mass produce. A number of inventors contributed to the development of the strobe, including Etienne Oehmichen who, with the *Société Anonyme des Automobiles et Cycles Peugeot*, received French and Swiss patents around 1920 for a strobe-like device that he specified would be used to examine motors in motion. In 1926 and 1927, Laurent and Augustin Seguin were awarded patents for their "flash-producing apparatus" by France, Switzerland, and the U.S. The brothers marketed this early strobe as the "Stroborama" and, by the early 1930s, the Seguins claimed industrial, government, and university laboratories around the world as their customers.

What made the stroboscope so useful was that the flash was renewable—as soon as the capacitor recharged, it was ready to go again; and since the flashing rate could be adjusted, it was possible to measure the speed of machinery in motion, as well as to spot any irregularities in the mechanism's operation, just by watching the machine in the rapidly-flashing light. However, devices like the Stroborama produced weak, long flashes that were not suitable for photography. Instead, it was the engineering and entrepreneurial talents of Harold Edgerton at the Massachusetts Institute of Technology (MIT) that made the strobe a commonplace device. Edgerton built on the existing work on strobes and experimented with ways to increase the flashing rate, make the flashes both brighter and shorter in duration, and produce light of the right color for photography.

Edgerton came to MIT as a graduate student in electrical engineering in 1926 to study the large motors used in power-generating plants. He understood the limitations of the neon strobes already in use, so he adapted a commercially available mercury tube, synchronized its short flashes to a motor's speed to make the spinning parts appear stationary, and shot his first stroboscopic photographs and motion pictures. Each brilliant flash lasted about 10 microseconds (1 microsecond $= 1 \times 10^{-6}$ seconds). From these first experiments, Edgerton created a variety of flash lamps and stroboscopes.

Throughout the late 1930s and 1940s, Edgerton concentrated on adapting the strobe for new applications. His images of stage shows and sporting events, illuminated by large strobes hung in theater and arena rafters, captured the public's imagination. These enormous flashes were later adapted for aerial nighttime reconnaissance photography during World War II. Edgerton's first camera-mounted, portable strobe ushered in a new era in photojournalism. Restless children became easier to photograph with the introduction of a strobe for studio photographers. To make high-speed motion pictures, Edgerton designed a camera without a shutter that rushed the film through continuously at speeds of around 23 meters per second. In a dark room, the regular flashing of the stroboscope acted like a shutter. Edgerton also created multiple-exposure still images of the essence of movement, such as the beat of a bird's wing. Working with explorer Jacques Cousteau, Edgerton developed strobes for underwater photography.

By the late 1940s, the demand for strobes supported more than 35 manufacturers in the U.S. During the 1950s, much of the design and manufacturing work moved first to Europe and then to Japan. Continual technical development has yielded shorter charging times, lighter-weight units, more automated functions, increased flash output, and improved circuitry, achieved by incorporating new semiconductors like transistors and later integrated circuits. The strobe has indeed become a standard tool in the photographer's kit.

See also **Cameras; Color Photography**

JOYCE E. BEDI

Further Reading

Bron, P. and Condax, P.L. *The Photographic Flash: A Concise Illustrated History*. Bron Elektronic, Allschwil, 1998.

Bruce, R.R., Ed. *Seeing the Unseen: Dr. Harold E. Edgerton and the Wonders of Strobe Alley*. George Eastman House, Rochester, NY, 1994.

Edgerton, H.E. *Electronic Flash, Strobe*. MIT Press, Cambridge, MA, 1979.

Edgerton, H.E. *Flash! Seeing the Unseen by Ultra High-Speed Photography*. Hale, Cushman & Flint, Boston, 1939.

Gus Kayafas, Ed. *Stopping Time: The Photographs of Harold Edgerton*. H.N. Abrams, New York, NY, 1987.

Talbot, W.H.F On the production of instantaneous photographic images. *Philosoph. Mag*., 3, 15, 73–77, 1852.

Submarines, Military

Although the first military submarine operated as early as 1775 and development continued throughout the 1800s, the submarine was really a creature of the twentieth century; it was the submarine and the aircraft carrier that defined naval warfare in that century.

The basic technology of the submarine is quite simple and has remained constant since its inception. The boat submerges by taking on water through vents to decrease its buoyancy and surfaces by expelling the water with compressed air. The outward appearance of the military submarine has remained remarkably constant throughout its modern development—a cigar-shaped hull topped by the immediately recognizable conning tower with a periscope for viewing the surface.

We can break submarine technology into five categories:

- Propulsion
- Hull design
- Weaponry
- Stealthiness
- Ancillary technologies.

The method of propulsion for the first half of the century was the diesel-electric system. Standard diesel engines were used for general operation on the surface but could not be used while submerged because of the enormous amount of air required for combustion. The submarine would only dive to attack or avoid detection, at which time the boat switched to power provided by electric batteries, charged from the diesels while on the surface. Most submarines were double hulled, with water filling the space between the two hulls while submerged.

The weapon that made the submarine useful was the self-propelled torpedo, powered by compressed air, which provided the sub with a deadly and stealthy attack. These technologies have been supported by countless others, each complex in its own right: atmosphere regeneration, escape and rescue techniques, underwater communications, weapons guidance, electronic countermeasure, and countless others, but all technologies unique to the military submarine revolved around one concept: avoiding detection by stealth.

The first significant use of submarines came in World War I when German U-boats attacked Allied shipping in the Atlantic Ocean. Losses of merchant ships increased and threatened to cripple the Allied war effort until a simple and ancient remedy was rediscovered—sailing the merchant ships in convoys. U-boats sank over 11 million tons of shipping but ceased to be a serious threat after the adoption of the convoy system.

In World War II, the critical Battle of the North Atlantic was a struggle defined by the technological accomplishments on each side as Germany's U-boat force sought to avoid detection from the eyes and ears of Great Britain's anti-submarine warfare (ASW) force. In the efforts at stealth and avoidance of detection, the most significant developments were the deployment of ASDIC (for Anti-Submarine Detection Investigation Committee), or sonar, followed closely by airborne radar. The most deadly enemy of the submarine turned out to be aircraft, which could detect surfaced submarines by means of radar allowing an attack with bombs or, as the submarine dived, depth charges. The U-boats countered British radar with a radar detector called Metox that warned of attack, but the British eventually deployed a new radar using a centimetric wavelength undetectable by Metox. The Germans did not discover the use of the new radar and were slow to develop an effective counter.

U-boat losses continued to rise. As a stopgap measure the Germans deployed the schnorkel, developed before the war by the Dutch but captured by the Germans upon the surrender of the Netherlands. The schnorkel, or snorkel to the Americans and snort to the British, was a simple device—a breathing tube that could be raised similar to a periscope that allowed the submarine to run its diesel engines while submerged. While technologically interesting, its practical deployment was a failure; Allied radar could soon detect even the schnorkel protruding from the water.

The war ended before the Germans could deploy their own next wave of technology embodied by the Type XXI "Walther" boat, powered by hydrogen peroxide fuel and much larger battery capacity that gave it a fast underwater speed.

A submarine is only as effective as its weaponry is reliable, and World War II saw examples of massive weapons systems failures. In the Norwegian campaign of 1940, the earth's magnetic field interfered with the operation of the U-boats' magnetically armed torpedoes. German U-boats operating off the Norwegian coast aimed torpedoes at unsuspecting British capital ships only to hear their duds clank off the sides of the targets.

In the Pacific, American submarines were armed with hopelessly defective torpedoes that rendered the American submarine fleet useless for many months until the flawed torpedo design was corrected. When effective torpedoes reached the American subs, their effect was devastating. The Japanese never developed an effective ASW force or doctrine, and U.S. subs ran wild, destroying over 60 percent of Japanese merchant shipping and paralyzing the import-constrained Japanese economy. While less well known than the great carrier battles and island invasions, the U.S. submarine force contributed at least as much to the defeat of Japan.

The most significant single development in submarine technology has undoubtedly been the use of nuclear propulsion. The first nuclear-powered boat was the *USS Nautilus*, launched in 1955. Nuclear power freed submarines from the need to surface or schnorkel. Subs could stay at sea for months longer than before and stay submerged indefinitely. The drawback of nuclear power was that it was relatively noisy, and a new generation of diesel submarines remained in use through the remainder of the century, particularly in short-range roles. The submarine also entered the area of strategic nuclear warfare, as it provided an ideal platform for long-range missiles tipped with nuclear warheads. The first launch of a ballistic missile from a submarine came from the *USS George Washington* in 1960.

With the security of a nation's entire population dependent on its defense against enemy ballistic missile submarines, antisubmarine and stealth technology became even more important. Both the U.S. and the Soviet Union devoted enormous amounts of resources in research in the race to detect the other's subs and protect their own. If the ASW-submarine contest of World War II was a battle of technology, the competition between Cold War fleets was even more so. Hull design improved tremendously; World War II subs could dive to 120 meters, and modern submarines can reach much greater depths. Hulls were also more streamlined, further increasing speed.

Weaponry diversified from earlier years. In addition to ballistic missiles and the traditional torpedo, subs began deploying sophisticated cruise missiles, first for attacking surface naval targets and, with the introduction of the American Tomahawk cruise missile, land targets.

The airplane, while still useful, gave way to the submarine itself as the most effective antisubmarine weapon with hunter-killer submarines on both sides patiently stalking the other's missile submarines lurking deep in the ocean as far as possible from enemy bases. The use of active emission sonar fell from favor except for targeting immediately before an attack, as its use gave away the position of the attacking sub. Passive listening sonar became the preferred method of tracking an enemy boat, and hence silence became the most important defense for the submarine.

A representative example of the many developments was when ASW forces began finding subs by the use of magnetic anomaly detection (MAD), the Soviet constructed their Alfa class with hulls of nonmagnetic titanium at great expense. Unfortunately for the Soviets, the machinery of the Alfas was so noisy they could be easily located by passive sonar. Such tradeoffs and competitions existed in all facets of submarine and ASW technology.

With the end of the Cold War, the usefulness and cost-effectiveness of the nuclear submarine began to be seriously questioned, but technological advances continued. The *USS Seawolf*, launched in 1997, boasted a nuclear reactor fueled with liquid sodium and the Virginia class promised to be even more advanced when deployed early in the twenty-first century.

The use of the submarine and its technological advance paralleled the changes of war in world society over the course of the century. World War I began with the U-boats operated by a strict law of maritime warfare, warning merchant ship crews of their presence and patiently waiting for the embarkation of the crew in lifeboats. Well before the end of the century, nuclear submarines lurked in the depths, each waiting to destroy dozens of enemy cities without warning.

See also **Radar Aboard Aircraft; Sonar; Submersibles**

FRANK WATSON

Further Reading

Clancy, T. *Submarine: A Guided Tour Inside a Nuclear Warship*. G.P. Putnam's Sons, New York, 1993.

Davies, R. *Nautilus: The Story of Man under the Sea*. Naval Institute Press, Annapolis, MD, 1995.

Miller, D. and Jordan, J. *Modern Submarine Warfare*. Military Press, New York, 1987.

Van der Vat, D. *Stealth at Sea: The Submarine at War*. Weidenfeld & Nicolson, London, 1994.

Submersibles

Beginning in the 1860s, and up to World War I, most submersible inventions were marketed for their military potential. After World War I, however, interests in oceanic exploration, for purposes ranging from scientific research to resource exploitation and wreckage recovery, prompted the development of several types of civilian submersibles. A further reason for the development of submersibles was an understanding of the limit of the human body to withstand great depths. Initially, diving bells were the basic means of underwater exploration, but by the twentieth century it was clear that no amount of improvement to the open diving bell (where the air became more compressed as the depth increased) could compete with enclosed habitats.

Enclosed bells were bathyspheres (from the Greek "deep sphere"), submersibles tethered to the surface for suspension, air, and power supply, first developed in the late nineteenth century. In 1930, naturalist William Beebe and geologist Otis Barton designed a stainless steel bathysphere that was 1.45 meters in diameter, with walls 3.8 centimeters thick, and three windows of quartz glass for observation purposes. The inside equipment was spartan, with no seat for the two occupants, internal oxygen tanks with 8 hours of supply, and soda lime and calcium chloride chemicals spread to absorb carbon dioxide (CO_2) and moisture. First diving to 244 meters in 1930, the two scientists eventually established a record of 923 meters in 1934. This record would hold until after World War II, partly because of limited interest in and therefore limited funding for oceanographic science.

In the meantime, a Swiss physicist, Auguste Piccard, had successfully navigated to 18 kilometers into the stratosphere in a sealed sphere, FNRS-1, attached to a balloon (the Belgian National Science Foundation had paid for the craft, hence the French language acronym that christened it). Based on his flight experience, Piccard conceived of a similar plan, this time to reach great oceanic depths. He felt, however, that the bathysphere was too vulnerable due to its dependency on the tether, which might snap. Thus, Piccard and his son, Jacques, decided to include a float that would have a function similar to that of a balloon. The float featured a gasoline tank (gasoline is incompressible and lighter than water, which would help raise the vehicle to the surface). Several tons of iron pellets were also in the float to provide negative buoyancy. Although the craft, known as the bathyscaphe ("deep boat") FNRS-2, was eventually launched to a depth of 4048 meters in 1954. Piccard was no longer involved in the project, after a falling out with the French Navy, which had taken over the project from Belgium. He therefore shifted his interest to a new project, a bona fide exploration submarine.

This new model, the Trieste (named in honor of the Italian city that had funded its construction), was divided into two sections, with the upper containing 106 cubic meters of gasoline, two ballast water tanks, and 9.1 tons of metal pellets (later 16 tons). The second section was a steel alloy sphere large enough for two people. The whole contraption weighed 50 tons. Although built for Auguste Piccard and his son Jean and modified in the 1950s, it was eventually transferred to the U.S. Navy, which helped Jean Piccard and US Navy Lt. Don Walsh set a depth exploration record in January 1960 by reaching a depth of 10.91 kilometers in the Mariana trench. Trieste's exploit, as well as its assistance in discovering the wreckage of the *USS Thresher* submarine off Massachusetts in 1963 reflected a shift in military and governmental circles in favor of oceanographic research and recovery. Allyn Vine of the Woods Hole Oceanographic Institute had recommended the purchase of Trieste by the U.S.. His contribution to the development of research submersibles would soon be acknowledged through the naming of a new type of deep research vessel (DRV) in his honor.

The Alvin (named for Allyn Vine) is one of the first and longest serving DRVs. Operated by the Woods Hole Oceanographic Institute since 1965 and weighing 17 tons, it uses six reversible electric thrusters, can carry a payload of 680 kilograms at a maximum speed of 2 knots. Its crew of three can remain submerged for up to 72 hours in the titanium pressure hull. Like other research submersibles of its kind, Alvin requires a support ship to remain in operation, as its maximum cruising range is limited to five kilometers. Alvin's early successes (it helped locate a lost H-bomb off the Palomares coast of Spain) also prompted a "golden age" of civilian submersible design in the 1960s, when several commercial firms took to designing DRVs. Westinghouse built the Deepstar 4000 for oceanographer Jacques Cousteau, while Grumman designed the PX-15 Ben Franklin, a mesoscaphe for use in exploring mid-depth phenomena such as the Gulf Stream. Although mostly of steel and titanium, some crafts came to include acrylic plastics that allowed crew to have a greater view outside the observation compartment.

The first submersible to be used for tourism was the Auguste Piccard, capable of carrying 40 passengers and used during the Swiss National Exhibition in Lausanne in 1964 to make dives in Lake Geneva. Later models, much smaller, are still used for underwater viewing. Canadian and French companies did attempt the construction of a long-range commercial nuclear-powered submarine, SAGA-1, capable of carrying 15 passengers for a month to investigate oil drilling and other oceanic resource exploitation. However, financial problems suspended the project in 1987.

DRVs also rely on the use of remotely piloted vehicles (RPVs). Although limited in their steering and recovery capacities, these machines began playing an important role in tracking underwater wrecks. RPVs usually feature a steel frame that includes the necessary instruments and an engine, but more advanced types feature tracks for crawling on the sea bottom. The most famous were the ones used by Dr. Robert Ballard to locate the wreck of the *Titanic* in 1985–1986, which included the Jason, piloted from Alvin.

Small civilian submarines are commonly used in industrial surveys and underwater tourism, but the infrastructure required for bigger ships makes their cost prohibitive to the extent that governments are either the main sponsors or even the main operators of these ships.

See also **Submarines, Military**

GUILLAUME DE SYON

Further Reading

Ballard, R.D. *The Discovery of the Titanic.* Warner/ Madison Press, New York, 1987.
Beebe, W. *Half Mile Down.* Harcourt Brace, New York, 1934.
Geyer, R.A. *Submersibles and Their Use in Oceanography and Ocean Engineering.* Elsevier, New York, 1977.
Piccard, J *The Sun Beneath the Sea.* Charles Scribner's Sons, New York, 1971.
Shenton, E.H. *Exploring the Ocean Depths.* W.W. Norton, New York, 1968.

Superconductivity, Applications

The 1986 Applied Superconductivity Conference proclaimed, "Applied superconductivity has come of age." The claim reflected only 25 years of development, but was justifiable due to significant worldwide interest and investment. For example, the 1976 annual budget for superconducting systems exceeded $30 million in the U.S., with similar efforts in Europe and Japan. By 1986 the technology had matured impressively into applications for the energy industry, the military, transportation, high-energy physics, electronics, and medicine. The announcement of high-temperature superconductivity just two months later brought about a new round of dramatic developments.

By 1986 the energy industry witnessed development of large superconducting projects for fusion power generation, magnetic energy storage, transmission lines, and industrial motors and generators. For instance, the international large coil test, begun in 1977 to demonstrate the capability of producing fields sufficiently strong for fusion power, utilized six D-shaped superconducting magnets, each 4.5 meters tall, 3.5 meters wide, and 1 meter thick. The coils, developed by Japan, Switzerland, Euratom (a European consortium) and three U.S. industries, were toroidally assembled at Oak Ridge National Laboratory where they successfully produced a 9-tesla field in 1987. Enormous superconducting solenoids were designed in the U.S. and Japan to store energy for diurnal load leveling by electric utilities. The designs for 5000 megawatt-hours storage proposed magnets 1 kilometer in diameter by 20 meter tall, supported by an underground earth-based structure. Superconducting transmission lines such as the 1000 MVA prototype developed from 1971–1986 at Brookhaven National Lab were designed for underground service to mitigate power congestion in large cities.

The British and U.S. Navies utilized superconducting magnets to build homopolar motors for ship propulsion. Although only 2 horsepower in 1965, these motors grew to 3250 horsepower by 1970 in the U.K. and to similar size by 1983 in the U.S. High-speed trains levitated with superconducting magnets exceeded 500 kilometers per hour by 1980 on the Japanese Miyazaki test track. In the field of high-energy physics, superconducting dipole magnets were built to accelerate beams of subatomic particles around multi-kilometer long rings, while large superconducting solenoid magnets were constructed to detect the particle spray from collisions of the accelerated beams. The "energy-doubler" accelerator built at Fermilab in Batavia, Illinois between 1971 and 1983 utilized 744 6-meter-long dipoles, and 244 2-meter-long quadrapole magnets.

Magnetic resonance imaging (MRI) provided the largest commercial success for superconductivity. Introduced in 1980, by 1986 MRI systems were used in over 700 hospitals and represented a $1 billion per year industry. The very stable 1–2 tesla field produced by the superconducting MRI solenoids coupled with sophisticated computer systems enabled a noninvasive technique for seeing inside the body (see entry on Nuclear Magnetic Resonance). The second major medical application of superconductivity utilized their ability to detect magnetic fields as small as 10^{-14} tesla. Medical devices using superconducting quantum interference devices (SQUIDs) were developed during the 1980s to detect the fields arising from miniscule currents within the human brain and heart, allowing nonsurgical diagnoses of irregularities. Josephson junctions, the building block for SQUIDS, were also utilized to develop high-precision high-speed electronics. Projects in the U.S. and Japan explored the possibility of super-fast computers, developing logic circuits by the mid-1980s with switching times of only 2.5 picoseconds.

Numerous technical challenges were overcome to achieve the remarkable progress reported above. Constructing reliable superconducting magnets required breakthroughs in five different areas: structural support, conductor stability, protection systems, AC losses, and optimized material microstructures. Structures were developed to support pressures exceeding 100 atmospheres produced inside the coils from the combination of high

Figure 29. Enlargement of 0.8-millimeter diameter NbTi composite superconductor fabricated in 1992 by the Advanced Superconductor division of Intermagnetics General Corporation. The cross section of the conductor is superimposed on an enlarged detail of the 6-micrometer diameter filaments. The NbTi filaments are surrounded by a high-purity copper matrix. [*Source: Image by Peter J. Lee. Used with permission.*]

currents and high fields. Conductor design principles were established so that thermal energy, released when large forces caused conductor motion, would not cause permanent loss of superconductivity. Magnet protection systems were devised to avoid permanent damage to the coils if large portions became nonsuperconducting. Conductor cables were twisted to minimize losses associated with changing magnetic fields. Multistage conductor fabrication processes involving bundling, swaging, and heat treatments were developed to maximize the current-carrying capacities. Maximum current densities for niobium–titanium (NbTi) conductors rose from 2000 amps per square millimeter at 4.2°Kelvin and 5 tesla in 1980 to 5000 amps per square millimeter by 1991. As shown in Figure 29, an optimized NbTi conductor was made up of thousands of 10-micrometer-sized filaments embedded in copper.

The discovery of high-temperature superconductivity (HTS) in 1986–1987 presented the exciting possibility of simplified cooling requirements, but introduced an entirely new set of technical challenges. Foremost among these was the materials problem of converting ceramic superconductors into practical wires, tapes, or films. Although large superconducting currents were measured within the plane of single grains, by 1991 researchers identified that intergranular current flow was very sensitive to grain alignment and grain-boundary cleanliness. Successful processing techniques required very pure original ingredients, and the means to produce flattened gains with large intergranular contact surfaces. Among the many HTS materials, bismuth–strontium–calcium–copper–oxide BSCCO) and yttrium–barium–copper–oxide (YBCO) have demonstrated the most promising properties for applications. The powder-in-tube approach, introduced by the Japanese in 1989 demonstrated a reliable method to produce silver-clad BSCCO conductors. Here, superconductor-precursor powders are packed into silver tubes and subsequently subjected to a sequence of heat treatments and deformation steps. Multifilamentary versions of these conductors, first developed by Sumitomo Electric in 1991 are now widely used in emerging applications. In 1995 Los Alamos National Lab reported critical current densities exceeding 1 million amps per square centimeter at 75°Kelvin in thin-film tapes of YBCO. Because it can support higher current densities than any other superconductor at 75°Kelvin and at fields larger than 20 tesla, considerable effort is underway to produce useful conductor lengths with this material.

Since 1987 worldwide development of HTS applications has been increasing rapidly. Examples of 5000-horsepower motors, 1.2 MVA transmission lines, and MRI coils mirror the low-temperature superconducting (LTS) applications. Although the number of large-scale LTS projects has been dwindling since 1986, new HTS applications have emerged. The levitating strength of HTS materials is being utilized in flywheel energy storage systems, telescope tracking structures, and maglev transportation vehicles. Combined with compact refrigerators, HTS filters are being utilized in cellular-phone base stations and satellite communication systems.

See also **Josephson Junction Devices**

JOHN PFOTENHAUER

Further Reading

Birmingham, B.W. and Smith, C.N. A survey of large scale applications of superconductivity in the US. *Cryogenics*, 16, 59–71, 1976.

Dahl, P.F. *Superconductivity: Its Historical Roots and Development from Mercury to the Ceramic Oxides.* American Institute of Physics, New York, 1992.
Hulm, J.K. International Cooperative—Collaborative Perspectives: Superconductive Science and Technology. *IEEE Trans. Magn.*, MAG-23, 2, 423–426, 1987.
Sheahen, T.P. *Introduction to High Temperature Superconductivity*. Plenum Press, New York, 1994.
Simon, R. and Smith, A. *Superconductors: Conquering Technology's New Frontier*. Plenum Press, New York, 1988.
Weinstock, H. and Nisenoff, M., Eds. *Microwave Superconductivity*. Kluwer, Dordrecht, 2001.
Wilson, M.N. *Superconducting Magnets*. Oxford University Press, Oxford, 1983.

Superconductivity, Discovery

As the twenty-first century began, an array of superconducting applications in high-speed electronics, medical imaging, levitated transportation, and electric power systems are either having, or will soon have, an impact on the daily life of millions. Surprisingly, at the beginning of the twentieth century, the discovery of superconductivity was completely unanticipated and unimagined.

In 1911, three years after liquefying helium, H. Kammerlingh Onnes of the University of Leiden discovered superconductivity while investigating the temperature-dependent resistance of metals below 4.2°Kelvin. Later reporting on experiments conducted in 1911, he described the disappearance of the resistance of mercury, stating, "Within some hundredths of a degree came a sudden fall, not foreseen [by existing theories of resistance]. Mercury has passed into a new state, which ... may be called the superconductive state." By February of 1914 Onnes discovered that tin and lead were also superconductors, and that all three elements remained superconducting only below their threshold temperature, current, and magnetic field. In 1913 Onnes proposed production of intense magnetic fields, 200,000 times larger than the earth's magnetic field of 0.5 gauss, by winding a 30 centimeter diameter coil out of superconducting lead wire. Unfortunately the idea was not to be realized at that time. As the current, and resultant magnetic field were increased, the wire became resistive at a much lower current than the threshold current in zero magnetic field. In 1917 Silsbee explained the reduced performance by relating the maximum allowable current to the presence of the superconductor's threshold magnetic field at the conductor surface. Of the many superconductors known today, the 24 that are pure elements all lose their superconductivity at magnetic fields of less than 0.2 tesla (2000 gauss); far below the multitesla fields proposed by Onnes.

Another surprise in the field of superconductivity came in 1933 when Meissner and Ochsenfeld found superconductors to behave differently than perfect conductors with respect to their interactions with magnetic fields. As shown in Figure 30, when a perfect conductor (here, a solid sphere) is cooled in the presence of a magnetic field, and the field is subsequently removed, surface currents are established to maintain the magnetic field in the sphere. However, cooling a superconductor below its transition temperature in the presence of a magnetic field, causes the field to be immediately expelled from the superconductor. Such behavior motivated Gorter and Casimir in 1934 to describe superconductivity as a separate thermodynamic state, in a similar sense that ice is a separate thermodynamic state of water.

The investigation of alloy superconductors between 1930 and 1935 presented further perplexities. For these materials, two critical magnetic fields H_{c1} and H_{c2} were identified. Below H_{c1} the alloys behaved as the elemental superconductors, expelling the magnetic field. However, between H_{c1} and H_{c2}, magnetic flux, and associated nonsuperconducting or "normal" regions, increasingly penetrated the superconductors until at H_{c2} superconductivity was completely eliminated. In 1950, Pippard at Cambridge explained why supercon-

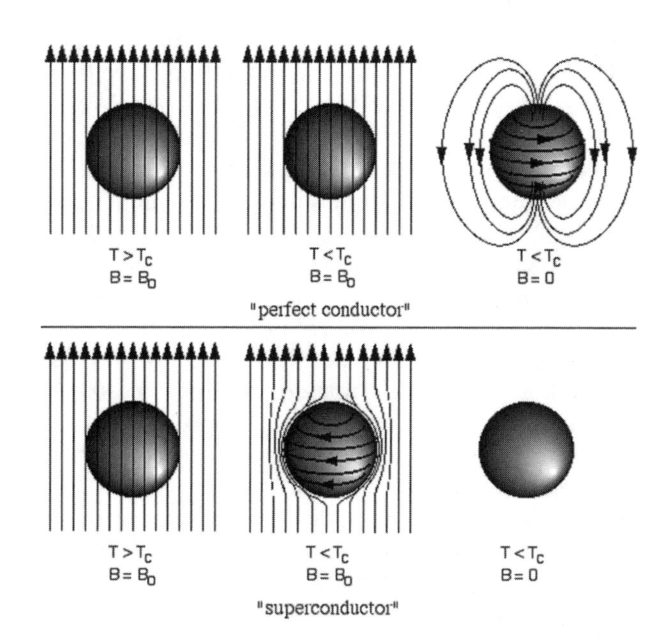

Figure 30. The response of a perfect conductor (above) and a superconductor (below) to the sequence: apply magnetic field, cool below the critical temperature, remove magnetic field.

ducting elements and alloys behave differently basing this on the surface energy between the superconducting and normal regions. For reasons similar to those dictating whether water will spread-out or bead-up on various surfaces, Pippard showed that the elements favored a minimum surface between the superconducting and normal regions, thereby expelling all magnetic flux, while the alloys favored maximum interfacial surface, thereby enabling flux penetration. In 1957 Abrikosov at Moscow accurately predicted that the penetration of magnetic flux would take the form of a regular array of flux lines. Also in 1957, a microscopic explanation for the existence of superconductivity was provided. The BCS theory, named for Bardeen, Cooper, and Schrieffer, of the University of Chicago, accounted for the absence of resistance by identifying a coupling of electrons in a low-energy state. In the coupled state, usually dissipative collisions with the vibrating atomic lattice became shared exchanges of momentum.

During the 1950s a foundation was quietly being laid for an explosion of applications in the 1960s. In 1955, Yntema at the University of Illinois utilized an advantage of current density in a strained sample of Niobium wire to wind a small magnet that produced an unexpectedly high field of 0.7 tesla. Hulm at Westinghouse, followed quickly by others, recognized the importance of the strained, or cold-worked, material. By 1961 magnets were developed using NbZr, NbTi, and Nb_3Sn that produced fields as high as 7 tesla. All were surprised that the conductors were able to carry such large currents. An understanding developed gradually through the 1960s that for the alloy, or type II, superconductors the maximum current density was not intrinsically linked to the magnetic field. Rather, the maximum current was dependent on how well the flux lines were pinned by microstructural imperfections in the material. Following this understanding, numerous magnets were developed that produced the fields envisioned by Onnes in 1913. Two developments in 1956 also birthed a variety of superconducting electronics applications. In the first, a layered structure of tin, insulator, and lead was utilized as a fast-switching element in a digital computer. The second, an observation by Josephson of superconductive "tunneling" through insulators, enabled alternative fast-switching devices, and superconducting quantum interference devices (SQUIDs) that allow extremely sensitive measurements of magnet fields.

A final explosion in superconductivity occurred in 1986. Convinced the search for superconductivity in the intermetallic compounds should not be further pursued, Bednorz and Müller at IBM in Zurich conducted a four-year search through a class of ceramic materials called perovskites. Their cautious announcement in September of 1986 entitled "Possible High T_c Superconductivity in the Ba–La–Cu–O System" described a precipitous drop in resistance at 35°Kelvin, more than 10°Kelvin higher than any previous superconductor. A flurry of reports during the next two years from the U.S., Japan, and China revealed additional high-T_c superconductors with critical temperatures ranging up to 150°Kelvin. The difficult task of converting the ceramic materials into reliable electronics devices and wires for magnets occupied engineers and scientists for the remainder of the twentieth century. Their determination is now providing an exciting array of commercial applications.

See also **Superconductivity, Applications**

JOHN PFOTENHAUER

Further Reading

Dahl, P.F. *Superconductivity: Its Historical Roots and Development from Mercury to the Ceramic Oxides*. American Institute of Physics, New York, 1992.

Hazen, R.M. *The Breakthrough: The Race for the Superconductor*. Simon & Schuster, New York, 1988.

Lynton, E.A. *Superconductivity*. Methuen, London, 1969.

Mendelssohn, K. *The Quest for Absolute Zero: the Meaning of Low Temperature Physics*. Taylor & Francis, London, 1977.

Schooley, J.F., Ed. Applied Superconductivity Conference, 1986. *IEEE Trans. Magnet*., MAG-23, 2, 1987. The first nine articles in this volume comprise a session celebrating the seventy-fifth anniversary of the discovery of superconductivity and recounting its history.

Sheahen, T.P. *Introduction to High Temperature Superconductivity*. Plenum Press, New York, 1994.

Simon, R. and Smith, A. *Superconductors: Conquering Technology's New Frontier*. Plenum Press, New York, 1988.

Surgery, Plastic and Reconstructive

As early as 1500 BC, surgeons in India used leaves to rebuild the amputated noses of criminals. By the Renaissance, Tagliacozzi devised a flap and graft using the patient's skin. A hiatus then prevented further development of plastic surgery technologies because of a belief that restoring and remolding was disreputable in that it interfered with the artistry and will of God. This attitude lasted until after World War I when plastic surgery came of age as an established specialty. The dire need for reconstruction and cosmesis from the mutilating defects of war spurred the research of new technologies.

In 1917, the tubed pedicle flap was devised simultaneously by Sir Harold Gillies (London) and W. Filatov (Ukraine). The graft could now be freed to grow independently. The dermatome, a surgical device that allowed the operator to excise tissue of uniform and predetermined thickness, was invented by Padgett in 1939 and used extensively in plastic surgery.

Wartime morbidity and industrial accidents challenged surgeons to restore circulation to damaged body parts. The first successful reattachment of an arm was reported in 1962 and of a finger in 1968. The 1980s saw the rise of microsurgery where small vessels, muscles, bone, nerves, and veins could be repaired because microscopy allowed adequate visualization for manipulation of tissues without damage to tiny structures. Repair of cleft lip and cleft palate and other congenital anomalies, including the genitalia, were improved with microsurgery. Chemical and mechanical technology contributed to improved healing through the development of synthetic antibiotics and safer and longer lasting anesthetics. Early antibiotic use hinged on Fleming's discovery of *Penicillium notatum* in London (1928), which was followed by tyrothricin (Rene DuBos) and Waksman's development of streptomycin in 1944.

After World War II, small-breasted Japanese women, emulating the images of American women, underwent breast augmentation surgery using silicone injection. Failure of the procedure resulted in granulomas and silicone migration. This led to the development of a gel prosthesis in 1963 by Cronin and Gerow with a Dacron patch for attachment to the chest wall. A variety of breast implants continued to be used, some with smooth outer envelopes, others with a fuzzy polyurethane shell believed to stimulate tissue retention. Internal materials ranged from dimethylsiloxane nonliquid gel to saline. One technology used tissue expanders for the gradual introduction of saline over a period of time. The DIEP procedure is a microsurgical approach to breast reconstruction using the patient's own skin and fat. After mastectomy, a flap is made from abdominal tissue, and after blood flow is established it is transplanted to the chest wall site.

Cosmetic facial surgery gained in acceptance for the general public as the media promoted stars who appeared to defy ageing. The rhinoplasty, originally developed in Europe, gained popularity as second and third generation Jewish and Italian women in the U.S., desirous of assimilation, underwent this surgery. Autogenous graft materials were taken from ear or rib cartilage, septal tissue, and rib or iliac bone. Gore-Tex, a synthetic material, has been used to augment facial soft tissue.

During the last quarter of the twentieth century, tissue expanders and endoscopy (1993) contributed to less operative time and trauma, faster healing, and reduced intraoperative bleeding. "Facelifts" consist of many procedures. Brow and forehead lifts remove wrinkles from the upper face. Endoscopy removes excess skin and fat from around the upper and lower eyelids. Implants of collagen are used to enhance cheekbones. In the 1990s came the startling news that *Clostridium botulinum*, the same organism known to cause botulism, was being used in cosmetic surgery. The organism produces two toxins, type A and type B. Tiny amounts of these as Botox are injected into selected small facial muscles to paralyze them so that frowning will not be possible. Some doctors use electromyograms to guide the needle in the procedure. A self-inflating expander that contains salt, which gradually absorbs fluid, was also being researched at the turn of the twenty-first century.

Perhaps the most ubiquitous technology has been the laser, beginning in 1961 for ophthalmology, then introduced to plastic surgery in the 1990s. The original ruby crystal was replaced with argon and then the pulsed-dye laser. In 1995, laser resurfacing using carbon dioxide (CO_2) was introduced. The tremendous use of lasers within the field of plastic surgery has resulted in removal of hemangiomas, tattoos, vascular lesions, and wrinkles. Endoscopic surgery combined with the erbium: yttrium–aluminum–garnet laser is another late-twentieth century technology.

Between 1930 and 1990, chemicals such as resorcinol, salicylic acid, phenol, and trichloracetic acid (TCA) peels were used to smooth skin. Dermabrasion, a mechanical process, was also employed. Dressings, adhesives and masks to aid healing developed with each of these technologies.

Liposuction has been used since the late 1970s. This technique uses a cannula and suction equipment to remove fat from the thorax, abdomen, and extremities. Internal ultrasound is combined with suctioning so the operator can better view tissue planes.

For most of the twentieth century, cutting tools were made from metal and ceramic materials. The latest development is the argon gas torch, which both creates an incision and coagulates the blood to limit bleeding. Technologies that have greatly enhanced the ability to stop capillary bleeding during cosmetic surgery are platelet gel and fibrin glue, which seal wound surfaces and stop bleeding during surgery and the formation of postoperative hematomas. Superior to commercial products, the patient's own blood cells can be collected and used to preclude disease transmission and problems of tissue incompatibility. Suture materials developed from silk and gut have been replaced with vicryl and absorbable synthetic materials; needles are preloaded and packaged in sterile containers.

Body fashion, like other cultural phenomena, changes with time, place, and value. It appears that the more the human body is exposed to display, the greater a variety of technologies will develop to mold it into culturally pleasing icons.

LANA THOMPSON

Further Reading

Biggs, T.M. and Humphreys, D.H. Augmentation mammaplasty, in *Grabb and Smith's Plastic Surgery*, Smith, J.W. and Aston, S.J. Eds. Little, Brown & Company. Boston, 1991.

Castiglioni, A. *A History of Medicine*. Alfred A. Knopf, New York, 1958.

Haeger, K. *History of Surgery*. Bell, New York, 1988.

Jankaukas, S., Kelman, C.I. and Grabb, W.C. Basic technique of plastic, in *Grabb and Smith's Plastic Surgery*, Smith, J.W. and Aston, S.J., Eds. Little, Brown & Company. Boston, 1991.

Man, D., Plosker, H. and Winland-Brown, J. The use of autologous platelet-rich plasma (platelet gel) and autologous platelet-poor plasma (fibrin glue) in cosmetic surgery. *Plast. Reconstr. Surg.*, 107, 229–239, 2001.

Margotta, R. *The History of Medicine*. Smithmark, 1996.

Roberts, T. and Pozner, J.N. Aesthetic laser surgery in *Plastic Surgery: Indications, Operations and Outcomes*, Guyuron, B., Ed. Mosby, St. Louis, 2000.

Synthetic Foods, Mycoprotein and Hydrogenated Fats

Food technologists developed synthetic foods to meet specific nutritional and cultural demands. Also referred to as artificial foods, synthetic foods are meat-free and are designed to provide essential fiber and nutrients such as proteins found in meats while having low saturated fat and lacking animal fat and cholesterol. These foodstuffs are manufactured completely from organic material. They have been manipulated to be tasty, nutritionally sound with major vitamins and minerals, have appealing textures, and safe for consumption. Synthetic foods offer people healthy dietary choices, variety, and convenience.

Mycoprotein

Mycoprotein is created from *Fusarium venenatum* (also known as *Fusarium graminearum*), a small edible fungi related to mushrooms and truffles that was initially found in the soil of a pasture outside Marlow in Buckinghamshire, England. Concerned about possible food shortages such as those experienced in World War II Europe; as global populations swelled postwar, scientists began investigating possible applications for this organism as a widely available, affordable protein source. Scientists at one of Britain's leading food manufacturers, Rank Hovis McDougall, focused on mycoprotein from 1964. At first, they were unable to cultivate fungus to produce mycoprotein in sufficient quantities for the envisioned scale of food production. Food technologists devoted several years to establishing procedures for growing desired amounts of mycoprotein. They chose a fermentation process involving microorganisms, somewhat like those historically used to create yogurt, wine, and beer.

A controlled fermentation process permits consistent production and harvesting of mycoprotein. Unlike alcohol fermentation, mycoprotein fermentation retains the microorganism cells instead of discarding them when the process is completed. The process begins with the sterilization of a fermenter container to protect microorganisms from harmful contaminants. Technologists provide the microorganism culture with glucose to sustain growth and expansion. Ideal fermenting conditions for mycoprotein production require sufficient oxygen and nitrogen, temperature monitoring, and adequate glucose, biotin, and mineral nutrient supply. Steady, sustained mixing ensures suitable oxygen and food levels. For small fermenters, magnetic stirrers or paddles are used to achieve mechanical stirring of the culture. Large-pressure cycle fermenters function by appropriating air movement inside the fermenter to mix microorganisms. When the desired amount of cells has been generated, technologists remove them from the fermenter. Harvested mycoprotein resembles soft bread dough and is bland tasting. Consisting of fine fibers, fermented mycoprotein has a texture reminiscent of lean meats. Approximately 50 percent of mycoprotein is protein.

British nutritionists selected mycoprotein as the main ingredient to create food that was eventually given the brand name Quorn™. For a decade, they evaluated mycoprotein's safety with such studies as feeding it to a succession of test subjects, beginning with rats and dogs, and finally humans. Several generations of animals that were fed mycoprotein diets flourished. Human digestibility trials were also successful, and the U.K. Ministry of Agriculture, Fisheries and Food designated mycoprotein safe for nutritional consumption. Beginning in 1985,

Marlow Foods Ltd, a subsidiary of the pharmaceutical manufacturer AstraZeneca, used mycoprotein to manufacture a variety of Quorn foodstuffs flavored to substitute for chicken and beef.

The manufacturing process to create specific meat substitutes requires binding mycoprotein cells with other ingredients so that the muscle tissue structure of meat is simulated. Food technologists mix vegetable flavorings and egg white with mycoprotein to manufacture food products that appeal to consumers. They designed the mycoprotein-binder mix to resemble cuts of meat, patties, or nuggets. Those shapes are steamed to set the binder, then frozen and packaged for distribution. Marlow Foods Ltd produces Quorn wedges by fermenting wheat and corn sugars and mixing them with thickening agents and egg whites. The porous white substance can absorb spices, sauces, and flavorings for specific applications. Quorn Foods Inc. markets mycoprotein products as ingredients for cooking, including stir fry tenders, or complete entrées such as lasagna. These items can be grilled, cooked, or baked in regular kitchen appliances, including microwaves. They are intended to make food preparation convenient and quick.

Although protein shortages that had been projected in the 1960s did not develop, the early twenty-first century bovine spongiform encephalopathy (BSE) epidemic in Europe raised concerns about meat consumption and increased demand for alternative foods. For the most part, consumers interested in meat substitutes have accepted Quorn, and it became the leading synthetic food in Britain and international markets. Both health food stores and mainstream groceries in Europe and the Americas stock Quorn products. Approximately, one billion units of Quorn food were sold to an estimated 20 million people by the early twenty-first century. Besides its nutritional benefits, Quorn has been proven to reduce cholesterol levels and help dieters because people feel full after consuming Quorn products and are less inclined to indulge in excessive calories. Initially Quorn was expensive, but manufacturers have reduced prices to be similar to those of meat-substitute competitors.

When Quorn foods became available in the U.S. in January 2002, the Center for Science in the Public Interest (CSPI) demanded that the Food and Drug Administration (FDA) ban it. CSPI spokesman Michael F. Jacobson, PhD, blamed Quorn for U.S. consumers' digestive problems and attempted to brand it as a controversial new food. European consumers had not experienced similar Quorn-related health issues. Journalists reported that the CSPI had exaggerated and manipulated statistics, attributing symptoms from other disorders and causes to Quorn in an attempt to discredit the product. Resarchers initiated studies, including one at London's Royal Brompton Hospital, to evaluate Quorn and determined that it posed no significant allergen and health risks. Scientists consider mycoprotein among the safest proteins for dietary use. Because of CSPI pressure, the FDA did insist that Marlow Foods Ltd change labels inaccurately stating Quorn was derived from a mushroom family member because the company was afraid the term fungus would repulse consumers. Using chemostats, microbiologists continue to study how the mycoprotein fungus has evolved and mutated, and they have isolated variation strains to evaluate their impact on mycoproteins. Although Quorn had been economically profitable, AstraZeneca divested Marlow Foods Ltd to Montagu Private Equity in spring 2003.

Hydrogenated Fats

Food technologists create hydrogenated fats by processing vegetable oils, consisting of glycerides and fatty acids, with chemicals to achieve certain degrees of hardening. Partial hydrogenation stiffens oils, while full hydrogenation converts liquid oils into solid fat. The hydrogenation process involves moving hydrogen gas through heated oils in vats containing metals, usually copper, nickel, or zinc. When the metal reacts to the gas, it acts as a catalyst to relocate hydrogen molecules in the oil to create different, stiffer molecular shapes. This chemical reaction creates *trans* fats. Saturation of fats in these synthetic molecules increases according to the degree of hydrogenation achieved.

Hydrogenation makes oil-based food more manageable; for example, margarine can be made firm enough to form sticks, and it also enhances flavor and extends freshness and shelf life. Hydrogenated fats are frequently used as ingredients for doughnuts, chips, crackers, cookies, french fries, and candy bars, which are popularly known as junk food and have minimal nutritional value.

Nutritional investigations have deemed hydrogenated fats unhealthy if consumed excessively because of their saturated fat content. Medical studies have linked hydrogenated fats to interference with essential physiological chemical processes such as metabolism and lipoprotein receptor functioning. Hydrogenated fats increase risks of cancer, heart disease, and obesity. Most experts agree that eating unsaturated plant and fish fats in moderation is safer than consuming hydrogenated fats and advise people to avoid products containing *trans* fats, which are unnecessary for bodily processes.

See also **Food, Processed and Fast; Food Preparation and Cooking**

ELIZABETH D. SCHAFER

Further Reading

Akoh, C.C. and Min, D.B., Eds. *Food Lipids: Chemistry, Nutrition, and Biotechnology*. Marcel Dekker, New York, 1998.
AstraZeneca shucks Quorn. *Chem. Eng. News*, 81, 14, 2 June 2003.
Avoid the fat trap. *Better Nutrition*, 64, 27 August 2002.
Fumento, M. Food fight. *Forbes*, 170, 56, 11 November 2002.
Grady, D. What's in those nuggets? Meat substitute stirs debate. New York Times, 14 May 2002, F1.
Griffen, A.M. and Wiebe, M.G. Extracellular protease produced by the Quorn myco-protein fungus *Fusarium graminearum*, in *Batch Chemostat Cult. Microbiol.*, 143, 3007, September 1997.
Hirsberg, C. Quorntroversy. *Vegetarian Times*, February 2003, 13.
Hunter, B.T. *The Great Nutrition Robbery*. Scribner, New York, 1978.
Hunter, B.T. Make way for mycoprotein in U.S. food supply. *Consum. Res. Mag.*, 84, 24, 2001.
Jacobson, M.F. New foods, new questions. *Nutrition Action Health Letter*, 29, 2, May 2002.
Knight, N. Roberts, G. and Shelton, D. The thermal stability of Quorn pieces. *Int. J. Food Sci. Technol.*, 36, 47, 2001.
Mann, G.V. Metabolic consequences of dietary trans fatty acids. *Lancet*, 343, 1268, 1994.
Mossoba, M.M., *et al. Official Methods for the Determination of Trans Fat*. American Oil Chemists' Society Press, Champaign, IL, 2003.
Pyke, M. *Synthetic Food*. J. Murray, London, 1970.
Rapoport, C. The Brits invent a new food. *Fortune*, 129, 16, 1994.
Sharp, T. Quorn mycoprotein: a new food for better nutrition, in *Implementing Dietary Guidelines for Healthy Eating*, Wheelock, V., Ed. Blackie Academic & Professional, London, 1997.
Spiller, G.A., Ed. *Handbook of Lipids in Human Nutrition*. CRC Press, Boca Raton, FL, 1996.
Underwood, A. Bad crop of Quorn? *Newsweek*, 140, 59, 2002.
Utley, A. "Queen for Quorn probes novel foods. *Times Higher Education Supplement*, 8, 21 November 1997.

Useful Websites

Marlow Foods Ltd: http:/www.quorn.com
Center for Science in the Public Interest (CSPI): http://www.quorncomplaints.com

Synthetic Resins

Chemistry became particularly conspicuous in the twentieth century through synthetic polymers. They include resinous products that are converted into plastics, laminates, surface coatings, and adhesives. Polymers exist because carbon has the property of forming single and multiple bonds with other carbons. In 1922 Hermann Staudinger suggested that polymers were macromolecules. Despite initial opposition, his ideas were accepted from around 1930 and had a considerable impact on industrial developments. The theory and mechanism of the processes whereby small molecules, the monomers, join in repeating units to create giant molecules, the polymers, was established around 1930, following the studies of Wallace Hume Carothers at the DuPont Company in the U.S. He identified two processes, condensation and addition, that distinguish between the main types of products. This provides a useful means for understanding historical developments.

From around 1900, chemists, electrical engineers, and inventors sought out novel products to replace or supplement natural rubber and gutta percha. Most promising was the chemical reaction between phenol and formaldehyde. Leo Hendrik Baekeland, a Belgian who had emigrated to the U.S., carefully controlled the conditions and recognized the catalytic action of acids and bases. He perfected the process in 1907. His main product was a resin readily converted into the first of the thermoplastics, those that set hard and rigid. Baekeland set up the General Bakelite Company in Perth Amboy, New Jersey, in 1910. Another early inventor was Sir James Swinburne, in England, but his process was covered by Baekeland's patent.

The availability of rapid hot molding from 1930 was a boon to growth. Bakelite was used extensively in the electrical and automobile industries, and also for cigarette holders, teapot and umbrella handles, telephones, and cabinets for radio sets. Bakelite had one disadvantage—its color, a dark brown, which at best appeared as black. It was followed with condensation products that could be pigmented, the amino resins. Almost transparent, they were made from urea and formaldehyde, and were discovered by Edmund Charles Rossiter in England at British Cyanides in Oldbury, Gloucestershire, during the early 1920s. They caused quite a stir when displayed during December 1926 in the form of molded colored tableware for the Christmas season at Harrods, in London. Melamine resins, made by a similar process, followed in 1939, and were mainly the work of the American Cyanamid Company in New Jersey. Melamine resins found considerable use in wartime as laminates for maps, molding powders for trays and so on, and after 1945, as kitchen and bar tops, associated with Formica.

Alkyd resins are condensation products from polyols, such as glycerol, and dibasic acids, such as phthalic anhydride. First known as glyptals, they were introduced commercially in 1929 by General Electric in the U.S., mainly as a result of research by Roy H. Kienle. By 1938, production of alkyd resins exceeded the production of phenolic resins, since the greatest demand for resins was as coating materials.

A variety of resin formulations in which monomers containing the vinyl grouping, $CH_2{=}CH{-}$, became available from the late 1920s. These addition polymers include poly- vinyl acetate, -styrene, -vinyl chloride, -methyl acrylate, and -methyl methacrylate. In contrast to Bakelite and the amino plastics, their products are thermoplastic, which means that they can be softened and hardened by heating and cooling. Vinyl acetate resins were introduced in 1928 in the U.S. and Canada. Next was polystyrene, following the research of Hermann F. Mark and colleagues at IG Farben in Germany, and introduced commercially in 1932. Polyvinyl chloride (PVC), had been discovered at Griesheim-Elektron, in Germany, in 1912, but commercialization was delayed due to the problem of dealing with the hydrogen chloride gas that was evolved in the process. In the 1920s, B.F. Goodrich in the U.S. overcame this difficulty through the addition of plasticizers, though these were not required in the processes developed in the U.S. and Germany in the early 1930s. PVC was successfully introduced during 1932–1934, and was the most important vinyl product.

Acrylic is the generic name for polymers made from methyl acrylate, acrylonitrile, etc., though it is normally used with reference to methacrylates. The latter range from soft semifluids, used for adhesives and finishing of textiles, to hard, tough resins, for molded items and clear sheets. The possibility of an organic glass arose from studies at McGill University, Montreal, by William Chalmer. During 1928, the German company Röhm & Haas introduced a safety glass from methyl acrylate. In 1931, methyl methacrylate was found to afford a polymer that could be processed with greater ease. Polymethylmethacrylate (PMMA) became viable commercially in 1932, when John W. C. Crawford at ICI in England discovered an economical way of manufacturing the monomer. Production began in 1934, and the technology was licensed to Röhm & Haas, whose product was known as Plexiglas, in return for a license to manufacture cast sheet, ICI's Perspex, in Britain. The bulk of production until 1945 was employed in aircraft construction. Some was used to prepare false teeth, as a substitute for vulcanized rubber.

Carothers and colleagues at DuPont during 1928–1932 developed important routes to new polymers, including neoprene in 1929. Carothers discovered the condensation product nylon in 1934–1935, in part because he placed a strong emphasis on the structure of natural polymers. Nylon is a polyamide, as are proteins, made by condensation between 6–6 hexamethylene diamine and adipic acid. Introduced in 1938, nylon was the first commercial fully synthetic fiber. Famous for nylon stockings, its first use was in toothbrushes. Though Carothers' group also investigated the structurally similar polyester resins, prepared from an acid and an alkali, such as ethylene glycol and adipic acid, the first commercial product was discovered at ICI in 1942. This was the basis of the fiber Terylene, a linear polyester, polyethylene terephthalate. In 1942, glass-reinforced polyester was used in the manufacture of boats.

The discovery of polythene, however, took place in England. In December 1935 chemists at ICI investigated a white, waxy solid obtained when ethylene (ethene) was subjected to high pressure. It was found to be a polymer with excellent electrical insulating properties. The reaction took place because there was a leak in the apparatus, which allowed oxygen to enter, which then acted as a catalyst. By the end of 1938, one ton of polythene had been manufactured. A new plant was opened at the beginning of September 1939, just as World War II broke out. Polythene was used in high-frequency radio transmitters since, unlike other insulators, it absorbed little energy, due to its inert structure. From 1948 the flexible polythene became an important product used in consumer goods: bowls, buckets, wrapping film, carrier bags, soft drink, and milk cartons. In 1955 an underwater cable with polythene as the insulating material was laid down between Britain and the U.S. Polyethylene, normally familiar as a plastic, was later used in textile coating resins and emulsions.

The original process was worked at around 2000 atmospheres. In 1953, the German chemist Karl Ziegler at the Max Planck Institute discovered that polythene could also be synthesized at low pressures in the presence of somewhat expensive organo-metallic catalysts suspended in organic solvent. The process gave high-density polythene (HDPE), which was stiffer and well suited to the manufacture of crates. This process was subsequently adopted by Hoecht in Germany. Similar processes were investigated by Robert L. Banks and J. Paul Hogan at Phillips Petroleum in Bartlesville, Oklahama, who worked on polymerization of ethylene (1951), and propylene (1951–1953). There was considerable litigation over priority but the key patent was eventually awarded to the two Phillips' inventors. The Phillips product, called Marlex, was introduced in 1954 and later in the decade was used for hula hoop, a toy for children. Polymerization of propylene in the presence of a similar active catalyst, as discovered by Guilio Natta and co-workers at Milan Polytechnic 1954, led to the introduction of polypropylene in 1956. For their contributions, Zeigler and Natta jointly received the Nobel Prize in 1963.

Polyurethane contains the repeating urethane group, –NHCOO–. It was discovered by Otto Bayer, at Bayer in Germany in 1937, and developed commercially during the early 1950s in England and the U.S. It is a linear condensation polymer made from diisocyanate and a dihydric alcohol such as ethylene glycol, used in flexible and rigid foams, especially as insulation in fridges and freezers. Significantly, this is a condensation polymerization that, unlike most others, takes place without the loss of a small molecule.

Epoxy resins, such as the Swiss CIBA's Araldite, were discovered in the late 1930s, and introduced in 1946. They are made by condensation of epichlorohydrin and bisphenol A. The resins in the uncured state are thermoplastic, ranging from low-viscosity liquids to high melting-point solids. They are cured, or hardened, with polyamides, and used in surface coatings, adhesives, castings, and tooling applications.

See also **Adhesives; Composite Materials; Fibers, Synthetic and Semi-Synthetic; Plastics, Thermoplastics; Synthetic Rubbers**

ANTHONY S. TRAVIS

Further Reading

Bijker, W.E. *Of Bicycles, Bakelite, and Bulbs*. MIT Press, Cambridge, MA, 1995, pp. 101–197.

Blakey, W. The History Of Aminoplastics: The Sixth Chance Memorial Lecture of the Society of Chemical Industry. *Chem. Ind.*, July 1349–1357, 1964.

Fenichell, S. *Plastics: the Making of a Synthetic Century*. New York, Harper Collins, 1996.

Friedel, R. *Pioneer Plastics*. Wisconsin, Wisconsin University Press, 1983.

Furukawa, Y. Inventing Polymer Science: Staudinger, Carothers, and the Emergence of Macromolecular Chemistry. University of Pennsylvania Press, Philadelphia, 1998.

Hochheiser, S. *Rohm and Haas: A History of a Chemical Company*. University of Pennsylvania Press, Philadelphia, 1986. Methylmethacrylate resins.

Kaufman, M. *The History of Polyvinyl Chloride*. Maclaren, London, 1969.

Miekle, J.L. *American Plastics: A Cultural History*. Rutgers University Press, New Brunswick, 1996.

Morris, P.J.T. *Polymer Pioneers*. Center for the History of Chemistry, Philadelphia, 1986.

Mossman, S.T.I. and Morris, P.J.T., Eds. *The Development of Plastics*. Royal Society of Chemistry, Cambridge, Special Publication No. 141, 1994.

Seymour, R. B. *History of Polymer Science and Technology*. Marcel Dekker, New York, 1980.

Travis, A.S. Modernizing organic chemistry: Great Britain between two World Wars, in *Determinants in the Development of the European Chemical Industry, 1900–1939: New Technologies, Political Frameworks, Markets and Companies*, Travis, A.S., Homburg, E., Schröter, H. and Morris, P.J.T., Eds. Kluwer, Dordrecht, 1998, pp. 171–198.

Travis, A.S. Amino plastics and the melamine story. *Educ. Chem.*, 1, 16–19, 2000.

Vale, C.P. and Taylor, W.G.K. *Aminoplastics*. Iliffe Books, London, 1964. Includes a historical introduction, and a portrait of Hanns John who in 1918 patented the first industrial use of urea-formaldehyde resins.

Synthetic Rubber

Rubber is a ubiquitous material in modern society, enhancing the quality of life in a myriad of applications. It was unknown in the Western world until the Spanish began their explorations of America, where they found Indian tribes playing games with a ball made from the milky sap obtained by cutting the bark of local trees. In France this sap, or latex, was called caoutchouc after the native name for the "weeping tree" that produced it, while the English called it Indian rubber because it was useful for removing pencil marks from paper. A number of trees and shrubs produced such latex, including the orders Euphorbiaceae, Urticaceae, Apocynaceae, and Asclepiadaceae, but only two natural sources became commercially important (*Hevea brasiliensis* and *Parthenium argentatum*).

The use of rubber for practical purposes was slow to develop because the tree latex coagulated quickly and was difficult to process in the solid form. After solvents were discovered that would dissolve the solid rubber, products were made that took advantage of rubber's elasticity and waterproofing capability, but these crude materials suffered from an inherent stickiness and a form that changed depending on the temperature. With the discovery of the vulcanization process by Charles Goodyear in 1839, the rubber industry had a technique for eliminating these difficulties and better consumer products soon appeared on the market.

Although plantations of rubber trees had been developed primarily in Southeast Asia, the demand for rubber became greater than the supply by the early twentieth century, and the rubber industry began a search for a viable synthetic rubber that would have the same desirable properties as the natural material. From the 1860s, chemists had been attacking the problem of the chemical composition of rubber, with the idea that once they could break natural rubber down into its chemical components, it would be possible to synthetically reconstruct rubber from those components and thus eliminate the need for the rubber tree.

Although this process has never been exactly achieved, it has been possible to prepare a large number of synthetic materials (mainly derived from crude oil) that have rubber-like or elastic properties. While they may be called "synthetic rubbers," they are not completely chemically equivalent to natural rubber, and are often called "elastomers" to reflect this difference. Thus, there is only one type of natural rubber, but there are many kinds of synthetic rubbers.

In 1860, Greville Williams in England isolated from natural rubber a relatively simple molecule named isoprene, a liquid hydrocarbon with the formula C_5H_8. On standing, pure liquid isoprene became viscous and formed a rubber-like material, leading to the exploration of catalysts that would enhance this process of polymerization, in which a simple monomer unit combined with itself many times to form a much larger molecule, or polymer. Chemists also examined the polymerization of other small monomer molecules with structures similar to isoprene, and were able to make a number of different elastomers, each with its own unique set of properties and its own advantages and disadvantages.

For example, one of the first commercial synthetic rubbers was methyl rubber, first produced by Bayer in Germany in 1910 from methyl isoprene and commercially produced during World War I. This was expensive and inferior to natural rubber, degrading when exposed to oxygen. Another synthetic rubber introduced in 1931 by Wallace Carothers at DuPont in the U.S. was neoprene, a polymer made from the monomer chloroprene, a derivative of isoprene that contained a chlorine atom. Neoprene was far superior to natural rubber in its resistance to gasoline and sunlight, and quickly found a market in the budding automobile industry. The question remained, however, if it would be possible to synthetically produce a general-purpose rubber with many uses, rather than specialty synthetics with limited applications.

The answer came in the development of copolymerization, a technique in which two different monomers combined in the polymerization process. Throughout the 1930s, several countries were seeking to reduce their dependence on natural rubber. In Germany copolymerization of styrene with butadiene (Buna S) or acrylonitrile (Buna N) had already been achieved by 1930 by IG Farbenindustrie in the laboratory. When World War II cut off the natural rubber supplies from the East Indies to the U.S., the U.S. embarked on a massive synthetic rubber program, deciding that the reaction of styrene with butadiene to produce a synthetic rubber similar to Buna S that the U.S. called GR-S would provide the best all-purpose rubber for the war effort. Since both of the monomers could be obtained from domestic sources, the U.S. began the construction of a number of plants for monomer and rubber production, and a new synthetic rubber industry was created where none had existed before. In 1939, U.S. synthetic rubber production was a meager 1700 tons. By 1943 production was 230,000 tons, and it reached 1,000,000 tons by the end of the war. In scope and complexity, this program rivaled construction of the atomic bomb in the Manhattan project.

After World War II, synthetic rubber plants were built worldwide, and by 1960 the use of synthetic rubber surpassed that of natural rubber for the first time. According to the International Institute of Synthetic Rubber Producers, by the end of 2001, "The yearly capacity of synthetic rubber manufacturing plants around the globe totals about 12 million metric tons and the capacity of tree-grown natural rubber produced on rubber plantations is approximately 8 million metric tons."

The Rubber Manufacturers Association indicates that about 70 percent of all rubber used today is synthetic in origin. There are about twenty different types of synthetic rubber currently in use, made from materials derived from petroleum, coal, oil, natural gas, and acetylene. However, there are a plethora of variations for each type (see Table 2). Where copolymerization is involved, both environmental conditions and amounts of the different monomers can be varied. It is also possible to blend different types of rubbers with each other and with other materials. Thus, the rubber industry can produce a vast number of

Table 2 Types of synthetic rubbers and principal uses.

Type	Principal use(s)
Styrene–butadiene (SBR)	Tires
Isoprene (IR)	Footwear, tires,
Polybutadiene (BR)	Blending with other rubbers in tires
Nitrile–butadiene (NBR)	Hoses, artificial leather
Acrylate (ACM)	Gaskets, textile coating, paper making
Chloroprene (CR)	Hoses, sealants
Chlorsulfonyl polyethylene (CSM)	Anti-corrosive coatings
Fluorocarbon (CFM)	High temperature seals and hoses
Isobutene–isoprene (IIR)	Tires
Ethylene–propylene (EPDM)	Automobile components
Ethylene-Vinyl Acetate (EVAC)	Cable coverings, textile proofing
Silicone (SI)	Aerospace, food processing, medical
Polyurethane (Ue)	Fabric coatings, insulation, foams
Thermoplastic Rubbers (TR)	Footwear and adhesives
Polysulfide (T)	Sealants for runways, bridges, structures

rubber products, each with its own set of unique properties that are required for a specific use. For example, the GR-S rubber of World War II, now called SBR rubber, is still the major synthetic being produced today, primarily because its properties are close to that of natural rubber; but there are over 500 different grades of SBR, which accounts for approximately 60 percent of total synthetic rubber production.

See also **Plastics, Thermoplastics; Plastics, Thermosetting; Synthetic Resins**

JAMES J. BOHNING

Further Reading

Bisio, A. *Synthetic Rubber: The Story of An Industry*. International Institute of Synthetic Rubber Producers, Houston, Texas, 1990.

Herbert, V. and Attilio B. *Synthetic Rubber: A Project That Had to Succeed*. Greenwood Press, Westport, CT, 1985.

Morris, P.J. T. *The American Synthetic Rubber Research Program*. University of Pennsylvania Press, Philadelphia, 1989.

Stern, H. J. *Rubber: Natural and Synthetic*. Maclaren, London, 1967.

Whitby, G. S., Ed. *Synthetic Rubber*. Wiley, New York, 1954.

Useful Websites

International Institute of Synthetic Rubber Producers: http://www.iisrp.com/synthetic-rubber.html

The Story of Rubber: A Self-Guided Polymer Expedition: http://www.psrc.usm.edu/macrog/exp/rubber/menu.htm

Systems Programs

The operating systems used in all computers today are a result of the development and organization of early systems programs designed to control and regulate the operations of computer hardware. The early computing machines such as the ENIAC of 1945 were "programmed" manually with connecting cables and setting switches for each new calculation. With the advent of the stored program computer of the late 1940s (the Manchester Mark I, EDVAC, EDSAC (electronic delay storage automatic calculator), the first system programs such as assemblers and compilers were developed and installed. These programs performed oft repeated and basic operations for computer use including converting programs into machine code, storing and retrieving files, managing computer resources and peripherals, and aiding in the compilation of new programs. With the advent of programming languages, and the dissemination of more computers in research centers, universities, and businesses during the late 1950s and 1960s, a large group of users began developing programs, improving usability, and organizing system programs into operating systems.

Systems programs were developed for several early computers in the late 1940s and early 1950s. They developed along different lines but were responses to similar challenges that all early computer operators faced, namely, how to automate and make more efficient the programming of a computer. In 1948, David Wheeler, a doctoral student studying and working on EDSAC at Cambridge University, wrote one of the earliest but most vital programs called the "Initial Orders." This program allowed the computer, rather than a technician, to complete the tedious task of converting a symbolic sequence of instructions into the long strings of binary code that the machine could execute. Initial Orders was hardwired into EDSAC on rotary telephone switches before the main program was loaded. Later developments of the Initial Orders included a new innovation in system software called the Wheeler Jump. The Wheeler Jump facilitated the processing of short instruction sequences called subroutines during the operation of a main program. The EDSAC group at Cambridge continued this development of systems programs and in 1951 published the results of their research in their influential book, *The Preparation of Programs for an Electronic Digital Computer*.

In the early 1950s, Grace Hopper, one of the earliest American computer programmers, developed a program called the A–O compiler for the commercial computer manufacturer UNIVAC. The A–O compiler also attempted to automate the process of programming. Grace Hopper and her colleagues at UNIVAC, whose work paralleled that of the EDSAC programmers, also wrote, tested, and stored a library of frequently used sequences. The automatic programming concepts behind Hoppers' A–O compiler and the sequence library became integral components of systems software and allowed programmers to efficiently reuse code. Another notable early system program was developed in 1954. The program, a type of compiler that could convert user instructions into machine code, was installed on the Massachusetts Institute of Technology's Whirlwind computer. Although built specifically to handle algebraic problems, the system programs provided basic operating abilities such as converting and executing instructions and providing facilities for data storage and address location.

In the 1950s and 1960s, developments in systems programs also occurred among the users of computers. Assisted by the development of programs that aided programming, users had a greater range of tools to develop their own programs and modify their systems. One such user group, formed in 1955, was made up of users of the IBM 704 computer. The group, named SHARE, shared tips, developed programs, and had a significant influence on the development of several components for IBM operating systems. Also during the 1950s, the computer users at General Motors Research Laboratories devised a program to handle batch processing. This program, sometimes called a monitor, represented an early but successful system to manage and schedule a computer's resources.

As the 1960s progressed, user needs from the commercial, military, and academic sectors became more diverse and system software became increasingly complex, occupying large portions of the computer's memory. The large computer manufacturers such as IBM, Honeywell, and RAND focused their software development on systems programs. Separately sold application programs, such as word processing and spreadsheet software, did not enter the commercial market until 1965. The so-called software crisis of the 1960s was a recognition of the difficulty of organizing and coordinating system programs into a workable operating system. The development of IBM's OS/360 was an example of this crisis; released in 1966, OS/360 was 18 months behind schedule and $125 million over budget.

The 1970s and 1980s saw a turn away from some of the complications of system software, an interweaving of features from different operating systems, and the development of systems programs for the personal computer. In the early 1970s, two programmers from Bell Laboratories, Ken Thompson and Dennis Ritchie, developed a smaller, simpler operating system called UNIX. Unlike past system software, UNIX was portable and could be run on different computer systems. Due in part to low licensing fees and simplicity of design, UNIX increased in popularity throughout the 1970s. At the Xerox Palo Alto Research Center, research during the 1970s led to the development of system software for the Apple Macintosh computer that included a GUI (graphical user interface). This type of system software filtered the user's interaction with the computer through the use of graphics or icons representing computer processes. In 1985, a year after the release of the Apple Macintosh computer, a GUI was overlaid on Microsoft's then dominant operating system, MS-DOS, to produce Microsoft Windows. The Microsoft Windows series of operating systems became and remains the dominant operating system on personal computers.

See also **Computers, Mainframe; Software Application Programs; Software Engineering**

JESSICA R. SCHAAP

Further Reading

Brinch Hansen, P., Ed. *Classic Operating Systems*. Springer-Verlag, New York, 2001.

Campbell-Kelly, M. and Aspray, W. *Computer A History of the Information Machine*. Harper Collins, New York, 1996.

Campbell-Kelly, M. Programming the EDSAC: early programming activity of the University of Cambridge. *Ann. Hist. Comp.*, 2, 7–36, 1980.

Ceruzzi, P. *History of Modern Computing*. MIT Press, Cambridge, MA, 1998.

Rosen, S., Ed. *Programming Systems and Languages*. McGraw-Hill, New York, 1967.

Weizer, N. A history of operating systems, *Datamation*, 27, 118–134, 1981.

T

Tankers, *see* **Ships: Bulk Carriers and Tankers**

Tanks

Despite some curiosity on the part of a handful of other nations, the development of the tank in the twentieth century was largely a British affair. Yet even Britain did not intentionally set out to develop it. Like many things, the tank was the result of other technologies being developed as well as a response to the dangers some of those very technologies presented.

The three things needed in order for the tank to become a reality were power, protection, and vision. The first of these, power, evolved slowly. James Watt's improved steam engine in the latter half of the nineteenth century seemed to offer the as of yet unrealized promise of armored travel. In 1883, Germany's Gottlieb Daimler's four-stroke gas-powered engine, mounted on a bicycle, moved armored vehicles further along. In 1899 British engineer F.R. Simms created the "War Car," a Daimler-powered motorcycle, boasting a machine-gun mounted behind an armor shield. A year later Simms showed off a more deluxe version of the "War Car," this one encased in armor, bristling with a cannon and machine-guns and mounted on four steel-tired wheels. The age of armored, mechanized warfare had begun.

Once the need for power had been at least temporarily satisfied, protection became the next concern. Advancement in weapons systems such as high explosives and the machine gun made survival on the battlefield near impossible without armored protection. In England, Henry Bessemer's steel production process quickly replaced dependence on iron. Manganese steel proved to be better than ordinary steel, but it soon gave way to nickel steel. Both manganese and nickel steel seemed to offer an immediate, albeit expensive response to the development of metal-piercing munitions.

However, solutions to the problems of power and protection only proved that without vision, the tank would never become a staple of modern warfare. Despite the encouraging developments, warfare in the early twentieth century seemed to remain dependent on the type of horsepower that required oats. World War I began a shift in thinking. The first modern war, with long-range artillery, machine guns, air power, and trench fortifications proved deadly to advancing troops. The result was either a horrendous loss of life or a frustrating stalemate. In an attempt to break the logjam, Allied commanders who had once doubted the value of armored vehicles began to use them. Initially, armored cars were used, but the poor roads and muddy ground which became known as no-man's land proved to be the downfall of these wheeled vehicles. The British Navy then stepped in, and with the help of the Royal Engineers produced the next generation of armored vehicles. The result was at that point an unimaginable contraption. It was a machine of geometric proportion with steel tracks on each side, coated with armor and loaded with cannon and machine guns. At first it was called "Centipede," then "Big Willie," and finally "Mother." The evolution from F.R. Simms' "War Car" was complete—the tank was born. In September 1916, British M-1 tanks went into battle. Weighing in at 28 tons, the tanks carried a crew of eight, half of whom fired the two 6-pound (2.7-kilogram) cannon and four machine guns. They spanned the 12-foot (3.5-meter) enemy trenches without mishap, negotiated the barbed wire and shell holes, and terrified the Germans, who quickly surrendered. Tracked vehicles would now be a new and necessary weapon. By war's end, France, Germany and the U.S., as well as Britain, were using tanks. The U.S. bought hundreds of French-built Renaults, and placed many of them under the command of Lt. Colonel George S. Patton, Jr. Patton, a former cavalryman, proved the tank's value not only as a supporter of advancing ground troops, but also as an independent assault weapon.

In the years following World War I, tank development varied. Nations, tired of war and the financial obligations it implied, either halted or slowed research and development. Germany, barred from any military development as a condition of the Versailles Treaty, could only experiment with tanks captured from the British. The U.S. Army did away with any plans to create an independent Tank Corps, relegating it to infantry support. The same lack of vision was evident in France and Britain, which also handed the tank over to the infantry. Nothing epitomized the lack of interest in the tank better than the experience of American engineer J. Walter Christie. Christie had studied World War I tanks and subsequently created a more modern fighting machine. His most important innovations were a suspension system, which made them more mobile, sloped armor to minimize damage from anti-tank weapons, a more powerful engine, and a larger gun mounted on a rotating turret. The U.S. was not interested, so in 1931 he sold his design to the Soviet Union, who modified it and created the "T" series of tanks, the first wave of armor to be designated as main battle tanks (MBT), which could combine mobility, lethality, and survivability. The Soviet's first effort, the T-34, equipped with a 76 millimeter gun proved to be up to the challenge of German tanks when Germany invaded the Soviet Union in 1941. Despite the restrictions imposed upon it after World War I, German military thinkers realized that the tank could make their armies more effective. They created entire divisions of tanks, with motorized infantry in supporting roles. It would take several years of fighting and dying before the Allied powers developed tank technology to challenge Germany's Tiger, Mark, and Panther series. Each sported heavier armor and firepower than anything America, France or Britain could put into action. The Panther was widely believed to be the best tank of World War II. The experiences of American and British tankers facing and surviving superior German technology contributed to the campaign in those two countries to pursue advanced tank technology in the postwar world.

That pursuit continued despite the belief that tanks had become outmoded in the nuclear age. The Soviet Union and France, as well as Britain and the U.S. took the lead in producing main battle tanks with greater speed, better protection and deadlier firepower. In particular the addition of infrared technologies on Russian and British tanks made them more effective in night engagements. Britain's Centurion series became an extremely effective armored vehicle when it was introduced after World War II, and as it was phased out in the 1960s it was snapped up in large numbers by other countries. The Centurion was replaced by the Chieftain, which in turn gave way to the Challenger, which featured a 120-millimeter gun, laser range-finding, thermal imaging, and night vision. The U.S. would ultimately become a leader in main battle tank innovation, but progress was slow. The wars America found itself enmeshed in, such as Korea and Vietnam, featured terrain inappropriate for extensive mechanized warfare. Additionally, air power had become the key piece in the U.S. defense puzzle. The result was a series of U.S. tanks, the "M" series, which were big, slow, underequipped in terms of firepower, and fueled with gasoline, making them instant fireballs if hit in the right spot. The M-60 was the last in the line, described by one American officer as "an inferior tank, part of a tired, old, second rate series." Meanwhile, the Soviet Union modified the T-34, and eventually spun it off into the T-72. Despite its cramped crew space, which occasionally exposed them to dangerous moving parts, Soviet tanks were formidable vehicles, with sloped armor, low silhouettes, and an accurate weapons system. In Britain, engineers found that angling the layers of armor on a tank could deflect anti-tank rounds. That development, along with breakthroughs in targeting and firepower produced new generations of tanks in Sweden, Germany and Israel. In the early 1980s the U.S. introduced the M1A1 Abrams, which received its first battlefield test during the 1991 Gulf War, and proved equal to the challenge. It cruised smoothly at 66 kilometers per hour, supported by torsion bar suspension and extra long tracks. The air-cooled gas turbine engine ran on a number of fuels, and produced less smoke and noise. It was also under 3 meterstall, which made it a hard target to hit. Inside, the four-person crew had a thermal imaging system to find and track targets, a 120 millimeter main gun with a range of 2500 meters, and a ventilation system to counteract nuclear, biological or chemical warfare.

Although tanks were plagued with problems of power, protection, and a lack of vision about their use at the beginning of the twentieth century, and despite the massive advances in weapons of all kinds during the century, at the end the tank remained a vital instrument of warfare.

See also **Warfare**

JOHN MORELLO

Further Reading

Fleming, T. Tanks. *Invention Technol.*, 10, 3, 54–63, 1995.
Macksey, K. *Tank Versus Tank: The Illustrated History of Armored Battlefield Conflict in The Twentieth Century*. Grub Street, London, 1999.

Technology and Ethics

Technology is manifest not simply in the mechanical, chemical, and electrical achievements from the first third of the twentieth century (e.g., automobiles, airplanes, synthetic materials, drugs, radios, motion pictures), the creations of physics that dominated in mid-century (e.g., nuclear weapons and space flight), and the electronic and biological inventions of the last third of the century (e.g., computers and genetic engineering) but also in the manifold social influences and impacts of these and related products, process, and systems. Such implications range from the economic and cultural to the legal and political—from new forms of production and consumption to the development of distinct governmental regulations and agencies. Yet underlying all such responses—and, indeed, originally calling forth the technologies themselves—are cultural and ethical commitments. As Aristotle argues in the opening pages of his *Ethics*, all human actions arise from some vision of the good; the critical examination of such visions as are present in the Greek "ethos," which we now call culture, is what constitutes "ethics."

The Ethos of Technology

From its sixteenth century beginnings the momentous cultural commitment to distinctly modern forms of making and using has rested on a vision of technology as a way to achieve the good if not a form of it. At the beginning of the twentieth century, Leo Baekeland, the inventor of Bakelite—the archetype for a whole family of plastics whose protean slickness contributed a new molecular level of artifice to such devices as the telephone and record player—summed up the prevailing moral assessment of technology as doing:

> "more for the betterment of the race than all the art, all the civilizing efforts, all the so-called classical literature" [*Baekeland 1910, p. 40*].

He continues:

> "The modern engineer, in partnership with the scientist, is asserting the possibilities of our race to a degree never dreamt of before: instead of cowering in wonder or fear like a savage before the forces of nature, instead of finding in them merely an inspiration for literary or artistic effort, . . . he fulfills the mission of the elect and sets himself to the task of applying his knowledge for the benefit of the whole race" [*Baekeland 1910, pp. 38–39*].

Taking hold of a baton passed from the sixteenth through the nineteenth centuries, the first half of the twentieth sought to extend what has been called "a second creation of the world," technologically transforming reality from something to which human beings adapted into that which they designed in their own image. This view of technology as justified by its pursuit and realization of material welfare and human freedom was reiterated fifty years later by C.P. Snow (1959) in his famous "two cultures" lecture, where he castigated literary intellectuals as "natural Luddites" for their worries about the materialistic values inherent in the great, ongoing achievements of the scientific and industrial revolutions: improved health care, increased food production, and the democratization of education. The primary ethical justification of modern technology across the century was the conquest of nature and the promotion of humanization as the pursuit of freedom.

During the second half of the century, however, the new technological world itself came to be recognized as requiring its own adaptations. Freedoms were not themselves always free. Thus, although the 1900s opened with a confidence in and remained throughout deeply sympathetic to an almost unqualified faith in the moral probity of technology, the later twentieth century witnessed the emergence, even within the techno-scientific community, of a series of critical ethical questions addressed to technological humanism.

The Question of Dehumanization

One powerful statement of this criticism argued that technological change fostered a form of dehumanization insofar as it separated human begins from nature and tradition, and subordinated the rich variety of human experience to judgments of instrumental rationalism. An especially influential articulation of technology as dehumanizing, inherited as well from the nineteenth century, focused on the issue of alienation in manufacture. For Karl Marx, alienation was narrowly defined in terms of the loss of control by workers over the processes and products of their labor, a loss sponsored by rationalist divisions of labor and large-scale industrialization. Division of labor is, however, but a special case of the more general phenomena of technology being disembedded or torn asunder from culture. Prior to the modern period, traditional technologies were characteristically embedded in a human life world; that is, hedged about with mores and counter-mores institutions. In agriculture, the major sector for employment before the twentieth century, the planting and harvesting of crops, the slaughtering of animals, and the consuming of foods were embedded in, that is, part and parcel of, age-old religious and cultural rituals and taboos. In many so-called pagan cultures the practice of agriculture required sacrifices to the gods; Hinduism, Buddhism, Judaism, and Islam all limit the partaking of certain foods, from animals to alcoholic beverages, and dignified seasons and celebrations with particular food associations.

It is not even correct to describe the traditional relationship as one of cultural ends guiding technical means, because our contemporary means–ends distinction was conspicuous by its absence in the premodern web of life. Each human activity was folded over or implicit in others. With industrial production, however, the web was unraveled so that means–ends distinctions came to the fore, and technology as means was cut loose from any particular ends in order to be pursued and developed on its own to an unprecedented degree. The result is what sociologist William Fielding Ogburn termed "culture lag," as life-ways ran to catch up with the explosion of new products and processes being introduced into human experience. The often-felt "loss of control" of our attempts to control nature are a further expression of this disembedding, and the root impetus calling for ethical reflection on the new means being placed at the disposal of a plethora of human desires and intentions cut free from traditional constraints.

Take some examples from the fields of transportation and communication. The automobile and the airplane turned the geographic niches of culture into porous boundaries, first nationally and then internationally, while introducing into ethos itself a manifold of unexpected consequences. The deployment of the assembly line by Henry Ford in 1914 transformed a toy of the wealthy into a consumer product for the many, facilitating the mid-century phenomenon of freeway suburbanization, while contributing to the sexual revolution in the form of a dating bedroom on wheels and to a geologically significant build up of atmospheric carbon. The invention of the airplane by the Wright Brothers in 1903 likewise transformed warfare, diplomacy, business, and tourism in diverse and sometimes conflicting ways that continue to play themselves out on the global stage. The communications technologies of telephone, radio, motion pictures, television, and computers further reduced the experiential importance of that place and history in which culture was once exclusively embodied. The displacement of ethos from localized initiation rituals of birth and death or forms of dress and speech into communications media styles and mixes supporting a cafeteria of entertainment programming is a uniquely twentieth century phenomenon that prolongs the disembedding trajectory. One correlate of such broadscale disembedding was what conservative culture critics of technology such as Lewis Mumford described as a lowering of the standards of taste in material culture and the replacement of spiritual discipline with the pleasure principle.

Questions of Ethics

Ethical criticism of such massive cultural dislocations remained more or less marginal until the invention of nuclear weapons took the vague unease characteristic of conservative intellectuals and placed the ethics of one specific technology in the forefront of public discourse. After Hiroshima and Nagasaki, many scientists and engineers found their gut intuitions given voice by J. Robert Oppenheimer, the engineer–manager of the atomic weapons program, when he said "In some sort of crude sense which no vulgarity, no humor, no overstatement can quite extinguish, the physicists have known sin." (Others violently complained that Oppenheimer had no right to beat his breast in public for them.) As Albert Einstein summarized the situation in less religious but nonetheless equally dramatic words, "The bomb ... and other discoveries present us with ... a problem not of physics but of ethics."

World War II likewise confronted the human community with instances in which even the most humane techno-sciences, those associated with medicine and its professional ethos of care, had been bent and corrupted through unthinking subordination to base political agendas. German and Japanese physicians performed medical research on patients that amounted to forms of torture, while developing chemical and biological weapons for use on noncombatants and combatants alike. As a result, the Nuremberg War Crimes Tribunal sought to establish new and stricter guidelines for the conduct of medical experimentation on human subjects, making free and informed consent a paramount requirement and applying the principle of distributive justice to the apportioning of any benefits from such research. Subsequent investigations disclosed immoral medical experiments not only among the enemies of democracy but within democratic regimes themselves, with medical treatments being withheld from minorities, as in the Tuskegee experiments on African Americans suffering from syphilis, and with soldiers and citizens being exposed to harmful doses of radiation, as in the nuclear tests at Nevada and in the South Pacific, all in the name of techno-scientific knowledge production or national defense.

Indeed, in the five post-World War II decades one can plot a series of ethical discussions, often initiated by techno-scientists attempting to create appropriate cultures of containment for new technological powers:

1. In the 1950s it was the ethics of testing nuclear weapons, leading to the Limited Nuclear Testban Treaty (1963), as well as debates about the ethics of nuclear deterrence policies.
2. In the 1960s developments in artificial intelligence began to challenge traditional views about the uniqueness of human thinking, and biologist Rachel Carson's *Silent Spring* (1962) exposed the ecological destructiveness of excessive pesticide use—the latter of which lead to establishment in the U.S. of what became an internationally influential government institution, the Environmental Protection Agency (1970). Subsequent arguments for renewed appreciations of the natural world on both anthropocentric and non-anthropocentric grounds inspired a whole new disciplinary discourse of environmental ethics, and eventually the idea of a World Charter for Nature (1984).
3. During the 1970s issues of environmental health merged with questions about how to allocate equitably costly high-tech medical devices and treatments to engender, in a dialogue between biomedical practitioners and ethicists, the field of biomedical ethics or bioethics. Questions about the safety of the first genetically engineered organisms prompted genetic engineers in the early 1970s to adopt a voluntary, worldwide moratorium on this technology, in order to establish appropriate protocols for its safe pursuit.
5. In the 1990s it was topics such as the loss of biodiversity, global climate change, and reproductive cloning that became major foci for ethical discussion and debate.

Practical Responses

Practical responses to this spectrum of techno-ethical issues can be observed at the level of the professional scientific and technical community and at the level of public decision making. One of the most remarkable features of twentieth century technical professions is the effort to formulate codes of ethics to guide their members in dealing with a host of potential ethical dilemmas. Not only has the medical profession progressively refined its ethical guidelines for the treatment of patients, but engineers have formulated codes that go well beyond the promotion of corporate loyalty or professional interests. As the century opened, there were no explicit codes of engineering ethics. When such codes were initially formulated in the 1910s they emphasized responsibilities to employers and clients. By the end of the twentieth century, however, it was customary for engineering ethics codes to inspire their members to hold paramount "the safety, health, and welfare of the public" in the performance of their technical tasks and, indeed, to educate the public about the risks as well as the benefits of engineering projects.

Stimulated in part by a number of high-profile engineering disasters such as the nuclear meltdowns at Three Mile Island (1979) and Chernobyl (1986), the Union Carbide chemical plant explosion in Bhopal, India (1984), and the loss of the space shuttle Challenger (1986) engineering professionals also sought creative ways to educate and enforce their new codes, to support whistle blowers, and to engage the public in establishing appropriate institutions for the monitoring and regulation of technology. As a result, engineering curricula in many universities now teach engineering ethics, and the Institute for Electrical and Electronic Engineers (IEEE), the largest professional engineering society in the world, gives an occasional award for Outstanding Service in the Public Interest. The American Association for the Advancement of Science (AAAS), the largest interdisciplinary scientific society in the world, likewise has a standing committee on "Scientific Freedom and Responsibility" which gives an annual award and works to engage science in the protection of human rights.

The 1980s also witnessed the exposure of a number of cases of misconduct, particularly in publicly funded biomedical research. This resulted in explicit legislative and institutional efforts to develop clearer guidelines for the responsible conduct of research in areas such as scientific record keeping, authorship and peer review, the treatment of laboratory animals, conflict of interest, and intellectual property rights. In the U.S. the National Institutes of Health, the largest funder of biomedical research, started to require ethics education for all graduate students and special boards of scientists and nonscientists to approve all research protocols involving human participants.

One convergence across these levels of professional and political response to the ethical challenges of technology was to abandon any strict laissez faire attitude toward whether or not, and how, the new powers of technological medicine and scientific technology might properly be deployed. Within the technical community a consensus emerged to try to avoid either a techno-scientific or market determinism in which things were made and used simply because they could be made and used. The basic belief was that enhanced and extended technological power called for an enhanced and extended practice of informed democratic consent in a world that was in effect becoming one gigantic socio-technical experiment. As Kristin Shrader-Frechette (1991) argued at length, insofar as the deployment of technologies constituted social experimentation, public participation is required. Yet public participation alone is not enough. Democratic intelligence depends on more than effective linkage of technological development to public desires or values; it also requires ethical insights to help it make informed and intelligent decisions about which technologies to accept, which to modify, and which to reject.

Theoretical Responses

The philosophical response that aimed to increase the kind of insight needed when faced with a thicket of ethical challenges centering around nuclear weapons, chemical engineering, high-tech medicine, computers, climate change, and biotechnology proceeded along two paths. One path has been to attempt a global or holistic assessment of modern technology as a transformation of the human condition. Here the work of such twentieth century thinkers as José Ortega y Gasset, Martin Heidegger, and Hans Jonas may serve as leading examples. Jonas, for instance, argues:

> "Modern technology has introduced actions of such novel scale, objects, and consequences that the framework of former ethics can no longer contain them... No previous ethics had to consider the global condition of human life and the far-off future, even existence of the race. These now being an issue demands...a new conception of duties and rights, for which previous ethics...provided, not even the principles"

In response, Jonas proposes a "heuristics of fear" to heighten the imagination of worst-case scenarios and thus introduce into the dynamism of modern technology a cautionary modesty.

One policy instantiation of such modesty is illustrated by the European Union adoption of the precautionary principle: essentially the view that a new technology should no longer be considered innocent until proven guilty (the classic modernist stance), but dangerous until proven safe. The weakness of such global stance, however, is that it remains at odds with the prevailing ethos of enthusiasm for technology still emergent elsewhere in the global marketplace, is difficult to implement in a pluralistic society, and abstract from any residues of those traditional forms of life that might actually support it. Besides, in concrete policy debates it is difficult to know how safe is safe enough.

Another path has been to take on problems one at a time, responding in a more piecemeal and pragmatic fashion, adapting received forms of ethical analysis and reflection. The two major modern ethical frameworks are known as consequentialism and deontologism. In consequentialist ethics, the rightness or wrongness of an action is dependent on the goodness or badness of its consequences or results (John Stuart Mills' utilitarianism is an example), whereas in deontological ethics rightness and wrongness are perceived as independent values of certain actions (Immanuel Kant's categorical imperative is the archetype). For instance, a deontologist might argue that respect for human autonomy and dignity demands without exception free and informed consent from all human research participants, even if this might compromise research that has a good chance of developing beneficial therapies. By contrast, a consequentialist may want to look at particular cases and insist that informed consent be justified on the basis of good results. In both frameworks, attempts have been made to understanding consequences and articulate rights so as to better encompass the powers of advancing technology—but not always with success.

The most common late-twentieth century proposal to enhance consequentialism centered on risk cost-benefit analysis. The problem for consequentialism is the prevalence of unintended consequences and complex risks, especially those of low probability and high magnitude of harm (such as nuclear disasters) or epistemological uncertainties (such as the anthropogenic dimensions of global climate change). The existence of such cognitive debilities led David Collingridge (1980) to describe what he termed the paradox of the social control of technology: in the early stages of a technology, when it would be relatively easy to modify its development, we seldom possess the knowledge to make rational decisions; by the time we have more experience and a better understanding of its consequences and risks, technological momentum has made control difficult, if not impossible. This paradox suggests the need to develop social institutions dedicated to proactive technology assessment and, whenever possible, the choice of more over less flexible technologies.

With regard to deontologism, there has been a consistent effort to isolate a few firm boundary principles, as with the obligation to seek informed consent for any human subject experimentation. The problem is that in a techno-scientific, pluralist democracy, all fundamental principles tend to rest on a minimalist public consensus more than on reasoned insight—a fact that tends to promote advocacy lobbying. Under such circumstances deontological boundaries get restricted to procedural rather than substantive issues; and even there, subject to modification by strong public opinion. In the last quarter of the twentieth century, for instance, a deontological prohibition on the use of unproven drugs was undermined by AIDS activists demanding treatment.

Conclusion

By the end of the twentieth century the enthusiastic commitment to technology as being good under virtually all circumstances had been qualified by a more nuanced faith and by diverse efforts to bring critical ethical reflection to bear on the opportunities and threats associated with the most rapid and expansive period of technological changes in human history. Two other complementary developments were efforts to model complex phenomena and the rise of interdisciplinary and social networks, as manifest most conspicuously in both cases with computers and the Internet. To some extent, ethical reflection coupled with modeling and communications networks can be read as responses to the cultural disembedding that has been the hallmark of modern technology. But insofar as technology has been disembedded not simply from culture but from nature as well, the long-term viability of this solution is unclear. The rich diversity of the planet, the product of eons of evolutionary interaction across multiple scales may simply be impossible to model in its fullness. Models necessarily simplify and leave things out and are not terribly good persuaders of public action. Indeed, the century ended with the emergence of ethical questions about scientific and technological modeling.

See also **Technology, Society and the Environment**

Carl Mitcham

Further Reading

Baekeland, L.H. Science and industry. *Trans. Am. Electrochem. Soc.*, 17, 37–53, 1910.

Harremoës, P. *et al.* Eds., *The Precautionary Principle in the 20th Century: Late Lessons from Early Warnings.* Earthscan, London, 2002. Fourteen case studies of such issues as radiation, asbestos, CFCs, and "mad cow disease."

Jonas, H. *The Imperative of Responsibility: In Search of an Ethics for the Technological Age.* Translated by Jonas, H. and Herr, D. University of Chicago Press, Chicago, 1984.

Katz, E., Light, A. and Thompson, W.B., Eds. *Controlling Technology: Contemporary Issues*, 2nd edn. Prometheus Books, Amherst, NY, 2003.

Mitcham, C. and Mackey, R., Eds. *Philosophy and Technology: Readings in the Philosophical Problems of Technology.* Free Press, New York, 1972; paperback 1983. Section two highlights "ethical and political critiques." Includes a select, annotated bibliography.

Shrader-Frechette, K.S. *Risk and Rationality: Philosophical Foundations for Populist Reforms.* University of California Press, Berkeley, 1991.

Technology and Leisure

Leisure, as defined by the *Encyclopedia of Social Sciences* (1933), is the "opportunity for disinterested activity," that is, activity that is not defined by its utility. Particularly in the industrialized world, leisure is usually contrasted to work; "free time" outside gainful employment is leisure time. The other important dimension of leisure activity is that it is voluntary, freed not only from the constraints of employment, but also open to the whims of the individual. In this sense, leisure is closely related to recreation and entertainment.

The contrast of leisure to work emerged around the beginning of the twentieth century. Mass industrialization and the factory system had forever altered the conditions of labor, dislodged the personal identification of many laborers with their work, and associated the workplace with drudgery. The Industrial Revolution also revised the pace and schedule of this work; by 1930, labor laws had significantly reduced the length of the typical working week in Europe and the U.S. Reactions to these changes were both optimistic and pessimistic. Some, such as the American writer Archibald MacLeish sang the praises of mechanization, believing it would lead to a "civilization in which all men would work less and enjoy more" (*The Nation*, 8 Feb. 1933). Others focused on the "leisure problem," the notion that idleness of the working classes—whether due to leisure time or mass unemployment—would lead to social unrest and immorality. International attention to the complex of issues raised by the transformation of labor came to a head during the 1920s and 1930s through organizations such as the International Labor Organization (founded in 1919) and international conferences that focused attention on workers' spare time.

The concept of leisure in contrast to work time is thus historically conditioned. It has only prevailed since the beginning of the twentieth century. Until then, leisure had been more typically depicted as a form of privilege, as a reflection of the status of free men; it conveyed the opportunity to devote oneself freely to pursuits such as the pursuit of knowledge. The tradition of considering leisure as the basis for contemplation and wisdom went back as far as Aristotle. In the late nineteenth century, writers such as Thorstein Veblen and Paul Lafargue insisted on the social acceptability and, in the case of Veblen, the benefits of idleness and leisure. Veblen's *Theory of the Leisure Class* (1899) depicted the "nonproductive consumption of time" in social terms, arguing that it was not mere idleness, but time spent privately in order to produce "immaterial" goods by those who could afford to do so. Examples of the results of leisure were "quasi-scholarly or quasi-artistic accomplishments," cultural observances, physical exploits, and the like (explained in the chapter entitled "Conspicuous Leisure,"). In short, leisure for Veblen was a "requisite of decency" in society.

The characterization of leisure as time away from work stimulated the creation of spaces for the "business of pleasure" (a term coined by Peter Bailey). These included sites that were specially constructed for the immediate delivery of entertainment, such as amusement parks, racetracks, and theaters. With the expansion of railroad networks, rise of the automobile, and introduction of airlines and cruise ships, travel to local, regional, and remote destinations for recreation and tourism also became feasible. In the last third of the twentieth century, the accelerating development of computer and network technology pushed the virtual worlds created through networked environments, video and computer games, and virtual reality, to the forefront of entertainment options, with a commercial impact rivaling that of older media such as motion pictures. In each phase of the expansion of available leisure spaces, technology played an important role both in expanding access to hitherto unavailable spaces and in creating compelling entertainment experiences within these spaces.

These notions of leisure and the activities associated with leisure time have thus been dramatically transformed over the course of the twentieth century. Generally speaking, the role of technology has been two-fold in this transformation. First, the deployment of technology has altered conditions of work and the efficiency of labor, and mechanization and industrialization have created a new context for leisure. Shortened working weeks, for example, stimulated discourse on the "problem" of free time. Religious, political, and social leaders considered the nature and importance of leisure activities, while industrial psychologists reconceived leisure as reducing the physical and psychological stresses suffered by the work force, particularly in the industrialized West. The notion of leisure was transformed, no longer a privilege of the "leisure class" but a space of possibilities for time away from work; something no longer to be repressed and feared, but cultivated. As Ida Craven asserted in an article on leisure for the *Encyclopedia of Social Sciences* in 1933, "the tone of any society is largely determined by the quality of its leisure." Technology has also played a second role in the shaping of modern leisure, less in the context than in the nature of leisure. Technological innovation and new systems of production have also transformed the practices associated with leisure time, creating for example a wealth of new recreational activities and entertainment media. At the end of the twentieth century, the appeal to consumers of free-time pursuits such as motion pictures, sports, computer games, and amusement parks depended more often than not upon the introduction of cutting-edge technologies into these realms. While a century earlier, technology, engineering, and factory production were opposed to the contemplative pursuits of the leisure class, by the end of the twentieth century such diverse concepts as Walt Disney's "imagineering," the "studio system" in movie production, or videogame design suggested that entertainment and leisure had been fully integrated into post-industrial patterns of production and consumption.

See also **Changing Nature of Work; Computer and Video Games; Entertainment in the Home; Sports Science and Technology; Technology, Arts and Entertainment**

HENRY LOWOOD

Further Reading

Bailey, P., Ed. *Music Hall: The Business of Pleasure*. Open University Press, Milton Keynes, 1986.

Biggart, N. Labor and leisure, in *The Handbook of Economic Sociology*, Smelser, N.J. and Swedberg, R., Eds. Princeton University Press, Princeton, 1994.

Craven, I. Leisure, in *Encyclopedia of Social Sciences*, Seligman, E.R.A., Ed. Macmillan, New York, 1933.

Cunningham, H. *Leisure in the Industrial Revolution c.1780–1880*. Croom Helm, New York, 1980.

MacLeish, A. Machines and the future. *The Nation*, 8 February, 1933.

Veblen, T.B. *The Theory of the Leisure Class*. Macmillan, New York, 1899. Reprinted with an introduction by Lekachman, R. Penguin, London, 1994.

Technology, Arts and Entertainment

Introduction and Survey

Technology plays an important role in many cultural activities, especially in entertainment and in the fine and popular arts.

In cultural activities related to technical media several transformations—"mediamorphoses"—can be distinguished. In the graphic mediamorphosis of early modern times, messages and communications were transformed into print and music into notations; later, in the second half of the eighteenth century, chemical and mechanical mediamorphosis stored visual or aural realities. With photography, film, the wax cylinder, and audio disks, only recording and playback were required. It was no longer necessary to add a symbolic intermediate stage as in writing, printing, or musical notation. Laborious "true to nature" painting was largely replaced by the camera.

In the next stage, the electronic mediamorphosis of the first half of the twentieth century, different codes of communication were translated into electronic codes; the storage and playback of visual images and of sound played a dominant role with electronic recording replacing acoustical recording. With digitization in the 1980s, this development gained fresh momentum in a digital mediamorphosis. Electronic media for storing information were combined with digital computers, which provided increased storage and networking. With digital codification it became easier not only to reproduce "reality" but also to create "new realities"—computer-animated images for example. As a consequence, the difference between "pure" and "applied" art has become smaller and a new definition of art will soon be required. Whereas in the entertainment industry there has been a trend towards mergers and concentration, self-employment has likewise increased since the early 1980s. Together with large, expensive sound, film or TV studios, digitization has made the rise of small "bedroom productions" studios possible in which inexpensive audio or video productions can be made in a professional manner.

These activities described previously often interfere with copyright laws. Copyright has always been at war with technology because technical change often challenged intellectual property regulations and the accepted notions of authorship. New technologies of reproduction such as the photocopier or the video cassette have made it increasingly difficult for copyright holders to control the copying and distribution of their intellectual property; digital technology has added another dimension to this. Works in digital form are easy to copy and to distribute via the Internet; illegal use is difficult to detect and to prove. In line with this, the self-understanding of many modern artists equipped with scanners and samplers has changed significantly. They do not regard themselves any longer as "original geniuses" but as processors of information or manipulators of discovered material; a DJ's criterion for expertise is his skill in collating existing materials. There is a host of lawsuits pending on breaches of copyright, and in many cases pragmatic compromise arrangements have been made by giving licenses at the cost of a share of royalties.

Visual Arts and Technology

The history of art is, to a large extent, also a history of technology. Every art genre employs technology of some kind, the materials of visual art or the instruments of music making. There has been, and still is, continuous research on chemical processes for the development of paint varnishing that has affected painting in complex ways. After the introduction of monocular perspective painting by the fifteenth century Italian painter Filippo Brunelleschi the invention of the camera obscura also played an important role. Similar to this significant change in the style of visual representation and the theory of visual perception, the invention of rapid serial photography by Etienne-Jules Marey and Edweard Muybridge in the 1870s and 1880s brought about a change in the way we perceive elements of visual motion and their representation.

In the early twentieth century the Italian futurists were fascinated by the machine and the motion of speed. Futurist esthetics valued the dynamic over the static and technology over nature. The cities with their noise and industrial rhythms were their preferred new material of art. Like the Italian futurists the Russian constructivists of the decade following the October Revolution of 1917 had an almost unlimited faith in the powers of industrial technology. Unlike the futurists, however, some of whom sympathized with Italian fascism, they dedicated their art to the service of the (Soviet) State. For the constructivists, the artist was an artist–engineer who made use of art in a radical reconstruction of society; but not all art movements of the early twentieth century were so fascinated by technology. Partly due to the terrors of World War I, which was also called the "war of the engineers," many artists realized technology's destructive power. The dadaists greeted technology with irony and the surrealists alluded to the disquieting and frightening aspects of the machine.

The introduction of electrical engineering and applications such as electronic means of controlling movement evoked the interest of artists. This gave rise to the "Kinetic Art" movement which investigated the dynamics of motion and the way, motion affects visual perception. Already in the late 1920s the theorist–artist Lászlo Moholy-Nagy, associated with the Bauhaus movement, which developed far-reaching insights into the relationship between technology and the arts, advocated a new perception of art in which dynamic values replaced static ones.

The Bauhaus movement in Germany in the 1920s proved enormously influential in all parts of the world. In a way it was a model for the art and technology movement in the United States in the 1960s. In the early 1960s, Billy Klüver, a research engineer and specialist in laser technology, worked together with artists like Merce Cunningham, Claes Oldenburg, Jasper Johns and Robert Rauschenberg, not only by providing technical assistance but as equal partner in the creative process. Collaborations with artists also led other scientists and engineers to make discoveries and achieve technical breakthroughs. However, the success of this cooperation was rather limited; many of these collaborations were marked by tensions and misunderstandings. For most of the artists involved, redefining the function of art was probably too radical a step. The same applies for the scientists and engineers, because members of those communities were rather wary of artists in general. Still, since the 1960s many industrial corporations with artists-in-residency schemes have provided artists with an opportunity to become acquainted with research and development activities in industrial companies, an arrangement that has proved beneficial to both parties.

Consumers and media saturation are themes highlighted by the pop art movement of the 1960s in which artists like Robert Rauschenberg (who also participated in the art and technology movement) and Andy Warhol played a large role. Pop art abandoned the artistic principles of modernism and related directly to the image-making technologies of mass culture. In Andy Warhols's machine esthetic there was a subversive effect. His technique featured the repetition, banality and boredom of automated production. This subversive complicity may also be observed in the works of many postmodern artists who make use of media technology.

In its fragmented, multiple narratives, multimedia performance exemplifies postmodern esthetics. It draws on a variety of different mixed media movements: the *avant garde* events of futurism and dadaism, the mixed media works of the 1960s fluxus movement; for example, Nam June Paik's sculptures made from television sets, as well the media-based interventions of the situationist movement. From the 1980s onwards, artists such as Laurie Anderson have incorporated digital technologies into their performances. Anderson's use of sampling and other sound-treating effects have to be seen in the context of her exploring technology's effect on gender and subjectivity. The postmodern performance work of Heiner Mueller deliberately dehumanized characters—technology itself becomes a character in his productions. In Peter Wilson's work, video and projected images are involved in a staged universe in which actors carry out repetitive tasks.

An art form that relies to a large extent on modern technological media is holography, a technique using light waves to record an image of a three-dimensional object on a two-dimensional photosensitive plate. In 1968, Stephen Benton of the Polaroid Corporation introduced a new form of whole light transmission hologram; both the vivid coloring and the clarity of the holograms had a great impact on artists interested in the new medium. By the mid-1970s the gradual development of holography resulted in different holographic phenomena; for example, animated portraits using pulsed laser, motion pictures ("multiplex") holograms, acoustical holograms, and facilities to mass-produce embossed holograms. However, the tension between highly developed technical skills and the artistic content has always been a problem in holographic art. Recently though, artists have managed to successfully explore its potential for communicating human and esthetic values. By creating a new, esthetically meaningful paradigm for holography the new generation of holographic artists abolished old stereotypes.

Music and Technology

Technology has always been inseparable from music. The moment man ceased to make music solely with his voice, technology entered the scene. Some social philosophers argue that technology and musical techniques, content and meaning, develop together dialectically. Although technology has always played some role in music, an increasing technologization took place during the twentieth century, and some genres of contemporary music seem to be completely dominated by technology.

Already in the early twentieth century, an acceleration of the role of technology in music can be observed—a new "machine music" came into existence, electronic musical instruments were developed, and music composers seemed to turn into sound researchers. Recording engineers acquired increasing importance and the rise of studio esthetics had a significant effect on the listeners' expectations in concert halls. Shortly before World War I the Italians futurists demanded the rejection of traditional musical principles and their substitution by free tonal expression. This led to the design of noise instruments, the *intonarumori*. With the futurists and others, "electricity, the liberator" became the slogan of the day. Instruments like Thaddeus Cahill's "telharmonium" (1906) tried to imitate the sound of a symphony orchestra, insinuating that the orchestra players might soon be made redundant. The player piano lifted musical performance out of concert halls and transferred it to private homes. In the early 1920s Leon Termen built his "theremin," an electrical instrument helping which the human hand seemed to conjure sound from the air. This process was based on obtaining audible frequency beats formed by the interference of inaudible high-frequency oscillations. Many inventors of that time tried to design an electronic organ. In 1934 the American inventor Laurens Hammond developed the Hammond organ, an instrument with ninety-one small tone wheel generators with harmonic drawbars placed above the keyboard to permit the mixture of different tones. The instrument—easier to play than a conventional organ—proved immensely popular, as did the electric guitar, which in the 1950s and 1960s was to become the most important instrument of pop music. The guitar's amplification started in the 1930s in response to guitarist's demands for their solos to be heard over the sound of big bands. It facilitated an expansion of traditional guitar solo techniques and allowed the implementation of new techniques resulting in new effects like sustained tone.

Shortly after World War II sound recordings, disks and audio tapes were essential for the origins of "concrete music" composed from altered and rearranged sounds from the environment. Many composers in the (analog) electronic studios were not satisfied with this, however. They aimed at producing new sounds by applying simple oscillators to generate electromagnetic waves, which could then be translated into pure sound. Already in the mid-1930s the Russian physicist Evgenij Sholpo had applied the principle of artificially synthesizing an optical phonogram to his "variophone." In 1945 the U.S. inventor John Hanert with his "electrical orchestra" attempted to give the composer control over the complete fabric of musical composition. A tone was broken down into its characteristics—frequency, intensity, duration or timbre. Hanert thus reduced music to its constituent elements and reassembled it into coherent musical structures. In the 1950s, the RCA engineers Harry Olson and Herbert Belar made great efforts to synthesize sound. The apparatus they designed was, however, cumbersome and expensive. In 1966, Robert A. Moog started producing his Moog synthesizer using transistors and the technique of voltage control. He devised oscillators controlled by the amount of voltage that would alter the volume, pitch or overtones of the sound.

Although voltage-controlled synthesizers could produce a large variety of sounds and were immensely popular, their timbral capabilities remained limited. Computers and the "digital revolution" of the 1980s remedied this. Already in 1956 Lejaren Hiller and Leonard Isaacson at the University of Illinois had experimented with computer music and used calculated procedures to generate musical scores. A year later, Max Mathews at the Bell Telephone Laboratories in Murray Hill, New Jersey, produced the first computer-generated sounds. The first experiments in digital synthesis were made in the mid-1970s; the "synclavier," invented in 1977, constructed every sound from scratch. The sampling techniques of the 1980s enabled musicians to treat all sound as data: once sampled, anything could be reproduced and reshaped. In 1983, the establishment of MIDI, or musical instrument digital interface, enabled musicians to easily transfer digital information between different electronic instruments as well as between instruments and computers.

Does all this mean that the introduction of electronics and computers into music making during the twentieth century has led to an ever-increasing musical perfection and to an opening up of new creative fields for professionals and amateurs? Some technological optimists are of that opinion and many "art music" composers, too, regard the computer as a useful tool for enhancing their creative abilities and relieving them from routine work. Others come to a negative conclusion: they argue that the computer stifles artistic creativity, produces a trend towards uniformity, devalues intellectual and artistic skills and has brought about technological dehumanization.

All these controversies aside, the revolution in music making as a consequence of electrification and electronics has taken place in only a few music genres. "Art music" making has proved remarkably resistant to electronics and computers; many composers of "minimal music" even feel disturbed by electronic sounds and prefer their music "unplugged."

Apart from music making, recording has been of great interest for the development of music in the twentieth century. Although many music enthusiasts were fascinated by it, conductors often declined to make recordings because they objected to their poor quality and resented the cold atmosphere in recording studios inimical to artistic inspirations. With improved recording facilities, however, the situation changed. In the 1960s Glenn Gould, the Canadian pianist, regarded the recording studio as the center of music making, relegating live performance to the fringe. Indeed, popular music from the 1960s onwards is unimaginable without the vast array of electronic studio equipment in existence; and other forms such as jazz would have developed differently without it. This is because recordings captured improvisations, which are extremely difficult to write down; but recording has also influenced music making. Before the rise of sound recording, for example, most violinists used vibrato sparingly. Once sound recording had been introduced, vibrato adopted a compensatory role. It made it possible for violinists to overcome the limitations of early recording equipment, served to mask imperfect intonation and also helped project a greater sense of the artist's presence.

Technology, particularly means of transportation like trains, cars and airplanes, have been an often recurring theme in both "art music" and "popular music." Many composers of the early twentieth century wrote music to reflect a changing world. In the 1920s railways and particularly railway engines aroused the interest of many composers, as is documented in Arthur Honeggers *Pacific 231*, named after one of the fastest American railway engines of its time. In *Pacific 231* the composer successfully transformed features like speed, dynamics, and energy into the language of music.

At the turn of the twentieth century artists also perceived the recently invented airplane as an esthetic event with wide-ranging implications for artistic and moral sensibility. Even more than in railways, artists and musicians transformed the airplane into a spiritual creation. The age of the airplanes was supposed to bring about unlimited individual mobility, peace, and harmony, but already before World War I it became clear that the airplane could also be used for destructive purposes. The utopia of peaceful internationalism gave way to aggressive nationalism and the two world wars bear witness to the misuses of flight. All these different feelings are reflected in paintings and literature, but also in twentieth century musical compositions. The airplane in the American composer George Antheil's Airplane Sonata (1921) manifests itself in machine-like driving rhythms and insistent ostinatos. The German composers Kurt Weill und Paul Hindemith transformed Charles Lindbergh's first transatlantic solo flight of 1927 into a dramatic tone poem (1929) and in 1945 the Czechoslovak composer Bohuslav Martini wrote his *Thunderbolt P-47*, a spectacular piece of program music, as an homage to the victorious U.S. Air Force. At the same time the American composer Marc Blitzstein wrote his symphony *The Airborne* about the glory but also about the terror inherent in aviation. In the second half of the twentieth century space flight also became a theme in music with generally positive connotations.

Performing Arts and Technology

Technology has always affected the theater. The Ancient Greeks introduced scenic painting but also the *deus ex machina*, a wooden elevator that brought a character onto the stage straight from heaven. In the eighteenth century, with the construction of the National Opera in Paris, there were wing-supporting chariots, overhead rigging sets, and an elaborate system of wing-operated floor traps set between the chariot slots. Those operations were mainly operated by "manpower"—a sizable number of stagehands—as well as by horses. In the nineteenth century, powered prime movers took over this task.

In the modern theater there are a variety of mechanisms used to erect, position, and manipulate scenery: hoists, lifts and horizontal drives. Hoists raise or lower scenery from the stage penthouse, lifts bring up scenic elements beneath the stage floor, and horizontal drives slide platforms, flats, and large properties from the wings onto the visible part of the stage or from one area of the stage to another.

Lighting and lighting effects have been of special interest in the performing arts for a long time. The modern era of stage lighting started with the invention of a practical electric lamp by Thomas A. Edison in 1880; gas lighting, which had been used before, was quickly discarded. At the turn of the twentieth century, incandescent lamps were in general use for stage lighting. Shortly before World War I concentrated oil filaments made incandescent spotlight possible, which in turn gave rise to the further development of stagecraft. Much lighting equipment was developed for special effects, moving clouds, for rain or snow or for fire effects. The oldest effects projector, the "Linnebach lantern," dates from the World War I era. In this "scene projector," as the Linnebach lantern was also called, a concentrated light source is placed in a deep black box; a painted slide is placed on the side of the box left open. The design painted on the glass is projected against a drop onstage, greatly enlarged, at a relatively short distance. As no lens is used in a Linnebach lantern, the light source must be powerful and concentrated.

Around the mid-twentieth century new interest arose in the use of projections. At a Wagner music festival at Bayreuth, Germany in the 1950s, Richard Wagner's grandson Wieland reduced the three-dimensional scenic elements to their essentials and then flooded the stage with multiple, overlapping projected patterns. Shortly afterwards the Czechoslovakian designer Josef Svobada developed his concept of "visions on space." He massed three-dimensional screens and, with slide and films, created a montage effect. In the early twentieth century motor-driven dimmers controlled auditorium lighting, but they were cumbersome and expensive for stage lighting. In 1890 the first remote-controlled dimmer was used, beset with problems. A good half-century later, remote control worked satisfactorily as the result of the work by George Izenour from Yale University, who in 1948 developed a dimmer with a thyratron, a type of electron tube. A few decades later, computerization increased the parameters of what can be done to enhance theatrical illusion and to come close the Wagnerian ideal of "total theater." With computerization hardly anything has to be left to chance.

Remote control is a key innovation in twentieth century theater technology. Computers are part of every operation. They seem to be best at repetitive or standardized operations and repetition, which is necessary for grand spectaculars requiring the complex coordination of various scenic elements. The director, as the main creative artist in the theater, is in charge of all this; interpreting the script, guiding the designers, and also manipulating the performers in order to realize his artistic objectives. Today many actors and critics argue that computer-controlled theater has established conditions in which the director, as puppet-master, is in total command of what is happening. He or she sits at the console and coordinates the sound, adjusts the lightening, and shifts the scenery. The director manipulates the performer, who is bereft of spontaneity and required to conform rather than to create. This pessimistic assessment plays down the important role that some directors already had in precomputer times and also the amount of artistic freedom famous actors still enjoy today, but the problem hinted at above certainly exists.

See also **Audio Recording; Film and Cinema**

HANS-JOACHIM BRAUN

Further Reading

Benjamin, W. *Illuminations*. Fontana, London, 1970.

Bettig, R.V. *Copyrighting Culture: The Political Economy of Intellectual Property*. Westview Press, Boulder, CO, 1996.

Bijvoet, M. How intimate can art and technology really be? A survey of the art and technology movement of the sixties, in *Culture, Technology & Creativity: The Late Twentieth Century*, Hayward, P., Ed. John Libbey, London, 1991, pp. 15–37.

Braun, H.-J. Introduction: technology and the production and reproduction of music in the twentieth century, in *Music and Technology in the Twentieth Century*, Braun, H.-J., Ed. Johns Hopkins University Press, Baltimore, 2002, pp. 9–32.
Braun, H.-J. Movin' on: trains and planes as a theme in music, in *Music and Technology in the Twentieth Century*, Braun, H.-J., Ed. Johns Hopkins University Press, Baltimore, 2002, pp. 9–32.
Carson, C. Theatre and technology: battling with the box. *Digital Creativity*, 10, 3, 129–134, 1999.
Chadabe, J. *Electronic Sound: The Past and Promise of Electronic Music*. Prentice Hall, Upper Saddle River, NJ, 1997.
Coyle, R. Holography—art in the space of technology: Margaret Benyon, Paula Dawson and the development of holographic art practice, in *Culture, Technology & Creativity: The Late Twentieth Century*, Hayward, P., Ed. John Libbey, London, 1991, pp. 65–88.
Douglas, D. *Art and the Future: A History-Prophecy of the Collaboration Between Art, Science, and Technology*. Thames & Hudson, London, 1973.
Duckrey, T., Ed. *Ars Electronica. Facing the Future: A Survey of Two Decades*. MIT Press, Cambridge, MA, 1999.
Elm, T. and Hiebel, H.H., Eds. *Medien und Maschinen. Literatur im technischen Zeitalter [Media and Machines. Literature in the Technological Age]*. Rombach, Freiburg, 1991.
Fidler, R. *Mediamorphosis: Understanding New Media*. Pine Forge Press, Thousand Oaks, 1997.
Flender, R. and Lampson, E., Eds. *Copyright. Musik im Internet*. Kadmos, Berlin, 2001.
Guderian, D., Ed. *Technik und Kunst [Technology and the Arts]*. VDI-Verlag, Düsseldorf, 1994.
Hayward, P. Introduction: technology and the (trans)formation of culture, in *Culture, Technology & Creativity: The Late Twentieth Century*, Hayward, P., Ed. John Libbey, London, 1991, pp. 1–12.
Hughes, R. *The Shock of the Modern*. Thames & Hudson, London, 1991.
Izenour, George C. *Theater Technology*, 2nd edn. Yale University Press, New Haven, 1988.
Katz, M. Aesthetics out of exigency: violin vibrato and the phonograph, in *Music and Technology in the Twentieth Century*, H.-J. Braun, Ed. Johns Hopkins University Press, Baltimore, 2002, pp. 174–185.
Kurzweil, R. *The Age of Spiritual Machines*. Viking Penguin, New York, 1999.
Lessig, L. *Code and Other Laws of Cyberspace*. Basic Books, New York, 1999.
Lunenfeld, P. *Snap to Grid: A User's Guide to Digital Arts, Media and Cultures*. MIT Press, Cambridge, MA, 2000.
McSwain, R. The social reconstruction of a reverse salient in electrical guitar technology. noise, the solid body, and Jimi Hendrix, in *Music and Technology in the Twentieth Century*, H.-J. Braun, Ed. Johns Hopkins University Press, Baltimore, 2002, pp. 186–198.
Manning, P. *Electronic and Computer Music*, 2nd edn. Oxford University Press, Oxford, 1993.
Millard, A. *America on Record: A History of Recorded Sound*. Cambridge University Press, Cambridge, 1995.
Murphie, A. negotiating presence—performance and new technologies, in *Culture, Technology & Creativity: The Late Twentieth Century*, Hayward, P., Ed. John Libbey, London, 1991, pp. 209–226.
Murphie, A. and Potts, J. *Culture and Technology*. Macmillan, Houndmills, 2003.
Pinch, T. and Frank, T. *Analog Days: The Invention and Impact of the Moog Synthesizer*. Harvard University Press, Cambridge, MA, 2002.
Rötzer, F., Ed. *Digitaler Schein: Ästhetik der elektronischen Medien [Digital Appearances: Aesthetics of Electronic Media]*. Suhrkamp, Frankfurt, 1991.
Schmidt Horning, S. From polka to punk: growth of an independent recording studio, 1934–1977, in *Music and Technology in the Twentieth Century*, H.-J. Braun, Ed. Johns Hopkins University Press, Baltimore, 2002, pp. 136–147.
Smudits, A. *Mediamorphosen des Kulturschaffens: Kunst und Kommunikationstechnologien im Wandel [Mediamorphoses of Cultural Creation:Change in Art and Communication Technologies]*. Wilhelm Braumüller, Vienna, 2002.
Theatrical Production, in *The New Encyclopaedia Britannica, Macropaedia*, 15th edition, vol. 28. Encyclopedia Britannica Inc., Chicago, 2003, pp. 561–615.
Théberge, P. *Any Sound You Can Imagine: Making Music/Consuming Technology*. Wesleyan University Press, Hanover, 1997.
Tisdall, C. and Bozzola, A. *Futurism*. Thames & Hudson, London, 1993.
Walker, J.A. *Art in the Age of Mass Media*. Pluto, London, 1983.

Technology, Society, and the Environment

The technology of the twentieth century significantly shaped not only society, but also environmental processes, in ways and on a scale unprecedented in human history. While the twentieth century "partnership" between technology and society in many respects continued nineteenth century trends, at least until the final decades of the twentieth century, the relationship with the environment changed fundamentally over the course of the century. Society for the first time significantly influenced, both advertently and inadvertently, not only the form but also the large-scale function of external nature. By the end of the century a thoroughly "technologized" society was uncertainly confronting the resultant challenges. Some of these, such as the risk of climate change and the resultant imperative to reduce greenhouse gas emissions, threatened to change the nature of industrial society itself (reasserting perhaps the more traditional tendency of the environment to shape society), while others, such as the challenge of regulating genetic technology, presented unparalleled ethical dilemmas. These developments are examined below through the lens of three overlapping clusters that exemplify the co-evolving relationship between technology, society, and environment over the course of the twentieth century.

While many environmental developments, such as climate change, were global phenomena, the spread of twentieth century "technologized" society was not uniform. Centered in the leading industrial regions of the U.S., Europe, and Japan, twentieth century technology spread unevenly and at different rates across the rest of the globe. This account will concentrate on the leading industrial regions but within the context of this differential development. First a brief review of the emergence of significant interdependence between technology, society and environment during the twentieth century.

While nineteenth century railroads had significant social effects; for example, generating markets, helping impose rigorous conceptions of time and timekeeping, and facilitating the generation of national self-identity, many twentieth century technologies took such tendencies much further. Many technologies, including the automobile, electric light and telephone, deeply fashioned lifestyles and the twentieth century social order, influencing mental landscapes as much they did physical ones. By the end of the twentieth century it was common to think of technologies as "iconic," or emblematic of culture more generally. This increasing integration of society and technology mediated an unparalleled mutual dependence between society and the environment. By the end of the century social practices and behaviors, mediated by technology, were a significant factor in the world's climate and in the ozone concentration of the stratosphere, for example. Many related developments over the course of the century underline this interdependence. The role and scale of government grew with the necessity to support and regulate large-scale technological systems, such as those of transport and telecommunication, and also from the 1970s their environmental impacts. Trade and commerce, facilitated by advanced computing and communication technologies, took on a progressively more global character, reflecting the scale of environmental impacts; and dominant environmental attitudes evolved from regarding external nature as little more than a source of resources and sink for wastes, to a widespread concern with resource depletion and environmental pollution.

The first cluster to be considered, the "coketown cluster," emerged in the second half of the nineteenth century and centered on the coal, iron, steel, and railroad industries. These industries were concentrated in "smokestack" cities that bore the brunt of the urban air pollution that provided the environmental signature of this cluster. Originally centered in northwestern Europe and the U.S., it spread to Japan early in the twentieth century and the USSR in the 1930s, while later incarnations of it were still emerging in Soviet satellite countries and in China in the 1950s. This cluster was associated with a period of intense integration and consolidation in the world economy running from about 1870 until abruptly terminated by World War I. Facilitated and reinforced by emerging transport and communication technologies—the railroad, steamship and telegraph—and also by colonialism, it demonstrated many analogies with the globalization that marked the closing two decades of the twentieth century.

The dynamism associated with this burst in international trade and economic integration were enabled by contemporary technological developments such as refrigerated shipping, and bore witness to the voracity of the emerging urban markets of the smokestack cities. The environmental impacts of these developments started, for the first time, to become significantly disengaged in time and space from the activities driving them. For example the conversion of Argentinean pampas to beef production, New Zealand pasture to that of sheep, and Brazilian agricultural land to coffee production was driven by the pull of distant markets, whose environmental impacts became increasingly global. While consumer society, as it came to be known, was most intimately associated with the ensuing cluster, its essential ingredients were developing rapidly in the first decades of the twentieth century. By the beginning of the century even working class families throughout much of Europe and the U.S. could afford and had access to a growing range of consumer goods. Ready-to-wear clothing, bicycles, daily newspapers and comics, and standard forms of entertainment such as that of the earliest cinemas, signaled what was to become, in the following decades, a deluge (see below).

The growth of smokestack cities was facilitated, in part, by the extended reach of cities afforded by railroads and improved levels of urban sanitation resulting from the application of better wastewater and urban waste management technologies. The first decades of the century also witnessed the emergence of the modern city, driven in particular by innovations in electrical technology. One of these—the electric lightbulb—came to dominate street lighting throughout the industrialized world by the first decade of the century. The second was the advent of electrified tramways or trolleybuses.

Until these arrived (from 1890 onwards but increasingly from 1900) the spread of cities, facilitated by railroads, was constrained by transport from suburban stations being constricted to horse- and steam-powered vehicles. Electrified tramways or trolleybuses were faster, cheaper, cleaner, and more convenient and significantly encouraged the expansion of city boundaries. World War I, the first mechanized war in which tanks and aircraft played a significant part, signaled the end of the period of intense economic integration with which this cluster was initially associated. These various developments, however, heralded the emergence of the ensuing cluster that was to dominate the twentieth century, and one of whose signature technologies—the car—was to ensure that the age of electrified tramways and trolleybuses was short lived.

This next cluster, the "motown" cluster, was to dominate the twentieth century. At its core were the oil, electricity, automobile, aircraft, chemicals, plastic and fertilizer industries; and the powerful assembly lines whose facilitation of mass production paved the way for mass consumption, was one of its chief hallmarks. This cluster first took shape in the U.S. from about the 1920s, and quickly took hold in Canada, Western Europe, Japan, Australia and New Zealand and predominated from the 1940s. Its take-up was constrained by state ideology in the USSR and China, and by power and economic differentials in much of the rest of the globe, although by the end of the century this was changing with countries such as China rapidly superimposing motown characteristics onto still expanding coketown economies. While the motown cluster is complex, many of its essential characteristics are captured by the conjunction of Fordism, the name given to the principles of mass production, and the mass consumption that accompanied it. Henry Ford was pivotal to these developments. His first moving electrified assembly line (1912) and his seminal insight that wages should be set at a level that enabled workers to purchase the fruits of their labor; together with Taylorism (the idea that workers movements could be scientifically managed to optimize productivity), paved the way for consumer society. By 1923 Ford's workers could afford to buy a Model-T Ford with 58 days' wages. Fordism soon spread this principle to many consumer technologies, such as radios, washing machines and refrigerators, that became a hallmark of twentieth century industrial lifestyles.

One of the pivots upon which the growth of twentieth century consumer capitalism depended was marketing. As electricity grids extended rapidly in the early decades of the century it became clear that optimal use of the thermal generating plant that dominated supply provision required major loads additional to those of industry. Aggressive marketing, particularly of newly developed electric appliances, aided the rapid establishment of a significant domestic electricity market. Marketing was similarly a key feature of the rapid deployment of many seminal twentieth century technologies such as the phone and the car, upon which twentieth century lifestyles progressively depended.

The car exemplified these developments and was particularly influential in conditioning the physical form of cities, as well as lifestyles more generally. Most U.S. cities built after 1920 were shaped by the mass ownership of cars, a trend epitomized by Los Angeles where, similar to other U.S. cities, public trains were dismantled in the 1940s to make way for cars. It was at just this time that smog first became a political issue in Los Angeles. Photochemical smog, created by the action of sunlight upon the mixture of pollutants emanating mainly from cars, became a significant problem in many twentieth century cities, and is linked to a number of, principally respiratory, health problems. This, however, is only a fraction of the environmental impact of cars. Discussing these matters McNeill states, "In Germany in the 1990s [making a car] generated about 29 tons of waste for every ton of car. Making a car emitted as much air pollution as driving a car for 10 years. American motor vehicles (c.1990) required about 10 to 30 percent of the metals—mainly steel, iron and aluminum—used in the American economy. Half to two-thirds of the world's rubber went into autos." Cars also took up a lot of space and killed a lot of people (according to McNeill, from 25,000 to 50,000 per annum after 1925 in the U.S. alone) but remained perhaps the most coveted of twentieth century technologies.

Similar impacts were multiplied across the economy as a whole. The phenomenal scale-up in production and consumption facilitated by Fordism gave rise to immense increases in resource use and pollution (although improved technology made cleaner and less material and energy intensive production possible as the century progressed). Fordism's appetite for resources can be graphically illustrated with relevant statistics. Between 1900 and 1995, global coal output rose by a factor of over 6.5, while by about 1980, quarrying moved more earth than did natural erosion, for the first time. In simple physical terms this resulted in major environmental impacts, many of the most significant and hazardous of which result from the wastes from these activities, in the form of stored wastes, slag, and tailings. Among other technological activities whose environmental impacts became particularly marked in this period was large-scale hydraulic engineering. Although dams had been used for electricity generation from the late nineteenth century the dynamics of the motown cluster inspired multipurpose (i.e., irrigation and power generation) hydro developments on a far greater scale. Pioneered by the Tennessee Valley Authority in the 1930s, and by the Boulder (later Hoover) dam on the Colorado river, the world's largest when built in 1935, large-scale dams came to appeal to twentieth century political leaders as emblematic of industrial progress. However, the enormous environmental and social dislocation brought about by large dams had made for a major rethink by the end of the century, except in China which was proceeding with the controversial Three Gorges Dam project, which if finished to plan will be the largest in history.

Many other technologies were significant in shaping technology, society, and environment relations throughout the twentieth century. Agriculture changed beyond recognition, not only via mechanization and motorization, but also crucially via the large-scale deployment of agrichemicals and of newly developed high-yielding crop strains. It was developments such as these that enabled previously marginal land, such as Australia's wheat belt, to be profitably cultivated so as to feed the vigorously expanding appetites of twentieth century cities. By the end of the century some of the environmental costs of such developments, such as the increasing salinization of agricultural land, which by the 1990s seriously affected about 10 percent of irrigated lands worldwide, were causing many to reflect on their wisdom. Also of particular significance was the mid-century development of nuclear technology that irrevocably changed the nature of warfare and whose civilian implications were still being calculated at the end of the century.

Other significant repercussions of the motown cluster included its effects on social arrangements, including those of a familial, gender, and economic nature. For example, while the traditional burden of housework was lightened considerably by the advent of cheap, readily accessable household appliances, women also became valued for many production line tasks in which precision and endurance counted for more than strength and skill. The renegotiation of the social contract implied by Fordism resulted in a significant reshaping of traditional class distinctions with a relatively leisured and affluent middle class emerging on a large scale in the U.S. from the 1940s, and a decade or so later in Europe. A paradoxical result of this emerging affluence was that it afforded the leisure to critically contemplate the status quo. One result of this was the rise of large-scale environmental concern in the 1960s and regulatory responses to it, most notably from the 1970s. Triggered by a concern over the impact of widespread pesticide use, environmentalism evolved, alongside fears of an energy crisis in the 1970s, into the emergence of ideas whose target became the form of industrial society itself. This emergence of environmental politics, a creature of the last three decades of the century, paved the way for and fed into the development of a new cluster.

The "postindustrial" cluster emerged in the 1980s and matured rapidly through the 1990s. It centered on the rise of sophisticated computing and telecommunications technologies (also called information and communication technologies, or ICT), biotechnologies, and, facilitated by ICT, a turn-away from primary manufacturing to a newly dynamic service sector. These both exemplified a technological trend to move away from the manipulation of material power to a focus on the manipulation of knowledge and information. The spread of this cluster was particularly uneven, both geographically and in terms of the adoption of the technologies it embodied. While centered in the traditional industrial heartlands of the U.S., Europe, and Japan, the advent of small, cheap, powerful distributed computing facilitated by the rapid extension of the Internet in the 1990s allowed many other countries to become involved, particularly in ICT. Malaysia, for example, while still industrializing (in both coketown and motown terms) had an ambitious national plan that placed it at the forefront of nations installing ICT infrastructure. Similarly in the 1990s, India became a center of software production, on which the rapidly expanding ICT depended. This was facilitated by the real-time access ICT granted overseas companies, such as those in the U.S., to the products of Indian software engineers, and by the relatively low level of Indian wages.

Contemporary commentators discussed how the scope and scale of the emerging ICT and the way they changed how time and space were experienced and conceived, heralded a new form of information or postindustrial society. The implications of these changes were unclear however by the end of the century. Many of the changes that were evident, such as further deskilling of the workplace, simply intensified trends evident throughout the course of the twentieth century.

If anything, far greater uncertainties surrounded the biotechnologies emergent at this time. While promoted as having the potential to feed the worlds starving, the actual gains delivered by genetically engineered crops and foods by the end of the century, appeared to amount to little more than a continuation of those gains granted over the course of the twentieth century by more traditional selective-breeding techniques. The application of genetic engineering also presented both many ethical and moral dilemmas, and threatened a variety of environmental impacts that made many question its continuing expansion.

The tendency to critical evaluation marked by the emergence of environmental politics and ambivalence to emerging technologies was a hallmark of the final two decades of the century. From its beginnings in the 1960s, environmental politics was focused on the systemic environmental effects of industrial society, which both marked it out from an earlier conservation movement and made for a departure from the technological optimism that had dominated the century. Events only reinforced and confirmed these trends. Notable among these were the emergence of the problems of acid rain (see Electricity Generation and the Environment), ozone depletion and climate change. While the Montreal Protocol (1987), and subsequent amendments to it, most notably London (1990) and Copenhagen (1992), are widely regarded as having brought ozone depletion under control, climate change was another matter. The push to decarbonize the energy economy its control necessitated was a challenge many governments baulked at and by the end of the century the implementation of the Kyoto Protocol (1997) was uncertain and disputed (see Electricity Generation and the Environment).

Decarbonization was part of a broader reaction to the polluting and resource-intensive technologies that had dominated the century. A push to dematerialize industrial economies by reinforcing trends to greater efficiencies in production and further reduce the intensity of energy and resource use marked this reaction. Operating under the rubric of sustainable development, first popularized by the 1987 report of the UN Commission on Environment and Development entitled "Our Common Future" and legitimated by the 1992 Earth Summit in Rio de Janeiro, the 1990s were marked by a variety of approaches to these matters. While many of these, such as notions of clean (or sometimes cleaner) technology and that of industrial metabolism (which aimed to mimic nature by reusing and recycling waste streams) held great promise, there was some incoherence about their development.

While much of the reduction in the energy and resource intensity of leading industrial economies resulted from improved industrial efficiencies, a great deal of it, from about the 1970s, also resulted from the rapid emergence of a service sector. This occurred as traditional smokestack industries moved to less developed nations. The service sector, which is far less energy- and resource-intensive than traditional smokestack industries, centered on activities such as finance, tourism and the retail sector, and was given a boost by the popularity of economic deregulation during the 1980s and 1990s. In the mid-1990s the World Trade Organization was formed to extend economic deregulation to the sphere of global trade. This period in many ways picked up the drive to integrate the world economy with which the century started while also reflecting its laissez-faire economic outlook, but this time, aided by ICT, it was far more successful. Unfortunately, however, this expansionary success ensured that the push to dematerialize industrial economies was marginal to continuing global motown trends in resource and energy use.

It was with such tensions that the century closed. Underpinning these tensions were emergent signs, signaled for example by the Kyoto Protocol, of recognition of the need for a new contract between technology, society and environment; one in which technological development, informed by an increasing awareness of the intimate interconnections between technology, society, and environment, would be driven by debate and consent rather than simply by profit or serendipity. These, however, were matters for a new century.

See also **Electricity Generation and the Environment; Environmental Monitoring**

STEPHEN HEALY

Further Reading

Blunden, J. and Reddish, A., Eds *Energy, Resources and Environment*. Hodder & Stoughton, London, 1991.

Elliot, D. *Energy, Society and Environment*. Routledge, London, 1997.

Fichman, M. *Science, Technology and Society: A Historical Perspective*. Kendall/Hunt Publishing, Dubuque, IA, 1993.

McNeill, J. *Something New Under the Sun: An Environmental History of the Twentieth Century World*. Penguin, London, 2000.

Melosi, M.V., Ed. *Pollution and Reform in American Cities, 1870–1930*. University of Texas Press, Austin, 1980.
Roberts, G., Ed. *The American Cities and Technology Reader: Wilderness to Wired City*. Routledge, London, 1999.
Silverton, J. and Sarre, P., Ed. *Environment and Society*. Hodder & Stoughton, London, 1990.
Smith, D.A. M*ining America: The Industry and the Environment, 1800–1980*. University Press of Kansas, Lawrence, 1987.
Stradling, D. *Smokestacks and Progressives: Environmentalists, Engineers, and Air Quality in America 1881–1951*. John Hopkins University Press, Baltimore, 1999.
Tarr, Joel A. *The Search for the Ultimate Sink: Urban Pollution in Historical Perspective*. University of Akron Press, Akron, OH, 1996.

Telecommunications

The history of two-way communication is embedded in a series of much deeper histories: technology, imperialism, and the rise of the nation state, to name but a few. These, in turn, are embedded in a discourse that returns theory to a central place in non-Marxist histories. Several theoretical enterprises emerging from the academic disciplines of economics and political science look to technology as one of the major engines that drive history and to telecommunications technology as the most significant technology in the current world economy.

Tilly notes that there are essentially two types of states: trading states and territorial states.

Trading states are maritime in nature, exert weak geopolitical control over long distances, and excel at the long-distance communications needed to manage their economies. To exchange goods profitably, trading states have needed a great deal of mutual trust and sophisticated commercial arrangements among a community of spatially distant merchants. Updating Tilly's argument, one would add managers and consumers to merchants. Trading states have fairly fluid social structures and a mutually supportive relationship between commerce and government. In the political realm they tend to relative democracy. The archetypes have been Holland and Britain.

Territorial states are continental in nature, exert strong geopolitical control over short distances, but have very little reach beyond that. The economies of territorial states tend to be command economies, with the government in charge, and with a one-way information flow down a strict social hierarchy. In the political realm they tend to relative autocracy. The archetypes have been China and Russia. Some states such as the U.S., France, Germany, Japan, and Spain have caused untold chaos at different times in world history shifting between different aspects of the two types or, worse, attempting to be both at the same time.

From the1400s onward, wealth generation has been driven by the remarkable success of trading states and their ever-increasing domination of the world economy. During the 1900s, this proved to be even more the case, although it did not seem that way in 1904 when British geopolitician Mackinder suggested that nineteenth century improvements in communications, especially terrestrial telegraphy and railroads, meant the commercial domination of the world economy by trading states was ending. In the early 1900s, however, wireless telecommunications began to shift the balance back toward trading states. The three main wars of the twentieth century (World Wars I and II, and the Cold War) were about the containment of the two main territorial states of the period, Germany and Russia, by the two main trading states, Britain and the U.S. Much of this containment was through superior telecommunications.

Although the global telecommunications system was shaped by the spectacular growth of the British submarine cable system after the first successful transatlantic cable opened for business in 1866, the 1900s saw considerable improvement of a system that had reached much of its modern shape by the late 1800s. Seven innovations stand out:

1. Low-frequency wireless telegraphy, starting around 1900
2. High-frequency wireless, which allowed telephony as well as telegraphy, in the 1920s
3. Microwave wireless, which led to radar, in the 1930s
4. Submarine telephone cables in the 1950s
5. Satellites in the 1960s
6. Fiber-optic submarine cables in the 1980s
7. Cellular wireless telephones in the 1990s.

These seven innovations have increased the capacity or convenience of the global communications system and have trended towards radically lowered costs. Developments in two-way communications often encouraged one-way communications, such as the development of broadcast radio entertainment out of wireless telecommunications. Occasionally the flow of innovation has reversed, as in the importance of electronic television technology to the development of radar.

Two areas of interaction between commercial and state power have been driven by telecommunications. First, states have sought to control communications systems, to prevent other states controlling them, or both. Most states, America being a rare exception, developed post, telegraph, and telephone systems (PTTs) as government monopolies in the 1800s, and frequently added such new technologies as broadcasting as they arose in the 1900s. Second, wireless telecommunication allows real-time connections without people or vehicles being tied to a spatially fixed infrastructure. The advantages of the cellular telephone to individuals are obvious. Commercially, the global positioning system (GPS) allows us to operate the world economy in an efficient, "just-in-time" way. Militarily, GPS allows considerable geopolitical control at very great distance, with the extreme accuracy and concentration of force afforded by "smart" bombs and cruise missiles, and without committing large numbers of ground troops.

Telecommunications are inextricably bound up with geopolitical power. In 1866 British capital began the successful installation of a global network of submarine cables that was never surpassed, investing as freely in Buenos Aires as in Birmingham. The geography of the network was such that most of the world's telegraphic messages passed through Britain. The cable companies involved cooperated closely with the British government. In World War I, Britain's control over global telecommunications and the skill of British code breakers was so complete that Room 40 in the British Admiralty was able to intercept and decode the notorious Zimmerman telegram. In 1917 German foreign minister Zimmerman offered Mexico the return of historic Spanish possessions in the American Southwest for attacking America, helping bring America into World War I.

The technological history of telegraphy is in three parts: terrestrial, submarine, and wireless, with only the last being a technology of the 1900s. The first great terrestrial companies were established in America after 1844. The first global telecommunications system was the British submarine cable system that grew after 1866, but required "repeater" stations on islands under British control dotted around the world's oceans, hence some of the odder territorial acquisitions of the Empire. Of the seven innovations in telecommunications of the 1900s, the first—low-frequency wireless—built on this base of global control.

Low-Frequency, Long-Wave Wireless

The possibility of wireless telegraphy was demonstrated by Hertz's empirical verification of Maxwell's equations in 1888. Eight years later, Marconi patented the first wireless telecommunications system. In 1901 Lloyd's gave Marconi's a monopoly on ship-to-shore wireless and required it for a ship to achieve their highest insurance rating. Marconi's commercial success was assured and the *Titanic* disaster of 1912 made Marconi a household name.

Wireless quickly became one of the most contested technological arenas of the early twentieth century. Nonmilitary researchers and entrepreneurs of this pioneer period dreamed of something like cellular telephones, well beyond the available technology, but it was in the geopolitical and military arenas that wireless took deepest hold. Several countries, the U.S. and Germany in particular, saw it as a way of challenging Britain's telecommunications dominance with all that that implied for British global hegemony. At the military level, navies quickly realized that wireless offered centralized command and control of ships at sea and began to force the ship-to-shore technology Marconi pioneered. Afraid of depending on a British supplier, the U.S. and German navies made major investments in wireless before 1910.

World War I caused even more rapid evolution of wireless than the naval armaments race before the war. By 1915, the British Army was using low-frequency wireless telegraphy to allow airplanes to spot for guns. By late 1918, Allied airplanes were being fitted with higher frequency, wireless telephone systems to allow centralized command and control of the air war. "Plan 1919" called for Allied wireless telephony to coordinate air and ground advance using attack airplanes and tanks. In World War II, Germany, laggard in 1918, demonstrated such *blitzkrieg* to perfection. The huge numbers of vacuum tubes produced for "Plan 1919" allowed commercial broadcasting to develop in the early 1920s, operating at the same frequencies as the radios built for mobile war, today's medium waves or AM band.

High-Frequency, Short-Wave Wireless

Vacuum tubes also made possible much higher frequencies for telecommunication. Marconi experimented successfully in the 1920s with high-frequency short waves, and "beam" antennae for global wireless telegraphy and telephony. The resulting imperial network of short-wave beam wireless stations was extremely cheap to build and operate and effectively renewed Britain's control of global communications by the mid-1920s. The submarine cable companies, seeing their profits evaporate, persuaded the British government to save them in 1929 by forcing Marconi to merge with them. The resulting company became Cable and Wireless in 1934.

Microwave Wireless

Although microwave wireless had its origins in the commercial world when Standard Telephones and Cables, based on work in their Paris laboratories, installed a "micro-ray" telephone relay across the English Channel in 1931 operating in the 17.6-centimeter band, microwave technology matured in war, not peace, and with radar, not telephony. In 1908 Wells' novel, *The War in the Air*, suggested an irresistible fleet of German zeppelins could cross the Channel, destroy the British fleet, bomb Britain into submission, then continue across the Atlantic to destroy the American fleet and bomb New York. Zeppelins fared poorly in World War I, but German airplane raids scared Britain into developing the world's first independent air force to both attempt defense against bombers and to bomb back. Three strands of reasoning developed: in Italy, Douhet argued that bombing civilians into submission would be the way to win future wars; in America, Mitchell argued for precise bombing of strategic targets; and in Britain, Trenchard argued that air forces could control fractious provinces of the Empire more cheaply and with less loss of (British) lives than could occupying armies, a policy known as "control without occupation."

Despite arguments that "the bomber will always get through," defense began to seem possible when Watson-Watt suggested to Britain's Committee for Imperial Defence that radio might locate incoming bombers. Watson-Watt routinely transmitted bursts of high-frequency radio to examine the ionosphere and inform the imperial short-wave beam wireless stations what the best frequencies were that day. He noted early returns when aircraft flew overhead during transmissions. Radar matured very rapidly under renewed German aerial threat after 1934. Radar was not really new, being patented in Germany in 1904 to help ships enter harbors in fog. STC's "micro-ray" telephone link across the Channel led the French to equip Atlantic liners with experimental (and unsuccessful) collision-avoidance radar in the late 1930s. At the military level, after World War I all the major navies experimented secretly with radar to solve the problems of gunnery in poor visibility; by World War II all had implemented radar systems.

It was air war did the most to improve radar. In ships or onshore radar could be heavy, bulky, and use lots of energy. Britain's Committee for Imperial Defence decided by 1936 that a day battle for Britain could be won using ground radar-directed fighters, forcing Germany into night bombing and requiring lighter, smaller, more energy-efficient microwave systems in fighters themselves. Massive technology forcing ensued and the British deployed Mark IV airborne intercept radars operating at 1.5 meters in early 1941, during the nighttime blitz on London. In part, this technology succeeded by drawing on Britain's pioneering electronic television, commercially deployed in 1936. Television engineers maintained the crucial radar systems and, at war's end, radar engineers found employment in television. Blumlein, Britain's greatest electronics engineer, helped develop electronic television, Mark IV airborne intercept radar, and H2S ground-imaging, bomb-aiming radar before his untimely death in June 1942. H2S radar was crucial for Britain's commitment to strategic bombing, allowing bombing by night or through thick cloud or smoke. It and its counterpart, Airborne Intercept Mark IX radar, operated in the 10-centimeter microwave band, later moving to 3 centimeters.

After World War II was over, microwave technology returned to its peacetime origins in microray telephony. Because little real estate had to be acquired, just hilltop sites for line-of-sight relay towers, microwave telephony was cheap, offered high-capacity communications, and was rapidly installed in countries such as the U.S. and Canada where sheer distance made co-axial cable expensive. By the late 1940s such microwave relays were also planned to carry television signals coast-to-coast in North America.

Submarine Telephone Cables

In 1956 Bell Telephone laid the first transatlantic telephone cable, TAT-1, using analog transmission technology. Initial capacity was only 36 simultaneous calls but it also carried more telegraphic traffic than the then total capacity of all existing submarine telegraph cables, rendering them obsolete. Unlike the submarine telegraph cables, telephone lines need a great deal of power to drive speech long distances. Bell devised vacuum tube amplifiers compact enough to be installed within the cable and reliable enough for 25 years service on the sea floor. TAT-1 needed 51 amplifiers in each direction to push signals across the Atlantic.

Although such telephone service was well beyond the pockets of the average person, TAT-1, its successors through TAT-7, and its competitors, the Anglo-Canadian CANTATs-1 and -2, ushered in a revolution in business communications across the Atlantic. Capacity mushroomed, from 36 simultaneous calls in 1956 to 11,173 when TAT-7 opened in 1983. Together with jets, which entered transatlantic service in 1959, the TATs made possible effective multinational corporations. The large American automobile manufacturers, Ford and General Motors, became proto-multinationals before World War II, but operated their American, European, and Japanese factories as independent fiefdoms with independent management structures. The communications and transport revolutions of the 1950s made centralized management possible from America through instant voice communication and rapid site visits to remote factories. As the European economy recovered following World War II, European companies such as Nestlé followed suit.

Pacific telephone cables and the development of jets able to reach Tokyo from London or New York nonstop have ushered in the era of global companies, especially once the Cold War ended and Russian airspace was opened to Western jets. Telecommunications capacity grew more slowly across the Pacific in the analog era. The first Pacific telephone cable was Transpac-1, installed in 1964 with 142 lines. By 1977 there were only 987 lines from America to Asia with a further 1380 from Canada to Australia and New Zealand.

Submarine telephone cables had severe limitations. TAT-1, at approximately $1 million per voice circuit, made calls expensive. Even TAT-6, at $179 million for 4,000 circuits, was stunningly expensive, as was TAT-7. AT&T proposed a 16,000-circuit analog TAT-8 that would have cost close to $1 billion and required a thousand built-in amplifiers. A cheaper solution was needed.

Satellites

Satellites were seen as the first alternative to the expensive, low-capacity submarine telephone cables of the 1950s. The first communications satellite, Telstar, was designed by Bell Telephone and launched into low-earth orbit in 1962. Satellites were relatively cheap, especially when launched into geosynchronous orbits so they appeared stationary above the earth's surface. They promised much higher capacity than analog submarine cables. The American International Telecommunications Satellite Organization (Intelsat) was created in 1964 to manage the satellite system, although with geopolitical as well as commercial aims. Although the experimental satellites of the 1960s and 1970s had limited capacities Intelsat-V, launched in 1980, and -VI, launched in 1981, added 45,000 transatlantic telephone circuits between them.

Until the early 1990s Intelsat had as much of a telecommunications monopoly as Britain's Eastern Companies before World War I or Cable and Wireless in the late 1930s. As signals passed through Intelsat they were decrypted by the National Security Agency, giving America the same geopolitical advantages that Britain had previously accrued. Four events have reduced the commercial and geopolitical advantages of satellites. First, the 1986 failure of the American space shuttle, Challenger-7, threw the American launch program into disarray and resulted in most commercial satellite launches moving to the French Ariane program, although they would have likely done so anyway because the shuttle's cargo bay restricted satellite diameter. Second, the success of the noise- and delay-free fiber-optic cables after TAT-8 was laid in 1989, markedly reduced the utility of satellites for communications. Third, the technology of satellites is such that there are relatively few parking slots in geosynchronous orbit, they last only about ten years, and their stability is controlled by their diameter, which in turn is controlled by the launch vehicle. Fourth, rapidly improving encryption capabilities have made decryption impossible without acquiring the cryptography of other states through intelligence assets. Satellites have therefore had to find other uses: gathering remotely sensed data of human and natural activities on the earth's surface; for GPS location of mobile assets; and for digital television transmission. By the mid-1990s there were 30 satellites in geosynchronous orbit, most of which were launched aboard Arianes, 22 owned by Intelsat, 4 by Inmarsat of London, and 4 by PanAmSat of Greenwich, Connecticut.

Fiber-optics

In a talk given in 1969 Alec Reeves, the father of pulse code modulation (PCM), the digital encoding system universally used in telecommunications, forecast submarine fiber-optic cable using PCM within twenty years. The first, TAT-8, entered service in late 1988 carrying some 11,500 circuits and operating at 280 million bits per second with amplifiers about every 65 kilometers. Despite considerable problems developing an optically pure glass with low transmission losses and optical amplifiers that could be installed in the cables, fiber-optic technology has exploded since 1988. Capacity has skyrocketed and amplifier spacing has lengthened. By the late twentieth century, 20 billion bits per second was possible in the laboratory. The target is a trillion bits per second through an individual fiber no bigger than a human hair, the equivalent of ten million simultaneous telephone calls.

Replacing analog with fiber-optic cables has returned emphasis to submarine transmission. Fiber-optic cables not only have very much higher capacity than analog ones, but also have the vanishingly low sampling errors important to PCM transmission. Laid over true terrestrial distances, they have none of the conversation interrupting transmission delays that occur in sending signals to a geosynchronous satellite and back to earth again, a fraction of a second even at the speed of light. Their low error rate makes them ideal for business data transmission, which is where most of their capacity is being used, although their vast overcapacity has led to extremely low cost for international telephony. Without them, the Internet would be impossible.

Cellular Telephones

The final great innovation in twentieth century communications has been the development of the mobile, or cell, phone. Using cellular technology this allows individuals to finally be free of the landlines that have connected them since the development of telegraphy in the 1840s. Cellular technology requires sophisticated computers to switch phones between cells and a network of fiber-optic landlines to link originating cells to the call's destination. The capacity of phones to carry data is still relatively limited, but visionaries propose cellular technology that would allow much more than mere telephony, up to and including individual mobile computing using "heads-up" interactive visual displays worn like eyeglasses.

See also **Electronic Communications; Fax Machine; Mobile (Cell) Telephones; Radio-Frequency Electronics; Satellites, Communications; Telephony**

PETER J. HUGILL

Further Reading

Aitken, H.G.J. *The Continuous Wave: Technology and American Radio, 1900–1932*. Princeton University Press, Princeton, 1985.
Beauchamp, K. *History of Telegraphy*. Institution of Electrical Engineers, London, 2001.
Burns, R.W. *The Life and Times of A D Blumlein*. Institution of Electrical Engineers, London, 2000.
Callick, E.B. *Metres to Microwaves: British Development of Active Components for Radar Systems, 1937–1944*. Peter Perigrinus, London, 1990.
Crowther, J. G. and Whiddington, R. *Science at War*. Philosophical Library, New York, 1948.
Douglas, S.J. *Inventing American Broadcasting 1899–1922*. Johns Hopkins University Press, Baltimore, 1987.
Edgerton, D. *England and the Aeroplane: An Essay on a Militant and Technological Nation*. Macmillan, London, 1991.
Guerlac, H.E. *Radar in World War II, Vol. 8: The History of Modern Physics, 1800–1950*. Tomash/American Institute of Physics, 1987.
Hecht, J. *City of Light. The Story of Fiber Optics*. Oxford University Press, New York, 1999.
Headrick, D.R. *The Invisible Weapon: Telecommunications and International Politics, 1851–1945*. Oxford University Press, New York, 1991.
Hook, S. *Wireless. From Marconi's Black-Box to the Audion*. MIT Press, Cambridge, MA, 2001.
Hugill, P.J. *Global Communications since 1844: Geopolitics and Technology*. Johns Hopkins University Press, Baltimore, 1999.
Jones, R.V. *The Wizard War: British Scientific Intelligence, 1939–1945*. Coward McCann, New York, 1978.
Mackinder, H.J. The geographical pivot of history. *Geographic. J.*, 23, 421–437, 1904.
O'Neill, E.E. *A History of Engineering and Science in the Bell System: Transmission Technology (1925–1975)*. AT&T Bell Laboratories, Indianapolis, IN, 1959.
Tilly, C. *Coercion, Capital, and European States, AD 990–1990*. Blackwell, Oxford, 1990.

Telephony, Automatic Systems

Fundamental to the expansion of telephone service was the widespread use of automatic telephone switches. Initial telephone service sought to connect every telephone directly with every other instrument in the same market. Soon manual switchboards were installed for more efficient operation. As more telephones entered service, switchboards became larger and more operators were required. Replacing often-slow manual operators, automatic devices were electromechanical for most of the first century of telephone usage, slowly being replaced by more efficient electronic (and eventually digital) systems.

Strowger Electromechanical Systems

The idea of an automatic telephone switch was first developed by Daniel and Thomas Connelly. Along with Thomas J. McTighe, they obtained the first patent on automatic or machine-switched telephony in 1879 (just a year after the world's first manual exchange had opened) and continued to develop their ideas, though their system was not commercially successful. More than 25 other related patents followed from a variety of inventors, though no operating systems resulted.

The first practical automatic switch originated with the ideas of Almon B. Strowger, a Kansas City undertaker. Strowger was losing business to a competitor whose wife operated the local telephone manual exchange and could thus route business calls away from Strowger.

Seeking a way to eliminate the human operator, Strowger developed a stepped (or step-by-step, as the switch moved with every number entered) switch in 1886–1887, applied for his initial patent two years later while continuing to improve his system with the help of several collaborators, and formed a manufacturing company in 1890.

The first Strowger automatic step-by-step switch was installed in La Porte, Indiana (near Chicago) in 1892, serving about 75 subscribers and with a capacity of 100. The related telephone rotary dial instrument with its timed pulse was invented by a Strowger associate in 1896 (and introduced in a Milwaukee exchange) while an improved version appeared from the Ericsson firm that same year. By 1898 there were 22 automatic exchanges operating in the U.S. Other electromechanical systems included the semiautomatic system developed by Dane Sinclair in Britain (1883; installed in Glasgow, 1886), and the Lorimer system in Canada (1900), which although not successful, helped contribute to the later panel systems.

The Strowger system was "direct" in that the switch mechanically called up the telephone number as it received electrical pulses from the customer's dial, and is thus fast or slow depending on customer speed of use. Strowger switches were "progressive" in that the equipment does not "know" the next step (or number) until it is dialed. It utilized electromagnets to open and close mechanically linked connections. A metal selector or "wiper" moved both vertically and in a rotary horizontal fashion to reach the ten contact positions that made up each of the ten vertical levels. These were combined to create ever-larger switches. Strowger switches are often called "lost call" systems in that the constantly moving wiper seeks an unused line, feeding back a busy signal if one is not available.

Such systems were best suited to small or medium-sized markets (and indeed remained in use in some smaller towns until the end of the century). Large 10,000-line capacity Strowger switches had been developed by 1900 (the first was installed in New Bedford, MA). A decade later, more than 130 exchanges used Strowger equipment and served more than 200,000 customers. Use of such equipment was confined to independent (i.e., non-Bell System) companies. Versions of Strowger devices were also slowly taken up (and sometimes imported) by Canadian carriers and the major European powers (the first British public automatic exchange opened in Epsom, outside of London, in 1912, just as the British Post Office took over from private operators). Expansion of their use was often slow to develop due to a readily available and inexpensive female labor pool. In the U.S., however, World War I labor demands helped to push wider usage.

By the 1920s only about 15 percent of the world's telephones used automatic switches of some kind, although most major American cities were served by such switches by 1939. Half of German telephones were served by automatic switches by 1930; Britain achieved that mark six years later.

Rotary, Panel, and Crossbar Electromechanical Systems

The short-lived Lorimer switch gave rise to two systems, both of which originated with Western Electric. The "rotary" automatic system utilized a cylindrical wiper that could connect with 30 different positions for a total of 300 lines per switch. Though AT&T decided not to use it, the rotary switch was widely adopted after World War I in Europe and other countries and successive developments remained in use from about 1910 into the 1970s.

"Panel"-style automatic switches were first experimented with by Bell Systems in 1912 and after considerable research, an improved variation was introduced in 1921in Omaha, Nebraska. These were more complex indirect-control devices (speed of dialing had no impact on how the switch worked) that used moveable vertical rods to switch between 500 lines (five times greater than contemporary Strowger gear) arranged across a wide panel. They were expensive to maintain and were used only by the Bell System (in 26 large cities by the late 1920s), slowly giving way to the crossbar system after the 1950s.

Based in part on the panel system, and on Swedish inventions early in the century as well as the crossbar switch of 1913 developed by John N. Reynolds of AT&T's Western Electric, the first experimental "crossbar" exchange opened in Sweden in 1926. Convinced of its merits, AT&T introduced an improved version in early 1938, its No. 1 crossbar switching system, designed for large urban centers and first installed in Brooklyn, NY.

The crossbar switch used common control (call handling was totally separated from interconnection with the network), with switches making only small motions (allowing greater speed), thus suffering less vibration than the Strowger, and allowing greater flexibility. It consisted of relays, complex link-trunking, and allowed alternative routing of calls. The No. 5 crossbar switch eventually achieved a capacity of 35,000 lines, but could easily be adapted for smaller exchanges.

Their higher costs (compared to existing panel switches) delayed installation of improved crossbar systems, but eventually more efficient crossbars (the No. 4 and No. 5, introduced in the Philadelphia area in 1943 and 1948, respectively) were developed. In the 1950s similar crossbar systems also began to take hold in Europe and Japan. By this time, more than three quarters of the world's telephones were served by automatic switches of one kind or another.

Toll or long-distance automatic switching developed more slowly, however, and required development of universal telephone numbering plans (including the use of area codes) for different parts of the world. Direct distance dialing was introduced in the U.S. in 1951, using the capable No. 5 crossbar switch. While the first international automatic service was introduced between Brussels and Paris in mid-1956, initial customer direct-dialed international calls (New York to Britain) became possible only in 1970. Crossbar switches manufactured by several firms were soon the most widely used automatic switches in the world, and continued to appear in improved versions into the 1970s. Many remained in use beyond the end of the century, proving them to be highly reliable and efficient over time.

Electronic Systems

Based on postwar computer and transistor developments, experiments with fully electronic switches began at Bell Laboratories in the early 1950s. Offering the potential of much higher speeds and lower power requirements (and thus smaller equipment designs), the possibility of such switches became more viable with innovations in solid-state electronics and the development of computer stored-program control (SPC) after 1960. After 500 million dollars of research and experimentation expenditure and considerable delay, AT&T's No. 1 ESS (electronic switching system) was first installed in a New Jersey suburb of New York in 1965.

With ESS, electronic scanning of telephone lines replaced use of mechanical relays and switches. ESS systems utilized complex computer programming and parallel processing both to determine system faults and to provide service redundancy and thus reduce downtime. Reliability was said to allow no more than two hours of downtime over a 40-year period. Within a decade, more than 500 offices were served by ESS installations. The No. 2 ESS for smaller markets followed, the first of which entered service in 1970. By the 1980s more than 700 offices were served by No. 2 ESS equipment.

The first time-division all-digital (or second-generation) electronic switching system was the No. 4 ESS, introduced in Chicago and other major American cities in 1976, which could handle nearly 54,000 two-way voice circuits (and up to 500,000 call attempts) per hour. The No. 5 ESS was delayed by developmental problems and was first introduced in 1982 in Seneca, Illinois. It was also the first switch that AT&T had to sell in a competitive market, because after 1984 its operating companies were independent. After 1996, Lucent Technologies continued development and marketing of the system by which time an improved No. 5 ESS could handle up to a million call attempts per hour.

See also **Packet Switching; Telephony, Digital**

CHRISTOPHER H. STERLING

Further Reading

Aitken, W. *Automatic Telephone Systems*, 3 vols. Benn Brothers, London 1921–1924.

Baldwin, F.G.C. Automatic or machine switching systems, 1879–1922, in *The History of The Telephone in the United Kingdom*. Chapman & Hall, London, 1925.

Chapuis, R.J. Manual and electromechanical switching (1878–1960s), in *100 Years of Telephone Switching (1878–1978)*. North Holland, Amsterdam, 1982.

Chapuis, R.J. and Amos Jr., E.J. Electronics, computers and telephone switching: a book of technological history, in *100 Years of Telephone Switching, Part 2: 1960–1985*. North Holland, Amsterdam, 1990.

Hanscom, C.D. *Dates in American Telephone Technology*. Bell Telephone Laboratories, New York, 1961.

Joel Jr., A.E. and Schindler Jr., G.E., Eds. *A History of Engineering and Science in the Bell System: Switching Technology (1925–1975)*. Bell Telephone Laboratories, New York, 1982.

Moudon, R. *The Strowger Automatic Telephone Exchange*. Spon & Chamberlain, New York, 1919.

Noll, A.M. *Introduction to Telephones and Telephone Systems*, 3rd edn. Artech House, Norwood, MA, 1998.

Ryan, J.S. Signalling and switching as we enter the second century. *Telecomm. J.*, 43, 206–219, 1976.

Smith, A.B. and Campbell, W.L. *Automatic Telephony: A Comprehensive Treatise on Automatic and Semi-Automatic Systems*, 2nd edn. McGraw-Hill, New York, 1921.

Switching and signalling. *Post Office Electr. Eng. J.*, 74, 187–222, 1981.
Various. The development of the automatic telephone, 1889 through 1918. *Telecom History*, 2, 1–112, 1995.

Telephony, Digital

Digital telephony is the digital transmission of voice over a communications network. This technology entails digital encoding of voice and digital transmission with regenerating repeaters, switches, and handsets. Using digitally enabled multiplexing, voice is integrated with computer data and digitized images, video, and other information in current telephone and computer networks.

The Bell Telephone Company was formed by Alexander Graham Bell in 1878 as the first U.S. telephone system. The public switched telephone network (PSTN) evolved into an analog, circuit-switched, internationally connected telephone system. In analog transmission, signals proportional to voice input are propagated throughout the length of the channel. In circuit-switched networks, a connection is established and bandwidth is dedicated for the entire connection. Although human voice extends over a 20-kilohertz range, voice grade lines have been limited to a 4 kilohertz band in order to decrease resource use.

In 1939, pulse code modulation was developed by Alex H. Reeves as a method for encoding an analog signal such as voice with digital values. A coder–decoder (codec) samples the analog, 4-kilohertz voice band at 8000 times per second, which is sufficient (according to Nyquist's theorem, 1924), to capture the entire information in that band. Each sample is assigned one of 128 values (North America) or 256 values (Europe) and encoded in seven or eight bits, requiring a data rate of 64 kilobits per second (Kbps), including control information. Approximating continuous analog values by discrete levels, called quantizing, results in some information loss. For high-quality voice in audio compact disks, a 22-kilohertz band is sampled at 44,100 samples per second, with 16-bit samples encoded over 65 kilohertz levels, requiring a 705.6 Kbps (double for stereo) data rate without compression if this data is transmitted. Differential compression schemes, such as differential pulse code modulation and delta modulation, as well as predictive schemes, such as adaptive differential pulse code modulation, code excited linear prediction and multi-pulse excited linear predictive coding, allow lower transmission rates, but at decreased signal quality.

Once voice was digitized, it became feasible to combine several input lines into a round-robin output stream. In 1957, Bell Laboratories developed the T-1 carrier system, which was implemented by 1962 in a Chicago area exchange. This digital timed-based multiplexing system takes turns allocating 8 bits of each of 24 input lines to the output stream. The total output transmission rate of 1.544 megabits per second (Mbps) includes 8 Kbps for clock synchronization. In Europe, an equivalent E-1 carrier multiplexes 32 input lines, each at 64 Kbps, for a 2.048 Mbps output. T-1 lines are combined to form T-2, T-3 and T-4 lines for higher transmission rates; E-1 lines are similarly bundled together.

"T" carrier systems were soon directly connected to digital switches. AT&T's electronic switching systems were installed in private computerized branch exchanges (PCBX) by 1963 and in public network telephony by 1965. Manual functions were computerized in large digital switches, which were able to combine the functions of several analog switches. Very large-scale integration (VSLI), spearheaded by Intel in 1971 for the manufacture of computers, made these systems cost effective.

The PSTN's backbone system was mainly based on analog radio trunks in the 1970s, with some reliance on coaxial cable. A major impetus for the transition from analog to digital transmission was the development of optical fibers. Prototypes of optical fibers that were suitable for transmission were developed at Corning Glass Works by 1970. By 1977, General Telephone and Electric (GTE) had installed optical-fiber commercial systems. Fiber's high bandwidth, low attenuation, small diameter, and immunity to electromagnetic interference spearheaded the replacement of analog trunks with digital fiber trunks.

By the 1980s, AT&T had replaced its entire analog backbone, implementing Integrated Digital Networks (IDN), the digital transmission of voice and data throughout its backbone network. By the 1990s, Integrated Services Digital Network (ISDN) extended digital telephony to the local loop, providing digital services directly to customers in many locations. Many new functions were enabled, including automated call tracing, automatic call back, call waiting, and call forwarding. Motivated by the high bandwidth demands of the Internet, other end-to-end digital access technologies were developed in the 1990s, notably digital subscriber lines (xDSL) and cable modems.

Digital telephony has evolved in wireless systems as well. As mobile telephony was being developed during the 1940s to 1960s, only a limited number of channels (initially 11 in the 40-megahertz band) were available. Blocking and high costs were common. Each (very heavy) handset searched the frequency spectrum, looking for an idle channel. In 1981, the first modern cellular (analog) system, the Nordic Mobile Telephone System, went into commercial operation in Sweden. Advanced Mobile Phone System (AMPS), a cellular analog system developed at Bell Labs, became commercially available by 1983. More recently, Digital Advanced Mobile Phone System (D-AMPS) has become available in the U.S., as well as Global System for Mobile (GSM) communications, a digital cellular system that was developed in Europe by 1982, and other systems. A digital wireless metropolitan area network may be used for the local loop, connecting customers to the PSTN, following the standards published by Institution of Electrical and Electronics Engineers (IEEE 802.16, in 2002. Satellite systems, such as Iridium (1998), Globalstar (2000), and Teledesic (under development), on the other hand, promise end-to-end digital telephony.

Since telephony software became commercially available in 1995, Internet telephony (VoIP) has connected the Internet to the PSTN using speakers and microphones on personal computers. The Internet, which is a connection of computer networks all over the world, is based on packet-switching technology. Packet switching (see separate entry) breaks the data stream into small units and adds control fields to each unit, for functions such as addressing and error correction. Packet switching allocates bandwidth on a demand basis, making it difficult to guarantee the timely and predictable service that is necessary for good quality voice reception. PacketCable, designed for Internet protocol (IP) technology over cable modems, includes quality of service options for better reception. Several application programming interfaces have been defined to connect computer systems to telephone services, such as TAPI, TSAPI, and JTAPI.

Digital telephony has rapidly replaced analog telephony due to many factors: VSLI technology provides lower costs; discrete values sampled by receivers and regenerating repeaters enable recovery from most transmission noise; digital methods of multiplexing, such as time division multiplexing (TDM) and code division multiple access (CDMA), afford higher throughput. As VoIP, the PSTN, and satellite telephony continue to develop, will packet switching and circuit switching continue to coexist and will a single technology dominate in digital telephony?

See also **Mobile (Cell) Telephones; Packet Switching; Optoelectronics, Dense Wavelength Division Multiplexing**

TRUDY LEVINE

Further Reading

Bellamy, J. *Digital Telephony* 3rd edn. Wiley, New York, 2000.
Griffiths, J.M. *ISDN Explained.* Wiley, New York, 1990.
Tanenbaum, A. *Computer Networks*, 4th edn. Prentice Hall, Upper Saddle River, NJ, 2003.
Wright, D.J. *Voice Over Packet Networks.* Wiley, New York, 2001.

Telephony, Long Distance

Originally a means of communication suited only for small or medium distances, from about 1920, cable and wire telephony—soon to be supplemented by radio telephony—began to conquer the longer distances as well. Advances in telephone telephony, such as the post-World War II transoceanic cables, became a key factor in the trend towards the global information society. Much of the progress was based in new scientific knowledge or transformation of such knowledge into technological methods and artifacts.

Shortly after Alexander Graham Bell's invention of the telephone in 1876, it became clear that it was technically difficult to transmit speech signals over long distances. By using gutta percha-insulated cables and thick copper wires the distance was increased, but this traditional technology did not allow economically feasible lines much longer than 1000 kilometers. The first major innovation was based on the theory of telephone current transmission independently proposed in 1887 by Oliver Heaviside in England and Aimé Vaschy in France. According to this theory, the attenuation would be minimized, hence the speaking distance increased, if the line was "loaded" with self-inductance (the counter-electromotive force arising from the rapid variation of currents). Only around 1902 was the theoretical insight turned into a practical method of how to add self-inductance to wires and cables. In the U.S., George Campbell and Michael Pupin devised methods for discrete loading by inserting induction coils in the line. The Danish engineer Carl Krarup invented an alternative method of continuous loading by winding the copper conductors with thin wires of iron. The loading method was highly successful and soon used for almost all submarine cables and long landlines. Loading technology increased the speaking distance, but not dramatically. The big step forward in long-distance telephony proved to be the introduction of the vacuum tube as an amplifier or "repeater" of the weak speech currents. In 1915, Harold Arnold and his staff at American Telephone and Telegraph Company (AT&T) had developed a tube repeater that, in combination with coil loading, made it possible to telephone from New York to San Francisco. After the end of World War I, tube-amplified lines formed the basis of the first long-distance telephone networks in Europe and the U.S. By the late 1920s, almost all European countries were connected by telephone lines.

In no area of telephony was the power of the vacuum-tube repeater more impressive than in long-distance submarine telephony. Loading technology increased the efficiency and range of submarine telephone cables, and in 1928 AT&T engineers made a proposal of designing a transatlantic cable based on continuous loading. Nothing came of the plan, however, and with the development during World War II of submerged tube repeaters it was realized that this was the method to be used in a future transatlantic cable. The first cable crossing the Atlantic, named TAT-1, was completed in 1956 as a joint project of AT&T, the British Post Office, and the Canadian Overseas Telecommunication Corporation. The heart of the 4,244-kilometer cable was the 51 submerged tube repeater units. Subsequent ocean telephone cables were even longer. In 1963 a distance record was achieved with the laying of a transpacific cable between Canada and Australia over a distance of nearly 11,000 kilometers. Whereas the first generation of transoceanic cables used vacuum tubes in the repeater units, from the mid-1960s tubes were increasingly replaced by transistors.

TAT-1 made crucial use of another important innovation in telephone technology—the frequency-division-multiplex (FDM) coaxial system, an advanced form of carrier-wave telephony in which many messages can be sent simultaneously through a single wire. In 1918, AT&T inaugurated the first commercial carrier system, and in 1931 the company had developed the first submarine coaxial carrier cable for use between Havana and Key West. The TAT-1 coaxial cable was insulated with polyethylene and included 36 voice channels, a number that would increase drastically in later transatlantic cables. For example, the transistorized TAT-6 cable of 1976 had a capacity of 4,000 channels. The largest cable of the traditional coaxial type was the 15,000-kilometer-long ANZCAN cable between Canada and Australia that started service in 1984. Investments amounted to $500 million and each of the 1,124 transistorized repeaters cost nearly $100 000. With the invention of the laser in 1960 and the development of low-attenuation glass fibers about 1970, fiber-optical telephone communications became a possibility. The new cables were immediately used for submarine cables, as in 1988 when TAT-8 was opened for service between England and North America. This project, a collaboration between AT&T, British Telecom, and France Telecom, had a capacity that doubled the existing cable capacity across the Atlantic. In 1991 it was followed by a fiber-optic link between the U.S. and Japan.

With the development of vacuum tubes, wireless telephony became a possibility for communication over long distances. Experimental radio telephony across the Atlantic was first achieved in 1915 by AT&T, and six years later the same company developed a hybrid communication system that combined inductively loaded cables and wires with radio links. The breakthrough of wireless telephony occurred a few years later, with the introduction of short-wave radio transmission. In 1927 the first commercial transatlantic radiotelephone service started between London and New York, and within a few years intercontinental telephony over the combined radio and cable network was a reality. At that time, truly intercontinental telephony, defined as circuits joining different continents over a distance of more than 1,000 kilometers, was limited to radio links. For example, the length of the 1930 London to Sydney link was 17,000 kilometers, and in 1934 AT&T performed the first round-the-world telephone conversation through combined wired and wireless circuits.

During the last three decades of the century, an increasing part of intercontinental telephony was transmitted wirelessly through satellites rather than through cables. Many experts believe that future long-distance communications will rely on a mixture of optical-fiber cables and satellites, with a major part of the telephone traffic going through the satellite service.

See also **Mobile (Cell) Telephones; Radio Transmitters; Satellites, Communications; Telecommunications; Telephony, Digital**

Helge Kragh

Further Reading

Bray, J. *The Communications Miracle: The Telecommunication Pioneers from Morse to the*

Information Superhighways. Plenum Press, New York, 1995.
Brittain, J.E. The introduction of the loading coil: George A. Campbell and Michael I. Pupin. *Technol. Cult.*, 11, 36–57, 1970.
Fagen, M.D. *A History of Engineering and Science in the Bell System: The Early Years, 1875–1925*. Bell Telephone Laboratories, Whippany, NJ, 1975.
Faltas, S. The invention of fibre-optic communications. *Hist. Technol.*, 5, 31–49, 1988.
Headrick, D.R. Shortwave radio and its impact on international telecommunications between the wars. *Hist. Technol*, 11, 21–32, 1994.
Kragh, H The Krarup cable: invention and early development. *Technol. and Cult.*, 35, 129–157, 1994.
Kragh, H. History and prehistory of the first transatlantic telephone cable. *Polhem*, 13, 246–271, 1995.
Solymar, L. *Getting the Message: A History of Communications*. Oxford University Press, Oxford, 1999.
Wasserman, N.H. *From Invention to Innovation: Long-Distance Telephone Transmission at the Turn of the Century*. Johns Hopkins University Press, Baltimore, 1985.

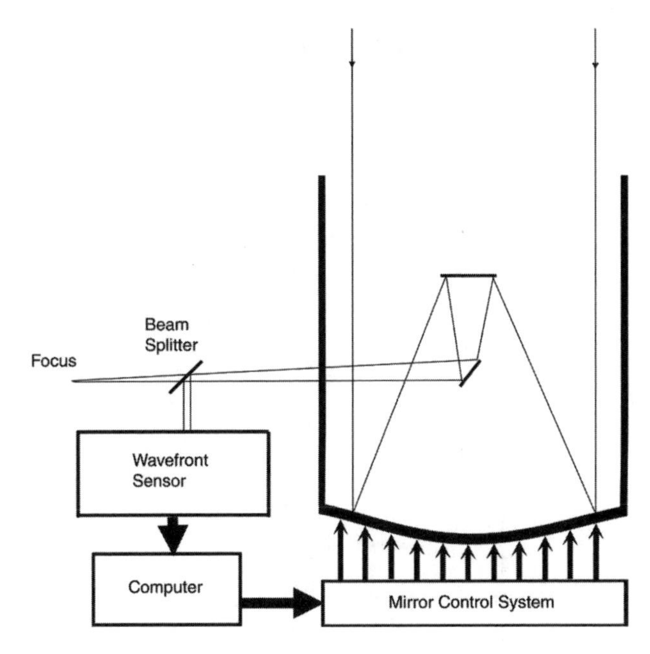

Figure 1. Active optics system.

Telescopes, Computer-Controlled Mirrors

The first half of the twentieth century saw a phenomenal increase in the size of astronomical telescopes, culminating with the 5-meter Hale instrument on Mount Palomar. However by mid-century it appeared that the ultimate size of earth-bound telescopes was inherently limited. Despite the use of new materials and advanced engineering, increasing the size of telescope mirrors seemed to make them inherently more sensitive to flexure and vibration. Moreover, the resolving power of earth-bound telescopes seemed to be ultimately limited by the distorting effects of atmospheric turbulence. One of the primary justifications for funding the Hubble space telescope was to avoid these difficulties.

In the second half of the twentieth century astronomers explored new technological approaches to solving these problems, many of which concentrated on developing increasingly sophisticated sensing devices. But there was also an impetus to rethink the telescope from its traditionally passive role and to give it some capacity to react to changes in viewing conditions. Eventually two separate, but complementary, systems emerged: active optics and adaptive optics.

Active optics involves changing the shape of the telescope's primary (and in some cases, secondary) mirror to optimize the instruments' resolving power (see Figure 1). Active optic systems analyze the image of a reference object (typically a bright star located near the object being studied) to generate a series of changes in the telescope's optical system, optimizing the image of the object being viewed. These changes may be performed relatively slowly and can correct for a variety of problems, including low-frequency wind vibrations, uneven thermal expansion, gravitational distortion and even the instrument's permanent figuring errors. Instruments with active optical systems can be made significantly lighter than conventional telescopes and are well suited to use on mirrors made up of multiple segments.

Adaptive optics involves changing the shape of a much smaller mirror, typically positioned within the telescope's optical train, with the goal of correcting wave-front distortions that occur as light passes through the earth's atmosphere (see Figure 2). Adaptive optics uses a guide-star feedback loop similar to that used in active optics but with much smaller corrections that are ideally made in the space of a few milliseconds.

Although suggested as early as 1953, practical implementation of these systems required the development of a variety of supporting devices, including wave-front sensors, deformable mirrors, actuators, computers, and computer software. While active optics is now considered a mature technology, the routine employment of adaptive optics to the entire visual spectrum is still under development. Significant correction of the atmospheric distortion of infrared light has been

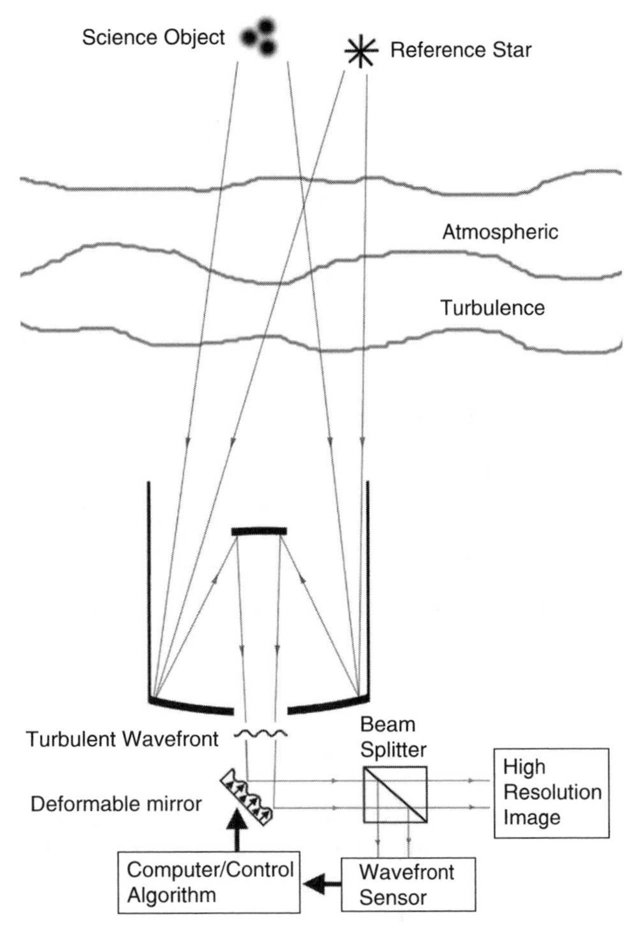

Figure 2. Adaptive optics system.

successfully achieved, but correction of distortion to higher frequencies remains a challenge.

Another important challenge has involved finding suitable reference stars for image and wavefront correction. The ideal reference is a bright star located fairly close to the object being studied but far enough away from the telescope's optical axis that its light is not part of the observation. The source needs to be bright enough to overcome noise in the wave-front detectors and it needs to be close enough to the science object to have roughly the same light-path through the atmosphere. Finding a suitable natural guide star (NGS) for every observation has proved to be a problem and the possibility of generating artificial reference stars was an early area of investigation.

This work received an unexpected boost from the U.S.'s Strategic Defense Initiative (SDI) in the 1980s. SDI planned to use a form of adaptive optics to focus a destructive laser beam on incoming missiles. To permit rapid response, the system used an artificial reference star created by shining a laser on a naturally occurring layer of sodium atoms approximately 90 kilometers above the earth's surface. When the SDI program was cancelled in 1991 much of this technology was turned over to astronomers. The use of laser guide stars (LGS) in astronomy dates from this time.

While the use of LGS appears to solve the reference star problem, the relative nearness of these objects introduces the problem of conical anisoplanatism. This refers to the fact that light from the LGS arrives at the telescope in a path that is cone shaped. This can cause the LGS wave-front distortion to differ significantly from that of astronomical object being studied. LGS systems can also be expensive to install and operate. A number of alternatives have been proposed, including the use of multiple guide stars (either natural or artificial) to produce a tomographic AO correction for entire sections of the sky.

Progress in both active and adaptive optical systems continues to accelerate and it is clear that future earth-bound astronomical telescopes will depend heavily on them. While complete correction of observing errors is not possible, even partial correction can produce dramatically improved images. The image quality of modern astronomical telescopes is commonly expressed in terms of the strehl ratio, which is defined as the ratio of the peak intensity of an image divided by the peak intensity of a diffraction-limited (ideal) image. A perfect strehl ratio is 1.0; for reference, the Hubble space telescope (HST) is estimated to have a strehl ratio of 0.97.

For earth-bound telescopes, the highest achievable strehl may be around 0.8 and day-to-day observations may fall short of that goal. However it has already been demonstrated that, using adaptive optics, a 10-meter surface telescope can produce infrared images that surpass those produced by Hubble. This is because of the much greater light-gathering area of the surface telescope. Moreover, the cost of building and operating large earth-based instruments can be significantly lower than launching telescopes into space. With the size of launchable space telescopes currently limited, attention is being increasingly directed towards the construction of large AO earth based observatories. The next generation of visual telescopes may be in the 20 to 30 meter range and instruments as large as 100 meters have been proposed.

See also **Telescopes, Ground; Telescopes, Space**

STEVEN TURNER

Further Reading

Babcock, H. The possibility of compensating astronomical seeing. *Public. Astronom. Soc. Pacific*, 65, 229–236, 1953.
Babcock, H. Adaptive optics revisited. *Science*, 249, 253–257, 1990.
Busher, D. Love, G. and Myers, R. *Adaptive Optics*. Wiley, New York, 2003.
Le Louarn, M., Hubin, N., Sarazin, M. and Tokovinin, A. New challenges for adaptive optics: extremely large telescopes. *Mon. Notices R. Astronom. Soc.*, 317, 535–546, 2000.
Marshall, E. Researchers sift the ashes of SDI. *Science*, 263, 620–623, 1994.
Ragazzoni, R., Marchetti, E. and Valente, G. Adaptive-optics corrections available for the whole sky. *Nature*, 403, 54–56, 2000.
Roddier, F., Ed. *Adaptive Optics in Astronomy*. Cambridge University Press, Cambridge, 1999.
Taroy, M., Ed. *Adaptive Optics: Current Abstracts With Indexes*. Nova Science, New York, 2002.

Useful Websites

Center for Astronomical Adaptive Optics, Steward Observatory, University of Arizona: http://caao.as.arizona.edu/publications/. Provides an extensive bibliography.

Telescopes, Ground

A telescope collects radiation from distant sources and converges it onto a detector such as the human eye or photographic plate. Historically astronomers have desired bigger telescopes that collect more light and enable scientists to observe fainter, more distant objects. Throughout the twentieth century, scientists and engineers have proposed novel designs for telescopes with increased light-collecting power. Choosing and implementing these has required a combination of community support, technological capability, and financial resources.

In 1900 astronomers still used two basic types of telescopes: refracting and reflecting. Refracting telescopes, which use a system of lenses to form an image, were precision instruments preferred by many professional astronomers for tasks such as measuring the positions of stars. Reflecting telescopes use glass mirrors coated with a thin layer of reflective metal to collect and focus light. Reflectors did not suffer from intrinsic optical defects that limited the usefulness of refracting telescopes. Somewhat easier to use and enclosed in smaller and less expensive domes, reflecting telescopes became the central instrument of the modern observatory as they were better suited for astrophysical studies of celestial objects via spectroscopy and photography.

The person perhaps most responsible for the development of large telescopes in the first half of the twentieth century was George Ellery Hale. Scion of a prominent Chicago family, Hale moved to California in 1903 and established the Mount Wilson Observatory overlooking Pasadena. Generous funding from the Carnegie Institution of Washington enabled Hale to commission specially designed telescopes to observe the sun. Hale later obtained a 60-inch (152-centimeter) glass disk which became the heart of the first large reflecting telescope at Mount Wilson. Even before workmen finished the 60-inch in 1908, Hale already had plans for a bigger telescope. Built with funding from a wealthy merchant and the Carnegie Institution, the 100-inch (254-centimeter) telescope entered service in 1919. It collected three times more light than the 60-inch and played a major role in reshaping cosmological theory. Astronomers like Edwin P. Hubble, for example, used it to establish the nature and distance of nebulae and provide evidence the universe was expanding.

In 1928 Hale began campaigning for a 200-inch (508-centimeter) telescope. He secured $6,000,000 from the Rockefeller Foundation and began plans for the world's largest telescope to be jointly run by Mount Wilson and the California Institute of Technology. The telescope was built on Palomar Mountain in southern California. Building the 200-inch telescope, especially its 20-ton mirror, was a Herculean effort and it was not dedicated until 1948, ten years after Hale's death.

The 200-inch defined what a large telescope should look like for the next three decades. Many slightly smaller telescopes; for example, the Kitt Peak 4-meter telescope and the Anglo–Australian Telescope, were based on its design. They featured tremendous dome enclosures, primary mirrors made of a single massive piece of glass, an immense horseshoe-shaped bearing, and traditional equatorial mounts. It was not until the Soviet Union completed its 6-meter *Bolshoi Telescop Azimutal'ny* in 1976 that astronomers had a telescope larger than the 200-inch. Despite the BTA's innovative altitude-azimuth mount, its capabilities were limited for years by a poorly performing mirror while Cold War politics limited its usefulness to Western astronomers. An altitude-azimuth mount allows the telescope to pivot simultaneously on up–down around one axis (in altitude) and horizontally (in azimuth) around the other. This was made possible by the availability of computer control in the 1970s.

All of the telescopes described thus far gather visible light. Telescopes can also collect other types of radiation. For example, astronomers and engineers built telescopes such as the United Kingdom Infrared Telescope (finished in 1979 on Mauna Kea in Hawaii) that were optimized for infrared observing. Telescopes can also capture gamma rays for high-energy astrophysics. When gamma rays pass through the earth's atmosphere they generate Cherenkov radiation, which specially designed telescopes can detect on the ground as flashes of blue light. Another variation are the "neutrino telescopes" located underground to detect elusive subatomic particles produced by reactions deep inside the sun.

While the 200-inch remained the largest optical telescope in the Western world for over 40 years, postwar astronomers devised many ways to collect more photons. One solution was electronic detectors such as image tubes and photomultipliers. These electronic devices were many more times efficient than photographic techniques, effectively giving scientists greater light-collecting power by recording more photons collected by a telescope. Electronic technologies refined after 1950 encouraged astronomers to build more sophisticated light detectors instead of bigger telescopes, which were far more expensive. Astronomers benefited from this phase of technological development well into the 1970s with the introduction of even more sensitive CCD (charge-coupled devices) detectors.

In 1980 a new telescope was introduced that broke from the traditional design established by the 200-inch. The Multiple Mirror Telescope (on Mount Hopkins in southern Arizona) featured six 1.8-meter mirrors on a common mount. These combined light collected at a common focus, making the MMT equivalent to a 4.5-meter telescope. The MMT's optical system—now converted to a single mirror—was innovative as was its use of a compact altitude-azimuth mount, extensive computer control, and inexpensive and efficient shed-like enclosure that rotated with the telescope. With the MMT astronomers showed they could collect more light at a reasonable cost.

From the 1980s astronomers and engineers developed several techniques for making large, lightweight, yet relatively inexpensive mirrors around which new telescopes were built. Jerry Nelson and his colleagues at the University of California based their design for the two 10-meter Keck telescopes on Mauna Kea around a mirror made of 36 smaller glass segments; scientists at the University of Arizona led by Roger Angel developed an innovative way to spin-cast large mirrors; and commercial firms such as Corning introduced thin "meniscus" mirrors. Engineers have used each of these solutions in a number of large telescope projects. Another potentially important development in the late 1980s was adaptive optics. Derived from military systems built for missile defense and satellite tracking, adaptive optics uses a complex optoelectonic system which can be adjusted many times a second. This offers astronomers the means to remove distortion caused by atmospheric turbulence and produce sharper pictures.

In 1980 the light collecting area of optical telescopes worldwide was about 150 square meters; in 2002 it was over 900. Astronomers use new telescopes differently than they did when the venerable 200-inch entered service. In Edwin Hubble's time astronomy was a mostly a solitary pursuit as the scientist worked alone at the telescope collecting data on photographic plates. Fifty years later ground-based optical astronomers sit in a warm, well-lit control room and monitor their data on computers. Astronomers can also operate some telescopes remotely from thousands of kilometers away while electronic data archiving may enable scientists to access far greater amounts of information. The increased size and sophistication of new telescopes have helped alter what it means to be an astronomer. As astronomers discuss plans for even larger telescopes—30 to 100 meters in size—and court potential patrons, the sociological effects of new and more complex telescope technology remains to be seen.

See also **Telescopes, Radio; Telescopes, Computer-Controlled Mirrors**

W. PATRICK MCCRAY

Further Reading

Florence, R. *The Perfect Machine: Building the Palomar Telescope*. Harper Perennial, New York, 1994.
King, H. *The History of the Telescope*. Sky Publishing Corporation, Cambridge, 1955.
Learner, R. *Astronomy Through the Telescope*. Van Nostrand Reinhold, New York, 1981.
Maran, S. A new generation of giant eyes gets ready to probe the universe. *Smithsonian*, 1987, 41–53.
Mountain, M. and Gillet, F. The revolution in telescope aperture. *Nature* 395, A23–29, 1998.
Osterbrock, D. *Pauper and Prince: Ritchey, Hale, and Big American Telescopes*. University of Arizona Press, Tucson, 1993.
Overbye, D. *Lonely Hearts of the Cosmos: The Story of the Scientific Quest for the Secrets of the Universe*. Little, Brown & Company, New York, 1991.
Smith, R. Engines of discovery: scientific instruments and the history of astronomy and planetary science in the

United States in the 20th century. *J. Hist. Astron.*, 28, 49–77, 1997.
van Helden, A. Building large telescopes, 1900–1950, in *Astrophysics and 20th Century Astronomy to 1950: Part A*. Gingerich, O., Ed. Cambridge University Press, Cambridge, 1984, pp. 134–152.

Telescopes, Radio

Radio telescopes are instruments for the study of cosmic radio emissions. In a strict sense, a radio telescope is any apparatus designed to detect radio waves—waves with wavelengths in the range between 1 millimeter and 30 meters—of extraterrestrial origin. The first designs of antennae and receptors to detect these waves were made in the late nineteenth century by scientists such as Americans Thomas A. Edison and Arthur E. Kennelly, the British Oliver J. Lodge, the Germans Johannes Wilsing and Julius Scheiner, and the French Charles Nordman. The purpose of these designs was the study of the sun as a source of radio waves, but no results were obtained. It was in 1942 when the British J. Stanley Hey for the first time successfully detected radio waves from the sun while studying the interference of British radar during World War II. Radio astronomy became a recognized part of astronomy during the 1950s, although successful experiments of that kind had been done as early as in 1932. This was the year in which the American radio engineer Karl G. Jansky, studying interference in transatlantic communications for the Bell Telephone Laboratories, identified short-wave noise resulting not only from thunderstorms, but also from an extraterrestrial source. As we now know, the signal he detected was synchrotron emission associated with energetic electrons accelerated in the magnetic field of the Milky Way galaxy. However, astronomers did not follow up Jansky's results. Other work, such as that of the American Grote Reber, who mapped the sky during the 1940s using contours of high and equal radio-wave intensity, required the design of new parabolic dish antennae suitable for detection of higher frequencies. The 10-meter parabolic reflector dish that Reber built in 1937 in his backyard to seek cosmic radio emissions is generally considered the first radio telescope used for astronomical research.

It was not until the end of World War II that radio astronomy received firm support. The equipment that was now freely available for use in this new field of astronomy and the knowledge acquired in the area of radar and radio communications made this possible: the great advances in microwave technology that had led to radar became available to astronomers. In fact, the instruments that were used during that period were not very different from the ones used nowadays, at least as far as the basic structure is concerned. In principle, a radio system consists of two basic elements: the large radio antenna and the radiometer or radio receiver. The antenna controls the direction of observation and then collects the radiation and converts it into an electric (alternating current) signal. This part of the radio system is very sensitive to polarization of the incident radiation, which is why crossed antennae are usually used to record the whole signal of the radio source. Antennae must be carefully designed in order to minimize the effects produced by different responses at various angles. On the other hand, it is the receiver that picks out from the signal a particular frequency and bandwidth to work with, which is subsequently processed and recorded. One of the problems radio astronomers have to deal with has been that for longer wavelength, larger apertures of telescopes are needed to achieve good resolution. Radio telescopes are limited by the largest practicable aperture, as well as by the diffraction produced by the receiver itself. A solution adopted in the second half of the 1940s emerged from working on solar eclipses. However, still better solutions were required in order to work regularly with radio telescopes without obtaining distorted images.

While the sensitivity of both radio and optical telescopes depends on the total surface area of the collector of radiation, the angular resolution of a radio telescope essentially depends on the largest distance between two points on the antennae system, which pick up signals to be processed simultaneously in the receiver. This distance is referred to as the aperture width. The larger the aperture width, the better the resolution. The aperture width is essentially equal to the diameter in what are called "filled aperture radio telescopes." These are frequently single-dish and their design is similar to optical telescopes. Another way of increasing the effective aperture width is using an array of radio telescopes. In other words, single telescopes at different places pick up the signal to be processed simultaneously, increasing the angular resolution. Martin Ryle and D. Vonberg at Cambridge, U.K., first employed this technique, known as radiointerferometry, in the 1940s. The particular technique used, still known as earth rotation aperture synthesis, was a two antennae radio interferometer with a maximum spacing of about 0.5 kilometers. Interferometric techniques at radio wavelength were developed during the 1950s and the 1960s. The highest resolution results were obtained with arrays composed of big single radio telescopes separated by very large distances. In these cases, cables, of course, do not link the components. Two typical examples of how these radio telescopes work are the British MERLIN (multi-element radio-linked interferometer network) begun in 1976 and extended in the 1990s, which works with radio links, and the VLA (very large array) in New Mexico dedicated in 1980, which uses a technique called VLBI (very long-baseline interferometry) where signals are recorded separately and combined later by computer. The latter currently represents the largest improvement in angular resolutions of radio telescopes, far superior to those achievable with optical observations. The development and use of self-calibration techniques in the 1970s—a decade after the development of the VLBI, with a base line of thousands of kilometers—allowed the construction of images with a milliarcsecond resolution.

During the twentieth century the use of radio telescopes led to many interesting results and important discoveries not only in astrophysics, with the discovery of superluminal radio sources moving at speeds apparently faster than the speed of light and the observation of pulsars and quasi-stellar radio sources (commonly known as quasars); but also with the possibility of addressing cosmological questions, such as studying the cosmic microwave background to investigate the age and evolution of the universe. Current projects that radio astronomers are working on, are in the areas of millimeter and submillimeter interferometry and water vapor radiometry (for dealing with atmospheric phase fluctuations). These, combined with advanced digital processing techniques, are expected to yield results on the bases of which future goals in astronomy research could be formulated.

PEDRO RUIZ CASTELL

Further Reading

Hanbury-Brown, R. *Boffin: A Personal Story of the Early Days of Radar, Radio Astronomy and Quantum Optics*. Institute of Physics Publishing, Bristol, 1991.

Kellermann, K.I. and Moran, J.M. The development of high-resolution imaging in radio astronomy. *Ann. Rev. Astron. Astrophys.*, 39, 457–509, 2001.

Malphrus, B.K. *The History of Radio Astronomy and the National Radio Astronomy Observatory: Evolution Toward Big Science*, 1996.

Pawsey, J.L. and Bracewell, R.N. *Radio Astronomy*. Clarendon Press, Oxford, 1955.

Smith, R.C. *Observational Astrophysics*. Cambridge University Press, Cambridge, 1995.

Sullivan III, W.T. *The Early Years of Radio Astronomy*. Cambridge University Press, Cambridge, 1984.

Verschuur, G.L. *The Invisible Universe Revealed: the Story of Radio Astronomy*. Springer Verlag, New York, 1987.

Telescopes, Space

The ability to place astronomical telescopes and other detectors in space has given rise to the term "space astronomy," which describes astronomy performed from space rather than earthbound observatories. Its main advantage over terrestrial astronomy is its elimination of the deleterious effects of the earth's atmosphere, which not only absorbs some wavelengths of radiation but also distorts the images of stars and other objects of interest. Indeed, astronomy at some wavelengths (e.g., far infrared, submillimeter and x-ray astronomy) has advanced only through space astronomy, simply because the radiation is absorbed by the earth's atmosphere.

Although the term "space telescope" may refer to a module or payload on a space station, it is more commonly applied to a dedicated astronomical satellite or space observatory. There have been literally hundreds of space astronomy missions, observing in a wide variety of wavelengths from the radio end of the spectrum to x-rays and gamma rays.

The first space-based observatory was the National Aeronautics and Space Administration's (NASA's) Orbiting Solar Observatory, OSO-1, which was launched in March 1962 to expand on the work of the Explorer spacecraft series. OSO-1 operated for almost two years, during which time it transmitted data on more than 140 solar flares. The data collected from its 13 experiments was tape-recorded on board the satellite and transmitted to earth in a 5-minute period on each orbit (a communications method known as store-and-forward).

Previous science spacecraft had been designed as individual instrument carriers with no attempt at standardization, but the nine OSO spacecraft were based on a standardized observatory platform on which a variety of different payloads could be mounted—a design method that has become the norm for most types of spacecraft. Thus, although OSO-1 weighed only 206 kilograms at launch, by the time OSO-8 was launched in 1975, the standardized platform was supporting a spacecraft weighing 1052 kilograms.

The OSO platform was spin-stabilized by rotating the base section, using nitrogen thrusters known as "gas jets," while the upper section, which contained the pointing-dependent part of the astronomical payload, remained pointing towards the sun. The spacecraft's spin axis was kept perpendicular to the solar vector by magneto-torquer coils in the base section, which aligned themselves, and thus the spacecraft, with the earth's magnetic field. Meanwhile, a gyroscope in the upper section acted as a memory to ensure that the spacecraft acquired the sun's light quickly on each orbit after emerging from the earth's shadow.

The early OSO launches were followed in 1968 by NASA's Orbiting Astronomical Observatory (OAO), which carried no less than eleven telescopes, enabling it to observe stars in the infrared, ultraviolet, x-ray and gamma-ray parts of the spectrum. Telescope mirrors up to 96 centimeter in diameter could be mounted inside the satellite's cylindrical core. Having realized that a spin-stabilized platform could not provide the stability required for accurate astronomical observations, a three-axis stabilized platform was designed for OAO. Its attitude was controlled by a number of rotating wheels mounted inside the spacecraft and aligned with each of its three axes, a type of stabilization system now used for the majority of manned and unmanned spacecraft.

The best-known space observatory is NASA's Hubble space telescope (HST), an optical telescope of the Cassegrain type named after the American astronomer Edwin Powell Hubble. It incorporates a number of camera and spectrometer payloads tuned to various frequency bands, and was designed to be launched to, serviced in, and eventually retrieved from, low-earth orbit by the Space Shuttle. Some indication of the progress made between the OSO series and the HST is provided by the increase in mass: the HST weighed some 11,250 kilograms at the time of its launch in April 1990.

The pointing accuracy of space telescopes has also increased markedly since OSO-1, which was accurate only to about 1 arc-minute. OSO-8, for example, equaled the arc-second accuracy of terrestrial telescopes and the three-axis stabilized OAO series reached 0.03 arcsecond. For comparison, the HST has a pointing stability of 0.007 arcseconds and the European Space Agency's (ESA's) Hipparcos astrometry satellite has an incredible 0.001 arcsecond stability.

Prior to launch, the capabilities of the HST's revolutionary payload were well publicized. Its 800-kilogram, 2.4-meter-diameter primary mirror, for example, would enable the telescope to resolve something the size of a small coin from a distance of 20 kilometers and detect the light of a firefly from about 16,000 kilometers. Unfortunately, it was discovered after launch that the mirror had been ground incorrectly as a result of a measurement error, to the extent that it was 2 micrometers (0.002 millimeters) too flat at the edges.

A servicing mission to correct the mirror's spherical aberration was conducted in December 1993, replacing one of HST's five main payloads with an optical instrument known as COSTAR (corrective optics space telescope axial replacement), which deployed five pairs of small corrective mirrors between the primary mirror and three of the remaining payloads. Subsequent servicing missions have changed or upgraded the original payloads and HST has made many successful observations and discoveries, including the derivation of a new value for the Hubble constant (the constant of proportionality between the recession speeds of galaxies and their distances from each other) and thus an improved estimate of the age of the universe. In fact, the HST is so popular among astronomers that observing time, which has to be booked in advance, is many times oversubscribed.

A selection of some of the more recent space-based telescopes is provided in Table 1. At the time of writing, several advanced missions are being planned by space agencies, including NASA's Next Generation Space Telescope, the European Space Agency's Integral gamma ray source mapper, and a number of infrared telescopes.

See also **Space Shuttle; Space Stations; Telescopes, Radio; Telescopes, Computer-Controlled Mirrors**

MARK WILLIAMSON

Further Reading

Cornell, J. and Gorenstein, P., Ed. *Astronomy from Space: Sputnik to Space Telescope*. MIT Press, Cambridge, MA, 1983.

Field, G. and Goldsmith, D. *The Space Telescope: Eyes Above the Atmosphere*. Contemporary Books, Chicago, 1989.

Goodwin, S. *Hubble's Universe: A New Picture of Space*. Constable, London, 1998.

Kopal, Z. *Telescopes in Space*. Faber & Faber, London, 1968.

Lawton, A.T. *A Window in the Sky: Astronomy from Beyond the Earth's Atmosphere*. Pergamon Press, New York, 1979.

Leverington, D. *New Cosmic Horizons: Space Astronomy from the V2 to the Hubble Space Telescope*. Cambridge University Press, Cambridge, 2000.

Peterson, C.C. and Brandt, J.C. *Hubble Vision: Astronomy with the Hubble Space Telescope*. Cambridge University Press, Cambridge, 1995.

Table 1 Selected Space Telescope Missions.

Spacecraft nation/agency	Launch date	Waveband(s)/mission
Hipparcos ESA	Aug 1989	Accurate measurement of star positions
COBE NASA	Nov 1989	Detect and map cosmic background radiation
HST NASA	Apr 1990	Visible, IR, UV imaging telescope
ROSAT Germany	Jun 1990	X-ray and extreme UV all-sky survey
GRO/Compton NASA	Apr 1991	Gamma ray observatory
ISO ESA	Nov 1995	Infrared observatory
SOHO ESA	Dec 1995	Solar and heliospheric observatory
SAX Italy	Apr 1996	X-ray observatory
Muses B/Haruka Japan	Feb 1997	Galactic radio source mapping
Chandra NASA	Jul 1999	X-ray observatory
XMM/Newton ESA	Dec 1999	X-ray spectroscopy & mapping

Television, Beginning Ideas (late Nineteenth and Early Twentieth Century)

In 1873 Willoughby Smith discovered the photoconductive property of selenium, the changing of electrical conductance with light falling on its surface. This property was utilized in many of the early schemes for television (the word dates from 1900) until the development of suitable amplifiers and photoemissive cells made selenium cells obsolete in the 1920s.

The photoconductive property of selenium was easily demonstrated and, in the following decade there was an expectation that "distant vision" would soon be a reality. This expectation was probably encouraged by the work in 1880 of Alexander Graham Bell and Charles Tainter on the photophone, a communication device that permitted sounds to be transmitted over a distance by means of a modulated beam of sunlight aimed at a selenium cell. The simplicity of the photophone and the lack of effort involved in its development, together with Bell's invention of the telephone in 1876, which enabled "hearing by electricity" to be readily implemented, stimulated workers in their quest to achieve "seeing by electricity." Several suggestions for "telectroscopes" were put forward in the 10-year period following Willoughby Smith's 1873 discovery.

The earliest ideas for these were based on notions that had been advanced from 1843 for picture telegraphy systems, but from around 1883 more appropriate ideas began to be proposed. In a basic system of television the light values of each elementary unit area of an illuminated scene or object are determined by an analyzing scanner-photoelectric cell arrangement and the amplified varying electrical signal is transmitted to the receiving apparatus. Here, the signal is again amplified and applied to an electrically controlled varying light source system so that, by means of a synthesizing scanner, an image of the original scene or object can be reproduced.

The implementation of such a system was not easy and more than 50 years would elapse, before John Logie Baird demonstrated a rudimentary form of television. Nevertheless, the work of the nineteenth century television pioneers was not wholly unproductive and, by the end of the century, some of the basic system components needed to implement a television scheme had been put forward. The ideas of Paul Nipkow (1883), Lazare Weiller (1889), and Marcel Brillouin (1894) led to television scanners—the apertured disk, the mirror drum, and the lens disk, respectively—which were widely utilized in the 1920s and 1930s. In addition Wilhelm Hallwachs' work on the photoelectric effect in 1888 followed by the detailed investigations from 1889 of Julius Elster and Hans Geitel on photoelectricity, together with the fundamental researches which were being undertaken, contemporaneously, on the conduction of electricity in gases and in vacuums were important contributions that were to play a vital part in the progress of television. The latter work led to the invention, by Karl Ferdinand Braun in 1897, of the cathode ray tube (CRT) as a practical laboratory instrument.

Following Braun's publication, the development and use of cathode ray tubes was pursued by several investigators, and so it was perhaps inevitable that it would be incorporated into a television system. Boris Rosing, in Russia in 1908, was the first person to engage in work on television using a CRT receiver, although prior to this date Max Dieckmann and Gustav Glage, in 1906, had developed a "method for the transmission of written material and line drawings by means of cathode ray tubes," In his work, Rosing was assisted by his student Vladimir Zworykin who later, from 1923, and in the U.S., tried to evolve an all-electronic television system. He described his "kinescope" (CRT) display tube in 1929 and his "iconoscope" (CRT) camera tube in 1933.

On 9th May 1911 Rosing recorded in his notebook, "... A distinct image was seen [on the screen of the CRT] for the first time consisting of four luminous bands." In the same year Alan Archibald Campbell-Swinton elaborated on his earlier (1908) ideas, founded on a cathode ray camera tube and a cathode ray display tube, for an all-electronic television system. These ideas influenced James McGee who, from 1932, led the team at Electric and Musical Industries (EMI), which developed the emitron camera tube, the British equivalent of the iconoscope.

During the period 1912–1922 only a few new schemes for television were advanced. This situation changed greatly from 1923 because of technical advances in electronics. In 1904 J. Ambrose Fleming invented the diode valve. Two years later Lee de Forest invented the Audion (triode) valve and in 1912 discovered that the valve, in addition to having applications in detecting and amplifying circuits, could be utilized in an oscillator to generate electromagnetic waves.

World War I gave an impetus to the use of valves in signaling systems, and so stimulated developments in circuit and radio techniques, so that by 1918 triodes could be manufactured to cover wide power and frequency ranges and were suitable for both receiving and transmitting purposes. Consequently, by 1920 the time was opportune for the establishment of sound broadcasting: the radio systems were available and public demand was growing. The growth of commercial radio telephony, domestic broadcasting, and facsimile transmission influenced the progress of television. By the early 1920s, all the basic components of a rudimentary television system appeared to be at hand; and so, from around 1923 determined efforts to advance television were being made in the U.K., the U.S., France, Germany and elsewhere. At first these endeavors were mainly those of individuals—Baird in the U.K., Charles Jenkins in the U.S., Edwin Belin of France and Denes von Mihaly, a Hungarian working in Germany—working in isolation from others, but from 1925 this situation changed. Bell Telephone Laboratories in New York began an ambitious program that led to an impressive demonstration of the first U.S. television in April 1927, by wire from Washington D.C. to New York. Later, General Electric, Westinghouse Electric and Manufacturing Company, and the Radio Corporation of America (RCA), of the U.S.; Fernseh AG and Telefunken of Germany; the Marconi Wireless Telegraph Company, the Gramophone Company, Electric and Musical Industries (EMI), in addition to Baird's company in the U.K. and its subsidiaries, and others, all carried out experimental investigations in the television field. Baird made transatlantic transmissions of television signals by short-wave radio from London to New York in February 1928, and began experimental broadcasts in collaboration with the British Broadcasting Corporation (BBC) in September 1929.

In the U.K., the determined efforts of EMI led, on November 2, 1936, to the inauguration of the world's first, public, regular high-definition television service from studios and transmitters at Alexandra Palace, north London.

See also **Iconoscope; Television, Electromechanical Systems**

Russell W. Burns

Further Reading

Burns, R.W. *Television: An International History of the Formative Years.* Institute of Electrical Engineers, London, 1998.

Abramson, A. *The History of Television, 1880 to 1941.* McFarland, Jefferson, NC, 1987.

Burns, R.W. *British Television, the Formative Years.* Institute of Electrical Engineers, London, 1986.

Burns, R.W. *John Logie Baird, Television Pioneer.* Institute of Electrical Engineers, London, 2000.

Shiers, G. *Technical Development of Television.* Arno, New York, 1977.

Wilson, J.C. *Television Engineering*. Pitman, London, 1937.
Zworykin, V.K., and Morton, G.A. *Television: The Electronics of Image Transmission*. Wiley, New York, 1940.

Television, Cable and Satellite

In the early days of television experiments, transmission of signal was by wire connections and transmissions over phone lines. Radio communications came later. For example, the demonstration in April 1927 by Bell Telephone Labs between Washington, D.C. and New York City was transmitted by both wire and radio, and John Logie Baird also transmitted television between London and Glasgow using conventional telephone lines in 1927. All prewar British television outside broadcasts used post office telephone lines to get the video signal back to Alexandra Palace prior to radio broadcast. However, consumers received their signal only by radio, and television distribution by cable did not start until the 1950s, and even then in only a limited way.

During World War II, television broadcasts halted in Europe, but they carried on in the U.S., although in a reduced sense. After the war, television broadcasts in England restarted and by 1950 there were early attempts in London to use existing radio relay cables to distribute television. Radio relay was used in England, Germany, and Russia to distribute radio from a master antenna via cable to blocks of houses. By 1953, cable systems were being designed and installed elsewhere in England, still on the radio relay principle.

While there is dispute over who built the first U.S. system, the National Cable Telecommunications Association (NCTA) has given credit to Ed Parsons, who set up a local cable system in Astoria, Oregon. Seattle radio station KRSC had announced it would build television station KRSC-TV, now KING-TV. Parsons' wife had seen television in 1947 at a convention of the National Association of Broadcasters (NAB), and remarked that she would like to have television at home. When KRSC-TV went on the air November 25, 1948, Parsons had an antenna and booster installed on the roof of the eight-story John Jacob Astor Hotel, connected with twin lead to his penthouse apartment a short distance down Commercial Street. By New Year's Day 1949 he had run a feed to Cliff Poole's music store across the street from the hotel, using coaxial cable, and later to the whole town. Early cable transmission was by ordinary cable, but coaxial cable began to be used from 1941 in the U.S. and the 1950s in the U.K.

Cable TV developed as a medium for distributing or relaying broadcast signals to outlying communities which could not get good reception of broadcast signals due to being out of range of the transmitter. Early cable operators such as those in Pennsylvania in 1949 were mostly local entrepreneurs operating community antenna television or CATV, which consisted of a tall antenna with a repeater station and amplifier, connected by wire to a few homes. From about 1953, operators began to build microwave relays to bring in distant television signals. By 1961 there were 700 CATV systems in the U.S., though the industry soon coalesced into a few large operators called MSOs, or multiple system operators, as entrepreneurs began to consolidate small networks into larger operations. The industry continued to be primarily a master antenna service operating outside of major metropolitan areas until 1975, when Home Box Office (HBO) inaugurated satellite distribution of its pay movie service. The inaugural broadcast occurred on 1 October 1975 (Philippine time), with the transmission of the Ali–Frazier heavyweight prize fight from the Philippine Islands (the "Thrilla from Manila") to cable television systems in Florida and Mississippi. HBO then began to sell its satellite-delivered movie service to cable operators, who in turn sold it to subscribers. Shortly after, a number of cable networks, both pay and advertiser-supported, began to appear. With these new sources of programming not available off the air, cable television began to penetrate larger cities.

Communications satellites receive television signals from a ground station, amplify them, and relay them back to earth. Satellite distribution of television began in 1962. On 11 July 1962 satellite dishes at the Radome in Pleumeur-Bodou, France and British Telecom's Goonhilly Earth Station, in Cornwall, U.K. received the first transatlantic transmission of a television signal from a twin station in Andover, Maine, in the U.S. via the TELSTAR satellite. TELSTAR had a low, elliptic orbit and was only usable for three or four 40-minute periods in each 24 hours. It delivered satellite television during its seven months in orbit.

The satellites used today for cable and broadcast program distribution are in geosynchronous orbit, so that they appear to be stationary in space, affording the use of relatively low-cost fixed receiving antennas. The early antennas were 10 meters in diameter, but antennas smaller than 3 meters can be used today. The original downlink (satellite-to-ground) was the so-called C band, 3.7 to 4.2 gigahertz. Analog frequency modulation was used, and remains in declining use today. Some newer satellites use downlinks in the Ku band, about 11.7 to 12.2 gighertz. Most newer satellite links use digital transmission of MPEG-2 (Motion Picture Experts Group) compressed video. This permits more programs to be transmitted in the same bandwidth, and permits use of smaller earth station antennas.

Cable television systems start at "headends," where signals are brought together from a number of sources. Headends supply signals directly to subscribers located close by, and supply signals to more distant subscribers using fiber-optic cable. Figure 3 illustrates a modern cable television system distributing signals to a large metropolitan area. The headend supplies signals to a number of "hubs." Radiating from each hub are a number of fiber-optic cables, each connecting to one or more "nodes." The nodes convert signals from optical to electronic form, where they are distributed via coaxial cable (coax) and radio-frequency amplifiers, to individual homes. A portion of the signal is removed to send to a group of homes, at a "tap." The tap is a passive (nonpowered) device that draws a predefined portion of the signal from the cable and sends it to one or more homes. Homes are attached to the tap by way of "drop" cable, smaller, flexible coaxial cables.

Signal strength is reduced as the signals travel through the coaxial cable. The signal strength reduction is due to two phenomena: (1) every time some of the signal is taken out of the cable at a tap, the remaining signal going downstream is weakened by the signal removed to supply the

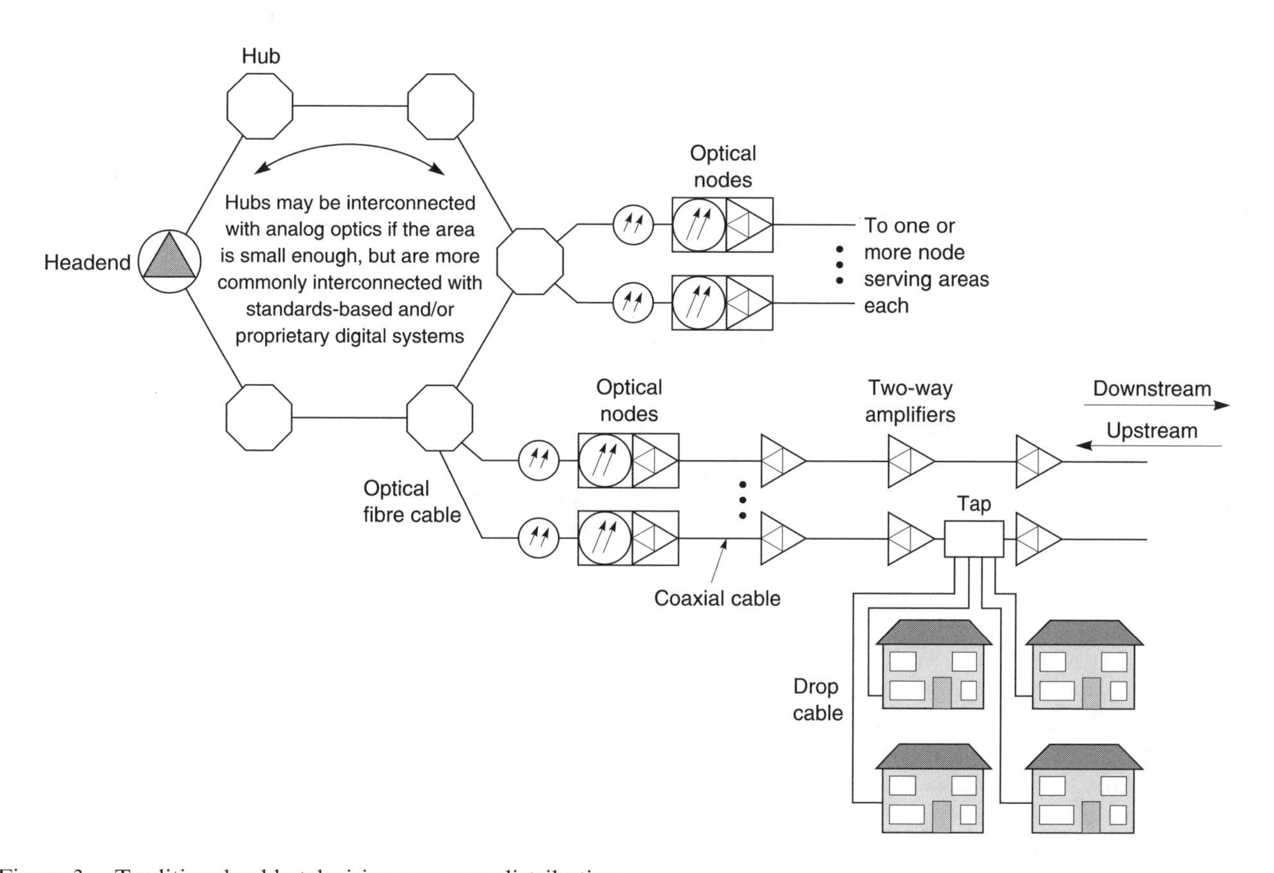

Figure 3. Traditional cable television program distribution.

homes attached to that tap; and (2) the coaxial cable itself causes signal loss due to resistance in the conductors and losses in the dielectric insulator between the conductors. Because of these losses, amplifiers must be used periodically to increase signal strength.

Traditional cable television program distribution is downstream only (see Figure 3). However, many modern services require two-way communications. These services include cable modem and telephone service, as well as interactive video. Bidirectional amplifiers facilitate this two-way communications by amplifying signals flowing in both directions. In order to separate the upstream and downstream signals, it is necessary to transmit them in different frequency bands. North American practice is to use the band from 54 megahertz to as high as 870 megahertz for downstream transmissions, and the frequency band from 5 to 42 megahertz for upstream transmission. Other parts of the world use slightly different frequency plans. Each amplifier uses "diplex filters" to separate these bands at both the input and output of the amplifier, then uses separate amplification circuits for the two frequency bands. Separate fiber strands usually carry different direction signals between the hub and the optical node.

Each amplifier adds distortion and noise to the signal, so the number of amplifiers used in any one signal path (called a cascade) must be limited. Also, since the amplifiers tend to be somewhat trouble-prone due to their complexity, reliability suffers as the cascade is lengthened. In early systems, amplifier cascades of 20 were common, and cascades of 50 amplifiers were sometimes used. Today, with the introduction of fiber optics, the number of amplifiers in a cascade has been reduced to typically six, with a strong trend toward fewer amplifiers.

Fiber-optic cables are being brought deeper into the cable plant to facilitate this reduction in the number of amplifiers in cascade. Long distances can be achieved with fiber-optic transmission (Figure 4). This figure plots the loss per kilometer for two sizes of coax, and also for fiber-optic cable. The loss on the vertical axis is measured in decibels (dB), with lower numbers (less loss) being better. This lower loss of fiber-optic cable permits long distances between the hub and the node, with no amplification being needed in most cases. Optical amplification is practical and is used where needed.

The systems described so far are called hybrid fiber-coax (HFC) systems. The latest trend is toward either fiber-to-the-curb (FTTC) or fiber-to-the-home (FTTH) systems, neither of which use

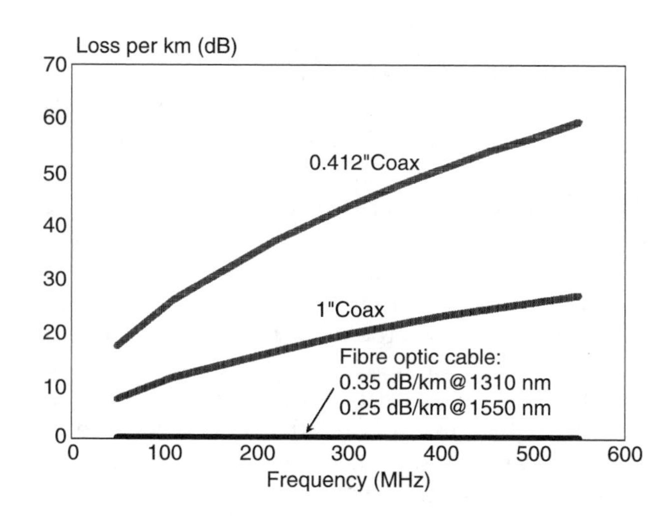

Figure 4. Signal loss per kilometer measured in decibels in fiber-optic cable and two sizes of coaxial cable. Lower numbers indicate less loss and therefore permit long distances between hub and node with no amplification needed in most cases.

RF amplifiers. In FTTC systems, the fiber is brought to a location that serves between four and sixteen homes, and broadcast services are distributed over coax from there. Telephone service is distributed from the end of the fiber to homes, using conventional telephone cable. Data is distributed on data cables, or in some cases may use the same cable as the phone service uses.

FTTH systems bring a fiber cable to the side of the home, where it terminates in a conversion device that supplies video, voice, and data signals for use in that one home. FTTH systems can be divided into passive systems, which have no active (amplification and signal processing) devices at all between the hub and the home (PONs, or passive optical networks); and active systems, which have one active device. The active systems can achieve much longer reach between the hub and the home, and can use lower cost optical components. These systems are beginning to be deployed at the time of writing, in competition with conventional HFC plants. These new architectures offer superior reliability, higher quality signals, and much higher data rates than are practical with HFC networks.

Signal distribution employs "frequency division multiplexing," or FDM, in which each signal is assigned a frequency band at which it is transmitted. The frequency band is called a channel. The signal is modulated, or impressed on, a carrier frequency in that assigned channel. For analog transmissions, one program is assigned to each channel, whereas for digital transmissions, several programs may be transmitted simultaneously on one channel. Simultaneous transmission of multiple programs is done using "time division multiplexing," or TDM. In TDM, a portion of one program is transmitted, followed by a portion of another and so on. The transmission is fast enough that all programs appear to the user to be transmitted simultaneously. At the television, when the viewer selects a channel, he or she is tuning the television to that carrier frequency, and the television is recovering, or demodulating, the channel.

In HFC networks, data and telephone traffic are also transmitted using a similar combination of FDM and TDM. In FTTC and FTTH systems, television programs may be transmitted using FDM as on HFC networks, but data and telephone traffic normally use base-band TDM, in which they are not modulated onto carriers, but rather are time division multiplexed and used to turn optical transmitters on and off at a fast rate. Video programs may also be transmitted this way, although it remains more common to use FDM to transmit video programs.

See also **Optoelectronics; Satellites, Communication**

James O. Farmer

Further Reading

Adams, M. *Open Cable Architecture*. Cisco Press, Indianapolis, 2000.

Burns, R.W. *Television: An International History of the Formative Years*. Institute of Electrical Engineers, London, 1998.

Ciciora, W., Farmer, J. and Large, D. *Modern Cable Television Technology: Video, Voice and Data Communications*. Morgan Kaufmann, San Francisco, 1999.

Easton, K.J. *Thirty Years in Cable TV: Reminiscences of a Pioneer*, Pioneer Publications, Mississauga, Ontario.

Farmer, J. *Fiber-to-the-Home: Why it's Inevitable*. PBIMedia, Potomac, MD, 2001.

Hecht, J. *City of Light: History of Fiber Optics*. Oxford University Press, New York, 1999.

Junkus, J.J. and Sawyer, M.J. *Digipoints*, vols. 1 and 2. Society of Cable Telecommunications Engineers, Exton, 1998 and 2000.

Laubach, M., Dukes, S. and Farber, D. *Breaking the Access Barrier: Delivering Internet Connections over Cable*. Wiley, New York, 2001.

Raskin, D. and Stoneback, D. *Return Systems for Hybrid Fiber/Coax Cable TV Networks*. Prentice Hall,. Upper Saddle River, NJ, 1997.

Taylor, A.S. *History Between Their Ears*. Cable Center, Denver, 2000.

Walker, P., Ed. *Milestones: A History Of Cable Television*. National Cable Television Center And Museum, Denver, 1998.

Useful Websites

The Cable Center, Cable History: http://www.cablecenter.org/history/index.cfm

Television, Color, Electromechanical

Prior to 1861 photographs in color could only be reproduced by hand painting daguerreotypes, which were direct-image photographs produced on a silver-coated copper plate. In 1855 James Clerk Maxwell suggested a method for creating a color image. If a scene or object were photographed through red, green, and blue filters separately to obtain three negatives and the three positives prepared from them were projected with the images in alignment onto a screen by means of three lanterns, fitted with the appropriate red, green, and blue filter in front of the projection lens, a colored image would result. Maxwell demonstrated the correctness of his notion at a meeting of the Royal Institution in London in May 1861.

John Logie Baird adapted Maxwell's concept to display colored television images on 3 July 1928, for the first time anywhere. His sending and receiving apparatus utilized Nipkow disk scanners, each having three spiral sets of apertures. The three sets were covered separately by red, green, and blue filters. At the sending end of the television link the scene or object was analyzed, sequentially, into its three primary color components, and at the receiving end the video signals corresponding to these were used to reconstitute three primary color images in register. Because of the persistence of vision, the final image was colored.

On 27 June 1929 Bell Telephone Laboratories (BTL) gave a demonstration in New York of color television using Nipkow disk scanners and transmitting stills. However, whereas Baird used a single scanner having three spirals, at each end of his television link and a single transmission channel, BTL employed a single scanner, having a single spiral of apertures, at each end of the television link, and three transmission channels. The red, green, and blue color contents of the scene or object being televised were transmitted simultaneously and not sequentially. An advantage of this method was that the same scanning disks and motors, synchronizing equipment, and circuits were applicable as in the monochrome television scheme. Neither scheme led to color television broadcasting services.

On 4 February 1938 Baird gave a public demonstration of television in the Dominion Theatre, London, at which high-definition images of about 3 by 2.7 meters, were shown in color, the television signals being received by radio, using a wavelength of 8.3 meters, from the Crystal Palace transmitter, about 16 kilometers away

The transmitting apparatus consisted of a 20.3-centimeter-diameter mirror drum, provided with 20 mirrors rotating at 6000 revolutions per minute (rpm). These mirrors reflected the scene to be transmitted, through a lens, onto a 500-rpm Nipkow type disk provided with 12 concentric slots positioned at different distances from the disk's axis. Each of the slots was covered by a color filter, blue–green and red being used alternately. By these means the fields given by the 20-mirror drum were interlaced six times to give 120-line picture signals repeated twice for each revolution of the disk.

At the receiver a similar system of rotating mirror drum and disk was employed together with a high-intensity arc lamp source. The light intensity was modulated; that is, made to fluctuate, in exact conformation to the variations from the transmitting end, as it passed through a modified Kerr cell connected to the television receiver.

Another demonstration was given by Baird, in his private London laboratory, on 27 July 1939. However, Baird utilized a cathode ray tube (CRT) at the receiver in place of the mirror drum and slotted disk. A rotating disk having 12 circular filters, alternately red and blue–green, was positioned in front of the screen of the CRT.

The following year Baird, on 17 September, patented a method of color television that enabled him to demonstrate, in April 1941, 600-line color television. Again the two-color principle was adopted but the sending-end scanner was now a CRT of the type that had been employed as a projection unit for cinema television. Both the sending-end and receiving-end CRTs had two-color filter disks rotating in front of their screens.

Further development of this system allowed Baird to demonstrate colored stereoscopic images to the press on 18 December 1941 (Figure 5). At the transmitter, the primary scanning beam of light, after having passed through one of the color filters and the projection lens, was divided by a system of two pairs of parallel mirrors into two secondary beams spaced apart by a distance equal to the average separation of an observer's eyes. By means of the revolving shutter disk, the scene was scanned alternately by each secondary beam.

The receiver included a color disk identical to that of the transmitter and a revolving shutter, both synchronized to the corresponding disks at the sending end of the system. Hence each eye of the viewer alternately observed red, green, and blue images. The shutters used differed at the transmitting and receiving ends to minimize flicker.

Baird's only competitor in the color television field by 1940–1941 was the Columbia Broadcasting System (CBS). On September 4, 1940 in New York, CBS engineer Peter Goldmark demonstrated equipment that comprised a Farnsworth image dissector tube camera and associated three-color filter disk rotating at 1200 rpm at the sending end; and a 23-centimeter CRT and another identical rotating filter disk at the receiving end. A 343-line image 14 by 18.5 centimeters was displayed.

From early 1942 Baird began to consider nonmechanical methods of color television and patented one version on 13 May 1942. He adapted the Thomas system of color cinematography in which an optical unit in the camera automatically produces images on a black and white film of the red, green, and blue components of a scene. At the receiver an identical optical unit combines the three cine film images to display a colored image. In the television version of this system the film camera and projector were replaced by an electronic camera and a projection CRT.

Two years later, on 16 August 1944, Baird gave a demonstration to the press of his telechrome tube—the world's first multigun, color television display tube. The tube employed either double or triple, separate and independent, electron guns and multiple fluorescent screens depending upon whether two- or three-color reproduction was required. Only the two-gun version of the telechrome tube was demonstrated.

In 1944 it was clear that electromechanical methods of color television would not be viable in a domestic situation—although CBS persisted with its scheme—and by 1950 all electronic frame-, line-, and dot-sequential color television systems were being considered.

See also **Television, Electromechanical Systems; Television: Color, Electronic**

Russell W. Burns

Further Reading

Abramson, A. *The History of Television, 1880 to 1941*. McFarland, London, 1987.

Burns, R.W. *Television: An International History of the Formative Years*. Institution of Electrical Engineers, London, 1998.

Burns, R.W. *John Logie Baird, Television Pioneer*. Institution of Electrical Engineers, London, 2001.

Ives, H.E. Television in color. *Bell Laboratory Record*, 7, 439–444, July 1929.

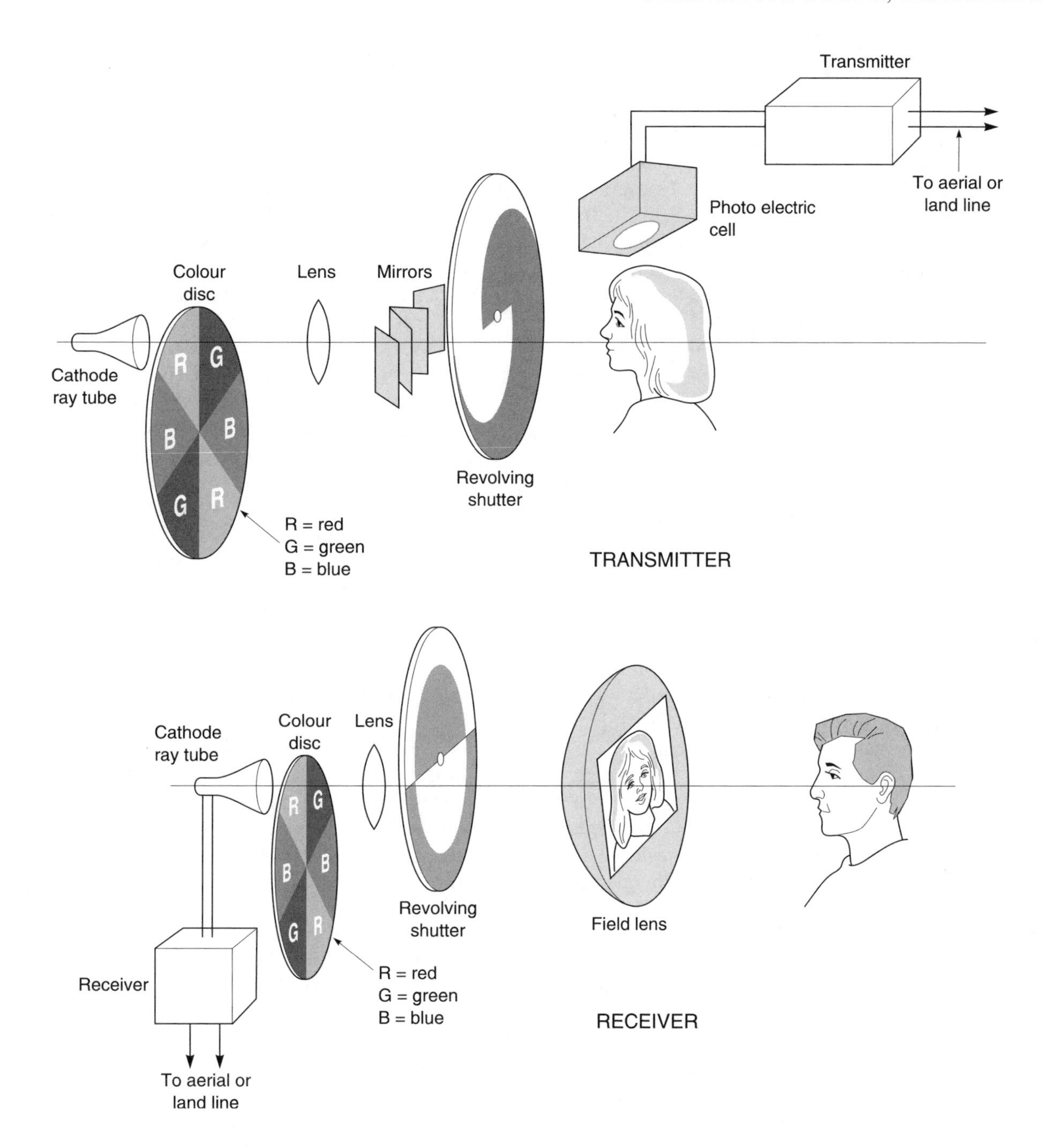

Figure 5. Schematic diagrams showing the layout of the apparatus that Baird used to show stereoscopic color television.
[*Source: Electrician, December 1941, p. 359.*]

McLean, D. F. *Restoring Baird's Image*. Institution of Electrical Engineers, London, 2000.

Television: Color, Electronic

At the end of the twentieth century, some 900 million people watched electronic color television around the world. Americans owned almost as many color television receivers as indoor toilets. In a generation the medium supplanted wireless broadcasting and cinema as the world's premier source of news and entertainment.

Although John Logie Baird had worked on an all-electronic color television from 1942, the development of this technology after World War II can be assigned to one organization. Under the sponsorship of chief executive David Sarnoff, the Radio Corporation of America (RCA)

Laboratories undertook the research and development necessary to make color television a reality between 1945 and 1953. The staff also made the transmission standards and hardware compatible with those in place for monochrome television. After the Federal Communications Commission (FCC) approved standards based on this effort, RCA underwrote production, programming, and marketing efforts until American consumers began returning RCA's $100 million investment in the early 1960s. Other countries adopted this standard or adapted it with various modifications.

Until RCA's researchers took up the challenge after World War II, it appeared that the FCC would approve the Columbia Broadcasting System's (CBS) electromechanical field-sequential system of color television, which was incompatible with the monochrome system already in place. Supported by other RCA engineers, the staff at RCA Labs proposed the conceptual framework, established most of the principles and techniques, and demonstrated the technologies necessary for the electronic pickup, broadcast, and reception of color television in the home. This system was the foundation for analog broadcast standards internationally.

The laboratories' researchers had experimented with field-sequential color systems during the early 1940s. This research demonstrated the inherent limitations of the technology in terms of picture brightness, monochrome compatibility, and image scalability. RCA then committed to developing an all-electronic system. In 1947 the Labs demonstrated a color system using Alda Bedford's principle of "mixed highs." Because the human eye distinguishes changes in brightness but not color at high levels of detail, high-frequency components of the primary additive colors—red, green, and blue—could be blended into one signal. This reduced the amount of bandwidth used.

By the end of 1948 technical and regulatory pressures forced RCA's researchers to compress the bandwidth to the 6 megahertz used for monochrome broadcasting. In January 1949, the Labs' leadership committed to meeting that limit while maintaining equivalent resolution, brightness, and flicker. Monochrome signals would be displayed on color receivers, while color signals would be received without adjustment on monochrome receivers.

That September, the FCC began hearings on proposed formats for color television. For the next eight months, under the pressure of attention by the government, the media, and the competition, RCA Labs staff turned their concepts into demonstrations of principle. Mixed highs and Clarence Hansell's application of time-division multiplexing to color transmission permitted the compression of the televised signal into two components. One carried the luminance, or brightness, and the other the chrominance, or color information. A monochrome set would simply ignore the color component. John Evans and Randall Ballard's use of dot interlacing to scan the colors was combined with Bedford's color-sampling reference burst to synchronize frequencies between transmitter and receiver. Harry Kihn's "Kolor Killer" circuit enabled color receivers to show monochrome signals.

As for hardware, Richard Webb built color cameras using three image orthicons, the standard monochrome image tube developed at the labs. Each tube scanned an image in a primary color. Dichroic mirrors (which transmit certain wavelengths or colors of light while reflecting others) and electronic circuitry then blended the three primaries and sent the image to the amplifier. RCA began producing a commercial camera based on this design in 1952.

The weak point, technically and socially, was the receiver. RCA's "triniscope" used three cathode-ray tubes (CRTs) and dichroic mirrors to combine the primary colors and project the image on a screen. The sheer bulk of this design required that RCA develop a practical household alternative.

Harold Law's refinement of Al Schroeder's shadow-mask tube offered the best solution. Like a monochrome CRT, the screen's glow was based on the intensity of the electron beam and signal. But the color tube contained three electron guns, one for each primary color. A perforated mask next to the screen enabled the beams to strike the appropriate red, green, or blue phosphors clustered in thousands of triads. The intensity of the beam and the relative brightness of the phosphors determined the colors seen by the viewer. More than any other component of the system, the shadow-mask CRT made color television (and computers) a household technology.

In July 1950, the National Bureau of Standards (NBS) reported that RCA's system was demonstrably superior to CBS's system and offered the most "opportunity for improvement." Nonetheless the FCC rejected RCA's system in September 1950. The rest of the television industry had already started refining RCA's format and technologies. The second National Television System Committee (NTSC) was organized in January 1950. Members of the industry and the FCC agreed to develop with RCA a standard for electronic color television broadcasting and reception. The NTSC adopted RCA's system with one significant exception.

Hazeltine Corporation, an independent laboratory with a cross-licensing agreement with RCA, developed a different approach to analyzing the color signal. While RCA's staff understood this process as one of sampling the combination of brightness and color, Arthur Loughren and Bernard Loughlin represented the dot-sequential signal as a color subcarrier added to a monochrome-carrier wave. This change in perspective enabled engineers to treat color television with the traditional tools of monochrome video and sine-wave mathematics. In April 1950 Hazeltine demonstrated its concepts of constant luminance and "shunted monochrome" to RCA Labs, which adopted them to resolve outstanding display problems.

RCA unveiled its version incorporating this approach in June 1951; field testing began in February 1952. During this time Schroeder and Ray Kell reduced color fringing by adapting quadrature-amplitude modulation. Color was now transmitted on a wideband orange–cyan I axis and narrowband green–purple Q axis.

The FCC approved the NTSC standard on December 17, 1953. Japan, Canada, Central America, and some South American countries followed suit. In the 1960s, all of Western Europe except France adopted the German company Telefunken's NTSC variant, phase alternating lines (PAL); France persuaded Russia and Eastern Europe to use sequentiel colour avec mémoire (SECAM). PAL, adapted from NTSC, has phase reversal that avoids color errors resulting from amplitude and phase distortion of the color modulation sidebands during transmission. For many European television engineers, NTSC's receivers, which have manual tint and intensity controls, have such large consumer color control that NTSC is said to stand for "never the same color twice," while American engineers dub SECAM "system essentially contrary to the American method." Today, the North American NTSC system is still incompatible with France and Eastern Europe's SECAM and Western Europe's PAL, with different receivers required for each.

Until the mid-1960s color cameras continued to use image orthicons, when Philips's plumbicon tube supplanted them; RCA introduced solid-state charge-coupled devices (CCDs) to cameras in 1984. Color receivers continue to feature shadow-mask variants with Sony's 1968 Trinitron tube being the notable improvement.

See also **Television: Color, Electromechanical**

ALEXANDER B. MAGOUN

Further Reading

Barnouw, E. *Tube of Plenty: The Evolution of American Television*, 2nd revised edn. Oxford University Press, New York, 1990. Best history of programming and its relationship with technology and regulation.

Brown, G.H. *And Part of Which I Was: Recollections of a Research Engineer*. Angus Cupar, Princeton, 1982. Thoughtful and humorous account of color standards competition by head of RCA's color program.

Fisher, D.E. and Fisher, M.J. *Tube: The Invention of Television*. Harcourt, New York, 1997; Counterpoint, Washington D.C., 1996. Relies on secondary sources for history of color.

Inglis, A.F. *Behind the Tube: A History of Broadcasting Technology and Business*. Focus Books, Boston, 1990. Best analysis of competing color systems and corporate motives for their support.

Schreiber, W.F. and Buckley, R.R. Introduction to color television—part I. *IEEE Proc.*, 87, 1, 173–179, January 1999. Argues that field-sequential color system could have worked with color tube and monochrome compatibility was a mistake.

Slotten, H. *Radio and Television Regulation: Broadcast Technology in the United States, 1920–1960*. Johns Hopkins University Press, Baltimore, 2000. Study of standards debates, including that over color, as struggle between those who understood and designed the systems and the political bodies that had to regulate it.

Useful Websites

IEEE History Center RCA Oral History Collection: http://www.ieee.org/organizations/history_center/oral_histories/oh_rca_menu.html

Interviews with RCA's Harold Law, Humbolt Leverenz, and Paul Weimer, who had basic roles in developing RCA's system.

Television, Digital and High-Definition Systems

The term "high definition" was first applied to television in the 1930s to compare all-electronic analog television (TV) systems with the older and partially mechanical systems then still used experimentally. Regular television service as inaugurated by the British Broadcasting Corporation (BBC) in 1936 (using 405 scanning lines) and in the U.S. in 1941 (525 lines) was often termed "high definition," as was a Dumont experimental system of 1939–1940 that briefly achieved 800 lines. Decades later, however, the term came to mean something quite different.

Originally restricted to analog technology, high-definition (usually defined as greater than 1,000 scanning lines, roughly equivalent to 35-millimeter film quality) television development focused on digital methods after 1990. By the early twenty-first century, however, a multiplicity of digital television technical standards appeared more likely than world agreement on any single approach. Digital television was thus following in the footsteps of analog, where in the 1960s three different color TV systems developed more for economic and political than technical reasons: NTSC, National Television System Committee, from the U.S.; PAL, phase alternative (or alternating) line, from Germany; and SECAM, sequential colour avec mémoire, from France.

HDTV Origins

Modern era high-definition television research began in Japan in the late 1960s as Nippon Hoso Kyokai (Japan Broadcasting Corp.) engineers sought to improve on the NTSC 525-line standard. This effort expanded into full-time research into both video and audio aspects of an improved system about 1970. The resulting 1125 scanning-line analog "Hi-Vision" system was first demonstrated to American policymakers in early 1981, four years after American engineers had also begun to investigate HDTV's potential. Interlaced scanning and field or frame standards matched existing NTSC practice, but the picture's aspect ratio was widened to 16:9 rather than NTSC's 4:3 (to better approximate telecasting of widescreen motion pictures), and multiple sound channels were digital. Because the highly complex HDTV signal would require the equivalent of six (later four) NTSC 6-megahertz channels to transmit, however, Hi-Vision was designated for production and not transmission.

Concern over the likely high cost of any high-definition system led to development of several analog alternatives—dubbed "advanced" or "enhanced"—which, providing far fewer than 1,000 scanning lines, would require less spectrum space while still offering improved picture quality. Major U.S. manufacturers and trade associations established an Advanced Television Systems Committee (ATSC) in 1983 to coordinate American research efforts and conduct comparative tests. Some proponents argued for a gradual transition to HDTV through these intermediate systems. Research continued into developing HDTV systems that could be "downward compatible" (receivable in lower-quality analog form) on existing NTSC sets to ease the introduction of HDTV.

Several European countries had by the mid-1980s begun cooperative development of analog HDTV technology based on their 50 hertz electrical standard, seeking to avoid acceptance of a Japanese system, imports of which might wipe out their domestic consumer electronics industry as had happened earlier in the U.S. Europe soon focused on perfecting the MAC (multiplexed analog components) family of transmission standards. The Japanese concentrated increasingly on their MUSE (multiple sub-Nyquist sampling encoding) transmission standard which applied signal compression to force HDTV into only 8.1 (later 6) megahertz of bandwidth, thus making a broadcast service more likely. Both European and Japanese efforts focused on satellite-delivered HDTV, bypassing bandwidth-limited broadcast stations. Faced with substantial industry pressure not to pursue a satellite option, however, the U.S. focused more on a terrestrial broadcast service when the U.S. Federal Communications Commission issued its first inquiry concerning advanced modes of television in mid-1987. The FCC also soon made clear its preference for a true HDTV system, bypassing intermediate "advanced" or "enhanced" stages. What then appeared imminent, however, took more than a decade to even begin to achieve.

Digital Breakthrough

In mid-1990, General Instrument transformed the HDTV picture by announcing computer models of a proposed fully digital system ("DigiCipher") of high-definition television. Practical models were soon being tested in laboratories. Under industry and FCC pressure, several competing digital HDTV proponents merged the best parts of their systems into a so-called "Grand Alliance" in 1993 as laboratory and field testing continued. The FCC allotted an additional channel to each on-air station to encourage development of HDTV parallel to continuing analog transmissions. In late 1996 the commission adopted the Grand Alliance system as a formal set of standards. No less than 13 different video scanning modes were included, ranging from 480 to 1080 lines, and allowing either interlaced or progressive scanning. Agreement was reached on use of the MPEG-2 (Motion Picture Experts Group-2) set of compression tools to condense the HDTV signal by a ratio of 55 to 1, allowing it to fit into the 6-megahertz (8-megahertz in Europe) channels presently used for analog service.

In Europe, cooperation on what had become known as the D2-MAC standard collapsed early in 1992 for lack of sufficient demand and the expense of introducing the system. Growing European satellite television success was based on existing analog formats, further undermining the attempt at a continent-wide digital standard. Faced with the digital transition elsewhere, Japan reduced its backing of analog MUSE, despite the fact the system was in regular NHK operation (though to few receivers given their very high cost) and commenced active work on a digital system.

Japan's introduction in 1991 of regular analog HDTV service (transmitted eight hours daily from a domestic satellite) was the world's first. Scheduled HDTV first aired in the U.S. when WRAL-TV (Raleigh, NC), began a limited but regular digital HDTV service five years later. Other stations slowly followed, most of them transmitting only a few hours a week. Costs of converting a station's facilities to full digital HDTV operation ranged up to $10 million or more. Stations in the ten largest markets began offering a few HDTV hours weekly in November 1988 and the system's use slowly expanded to smaller markets. By FCC rules, all stations were to be providing at least some HDTV service by May 2002.

Complicating the digital TV picture, American broadcasters by the turn of the century had become increasingly interested in an application of digital television that promised more immediate revenue—the ability to transmit four or more "standard" (or slightly degraded) NTSC channels rather than a single high-definition signal. They argued that this would avoid the huge expense of HDTV conversion (which showed little promise of creating additional industry revenue) while allowing broadcasters to better compete in a multichannel environment. In Europe and Japan, "enhanced" definition TV services were available by the late 1990s. Still, by 2002 it was increasingly clear that widespread use of HDTV would take far longer than proponents had earlier projected—easily another decade or more would pass before existing analog systems could be turned off.

Part of the delay and confusion is because, contrary to widespread opinion even within the industry, there is no single digital television standard for either production (there are actually thirteen) or transmission (where there are five, two of them for high-definition service). Another problem is confusion over nomenclature—when does high definition really mean *high* definition. Different players are pursuing different HDTV strategies using different standards, helping to confuse the marketplace. Finally, the cost of receivers (in the aggregate, far more costly than the industry's conversion costs) are off-putting to many would-be set buyers. Prices still averaged $3000 early in 2002 (though down from ten times that level a few years earlier). This last point means that "downconversion," or the ability of existing analog receivers to receive (with a conversion box) at least a semblance of "lower" definition television, is a hugely important aspect of what is now clearly going to be a slow transition.

Still, the promise of digital high definition remains strong. While the analog systems of today provide a video image made up of about 210,000 pixels (individual picture elements), digital HDTV images provide ten times that number. That image can be formatted in either a standard (three units high by four units wide) or wide-screen (nine high by sixteen wide) frame. The latter is closer to theatrical film's format, and thus avoids having to use either letterbox (leaving gray bars above and below the film) or pan-and-screen (editing the picture to better "fit" the television screen) techniques when showing films on television. A major controversy in developing digital standards has been whether to continue the use of interlaced picture scanning (standard in analog systems), or adopt the progressive scanning used on computer screens. The latter is better in many ways (smoother transitions and generally more capable), but requires more bandwidth. U.S. policymakers have allowed the use of either mode, leaving the eventual decision to individual broadcasters. This may force the manufacture of television receivers able to do both, at greater expense than a clear FCC standards decision one way or the other. Finally, the approved audio standard for digital television—the five-channel Dolby Digital (AC-3) surround sound system—compresses 5.1 channels of audio to a 640 kilobytes per second stream. While sophisticated, most current television production and transmission equipment cannot adequately handle it and some time will pass before consumers can receive the full advantage of this system.

CHRISTOPHER H. STERLING

Further Reading

Abramson, A. *The History of Television, Volume 2: 1941–2000*. McFarland, Jefferson, NC, 2002.

Benson, K.B., and Fink, D.G. *HDTV: Advanced Television for the 1990s*. McGraw-Hill, New York, 1990.

Brinkley, J. *Defining Vision: The Battle for the Future of Television*. Harcourt Brace, New York, 1997.

de Bruin, R. and Smits, J. *Digital Video Broadcasting: Technology, Standards, and Regulations*. Artech House, Norwood, MA, 1999.

Dupagne, M., and Seel, P.B. *HDTV: High-Definition Television-A Global Perspective*. Iowa State University Press, Ames, 1998.

Evans, B. *Understanding Digital TV: The Road to HDTV*. IEEE Press, New York, 1995.

Forrester, C. *The Business of Digital Television*. Focal Press, Boston, 2000.
Hartwig, R.L. *Basic TV Technology: Digital and Analog*. Focal Press, Boston, 2000.
Sudalnik, J.E. and Kuhl, V.A. *High-Definition Television: An Annotated Multidisciplinary Bibliography, 1981–1992*. Greenwood, Westport, CT, 1994.

Useful Websites

Advanced Television Systems Committee (ATSC) website: http://www.howstuffworks.com/framed.htm?parent=hdtv.htm&url=http://www.atsc.org/

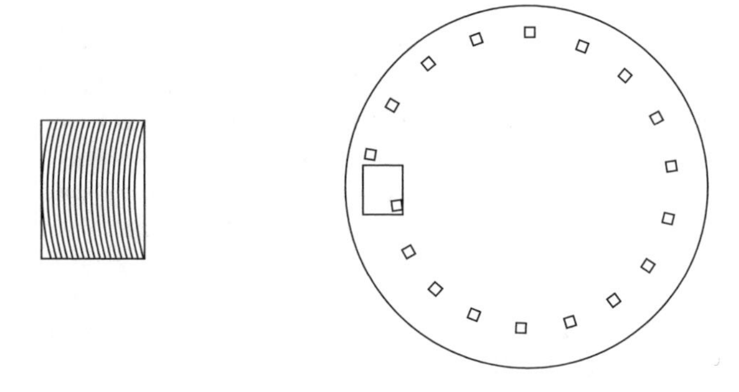

Figure 6. A Nipkow disk scanner and the paths of the apertures across the image plane.

Television, Electromechanical Systems

Television is a form of telecommunication for the transmission of signals representing scenes—the images of the scenes being reproduced on a screen as they are received or recorded for subsequent use. In monochrome television the luminance, but not the color, of an object is reproduced; in color television the reproduced picture simulates both the color and the luminance of the object.

As with a cine film, television consists of a series of successive images that are manifested by the brain as a continuous picture, because of the persistence of vision. Unlike a cine film, where the light values of each picture element of a scene are simultaneously recorded on a film, the light values of a televised scene are scanned in portions that are transmitted sequentially to the recording or display device, since in practice, only a single transmission link is used for a given television camera-display system. This restriction necessitates the utilization of a scanning device and a photosensitive cell at the transmitter to sample a two-dimensional image and convert it into an electrical signal, and another scanner and a display device at the receiver to synthesize or reconstitute the reproduced image from the transmitted electrical signals. Usually, the scanning process follows a left-to-right, and a top-to-bottom sequence as in the reading of a printed page.

Electromechanical television scanners date from c.1880. Although very many suggestions were advanced during the period 1880 to 1930, only the apertured disk scanner of Paul Nipkow (1884), the mirror drum scanner of Lazare Weiller (1889), and the lensed disk scanner of Marcel Brillouin (1891) were subsequently extensively used by experimenters. Of these types, Nipkow's disk was the simplest and the most versatile.

Figure 6 shows a scanning disk (for picture analysis) pierced by 24 apertures arranged at equal angular displacements along a spiral line. Each of the apertures has the shape of a picture element (pixel) and allows the light flux corresponding to the brightness of a picture element of the scene being televised to be incident, via a lens, onto a photoelectric cell. When the disk rotates, each aperture scans a given path (line) of the image field, and after one rotation of the disk the whole of the image has been scanned by 24 paths (lines). The process is then repeated, the number of scans of the complete image (rotations of the disk) per second being the frame rate. With a Nipkow disk the commencement of the line scanning, and also the frame scanning, is carried out automatically. Despite his patent, Nipkow never built a working prototype, as he had no way of amplifying the weak signal from the photocell.

John Logie Baird (1888–1946) employed various types of Nipkow disk type scanners in his early work on television and on 26 January 1926 demonstrated in London, for the first time anywhere, a rudimentary television system. Subsequently his basic scheme was adopted and adapted by many workers. Initially Baird used a "floodlight" method of illuminating his subjects (Figure 7(a)), but when this caused discomfort to his subjects he inverted the positions of the light sources and the photocells to produce a "spotlight" method of scanning (Figure 7(b)). Baird patented the method, as did Bell Telephone Laboratories (BTL) in New York, but unknown to Baird or BTL the method had been patented in 1910.

Both Baird and BTL adapted their systems to demonstrate color television, stereoscopic television (Baird), large screen television, multichannel television, and two-way television. BTL's experimental two-way system was in operation in New York for approximately one year from April 1930 and was used by around 17,000 persons. A novel application was observed when two deaf persons

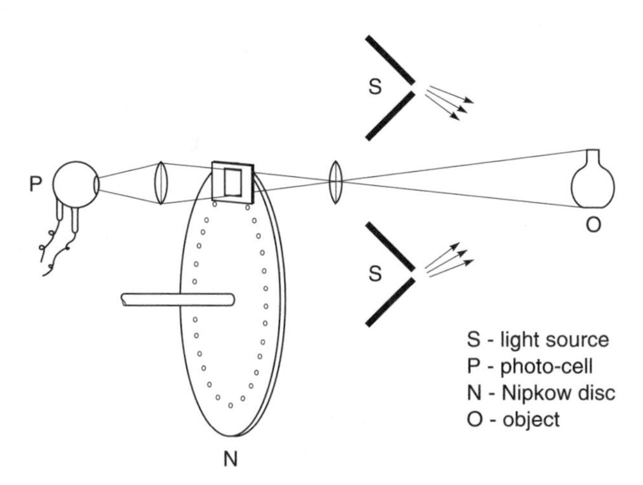

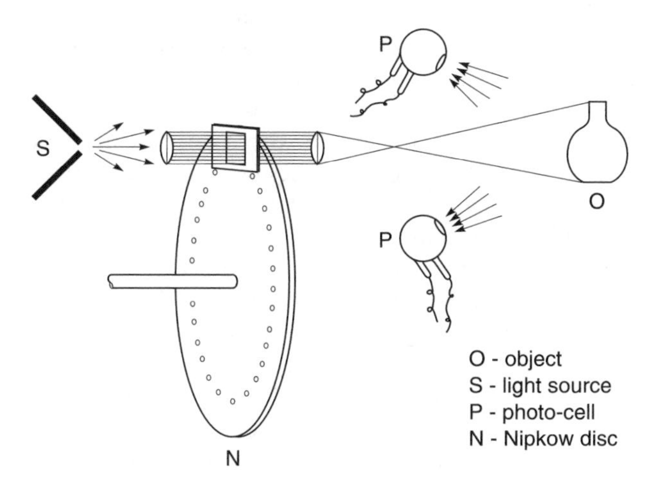

Figure 7. (a) Floodlight scanning system; (b) spotlight scanning system.

carried on a telephone conversation by reading each other's lips.

From 1880 to 1930 progress in television proceeded in an empirical manner. However, in a 1930 issue of the periodical *Fernsehen*, Moller and Kirschtein, in two separate papers, showed how the optical efficiencies of scanners could be calculated according to the principles of optics and photometry. Analyses of the relative efficiencies of the Nipkow disk and the Weiller mirror drum led to the conclusion that the aperture disk should be used at the transmitting end of a low-definition system of more than 40 lines or so. Below 40 lines it was advisable and more efficient to employ the mirror drum. Since Baird's low-definition system was based on 30 lines per picture, at 12.5 pictures per second, the British Broadcasting Corporation (BBC) utilized mirror drum scanners in its low-definition service. Further analysis showed that there was no optical advantage in using a Brillouin multilensed disk rather than a single lens-aperture disk combination for either floodlight or spotlight scanning. Again, analysis indicated that the Nipkow disk was unsuitable for image synthesis and from about 1932 its use declined. In some systems it was replaced by the mirror screw, as advocated by Hatzinger (1930). Essentially, this scanning element comprised a number of mirrors arranged like a spiral staircase on a central shaft so that the mirrors made one complete spiral. The mirror length was the same as the picture width the number of mirrors was equal to the number of scanned lines. For the 84-line images, first displayed in 1931, the mirror screw used 84 mirrors, each 100 millimeters long and 1 millimeter deep to produce a picture 84 by 10.0 millimeters in size. This image size was a substantial improvement over the display size achievable with the Nipkow disk.

By the early 1930s, the Radio Corporation of America (RCA), Farnsworth Television Inc., and Electric and Musical Industries (EMI) were undertaking much research and development work on the evolution of an all-electronic, high-definition television system. Their electronic cameras were

Figure 8. TV Dumont 180.

called the iconoscope, the image dissector, and the emitron respectively. Of these, the iconoscope and the emitron embraced the very important principle of charge storage. Philo Farnsworth's image dissector, basically, was the electronic equivalent of the Nipkow disk–photocell combination and was relatively insensitive compared to tubes of the iconscope and emitron type.

On 2 November 1936 the London BBC television station was inaugurated at Alexandra Palace. It operated with both electronic and mechanical scanners on an alternate basis. By 19 December 1936 it was apparent that the emitron cameras were greatly superior to those that used Nipkow disks. As a consequence, after 2 January 2 1937, only the 405-line, all-electronic system of EMI was operational.

See also **Iconoscope; Television: Color, Electromechanical; Television: Color, Electronic; Television, Beginning Ideas (Late nineteenth and Early twentieth Century)**

RUSSELL W. BURNS

Further Reading

Abramson, A. *The History of Television, 1880 to 1941.* McFarland, London, 1987.

Burns, R.W. *Television: An International History of the Formative Years.* Institution of Electrical Engineers, London, 1998.

Burns, R.W. *British Television: The Formative Years.* Institution of Electrical Engineers, London, 1986.

Burns, R. W. *John Logie Baird, Television Pioneer.* Institution of Electrical Engineers, London, 2001.

Television, Iconoscope, *see* **Iconoscope**

Television Recording, Disk

Recording television signals onto a disk has been a goal of engineers since the inception of television broadcasting, and in recent decades the effort to perfect a disk television recorder has resulted in many competing systems. The disk phonograph was established as the medium for home audio listening by the time television was tentatively introduced in the 1920s and 1930s, so it was natural that experimenters imagined a disk-based accompaniment to future home television sets. The earliest such system was probably that of Scottish inventor John Logie Baird. His "phonovision" disks, developed in 1926, recorded low-resolution television signals in a spiral groove on a phonograph disk. The output of the Baird camera, consisting of electrical pulses, was simply fed to an ordinary electromagnetic phonograph recording head, where the pulses modulated the movement of the cutting stylus.

Despite Baird's progress, the introduction of television in the U.S. and Europe in the late 1930s would be based on the live broadcast of programs, instead of the distribution of recorded programs as in the case of the phonograph. The invention of a successful videotape recording system by the Ampex Corporation (U.S.) in the 1950s led to predictions of home video recorders, and these began to appear in the 1960s (see Television Recording, Tape). The disk format was universally abandoned for commercial video recording, in part because the considerable bandwidth required for television signal recording demanded large areas of recording surface. This surface could be conveniently wound onto a spool in videotape systems, but would have required a very large disk to record even the 15- to 20-minute television shows of the day. However, for certain types of short-duration recording, the fast access possible only with a disk led to special-purpose devices such as the Ampex "instant replay" disk recorder, introduced in 1966 and widely used in the broadcast of sporting events. The Ampex instant replay recorder utilized thirty concentric tracks on a large magnetizable disk, each track representing a single frame of television. Thirty in-line heads read these tracks and allowed the broadcast of slow motion, fast motion, or still images of up to thirty seconds of video.

Improvements in technology also led to the revival of video recording on phonograph-like disk with the RCA Corporation's (U.S.) "Selectavision" system introduced in early 1981. Using the so-called capacitance electronic disk, Selectavision employed a very fine groove and stylus to achieve a recording time of over 60 minutes per side on 12-inch (300-millimeters) diameter plastic disk. The walls of the exceedingly fine groove interacted with a special stylus to establish a varying capacitance between the stylus and the disk. This varying capacitance contained the video information that was then processed and displayed on an ordinary television receiver. It was not possible for the consumer to make an original recording on the home players. Although considered a technical success, Selectavision was introduced at almost the same time as the soon-to-be popular Betamax and VHS home videotape systems. This competition, along with quality control problems and the lack of recording capability, caused RCA to discontinue Selectavision in 1986.

Meanwhile, various electronics manufacturers were experimenting with a video recording technology of a very different sort. These systems utilized the relatively new technologies of digital recording and optical reproduction by laser. Telefunken (Western Germany), RCA, MCA (U.S.), Thomson (France), Sony (Japan), and Phillips (Netherlands) all demonstrated such videodisc systems by 1982, with the first commercial product being the Philips/MCA "DiscoVision," first sold in Atlanta, Georgia in December, 1978. The DiscoVision system recorded video information as pits on an aluminum-coated disk, and the player read the information using a reflected laser beam. However, it was not a digital recording, but rather a frequency-modulated recording system reminiscent of earlier videotape recording technology. DiscoVision failed almost immediately, as did most of the competing systems of the 1980s. A similar Pioneer "Laserdisc" format on 12-inch disks was one of the one or two videodisc formats to survive into the 1990s, although it sold in numbers that were dwarfed by VHS sales.

While the compact audio disc, the audio-only variation of this technology developed by Sony and Philips, proved to be a commercial success, they also provided the basis for a new type of laser videodisc technology employing digital data recording. Pioneer Corporation was among the first to offer a digital videodisc, and employed the same sort of pulse code modulation circuitry already in use in long-distance telephony, data recording on magnetic tape, and other applications. The shift to digital video recording was spurred by the introduction of the personal computer, which utterly changed the context of videodisc development. The CD-ROM, which became a popular way for software companies to deliver their products, was also once considered a competitor to VHS. While much smaller than the original digital videodiscs, it was in most other ways similar in operation. It was not common, however, to watch CD-ROM video presentations on home television sets, but rather on personal computers. However, some record companies began to include short video clips on a CD-ROM disk in 1991, calling this product the "interactive" CD, or CD-I. CD-I players were intended to be connected to both the home audio system and to a television receiver. There were several other similar formats during the 1990s, such as the Eastman Kodak Company's (U.S.) Photo CD, intended to compete with photographic prints for the storage of still images.

Video CD-ROMs achieved considerable commercial success, but did not have quite enough recording capacity to compete with videotape for the presentation of feature motion films, and at the peak of the CD-ROM's popularity it did not seriously threaten the videotape market. It was, however, a more flexible video format than VHS, capable of conveniently storing moving and still images and providing multiple grades of quality to suit the user's needs. Like all disk formats, the CD-ROM allows faster access to data located anywhere on the disk as compared to a linear tape format. Further, by 1996, a user-recordable disk called the CD-R was introduced, with a storage capacity comparable to a CD-ROM.

A refinement of the CD-ROM and CD-R is the Digital Video Disc (DVD, sometimes referred to as the Digital Versatile Disc) first offered in 1996. This format (which, like its predecessors, stores information as pits on a small laser-read disk) packs more information onto a CD-ROM size medium, and for the first time beats VHS tape in terms of both picture quality and recording time. Like the CD-ROM, it is equally applicable to both home video and personal computer applications. The DVD began to sell in large enough numbers by 1999 to garner both larger amounts of shelf space in video stores as well as inclusion as original equipment in many new personal computer systems. A short time later, recordable DVD drives became available for use in conjunction with personal computers. While the DVD may itself be superceded, it seems likely that the place of the laser-read disk format in television recording is assured for the near future.

See also **Audio Recording, Mechanical; Audio Recording, Tape; Television Recording, Tape**

DAVID MORTON

Further Reading

Burns, R. *John Logie Baird, Television Pioneer*. Institution of Electrical Engineers, London, 2000.

Daniel, E.D. *et al.* Eds. *Magnetic Recording: The First 100 Years*. IEEE Press, Piscataway, NJ, 1999.

Graham, M.B.W. *RCA and the Videodisc*. Cambridge University Press, Cambridge, 1986.

McLean, D.F. *Restoring Baird's Image*. IEEE Press, London, 2000.

Television Recording, Tape

The development of video tape recording is closely associated with the development of television. The rapid expansion of television broadcasting in the post-World War II period led to a demand for video recording to facilitate production of programming. Initially, this demand was satisfied by the use of motion picture camera to photograph images produced by a high-brightness cathode-ray tube called a kinescope tube. Such kinescope recordings (the word came to describe a filmed recording although the word actually refers to the cathode-ray tube) were less than ideal, as the process introduces a variety of image defects due to poor resolution, compressed brightness range, nonlinearity, and film grain and video processing artifacts. Given that the resulting signal to noise ratio was often less than 40 decibels (dB), it is not surprising that television producers sought an alternative recording method to make a permanent document for rebroadcast or archiving.

Initial experiments with video recording in the early 1950s were based on magnetic tape technology developed for audio tape recording, a technology that had reached maturity in the late 1940s. However, video recording proved to be a much more difficult technical challenge than audio recording, due to the higher bandwidth required. In contrast to audio signals, which range from about 20 hertz to 20 kilohertz, or roughly 10 octaves, video signals require a recording range from about 30 hertz up to 5 megahertz, more than 17 octaves. With contemporary technology, this required very high tape speeds, which consumed enormous amounts tape. Moreover, these high speeds led to unacceptable wear on tape recording and reproducing heads.

Early experiments, which used a fixed recording head, attempted to solve the problem by using multiple tracks on a single piece of tape, either by running the tape past the head back and forth, or electronically splitting the incoming signal into several frequency bands and recording on separate parallel tracks on the tape. Attempts to solve the mechanical and electronic issues associated with these systems were unsuccessful, and no such machine was ever commercialized.

Led by Charles Ginsburg, the Ampex Corporation solved the problem of video recording by adopting a new type of recording head. The VRX-1000, introduced in 1956, was the first practical videotape recorder. At $50,000, it was used by television networks for videotape delayed broadcast, rather than home recording. Referred to as the quadraplex system, the Ampex machine moved tape past a thin wheel on which were mounted four recording heads that rotated at right angles to the tape motion. This system allowed writing on thin parallel tracks on the tape at very high effective speeds, thereby allowing the capture of television pictures. Initial quadraplex machines were capable of monochrome recording and subsequent improvements allowed the recording of color video. Quadraplex machines were used until the mid-1970s, when they were supplanted by helical scan recorders.

Quadraplex video recorders were in widespread use in television production by the late 1950s. However, these machines were expensive and required constant adjustment to perform well, and manufacturers sought to develop less expensive machines based on emerging transistor technology. In 1958 Ampex developed the helical scan recorder, which wrapped the recording tape in a spiral path around a rotating cylindrical recording head. This allowed the recording of a complete TV field on a single track, and considerably simplified the electronics needed to process the image. However, due to problems with timebase correction, helical scan systems did not match the performance of quadraplex recorders until the early 1970s. After that time, advances in large-scale integration of silicon devices improved the quality and lowered the cost of helical scan recorders so that they replaced quadraplex machines for studio use by the late 1970s.

Helical scan machines were marketed and used in a number of other applications during the 1960s. Television stations in smaller markets or that were not affiliated with the large networks were willing to use helical scan machines with lower levels of performance due to their lower cost. Portable helical scan machines were also developed during the 1960s allowing taping of news and sports recording, though these units were rather heavy and required carts to move them.

Continued development led to the first successful video cassette recorder, the U-Matic system, jointly developed by the Japanese firms Sony, Matsushita, and JVC, and introduced in 1969. The tape cassettes were expensive (almost $100 for one hour), but the machines were widely adopted by institutional users to make training films, replacing 16-millimeter film. In 1974 Sony developed a portable camera system for CBS, which was very successful, but still heavy—one person was needed to carry the recorder, the other the camera. Subsequent competition led to the introduction of smaller recording camera units.

The first home video recorders were introduced by Sony and Matsushita in the mid-1960s, but these reel-to-reel machines were very expensive, had poor picture quality and could only record for a short period of time (less than one hour). With the introduction of chromium dioxide tape in the late 1960s, higher recording densities became possible, and home machines based on the U-Matic format were introduced. Their high cost largely limited their use to institutional settings, and it was not until the introduction of the VHS and Beta cassette formats in the mid-1970s that prices declined to a level that led to widespread consumer acceptance. The key technical innovation used in both VHS and Beta was azimuth recording, which allowed tracks to be recorded much closer together, reducing the amount of tape needed for a given recording time. VHS and Beta machines offered similar levels of performance, but the earlier introduction of longer recording times by VHS machine makers led to their eventual domination of the market by the late 1980s. Subsequently, smaller size cassettes, such as the 8 mm and VHS-C formats, were developed for use in portable video cameras for consumer use.

Although digital recording is an idea that dates to the late 1930s, the digital recording of video signals was delayed for many years by the difficulty of developing adequate analog to digital converters. The rise of large-scale integrated circuits in the 1970s solved this problem, and the first digital video recorders were operating in laboratories by the mid-1970s. Negotiations over a common digital recording format extended into the early 1980s, and it was not until 1986 that the first generation of digital video recorders were marketed by Sony of Japan and Bosch-Fernseh of Germany. By the end of the twentieth century, digital video recording dominated broadcast recording.

See also **Audio Recording, Tape; Television Recording, Disk**

MARK H. CLARK

Further Reading

Daniel, E.D., Mee, C.D. and Clark, M.H. *Magnetic Recording: the First 100 Years*. IEEE Press, 1999.

Daniel, E.D. and Mee, C.D. *Magnetic Storage Technology*, 2nd edn. McGraw-Hill, 1996.

Marlow, E. and Secunda, E. *Shifting Time and Space: the Story of Videotape*. Praeger, New York, 1991.

Nmungwun, A.F. *Video Recording Technology: Its Impact on Media and Home Entertainment*. L. Erlbaum Associates, Hillsdale, NJ, 1989.

Thin Film Materials and Technology

Thin film technology—the growth or deposition of mechanical strengthening, optical, electronic, magnetic, or semiconductor materials in an ultrathin layer—resulted from rapid development of materials science and technology in the late twentieth century. Thin film technology aided the development of devices such as transistors and microelectronic components such as diodes, capacitors, sensors, and resistors, microelectromechanical systems (MEMs), and solar cells. Thin films enable transparent conductive coatings for touch screens and thin conductive films on magnetic read–write heads that allow increased magnetic data storage capacity. In aerospace engineering thin films contribute to strengthening against wear friction and corrosion. They are also used to coat microcircuits and optical lenses to withstand stress and extreme temperatures, protect them from damage and wear, give antireflection or polarizing coatings, and improve durability and performance.

Chemists create thin films, usually only a few micrometers or less thick but potentially nanometers for monolayers, through some method of depositing atoms from a source material target onto a foundation called a substrate (often a silicon wafer, but could be metal or glass). Deposition techniques and material sources can be adjusted to design thin films with tailored properties or thickness to meet specific industrial needs and conditions.

Thin films technology is based on research and processes related to vacuum, gas diffusion, and thin material layers that had gradually developed in the nineteenth century. In 1852, Sir William Robert Grove observed in experiments involving electric discharges between electrodes in a low-pressure atmosphere that metal from one of the electrodes was deposited on the glass walls containing the electrodes and the gas. This "vapor deposition" of metal films, also reported by Michael Faraday in 1854 and Julius Plücker in 1858, is now known to be caused by sputtering (see below). By 1887, Robert Nahrwold heated platinum wires inside a vacuum to deposit material for thin films.

The first commercial use of thin films was in 1901, when mass production of Edison's "gold molded" cylinder phonograph records was enabled by a vacuum coating process that deposited gold vapor from gold electrodes. With awareness of possible industrial uses for vacuums, research into vacuum equipment accelerated, and then shifted to applications by the mid-twentieth century. Innovation of existing thin film processes increased in the mid-twentieth century, quickly escalating from the 1960s through the turn of the century, with chemists adapting processes and materials to fabricate new thin films to create desired structures. Developments paralleled microelectronics demands, particularly the need for transistors and integrated circuits to have pure layers with no microstructural defects that could affect electronic properties. Thin film technology also progressed as processes and underlying knowledge of materials such as electroceramics advanced. For example, discovery of high-temperature superconducting oxides (ceramics) in the late 1980s stimulated development of new thin film deposition techniques, owing to potential applications in superconducting electronics.

Technologists consider vacuum evaporation to be the most efficient and productive thin film deposition method. Physical vapor deposition (PVD) processes usually involve depositing material from a vaporized solid or liquid target source by moving atoms through a vacuum or low-pressure gas or plasma to condense onto a substrate.

Vacuum process technology relies on vacuums that are as empty as possible of any particles and gases that might interfere with materials being deposited within the vacuum. Technologists heat a selected source material in a vacuum so that sublimation or evaporation, by thermal or electron beam heating, results in atoms or molecules forming a film on a substrate. Substrate materials are often metals or glass composed of aluminum, silicon, or beryllium that are smooth, mechanically strong, chemically stable, and have desired thermal qualities. The quality of thin films is diminished with exposure to any moisture or contaminants that enter the vacuum if seals leak, pressure is not maintained, or the chamber is not cleaned.

Molecular beam epitaxy (MBE) methods in which the film crystal structure is "ordered" as it is deposited (growth of the deposited crystal is oriented by the lattice structure of the substrate) were developed by several chemists in the U.S. (including Alfred Cho and John Arthur at Bell Labs), Europe, and Asia during the 1960s and 1970s. Vapors from heated sources form beams that condense onto a heated substrate to create thin crystalline films just as in vacuum evaporation. However the timing and content of these beams can be carefully controlled by shutters in front of the heated sources that can close and block the atomic beams. In the 1960s, this assisted the subsequent development of integrated circuitry. The molecules are placed one layer at a time on the substrate, producing monolayer films that are more suitable for tiny electronics applications than films in which the substrate and polycrystalline film are not so evenly matched and the film's grain size is larger than the circuits for which the films will be used. This process occurs in an ultrahigh vacuum compared to other thin film deposition techniques, and as a result, films tend to be cleaner than those created by other methods. Because MBE relies on vacuum pressures and hygienic measures often too unstable to sustain, technologists devised technology in which various aspects of MBE deposition occur separately in a series of connecting chambers which process crystalline wafer substrates. After being decontaminated and prepared, a wafer moves on a platform along a track and is heated from 500 to 900°C before entering the growth chamber where films are formed.

Sputtering deposition differs from high-vacuum techniques because a rare gas such as argon is always moving in the vacuum chamber. Gas flow and a throttle valve determine gas pressure in the chamber. The rare gas is ionized in the electrical field formed by the substrate and diode or magnetron target in the chamber, creating an ionized plasma with an overall neutral charge. The accelerated ions strike the target causing its atoms to be ablated, and the target atoms are directed towards the substrate. Technologists often choose sputtering to create thin films from refractory sources including tungsten because deposition can occur at less than the materials' normal melting point. Sputtering also appeals because technologists can design films similar to alloy materials or compound sources. This method efficiently shifts atoms from targets to substrates to produce a film almost identical to its source. Evaporation deposition techniques are not as consistent because fluctuating vapor pressure affects how source material is deposited.

Unlike PVD methods where material is removed from a solid target, chemical vapor deposition (CVD) uses reactive carrier gases to form new material on a heated substrate. These gases either break down into reactive precursors or interact, and the resulting materials coat substrates. CVD offers chemists the capability to create a large variety of thin films suitable for numerous uses. CVD technologists use a reactor that provides the energy necessary to cause chemical reactions to deposit material to form films. They have innovated processes for specific material sources such as metal-organic CVD for organometallic materials used in semiconductors. Plasma is the energy source for reactions in plasma-enhanced CVD, in which vaporized compounds in the plasma trigger the coating of substrates. A vacuumless process, atmospheric pressure CVD moves substrates on a belt through the chamber to deposit materials for films used to coat silicate glasses. In this technique, coating colors can be achieved by the selection of specific compound sources. Diamond thin films that increase the surface hardness of cutting tools have been a popular product of CVD methods. Based on work patented by American William G.

Eversole in the 1950s and developed in England by John C. Angus, H.A. Will, and W.S. Stanko in the next decade, CVD methods have continued to improve throughout the late twentieth century for optical and electronic usages.

Ion beam deposition (IBD) refers to several interconnected processes that create predictable quality and characteristics in thin films. Kasturi L. Chopra and M. R. Randlett first used ion beam sputtering to make thin films in the mid-1960s. Chemists appropriated techniques for IBD, using ion beams that are usually broad and high energy to prepare substrates prior to deposition by removing contaminants. They then utilize ion beams to deposit materials by sputtering techniques as described above or as an aid for other methods to achieve optimal film production. Typically, the ion beam is aimed at a metal. As a result, target material sputters onto substrates, forming thin films. Often, an additional ion source known as the ion assist source (IAD) provides energy in ions that thicken and stabilize the films' surfaces and enhance their strength and capabilities. Engineers have appropriated IBD to reinforce magnetic heads with carbon films.

See also **Ceramic Materials, Coatings, Pigments and Paints; Crystals, Synthetic; Integrated Circuits, Fabrication, Materials and Industrial Processes, Nanotechnology, Materials and Applications, Optical Materials, Semiconductors, Crystal Growing and Purification; Transistors**

ELIZABETH D. SCHAFER

Further Reading

Elshabini-Riad, A.A.R., and Barlow III, F.D. *Thin Film Technology Handbook*. McGraw-Hill, New York, 1998.

Francombe, M.H. and Vossen, J.L., Eds. *Thin Films for Emerging Applications*. Academic Press, Boston, 1992.

George, J. *Preparation of Thin Films*. Marcel Dekker, New York, 1992.

Hopwood, J.A., Ed. *Ionized Physical Vapor Deposition*. Academic Press, San Diego, 2000.

Mahan, J.E. Physical Vapor Deposition of Thin Films. Wiley, New York, 2000.

Mattox, D. *Foundations of Vacuum Coating Technology*. Noyes, Norwich, NY, 2003.

Ohring, M. *The Materials Science of Thin Films*. Academic Press, Boston, 1992.

Ramesh, R. *Thin Film Ferroelectric Materials and Devices*. Kluwer, Boston, 1997.

Schuegraf, K.K., Ed. *Handbook of Thin Film Deposition Processes and Techniques: Principles, Methods, Equipment, and Applications*. Noyes, Park Ridge, NJ, 1988.

Smith, D.L. *Thin-Film Deposition: Principles and Practice*. McGraw-Hill, New York, 1995.

Venables, J. *Introduction to Surface and Thin Film Processes*. Cambridge University Press, Cambridge, 2000.

Wasa, K. and Hayakawa, S. *Handbook of Sputter Deposition Technology*. Noyes, Park Ridge, NJ, 1992.

Timber Engineering

Timber engineering is the technology of creating wood products not found in nature. Manufactured lumber has characteristics superior to those found in its individual components.

Glued layers of hardwoods or veneers were used for decoration by the ancient Egyptians. The first plywood made from layers of softwood was developed in the early twentieth century. In 1905, the directors of Portland, Oregon's Lewis & Clark Exposition asked the Portland Manufacturing Company to devise for display some new and unusual wood product. To bring attention to the region's rich timber resources, the company manufactured the first Douglas fir plywood.

Appeal for the product was immediate and worldwide in scope. Mills everywhere produced thin rectangular sheets of the lightweight wood product. Assembled so that the grain of each ply alternated direction by 90 degrees, it was strong, warp resistant, dimensionally stable, and did not split. It was useful in such applications as door panels, drawer bottoms, crates, trunks, and partitions. If the material had one shortcoming, it was the tendency to delaminate when exposed to dampness. Adhesives were not waterproof and early plywood was limited to interior or protected exterior use.

Contemporary to the development of softwood plywood was the refinement of glued laminated lumber or glulam technology. The process, which allowed the use of thinner (younger) trees, produced beams, columns, and lumber to serve as structural elements. Although glued timbers were first used in Britain during the 1860s, the technology developed no further. A building boom in Germany around 1900 and a shortage of large size timbers prompted lumber mill owner Otto Hetzer of Weimar, to investigate methods for fabricating long beams from the materials at hand. Hetzer eventually received five patents for methods of assembling or laminating wood segments into beams of great length.

The most practical system was a laminate of individual narrow planks no more than 5 centimeters thick. Beams were made thicker and wider by adding planks. Assembled with their grain running parallel, the slender "lams" were relatively pliable and could be bent into job-specific shapes during gluing. Although no metal fasteners were used to join the lams, finished lumber had great strength as well as length. Thus, glulam lumber was useful in extended clear span structures.

The first public demonstration of the technique was at the Brussels World's Fair in 1910. The technology spread quickly throughout Germany and was later adopted in Switzerland, Sweden, the Netherlands, and Italy. Although the U.S. investigated the system in 1920, it was 1934 before the first laminated lumber structure was erected there. Thereafter, the technology was embraced for public structures nationwide. Glulam had much in its favor: it demonstrated an economic use of materials and made possible new innovative building designs; prefabricated components meant faster assembly and cost saving. Over the years the technology changed little.

Plywood's inherent strength and lightness lent itself well to the construction of aircraft, which were made primarily of wood. In 1915 the LWF (laminated wood fuselage) Engineering Company of New York City fabricated one of the first molded wood aircraft fuselages. Aircraft builders experimented with plywood for airframes, wings, and tails throughout the 1920s and 1930s.

Adhesives, critical to the growing engineered lumber industry, varied little from those used in previous centuries. During the 1920s, glues were still derived from animal and vegetable matter. Bone and hide wastes, casein from cow's milk, and in 1926 soy beans were among the materials used to produce glue.

The 1930s brought several important breakthroughs in timber engineering technology. Heat along with pressure became part of the adhesive curing process. A significant turning point was reached with the introduction of two resin adhesives made from synthesized formaldehyde gas. Water-resistant urea formaldehyde and waterproof phenol formaldehyde adhesives dramatically increased the potential uses for engineered lumber.

Worldwide shortages of vital metals during World War II placed plywood in an important position in the military industries of several nations. Undoubtedly, the best known and most successful aircraft was the British Mosquito bomber. Designed with twin engines and capable of speeds of over 600 kilometers per hour, the fuselage was assembled from molded plywood halves. Plywood was used by the Allies in their rush to produce naval vessels. The coastal forces of Britain's Royal Navy employed a fleet of plywood motor-torpedo and motor-gun boats (MTB and MGB). American-built Patrol Torpedo (PT) boats while not made entirely of plywood, used much in their construction. Each steel-hulled American-built Liberty ship contained 2325 square meters of plywood. Landing craft used to deliver vehicles and personnel to the beaches of Normandy were part of a fleet of 20,000 such plywood vessels made by the Higgins Company of the U.S. The Axis powers used plywood as well, perhaps most notably in Japan's lightweight explosive laden plywood suicide boats.

Since the end of World War II plywood's use rose dramatically, primarily in construction. Plywood subflooring, decking, siding, and roofing revolutionized the construction of light frame buildings, especially in the U.S.

In the postwar era, particle board also became a component of the construction industry. Particle board was first produced early in the century, but its commercial production dates only from the 1940s. There was a definite economic benefit in boards made of wood chips, shavings, and trimmings as well as agricultural fibers. Bonded with water-resistant adhesive it was used most often for interior work.

The shortage of large timber first observed at the beginning of the century was even more acute at the end. What was formerly considered waste was turned into reliable, highly predictable dimensioned lumber of uniform composition.

Several new engineered lumbers appeared late in the century. Panels of oriented strand board (OSB) were manufactured from rectangularly shaped wood strands arranged in layers with alternating grain. Manufactured like plywood, they had common characteristics. Chipboard panels were yet another lumber made from what was a formerly waste. Particles of resin-coated softwood were bonded together by heat and high pressure. Because it was unable to tolerate dampness, it too was used primarily for interior work.

During the 1970s, new applications were devised for laminated veneer lumber (LVL). Made from plys or veneers, the laminations are glued and processed to form material with the dimensions of sawn lumber. While built up of thin plys as in plywood, it differs in that all grains are aligned. The segments can be assembled to produce material whose finished length is limited only by the size of the machines used to manufacture it and the means of transporting it.

With ever-dwindling supplies and an expanding market for lumber products, new and ingenious methods continue to be devised to utilize available wood to its fullest extent.

See also **Adhesives; Buildings, Prefabricated**

WILLIAM E. WORTHINGTON, JR.

Further Reading

Andrews, H.J. *An Introduction to Timber Engineering*. Pergamon Press, Oxford, 1967.
Baldwin, R.F. *Plywood Manufacturing Practices*. Miller Freeman Publications, 1975.
Bulkley Jr., R.J. *At Close Quarters: PT Boats in the United States Navy*. U.S. Government Printing Office, Washington D.C., 1962.
Cour, R.M. *The Plywood Age, A History of the Fir Plywood Industry's First Fifty Years*. Binsford & Mort, Portland, 1955.
Levin, E. Ed. *Wood in Building*. Architectural Press, London, 1971.
Mondey, D., Ed. *The Hamlyn Concise Guide to British Aircraft of World War II*. Hamlyn Aerospace, London, 1982.
Muller, C. *Holzleimbau, Laminated Timber Construction*. Birkhauser, Basel, 2000.
Perkins, N.S. *Plywood: Properties, Design, and Construction*. Douglas Fir Plywood Association, Tacoma, 1962.
Sellers Jr., T. *Plywood and Adhesive Technology*. Marcel Dekker, New York, 1985.

Tissue Culturing

The technique of tissue or cell culture, which relates to the growth of tissue or cells within a laboratory setting, underlies a phenomenal proportion of biomedical research. Though it has roots in the late nineteenth century, when numerous scientists tried to grow samples in alien environments, cell culture is credited as truly beginning with the first concrete evidence of successful growth *in vitro*, demonstrated by Johns Hopkins University embryologist Ross Harrison in 1907. Harrison took sections of spinal cord from a frog embryo, placed them on a glass cover slip and bathed the tissue in a nutrient media. The results of the experiment were startling—for the first time scientists visualized actual nerve growth as it would happen in a living organism—and many other scientists across the U.S. and Europe took up culture techniques. Rather unwittingly, for he was merely trying to settle a professional dispute regarding the origin of nerve fibers, Harrison fashioned a research tool that has since been designated by many as the greatest advance in medical science since the invention of the microscope.

Indeed, after this initial experiment, Harrison left it to others to take up the mantle of culturing tissue; and it is Alexis Carrel who became inextricably associated with the technique in its early history. Carrel, a French surgeon, worked at the Rockefeller Institute in New York, where he cultured *in vitro* the cells of many warm-blooded animals. Generally, from 1910 until 1920, research into tissue culture consisted primarily of simply determining how many tissues could survive in cultures and for how long. In 1912, Carrel claimed to have established an immortal strain of chick cells—a claim that attracted a huge amount of international press attention and alerted the public to the potential of this technique for the first time.

By contrast, many scientists continued to lament both the lack of medically relevant research derived from cells in culture and the difficulty of getting cultures to survive for any length of time *in vitro*; for many years opinion remained divided on the method's relevance and prospects. However, a number of technical advancements in the 1940s began to sway scientific opinion. A major breakthrough occurred in 1943 when Wilton Earle, working at the National Institutes of Health, developed the first permanent mammalian "cell line." This term applies to a culture of cells that either through innate or induced malignancy does not die *in vitro*, enabling scientists to work with a standardized model that can be easily distributed. After World War II widespread use of antibiotics prevented the hitherto disastrous problem of continual bacterial infection and made cell survival in culture far less difficult to attain. No less important were the development of optimal, artificial growth media in the U.S. in the early 1950s and improved methods such as nitrogen freezing for long-term storage of cultures.

Subsequently, cell culturing was used in researching most medical problems, replacing the traditional animal models in experiments to determine the cellular basis of cancer. The International Tissue Culture Association was founded in 1946, which aimed to foster an international community to disseminate findings and also train those lab workers who now chose to work on cultures. Thus, at the beginning of the 1950s, scientific unanimity had been reached on the technique's value.

In 1951, researchers at Johns Hopkins University in Baltimore, Maryland, achieved the first human cell line—HeLa—derived from the cervical tumor of an American woman named Henrietta Lacks. This cell line was very widely used, but neither Lacks nor her family learned of HeLa until well after her death, a distressing episode that the Clinton administration formally acknowledged in 1997. As well as their use in research on cancer, cultures began to be inoculated with viruses to produce vaccines against human diseases. Indeed, HeLa cells became the first to be produced on an industrial scale at Tuskegee, Alabama, for the purposes of producing polio antibodies. Such was the demand for mass-produced HeLa that Tuskegee scientists distributed over 600,000 cultures within two years. Other researchers followed suit, and cell cultures rapidly became the experimental model of choice for virologists. However, HeLa possessed a remarkable propensity to contaminate and colonize other cultures, which ultimately undermined millions of dollars of research on what transpired to be falsely labeled HeLa cells, including a much vaunted collaborative project between the U.S. and USSR in the 1960s as part of Richard Nixon's so-called war on cancer.

In 1962, a centralized store of cell cultures was established at the American Type Culture Collection in Washington in order to ensure ready access to a wide variety of cultures with (crucially, in the light of the HeLa fiasco) defined and documented characteristics. In 1975, the value of cultures increased further, with the production of specialized hybridoma cells by Milstein and Köhler, in Cambridge, England. These cells, the product of fusion between a cancer cell in culture and an antibody-producing lymphocyte cell, produce high quantities of a specified antibody—a process that is simultaneously of immense scientific and commercial value. Indeed, in some cases it has been estimated that the products secreted by certain cells in culture can be worth up to $3 billion, so it is little wonder that commercial firms value the technique as highly as academic researchers.

From the 1980s, cell culture has once again been brought to the forefront of cancer research in the isolation and identification of numerous cancer causing oncogenes. In addition, cell culturing continues to play a crucial role in fields such as cytology, embryology, radiology, and molecular genetics. In the future, its relevance to direct clinical treatment might be further increased by the growth in culture of stem cells and tissue replacement therapies that can be tailored for a particular individual. Indeed, as cell culture approaches its centenary, it appears that its importance to scientific, medical, and commercial research the world over will only increase in the twenty-first century.

DUNCAN L. WILSON

Further Reading

Brown, R.M. and Henderson, J.H. M. The mass production and distribution of HeLa cells at Tuskegee Institute, 1953–1955. *J. Hist. Med. Allied Sci.*, 38, 415–431, 1983.

Freshney, I.R. *Culture of Animal Cells: A Manual of Basic Technique*. Alan R. Liss, New York, 1983.

Gold, M. *A Conspiracy of Cells: One Woman's Immortal Legacy and the Medical Scandal It Caused.* State University of New York Press, Albany, 1986.

Landecker, H. Immortality *in vitro*: A history of the HeLa cell line, in *Biotechnology and Culture: Bodies, Anxieties, Ethics*, Brodwin, P., Ed. Indiana University Press, Bloomington, IA, 2000.

Landecker, H. New times for biology: nerve cultures and the advent of cellular life *in vitro*. *Stud. Hist. Phil. Boil. Biomed. Sci.*, 33C, 4, 667–694, 2002.

Masters, J.R. W. Human cancer cell lines: fact and fantasy. *Nat. Rev.: Mol. Biol.*, 1, 3 233–236, 2000.

Witkowski, J.A. Dr. Carrel's immortal cells. *Med. Hist.*, 24, 129–142, 1980.

Useful Websites

American Type Culture Collection: http://www.atcc.org

Tomography in Medicine

The word tomography derives from the Greek word *tomos* meaning a section, and the technique was proposed by Gustave Grossman in papers published in 1935 describing the experimental details of tomography and its clinical application. A tomographic image is essentially a picture of a slice of the patient's anatomy. Developments in x-ray tomography culminated at the end of the twentieth century with the invention of computed tomography (CT), which has found numerous applications in diagnostic medicine including scanning for cancer, subdural hematomas, ruptured disks and aneurysms. It substantially reduced the requirement for exploratory surgery. Tomographic techniques are also employed in other diagnostic imaging modalities such as ultrasonography. Early B-Mode ultrasound scanners were mounted a single transducer element on an articulated arm to produce an acoustic tomographic slice of the body. The arm was moved through a series of positions to allow the "slice" of data to be collected.

In standard x-ray imaging, three-dimensional patient information is projected onto a two-dimensional surface. Each point on the image is a projection of the attenuating properties of all the tissues along the x-ray beam's path. Tomographic techniques sought to examine a section through the body ignoring the tissue above and below the section of interest. In early x-ray tomography, the x-ray tube traveled in an arc above the patient while the film cassette in the table, beneath the patient, traveled in the opposite direction so that x-rays remain incident on the film throughout the exposure (Figure 9). The patient was positioned so

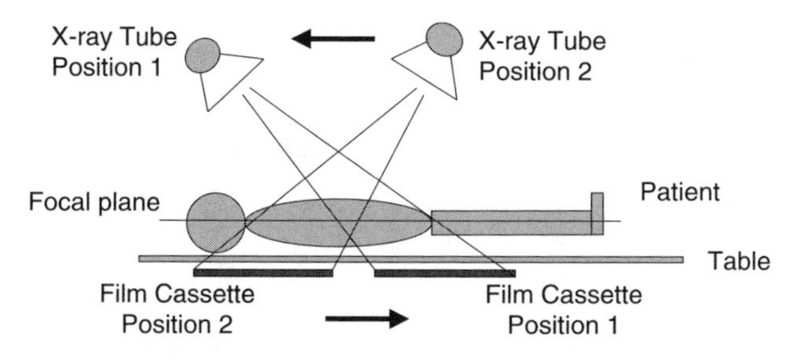

Figure 9. Motion of the x-ray tube and film cassette in x-ray tomography.

that the area of interest within the body was at the focal plane of the x-ray tube. When the x-ray tube moved through the tomographic angle, objects above and below the focal plane were blurred, thereby largely obliterating their contrast. As a result, the objects at the focal plane were seen more clearly than objects above and below the plane. The tomographic angle determines the slice (depth of objects in the focal plane) thickness. As the out-of-plane anatomy is not removed from the image, but blurred out, the effective image contrast is reduced. The clinical importance of tomography was quickly established, and the first atlas of clinical tomograms was compiled in 1940.

The development of CT followed from the work of numerous scientists who had built a variety of apparatus that aimed at true section imaging, which is achieved when the presence of planes above and below the image are of no consequence. Gabriel Frank (1940) showed how a transverse axial section can be sharply imaged without contribution from adjacent sections. Images were reconstructed by a form of optical back-projection, a method that emulates the image acquisition sequence in reverse. This was a form of CT without the computing, three decades before CT was developed. A Japanese radiologist, Shinji Takahashi, conducted a series of experiments in the 1940s and 1950s which also anticipated the later developments in CT. W.H. Oldendorf (1961) designed an apparatus to measure radiodensity at a point, uncoupling the effects of all other points in the same plane. This was achieved by arranging that the signals were at different frequencies. Oldendorf is often credited with producing the first laboratory CT of a "head" phantom.

Development of emission tomography methods overlapped with the development of transmission CT. Tomographic nuclear medicine was studied in the 1960s and developed into computerized tomography with the invention of single photon emission computed tomography (SPECT) in the 1970s. Experimental work leading to the development of modern positron emission tomography (PET) systems began in the 1950s.

In 1972 Godfrey Hounsfield, a British engineer employed by Electric and Musical Industries (EMI), developed the first CT system. CT technology quickly evolved through a series of generations with most of the initial developments focusing on increasing the scanning speed and thus reducing problems with patient motion. Developments in computing complemented these advances enabling rapid reconstruction of images from the huge data sets obtained. Hounsfield's scanners included what have become known as first, second, and third generation scanners developed between 1972 and 1977. Although numerous designs of scanners were patented (particularly in later generation scanners) the following broadly applies to the designs.

First generation systems used an x-ray tube, which emitted pencil beam x-rays. Transmitted x-rays were measured by a single detector (per slice). The tube and detector were translated across the object to acquire x-ray transmission data (Figure 10). The detector and x-ray tube were then tilted at an angle of 1 degree, and the translation process was repeated until the system had completed a 180-degree rotation and a complete set of CT data was obtained. Approximately 30,000 projections were taken with scanning times running to several minutes per slice.

Second generation CT scanners employed a linear array of 30 detectors to collect x-ray transmission data. The x-ray beam was collimated to a narrow fan. Second generation scanners did not have as good scatter rejection as first generation but slice acquisition time was as much as 15 times faster. Additional data detection capability was used to increase speed of acquisition and image quality.

By the mid-1970s third generation scanners had been developed, and these were the first fast medical scanners. More than 800 detectors were employed, arranged in an arc encompassing the width of the patient. This eliminated the need for translation motion. In this design both tube and detectors rotate together inside a gantry to acquire the necessary projections. Scan times per slice were reduced to approximately 5 seconds.

In the late 1970s, fourth generation CT scanners were built with approximately 5000 individual detectors arranged in a 360-degree ring around the patient (Figure 11). The detectors remain stationary while the tube is rotated.

Obtaining CT images requires taking x-ray transmission measurements at numerous angles

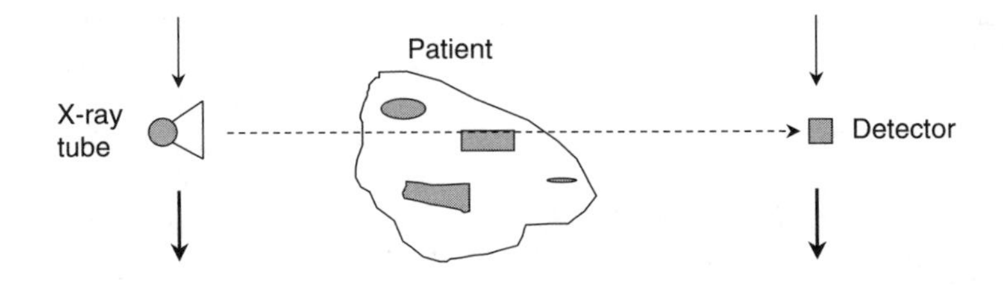

Step 1: A series of data is acquired as x-ray tube and detector translate along the object

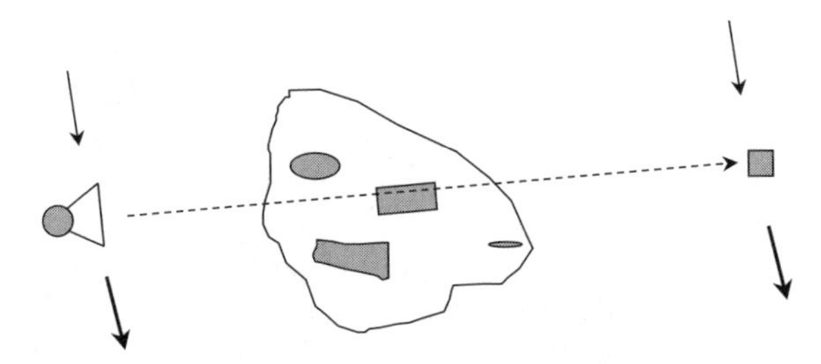

Step n: A rotation on n degrees and translation data acquisition repeated

Figure 10. First generation computed tomography (CT).

around the object. This is roughly analogous to precisely locating the seeds inside an orange by probing it from many points with a thin wire. The image may be built up through a back-projection technique, which is essentially a mathematical method for establishing the geometry of the object based on data collected from x-rays taken at multiple angles through the object.

The mathematical principles underlying CT were developed by an Austrian mathematician, Johann Radon, in 1917. Radon theoretically demonstrated that an image of an unknown object could be reproduced from an infinite number of projections through the object. Images are reconstructed from the raw data through sophisticated algorithms. This involves processing 800,000 or so projections for each picture element (pixel) displayed, as well as applying image-processing algorithms. Filtered back projection is a common algorithmic image reconstruction technique. The process emulates the acquisition procedure in reverse and convolves the solution with a filter of kernel to correct for known errors in the reconstruction process and to emphasize particular characteristics in the image. Different kernels are available for types of soft tissue and bone and may be selected by the operator.

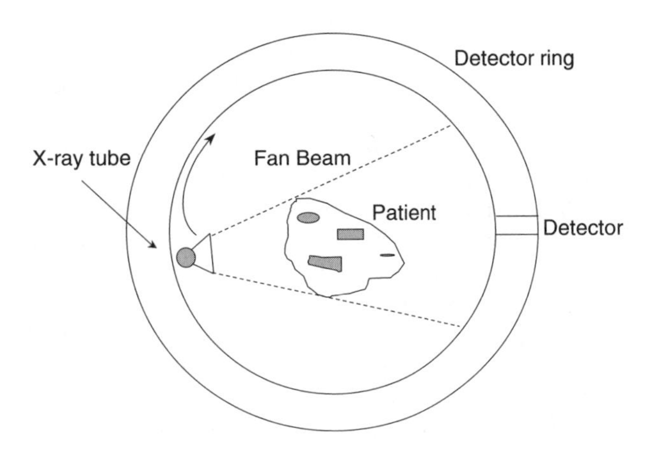

Figure 11. Fourth generation scanner. The detectors are fixed in a ring around the patient, and the x-ray tube rotates during acquisition.

Digital technology for display of images became prominent in the 1980s. Image displays for CT typically have a 512 by 512 display matrix with each pixel in the matrix having a distinct value (or CT number) based on the calculated attenuation value for the corresponding volume element (or voxel) in the patient. Images are usually presented as a series of two-dimensional images on a monitor, with each image representing a slice through the patient. Computer programs can also produce three-dimensional reconstruction of CT images.

The development of CT has been compared in significance to Wilhelm Roentgen's discovery of x-rays. Hounsfield won the Nobel Prize for Medicine rather than Physics, reflecting the considerable clinical benefits in producing high-contrast images of slices through a patient. Developments in CT technology through the 1980s included the helical CT scanners that acquire information while the table moves the patient through the ring of detectors. The effect of this linear translation through a rotating acquisition is to produce a helical or spiral motion of the x-ray tube around the patient. Helical scanners increase the efficiency of multislice acquisition and reduce overall scanning time.

The 1990s saw the development of multidetector arrays. In these systems slice width is determined by the way the detectors are grouped together rather than by beam collimation, as was the case in earlier generations. Multidetector arrays bring CT imaging systems closer to acquiring a full three-dimensional data set by reducing slice thickness to about 0.5 millimeters.

See also **Positron Emission Tomography (PET); Ultrasonography in Medicine; X-Rays in Diagnostic Medicine**

COLIN WALSH

Further Reading

Bushberg, J., Seibert, J.A., Leidholdt, E.M. and Boone, J.M. *The Essential Physics of Medical Imaging*, 2nd edn. Lippincott Williams & Wilkins, Philadelphia, 2002.

Curry, T.S., Dowdey, J.E. and Murry, R.C. *Christensen's Physics of Diagnostic Radiology*. 4th edn. Lea & Febiger, 1990.

Goodenough, D.J. Tomographic imaging, in *Handbook of Medical Imaging, Volume 1. Physics and Psychophysics*, Beutel, J., Kundel, H.L. and Van Metter, R.L., Eds. SPIE Press, 2000.

Grossman, G. Lung tomography. *Brit. J. Radiol.*, 8, 733, 1935.

Hounsfield, G. Computerized transverse axial scanning (tomography), part 1: description of system. *Brit. J. Radiol.*, 46, 1023, 1973.

Mould, R.F. *A Century of X-Rays and Radioactivity in Medicine*. Institute of Physics Publishing, Bristol, 1993.

Swindell, S. and Webb S. X-ray transmission computed tomography, in *The Physics of Medical Imaging*, Webb S., Ed. Institute of Physics Publishing, Bristol, 1988.

Webb, S. *From the Watching of Shadows: The Origins of Radiological Tomography*. Institute of Physics Publishing, Bristol, 1990.

Wegener, H.O., *Whole Body Computed Tomography*, 2nd edn. Blackwell Scientific, 1993.

Useful Websites

http://www.nobel.se/medicine/laureates/1979/hounsfield-autobio.html

Transistors

In 1906, the American inventor Lee de Forest developed a triode, a three-element vacuum tube (or "thermionic" valve). Dubbed the de Forest Audion, it was a device that could detect and electronically amplify radio and telephone signals. In 1909 the American company Bell American Telephone & Telegraph (AT&T) bought de Forest's patent and improved the tube so that it could be used to amplify signals in long-distance telephony. A practical problem was that the vacuum tubes were often unreliable, slow, used too much power, and produced too much heat. For years, researchers in Western countries tried to make a solid-state amplifier—what became the transistor—in an attempt to enable the creation of smaller, faster, less power-hungry electronics.

Early experiments in transistor technology were based on the analogy between the semiconductor and the vacuum tube: the ability to both amplify and effectively switch an electrical signal on or off (rectification). By 1940, Russell Ohl at Bell Telephone Laboratories, among others, had found that impure silicon had both positive (*p*-type material with holes) and negative (*n*-type) regions. When a junction is created between *n*-type material and *p*-type material, electrons on the *n*-type side are attracted across the junction to fill holes in the other layer. In this way, the *n*-type semiconductor becomes positively charged and the *p*-type becomes negatively charged. Holes move in the opposite direction, thus reinforcing the voltage built up at the junction (Figure 12). The key point is that current flows from one side to the other when a positive voltage is applied to the layers ("forward biased").

The transistor is also a solid-state device that can amplify electrical current. Transistor stands for transit resistor. Its development went along two lines: basic research in solid-state physics to replace the old vacuum tubes (such as de Forest's Audion), which failed to solve technological problems, and multidisciplinary research activities at several industrial and university research labs. It was in December 1947 that John Bardeen and Walter Brattain working at Bell Telephone Laboratories in New Jersey, in a research team headed by

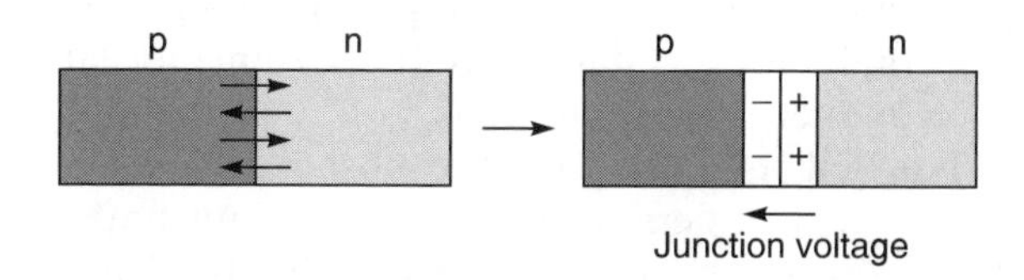

Figure 12. *p–n* junction between two semiconducting materials. Near the junction, electrons on the *n* side diffuse to the *p* material on the left and recombine with holes in that side, and holes diffuse from the *p* side to the *n* material on the right, recombining with electrons. At equilibrium a thin depletion region forms, with an overall negative charge in the *p* material and an overall positive charge in the *n* material. A junction voltage develops across the depletion region.

William Shockley, demonstrated the first transistor, a semiconductor device based on germanium. However, the German scientist Julius E. Lilienfeld from New York had patented the first field-effect transistor in 1926. It was a patent on a "method and apparatus for controlling electric currents." It was unlikely, however, that he ever got it to work. Nevertheless, in the early 1930s solid-state physics, and later semiconductor technology, was a promising research field for a broad range of researchers. Semiconductors were interesting to the radio and telephony industry because of their ability to rectify electrical current (allowing current to flow in one direction and not the other) and they were useful as electronic switches.

In the late 1930s, Bell's director of research Mervin Kelly recognized that a better amplifier was needed for the telephone business. He gave a group of researchers headed by Shockley the freedom to carry out scientific work in the field of solid-state physics. In 1939 Shockley further developed the principle of the field-effect transistor (FET), in which instead of the wire "grid" of de Forest's triode, an electric field controlled the stream of charge carriers between electrodes. This principle started a line of inquiry that led to new experiments. However, Shockley himself went off in other directions and was barely involved in the further experimental research of the Bell group. Meanwhile Robert W. Pohl and Rudolf Hilsch from Gottingen University made a solid-state amplifier in 1938 using salt as the semiconductor. It was a functioning device, but reacted to signals too slowly. Karl Lark-Horovitz and his research team at the physics department of Purdue University, Indiana also became involved in solid-state physics, working on improving the crystal rectifiers that were used as radar detectors in World War II. The team at Purdue worked with both silicon and germanium crystals. Such crystal detectors had no signal gain, but the work on germanium and techniques of growing and doping semiconductor crystals were important to later semiconductor researchers.

The Bell researchers did the most extensive work on crystal rectifiers in the radar program both in the U.S. and the U.K. In 1945 John Bardeen, a theoretical solid-state physicist, joined Shockley's group and a semiconductor subgroup was formed within Bell Laboratories. Shockley filled out his team with a mix of physicists, chemists, and engineers. In this subgroup Walter Brattain was an experimental physicist. Bardeen and Brattain continued the research on Shockley's earlier design sketches for the field-effect transistor. The substitution of the Fleming triode tube (developed by Ambrose Fleming in 1904) by a solid-state device—the transistor—formed the most important outcome of this semiconductor subgroup's research efforts.

In their experiments they placed the electric circuit contacts on two strips of golden foil, since Bardeen suggested that greater amplification could be obtained by placing the two point contacts closer to each other. Bardeen also suggested replacing silicon with high-purity germanium that made better rectifying contacts. Germanium is an *n*-type semiconductor (excess of electrons), and when current flowed in from the gold foil contact, holes were "injected" into the germanium surface. This created a *p–n* junction as described above. In the junction, current started to flow from one side to the other. In the case of their little construction, current flowed towards the second gold contact. The outcome was that a small current changed the nature of the semiconductor so that a larger, separate current started flowing across the germanium and out of the second contact. In other words, a small current was able to alter the flow of a much bigger one, thus effectively amplifying it. The first device was called a point-contact transistor because the wires stood directly in contact with the semiconducting material (Figure 13). Later Shockley developed the junction transistor (also called the sandwich transistor), of which there are two types, called *p–n–p* and *n–p–n* (depending on which material forms the inside layer). The field-effect transistor was not built until the 1960s, but today most transistors are field-effect transistors.

In 1956 Shockley and his colleagues shared the Nobel Prize for their invention of the point-contact transistor. Inner competition broke the Bell Lab team apart, but their invention was of great importance for Bell, of which numerous patents and licenses with amongst others General Electric, IBM, Texas Instruments, Philips, and later Sony Electronics bear witness.

Following Bells announcement in 1948 of the first working transistor, the transistor quickly became popular in industry as Bell licensed their transistor, and the first commercial product with transistors—a hearing aid—was sold by Raytheon in 1952. Military applications, as a replacement for the vacuum tube in communications and computing, swiftly followed. Transistors began to replace fragile vacuum tubes in consumer electronic devices, and the first U.S. transistor radio was sold by Texas Instruments in 1954. Sony Electronics especially was able to mass-produce miniaturized transistor radios from 1957. The

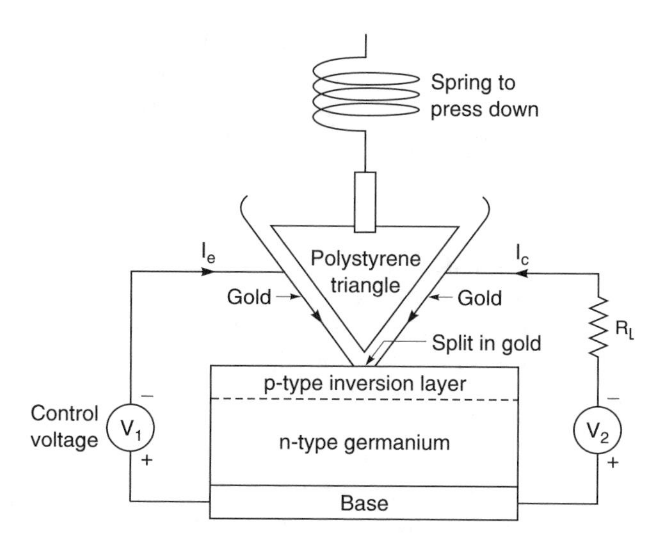

Figure 13. The first point-contact transistor, developed by Walter Brattain and John Bardeen in 1947. A single strip of gold foil over a plastic triangle is split to make two contacts. The triangle lightly touches the surface of a crystal of germanium, which sits on a metal plate attached to a voltage source. Current following from one gold contact to the other can be controlled by current introduced into the germanium. The actual device was about half an inch high. [*Reproduced with permission from Hoddeson, L. Innovation and basic research in the industrial laboratory: the repeater, transistor and Bell Telephone System, in Between Science and Technology, Sarlemijn, A. and Kroes, P. Eds. North Holland, Amsterdam, 1990.*]

transistor became the key to further developments in electronic technology and the consumer electronics industry. Considerable developments were made in the 1950s as a result of open sharing of technology between various industrial and university labs. These developments, like the means of introducing dopants (impurities) to very shallow depths using vapor phase diffusion, the use of silicon dioxide as a diffusion mask, and an all-diffused silicon transistor enclosed in oxide, led to a wide range of transistors.

A practical problem remained, however. Like the elements in the vacuum tubes, the electric components that formed the transistor needed to be soldered together. The more complex the electric circuits became, the more complicated the construction of the transistor. Computer technology in particular needed complex circuits. Because of this problem, practical application of transistors was slowed down. However, in 1958 Jack Clair Kilby of Texas Instruments developed the first integrated circuit or chip using some key achievements from the 1950s transistor research activities. His invention combined a collection of transistors arranged on a single chip of silicon in order to save space. This was the first step to integrated circuits that replaced individual transistors in computers—a refinement that led to the development of the modern microprocessor.

Figure 14. RCA transistor 1948.
[*Courtesy of the David Sarnoff Library.*]

See also **Integrated Circuits; Radio Receivers, Early; Radio Receivers, Valve and Transistor Circuits; Semiconductors; Valves/Vacuum Tubes**

KEES BOERSMA

Further Reading

Bardeen, J. The early days of the transistor. *Proc. Stocker Symp*., Raju, G.V.S., Ed., 1979, pp. 3–10.

Brattain, W.H. Genesis of the transistor. *Phys. Teacher*, 6, 108–114, 1968.

Braun, E. and Macdonald, S. *Revolution in Miniature: The History and Impact of Semiconductor Electronics*. Cambridge University Press, Cambridge, 1978.

Brinkman, W., Haggan, D. and Troutman, W. A history of the invention of the transistor and where it will lead us. *IEEE J. Solid-State Circuits*, 32, 12, 1997.

Hoddeson, L. Innovation and basic research in the industrial laboratory: the repeater, transistor and Bell Telephone System, in *Between Science and Technology*, Sarlemijn, A.,P., Ed. North Holland, Amsterdam, 1990.

Riordan, M. and Hoddeson, L. *Crystal Fire: The Birth of the Information Age*. W.W. Norton, London, 1997.

Shockley, W. The theory of *p–n* junctions in semiconductors and *p–n* junction transistors. *Bell Systems Tech. J.*, 27, 435–489, 1949.

Useful Websites

Brinkman, W., Haggan, D. and Troutman, W. A history of the invention of the transistor and where it will lead us: www.sscs.org/AdCom/transistorhistory.pdf

http://www.pbs.org/transistor

Transport

During the twentieth century, human beings witnessed a revolution in transportation technology. The way in which people and goods moved around the world changed dramatically. The nineteenth century, of course, also had its transportation revolution, as railroads and steam ships became dominant, and these forms of mass travel did not disappear during the twentieth century. Transport by sea continued to play a key role in the movement of people and cargoes, as ships became bigger and faster, culminating in giant super tankers used to transport oil around the world. Railroads also evolved, shifted to electric and diesel power. By the late twentieth century, high-speed trains whisked passengers to their destinations at speeds of up to 270 kilometers per hour.

The internal combustion engine led to widespread use of automobiles, buses, and trucks. Trucks, for example, would provide fierce competition for the railroads in the movement of goods across land. The development of aircraft also transformed travel around the globe. The use of airplanes to carry passengers and cargoes greatly reduced travel time. Journeys across the oceans were now measured in hours rather than in days and weeks.

Water Transport

Transportation over water continued to play an important role in the twentieth century. Intense competition among shipping lines had begun in the mid-nineteenth century as immigration from Europe to the Americas grew. Shipping lines sought to build bigger and more luxurious ocean liners in hopes of attracting first- and second-class passengers. By the early twentieth century, the Hamburg–America Line was the largest shipping company, covering 75 routes with 500 vessels. Rivalries among these shipping lines meant that the companies continually put bigger and faster ships into service, leading to much turnover in their fleets, as older vessels quickly became outdated. This competition also led to a price war in the years before World War I. Lower prices led to a notable increase in emigrant traffic in this period.

At the turn of the century, the U.S. sought to become more involved in the transatlantic shipping business. In 1902, J.P. Morgan formed the International Mercantile Marine Company. Morgan either bought or made cooperative agreements with numerous existing shipping lines. His biggest competitor was the Cunard Steamship Company, a British line. In order to combat Morgan, the Cunard line received subsidies and loans from the British government. It was then able to construct two 35,000-ton liners and in 1903 inaugurated direct service from Italy to New York. The Morgan–Cunard rivalry led to a rate war in 1904, which eventually led to the end of Morgan's venture in 1926.

Despite technological advances, sailing ships did not disappear immediately. However, by the early twentieth century, they had mostly been pushed to more marginal trade. There remained some important transoceanic routes for sailing ships, such as guano from South America, grain from Australia and Argentina, and timber from Africa and Asia. These sailing ships continued to operate where large cargo ships could fill their holds and find favorable winds. However, there were only a few holdovers that lasted until after World War II.

Most cargo was carried by tramp steamers, a business dominated by the British. These tramp steamers could carry cargoes of up to 10,000 tons while traveling at 13 knots. They consumed 70 tons of coal per day. They were some 140 meters long and 16 feet wide and were manned by a crew of 75.

Shipping in the early twentieth century was aided by the construction of the Panama Canal. This canal stretches 65 kilometers from shoreline to shoreline across the Isthmus of Panama, connecting the Atlantic and Pacific Oceans. Along with the Suez Canal, the Panama Canal became one of the two most strategic artificial waterways in the world. The canal opened to traffic in 1914 and was controlled by the U.S. until 2000 when control reverted to Panama. The great advantage of the Panama Canal was that ships no longer had to carry their cargoes and passengers around Cape Horn at the southern tip of South America, thus shortening the journey between the east and west coasts of the U.S. by some 8,000 nautical miles (14,800 kilometers).

Another technological development in the movement of goods around the world by sea was the increased use of tankers. The first tankers had been used in the late nineteenth century to transport oil. There was much concern over the use of tankers, particularly due to the threat of explosions. For this reason, tankers were not allowed through the Suez Canal until 1902. As petroleum became an increasingly important product in the early twentieth century, tanker freights increased dramatically. At first, most tankers were owned by the oil companies themselves. Later, private companies became involved, leasing their tankers to the large oil firms.

World War II saw important developments in the movement of people and goods across the seas. In particular, shipbuilding in the U.S. was a key factor in the Allied victory. U.S. shipyards produced an enormous fleet that kept the Allies supplied with constant convoys of materials and troops across both the Atlantic and Pacific Oceans. The most important of the ships produced during the war were the Liberty ships, of which 2650 were built between 1941 and 1944. Among the advances seen in the construction of the Liberty ships was that large sections were produced and then welded together. They had steel decks, and were powered by reciprocating steam engines fed by oil-fired boilers.

The postwar years saw an important evolution of shipping technology, as larger tankers and bulk carriers made their appearance. These developments were linked to a significant increase in world trade following the war. Ships became much faster, reaching speeds of 20 knots. There was also much concern over appearance as esthetically pleasing ships were built. Shipbuilders also tried to streamline the loading and unloading of cargo, introducing for example the loading pallet in 1958.

Perhaps the key development in the movement of goods by sea in the postwar era was the fact that oil became the key to the world economy. By the early 1950s, about 40 percent of sea trade was in petroleum products and the total amount of crude oil and gasoline shipped by tanker nearly tripled in the decade. There was already a plentiful supply of 16,000 deadweight ton (dwt) tankers built during the war. However, as oil consumption continued to grow, there was increased demand for higher capacity tankers. Soon, shipbuilders were constructing 50,000 dwt supertankers, which greatly reduced the cost of transport. These supertankers led to many other technological changes, including the need for larger building berths in shipyards, larger dry docks in repair yards, the construction of new terminal berths and storage facilities by oil companies. Changes in shipbuilding were often led by the Japanese, who pioneered the mass production of tankers in their shipyards, using large cranes to lift huge subassemblies and taking the lead in the use of computers in shipbuilding. At first the 50,000- to 60,000-dwt tankers were seen as the limit in size, as larger tankers could not pass through the Suez Canal. However by 1970, there were tankers of 200,000 to 300,000 dwt.

Railroads

By the early-twentieth century, railroads had reached maturity. At the time of World War I, there were about 1.6 million kilometers of railroad track around the world, with some 25 percent located in the U.S. During the first half of the twentieth century, new railroad construction slowed in most of the developed world. However, in other parts of the world, the building of new railways continued. Such was the case in Canada, China, Russia, and parts of Africa. A prominent example of this new construction was the completion of the Trans-Siberian Railroad in Russia. Work had begun on the railroad in 1891. In 1904, all of the sections between Moscow and Vladivostok had been completed. However, Russia had secured permission from China to build part of the line across Manchuria. After the Russo–Japanese War in 1904–1905, Russia feared Japan might take over Manchuria, and proceeded to build an alternative route across Russian territory known as the Amur Railway. This section was completed in 1916.

Another impressive railroad engineering feat of the early twentieth century was the completion of the Simplon Tunnel in the Alps, which despite many serious problems was opened to railroad traffic in 1906. One of world's longest railway tunnels at 20 kilometers long, it connected Iselle, Italy to Brig, Switzerland. The Simplon Pass had been an important European trade route for centuries. In the 1890s, a German engineering firm undertook construction of the Simplon tunnel using many new tunneling techniques. Other technological developments to aid railroad transportation in the twentieth century included electric signaling systems and the widespread adoption of diesel traction. By the late 1930s, the use of diesel power had occurred in many places as it proved more reliable and efficient than steam power. In most railroad networks, diesel locomotives are more cost effective than electric ones, which is generally only profitable in very high-traffic areas such as metropolitan commuter systems.

Electrification became more widespread in the second half of the twentieth century. The French pioneered electrification of locomotives with a direct supply of high-voltage alternating current at the industrial frequency. These French developments in turn led to subsequent large-scale electrification programs in countries such as China, Japan, South Korea, the Soviet Union, and India. Such conversions to electric traction were also encouraged by the high price of oil during the 1970s.

As was the case in the early twentieth century, some countries expanded their railroad networks in the latter part of the century. This was particularly true in countries that were undertaking major industrialization projects and where railroads were still the principal form of moving people and goods. Examples include China, the Soviet Union, and India. These countries built new rail lines in order to increase their capacity to carry raw materials to industrial areas of their countries. For example, China doubled the length of its rail network between 1950 and 1990, including numerous routes of up to 800 kilometers that connected coal fields in the west to the ports on the east coast. During the same period, the Soviet Union built 32,000 kilometers of new rail lines. This new construction included the more than 3000-kilometer-long Baikal–Amur Trans-Siberian Railroad that was started in the late 1970s. During the 1990s, India also undertook major railroad construction.

In other parts of the world, there was less new construction. Here, there was an emphasis on improving rail communications, especially in light of growing competition from highway construction and air travel. Railroad companies sought to improve passenger amenities, develop larger and more specialized freight cars, design more sophisticated signaling and traffic control, and improve motive power.

Two major railway tunnels built in the second half of the twentieth century were Seikan Tunnel in Japan and the Channel Tunnel connecting England and France. The Saikan Tunnel connects the islands of Honshu and Hokkaido and is the world's longest tunnel at 53.8 kilometers. The Japanese National Railways sponsored construction of the tunnel between 1964 and 1988. The tunnel has quickly become of limited use, as air travel between the islands is faster and almost as cheap. Nearly as long at 31 miles, the Channel Tunnel, popularly called the "Chunnel," provides freight and passenger railroad service on trains that can travel up to 160 kilometers per hour. Digging began in 1987 and the tunnel opened in 1994.

The most significant development in rail transportation was an emphasis on high-speed trains, and Japan pioneered this development. In 1957, the Japanese government concluded that the old Tokyo–Osaka line could not be upgraded to meet the needs of the heavily populated and industrialized Tokaido coastal belt between the two cities. In 1959, work began on a 500-kilometer-long line for electric passenger trains. The first *Shinkansen* (New Trunk Line) opened in 1964. Initially able to travel at speeds of 210 kilometers per hour, the high-speed line was a commercial success. From the 1970s through the 1990s, Japan extended its network of high-speed trains, some of which reached speeds of 274 kilometers per hour. In Europe, France took the lead in developing a system of high-speed trains. The first *Train à Grande Vitesse* (TGV) between Paris and Lyon went into service in 1981, the result of some twenty years of research. France built and planned additional lines with an ultimate goal of connecting all major cities with Paris as well as connecting with the high-speed systems in neighboring countries. Other European nations, including Italy, Germany, and Spain have also developed high-speed trains.

Trucks

Railroads faced competition from commercial trucking throughout the twentieth century. Gottlieb Daimler built the first motor truck in Germany in 1896. In the U.S., truck transportation began at the turn of the century. The Winston Company of Cleveland, Ohio began building trucks in 1898, making it one of the first truck manufacturers in the U.S. In 1904, there were a mere 700 trucks in use. This figure had increased to more than 150,000 by 1915. The first transcontinental coast-to-coast truck trip occurred in 1911, completed in 66 days. Early trucks were used mainly for making local deliveries and limited commerce between cities. Their usefulness was hindered by poor roads and the dominance of railroads in shipping goods over long distances.

However, before trucks could compete effectively with railroads, the country needed a system of well-paved roads. In 1916, the U.S. government passed the Federal Aid Road Act, which emphasized the construction of high-quality hard-surface pavements and highways for motor transportation, which helped intercity commerce utilizing trucks. As with other technologies, World War I played a key role in the development of motor trucks. In 1914, annual U.S. truck production was 24,900. By 1917, the country was producing 128,000 trucks. Many of these trucks went to Europe for the war effort to move soldiers and supplies and serve as ambulances. Because railroad arteries were often clogged during the war, large truck caravans were organized to drive the new trucks to East Coast ports from assembly plants in the Midwest. The trucks were also loaded with goods, demonstrating the possibilities for long-distance trucking and also calling further attention to the need for system of highways.

Air Transport

The use of aircraft would drastically alter the way in which people and goods traveled around the world during the twentieth century. Experiments with flight were not new in the twentieth century. The French had pioneered the use of balloons in the late-eighteenth century. In the late nineteenth century, the German Otto Lilenthal successfully flew gliders. Others had begun experiments with lighter-than-air dirigibles. The German Count Ferdinand von Zeppelin carried out the most important work, including the creation of a rigid but light frame that allowed for easier steering. Following Zeppelin's advancements, in 1910 the German company Deutsche Luftschiffahuts AG (Delag) was organized and took the lead in transporting passengers in its airships. Between 1910 and 1914, Delag carried more than 34,000 passengers. Such service continued after World War I, when the *Graf Zeppelin* flew more than 1.6 million kilometers in commercial service. However, commercial lighter-than-air service effectively came to end when the zeppelin *Hindenburg* exploded in 1937.

Inspired by Lilenthal, the young mechanics Wilbur and Orville Wright in the U.S. built the first aircraft that was heavier than air and able to maneuver in flight. In December 1903 the Wright brothers made the world's first successful flight, success based on the fact that it was a powered, sustained, and controlled flight that carried a human being.

Following the success of the Wright brothers, human beings would be fascinated with aviation, inspiring new designs and expanding the use of aircraft. The first commercial service was between Tampa and St. Petersburg, Florida, but expansion was further hastened by World War I and stories of World War I flying aces who captured the attention of the public. While commercial developments were interrupted by the war, there were many technical developments made during the conflict. During the 1920s, aviation was mainly a sport and form of entertainment. Amelia Earhart and Charles Lindburgh were among the most well-known pilots of this heroic age of flight.

At the same time, the first small commercial airlines began to operate, often to carry mail. The first were formed in 1919 in Germany, France, and Holland. These would be followed by many more during the 1920s and 1930s. Most of these early commercial airlines did not last long, either failing or merging with other airlines. In 1927, John Northrup and the Lockheed Aircraft Company developed the Vega aircraft, which served as model for modern commercial aircraft. The Vega used either 220- or 425-horsepower engines, carried a pilot and six passengers, could travel at speeds of up to 215 kilometers per hour, and had a range of between 800 and 1450 kilometers. Starting in the 1930s, airplanes were used increasingly for the transportation of people and goods. Aviation technology advanced, as monoplanes with all-metal fuselages and retractable undercarriages became widespread. In particular, three major airlines began to establish worldwide routes during the 1930s: Pan American, Imperial, and KLM.

These airlines often used seaplanes. Water provided potentially long runways at a time when most airport runways were only about 300 meters feet long. In turn, this meant that planes could be larger, use multiple engines, and have larger fuel tanks, allowing them to carry more passengers over longer distances. The 1930s also saw the introduction of first true modern commercial aircraft. In 1933, Boeing introduced its Boeing 247, to be followed by the Douglas Company's DC-2 and

DC-3. The DC-3 proved to be the first commercial aircraft to be profitable. It could fly at above 1500 meters, had a stressed aluminum sheathing that gave it considerable strength, and had a retractable landing gear. DC-3s carried most commercial traffic in the U.S. at the time, showing that commercial air travel could be safe, reliable and profitable.

The next technological advancement in the movement of people and goods through the air was the four-engine plane. Work on such aircraft had begun as early as 1913 and demand increased during the 1930s. Airlines sought to fly longer routes, such as across the Pacific Ocean, and only more powerful four-engine planes could lift enough fuel to fly such distances. Four-engine planes could also fly at higher altitudes, thus avoiding the "weather," making trips faster and more comfortable for passengers. They could also fly over mountains rather around them. Another technological advance that had to accompany the four-engine planes was the pressurized cabin. An early example of this new generation of commercial aircraft was the Boeing Stratoliner, introduced in 1940.

War once again played a role in developing aviation technology. During World War II, airplanes became bigger and faster. The war was also important as the first aircraft with jet engines were introduced. After the war, there was tremendous growth of commercial air travel, led by the Stratoliner, the DC-4, and the Lockheed Constellation, all aircraft that were faster and could travel greater distances than earlier planes. By the late 1950s, more people crossed the Atlantic Ocean by plane than by ship. In the 1950s, there was some use of turboprop planes, mostly in Europe. However, the turboprop planes would soon be surpassed by the widespread use of jet engines introduced during the war. By the late 1940s, jet engines were standard in most military aircraft and by the 1950s, they were used increasingly in commercial aviation. Due to the higher speeds and lower operating costs of jet aircraft, there was a major expansion of the commercial airline industry in the second half of the twentieth century. The potential of this technology could be seen in the 1970 introduction of the Boeing 747 "jumbo jet," capable of carrying up to 500 passengers.

Once jets became more widespread, the next technological advance was to produce supersonic aircraft. If successfully developed, airplanes traveling faster than the speed of sound would greatly reduce flight times around the world and revolutionize communications. Starting in 1962, France and Great Britain agreed to jointly develop a supersonic transport (SST0), known as the Concorde. The first passenger plane to break the sound barrier, however, was the Soviet Tu-144. Due to design problems, however, the Soviet craft was only in service briefly during the 1970s. The Concorde made its first flight in 1969 and entered service in 1976. It reduced the flight time between London and New York to about 3 hours. However, supersonic travel did not become profitable and also met resistance from environmental groups. In 2000, a Concorde crash outside Paris led many to reconsider the safety and value of the supersonic fleet. The Concordes were withdrawn from service because of high fuel and maintenance costs in late 2003.

See also **Aircraft Design; Bulk Carriers and Tankers; Rail; Urban Transportation**

RONALD YOUNG

Further Reading

Allen, G.F. *Railways of the Twentieth Century.* W.W. Norton, New York, 1983.

Allen, G.F. *Railways: Past, Present, and Future.* Morrow, New York, 1982.

Bilstein, R. *Flight in America: From the Wrights to the Astronauts*, revised edn. Johns Hopkins University Press, Baltimore, 1994.

Heppenheimer, T.A. *A Brief History of Flight: From Balloons to Mach 3 and Beyond.* Wiley, New York, 2001.

Heppenheimer, T.A. *Turbulent Skies: The History of Commercial Aviation.* Wiley, New York, 1995.

McCullough, D. *Path Between the Seas: The Creation of the Panama Canal, 1970–1914.* Simon & Schuster, New York, 1977.

Nersian, R.L. *Ships and Shipping: A Comprehensive Guide.* Penwell Books, Tulsa, 1981.

Rinman, T. and Brodefors, R. *The Commercial History of Shipping.* Translated by Hogg, P. Rinman and Linden AB, Goteborg, 1983.

Transport, Foodstuffs

Twentieth century foodstuffs were transported by land on vehicles and trains, by air on cargo planes, and by water on ships or barges. Based on innovations used in previous centuries, engineers developed agricultural technology such as refrigerated containers to ship perishable goods to distant markets. Technological advancements enabled food transportation to occur between countries and continents. International agreements outlined acceptable transportation modes and methods for shipping perishables. Such long-distance food transportation allowed people in different regions of the world to gain access to foodstuffs previously unavailable and incorporate new products they liked into their diets.

Land and ocean transportation methods are the most economical methods for food shippers. Because of expenses, aircraft have been used to transport food primarily when profits from sale of goods are expected to exceed shipment costs or urgency is involved such as the humanitarian Berlin Airlift. Since the mid-twentieth century, technology enabled foods to be safely preserved and packaged for transportation to serve as meals for passengers on aircraft and boats. Foods undergo such processes as being heated, cooled, or hydrated in order to be palatable. As a result, airplane passengers can dine on fresh fruit while flying 10 kilometers high above frozen lands in the winter. People on cruise ships can feast on fresh nonseafood meat and dairy products and edible vegetables when they are hundreds of kilometers from shore.

Preservation of the quality of foodstuffs en route from field to market is crucial to maintain flavor, texture, and appearance. Temperature and humidity regulation maintained by air circulation within shipping devices is essential for foodstuffs transportation. Most types of vegetables, fruits, meat, and dairy and poultry products require chilling or refrigeration. Humidity is maintained at 85 to 95 percent to prevent foods from dehydrating. Foodstuffs are usually loaded in large metal intermodal containers which are standardized 20- to 40-feet (6- to 12-meter) boxes that can be pulled as trailers by trucks on highways, carried by ships, or moved by railways as a trailer-on-flatcar (TOFC). In the early twentieth century, Mary Engle Pennington designed refrigerated rail cars that were adopted internationally. Container innovations included building a double hull to insulate products and minimize leaks and contamination.

Refrigerated trailers dominate road food transportation methods. This transportation mode minimizes food vulnerability to shipment damage from being harvested to placement on grocery shelves. Refrigerated transport enables fresh produce from milder climates to be shipped out-of-season to colder locations. Refrigeration is achieved by mechanical or cryogenic refrigeration or by packing or covering foods in ice. Ventilation keeps produce cool by absorbing heat created by food respiration and transferred through the walls and floor from the external air beneath and around the shipping trailer.

Engineers designed trailers to have one of two types of air circulation, bottom forced-air delivery or overhead top-air delivery. Most refrigerated trailers have overhead top-air delivery. A blower pushes cold air through ceiling ducts towards the rear of the trailer. The air moves horizontally through channels between packed items then downward, passing underneath the load before returning to the blower. Walls are grooved and floors have T rails or pallets so air flow is not obstructed. Vents remove heat from the circulating air.

Bottom forced-air delivery is most frequently used on ship containers although some over-the-road trucks have this type of air circulation. Air moves from a container's bottom, passing under and up through loads. Shipment boxes have vent holes in their tops and bottoms and are packed tightly so that the holes match up to facilitate air passage. Transportation vehicles are heated if the fresh foods being shipped risk frost damage when they pass through freezing weather conditions.

Tankers carry milk, oils, and other liquid foodstuffs. Live animals shipped for food consumption require specific transportation needs for humane and public health reasons. Special tanker trucks are designed to haul live fish between fisheries and markets. Other vehicles ventilated with air holes and equipped with ramps are built to transport cattle and poultry from farms to slaughterhouses.

Trailer-loading techniques aid ventilation, which keep foods at optimum hauling temperatures. Shippers prepare food for transit by precooling it according to industrial standards. Preferably, heat is removed immediately after perishables are harvested. Various methods are utilized, including application of liquid ice, hydrocooling, vacuum cooling, and refrigerated forced air chilling rooms. Foodstuffs can also be refrigerated by vehicles at loading docks. Prechilling results in foods retaining flavor, freshness, and appearance when they reach market and inhibits decay thus increasing the shelf life.

Food technologists design packaging materials for food transportation. Most produce is shipped in corrugated and fiberboard cardboard boxes that are sometimes coated with wax. Wooden and wire-bound crates are also used in addition to bushel hampers and bins. Mesh plastic, burlap, and paper bags hold produce. Meat is often vacuum packed on plastic trays that are placed in wooden lugs. Foods are occasionally wrapped in plastic liners or packed in ice to withstand damage in transit and limit evaporation.

Loading patterns are essential to achieve desired circulation to maintain desired shipping temperatures. Boxes are stacked to create horizontal and vertical air channels while ensuring that the load is stabilized to prevent damage from road vibrations and movements. Certain foods such as dairy products are lifted above the floor to prevent them melting due to road heat. Eggs are shipped in cartons placed in fiberboard boxes, which are packed tightly together to prevent breakage. Modular unitized metric (MUM) load standardizes loading unitized produce packaged together on pallets and reinforced with straps. Units are quicker to load and unload than individual cartons, and handling damage is minimized.

Approximately 5 percent of fresh foodstuffs are ruined during distribution. Some fruits and vegetables produce ethylene gas that can damage other foods. Genetic engineers have designed produce that is more tolerant of transportation stresses. Digital information and electronic alarms alert shippers to the temperature and humidity of shipments through modems and satellite links, and remote controls provide access to alter settings. Inside shipping containers, a controlled atmosphere is maintained. Engineers determined that removing oxygen prevents foods from decaying while in transit. Carbon dioxide or nitrogen gases are injected into the container to supplement temperature and humidity control. After produce is loaded, shippers seal the container's contents with plastic film. Air valves are shut, and gases are released. Carbon dioxide is used to inhibit mold in berries. Nitrogen protects green vegetables and fruits. Small loads are sealed in plastic bags with protective gases.

Ocean transportation of food is aided by international deep-water cargo ports connected to railroad yards. In 1966, SeaLand was the first company to utilize shipping containers to trade between Europe and the U.S. Transported in stacks on ships, containers are transferred to trains or trucks or stored in terminals. Globally, containers became accepted during the early 1970s, annually carrying millions of tons of cargo by the 1980s. Until refrigeration became economically feasible in the mid-twentieth century, fish were too perishable to ship beyond local markets. On refrigerated factory ships, filleting and processing occur at sea soon after fish are caught. Seawater mixed with tetracycline antibiotics are used to refrigerate fish and prevent bacterial growth. This reduces spoilage and lost profits in addition to expanding market possibilities.

See also **Food Preservation: Cooling and Freezing; Food Preservation: Freeze Drying, Irradiation, and Vacuum Packing**

ELIZABETH D. SCHAFER

Further Reading

Ashby, B., Hunt, R., Hinsch, T., Risse, L.A., Kindya, W.G., Craig Jr., W.L. and Turczyn, M.T. *Protecting Perishable Foods During Transport by Truck*. U.S. Department of Agriculture, Agricultural Marketing Service, Transportation and Marketing Division, Washington D.C., 1995.

Coyle, W. and Ballenger, N. *Technological Changes in the Transportation Sector: Effects on U.S. Food and Agricultural Trade*. U.S. Department of Agriculture, Economic Research Service, Washington D.C., 2000.

Friedland, K. and Koninckx, C., Eds. *Maritime Food Transport*. Böhlau Verlag, Köln, 1994.

Harris, T.C. *Inland Challenges Facing Deep Sea Carriers in Europe*. Intermodal, Rotterdam, 1998.

Heap, R., Kierstan, M. and Ford, G., Eds. *Food Transportation*. Blackie Academic & Professional, London, 1998.

Kochersperger, R.H. *Food Warehousing and Transportation*. Lebhar-Friedman Books, New York, 1978.

Kochersperger, R.H. *Traffic and Transportation: Servicing the Grocery Industry*. National American Wholesale Grocers' Association, Cornell University, Ithaca 1990.

Miles, G.H. *Alternative Food Delivery Systems: An Exploratory Assessment*. National Science Foundation, Washington D.C., 1977.

Walker, M.S. *Chilled Meat Technical Cold Chain Challenges*. Intermodal, Rotterdam, 1998.

Transport, Human Power

Human-powered transport in the form of bicycles and tricycles has played an important—though often unrecognized—role in developments since the nineteenth century in industrial practices, human mobility, and sporting achievement. It has stimulated engineering innovation far beyond its own boundaries. Henry Ford and the Wright Brothers all began their careers as bicycle mechanics, and bicycles continue to be used because of their simplicity of design as a testing ground for new innovations in materials and bonding methods.

The contemporary bicycle can be traced to the German Baron von Drais, who in 1817 invented a running machine or hobby horse which became popular in France and the U.K. The two-wheeled machine was propelled by straddling the frame and pushing one's feet against the ground. Over subsequent decades, a variety of hand-propelled or other models were built by engineers and craftsmen, but the birth of the bicycle industry is generally dated from when the Parisian carriage maker Pierre Michaux (or possibly his associate Pierre Lallement) added pedals to the front wheel of an old hobby horse in the mid-1860s.

Several different lines of innovation continued simultaneously through the 1870s and 1880s.

James Starley of the Coventry Machinists Company, transformed the heavy iron French "boneshaker" into the lighter steel machine now known as the "Penny Farthing." Starley increased the size of the front wheel to cater to the needs of racing cyclists wanting as much leg power as possible, though other potential users found the high-wheeled bicycle either too dangerous or—for ladies—too indecorous. Other innovations improved drive train efficiency, steering and rolling performance, whilst others focused on safety, compensating for the reduced power of a smaller front wheel by introducing various gearing devices, or by turning instead to tricycles. Innovations such as ball bearings and tangent spoking remain a feature of contemporary bicycles, whilst this period also introduced differential gearing, later transferred into automobile technology. The two key breakthroughs came with the Rover safety bicycle of John Kemp Starley (James's nephew), which attached gearing to the rear, rather than to the front, wheel and the pneumatic tire invented by John Boyd Dunlop from Dublin. During the 1890s, bicycle design stabilized around this low-wheeled, rear chain-driven, pneumatic tired configuration—this has remained the dominant bicycle design to the present day.

During the late nineteenth century, the bicycle was significant as a symbol of modernity, freedom, and mobility; it helped transform the urban landscape and it expanded the horizons of the poor, the rural population, and women. In the twentieth century this symbolism became attached to the automobile, and many cycle manufacturers moved into automobile production. Some, such as Starley's Rover Company, eventually abandoned bicycles altogether. Nevertheless, in Europe the bicycle remained a popular means of both mass transportation and mass leisure until after World War II. The remaining manufacturers adopted many of the mass production methods such as interchangeability of parts and vertical integration that had been advanced by the Pope Manufacturing Company and subsequently the Ford Motor Company. U.K. firms such as Raleigh, BSA and Hercules introduced sheet steel stamping, automatic machine tools and conveyor belt production lines during the first few decades of the twentieth century.

By the 1960s, the bicycle had become marginalized as a form of transport in the developed world due to increasing car ownership and car-based urban planning. Sports cycling took over from commuter and leisure uses, marked by a shift from the robust upright utility machines of the 1950s to the racing bike, characterized by drop handlebars, thin wheels and 10-speed derailleur gears, the latter replacing the once-popular three-speed Sturmey–Archer hub gear. From the 1960s, further innovations were introduced to counter lost markets. Alex Moulton's small-wheeled suspension bicycle inspired a rash of new everyday utility models using balloon tires instead of Moulton's more expensive rubber suspension device. The development of children's models such as the Raleigh Chopper and children's BMX (bicycle moto cross) sports bikes, followed this trend. In the mid-1970s, mountain bikes were developed by Californian downhill racing enthusiasts, based on children's bicycle frames from the 1950s and motorcycle and BMX components. Mountain bikes were sturdier than racing bikes, with thick tires and flat handlebars. By the late 1980s they had taken a major part of the bicycle market, accounting for over 60 percent of sales.

From the industry's perspective, mountain bikes are most significant in being central to a shift in world production to the Far East, especially Taiwan and Japan. Except for specialist models, bicycles sold in the developed world were generally, by the 1990s, constructed in Taiwanese factories and fitted with components made by the Japanese firm Shimano, whose highly innovative approach revolutionized much bicycle componentry. Indexed gearing became widespread, with gearshifters using ratchet mechanisms adapted from Shimano's fishing rod products, and integrated with the braking system. Other innovations focused on drivetrain transmission and integrated shoe-pedal systems. Shimano's American competitor SRAM introduced the "Gripshift" gearshifter in the mid-1990s, while both companies—as well as the revived Sturmey-Archer—developed in the late 1990s new hub gear mechanisms to attract the growing utility and leisure markets that arose as a result of concerns about fitness and the environment.

Most successful bicycle innovations have been proved in competition. The first bicycle races took place in the 1860s to test competing designs. In the 1930s, challenges to the hour distance record were made on "recumbent" bicycles, where the rider sits back with their legs pedaling in front of them (Figure 15). The greater speeds possible in this position, especially when partial or full fairings are added to reduce wind resistance—led to recumbents being banned from cycling competitions so as not to invalidate other models. In the 1970s, recumbent enthusiasts established their own sporting bodies, with speeds reaching as high as 128

Figure 15. The Kingcycle recumbent, marketed as "the fastest production road bike in the world."
[*Source: Neatwork distribution brochure.*]

kilometers per hour in race conditions. Also known as HPVs (human-powered vehicles), there are many competing designs—both bicycles and tricycles are common, and there is little standardization so far for wheel size, componentry, length or height.

The most well known HPV designer—and bicycle innovator more generally—is Mike Burrows, known principally for his very low tricycle design, the Windcheetah. Burrows also designed the Lotus Sport carbon fiber monocoque (single blade) bicycle on which Chris Boardman won the 1992 Olympic pursuit championship for Britain (Figure 16). This machine was designed to minimize aerodynamic drag, matching Boardman's "sports science" approach. A contrasting amateur approach came from the Scot Graeme Obree, who beat several records in the 1990s on a home-made bicycle using washing machine ball bearings, a combination of conventional and found components, and an innovative hunched riding position that was banned by the cycle sport regulators.

Future developments in bicycle design, production and performance are likely to include further improvements in materials and design, although past experience suggests that only in componentry will these filter through to the wider consumer market.

See also **Automobiles; Globalization; Sports Science and Technology; Transport; Urban Transportation**

PAUL ROSEN

Further Reading

Ballantine, R. *Richard's 21st Century Bicycle Book*. Pan Macmillan, London, 2000.

Burrows, M. *Bicycle Design*, Hadland, T., Ed. Open Road, York, 2000.

Dodge, P. *The Bicycle*. Flammarion, Paris, 1996.

Gaffron, P. Cycling special issue. *World Transport Policy and Practice*, 7, 3, 2001.

Hounshell, D. *From the American System to Mass Production, 1800–1932: The Development of Manufacturing Technology in the United States*. Johns Hopkins University Press, Baltimore, 1984.

Lowe, M. *The Bicycle: Vehicle for A Small Planet*. Paper 90, Worldwatch, Washington D.C., 1989.

McGurn, J. *On Your Bicycle: The Illustrated Story of Cycling*, 2nd edn. Open Road, York, 1999.

Norcliffe, G. *The Ride to Modernity: The Bicycle in Canada, 1869–1900*. University of Toronto Press, Toronto, 2001.

Rosen, P. *Framing Production: Technology, Culture and Change in the British Bicycle Industry*. MIT Press, Cambridge, MA, 2002.

Figure 16. Wind tunnel test of the Lotus Sport Superbike ridden by Chris Boardman.
[*Source: QED: The Bike, BBC Education, 1993.*]

Useful Websites

Gaffron, P. Cycling special issue. *World Transport Policy and Practice*: http://www.ecoplan.org/wtpp/.

Traveling Wave Tubes

One of the most important devices for the amplification of radio-frequency (RF) signals—which range in frequency from 3 kilohertz to 300 gigahertz—is the traveling wave tube (TWT). When matched with its power supply unit, or electronic power conditioner (EPC), the combination is known as a traveling wave tube amplifier (TWTA). The amplification of RF signals is important in many aspects of science and technology, since the ability to increase the strength of a very low-power input signal is fundamental to all types of long-range communications, radar and electronic warfare.

The traveling wave tube owes its name to its mode of operation: it is designed to cause an RF carrier wave to travel along its length in a carefully predetermined manner. The energy for amplification is derived from a high-powered electron beam, which is made to interact with the RF wave carried on a slow wave structure, usually in the form of a helix (see Figure 17). The helix, made of copper, or of tungsten or molybdenum wire, is supported by three or four ceramic rods to isolate the RF fields on the helix from the metallic walls of the surrounding vacuum envelope. The helix reduces the velocity of the RF wave so that it travels slightly slower than the electron beam. This allows an interaction between the electron beam and the RF wave, whereby electrons are, on average, decelerated by the electric fields of the RF wave and lose energy to it, thus amplifying the signal it carries. The remaining energy of the electron beam is dissipated as heat in a collector. An alternative type of slow wave structure—known as a coupled cavity design—is based on a series of accurately sized and shaped RF cavity sections, usually of copper, which are brazed together and coupled by a slot in the wall of each cavity.

The TWT was invented in 1943 by Rudolf Kompfner, an Austrian refugee working for the British Admiralty. He demonstrated the principle of traveling wave amplification at Birmingham University later that year and published the first results of his work in the November 1946 issue of *Wireless World* magazine. The first practical device was developed at Bell Telephone Labs (BTL) in 1945 by John R. Pierce and L.M. Field and a detailed theory of its operation was published by Pierce in 1947. Subsequent development work was done at BTL and Stanford University in the U.S., and by Standard Telephones and Cables (STL) in the U.K., with a particular eye to potential applications in the communications field. The first TWTs to enter operational service were built by STL for a television relay link between Manchester and Edinburgh, which was operated by the Post

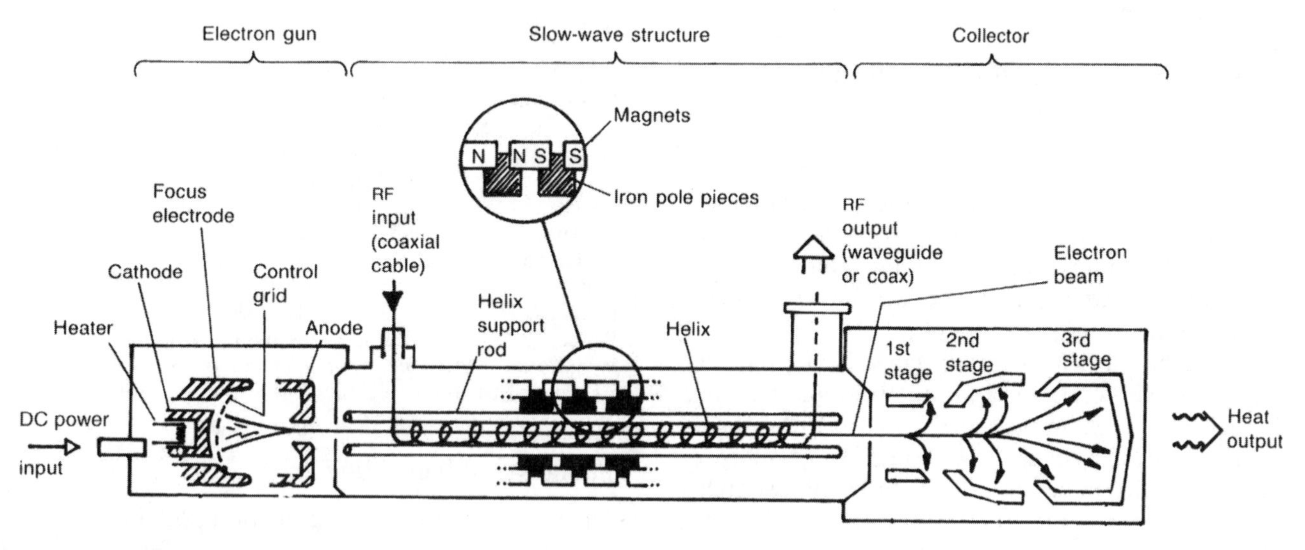

Figure 17. The component parts of a traveling wave tube (TWT). From left to right: electron gun, slow wave structure and collector. Note that the collector of the high-powered space tube depicted is designed to radiate directly into space.
[*Photo: Mark Williams.*]

Office and used by the BBC from 1952. These early tubes produced an output of about 2 watts across a band of frequencies centered on 4-gigahertz and had a gain of about 25 decibels.

Since then, TWTs have been developed to operate at a wide range of radio frequencies and at very high powers, both on earth and in space. In fact, TWTs have been used for space communications (in spacecraft and earth stations alike) since the first privately owned communications satellite, Telstar-1, was launched in 1962. It is historically significant that, without the TWT, the world would not have seen the early Olympic Games or the manned lunar landings "live via satellite."

Today, TWTs are broadband devices that can handle a signal bandwidth up to about 800 megahertz at Ku-band (12 to 18 gigahertz) and even wider bandwidths at higher frequency bands. Ground-based tubes deliver higher output powers than space-based tubes, typically up to 700 watts at Ku-band and 3 kilowatts at C-band (4 to 8 gigahertz) for helix tubes and up to 10 kilowatts for coupled-cavity tubes. Space-based devices, which tend to be of the helix type, can deliver up to about 300 watts at Ku-band. Individual TWTs have been built with power gains of more that 10 million (70 decibels).

In addition to exploiting the TWT for defense communications applications, the military services have developed its potential in the fields of radar and electronic counter-measures (ECM), for which its high gain and broad signal bandwidth are ideally suited. TWTs are also used for civilian and military space-based radars, weather and other remote sensing satellites, and all types of manned spacecraft.

Another type of electron tube, the klystron, was developed at Stanford University in the late 1930s by two brothers, Russell and Sigurd Varian. The klystron power amplifier (KPA) provides a useful alternative to the TWTA in radar systems and satellite earth stations. However, its narrower operational bandwidth and the fact that it cannot be easily retuned has made it unattractive for space applications.

Space-based TWTs are designed to be particularly reliable, since they cannot be repaired once launched. For example, the TWTs on the Voyager-1 spacecraft—at over 12 billion kilometers from the earth, the most distant example of twentieth century technology—were still working 23 years after its launch. Moreover, the ruggedness of the TWT was proved in the late 1970s when a satellite fell 9 kilometers into the Atlantic Ocean following a launch vehicle explosion: the recovered tubes were found not only to work, but to meet their original performance specification.

An important alternative to the TWTA, particularly for communications satellites, is the solid-state power amplifier (SSPA), an RF amplifier which uses semiconductor components—typically gallium arsenide field-effect transistors (GaAsFETs). Although the performance and reliability of SSPAs has been improved since their introduction in the mid-1980s, they are limited to relatively low-power applications at a given frequency since the solid-state medium is a much poorer conductor of heat than the materials used in a TWT. Typical late-1990s GaAsFETs could individually provide up to about 45 watts at C-band and 15 watts at Ku-band, while using power combination techniques, total output powers of 500 watts and 100 watts, at C and Ku-band respectively, were available for certain applications.

The fact that all three types of high-power RF amplifier—TWTA, KPA and SSPA—are still in use, and will remain so for the foreseeable future, indicates that there is often more than one engineering solution to a technological requirement.

See also **Radar, High Frequency and High Power; Radio-Frequency Electronics; Radionavigation; Satellites, Communications; Vacuum Tubes/Valves**

MARK WILLIAMSON

Further Reading

Elbert, B.R. *Introduction to Satellite Communication.* Artech House, Boston, 1999.

Elbert, B.R. *The Satellite Communication Ground Segment and Earth Station Handbook.* Artech House, Boston, 2001.

Evans, B.G. *Satellite Communication Systems.* Institution of Electrical Engineers, London, 1999.

Gittins, J.F. *Power Travelling Wave Tubes.* English Universities Press,. London, 1965.

Hansen, J.W. *Hughes TWT/TWTA Handbook.* Hughes Aircraft Company Electron Dynamics Division, Torrance, CA, 1993.

Kompfner, R. *Wireless World*, 52, 369–372, 1946.

Pierce, J. *IRE. Proc.*, 35, 111–123, 1947.

Pierce, J. *Traveling Wave Tubes.* Van Nostrand, 1950.

Williamson, M. *The Communications Satellite.* Adam Hilger/Institute of Physics Publishing, Bristol, 1990.

Tunnels and Tunneling

The history of tunnel construction goes back to the ancient civilizations of the Incas, Babylonians, Persians, and Egyptians, and therefore considerable experience in the construction of tunnels had already been gained worldwide by the beginning of the twentieth century. Tunnels were constructed to allow transportation through barriers (mountains, underground or underwater). In a country such as Switzerland or Canada, of which substantial parts are mountainous, tunnels were crucial for the development of a transportation infrastructure, and by the end of the nineteenth century the number of railway tunnels had greatly increased.

The optional methods for constructing tunnels increased in the twentieth century. The development of new methods and the improvement of existing ones were stimulated by the rapid increase of car traffic and the need for roads, for which new tunnels were needed. The choice for a particular way in a certain situation depends on the sort of material through which the tunnel is to be constructed. The most important difference is between hard rock and soft material. Besides that, the length and diameter of the tunnel has an influence on this choice. For allowing a sophisticated choice, geologic investigations into the behavior of the ground mass and the ground water are needed in an early stage of the tunnel project.

By the end of the twentieth century the following methods for tunnel construction could be distinguished:

1. Advance the heading in full face without lining. In this case, holes for explosives are drilled, and after blasting the debris is removed. This method is suitable when constructing tunnels through mountains and when the rock is stable enough to be self-supporting.
2. Classical methods. In this process the tunnel cross-section is gradually extended (contrary to the first-mentioned method) and temporary supports are constructed to prevent bursting of the tunnel wall.
3. Trenching or "cut and cover" methods. In this method a trench is excavated and, depending on the behavior of the groundwater, shields are constructed to prevent collapsing of the trench (Figure 18). In case of a stable ground, stiffening ribs may suffice. After constructing the tunnel, the trench has to be covered again.
4. Tunnel boring with a slurry shield. In this case a tunnel-boring machine (TBM; Figures 19 and 20) is used to excavate the tunnel and the walls are temporarily covered with bentonite in order to prevent collapsing, after which a series of tunnel segments is positioned.

Figure 18. Trenching method as used in the Rotherhithe tunnel. This tunnel was built under the river Thames in England from 1904 to 1908. Two trenches of about 4 meters wide were dug, and side walls built in them. The earth was then removed (letter A in photo indicates where it was dumped).

5. Immersed tunnels. This method can be applied when a tunnel is to be constructed underwater. A prefabricated tunnel segment is positioned at surface level and then sunk, after which the segments are joined.

Of these methods the first had already been fully developed by the beginning of the twentieth century. The second method was improved in the twentieth century by reducing the cost for temporary supports (e.g., by using sprayed concrete). The third method had also been developed in the second half of the nineteenth century, mainly for constructing underground railways in cities (starting with London). The last two methods are twentieth century accomplishments, although there are some roots in the nineteenth century. One of the first tunnels constructed by boring with a slurry shield was the Holland tunnel under the Hudson River in New York, which was opened in 1927. In 1825 the engineer M.I. Brunel had tried to construct a tunnel under the Thames using this method, but this effort had been so problematic that the method remained

Figure 19. Tunnel boring for the Rotherhithe tunnel. Part of this tunnel was bored with a rotary excavator, a hollow cylinder of 9.5 meters in diameter and 5.5 meters long. The front part (visible in the figure) was built up of cast-steel segments. Also shown are the two erectors for positioning the lining plates.

unpopular until the boring machines were improved. In 1970 for the first time more tunnels were constructed by boring than by blasting. In 1910 the first immersed tunnel was constructed: the Michigan Central Railroad tunnel under the Detroit River. The first immersed tunnel in Europe was built in the city of Rotterdam, the Netherlands (the Maas tunnel completed in 1942). Sometimes different methods were combined, such as in the case of the Detroit–Windsor tunnel that was constructed from 1928 to 1930 by using a combination of trenching and immersing.

The choice for a certain method is not only influenced by geologic considerations, but also by economic and social factors. Of course the methods mentioned above vary in cost and efficiency. Tunnel boring machines in general cost less time than other methods. In hard material, boring with a TBM can yield a progress of up to 15 meters per 8-hour shift, while methods using explosives usually do not yield a progress higher than 7 meters per shift. On the other hand the boring method requires an expensive TBM, which can undo the cost economic effect of the high speed of progress. The influence of the construction process at surface level also has an impact on the choice. Trenching in some cases may be cheaper than boring, but totally upsets the surface level situation, while boring can be done in such a way that it is hardly noticed at surface level (which of course is quite a social advantage). One of the most important social factors for the construction and use of tunnels is safety. Accidents such as a tunnel breakdown or a fire can totally isolate people from the outside world. Also, the air in the tunnel can be easily polluted unless special measures are taken to ensure constant refurbishment of fresh air. Lighting is another important safety issue. In the twentieth century the awareness of these nontechnical aspects grew when the number of tunnels increased and with that the number of accidents.

One of the major achievements in twentieth-century tunneling is the Simplon tunnel (Switzerland), the construction of which commenced in 1898 and was finished in 1905.

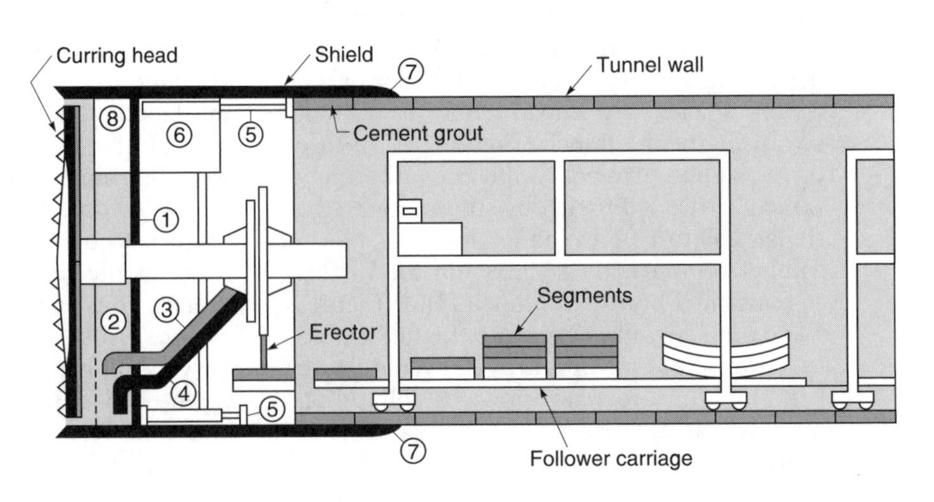

Figure 20. The tunnel-boring machine (TBM) with slurry shield method. (1) Pressure bulk head; (2) baffle; (3) bentonite supplies; (4) bentonite disposal; (5) press screw; (6) air lock; (7) tail seal; and (8) air cushion.

Hydraulic drills were used for removing the mountain rock. A serious delay was caused by a subterranean river that workers encountered in 1901. It was a critical time for the project because all the available materials for constructing girders seemed inefficient to resist the large forces. After only 45 meters, circumstances improved, but it had cost nearly £1000 per yard to overcome this distance. Hot springs that caused high temperatures in the tunnel were another big challenge for the engineers. On 24 February 24 1905 the Italian and the Swiss parts of the tunnel were connected with a heavy charge explosion. The excavating direction had been extremely accurate for that time (up to just centimeters). The tunnel was 19.2 kilometers long and was to be the longest for many years, and it also deserves mentioning that in the construction process only 39 lives were lost.

Some examples of other major tunnel projects in the twentieth century are the Channel tunnel that connects England and France, and the so-called Big Dig project in the city of Boston, Massachusetts, begun in the 1980s and scheduled for completion in 2004. The history of the Channel tunnel offers a nice illustration of political aspects of tunnels. The plans for this 50-kilometer tunnel for many decades were frustrated by the fear of the political effects of a connection between the two countries. The Big Dig project nicely illustrates the enormous impact tunnel constructions can have on life at surface level. By the end of the twentieth century the world's longest completed road tunnel was the Laerdal tunnel in Norway, which was opened on 27 November 2000 (24.5 kilometers in length). The longest railway tunnel was built in Japan (the Sei-kan tunnel, opened in 1988, nearly 54 kilometers length).

See also **Construction Equipment; Urban Transportation**

MARC J. DE VRIES

Further Reading

Beaver, P. *A History of Tunnels*. Peter Davies, London, 1972.

Bickel, J.O., Kuessel, T.R. and King, E.H., Eds. *Tunnel Engineering Handbook*, 2nd edn. Chapman & Hall, New York, 1996.

Harris, F. Ground Engineering Equipment and Methods. McGraw-Hill, 1983.

Turbines, Gas

During the twentieth century, the gas turbine was developed to fit many applications on land, sea and in the air. From early beginnings, the gas turbine came alongside, competed with, and often replaced the existing technologies of steam, water, and reciprocating internal combustion engines. Initial problems stemmed from a lack of knowledge and the techniques; the fundamentals were well enough understood, but what were lacking were the design techniques. Materials also held up developments; but after extensive experimentation, successful turbine designs were being constructed in the first ten years of the twentieth century.

The gas turbine has the advantage over traditional engines in that its combustion process is continuous and thus the equipment is less subject to cyclic heat stresses and its power is less limited—power is limited by combustion knock in spark ignition engines, but in diesel engines it is only limited by structural strength and maximum working pressures in the fuel injection systems. It also has fewer moving parts, so wear and tear is lessened. Despite these differences, the gas turbine still has the basic four functions of the four-stroke cycle but operates continuously: air is admitted, compressed, heated by burning fuel so that it expands and does work, and then the spent gases are expelled. However, unlike an ordinary engine, each of these processes takes place in a separate part of the engine and happens continuously; the oil engine has all processes within the cylinder and they follow on from each other.

Gas turbine engines can be classified according to the type of compressor used, namely radial or centrifugal flow, and axial flow. In the first of these, the air is compressed by accelerating it outward perpendicular to the longitudinal axis of the engine. Radial engines are divided into single-stage and two-stage compressors—the amount of thrust is limited because of the maximum compression ratio. Its advantages are its light weight, simplicity, and low cost.

Axial flow compressor engines may incorporate one, two or three spools (where a spool is defined as a group of compressor stages rotating at the same speed). In a two-spool engine, the two rotors operate independently of one another. The turbine assembly for the low-pressure compressor is the rear turbine unit. This set of turbines is connected to the forward low-pressure compressor by a shaft that passes through the hollow center of the high-pressure compressor and turbine drive shaft. This type of compressor can be found in the larger engines because of its ability to handle large volumes of airflow; it also has a high pressure ratio. However, it is more expensive to make, heavy in comparison with a radial engine, and more susceptible to damage by foreign matter being drawn in—which is why a "bird strike" is so damaging for an aircraft.

Gas turbines are limited by the resistance of the turbine blades to the intense heat generated within the machine, but sufficient experimentation had been undertaken to allow work to be done commercially by turbines by the outbreak of World War II. An early commercial producer of the machines was the Swiss firm of Brown Boveri, later ABB.

Initially, the engines produced were for industrial power applications; gas turbine-powered electricity-generating sets being a main use. The transport options were also being explored for this source of power. Ships and locomotives had been built that utilized steam turbines, and gas was seen as a further possibility. Gas turbine locomotives were built using Brown Boveri units for use in Switzerland, and Britain experimented with three gas turbine locomotives, two in 1947, of which one was by Brown Boveri, and one in 1959. The concept continued in trials up until the 1970s for railways, as Britain's experimental Advanced Passenger Train was driven by gas turbines and at one stage held a speed record for a rail vehicle thus powered. The most successful use of locomotives powered by gas turbines was in the U.S., though high fuel costs and competition from diesel locomotives meant that this use was over by 1969. The problem with simple gas turbines in transport is that without heat exchangers the efficiency is poor at part load; their use in the U.S. was on a particular Southern Pacific line where there was a steady gradient on either side of a mountain pass, where the engine was run at full power to the top and then shut down for the descent.

It was not long before thoughts of gas turbines to airborne transport were conceived—as had happened and been tried in earlier years on steam and internal combustion power. The course that events then took had a direct bearing on the diminution of the world and ultimately paved the way for the globalization of markets that was the position at the close of the twentieth century. Almost all modern applications have been developed from the construction and manufacturing techniques of aeroengines. The thin sections of high-temperature metals in aeroengine construction were initially used solely for weight reduction, and people who were developing land-based engines still thought in terms of thicker sections of traditional metals, except, of course, in turbine blades and disks. Then they realized that these thin sections reduced thermal stresses and allowed quick startup and shutdown compared to "old fashioned" engines. (It is the time taken for heat to "soak" through the thick materials without excessive temperature gradients that takes so long to start up and shut down steam turbines). By the time that this was realized, jet engines and turboprops had become big enough to be useful in marine and land applications, and the number of aeroengines being manufactured made it much cheaper to adapt them for other uses than to design special engines. Among the most popular land-based turbines today are Rolls-Royce Avons and RB211s.

The development of specialized materials contributed to the progress of the thrust to weight ratio of the turbine, especially in aircraft usage. Titanium and nickel alloys have been substituted for steel, while aluminum has virtually disappeared from use in aeroengines. The future suggests that there is great potential for composites of various types, including carbon, ceramic matrix, and metal matrix composites, depending of course on achieving low cost manufacture and cost effective exploitation. The range is necessary because of the temperature variation being so great within a turbine. Materials also supply the enabling technology for improvements in performance and reliability.

Serious development of the gas turbine and its application in jet engines began around 1935 in both Britain and Germany. In Britain the name of Sir Frank Whittle will always be linked with jet engines. He wrote a paper in 1930 outlining the principles of gas turbines and jet propulsion, and took out a patent on his ideas, but due to a lack of support for practical development of his theories he let the patent lapse in 1936. Eventually with some backing, he formed Power Jets Ltd. to experiment and develop full-size engines, resulting in the first jet-powered flight in Britain, which took place on 15 May 1941.

The Germans were, however, the first to actually fly a jet-powered craft in 1939 after experimentation dating from 1935, but it is believed that their first engine to run satisfactorily did so only after Whittle's. They were therefore able to use the technology in World War II, as did the British, flying Gloster Meteor aircraft fitted with jet engines. By that time, the course of the war was well advanced and the U.S. had yet to fly a craft in combat, but after the war, gas turbines became very widespread in both civil and military aviation fields. The jet engine was developed further and it became possible to travel at previously unheard of speeds for very long distances. As time passed, supersonic flight became possible for the first time and was offered commercially with the introduction of the Concorde aircraft in 1973.

On the maritime scene, gas turbines have worked well alongside the steam turbine. Their application on naval terms has been particularly successful, with the British Navy using only gas turbines to power its major warships since 1969. Gas turbine-powered cars have also been tried with some success, but nothing on a large commercial scale has been seen on the world's highways.

Gas turbines became an accepted form of power for transport and industry, and a journal, *The Oil Engine and Gas Turbine*, devoted its pages to them. Electric power generation has become the largest use of the turbine after air transport. The advantages of their application in this field are quick starting, the option of remote operation, low costs, and viable economics. Interest in power generation began in 1939, but the increase in oil fuel costs from 1973 alerted many to the possibilities of using the power in generating plants. This interest was also spurred by breakdowns with existing plant, often because high-level management decisions had pushed the introduction of very large sets too quickly and they were not adequately developed, resulting in delays in the commissioning of other generating stations.

The production of gas turbines continued into the twenty-first century for many of the uses outlined earlier, plus applications in powering offshore installations and compression on gas pipelines. They have not been taken as far commercially as once envisaged, due to limitations on their fuel supplies and the perceived advantages not meeting expectations. Many attempts have been made to overcome the poor efficiency at low power by using heat exchangers; these have mostly suffered from short life. However, gas turbines are an important source of power in the world, and constant development drives their improvement as a result of their use in the vast majority of large and fast aircraft in service.

See also **Turbines, Gas in Aircraft; Turbines, Gas in Land Vehicles**

ANTHONY COULLS

Further Reading

Kennedy, R. *Modern Engines and Power Generators*. Caxton Publishing, London, undated.

Kolin, I. *The Evolution of the Heat Engine*. Longman, London, 1972.

Miller, S. Advanced materials mean advanced engines. *Mater. World*, 4, 1996.

Robertson, K. *The Great Western Railway's Gas Turbines: A Myth Exposed*. Alan Sutton, Stroud, 1989.

Rolls-Royce. *Power for Air, Sea and Land*. Rolls-Royce, London, 1982 and 1989.

Whittle, F. The early history of the Whittle jet propulsion gas turbine. *Minutes Proc. Inst. Mech. Eng*., 152, 1945.

Turbines: Gas, in Aircraft

Based on the technology of the land-locked and heavy steam turbine, by the mid-1930s development began in earnest on the gas turbine for aircraft. In the end, the evolutionary development in technology had a revolutionary impact on transportation systems.

The turbojet engine for aircraft is a connected three-stage system. The compressor, in either the axial-flow or centrifugal-flow form (or a combination thereof), compresses entering air. The compressed air enters the next stage—single or multiple combustion chambers—where it is mixed with fuel and fired. The super-heated, expanded air flows over a turbine as it exits the system, producing thrust in the form of exhaust. The turbine is connected to the compressor on a rotating shaft (or shafts) that turns the entire system. The rotational forces of the turbojet system are not as strenuous as those in a conventional piston engine, and are therefore more mechanically efficient. Although the thrust in the form of exhaust is the most simplified use of the turbojet engine, the shaft horsepower can also be harnessed to drive propellers, drive gears, or applied to other uses (e.g., the 1950s American Chrysler Turbine automobile).

During the first three decades of the twentieth century, a number of developmental programs were considered around Europe and the U.S, but most were "paper projects." French, Swedish, and Swiss designers came up with ideas for turbine engines for use in aircraft, but materials, funding, and interest were practically nonexistent before the mid-1930s. By then, in the quest for speed in the air and active aerial rearmament programs, interest was rekindled in turbojet power for aircraft.

In two unrelated projects, one German and one Briton designed, built, and tested turbojet engines prior to the start of World War II. In Germany, the young physicist Dr. Hans von Ohain developed first the theory, then a working centrifugal-flow turbojet as an aircraft power plant. In Britain, a fundamentally similar, but noticeably different centrifugal-flow turbojet was designed and tested by Flight Lieutenant Frank Whittle (a Cambridge University mechanical engineering graduate) under the aegis of his new company Power Jets Inc.

Ohain's design was the first to reach fruition, with the financial backing of the German aircraft magnate Ernst Heinkel, director of the aircraft-manufacturing firm that bore his name. The first experimental turbojet aircraft, the Heinkel He-178, flew for the first time on 27 August 1939, four days before the outbreak of World War II. The British program, with Frank Whittle leading theoretical developments, was hampered by a lack of funding and materials constraints and required more time to put into operation. Whereas the Whittle WU (for Whittle Unit, his first engine) ran for the first time in April 1937, the modified W1 engine was not ready for flight-testing until 1941, when the experimental Gloster-E28/39 took off for the first time in April. Thus, the Germans had an apparent head-start in the race to develop jet engine technology.

However, the Germans during World War II lacked a coherent strategy, especially in technological development. The German emphasis on theoretical application of scientific principles encouraged them to seek out the most advanced designs possible in opposition to using practical applications. In the end, they turned to the potentially more powerful axial-flow turbojet engine to power its nascent jet force, while the British continued centrifugal-flow turbojet development. The end result was that although the Germans were able to commit jet-powered aircraft to combat operations, materials limitations and the Allied combined bomber offensive thwarted extensive production. The German equipment was revolutionary for the time, but production figures were relatively low, as was combat commitment; in addition, operations and tactics were a constant trial and error process with the revolutionary technology. The British, on the other hand, were able to commit turbojet-powered aircraft (in the form of the Gloster Meteor) to combat operations as Home Defense fighters.

During the war, two other jet programs developed directly from the German and British technology. The German's aided the Japanese, their Axis partners, in developing the Nakajima Kikka, a design inspired by the German Messerschmitt Me-262 and powered by BMW 003 axial-flow turbojets. However, the Kikka did not fly until the last days of the war and only as a prototype; it had no impact on the final outcome. The American jet program was inspired by the British Whittle developments. The Americans copied the Whittle engines directly, under license to General Electric, and used two to power the Bell XP-59a Airacomet. The Airacomet first flew on 1 October 1942. Although the Airacomet was superceded by the Lockheed P-80 Shooting Star before the end of the war, the Americans did not commit jet aircraft to combat operations.

Following the war, both military and civilian aircraft interests accepted the potential of the turbojet. Aircraft engine manufacturing firms continued development of turbojet engines for use as aircraft power plants. In addition to centrifugal-flow and axial-flow turbojets, the development of other related projects were initiated. Turboprops (turbojets that drive propellers), ramjets, and turbofans were all developed in order to explore the potential advantages of the engines with regards to speed, endurance, altitude, and efficiency.

Today, commercial aviation continues the development of high-flow bypass turbofans, which maximize efficiency in high-altitude sustained subsonic flight. Military applications include these engines, mainly for transport and bomber aircraft, while development continues on high-speed axial-flow turbojets with afterburners for fighter aircraft, to maximize speed and thrust at the expense of endurance.

Since 1937, the development of the turbojet has revolutionized air transportation. Initially conceived for military purposes with an emphasis on speed, the turbojet engine for aircraft has developed into a prime mover for various aircraft applications. Combining high efficiency and speed at high altitude, the turbojet engine has revolutionized air travel for the commercial passenger as well as the military pilot. Supersonic flight became possible for the first time and was offered commercially with the introduction of the Concorde aircraft in 1973.

See also **Civil Aircraft, Jet Driven; Turbines, Gas**

MIKE PAVELEC

Further Reading

Bathie, W. *Fundamentals of Gas Turbines*. New York, 1996.
Constant, E. *The Origins of the Turbojet Revolution*. Baltimore, 1980.
Driggs, I. *Gas Turbines for Aircraft*. New York, 1955.
General Electric Company. *Seven Decades of Progress: A Heritage of Aircraft Turbine Technology*. General Electric, Fallbrook, CA, 1979.
Gunston, B. *World Encyclopedia of Aero Engines*. Wellingborough, 1986.
Institution of Mechanical Engineers. *Development of the British Gas Turbine Jet Unit*. Institution of Mechanical Engineers, London, 1947.
Judge, A. *Gas Turbines for Aircraft*. New York, 1959.
Kerrebrock, J. *Aircraft Engines and Gas Turbines*. Cambridge, MA, 1977.
Leyes, R. *The History of North American Small Gas Turbine Aircraft Engines*. Reston, VA, 1999.

Neville, N. *Jet Propulsion Progress; the Development of Aircraft Gas Turbines*. New York, 1948.
Oates, G. *Aerothermodynamics of Gas Turbine and Rocket Propulsion*. Washington D.C., 1988.
Oates, E., Ed. *Aircraft Propulsion Systems Technology and Design*. Washington D.C., 1989.
Schlaifer R. and Heron, S. *The Development of Aircraft Engines and Fuels*. Chicago, 1950.
Treager, I. *Aircraft Gas Turbine Engine Technology*. New York, 1979.
Whittle, F. *Gas Turbine Aero-Thermodynamics: With Special Reference to Aircraft Propulsion*. Oxford, 1981.
Wilson, D. *The Design of High-Efficiency Turbomachinery and Gas Turbines*. Cambridge, MA, 1984.

Turbines: Gas, in Land Vehicles

The gas turbine has found widespread use in the aviation, marine, and stationary power areas. However the gas turbine has only seen limited use in land transportation.

As various companies began to experiment with gas turbines in the 1920s and 1930s some gave thought to using the turbine as a source of motive power for land vehicles. The turbine promised much higher power-to-weight ratios than conventional reciprocating engines and also had the capability of using cheaper fuels such as industrial heating oil, diesel fuel, and even powered coal. As with gas turbines in aviation, most development has occurred since World War II.

There have been no true production automobiles manufactured with gas turbine engines. However, there have been many experimental vehicles. Some of these vehicles utilized aviation engines and constructed vehicles around them. The exhaust blast was utilized for propulsion as in jet aircraft. These vehicles were not designed for general highway travel and were exclusively confined to high-speed, often record breaking, runs. Many individuals and groups have built exhaust-powered gas turbine vehicles for drag racing, automobile shows, and land speed record attempts. A more practical way of utilizing gas turbines to power a vehicle, tried by several manufacturers, was to connect the turbine to shafts for wheel-driven automobiles. Generally hydrostatic automatic transmissions were used to control the speed of the vehicle. Among the most successful of the gas turbine automobiles were Chrysler's several efforts but many other manufacturers, including Ford, General Motors, Rover, Volvo, Nissan, Toyota, Volkswagen, and Daimler Benz experimented with wheel-driven gas turbine automobiles. Rover produced the first true gas turbine-powered, wheel-driven automobile in 1950. Chrysler had been experimenting with turbines since the early 1950s and achieved its largest success in 1963–1964, when it produced 50 vehicles and distributed them on a test basis to typical consumers all over the U.S. General Motors experimented with geared gas turbines in heavy trucks and buses in addition to automobiles. Most of the turbines used in these vehicles were single-stage, centrifugal compressor designs. Some incorporated heat regenerators. While the vehicles worked fairly well, there were many issues to address. Fuel consumption was generally worse than conventional gasoline or diesel piston engines. There was also a lag in acceleration experienced in many of the test automobiles due to the spool-up time, or the time to reach operating speed, of the turbine. The consequences of catastrophic failure worried designers, and emissions were also a concern after the clean air act and the creation of the EPA in the United States. Despite these concerns, experimentation continued throughout the 1970s and 1980s.

While unsuccessful in civilian automotive use, the gas turbine was put to good use in the U.S. Army's M-1 Abrams main battle tank, built by General Dynamics and first delivered in 1980. The high power-to-weight ratio of the turbine as well as its smaller size and the capability for multiple fuels were the primary reasons that turbine power was chosen for the M-1. Many of the concerns raised in automobile testing, such as fuel consumption and emissions, were deemed to be of secondary importance in a military application. The M-1 has a 1120-kilowatt gas turbine coupled to a six-speed automatic hydrokinetic transmission.

The most successful gas turbine use in ground transportation has been in the railway arena. Early experiments began prior to World War II with full-scale locomotives entering production during and shortly after the war. The Swiss firm Brown-Boveri built a 1640 kilowatt locomotive for the Swiss federal railway in 1941. British Rail rostered three experimental locomotives during the 1950s and early 1960s. The first two, delivered in 1950 and 1952 respectively, were gas turbine electrics while the third, built in 1960, was a direct-drive geared locomotive. They were used mostly in the western region and were eventually withdrawn from service partly due to high fuel costs. In the U.S., both Westinghouse and General Electric (GE) adapted aviation-type gas turbines to railroad use. Westinghouse built one test locomotive in 1950 but did not follow with a full line of locomotives. General Electric was more successful. It built one test locomotive in 1948 for the Union Pacific railroad and followed with several different production models manufactured until 1961. There were three distinct types of gas turbine locomotive in service on the Union Pacific. The first was the single GE product of 1948. The second were a group of 24 3360-kilowatt GE turbines produced from 1952 to 1954. These were the first truly operational gas turbine locomotives in the U.S. The final group of gas turbines on the Union Pacific were the monstrous 6340-kilowatt "Big Blow" turbines built from 1958–1961. These were the largest and most powerful single locomotives in the world when produced. The Union Pacific was the only North American railroad to use gas turbine locomotives in revenue service. Both the Westinghouse and GE locomotives were gas turbine electrics. The multistage, axial flow gas turbine drove an electric generator, which powered traction motors located on the axles. Both were designed to burn bunker C oil, a heavy industrial fuel. The Union Pacific also experimented with a coal-burning gas turbine locomotive. This turbine had many more difficulties than the earlier turbines and was purely experimental. The Union Pacific gas turbines were very loud, and could only be operated in rural and isolated areas. Because of these concerns and the rising cost of fuel, all were withdrawn from service by the end of 1971.

Apart from locomotives, railways have also made use of gas turbines in self-propelled trains and railcars. SNCF, the French national railway, achieved success in the late 1960s and 1970s with the RTG gas turbine-powered high-speed train. Amtrak imported six in 1973 for use in the U.S. and several more were manufactured under license by the Rohr Corporation. The United Aircraft Corporation turbine train was used in both the U.S. and Canada in the 1970s and early 1980s. The British effort at a high-speed gas turbine train, the APT or Advanced Passenger Train, was not successful.

See also **Rail, Electric; Rail, High Speed; Tanks; Turbines, Gas**

JEFF SCHRAMM

Further Reading

Advanced Gas Turbine Systems for Automobiles. Society of Automotive Engineers, Warrendale, PA, 1980.

Alternative Sources of Railroad Motive Power. Association of American Railroads, Washington, D.C., 1985.

Carrol, W. *Automotive Gas Turbines*. Coda Publications, Studio City, CA, 1963.

History of Chrysler Corporation Gas Turbine Vehicles. Chrysler Corp., Detroit, MI, 1964.

Lee, T.R. *Turbines Westward*. T. Lee Publications, Manhattan, KS, 1975.

Simmons, J. and Biddle, G., Eds. *The Oxford Companion to British Railway History*. Oxford University Press, Oxford, 1997.

U.S. Passenger Rail Technologies. U.S. Congress Office of Technology Assessment, Washington D.C., 1983.

Turbines, Steam

The first steam turbine, of which there is any record, was made by Hero of Alexandria more than 2000 years ago. This simply demonstrated that a jet of steam, impinging on a paddle wheel, could convert heat energy into mechanical energy. In the late nineteenth century significant improvements in the efficiency of conversion were made by, among others, Sir Charles Parsons on Tyneside, U.K. and Charles G. Curtis in the U.S.

Early steam engines up to that time had involved very high rotational speed, which was difficult to utilize for many purposes unless speed-reducing gearboxes were employed. Parsons had deduced that moderate surface velocities and speeds of rotation were essential if the "turbine motor" was to receive general acceptance as a prime mover. His early designs arranged to divide the fall in pressure of the steam into small fractional expansions over a large number of turbine wheels in series so that the velocity of the steam over each wheel was not excessive.

At the close of the nineteenth century, many local power stations employed reciprocating steam engines to drive electric generators. Steam turbines had the advantage over reciprocating steam engines, which were based on the movement of a piston in a cylinder, of being lighter and more efficient. The Curtis multiple-stage steam turbine (patented in 1896, sold rights to General Electric in 1901) occupied a smaller space and cost much less than contemporary reciprocating steam engine-driven generators of the same output. The Curtis turbine was also shorter than the Parsons turbine, and was thus less susceptible to distortion of the central shaft.

The work that Curtis, Parsons, and others carried out in the development of steam turbines allowed large central power stations to be developed, providing electricity for the growing demand during the early 1900s. Early machines at the beginning of the twentieth century had ratings typically of less than 1 megawatt.

Operating Principle

Steam is supplied to a turbine at high pressure and temperature from a boiler and the energy in the steam is converted into mechanical work by expansion through the turbine. This expansion takes place through a series of fixed blades or "nozzles" that direct the steam flow onto moving blades on a turbine disk, which rotates under the action of the steam to produce mechanical power in the rotor shaft. A row of fixed blades and its associated moving blades are known as a "turbine stage." Many such stages in series form a turbine cylinder. The fixed blades are attached to the turbine casing, which contains the steam pressure. In large output turbines, the steam flow and expansion ratio are too great to be contained within a single turbine rotor and casing and several such turbine cylinders are combined to achieve the necessary output.

Steam turbine development in the U.K. progressed from 30-megawatt sets at the end of the 1930s, operating at 450°C and 41 bar, through to 100-megawatt sets in the mid-1950s, to 660-megawatt sets operating at 158 bar and 565°C in the late 1970s. Designs for 1300-megawatt electric sets existed at the end of the twentieth century.

Applications

The chief applications for steam turbines are:

- Electricity generation with machines ranging in size from comparatively low powers of the order of hundreds of kilowatts up to about 1300 megawatts in central power stations. Electric power generation on land is now produced almost exclusively by steam-driven turbine generators.
- As part of an industrial process to produce electricity but where the steam is used for other purposes after its exhaust from the turbine.
- Variable speed drives, such as boiler feed pumps and blast-furnace blower drives.
- Marine propulsion, such as large ocean liners of the early twentieth century. Parson's demonstration of *Turbinia* at the 1897 Spithead Naval Review was the world's first steam turbine-driven ship. Now largely superseded in conventional shipping, steam turbines are still used to drive nuclear-powered vessels.

Characteristics and Efficiency

The output characteristics of a steam turbine are based on the so-called "Rankine" cycle. Water is pumped into a boiler in which heat is supplied to convert the water into steam. This steam is then utilized in a steam turbine to do work (e.g., to drive an electric generator). The steam exhausting from the turbine is condensed and then pumped back into the boiler to complete the cycle.

The output of any heat engine such as a steam turbine is governed by the simple relationship that defines its efficiency:

$$\text{Efficiency}_{\text{max}} = 1 - (T_{\text{out}}/T_{\text{in}})$$

where T is the temperature of the steam (expressed in degrees Kelvin). This gives the so-called "Carnot efficiency," which is an ideal that cannot be achieved in practice, only approached. It will be seen that raising the input temperature of the steam as high as possible and reducing the temperature at which the steam is condensed back to water as low as possible are essential to achieve high heat-to-power efficiencies.

Efficiency Improvements

Improvements in the overall cycle efficiency are achieved by design refinements such as bled-steam feed heating, where some of the steam passing through the turbine is extracted to preheat the water entering the boiler. The improvement in efficiency obtained with feed-water heating increases with turbine inlet pressure.

Another refinement is "reheat" where the steam passing between the different stages of the turbine is raised in temperature by circulating it back through a section of the boiler. Increases in efficiency of the order of 3 to 5 percent can be gained in this way although, as operating steam temperatures are raised, there is a progressive reduction in the improvements that can be gained in this way.

Economies of Scale

In postwar years the trend toward larger and larger units in power stations enabled marked reductions in capital cost per kilowatt of capacity. It also enabled improvements in thermal efficiency to be made, partly as a direct result of the increases in size, and partly because the economic steam pressure increases with size. Steam temperatures have increased also over the period up to 565 to 593°C which is near the limit for ferritic steels in the turbines.

The highest efficiency steam turbines at the end of the twentieth century were to be found in Japan where improvements in manufacturing technology and materials, together with blades designed utilizing computer-based three-dimensional flow analysis and "super-critical" steam conditions of 250 bar, 600 to 610°C, have resulted in 1050-megawatt machines giving overall power station efficiencies of 49 percent.

See also **Electrical Energy Generation and Supply, Large Scale**

IAN BURDON

Further Reading

Hills, R.L. *Power from Steam: A History of the Stationary Steam Engine*. Cambridge University Press, Cambridge, 1993.

Ramage, J. *Energy: A Guidebook*, 2nd edn. Oxford University Press, Oxford, 1997, chapters 5–7

Useful Websites

Parsons, C.A. *The Steam Turbine*. Rede Lecture. Cambridge University Press, 1911: http://www.history.rochester.edu/steam/parsons/

U

Ultracentrifuge

As a scientific instrument, the ultracentrifuge is used primarily in biochemical research to sediment substances from a solution. A rotor with containers holding the solution is spun at high speeds, causing the solute to move towards the periphery of the container under influence of the generated gravitational fields. Historically, however, ultracentrifuges were first developed for analytical purposes to determine the size of colloidal particles.

Early High-Tech Machine

Beginning about 1910 and through the 1920s, Swedish chemist Theodor Svedberg analyzed the particle sizes of colloids. Under the influence of gravity, these particles gradually separated from a solvent at a speed depending upon their size. To improve the accuracy of his method, he began to use centrifuges to generate forces many times that of the earth's gravitational field. He developed this idea into a practical method and could subject the colloid particles to forces up to 5000 g (5000 times the force of gravity). In analogy with the common methods of ultrafiltration and the ultramicroscope, he called the apparatus an ultracentrifuge.

Svedberg used the new apparatus to determine particle sizes of both inorganic and organic colloids including proteins, expecting that these, like inorganic colloids, would show a wide distribution of particle sizes. However, when he analyzed hemoglobin, all particles seemed to have the same size. This suggested that this protein could be a well-defined molecule, which was a revolutionary thought at the time. To be sure, however, further analyses would be needed using fields of the order of 100,000 g.

This 20-fold improvement was difficult to achieve, but he succeeded after investing huge sums of money and conquering a variety of problems. The rotor, driven by oil turbines to facilitate lubrication of the bearings, spun in a hydrogen atmosphere to reduce heat production. The substance to be analyzed was placed in a special container and the sedimentation of the particles caused by the gravitational forces was recorded by illuminating the process with a special light source and taking photos at regular intervals. Some hours of calculation were needed to deduce the particle sizes from these photographs.

After experiments with this new ultracentrifuge Svedberg concluded that hemoglobin was indeed a monodisperse protein. After this surprising result, research at his laboratory became almost exclusively focused on proteins. In the following decade he tried to improve his ultracentrifuges, especially to increase the forces that could be generated. Almost all parts of the apparatus were optimized, including the shape of the turbines and turbine chambers, oil inlets, bearings, type of oil used, rotor balancing method, rotor size, and so on.

By 1937 Svedberg considered that he had reached the limit of modifications. He used his latest ultracentrifuge, which generated 400,000 g, until his retirement in 1949, after which the protein research was continued by his former colleagues. It was not until the mid-1970s, half a century after the first apparatus was developed, that the oil-turbine ultracentrifuge was taken out of use.

Vacuum Ultracentrifuge

American physicist Jesse Wakefield Beams studied optical phenomena in the 1920s. In his research he used rapidly rotating mirrors mounted on small conically shaped spinning tops (of the order of 1–2 cm in diameter) that were driven by compressed air. After 1930 he started to make these tops hollow, which allowed him to use them as small centrifuges. Because of the high speeds that could be achieved, Beams called them ultracentrifuges. He identified a wide variety of applications for the apparatus, including Svedberg's method of determining molecular weights.

Edward Greydon Pickels, one of Beams' students, developed the apparatus further for this application. This proved difficult because the high speeds heated the rotor and caused convection currents in the sedimentary solution. Pickels tried various solutions until in 1935 he produced a design in which the rotor spun in vacuum. A small wire, passing through a vacuum-tight gland, connected the rotor to the driving air turbine. This design solved all convection problems, allowing forces up to 1,000,000 g.

This design attracted the attention of scientists at the Rockefeller Institute for Medical Research in New York, and Pickels developed two types of ultracentrifuge further for them. The first type, the analytical ultracentrifuge, was used to determine particle sizes and was analogous to Svedberg's method. The second type, the preparative ultracentrifuge, could separate a substance from a solvent and was primarily used for concentrating viruses. Beams attempted to make his apparatuses as simple as possible to allow many scientists to use them. His instrument makers made some ultracentrifuges for others, which led to a limited distribution in the scientific world. Around 1937 his vacuum ultracentrifuge was marketed by an American company, but this resulted in commercial failure.

Svedberg was also prepared to sell his ultracentrifuges to others, but the apparatus was extremely expensive—in the order of $20 000—which was an enormous amount of money for scientists in those days. In the early 1940s the apparatus was marketed by a Stockholm company, but once again it was a commercial failure.

In 1946 Pickels was approached by a salesman who wanted to market an analytical ultracentrifuge based on Pickels' design. Together they formed Spinco, or Specialized Instruments Corporation. Pickels considered his design at the time too complicated and developed a more easily operated, foolproof version. Sales, however, remained low, and Spinco nearly went bankrupt. Subsequently, Pickels concentrated on developing the preparative ultracentrifuge, which seemed to sell reasonably well. This gave Spinco sufficient financial power to continue the production of the analytical ultracentrifuge as well, although in small numbers. Over the years, that number gradually rose.

Increasing numbers of scientists used ultracentrifuges for biochemical research, forming a research community that met at symposia and conferences at regular intervals. The ultracentrifuge was still not fully developed. As more scientists used them, each with their own approach and interest, new desires and problems in connection with the apparatus were articulated. Over the years a variety of ultracentrifuges have been developed to the point that it became a very common instrument in laboratories for many types of biochemical research.

BOELIE ELZEN

Further Reading

Beams, J.W. High speed centrifuging. *Rev. Mod. Phys.*, 10, 245–263, 1938.

Elzen, B. *Scientists and Rotors: The Development of Biochemical Ultracentrifuges*. Dissertation, University of Twente, Enschede, 1988.

Elzen, B. The failure of a successful artifact—the Svedberg ultracentrifuge, in *Center on the Periphery: Historical Aspects of 20th-Century Swedish Physics*, Lindqvist, S., Ed. Science History Publications, Canton, MA, 1993, pp. 347–377.

Pickels, E.G. High-speed centrifugation, in *Colloid Chemistry*, vol. 5, Alexander, J., Ed. Chemical Catalog Company, New York, 1944, pp. 411–434.

Rånby, B. Physical chemistry of colloids and macromolecules. *Proceedings of the Svedberg Symp*osium, Blackwell, London, 1987.

Schachman, H.K. *Ultracentrifugation in Biochemistry*. Academic Press, New York, 1959.

Svedberg, T. and Pedersen, K.O. *The Ultracentrifuge*. Clarendon Press, Oxford, 1940. Reprinted by The Johnson Reprint Corporation, New York, 1959. German edition: *Die Ultrazentrifuge—Theorie, Konstruktion und Ergebnisse*. Verlag von Theodor Steinkopff, Dresden, 1940.

Ultrasonography in Medicine

The word *ultrasound* refers to sound waves beyond the range of human hearing (oscillations above 20,000 cycles per second or 20 kHz). The dog whistle is an often-used example. Both sound and ultrasound waves are mechanical vibrations that can only propagate through matter (liquids, solids, and to some extent, gases). The composition and temperature of that matter determines the velocity of these waves. Whenever these waves encounter a boundary (such as an organ wall) or interface between two substances, there is a decrease in velocity, and some of the energy is reflected back as an echo, while the rest of the energy passes through to the next interface.

In clinical ultrasonography, a transmitter produces a short pulse (typically a few millionths of a second) of high-frequency electrical oscillations (1 to 10 MHz, or million cycles per second). The transducer, acting like a loudspeaker, converts this electric signal into a pulse of mechanical vibrations of about the same frequency and duration. With the transducer pressed firmly against the patient's body, and acoustically coupled to it with gel or oil, the pulse of ultrasound energy enters with little reflection at the skin and propagates inward through soft tissues and fluids. Then, quiescent and acting as a microphone, the same (or another) transducer senses any reflected, much weaker pulses of ultrasound, and transforms them back into electrical signals. Echoes from distinct organs, blood vessels, and other structures are amplified and processed by the receiver, and are sent to a computer, which keeps track of their return times and amplitudes. About a thousandth of a second later, another pulse is produced and sent off in a slightly different direction through the body, and the whole process begins anew. From echo data generated in this fashion, the computer can create a real-time image in which one sees the arms and legs of a fetus move around or a heart valve open and close.

Ultrasound is particularly useful in the study of soft tissues and organs that are too similar to provide adequate x-ray image contrast. Doppler ultrasound can detect and monitor the flow (or lack thereof) of blood in the arteries and veins. Ultrasound does not use ionizing radiation and usually costs less than other imaging modalities such as computed tomography (CT) or magnetic resonance imaging (MRI).

Ultrasonics originated in the domain of physics. The basic properties of sound were described in the classic work of Lord Rayleigh on *The Theory of Sound* (1877). The discovery of the piezoelectric effect by the French physicists Pierre and Jacques Curie in 1880 provided a way for generating ultrasound waves and became the principle of ultrasonic transducers. They found that certain ceramic materials, such as quartz, could convert electric signals to acoustic signals, such as ultrasound waves, and vice versa. The most common transducers are made of ceramic materials that have been processed to have these piezoelectric characteristics. It was neither possible to use this discovery for generating high-frequency sound for many years in the absence of suitable oscillators, nor to detect the low-amplitude electrical signals generated without suitable amplifiers.

While the early medical applications of ultrasound were diagnostic, those during the interwar years were therapeutic and very conventional. During the 1930s ultrasound was used for therapeutic heating, particularly in physical therapy, for the sterilization of biological preparations such as vaccines, and for cancer therapy, often in combination with x-rays. Some researchers during the late 1930s and early 1940s studied the biologic effects of ultrasound, and a few, such as the French physiotherapist Andre Dernier and the brothers Dussik (neurologist Karl and physicist Friedrich), suggested that ultrasound could be used to produce images of interior body structures such as the brain. However, the development of diagnostic medical imaging using ultrasound had to await the end of World War II and the postwar demobilization of expertise before the military sonar and radar techniques based on echo principles could be more fully applied to medical uses.

The emergence and development of diagnostic medical ultrasound was a worldwide phenomenon. During the decade 1948–1958, researchers in Japan, Europe, and North and South America worked mostly independently, with very little collaboration or exchange of information, to adapt military and industrial ultrasonic equipment to medical uses. In Japan, pioneering work was done by the physicist Rokuro Uchida at the research laboratories of Nihon Musen Company (now Aloka) and by physicians Kenji Tanaka and Toshio Wagai at Juntendo University School of Medicine in Tokyo. Using flaw detectors adapted for medical ultrasonic use they scanned the brain (to detect intracerebral hematomas and tumors), the gall bladder (to detect stones), and the breast (to detect tumors). These early machines displayed information (returning echoes of ultrasonic pulses) in A-mode (one-dimensional) as bright dots on an oscilloscope screen corresponding to points within the body. Another important group in Japan formed around Shigeo Satomura and Yasuharu Nimura at Osaka University. They discovered that the Doppler effect could be applied to ultrasonic energy and used it in their cardiovascular investigations, particularly of heart valves and blood flow, publishing their first results in 1956.

In the U.S., important early work in medical ultrasound was done by the internist George Ludwig, first at the Naval Medical Research

Institute in Bethesda, Maryland and then with the neurosurgeon Thomas Ballantine Jr. at the Massachusetts General Hospital and a group of physicists at the Massachusetts Institute of Technology's Acoustics Research Laboratory; by the surgeon John Wild and engineer John Reid in Minneapolis, Minnesota; by the radiologist Douglass Howry and nephrologist Joseph Holmes in Denver, Colorado; and, by the physicist William Fry and his group at the University of Illinois at Champaign.

Ludwig, using exclusively A-mode presentation, studied the velocity of ultrasound in various animal tissues, which became standards for later researchers, and worked on the detection of gallstones and foreign bodies embedded in tissues—in principle a flaw detection approach.

Wild began his work in 1949 using a Navy radar trainer operating at 15 MHz to measure the thickness of excised bowel tissue. The discovery that echoes from tumor-invaded tissue could be distinguished from those produced by normal tissue in the same sample led him to apply ultrasound to cancer detection, particularly of the breast. Together with Reid he built a B-mode contact scanner, which provided a cross-sectional, two-dimensional picture of the plane of the body scanned (and thus more accurate position information than A-mode). This permitted real-time scanning so that images appeared directly on the screen during the examination with no need for intervening film development.

Unlike Wild whose focus was "tissue characterization," Howry was primarily interested in producing accurate cross-sectional anatomical images as the basis of medical diagnosis. Using surplus U.S. Air Force radar equipment, he built a pulse-echo electronic scanner in his basement in 1949. In 1950 he recorded his first cross-sectional pictures obtained with ultrasound using a 35 mm camera. However, with only a horizontal scanning motion, it was not possible to make the kind of accurate anatomical pictures of living tissue he wanted. However, by 1951, working with engineers Roderick Bliss and Gerald Posakony, Howry built scanners that utilized a cattle watering tank and later the rotating ring gear from a B-29 gun turret as a water immersion tank system to introduce multiposition, or compound scanning, which resulted in the removal of "false" echoes and therefore better images. While one motor moved the transducer around the patient, another provided a second back-and-forth motion, resulting in compound scanning of the immersed subject. The problems inherent in this kind of water-bath coupling system for ill patients were obvious. Other engineers worked with Howry to build a portable scanner that utilized the principle of compound scanning, came into direct contact with the skin, and had an articulated arm to which the transducer was attached and could be easily manipulated and moved. This "Porta-arm" scanner, marketed by Physionics, enjoyed widespread clinical use beginning in 1964.

The work of Fry and his group was in the older tradition of ultrasound use in biophysical investigations and therapeutics, as opposed to diagnostic applications and soft tissue visualization. During World War II and until 1946, Fry worked on the design of piezoelectric transducers at the Naval Research Laboratory in Washington D.C. After establishing the Bioacoustics Research Laboratory at the University of Illinois in Champaign, he concentrated on the use of high-intensity ultrasound as a noninvasive surgical technique to treat neurological or other brain-related disorders such as Parkinson's disease and brain tumors.

The earliest diagnostic applications of ultrasound were made in the specialties of neurology, cardiology, gynecology, obstetrics, ophthalmology, and internal medicine. They were made in Europe, the U.S., Canada, Australia, and Japan. They were made by physicians, who could identify clinical needs, working closely with engineers and physicists, who could provide technical design and engineering skills. For example, in 1953, Inge Edler, a physician at the University of Lund in Sweden, began a collaboration with physicist C. Hellmuth Hertz that eventually launched clinical echocardiography using the pulse-echo technique. At the University of Glasgow's Department of Midwifery in the mid-1950s, physicians Ian Donald and John MacVicar worked with engineer Tom G. Brown and physicist Tom Duggan to introduce ultrasound as a diagnostic modality in obstetrics and gynecology. In the U.S., the physician Gilbert Baum worked with engineer Ivan Greenwood to pioneer the use of ultrasound in ophthalmology.

Steady advances in ultrasound technology occurred in the last half of the twentieth century. The A-mode and B-mode scanners of the 1950s and 1960s gave way in the 1970s to mechanical sector, linear, and phased array scanners, and to grayscale static imaging and real-time. In the 1980s real-time, two-dimensional ultrasound was standard, and computers were harnessed to produce even higher quality images. By the late 1980s color Doppler had been introduced. In the 1990s, the greatest advances were the development of three-dimensional grayscale and color Doppler and the introduction of ultrasound contrast agents to significantly improve diagnostic capabilities. Some of the major manufacturers of ultrasound machines have been: Acuson, Aloka, Diasonics, General Electric, Hewlett-Packard, Kretz, Medison, Philips, Picker, Siemens, and Toshiba.

Almost 25 percent of all imaging studies worldwide are ultrasound. Because of the wide accessibility and utility of ultrasonography, the World Health Organization now recommends its use after basic x-ray rather than more advanced imaging procedures such as CT and MRI.

See also **Nuclear Magnetic Resonance (NMR) and Magnetic Resonance Imaging (MRI); Tomography in Medicine**

RAMUNAS KONDRATAS

Further Reading

Blume, S.S. *Insight and Industry: On the Dynamics of Technological Change in Medicine*. MIT Press, Cambridge, MA, 1992. Chapter 3 deals with diagnostic ultrasound.

Hackmann, W. *Seek and Strike: Sonar, Anti-Submarine Warfare, and the Royal Navy*. Her Majesty's Stationery Office, London, 1984.

Koch, E.B. In the image of science? negotiating the development of diagnostic ultrasound in the cultures of surgery and radiology. *Technol. Cult.*, 34 858–893, 1993.

Tansey, E.M. and Christie, D.A., Eds. *Looking at the Unborn: Historical Aspects of Obstetric Ultrasound*, vol. 5 of the Wellcome Witnesses to Twentieth Century Medicine series. Wellcome Trust, London, 2000.

Wolbarst, A.B. *Looking Within: How X-Ray, CT, MRI, Ultrasound, and Other Medical Images Are Created and How They Help Physicians Save Lives*. University of California Press, Berkeley, 1999. Serious but non-specialized descriptions of how medical images are created. Chapter 7 deals with ultrasound.

White, D.N. Neurosonology pioneers. *Ultrasound Med. Biol.*, 14, 541–561, 1988.

Yoxen, E. Seeing with Sound: A Study of the Development of Medical Images, in *New Directions in the Sociology and History of Technology*, Bijker, W.E., Hughes, T.P. and Pinch, T., Eds. MIT Press, Cambridge, MA, 1987.

Urban Transportation

All the forms of urban transit popular in the twentieth century—buses, streetcars or trams, trolleybuses, cable cars, railroads, and subways—had their origins in the nineteenth century or before. In that century they merely served towns and cities, but in the twentieth century, enhanced mainly by electrical technology, they helped to create new and enlarged communities.

Buses and streetcars were a familiar sight in the nineteenth century. Buses developed from coaches and had their origin in Paris. There, in 1823, Stanislas Baudry, the owner of hot baths in a Nantes suburb, began running his horse-drawn vehicles to them from a stand outside a shop belonging to an M. Omnes, whose business had the slogan "Omnes omnibus"—Omnes for all. Baudry soon found people using his carriages to go to other places, so he renamed them "Omnibuses" and introduced other routes. George Shillibeer spread their use to England. He began a regular timetabled omnibus service in London on 4 July 1829. The fare was half that of a stagecoach, advanced ticket booking was unnecessary, and the omnibuses stopped wherever and whenever requested. Omnibus services were introduced across Europe and in the U.S., especially after the coming of railways.

Streetcars were pioneered in South Wales but developed in the U.S. in the late 1820s. Their use spread to Europe by way of England in the early 1860s. The lower friction between a metal wheel and track meant that a horse could pull over seven times the load of a horse pulling a bus, and new rails offered a much smoother ride than the poor roads. Thus it was to the streetcar that technology was applied, first to assist, then to supplant, the horse. Many forms of mechanical traction were tried, but only two succeeded—steam and cable—with only the latter surviving significantly into the twentieth century.

Cable Cars

Andrew Hallidie, an English-born wire rope manufacturer living in San Francisco in the late 1860s, largely devised cable car technology. After witnessing the cruel treatment of streetcar horses struggling up one of that city's famous hills, he devised a horse-free system using cable haulage. His first line ran up Clay Street and opened on 1 September 1873. It was 1584 meters long and rose 94 meters in that distance. A double track was laid. A steel wire rope of 3,353 meters was carried in an iron tube and supported on pulley wheels placed at 12-meter intervals. The cable was wound on a drum turned by a steam engine. The cars were powered by a constantly moving, endless loop of wire rope that traveled at a constant speed of 10 kilometers per hour. A grip mechanism protruded down from the car through a slot in between the running rails and into a second slot in the top of the tube carrying the cable. When the driver turned a hand wheel, the jaws of the grip opened or closed. In the closed position its jaws gripped onto the cable, and the car moved forward. Armed with only this gripper mechanism and a handbrake, the driver had to control his cable streetcar.

Worldwide, 81 towns and cities had cable streetcars. Their use was suited to places with hills, and they found most favor in the U.S. where 66 locations in 18 states built at least one line. There were also nine lines built in the U.K. (Birmingham, Douglas, Edinburgh, Glasgow, Liverpool, Llandudno, two in London, and Matlock), one in France (Paris), two in Portugal (Lisbon), two in Australia (Melbourne and Sydney), and four in New Zealand (Dunedin). However, the expense of extending existing lines or laying down new ones meant that cable streetcars were rarely modernized, as demonstrated by the fact that in those few places where they do survive, they are period pieces or curiosities of urban transit history, maintained for tourist purposes. Also, the technology did not allow cars from one route to pass over another one, each being a separate line. Thus the routes were inflexible, and it was their passengers who had to change, from line to line and car to car. As towns and cities developed rapidly in the early part of the twentieth century, they quickly outgrew their cable cars. As technology improved the electric streetcar, cable cars even lost their advantage for climbing hills.

Almost two thirds (41) of the U.S. cable cars had closed by the end of the nineteenth century, and there was only one new line opened anywhere in the world during the twentieth century—a route extension in Dunedin, New Zealand, which only lasted three years. Most of the remainder had closed by World War II, including the 41 kilometers of lines in Edinburgh, which ended on 23 June 1923, and the massive 75 kilometers of lines in Melbourne, which used 1200 cars and last ran on 26 October 1940. Both of these systems were replaced by electric traction.

Electric Streetcars

Experiments with electric traction on railways began in the 1830s, but it was not applied to urban transit until the late 1870s. On 31 May 1879 Dr Werner von Siemens demonstrated the first practicable application of electricity to the mass movement of people with a 320-meter line at the Berlin Industrial Exhibition. A small four-wheeled locomotive drew 150 volts from a third rail to power an electric motor, the current returning via the running rails. It hauled three passenger cars, each seating six people. Over the course of the exhibition, this carried over 80,000 passengers. On 12 May 1881, von Siemens opened a more permanent electric streetcar, linking the railway station at Lichterfelde outside Berlin with its Cadet School, some 2.5 kilometers, this time using only the running rails to conduct the 100 volt power. Other streetcars employing electrical conduct through their running or auxiliary rails opened in Ireland and the U.K. in the early 1880s, while in the U.S. there were experiments being made to conduct electricity thorough overhead wires. Leo Daft, John C Henry, Charles van Depoele, and others all used twin wire systems: one positive, one negative, along which ran a small four-wheeled "troller"—resembling a basic roller skate—attached to the streetcar by a cable. This system worked, but was not without its problems. The trollers were prone to come off the wires, and did not stop when the streetcar they were attached to did!

Frank Sprague rescued electric streetcars from becoming technological curiosities. Bound by a ridiculously tight contract to build an electric streetcar system in Richmond, Virginia, across steeply graded land over which trollers were useless, Sprague devised and perfected a single wire system, with a counter-sprung trolley pole and wheel, which conducted electricity through the streetcar, returning it via the track. He also refined the design of the electric motors and control equipment required and devised a means of mounting the motors in the streetcar's truck so that their performance was unaffected by rail joints and track undulations. When his Richmond system opened on 2 February 1888, Sprague had built the prototype for the majority of streetcar systems that would follow anywhere in the world. In essence it worked like this: the streetcar completed an electric circuit between the positive overhead wire and the negative track. Current from the wire passed down a cable in the trolley pole through a circuit breaker to a controller. This housed a series of electrical resistances, switched in and out of circuit by the action of the streetcar driver rotating a handle to preset, notched positions, which progressively allowed more and more current through. From the first few notches the current passed to the motors in series—one after the other—which produced the maximum torque to move the streetcar from stationary. Once moving, higher notches passed the current to the motors in parallel—at the same time—which kept them moving at the same speed. Advances in technology improved the performance of electric streetcars but this basic principle of their operation was unchanged.

Few technologies spread faster than the electric streetcar in the late nineteenth and early twentieth centuries. Their speed and power made them ideal for moving large numbers of people quickly. Streetcars could be loaded well beyond their stated capacity without any loss of performance, and they could be run in safety more closely together than trains. Few major towns and cities failed to have them, and some, such as Liverpool, planned new routes along with the building of new housing estates. In the U.S., electric streetcars were also built to link towns and cities. Called "interurbans," these were built in two main waves from 1901 to 1904 and from 1905 to 1908. By 1916 a maximum of 25,074 kilometers of lines had been built. Ohio had the most with 4503 kilometers; in the 1910s a journey of 1749 kilometers was possible using connected interurban systems.

Despite their obvious advantages, electric streetcars also had their drawbacks. The urban growth they fostered sometimes outpaced them, either requiring costly new routes and extensions, or providing a toehold for motorbuses. Streetcar rails also wore out, requiring constant repair and needing replacement every 15 to 20 years. In the U.K., crippling legislation placed huge financial burdens on operators for the upkeep of the roads, while in the U.S. a company called National City Lines was formed in 1936 to acquire local systems throughout the country and convert them to motorized bus operations. In 1949 it was revealed that National City Lines had ties with General Motors, Firestone, Standard Oil, and Phillips Petroleum, companies that had an interest in supplying buses, tires, and fuel. Elsewhere streetcar systems have proven remarkably resilient, and their value has been underscored by the fact that many European systems devastated during World War II were effectively rebuilt afterward. Others brought to the brink of closure, such as that in Melbourne, have been reequipped and given a secure future.

Motorbuses

Many applications were tried for the internal combustion engine in the late 1890s, including powering buses. The first regular service of gasoline-engine buses was introduced in London on 9 October 1899, using German Daimler engines. Seating capacities were low on these early motorbuses, typically 14 to 16 people, but the Peckham–Oxford Circus service, introduced on 30 September 1904 by Thomas Tilling, used 34-seater Milnes–Daimler buses. In 1905 the first motorbus services were introduced in Paris and on Fifth Avenue in New York, and they were introduced on lightly trafficked routes in Vienna in 1907. A measure of the success of these early motorbuses is given by figures for those in use in London: in 1905, 20 were in use; by 1908, 1066 were in use. The development of larger and more powerful engines allowed the seating capacity of motorbuses to rise, such that by 1919 the London General "K" type seated 46, and by 1926 Daimler–Benz of Germany produced a 60-seater double-decker, one of the first buses capable of taking a streetcar's load of passengers.

At first motorbuses were used on lightly trafficked routes, where there were insufficient passengers to justify the building of a streetcar line. They were also used to extend streetcar routes beyond their termini, to serve ever-growing suburbs. Once the seating capacity approached that of streetcars, buses posed a direct threat to the latter; and the introduction of pneumatic tires, improved suspension, and padded seats gave superior passenger comfort with which increasingly aged streetcar fleets could not compete. Finally, the wider availability of diesel engines from the late 1920s offered bus manufacturers and operators greater power at reduced cost, as they ran on a heavy oil produced at an early stage in the refining process, which was cheaply produced. Diesel engines were first used in buses in the U.K. in Sheffield in 1930 and very widely in the U.K. and elsewhere after 1933.

Trolleybuses

From 29 April 1882, streetcar pioneer Dr. Werner von Siemens began experimental operation of his 'Electromote', a four-seat, four-wheel dogcart electrically driven by power drawn from a pair of overhead wires via a troller device. His 550 meter line, which ran through the streets of Hallensee in Berlin, can be seen as the forerunner of the trolley bus. Essentially a hybrid of the streetcar and the bus, a trolleybus draws its power from a pair of overhead wires, but then technically works in much the same way as a streetcar, save for the fact that it is not bound to any rails. Therefore it benefits from the power of electric traction, but has the added ability to steer around obstacles that would otherwise block a streetcar's progress. A 6-kilometer trolleybus route served the 1900 Paris Exhibition, and the early 1900s witnessed experiments with trolleybuses in Austria, France, Germany, Italy, and Norway, where a 4-kilometer route opened in Drammen on 10 July 1902. Trolleybuses were introduced in Laurel Canyon in the U.S. in 1910, and in Chicago on 8 October 1914.

Most of the development work on the vehicles happened in the U.K. where, on 20 June 1911 the country's first trolleybus services were introduced in Leeds and Bradford. Up to 1926, trolleybuses developed as trackless streetcars; after 1926 they began to develop as electric buses. This change was largely due to the General Manager of Wolverhampton Corporation Transport—Charles Owen Silvers. Wolverhampton's streetcar system had been built using an unusual surface contact method of current collection for which spares became impossible to obtain by the early 1920s. A decision was taken to install overhead wire power distribution, followed almost immediately by a second decision—to scrap the streetcars. Thus, faced with an almost new power distribution system, trolleybus operation seemed to offer the best alternative to streetcars. The rough riding of the vehicles on the first route to Wednesfield, opened in July 1923, made Silvers examine their design. He experimented by putting trolleybus equipment in the latest design of motorbus body, creating a light, lively, and comfortable vehicle and a showcase transport system. By 1929 Wolverhampton had the largest trolleybus system in the world, and Owen Silvers played host to many delegations from overseas transport operators. Indeed, across the world, a number of trolleybus systems owe their existence to an initial visit to Wolverhampton.

Like motorbuses, trolleybus operation was first seen as an adjunct to that of streetcars, but they later became a direct competitor, enabling, as at Wolverhampton, the life of streetcar power systems to be extended. At the end of the twentieth century, trolleybuses could be found operating in many parts of the world, with the largest number of systems to be found in the former Soviet Union (185); followed by Eastern Europe (58); Western Europe (46); China (25); South and Central America (10); North America and Canada (9); East Asia (8); West Asia (2); and Australia and New Zealand (1).

Underground Railways (Subways, Metros)

Underground railways were a nineteenth-century solution to providing rapid urban transit to and from the heart of busy cities, without further clogging the streets. As with other forms of urban transit, they were transformed in the twentieth century by the application of electric traction to their operation. The world's first underground was the Metropolitan Railway, which was built to connect the center of London with its mainline railway termini. It was operated by steam locomotives and opened on 9 January 1863. Ten further railways, mostly built by separate companies between 1868 and 1907, expanded this line to produce the London Underground system, which has over 275 stations and 407 kilometers of underground tunnels, in some cases stacked three levels deep.

London had a monopoly of subways for almost 30 years, during which time it also opened the world's first electric underground railway—the City and South London—on 4 November 1890. This was also the world's first tube railway, but was only 3.01 meters in diameter and 12 meters below ground. Decidedly cramped, it was nicknamed the "sardine-box railway." The world's second tube system in Chicago demonstrated the best alternative to an underground—the elevated railway. Chicago's "El" was in fact four lines, opened between 6 June 1892 and 31 May 1900, the latter also being the first electric line. Elevated railways had been pioneered in New York by Charles T Harvey, who built an experimental single-track cable-powered elevated railway from Battery Place, at the south end of Manhattan Island, northward up Greenwich Street to Cortlandt Street. Dubbed the "one-legged railroad," the 800-meter single track was carried above the street on a row of single columns. Driven by a stationary engine, the cable was a loop that ran between the rails for propulsion of the cars, and then returned under the street. Harvey's railroad opened for business on 1 July 1868.

Whether under- or overground, the advantages of these urban transit systems are that they occupy dedicated, reserved, and often isolated tracks, which they do not have to share with trains belonging to other operators. Add the acceleration of electric traction, and this gives an urban transit system capable of moving millions of people in and out of cities daily. By the close of the twentieth century there were 124 metro systems worldwide, with 48 in European cities, 42 in Asia, 29 in the U.S., 3 in Africa and 2 in Australia. Their construction has also produced many heroic feats of engineering and outstanding examples of modern architecture. Notable examples are:

- The Liverpool Overhead Railway opened on 4 February 1893.
- Budapest's Metro opened in May 1896 between Vörösmartytér in the center and Széchenyifürdö. The second electric underground railway in the world and the first electric line in Europe.

- The Glasgow Subway opened for traffic on 14 December 1896, but closed the same day following an accident, the service not opening for regular working until 19 January 1897.
- The first line of the Paris Métro opened on 19 July 1900, between Porte de Vincennes and Porte Maillot, making it Europe's fourth oldest metro system, and probably the densest.
- The New York City subway system officially opened on 27 October 1904. It forms 1100 kilometers of New York City's 3200-kilometer public transit system, which is the largest in the world.
- The Moscow metro's first line opened on 15 May 1935 between Sokol'niki and Park Kul'tury, with a branch to Smolenskaya, which reached Kievskaya in April 1937, crossing the Moskva river on a bridge. On a normal weekday it carries 8–9 million passengers, some 3,000,000,000 a year, which is more than the London and New York systems combined.

These transit systems often played a vital role in the development of the cities they served. In the latter half of the twentieth century, major North American cities added subway systems to well established cities such as Washington D.C., San Francisco, and Montreal in an attempt to ease bus and automobile congestion by providing faster alternatives.

Suburban Electric Railways

The first railways usually served local areas by accident, as they passed through en route to larger towns and cities. Typically, suburban railways were built in response to urban growth, a generation or so after the first lines. As with other forms of urban transit, suburban railways were revitalized through electrification. A good example is the lines in the southeast of England serving London. These were electrified progressively from 1 December 1909, with most being converted between 1 April 1925 and 2 July 1939. Electric trains are not only more capable of moving people *en masse* than steam hauled ones, they also allow for a more responsive service. Extra trains can be pressed into service in minutes, as opposed to the 2 hours or so required to steam up a locomotive. Since the 1980s, rail operators in Europe and North America have also increased the capacity of their trains by using double-decked carriages. In Japan, employees help "pack" the subways and move as many people as possible on each train.

See also **Electric Motors; Rail, Diesel and Diesel Electric Locomotives; Rail, Electric Locomotives; Transport**

PAUL COLLINS

Further Reading

Aldridge, J. *British Buses Before 1945*. Ian Allan, Shepperton, 1995.
Clapper, C. *The Golden Age of Tramways*. Routledge & Kegan Paul, London, 1961.
Collins, P. *The Tram Book*. Ian Allan, Shepperton, 1995.
Dunbar, C.S. *Buses, Trolleys & Trams*. Paul Hamlyn, London, 1967. Integrated story of urban transit, along with Charles Clapper's book.
Halliday, S. *Underground to Everywhere*. Sutton Publishing, London, 2001.
Hanscom, W.W. *The Archaeology of the Cable Car*. Socio-Technical Books, Pasadena, CA, 1970.
Howson, H.F. *London's Underground*. Ian Allan, Shepperton, 1986.
Kinder, R.W. Revision of Dendy Marshall's *History of the Southern Railway*. Ian Allan, Shepperton, 1968.
Whipple, F. *Electric Railway*. Orange Empire Railway Museum, Perris, CA, reprint of 1889 publication.

Useful Websites

David Pirmann's encyclopedic site on the New York Subway is one of the oldest and most comprehensive websites, but many others provide information on individual transit undertakings: http://www.nycsubway.org.

V

Valves/Vacuum Tubes

The vacuum tube has its roots in the late nineteenth century when Thomas A. Edison conducted experiments with electric bulbs in 1883. Edison's light bulbs consisted of a conducting filament mounted in a glass bulb. Passing electricity through the filament caused it to heat up and radiate light. A vacuum in the tube prevented the filament from burning up. Edison noted that electric current would flow from the bulb filament to a positively charged metal plate inside the tube. This phenomenon, the one-way flow of current, was called the Edison Effect. Edison himself could not explain the filament's behavior. He felt this effect was interesting but unimportant and patented it as a matter of course. It was only fifteen years later that Joseph John Thomson, a physics professor at the Cavendish Laboratory at the University of Cambridge in the U.K., discovered the electron and understood the significance of what was occurring in the tube. He identified the filament rays as a stream of particles, now called electrons. In a range of papers from 1901 to 1916, O.W. Richardson explained the electron behavior. Today the Edison Effect is known as thermionic emission.

Two further experiments on the Edison Effect were significant contributions to the development of vacuum tubes. John Ambrose Fleming, a British student of James Clark Maxwell, working for the British Wireless Telegraphy Company, attached a bulb with two electrodes to a radio receiving system in order to improve the reception of wireless radio signals. Fleming patented the first electronic rectifier, the diode, or Fleming Valve in 1905. The vacuum tubes were indispensable to switch currents. For example, the Fleming Valve was incorporated in Guglielmo Marconi's wireless system in which communications could take place over great distances. The most serious limitation of the Fleming Valve, however, was that it was insensitive to changes in the concentration of occurrence in electromagnetic radiation. Moreover, the valves consumed large amounts of power.

In 1906 Lee de Forest invented the vacuum tube that became the most important electronic device in the first half of the twentieth century. To Fleming's tube he added a third electrode called a grid—a network of small wires surrounding the cathode to control the stream of electrons. This was the original "triode:" a glass tube containing a filament that heated a plate (the cathode) that emitted electrons, a collector plate (the anode) that collected the electrons, and a metal grid in between. The current flow from the cathode to the anode was highly dependent on the voltage of the grid and the fact that the current drawn by the grid was very low. The name triode came from the three active elements of the device—the anode, the cathode, and the grid. This tube was used as the controllable valve in electronic circuits; for example, to amplify music and voice and to make long-distance calling practical. De Forest called the tube the "Audion," which he patented in 1907 (see Figure 1). Fleming, who was aware of de Forest's activities, had disputed the American's claims to originality, which meant a lifelong enmity between the two inventors.

The first tubes were poorly constructed and able to handle only low levels of power. A further improvement of the vacuum tube came from Irving Langmuir who was working at the Research Laboratory of General Electric. Like some

Figure 1. Early DeForest Audion 1906. [*Courtesy of the David Sarnoff Library*].

other physicists at this time, de Forest believed that electron emission in high-vacuum conditions was impossible. Langmuir, however, studied the emission of electrons by hot filaments in a vacuum, because he realized that the Audion did not require the presence of gas to operate. This led to the invention of the high-vacuum electron tube in 1912 that could operate at much greater power levels. These tubes were used as current amplification devices for radio and telephone systems.

In addition to practical problems such as inadequate vacuum pumping, mechanical resonance, and poor welds, the vacuum tube had a major problem: the irregular arrival of electrons at the anode of thermionic tubes. Technologists were trying to solve this problem by adding extra grids to the tube. In 1919 Walter Schottky invented the first multiple grid vacuum tube, the Tetrode. He had found that the fundamental source of noise would be caused by the randomness of the emission from the cathode. Therefore he added a second grid, called the screen, to prevent the tube from producing unwanted oscillations. Bernard D. H. Tellegen, who worked at the Philips Research Laboratory in the Netherlands, made a further improvement to the multiple grid vacuum tube. He invented an advanced receiver tube known as the Pentode. His Pentode tube was a thermionic valve with five electrodes. After Tellegen developed the tube, the most common way of suppressing secondary emission was by using a suppressor grid, placed between a screen grid and the anode. The suppressor grid was maintained at the filament potential level, and it was able to hold back secondary electrons. In 1926 a patent was requested for this invention, and in the same year Pentodes were built into radios.

In 1920 Albert W. Hull of General Electric invented a special vacuum tube called the magnetron. The magnetron is a tube in which the electron beam is controlled by a magnetic field, making extremely high frequencies possible. It's invention led to the development of radar. A major invention was also made by two brothers—Sigurd and Russell Varian—at Stanford University, who applied the principle of velocity modulation to the tube. This high power microwave oscillator used a linear electron beam. It was called the klystron and was built into British and U.S. radar in World War II.

Although many improvements were made, the vacuum tubes still used a lot of electrical power, most of which ended up as heat, thus shortening the life of the tube. This became a big problem in the 1930s because the apparatus for radio, telephony, and computing needed receivers that required high power. Early digital computers like the ENIAC, which was built for the U.S. Army in the 1940s but only completed after the close of the war in late 1945, had about 19,000 vacuum tubes in them, which produced too much heat. The invention of the transistor meant that smaller, reliable, and less power-hungry devices could be produced. Although vacuum tubes are still used, the success of the transistor replaced the vacuum tube technology. After World War II, the transistor became the key to further developments in electronic technology.

See also **Radio Receivers, Early; Radio Receivers, Valve and Transistor Circuits; Transistors**

Kees Boersma

Further Reading

Aitken, H.G.J. *The Continuous Wave: Technology and American Radio 1900–1932*. Princeton University Press, Princeton, 1985.

De Forest, L. The Audion—Its Action and Some Recent Applications. J.Franklin Inst., 1920.

Dunsheath, P. *A History of Electrical Engineering*. Faber & Faber, London, 1962.

MacGregor-Morris, J.T. *The Inventor of the Valve: A Biography of Sir Ambrose Fleming*. London, 1954.
Tyne, G.F.J. *The Saga of the Vacuum Tube*. SAMS, Indianapolis, 1977.

Vertical Transportation (Elevators)

Despite the popular concept that the elevator was born at the Crystal Palace Exposition, it actually originated in New York City in 1853 when inventor Elisha Graves Otis first successfully demonstrated his revolutionary new concept—the elevator safety gear or break—which was to allow passengers to travel with safety. The true modern passenger elevator was conceived due to a catastrophic event in 1871 known as the Great Chicago Fire, when a three-day fire razed the city to a desolated wilderness on the plains of Illinois. This fateful day on the 8 October 1871 pinpoints exactly the beginning of the modern elevator.

From the ashes of the old Chicago, the City Fathers made the bold decision that the new City should be constructed from modern materials. The most critical of these forward-looking inventions was the rolled steel joist (RSJ), and by splicing, riveting, and welding together these steel sections, the new construction technique was unveiled. The population explosion in North America required that the new Chicago had to be rebuilt vertically, and with the need to build higher, a new form of elevator was required. The water hydraulic elevator concept had been used since the 1830s, but this design had its limitation on vertical rises due to the restrictions in pressure and the supply of high-pressure water mains. This restricted the vertical rise of the elevator and effectively limited the height of the building. The steam-driven elevator also had its origins in this period, with the first recorded use in 1835.

The first modern passenger elevator in a commercial building was installed in the ten-story Home Insurance Building in Chicago, completed in 1884. Some 13 years after the Great Fire, this building used a steel structure of half cast iron and half steel with supporting masonry walls. The modern elevator age was born with the advent of basic skyscraper technology.

Although the dynamics and control systems of these early elevators were only just evolving, the finishes, materials, craftsmanship, and designs of elevators were in the "Golden Age," when second best was not good enough. With the advent of more complicated control systems and additional safety equipment, components had to be fitted in the elevator shafts and on the elevator cars. This coincided with the beginning of the mass production era, when the large elevator manufacturers realized that this technique enabled more elevators to be built. In addition, new building fabrication techniques also required the elevator engineer to develop new vertical transportation techniques to move passengers higher, faster, and more safely.

Elevators in apartment blocks, banks, and commercial buildings with their imposing entrance lobbies were the symbols of wealth and power. Buildings became active, and they required vertical motion for commerce, bringing spaces alive, and moving people skyward not only rapidly but with grace. The elevator ride was becoming a travel experience to be enjoyed by all. In 1893 the five-story Bradbury Building in Los Angeles brought the elevator into the glass-roofed space spanning the gallery balconies. These elevator cars used open metal designs of the first true panoramic elevator in what John Portman would classify 60 years later as an atrium.

As the new quicker methods of solid elevator shaft construction were being adopted, the passenger's view in and out of the elevator was becoming more and more restricted and the use of glass and decorative materials in elevators was being quickly phased out. For half a century people were moved at ever-increasing speeds, with no views in or out, although the elevator finishes were in some instances very elaborate. The norm, however, was for metal panels (stainless steel, brass, or bronze) or painted finish with the occasional wood panel or mirror. A notable exception to this style was the 1936 Frank Lloyd Wright design for the Johnson Wax headquarters building in Racine, Wisconsin. There, circular observation elevators were used to move travelers through the gallery levels.

It was not until 1962 that architect John Portman had the inspiration to provide elevators in the atrium of a hotel he was designing, not in solid shafts but in an open environment within the building. This revolutionary elevator can be dated exactly to 1965 when the Regency Hyatt, Atlanta opened its doors to the first guests as well as the simply curious.

By the 1980s a new approach to wall-climbing elevators was being adopted. The "flying keyhole" and "birdcage" designs had run their design life, except in hotels, shopping malls, department stores, and speculative commercial developments where corporate and standardized design policies were used. The honest engineering approach was now in vogue with adventurous architects and designers, in which the elevator components and their function were highlighted and expressed.

From the mid-1980s, panoramic elevators had a very high percentage of glass in all visible surfaces to the elevator car. Old skills and craftsmanship were being rediscovered, and computer technology was being incorporated extensively in the control, drive, and design processes to provide not only a good elevator ride but also an experience for the passenger.

Architect Richard Rogers exploited this "total ride" concept in the heart of London in 1986 with his twelve-story neo-Gothic Lloyds of London, where the three elevator towers not only incorporate the elevators but are also fully external. Elevators had to open onto the building lobbies without any of the external environmental conditions penetrating the occupied space of the building. Elevator engineers had to develop new systems for air conditioning, water proofing, and designs to prevent ice and snow build-up on vital safety equipment, while still maintaining reliability, passenger confidence, and safety. This totally new design approach, which celebrated engineering components, had become an essential part of the design of the elevator.

Architects and engineers have always pushed the frontiers of design, none more so than I.M. Pei's Grand Louvre in Paris. The disabled-access elevator within the glass pyramid demonstrates that innovation and excellent design can be achieved on a two-stop low-rise elevator. This elevator is not only a moving sculpture wrapped around a spiral staircase but also, more importantly, a very functional elevator, offering visitors with a disability the opportunity to visit this world famous museum.

The elevator continues to bring new challenges and design opportunities that have been excluded from the elevator engineer's design brief since the advent of the skyscraper, when advanced engineering concepts of their day were required to build structures higher and make elevators travel faster than previously considered safe. Today we understand about height and speed; we now need to know and explore the experience, not in terms of being frightened but in terms of enhancing and experiencing the architecture and design of buildings. The kinesthetic experience (the experience of a body moving through space) will become a vital consideration in future elevator design. Elevator travel has become not just a necessity of the modern city and perhaps even the home but a totally pleasurable experience for the passenger's senses.

See also **Skyscrapers**

ROGER HOWKINS

Further Reading

Elevators. McGraw-Hill, 1927, 1935, and 1960.
Jallings, J.H. *Elevators*. 1918. Republished by Elevator World, 1995.
Strakosch, G. *The Vertical Transportation Handbook*. Wiley, 1983.
Installation Manual. NEMI, 1964.
Howkins, R.E. *Lift Modernisation Design Guide*. Elevator World, 1998.
Philips, R.D. *Electric Lifts*. Pitman, 1973.
Elevator World: U.S.-based monthly magazine.
Elevation: U.K.-based quarterly magazine.

Video Recording, *see* **Television Recording, Tape**

Vitamins

Vitamins are organic compounds essential in minute quantities for the maintenance of health and normal development in most animals and some plants. Many vitamins act as coenzymes or precursors of coenzymes to aid in regulating metabolic processes to produce body tissues or to store or release energy. Others are related to the sterols and hormones. Many vitamins are synthesized by plants and are present in the food intake of animals, though a few are synthesized in the animal body. The existence of such "special factors" in food has long been suspected; for example it has been known since the seventeenth century that fresh fruits and vegetables in the diet prevent scurvy. In 1890 the Dutch physician Christiaan Eijkman showed that beriberi resulted from eating "polished" rice, or rice from which the husks had been removed. When the husks were added to the diet, the disease disappeared, and it was recognized in 1901 that beriberi was due to a dietary deficiency.

Modern vitamin theory dates from 1912 when Sir Frederick Gowland Hopkins showed that animals did not thrive when fed on carefully purified fats, proteins, carbohydrates, mineral salts, and water; but that a very small addition of milk was enough to render the purified diet adequate. In the same year, Casimir Funk at the Lister Institute in London found that the anti-beriberi factor in rice-husks was an amine and propounded his "vitamine" (vital amine) theory, linking vitamin deficiency to other diseases including pellagra and rickets (rachitis). The term was applied to "accessory factors" generally, but as various chemical structures and functions were identified among these compounds and many were found not to be amines at all, they were called vitamins.

Table 1 Recognized vitamins and growth-factors.

Vitamin	Chemical Name	Biological Function	Occurrence
A fat-soluble (discovered 1909) (synthesized 1947)	Retinol	Absence causes night blindness and epithelial tissue becomes keratinous	Egg yolk, milk, fish liver oils
D fat-soluble D2 D3 (disc.1918) (synth. 1959)	Related to steroids ergocalciferol cholecalciferol	Essential for bone and tooth structure; prevents rickets in children and osteomalacia in adults	Egg yolk, milk, fish liver oils. Formed by UV irradiation of sterols
B water-soluble B1 (disc.1897) (synth. 1936) B2 (disc.1933) (synth. 1935) B3 (synth.1941) B6 (disc.1934) (synth. 1939) B12 (disc. 1948) (synth. 1972)	Thiamine Riboflavin Nicotinic acid (Niacin) Pyridoxine Complex cobalt compound Cyanocobalamin	Vertebrate nutrition Growth-promoting factor Pellagra preventive Absence causes skin lesions Anti pernicious-anemia factor. Essential for normal blood formation	Yeast, seed-germs, eggs, liver, flesh and some vegetables Yeast, vegetables, milk, liver Yeast, rice, heart, muscle and liver (nicotinamide) Liver; widely in animal tissues
C water-soluble (disc.1912; synth.1933)	Ascorbic acid	Preventive for scurvy; food antioxidant	Fruits; green leafy vegetables; made synthetically
E fat-soluble (disc.1922; synth.1939)	Tocopherols	Animal nutrition	
H (disc.1931; synth.1943)	Biotin	Protects against toxins in raw egg white	Egg yolk; liver; yeast
K fat-soluble K1 (disc.1929) (synth; 1939) K2	Phylloquinone Menaquinone	Produce prothrombin essential for blood-clotting	Synthesized by intestinal bacteria in animals
Pantothenic acid (disc. 1933) (synth. 1940)	Pantothenic acid	Important in animal nutrition; present in coenzyme A	Found widely in animal tissues
Folic acid water-sol. (disc.1941; synth.1946)	Folic acid	Prevents anemia	Green leaves; some animal tissues

In 1915, Americans E. V. McCollum and M. Davis showed that there were at least two kinds of vitamins, one soluble in fatty and the other in nonfatty foods. They were later called vitamin A and vitamin B, respectively, and each has since been found to be a group of compounds with related functions. The water-soluble group includes vitamins B and C. The vitamin B complex contains a large collection of compounds, all of which are essential constituents in various enzyme systems. Similarly the fat-soluble vitamin group is now known to include vitamins A, D, E, and K. The physiological functions of most of the vitamins have been examined in detail and their molecular structures have been determined, but their dietary importance has ensured the everyday use of the original alphabetical classification.

Water-soluble vitamins are absorbed in the intestine and carried in the blood to the tissues. They are distinguished from each other by the degree of solubility, a factor that influences their route in the body. In their free state, the B vitamins are inactive, and they must go through several chemical processes before they can perform their functions in the body. When combined with other substances, they are changed into their functional, or coenzyme, form and can then combine with proteins to form active enzymes that catalyze various metabolic and regulatory processes. The fact that vitamin C prevents and cures scurvy is well known, but the vitamin is also essential for the growth of bones and teeth, for the maintenance of subcutaneous tissues and the walls of blood vessels, and for the healing of wounds. A controversial theory suggests that the intake of large quantities of vitamin C can prevent or cure the common cold, but there is no clear evidence to support this claim. When the intake of water-soluble vitamins exceeds the bodily requirements, they are stored to a limited extent in body tissues, but most of the excess is excreted in the urine.

The fat-soluble vitamins are absorbed with the help of bile salts and are carried through the body in the lymphatic system. The body stores larger quantities of fat-soluble vitamins than of water-soluble ones. The liver provides the chief storage tissue for vitamins A and D; vitamin E is stored in body fat and in the reproductive organs. Vitamins of this group perform various functions. For example, vitamin A combines with proteins in the retina of the eye to aid night vision, though it may have other functions as yet uncertain. The anti-rachitic factor was labeled vitamin D; it is essential to growth, especially in calcium metabolism for bone formation and the avoidance of ricketts. Produced in the skin on exposure to sunlight, vitamin D is one of the few vitamins synthesized by animals. In 1922 a new factor called vitamin E, which facilitates animal growth and maintains fertility in rats and some other species, was identified. Vitamin K is essential for the enzymatic processes of blood clotting.

The number and variety of recognized nutritional factors is continually growing, though not all are essential to animal or human nutrition. Moreover, the definition of a vitamin is imprecise because the metabolic functions of enzymes, coenzymes, sterols, hormones and vitamins are closely related, and much detailed research is required to differentiate between these and other groups of biochemical molecules. Vitamins are not distributed equally throughout nature, nor do they perform the same functions in all species. Sometimes a compound that can act as a vitamin is present in combination with another compound that prevents its absorption and so destroys its activity. Both plants and animals are important natural sources of vitamins for human health, and the more restricted the diet the more likely it is that one or more vitamins will be lacking.

The vitamin requirements of most organisms are fairly well known, but there is not uniform agreement about the requirements for a healthy human diet. Differences arise due to the various ways requirements are determined and to the scanty data available for some of the vitamins. Studies in this field give rise to the subject of human nutrition in which the quantities of the various vitamins figure importantly, along with minerals and other essential trace elements in the diet. It is generally thought that a balanced diet supplies all the vitamins needed for a healthy lifestyle, but this is not necessarily true. Long storage of fruits and vegetables after harvesting may result in the loss of vitamin C due to oxidation. Washed vegetables lose water-soluble vitamins; heating and overcooking destroys them. Manufactured foods therefore often require added vitamins to replace such losses. Sometimes extra vitamins are added to raise the proportion above its natural level, and many people also take vitamin supplements. Milk fortified with vitamin D was first introduced by the Borden Company in the U.S. in 1933. As vitamins were isolated and then synthesized, they could be manufactured by pharmaceutical companies. Vitamin C was first to be synthesized in a laboratory in 1933, followed by vitamin B2 in 1935. By 1936, vitamin and iron supplement sales were widely available in the U.S. In Britain, though not in the U.S., vitamin supplements are classified as drugs.

The chemical structures of all the known vitamins have been determined and most can now be manufactured synthetically by chemical or biochemical processes. The quantity of each product is governed by economic considerations and wide variations in annual production occur. Thus, only about 10 tons of vitamin B_{12} are manufactured each year, but 50,000 tons of vitamin C (ascorbic acid) are made. Manufacturers have developed variations in the synthetic processes, most of which are the subject of trade secrets. Synthetic vitamins produced in bulk are used in the food industry, and large quantities are added to animal feed, but there are also other uses. For example, some vitamins act as antioxidants, and this property has been used in the manufacture of plastics. A small but growing market is also developing for vitamins that benefit the skin, and vitamins are beginning to appear in cosmetics, skin creams, lotions, and shampoos. Research to identify new vitamins and growth factors continues along with efforts to discover more economical methods of manufacture and new uses for the known vitamins.

N.G. Coley

Further Reading

Apple, R.D. *Vitamania: Vitamins In American Culture*. Rutgers University Press, New Brunswick, 1996.

Brewer, S. *The Daily Telegraph Encyclopaedia of Vitamins, Minerals and Herbal Supplements*. Robinson, London, 2002.

Diplock, A.T. *Fat-Soluble Vitamins: Their Biochemistry and Applications*. Heinemann, London, 1985.

Djerassi, D. *The Role of Vitamins in Skin Care*. Hoffmann–La Roche, Nutley, NJ, 1993.

Dyke, S.F. *The Chemistry of the Vitamins*. Wiley/Interscience, London, 1965.

Griffith, H.W. *Vitamins, Herbs, Minerals and Supplements*. Fisher, Tucson, AZ, 1998.

Harris, L.J. The discovery of vitamins, in *The Chemistry of Life: Lectures on the History of Biochemistry*, Needham J., Ed. Cambridge University Press, Cambridge, 1970, pp. 156–170.

Kiple, K.F. and Ornelas, K.C., Eds. *The Cambridge World History of Food*. Cambridge University Press, Cambridge, 2000, vol.1, pp. 741–784.

Marks, J. *A Guide to the Vitamins: Their Role in Health and Disease*. MTP Press, Lancaster, 1975.

Mervyn, L. *Thorson's Complete Guide to Vitamins and Minerals*, 3rd edn. Thorsons, London, 1986.

Reinhard, T. *The Vitamin Source Book*. Lowell House, Los Angeles, 1998.

Warfare

Twentieth-century warfare begins with World War I (1914–1918) even though this conflict had more in common with wars of the previous century than it did with those that followed. The Great War opened with maneuvers by huge field armies that culminated in frontal assaults by masses of infantry. After only a few months of mobile warfare, heavy casualties forced opposing armies to take shelter in trench systems that stretched all across France. Faced with the ensuing stalemate of the trenches, both sides adopted an attrition strategy that would defeat the opposition by bleeding his manpower and depleting his material resources. The strategy finally succeeded when an exhausted Germany surrendered in November 1918.

In spite of its similarity to nineteenth-century warfare, World War I witnessed several new developments, most notably the airplane, the tank, and the truck. Between 1919 and 1939, the implications of these new developments were worked out, producing new operational approaches that transformed warfare.

During World War II (1939–1945), European land warfare was dominated by mobile armored forces that swept back and forth across the continent. While armies fought on the ground, air forces contended for control of European skies. In this massive air war, Allied bombers devastated Germany's industrial base and population centers.

Meanwhile, in the Pacific region, the war centered on aircraft carrier task forces that battled each other and supported amphibious operations. The war started with Japan conquering much of the western Pacific, only to be pushed back by superior Allied arms and forced to surrender when American B-29 bombers dropped the only two atomic bombs ever used in war.

By the time World War II ended in the Pacific, Japan's military resources had been severely reduced by Allied military actions. The reduction of Japanese resources, along with the progressive weakening of Germany in the European theater, suggest that World War II, like the first, was an attrition war in which industrial capacity was as important as military forces.

Two of the most revolutionary developments of World War II were the atomic bomb and the long-range ballistic missile. When more fully developed and mated to each other during the Cold War (1946–1991), they became what is perhaps the most revolutionary weapon in military history, the nuclear-tipped, intercontinental-range ballistic missile (ICBM). In the end, the Cold War was another attrition conflict, ending with the economic exhaustion and collapse of the Soviet Empire.

The end of the Cold War reduced the tensions that had kept nuclear strike forces on hair-trigger alert since the 1950s. Although nuclear weapons still existed, relations between the U.S. and the Russian Federation that emerged from the defunct Soviet Union were no longer based on mutually assured destruction (MAD), as both sides reduced their nuclear forces and the U.S. continued developing missile defenses.

While there were a number of significant "limited" wars during the twentieth century, the five major episodes described above unleashed the greatest national energies. These energies were molded into major new military systems through the process of command technology that is rooted in England of the 1880s according to historian William McNeill. Before this time, weapons were either developed in government-owned arsenals or by private entrepreneur inventors. A major change began in 1886 when the British Admiralty, dissatisfied with the performance of the government arsenal at Woolwich, started contracting with private arms makers for the development of new weapons. Under this approach, the Admiralty established the specifications for a new weapon and effectively challenged the contractor to produce it. This contracting system marks the beginning of command technology. Tantamount to invention on demand, this process of state-sponsored research and development spread throughout the West, becoming the dominant paradigm for weapons acquisition by 1945.

One product of command technology during World War I was the tank, which was developed by the British to cross fire-swept terrain between the trenches and breech the German defenses. While the tank proved capable of completing its mission, its successes were limited due to technical limitations and a lack of understanding of how best to use the new weapon.

The principal enabling technology for the tank was the internal combustion engine, which also powered World War I trucks and airplanes. The former improved logistics by connecting troops in forward positions with railheads and supply depots in the rear. The latter opened an entirely new realm of warfare and, over the course of the war, suggested all the missions the airplane would perform in future wars.

Building on the lessons of World War I, air power advocates used the period between the two world wars to develop a rigorous body of air power doctrine. At the same time, the world's leading powers developed aircraft of increasing capabilities to execute the missions defined in their doctrines.

The U.S. emphasized long-range bombers to execute daylight, precision bombardment—the dominant doctrine in America's air force. England also developed bombers, but she also pursued fighter development because of the threat posed by the air force of a rearming Germany. In addition to bombers, Germany developed tactical aircraft to support its new approach to ground warfare—Blitzkrieg.

The basic ideas behind Blitzkrieg had emerged by the end of World War I, as the capabilities of tanks and aircraft improved. After the war, the Germans developed these ideas further and mated them to the panzer division, which included tanks, mechanized artillery, and motorized infantry. Through radio communications, these elements were integrated into coherent units that also used their radios to coordinate supporting air attacks by Germany's tactical fighters. Using their air support, the panzers would execute deep, penetrating attacks to unbalance opponents and keep them from shoring up their defenses once these had been breeched.

World War II in Europe opened with Blitzkrieg attacks that swiftly overran Poland in 1939 and France in 1940. It ended with Allied air forces supporting mechanized operations that pushed German forces out of their conquered territories prior to overrunning Germany itself.

While the Germans were perfecting Blitzkrieg, naval officers around the world were integrating aviation into naval operations. This entailed developing true aircraft carriers with landing decks that ran the full length of vessels, allowing aircraft to both take-off and land on the carrier. The advent of these carriers prompted a debate over which ship, the carrier or the battleship, would dominate the next war.

This question was settled decisively at Pearl Harbor on December 7, 1941, when Japanese carrier aircraft damaged or sank every American battleship in the harbor. Throughout the remainder of the war in the Pacific, the principal measure of naval power was the carrier task force in which battleships, cruisers, and destroyers used their firepower principally to protect their carriers from attack by enemy planes and submarines. The impact of the carrier on naval warfare is clearly illustrated by the May 1942 Battle of the Coral Sea, the first naval engagement in which the surface forces never sighted each other. Throughout the remainder of the century, the carrier task force dominated naval operations.

Three years before the Battle of the Coral Sea, physicist Albert Einstein alerted U.S. President Franklin Roosevelt to the potential of nuclear fission. After having this concept evaluated by a panel of scientists, Roosevelt launched the Manhattan District Project to develop an atomic bomb.

There were two major facets to this project: developing an industrial base to produce fissionable materials and designing the bomb itself. On July 16, 1945, less than three years after the project began, the world's first atomic bomb was detonated in New Mexico. Within a month, the U.S. had dropped two atomic bombs on Japan, forcing the Japanese to surrender. About a decade before the U.S. launched its atomic bomb project, the Germans began work on what would become the world's first long-range ballistic missile. In 1937, this program was greatly expanded with the establishment of a vast, new rocket development center at Peenemunde. The German program employed several hundred scientists and technicians who were supported by a large budget that could be coupled to Germany's industrial base and its university research facilities through a flexible contracting system.

This rocket program is a classic example of command technology. Guided by specifications established by the army's ordnance office, the Peenemunde team made rapid progress after 1937. In June 1942, the team completed the first successful test of the V-2 rocket, which became the world's first operational long-range missile when it began hitting allied cities in September 1944. This choice of targets, which was dictated by the missile's inaccuracy and the limited size of its warhead, meant that the V-2 was essentially a terror weapon with little real military value.

Immediately after World War II, both the U.S. and the Soviet Union absorbed German rocket developments and began working energetically to produce long-range missiles that could be used for military purposes. A major breakthrough came in the 1950s when both countries demonstrated the ability to produce thermonuclear bombs. This meant that warheads could be made that were light enough to be carried by a missile, yet powerful enough to compensate for missile inaccuracies. Moreover, the advent of the hydrogen bomb ushered in an era of "nuclear plenty," since fusion fuel is plentiful and inexpensive when compared to fission fuel.

At the same time, work was progressing on inertial guidance systems that would be much more accurate than the system used in the V-2. A major breakthrough here was the development of more sensitive inertial measuring units that were based on complex mechanical structures, computer advances, and improved electro-optical technologies.

The simultaneous resolution of guidance and warhead problems made the ICBM feasible. Paradoxically, because these weapons could destroy civilization, the doctrine governing their employment, mutual assured destruction (MAD), aimed to deter their use. MAD required each side to have enough nuclear weapons to absorb a nuclear attack and still be able to inflict unacceptable losses on the attacker.

America's first ICBM became operational in 1959. In developing this missile, the U.S. Air Force pioneered a new management discipline that was based on insights into the functioning of complex weapons.

Until well into the nineteenth century, weapons were largely simple, stand-alone devices. However, by World War I, they were often amalgams of complex components as in the case of the giant dreadnought class battleships that dominated naval warfare during the first two decades of the twentieth century.

During the World War II, air defenses and aircraft carriers raised the complexity of weaponry another order of magnitude. It was at this point that the pioneers of operational analysis made the point that optimizing the performance of complex weapons required a thorough understanding of how their components interacted with each other and with their operational environment. Assuring a proper "fit" between system components became the work of systems engineering. Bringing operational analysis and systems engineering together to create an effective weapon was the function of systems management, a discipline that was more fully developed and formalized in the U.S.'s huge ICBM program that was launched in the 1950s. The success of the ICBM program transformed systems management into the principal paradigm for managing major weapons programs, including those for self-guided and precision-guided munitions (PGMs).

A major inspiration for self-guided munitions was the airplane. Before the advent of artificial sensors, computers, and advanced servo motors, the presence of a pilot offered one means, beyond initial aiming, to guide a weapon to its target. Indeed, one of the best known early efforts to achieve precision guidance was the Japanese use of suicide pilots who attempted to fly their planes into U.S. ships during the World War II. Less well known are U.S. and German efforts to develop unmanned glide bombs and vertical bombs that could be controlled from the aircraft that dropped them.

Germany's desperate efforts to down Allied bombers near the end of the World War II spawned several innovative concepts in the area of precision-guided surface-to-air missiles or SAMs. Included here was the use of a simple infrared sensor to allow SAMs to home in on hot bomber engines. Another SAM was to have been guided by commands from the ground that reached the interceptor via a thin wire that played out as the missile flew toward its target. Fortunately for Allied bombers, these ideas came too late in the war to be implemented.

More fully developed after World War II, wire-guided missiles were used extensively in limited and regional wars such as the Vietnam War (1965–

1973) in which an American wire-guided missile achieved an 80 percent hit rate. Soviet wire-guided missiles were used extensively by the Egyptians to inflict heavy loses on Israeli armor during the early phase of the 1973 Egyptian–Israeli War.

Infrared heat-seeking technology was widely applied in missile guidance after 1945. By 1953, the U.S. had developed the world's first heat-seeking air-to-air missile. Widely used throughout the rest of the century, these missiles generally employed a small, nose-mounted infrared sensor to guide them to the hot engine tailpipe of enemy aircraft. Shoulder-held heat-seekers were also developed to protect soldiers against air attacks. By the end of the century, the spread of these small, portable missiles was causing concern that terrorists might use them against commercial jetliners.

Other precision-guided missiles used radar in their guidance systems. While some were designed for air-to-air combat, others were built to home in on the signal from air defense radars. Radar-guided SAMs also became central to effective air defenses.

Systematic efforts to develop defenses against aircraft began during World War I when the British tried to stop German bomber attacks on England. Twenty years later, with England facing the prospect of air attacks from a rearming Nazi Germany, Sir Robert Watson-Watt advised the British government that reflected radio waves could be used to locate attacking aircraft. This principal became the basis for a radar system that the British began deploying in the mid-1930s. By the time German planes attacked London in 1940, England had deployed an air defense system that used radar plots and radio communications to guide defensive fighters to attacking German planes.

The use of radar here is an important departure. The increasing speed and range of the airplane collapsed time and threatened to deprive the defender of adequate response time. Using instruments such as binoculars and listening devices to increase the power of human senses was no longer adequate for locating an attacking force. Radar marks the first effort in military affairs to extend human perception by using phenomena outside the normal range of man's five senses. The British Chain Home radar system could detect aircraft approaching at an altitude of 6000 meters at a range of 145 kilometers, providing a warning time of 15 minutes for planes flying at 580 kilometers per hour.

Faced with the threat of nuclear-armed Soviet bombers in the 1950, the US. developed a continental-wide air defense system with a forward-based radar system to provide the earliest possible warning of attack. Radar data were fed to computerized control centers that automated the manual process of vectoring interceptors to their targets. These centers could simultaneously track 200 attacking bombers while vectoring 200 interceptors to their intercept points.

As this system was becoming operational, both the U.S. and the Soviet Union began deploying ICBMs. Against these weapons, bomber defenses were essentially useless. The ICBM's speed allowed it to traverse thousands of kilometers in a matter of minutes, further compressing the time for defensive actions. Some way had to be found to recapture the lost response time.

Improved ground-based radars provided fifteen minutes of warning time in the case of an ICBM attack. An additional fifteen minutes were gained by deploying satellite-based, infrared sensors that surveilled enemy missile fields around the clock. More time was recovered by parsing time into picoseconds, providing billions of time units that could be effectively managed by high-speed computers to optimize defensive reactions.

Later missile defense concepts pursued under the Strategic Defense Initiative, which was launched in 1984 by the US., sought to improve the odds for a successful defense by placing interceptor missiles in space. Furthermore, the U.S. pursued various concepts for directed energy weapons, which promised a near instantaneous kill, since beam velocities approached the speed of light. Combining orbiting lasers with space-based interceptors would produce a defense capable of destroying enemy missiles during the boost phase before they released their multiple warheads and decoys.

As the twentieth century was ending, the U.S. was developing an airborne laser that could also destroy ballistic missiles during their boost phases. This weapon also promised to be effective against attacking aircraft.

The high-speed computer, so crucial to the prospects of missile defense, was also central to the development and proliferation of command and control systems after 1950. These systems formed an integrated "picture" of current situations based on information from a wide variety of sources. Included among these sources are battlefield sensors, overhead satellites, electronic intelligence, and units engaged in combat. This picture provided the basis for extending and tightening the control exerted of senior political and military leaders. Computerized systems also played a major role in managing military logistics, so essential to modern military forces.

Developments such as high-speed computers, lasers, radar, and infrared sensors point toward a fundamentally new departure in twentieth-century weaponry: the creation of advanced military capabilities based on esoteric scientific principles. These principles are generated through abstract, mathematical reasoning and are not readily discoverable through the traditional methods of careful observation and the manipulation of materials. Without the highly mathematical electromagnetic field theory of James Clerk Maxwell there would be no radio or radar. Without the work of scientists like J. J. Thompson, Ernest Rutherford, and Niels Bohr there would have been no atomic theory and no basis for conceiving nuclear fission.

Introducing scientists into the mix of engineers, technicians, and managers that was central to earlier forms of command technology greatly increased government's power to invent on command. As the century was ending, this enhanced form of command technology had created in military affairs a situation similar to what historian Walter McDougall described as a perpetual technological revolution.

Change had become one of the few constants in military affairs. Making effective "transformations" in force structures and doctrines to ensure success in future wars was more clearly than ever a core concern for military professionals and their civilian leaders.

See also **Battleships; Fission and Fusion Bombs; Military versus Civil Technologies; Tanks; Missiles; Radar; Sonar; Submarines, Military; Warfare, Biological; Warfare, Chemical; Warplanes**

DONALD R. BAUCOM

Further Reading

Brodie, B. and Fawn, W. *From Crossbow to H-Bomb: The Evolution of the Weapons and Tactics of Warfare,* revised edn. Indiana University Press, Bloomington, 1973.

Dupuy, T.N. *The Evolution of Weapons and Warfare*. The Bobbs-Merrill Company, New York, 1980.

Huges, T.P. *Rescuing Prometheus*. Pantheon, New York, 1998.

Johnson, S.B. *The United States Air Force and the Culture of Innovation: 1945–1965*. Air Force History and Museums Program, Washington D.C., 2002.

McDougall, W.A. *The Heavens and the Earth: A Political History of the Space Age*. Basic Books, New York, 1985.

McNeill, W.H. *The Pursuit of Power: Technology, Armed Force, and Society since A.D. 1000*. University of Chicago Press, Chicago, 1982.

Van Creveld, Martin. *Technology and War: From 2000 B.C. to the Present*. The Free Press, New York, 1989.

Warfare, Biological

In addition to the military use of natural or synthesized plant and animal toxins as poisons, biological warfare involves the use of disease-causing bacteria, viruses, rickettsia, or fungi to cause incapacitation or death in man, animals, or plants. Over the course of the twentieth century, biological weapons scientists, engineers, and physicians in various countries adopted existing technological and scientific practices, techniques, and instrumentation found in academic and industrial research to create a new weapon of mass destruction. Unlike the production of nuclear weapons, biological weapons research involves a synergistic relationship between the separate offensive and defensive components of each individual weapon system. Offensive research involves the identification, isolation, modification, and mass production of various pathogenic organisms and the creation of organismal delivery and storage systems. Offensive research is dependent in many cases upon the simultaneous success of a parallel defensive research program involving the creation of vaccines and protective health measures for researchers, military personnel, and civilians. In addition, defensive research involves the construction of accurate detection devices to indicate the existence of biological weapons whose presence can be masked during the initial phases of a natural epidemic.

Combining practices from medical bacteriology and public health such as pure culture technique and vaccination with standard industrial microbiological practices used in the mass production and storage of pharmaceuticals, foodstuffs, and chemicals, military researchers created industrially modified organisms that could be used against specific agricultural, civilian, and military targets. The dual-use nature of much of the technology used to create and store biological weapons, coupled with the availability of this technology on open world markets, has created difficulties for those nation states and international regulating bodies attempting to prohibit the further expansion of this military technology to other states and possibly even to terrorist organizations. The tactical and strategic utility of biological warfare has evolved as a function of the capabilities of specific biological weapons, and it was not until the perfection of airborne delivery systems using aerosolized dry matter by the U.S. and the Soviet Union during the 1960s that biological weapons revealed a destructive potential equal to the strategic threat of nuclear weapons.

Biological weapons were initially conceived for use against agricultural targets as a form of economic and psychological warfare. The creation of directed research programs in industrialized nations beginning during the 1930s indicated the possibility of using biological weapons as antipersonnel weapons and this research has continued to the present day. During World War I, against a backdrop of chemical weapons use by belligerents on both sides, the General Staff of the German Army carried out a secret campaign of sabotage involving injections of anthrax (*Bacilus anthracis*) and glanders (*Pseudomonus mallei*) into military livestock of neutral nations such as the U.S., Romania, Norway, and Spain that resulted in the deaths of numerous horses between 1915 and 1918. Following the war, the 1925 Geneva Protocol for the Prohibition of the Use in War of Asphyxiating, Poisonous or Other Gases, and of Bacteriological Methods of Warfare prohibited "the use of bacteriological methods of war." Unfortunately, many nations interpreted the treaty as reserving the right for retaliation in kind and top secret research was carried out in France, Germany, Great Britain, Japan, Poland, Germany and the Soviet Union during the interwar period.

The onset of World War II accelerated many of these research and development programs and initiated new programs in other countries, most significantly in the U.S. Although German scientists mass-produced nerve gases such as tabun, sarin, and soman, Nazi interest in biological warfare was constrained by Adolf Hitler's distaste for the weapon. Surprisingly little practical research was conducted and Germany was ill prepared for either carrying out or defending against biological warfare. While historical inquiry into the Soviet program during this and later periods continues, it now seems possible that Soviet military forces may have successfully used tularemia (*Francisella tularensis*) as a tactical biological weapon to thwart German advances during the pivotal battle of Stalingrad in 1942. Large-scale research was carried out by Japanese military scientists and physicians in Unit 731 under General Ishii Shiro in Japanese-occupied Manchuria from 1936–1945. Ishii tested the feasibility of anthrax, cholera (*Vibrio cholerae*), dysentery (*Shigella dysenteriae*) and plague (*Yersinia pestis*) through studies involving human experimentation and torture that resulted in the deaths of thousands of political prisoners, captured POWs, and civilians, including woman and children. Offered asylum by American officials after the war, Ishii's research results were added to the world's most advanced biological weapons research program. In cooperation with Great Britain and Canada, the U.S. program encompassed more than 23 different research projects involving specialized laboratory, pilot plant, and manufacturing equipment at top secret military facilities, and through research contracted to university, governmental, and industrial laboratories at a cost of $60 million. Significant research was conducted on a host of biological warfare agents including anthrax, brucellosis (*Brucella melitensis)*, dysentery, glanders, plague, tularemia and various chemical plant growth inhibitors. In addition projects were carried out on vaccination, mass production and storage, the aerosolization of pathogens, and the dynamics of airborne infection.

The U.S. and the Soviet Union continued research on biological warfare throughout the Cold War and during the 1950s and 1960s successfully "weaponized" anthrax, brucellosis, and tularemia as biological aerosols. During this period researchers moved beyond the liquid slurry form of these agents, and combining technologies of drying, freezing, and milling produced agents which could be stored for long periods and be disseminated from various delivery systems including aircraft and missiles to inflict mass casualties. Large-scale ecological testing in the Pacific Ocean involving laboratory animals verified the destructive capability of these weapons over thousands of square kilometers under specific meteorological and ultraviolet conditions. Although a signatory (like the U.S.) to the 1972 Convention on the Prohibition of the Development, Production, and Stockpiling of Bacteriological (Biological) and Toxin Weapons and on Their Destruction, the Soviet Union greatly expanded it's offensive biological weapons program under the Ministry of Defense during the 1970s and 1980s. Under Biopreparat, Soviet officials created the world's largest and most sophisticated research program involving more than 15,000 personnel, six research laboratories, and five large-scale production facilities that incorporated recent discoveries in recombinant DNA technology to create a new generation of biological warfare agents.

An anthrax outbreak in April 1979, in Sverdlovsk (now Ekaterinburg, Russia) resulted from the explosion of one such military production facility and raised questions about Soviet treaty compliance. The breakup of the Soviet Union since 1991 has greatly destabilized the security of these weapons facilities, and questions remain about the movement of former Soviet scientists and materials. In the U.S. during the fall of 2001 the mysterious use of letters containing domestic weapons grade anthrax contaminated citizens, postal workers, government officials, and members of the news media, resulting in five deaths. Although the difficulty and expense involved in standardizing airborne delivery systems has limited the use of biological weapons as a weapon of mass destruction, the inhalation fatalities resulting from the handling of anthrax contaminated mail indicates that biological warfare and biological terrorism pose a serious threat in the near future.

See also **Warfare, Chemical**

GERARD J. FITZGERALD

Further Reading

Carter, G. and Balmer, B. Chemical and Biological Warfare and Defense, 1945–1990, in *Cold War Hot Science: Applied Research in Britain's Defense Laboratories 1945–1990,* Bud, R. and Gummett, P, Eds. Harwood, Amsterdam, 1999.

Carter, G. and Pearson, G.S. North Atlantic chemical and biological research and collaboration: 1916–1995. *J. Strat. Stud.*, 19, 74–103, 1996.

Geissler, E. and van Courtland Moon, J. E., Eds. *Biological and Toxin Weapons: Research, Development and the Use from the Middle Ages to 1945*. SIPRI Chemical & Biological Warfare Studies, 18, Oxford University Press, Oxford, 1999.

Harris, S.H. *Factories of Death: Japanese Biological Warfare, 1932–45 and the American Cover Up*. Routledge, New York, 1994.

Lederberg, J., Ed. *Biological Weapons Limiting the Threat*. MIT Press, Cambridge, MA, 1999.

Wright, S., Ed. *Preventing a Biological Arms Race*. MIT Press, Cambridge, 1990.

Warfare, Chemical

Popular fiction forecast the use of poison gas in warfare from the 1890s. While an effort was made to ban the wartime use of gas at The Hague International Peace Conference in 1899, military strategists and tacticians dismissed chemical weapons as a fanciful notion. The stalemate of World War I changed this mindset. Under Fritz Haber, a chemist at the Kaiser Wilhelm Institute, Germany's chemical industry began making gas weapons. Compressed chlorine gas in 5730 cylinders was released against French Algerian and Canadian troops at Ypres, Belgium, on April 22, 1915. The gas attack resulted in approximately 3000 casualties, including some 800 deaths. Within months the British and French developed both gas agents of their own and protective gear, ensuring that chemical warfare would become a regular feature of the war.

A variety of lethal and nonlethal chemical agents were developed in World War I. Lethal agents included the asphyxiating gases such as chlorine, phosgene, and diphosgene that drown their victims in mucous, choking off the supply of oxygen from the lungs. A second type were blood gases like hydrogen cyanide, which block the body's ability to absorb oxygen from red corpuscles. Incapacitating gases included lachrymatorics (tear gases) and vesicants (blistering gases). The most notorious of these is mustard gas (*Bis*-[2-chloroethyl] sulphide), a blistering agent that produces horrible burns on the exposed skin and destroys mucous tissue and also persists on the soil for as long as 48 hours after its initial dispersion.

Ultimately chemical warfare had little decisive effect on the outcome of World War I. On the western front, approximately 500,000 casualties resulted from gas, of which less than 5 percent were fatalities. Defensive measures, including new gas helmets and masks, and effective decontamination procedures, kept pace with the appearance of new gases, reducing their long-term effectiveness. By 1917 gas shells were relegated to the status of auxiliary weapons, used to shut down enemy artillery and machine gun positions and disrupt communications.

After the war chemical weapons acquired a sinister reputation far out of proportion to their military effectiveness at the time. Much of this was due to the postwar writings of veterans, who projected their own experiences and fears onto a wider strategic venue. Military strategists like the Italian apostle of air power Guilio Douhet and Britain's Sir Basil Liddell Hart increased the negative perception of gas warfare by describing its use as a strategic terror weapon for use against civilian populations. The offensive use of chemical agents in wartime was outlawed, first by the 1922 Washington Disarmament Treaty, and then more broadly by the 1925 Geneva Protocols on gas and bacteriological warfare.

Despite the international agreements, most countries continued developing chemical weapons in the 1920s and 1930s, led by Germany, the U.S., and Great Britain. American and British scientists worked on improving existing agents, as well as creating new lethal agents using organic compounds such as ricin (the toxic distillate of the castor bean), BTX (botulinal toxin), and CN (chloroacetophenone), a lachrymatory agent. By 1936 German chemists, examining fluorine and phosphorus compounds for the insecticide industry, developed tabun (ethyl NN-dimethylphosphoramidocyanidate), the first nerve gas. This was followed in 1939 by sarin (isopropyl methylphosphonofluoridate). The German nerve gases, identified as G-agents after the end of World War II, were considerably more effective than the asphyxiating gases. Lethal in very small doses, the G-agents were absorbed through the skin and were odorless and colorless. Despite the work during the interwar years, the major combatants refrained from using chemical weapons against each other in World War II, fearing that battlefield use would soon translate into deployment against civilian populations.

After World War II, captured data and samples of the G-agents became the foundation for chemical warfare development programs by American and Soviet researchers. Sarin and VX, a second-generation reformulation of the G-agents, formed the core of the American chemical weapons arsenal until 1969. Following a series of embarrassing incidents, including the accidental release of gas at Dugway Proving Grounds in Utah in 1968, President Richard Nixon formally renounced American production, storage, and use of chemical weapons. The unilateral ban remained in effect until the Reagan administration in the 1980s. Facing a real and perceived Soviet advantage in conventional and strategic weapons, the U.S. resumed production of chemical weapons agents and delivery systems, which included a binary shell system that separated the chemical ordnance into two harmless components for storage, mixing upon explosion into the desired agent. Although the Soviet Union was recognized as maintaining the world's largest stockpile of chemical weapons, many details of the program remained hidden until after 1991, when the scope and extent of Soviet research into their own G-agent derivatives and biological toxin agents emerged. Since the collapse of the Soviet Union in 1991, however, the U.S. and Russia have moved toward chemical weapons disarmament and disposal.

Since World War I, chemical weapons have been used only sparingly, and then generally against weaker opponents without their own offensive or defensive capabilities and against helpless civilian populations. Fascist Italy acknowledged using chemical agents (later found to be mustard gas) during their 1935 invasion of Ethiopia. The Imperial Japanese Army regularly used a variety of agents, including phosgene, hydrogen cyanide, and mustard gas, against military forces and civilians in China from 1937 through 1945. A commercial compound of hydrogen cyanide known as Zyklon-B was also used to grim effect against Jews, Soviet prisoners, gypsies, and other victims in the Nazi death camps.

After World War II allegations about the use of chemical weapons by government or colonialist forces were often reported but rarely substantiated. Beginning in 1962 the U.S. used chemical defoliant agents to eliminate tactical cover for National Liberation Front and North Vietnamese Army forces operating in South Vietnam, as well as to destroy rice fields supporting them in Project Ranch Hand. Between 1962 and 1969, the United States Air Force sprayed 22,336 square kilometers with herbicides, the most common of which was Agent Orange (*n*-butyl 2,4-dichlorophenoxyacetate and *n*-butyl, 2,4,5-trichlorophenoxyacetate). While not deployed as an antipersonnel agent, since the end of the Vietnam War, Agent Orange has been linked to high rates of cancer, birth defects, and sterility among those exposed to it and their descendents. Refugees from Laos and Afghanistan throughout the 1980s described a "yellow rain" that brought illness and death to those exposed to it. The nature of the substance was the subject of debate for some years. One theory maintained the agent was really feces from a large swarm of honeybees, while Western defense analysts pointed to material evidence and eyewitness testimony that indicated the substance was a Soviet-produced chemical agent, T-2-trichothecene mycotoxin.

Iraq employed mustard gas, tabun, and other agents against Iran in the war from 1979 to 1988. Iran responded with its own gas weapons by 1986. Sarin and mustard gas were also used against Kurdish villages in northern Iraq during and after the Iran–Iraq War. The Iraqi chemical warfare program continued to develop agents and delivery systems, including warheads for *al-Hussein* intermediate ballistic missiles during and after the 1991 Persian Gulf War, despite international sanctions and the objections of the U.S. and its allies.

The Iraqi example indicates the trend of chemical weapons as strategic weapons "on the cheap" for developing nations. Relatively inexpensive to produce and adaptable to a variety of delivery systems, chemical weapons can provide a qualitative advantage previously unavailable. The proliferation of skilled technicians and scientists from the former Soviet Union after 1991 exacerbated this problem. By the turn of the twenty-first century, non-state actors in the form of terrorist organizations had arisen. In 1995 the Aum Shunrikyo religious cult released sarin on board commuter trains in Tokyo, killing seven and injuring another 144 people. This act highlighted the ease with which individuals and small groups with technical knowledge and access to chemicals could effectively "home-brew" sophisticated agents.

See also **Warfare, Biological**

BOB WINTERMUTE

Further Reading

Croddy, E., Perez-Armendariz, C. and Hart, J. *Chemical and Biological Warfare: A Comprehensive Guide for the Concerned Citizen*. Copernicus, New York, 2002.
Haber, L.F. *The Poisonous Cloud: Chemical Warfare in the First World War*. Clarendon Press, Oxford, 1986.
Spiers, E.M. *Chemical Warfare*. Macmillan, London, 1986.
Stockholm International Peace Research Institute (SIPRI). *The Problem of Chemical and Biological Warfare*, 5 vols. Almqvist & Wiksell, Stockholm. Humanities Press, New York, 1971.

Useful Websites

Fedorov, L.A. *Chemical Weapons in Russia: History, Ecology, and Politics* [*Khimicheskoye Oruzhiye V Rossii: Istoriya, Ekologiya, Politika*]. Federation of American Scientists, Washington D.C., July 27, 1994: http://www.fas.org/nuke/guide/russia/cbw/jptac008_l94001.htm

Warfare, High-Explosive Shells and Bombs

Among the most baleful of twentieth century technological accomplishments was the vast elaboration of the means for inflicting death and destruction in war. While nuclear and chemical weapons occasioned more revulsion, conventional high-explosive weapons wrought far wider harm.

A revolution began in the nineteenth century with the introduction of rifled cannon and effective explosive shells. This, in turn, brought an escalating contest between weapons and protection both for fortifications and ships. At the beginning of the twentieth century, shells were beginning to move from black powder fill to modern high explosives such as ammonium picrate and trinitrotoluene (TNT). High-explosive (HE) shells needed steel walls thick enough to withstand the shock of firing, limiting weights of bursting charges to no more than about 25 percent of the whole. Depending on the target, they might use either point-detonating or time fuses. The early time fuses continued, as they had in the nineteenth century, to depend on the time taken for a powder train of precut length to burn to its end.

Shells intended for penetration had very hard, rather blunt steel points to break the hard face of the armor, backed by a much softer and tougher body to hold the projectile together through the course of the penetration. A slightly softer cap over the point of the shell might serve to preload the armor's hardened face. Bursting charges generally were no more than 5 percent of projectile weight and were detonated by base fuses intended to delay until penetration had been achieved. Also widely

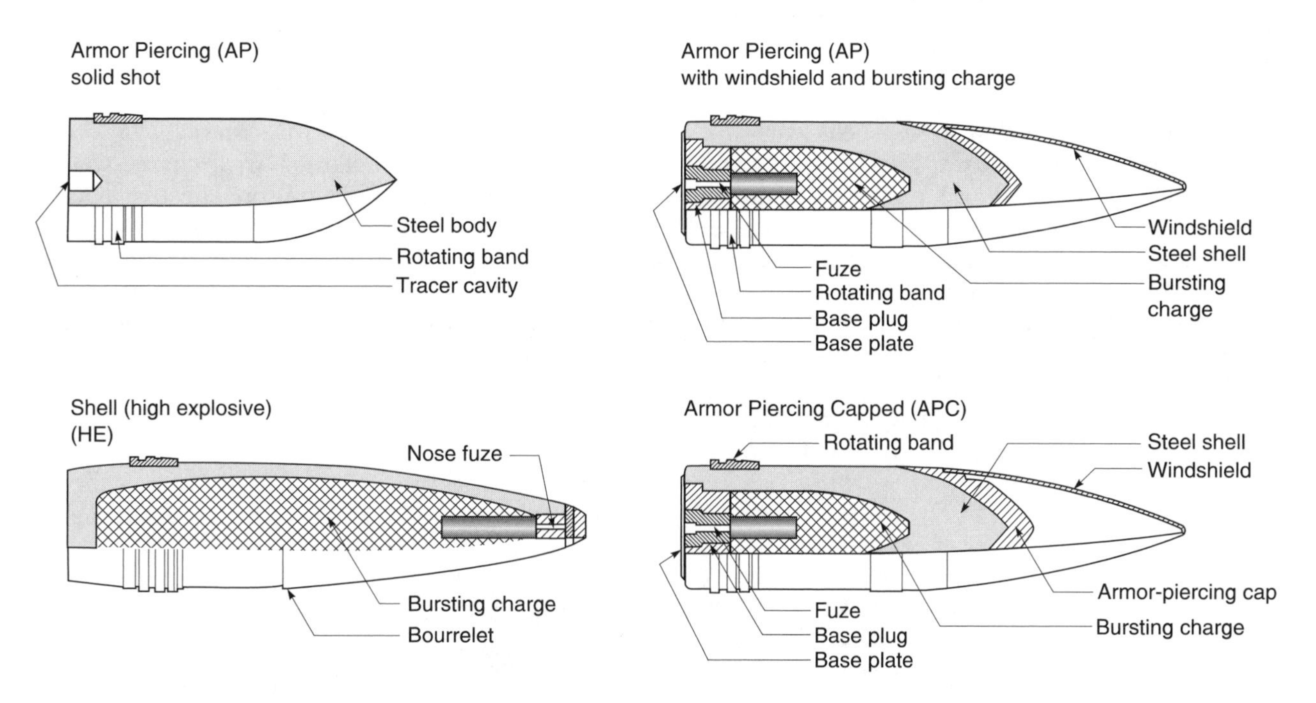

Figure 1. U.S. Army 155 mm World War II projectiles are broadly typical of types used by all nations throughout the century.

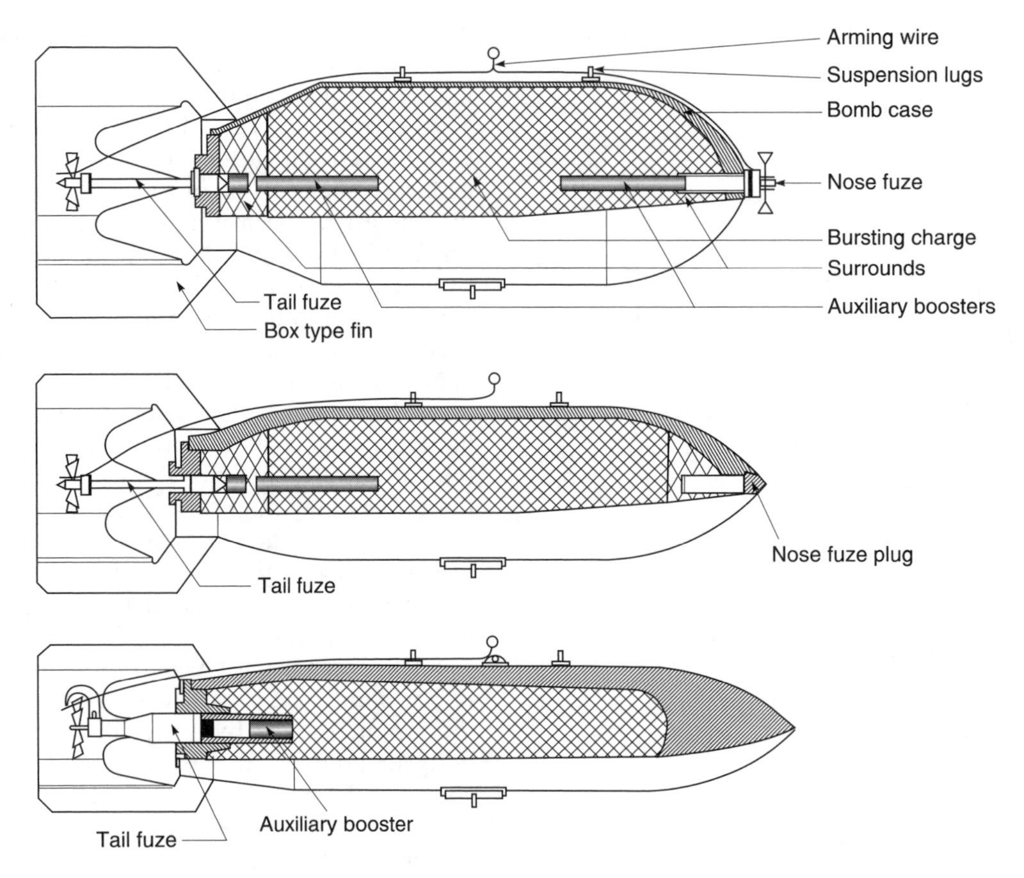

Figure 2. U.S. Army Air Force 1000 lb (450 kg) bombs from World War II also are typical, although later bombs became more streamlined to better suit jet speeds. From top to bottom they are general purpose (British medium charge), semi-armor piercing (British general purpose), and armor-piercing types, with ratios of charge to weight of ~57 percent, ~32 percent, and ~14 percent, respectively. The armor-piercing bomb is 185 cm long. Larger light case (British heavy charge) bombs with ~75 percent explosive weight were also employed for maximum blast effect.

employed in World War I were shells carrying chemical agents such as mustard gas.

Experience in World War I revealed a number of defects that impaired projectile performance, prompting a wide variety of refinements of detail. One innovation was a more reliable type of time fuse, based on mechanical clockwork. World War I also brought about interest in aerial bombs. Initially these were adapted from shells, but specially designed models soon were introduced. Owing to the much lower stresses, bombs could have lighter cases and explosive charges that could exceed 50 percent of weight. Larger charge weight fractions decreased fragmentation, effective against personnel and matériel, but increased blast effect against structures. Bomb weights increased as airplane capacities grew. Incendiary bombs filled with magnesium or jellied petrol (gasoline) were widely employed against flammable targets in World War II.

World War II brought accelerated improvements to existing types of weapons as well as

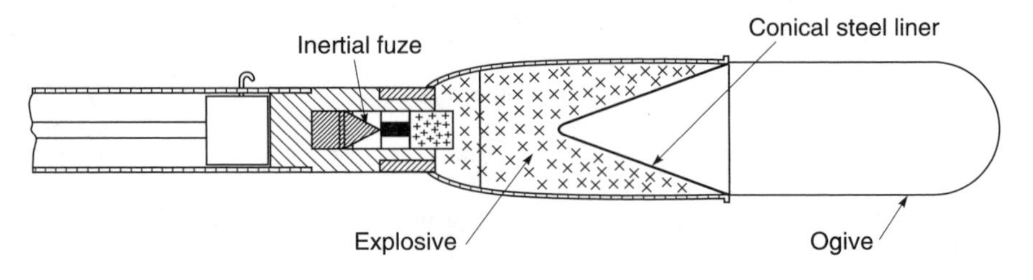

Figure 3. An early metal-lined shaped-charge munition, an American World War II Bazooka anti-tank rocket round of 2.75 in (7 cm) diameter.

innovation of new types. While TNT continued in very wide use, RDX (cyclotrimethylenetrinitramine) was introduced in quantity, frequently in mixtures with TNT. Often powdered aluminum was mixed with explosives. Both increased explosive power. An important development was introduction of metal-lined shaped charges. If a block of explosive has a conical cavity and if the detonation wave proceeds from the apex of the cone then the explosion products emerging from the cavity's walls are focused in a narrow stream along the axis of the cone. If the cavity is lined with metal, a narrow stream of metallic particles is projected. The tip of the stream reaches velocities of more than 5 kilometers per second (km/s), with bulk of metal mass following at more than 2 km/s. The pressure exerted by the jet is far beyond the yield strength of armor material, causing it to flow away like a fluid. In this manner the shaped charge digs a narrow hole in the armor approximately five to seven times as deep as its diameter, somewhat as a high-pressure jet of water in a block of gelatin.

Another wartime innovation was the proximity fuse, developed by the U.S. and Britain and not available to the Axis. This approximately trebled the effectiveness of antiaircraft fire and improved fragmentation and blast effects against ground targets. It worked using a tiny radar set carried in the nose of a shell or bomb. Cluster munitions emerged at this time as well. These were essentially grenades carried in bomb cases and scattered to increase the area covered by fragments.

The years following World War II brought the introduction of plastic-bonded explosives and further refinement of ammunition. Scientific research into the basic physics of weapons played an increasing role and led to development of a wide variety of munitions tailored to specific needs. Dueling tanks increasingly employed long-rod penetrators taking the form of slender darts fired to velocities approaching (by century's end) 2 km/s using light sabots to carry them down the bore of the large tank cannon. Made from a dense metal such as depleted uranium or tungsten, a rod can bore through steel armor of thickness approaching its length—approximately 60 cm for a 120-mm tank cannon, or nearly twice what could be achieved with conventional armor-piercing ammunition.

In the 1970s, explosively formed projectiles (EFPs) were developed for missile warheads. These consisted of shaped charges with shallow cones that ejected a compact mass of metal at velocities up to around 2.5 km/s. An EFP warhead detonated from a range of around 100 meters can deliver an effective armor-piercing blow. Another development of this period was the fuel–air explosive weapon (FAE), operating on the same principle as the detonation of fuel in a piston engine cylinder or of dust in a grain storage facility. Early FAEs employed a hydrocarbon-based fuel, which was dispersed in a cloud by one explosive charge and then ignited 100 milliseconds later by a second. The resulting deflagration (very rapid burning) typically produced a pressure wave less intense but lasting much longer than that of a high-explosive detonation. This long pressure pulse combined with the high energy content of the fuel (which, unlike an explosive, was able to take its oxygen from the air) to produce severe damage to structures over a wide area, as well as rupturing human organs. Effects were quite sensitive to environmental conditions, however, and efforts to develop FAEs with more predictability and flexibility were underway at century's end. Themobaric and enhanced blast weapons operated on similar principles; distinctions of terminology were not clearly settled. A variety of higher-energy explosives were investigated but considerations of cost as well as safety delayed their introduction.

See also **Battleships; Explosives, Commercial; Tanks; Warfare, Mines and Antipersonnel Devices; Warplanes, Bombers**

William D. O'Neil

Further Reading

Batchelor, J. and Hogg, I. *Artillery*. Charles Scribner's Sons, New York, 1972. General history of modern artillery.
Carleone, J, Ed. *Tactical Missile Warheads*. AIAA, Washington, 1993. Has introductory chapter on weapons history.
Frieden, D.R., Ed. *Principles of Naval Weapons Systems*. Naval Institute Press, Annapolis, 1985.
Padfield, P. *Guns at Sea*. St. Martin's Press, New York, 1974.
Wall, R. The devastating impact of sensor-fuzed weapons. *Air Force,* 81, March 1998.
White, M.P., Ed. *Effects of Impact and Explosion*. National Defense Research Committee, Washington, 1946. Particularly comprehensive on World War II bombs.

Warplanes, Bombers

Bombers apply aerospace technology to defeat an enemy through destruction of his will or ability to continue the conflict. In the twentieth century, the U.S. and the U.K found bombing particularly attractive because they were leaders in aerospace technology and disliked mobilizing large armies and suffering heavy casualties. Bombing requires aircraft that can carry sufficient bomb loads over great distances, penetrate enemy defenses, find targets in darkness and poor weather, and bomb accurately. Effective campaigns require adequate bases, trained personnel, fuel, munitions, replacement aircraft, spare parts, and the intelligence capability to select and assess damage to the proper targets.

Aircraft technology received a tremendous boost during World War I. Until 1914, aircraft were essentially hand-built, but in 1918 France, Britain, and Germany mass-produced sturdy airframes and powerful, reliable, lightweight engines. Strategic bombing emerged as a theoretical way to break the stalemate of trench warfare, but existing aircraft could not strike military or industrial targets effectively. Zeppelin and Gotha raids on Britain did not break British morale, but profoundly influenced postwar strategic thought.

From 1918–1939, rapidly advancing technology dramatically increased speeds, ranges, and payloads. Aircraft evolved from fabric-covered biplanes to aerodynamically efficient, metal-skinned monoplanes with enclosed cockpits, dependable instruments, and retractable landing gear. Interwar theorists asserted that airpower would win future wars, but failed to appreciate the difficulties of navigation, target identification, bomber self-defense, and the size and number of bombs needed to cripple an enemy economy. Thus, no air force was truly ready for war in 1939, and the Anglo–American bombing offensive only fully matured in 1944.

The 1930s Royal Air Force (RAF) failed to create technologies to enable bombers to survive in daylight or to find, hit, and destroy precise targets at night. The RAF had little alternative to night area attacks on cities until late 1943, but persisted in this approach long after night precision bombing became possible. Germany created formidable defenses against night bombers, and after smashing these in mid-1944, Britain achieved the prewar dream of relentless, devastating air attack. By then, the RAF had perfected navigation and bombing aids (*Gee*, *Oboe*, and *H2S*), radar countermeasures (chaff and jamming), and an elite Pathfinder Force to mark targets for the main bomber force. The late-war strategic force consisted largely of Lancasters, whose four Merlin engines delivered a 6000-kilogram payload 12,670 kilometers, and Mosquitoes, an innovative wooden design that used speed, maneuverability, and altitude to escape the enemy. Unarmed Mosquitoes precisely delivered 1800 kilogram bombs over 2300 kilometers. Mosquitoes also performed ground support, torpedo bombing, night fighter, photoreconnaissance, and electronic warfare missions.

The Boeing B-17 equipped with Norden bombsights could, in 1934, fly faster than any available fighter, and American theorists concluded that such bombers could conduct unescorted daylight precision attacks. In 1943, however, unescorted B-17s suffered severe losses, forcing America to develop long-range fighters. B-17s and B-24s attacked oil, transportation, and industrial targets; but smoke, bad weather, and enemy defenses frustrated the effort significantly. Nevertheless, bombing tied down enormous defensive forces and crippled Germany's economy in late 1944. B-29s—developed during the war—incorporated then-revolutionary designs for wing trusses, pressurized cabins, and remote-controlled guns. B-29s based in the Marianas incinerated Japanese cities from low altitude at night and mined Japanese waters. This ruined Japan's economy before atomic attacks ended the war.

In the 1930s, Germany developed a strategic bombing doctrine and a twin-engine bomber force equipped with excellent navigational and blind-bombing aids, but Hitler misused this force against both Britain and Russia. Before the war, lack of sufficiently powerful, efficient engines hindered heavy bomber development, but more importantly, Germany failed to develop an effective long-range escort fighter. Wartime research led to such major advances as swept wings (which improve handling, stability, and control), turbojets, rocket engines, ramjets, turboprops, and prototype intercontinental bombers like the Me-264 and Ju-290. However, Germany did not produce advanced designs like the Me-262 and Ar-234 in adequate numbers soon enough to affect the war's outcome. After the war, Americans and Soviet aircraft greatly benefited from German airframe, propulsion, and aerodynamic research.

During the Cold War, the Americans, British, and Soviets developed heavy bombers, because initially only bombers could deliver nuclear weapons to enemy territory. Bombers lost their primacy with the development of intercontinental-range ballistic missiles (ICBMs) and air defense missiles in the late 1950s, but remained a flexible, highly accurate force that—unlike ICBMs—could be recalled or retargeted after launch. Jet engines and improved aerodynamics greatly enhanced performance, and initially led to development of high altitude, high-speed bombers. Increasingly effective networks of radars, interceptors, and surface-to-air missiles (SAMs) soon forced bombers to operate at low altitudes and relatively low speeds. Bombers carried air-launched cruise missiles (ALCMs) to penetrate enemy defenses, and used electronic warfare to jam or deceive enemy sensors. Later bombers used "stealth" to reduce radar and infrared signatures. Each new generation of bombers was more survivable and lethal, but more expensive and purchased in smaller numbers.

Developed during World War II, Convair's B-36 never saw combat. After the war, the B-36 was the only American bomber with sufficient range and payload to strike Russia from North America with atomic weapons. Some favored canceling the B-36 and waiting for jet bombers, but the B-36 survived. With six propeller engines, B-36s flew too slowly to escape enemy fighters, and designers tested several concepts (such as "parasite" fighters carried in bomb bays) to reduce this vulnerability. The ultimate solution was a new bomber, but 385 B-36s served from 1948 to 1959.

The first American pure jet bomber, Boeing's B-47, was a revolutionary design. B-47s had six pylon-mounted jet engines, thin swept wings, and bicycle-type landing gear mounted in the fuselage. Some B-47s were equipped with rocket pods to shorten take-off length. B-47s were the first U.S. bombers that routinely used in-flight refueling, and also deployed at forward bases near the USSR. 2041 B-47s were built, serving from 1952 to 1966 in bomber and reconnaissance versions.

Boeing B-52s resembled larger, heavier B-47s, with swept wings, eight engines on pylons, and bicycle landing gear that pivoted for crosswind take-offs/landings. 744 B-52s were produced from 1954–1962 and subsequently modified extensively. Soviet SAMs required B-52s, originally designed for high altitudes, to fly low and carry up to twenty nuclear ALCMs. B-52s have dropped conventional unguided and precision munitions on Southeast Asia, Iraq, Yugoslavia, and Afghanistan. Eighty-five B-52s remain in the active force.

From 1946 until the late 1950s, the U.S. sought a delta-winged, high-altitude supersonic bomber. Supersonic flight created difficult problems of aerodynamic heating, structural fatigue, and flight control, which Convair's B-58 solved with composite materials and an innovative fuselage shape. The B-58 broke numerous speed records and won many aviation awards, but had unreliable bombing and navigation systems. More importantly, the B-58 was too vulnerable to Soviet SAMs, and was structurally unsuitable for low-altitude operations. Only 116 B-58s operated from 1961 to 1970. North American's XB-70 was a delta-winged Mach 3 nuclear bomber designed in the late 1950s that would have been the fastest and highest-flying bomber ever built. Like the B-58, the XB-70 would have been vulnerable to Soviet SAMs and unsuited to low-level operations. The XB-70 was cancelled after two prototypes were built.

Rockwell's B-1 emerged from the Advanced Manned Strategic Aircraft studies of the early 1960s. The B-1's variable geometry ("swing") wings enabled good high- and low-altitude performance and short take-offs/landings. Four turbofan engines permitted high subsonic and supersonic dash speeds. The B-1 had reduced radar signature and automatic high-speed terrain-following capability. Originally a nuclear bomber armed with bombs and up to 24 ALCMs, the B-1B was converted to a conventional precision bomber and fought in Iraq (in 1998), Yugoslavia, and Afghanistan. The B-1 entered service in 1985, and 93 remain in the inventory.

Northrop Grumman's B-2 "stealth" bomber employs special shapes, materials, and coatings to reduce visual, acoustic, infrared, and radar signatures. The B-2 resembles a "flying wing," and requires computer control for aerodynamic stability. America originally planned to acquire 132 B-2s for nuclear missions against the USSR, but scaled back to 21 aircraft when the Cold War ended. B-2s based in Missouri dropped precision conventional munitions in Yugoslavia and Afghanistan, and can reach anywhere on earth with one refueling from Missouri, Diego Garcia, or Guam.

The U.S. Air Force plans to upgrade its existing bombers until 2037. Future bombers will need sufficient range to strike anywhere on earth without refueling from U.S. bases. They may have "active stealth" and "quiet supersonic" capabilities. They will deliver nuclear, precision conventional, or directed-energy weapons to destroy fixed and mobile targets. Farther ahead, completely unmanned systems and "trans-atmospheric vehicles" (TAVs) may replace traditional bombers. TAVs would launch into low earth orbit, strike terrestrial targets with directed energy or hypervelocity kinetic projectiles, and land on a conventional runway.

Postwar British bombers emerged from Air Ministry specifications of December 1946, which called for a jet bomber able to carry a 4500-kilogram payload 6000 kilometers at 900 kilometers per hour and at 15,000 meters. Vickers, Handley Page, and Avro submitted designs, each featuring four jet engines mounted in the wing root, and from 1947 to 1964, Britain built 136 Vulcans, 86 Victors, and 107 Valiants, including prototypes.

Like the B-47, the Vickers Valiant had swept wings set high to permit a large bomb bay. The Valiant had a better payload and ceiling than the B-47, but the B-47 was faster. The Valiant became operational in 1955, and saw conventional action against Egypt in 1956. In 1962, the RAF realized that high-altitude bombers could not penetrate Soviet air defenses, and ordered the Valiants to fly at low level. The Valiant fleet quickly deteriorated due to the increased stress of low-level flying, forcing the RAF to withdraw them from service in January 1965.

The Handley Page Victor featured a T-tail, crescent wings, and the use of unique materials (a metal sandwich with a honeycomb filling) to reduce weight. The Victor became operational in 1958. In the early 1960s, Victors converted to low-level flight to penetrate Soviet defenses, and carried the Blue Steel nuclear ALCM. Due to the unexpected retirement of the Valiants, the RAF converted its Victors into tankers in the late 1960s. Victor tankers served until 1993, fighting in the Falklands and the Gulf War.

The Avro Vulcan was the world's first delta-wing bomber, and had a "kinked" leading edge to reduce buffeting. The Vulcan's delta wing, strong internal structure, and lightweight skin permitted a large payload and excellent, "fighter-like" handling at all speeds. Vulcans entered service in 1957, and gained improved electronic countermeasures (ECM), more powerful engines, larger wingspan, and in-flight refueling capability in 1960. In the early 1960s, the Vulcan converted to low-level flight and carried the Blue Steel ALCM. Vulcans incorporated terrain-following radar after 1966. Vulcans were intended to carry the American Skybolt air-launched ICBM, but America cancelled Skybolt in 1962. After British Polaris submarines entered service in 1969, Vulcans became conventional strike platforms, flying runway cratering and anti-radar missions during the Falklands War. The Vulcans retired in 1986.

Army officers dominated the 1930s Soviet military establishment, and many senior Air Force officers, aircraft designers, and engineers perished in Stalin's purges. During World War II, the Soviet Air Force focused on supporting the Red Army, and therefore in 1945 the Soviets lagged far behind Britain and the U.S. in strategic bombing theory, experience, and technology. Despite these obstacles, the Soviets created a strategic bomber force from scratch.

The Soviets initially copied the B-29, but the resulting Tu-4 lacked the range and survivability to attack the U.S. The twin-turbojet Badger, which replaced the Tu-4, was the first Soviet swept-wing bomber. Around 1500 Badgers were produced from 1953–1963 and some served as tankers and theater bombers until 1993.

Underpowered, inefficient jet engines hindered Soviet bomber development in the 1950s. The Bison, with four turbojets, could not reach the U.S. unrefuelled or carry ALCMs, so only 93 were built. The Tu-95, the world's only swept-wing turboprop bomber, had excellent range, but was slow and vulnerable. Nevertheless, roughly 100 Tu-95s comprised the primary element of the Soviet strategic bomber force from 1957 onwards. After 1959, Tu-95Ks carried one Kh-20 ALCM, upgraded to one or two Kh-22 ALCMs in 1982. In 1981, the Soviets began producing the Tu-95MS, and 63 (carrying 6 or 16 Kh-55 ALCMs) remain in service today.

In 1959, Khrushchev declared that ICBMs would be the primary strategic strike system, and transferred significant resources from bomber to missile production. In the 1960s and 1970s, the Soviets focused on developing ALCMs to penetrate Western air defenses, and also produced two theater bombers. The Soviets intended the Blinder to be a supersonic replacement for the Badger, but only built 300 Blinders due to unreliability and poor engine performance. The Backfire's variable-geometry swept wings and twin turbofans enabled supersonic dash speeds and low-altitude subsonic cruising. Some 500 Backfires were built after 1971, and saw combat in Afghanistan and Chechnya, carrying 3 Kh-55 ALCMs, 10 Kh-15 ALCMs, or conventional bombs. Backfires proved contentious during 1970s arms control negotiations. The Soviets agreed to remove the Backfire's refueling gear after the Americans insisted that Backfires, which could reach the U.S. with refueling, should be considered strategic bombers.

In 1970, the Soviets began designing the Blackjack, a variable-geometry supersonic intercontinental bomber carrying 12 Kh-55 ALCMs or conventional bombs. The Blackjack closely resembled the American B-1, and like the B-1, had reduced radar and infrared signatures. Fifteen Blackjacks entered service after 1987, and will probably remain in service until 2020. Russia recently cancelled rumored stealth bomber development for lack of funds, but continues to develop and deploy new precision ALCMs like the Kh-101 and Kh-555.

During the Cold War, the U.S. and Britain maintained strong navies to control the oceans, but the Soviets only needed the capability to deny NATO this control. Thus, maritime attack was a primary mission for Soviet bombers, unlike for British and American bombers. Soviet bombers carried long-range, supersonic antiship cruise missiles (ASCMs) to attack NATO shipping outside the range of carrier air defenses. Satellites, reconnaissance Tu-95s, and other sensors provided targeting information for ASCM-equipped Badgers, Blinders, and Backfires from the 1950s until today.

Bombers never became the "cheap" war-winning weapons that theory proposed, because bombers needed constant improvement to penetrate defenses that were themselves constantly improving to counter bombers. Bombers played a crucial role in maintaining nuclear deterrence from 1945–1991. The advent of stealth and precision-guided munitions give bombers their current great utility, but defenses will doubtless eventually evolve to counter these technologies.

See also **Aircraft Design; Warfare, High Explosive Shells and Bombs; Warplanes, Fighters and Fighter Bombers**

JAMES D. PERRY

Further Reading

Brookes, A. *V-Force: The History of Britain's Airborne Deterrent*. Jane's, London, 1982.

Dorr, R.F. and Peacock, L. *B-52 Stratofortress*. Osprey Aviation, Oxford, 2000.

Futrell, R.F. *Ideas, Concepts, Doctrine: Basic Thinking in the United States Air Force, 1907–1984*, 2 vols. U.S. Government Printing Office, Washington D.C., 1989.

Jenkins, D. *B-36 Peacemaker*. Airlife Publishing, Shrewsbury, 1999.

Knaack, M. *Post-World War II Bombers, 1945–1973*. U.S. Government Printing Office, Washington D.C., 1988.

Levine, A.J. *The Strategic Bombing of Germany, 1940–45*. Praeger, Westport, CT, 1992.

Maurer, M. *Aviation in the US Army, 1919–1939*. U.S. Government Printing Office, Washington D.C., 1987.

McFarland, S.L. *America's Pursuit of Precision Bombing, 1910–1945*. Smithsonian Institution Press, Washington D.C., 1995.

Miller, J. *Convair B-58 Hustler*. Aerofax, Hersham, 1985.

Pace, S. *B-1 Lancer*. Airlife Publishing, Shrewsbury, 1998.

Podvig, P. *Russian Strategic Nuclear Forces*. MIT Press, Cambridge, MA, 2001.

Warplanes, Fighters and Fighter Bombers

Although new as weapons, fighters played an important role in World War I. Early in the war, reconnaissance planes and bombers were joined by fighters whose task it was to engage the enemy in aerial combat. Light machine guns were synchronized to fire through aircraft propellers. It was the German firm of Fokker which developed the first effective synchronizing device; this gave the Fokker planes, agile monoplanes, superiority over the Allies comparatively slow and less maneuverable biplanes. Aircraft development was then marked by a continuous catching-up process between German fighters on the one hand and French and British fighters on the other. By mid-1916 the British DH-2 and FE-2b had outclassed the Fokkers in speed and rate of climb whilst the French Nieuport-11 and -17 and the Spad S. VII distinguished themselves by superior agility. Later in the year Germany managed to close the gap with biplanes of the Albatros-A series, but Britain rapidly caught up with the Sopwith F-1 "Camel" and France with the Spad S. XIII. Although the new Fokker D-VIII and D-VIII proved superior to the Camels and Spads they came too late and were too few. At that time another airplane with high potential did not play any military role: the Junkers D-1, built in 1918 and derived from the J-1 model of 1915, was an all-metal, duraluminum monoplane. It was fast and maneuverable but needed more time for development.

Compared with the biplanes of World War I, the all-metal piston engine monoplane fighters of World War II reached speeds of about 470 mph (750 km/h), thereby doubling the speed performance of First World War aircraft. The British Hawker Hurricane Mk-1, which left the assembly lines in October 1937, showed great versatility and was used as a night fighter, fighter–bomber and ground attack aircraft. Owing to its decisive role in the Battle of Britain the Supermarine Spitfire became even better known. More than most other airplanes of the time, the Spitfire (of which 40 different models were built) benefited from the application of recent aerodynamic research. Its greatest adversary was the German Messerschmitt Bf-109, which during its first years of service, can be considered the best fighter worldwide. It was complemented by the Focke Wulf-190 which went into service in July 1941, surpassed the Me-109 in several respects and, regarding speed and maneuverability, proved superior even to the Spitfire Mk-V. However, the best fighter of the period 1943–1944 was the North American P-51 Mustang. Equipped with British Rolls-Royce Merlin engines it was extremely advanced in structural aerodynamics. The U.S. Republic P-47 Thunderbolt, a heavy single-engine, single-seat fighter, was of equal stature with high speed and excellent rate of climb, combined with heavy firepower and great sturdiness.

During World War II the piston engines as used in fighters had reached their limits of technical capability. Therefore in Germany and Britain jet engines were developed which surpassed piston engines in performance. In Germany the experimental Heinkel He-178 flew in 1939; in Britain the experimental model of a Gloster, the E-28/39, followed in May 1941. Air engine designers had, among other difficulties, to cope with the lack of high performance materials for turbine engines. In the autumn of 1944 the German jet fighter Me-262 became operational and, in spite of various problems, proved to be an exceptional, fast aircraft. But its operational use came too late to have any significant impact on the outcome of the war.

Research during World War II, especially in Germany, had shown that swept-back wings eased shockwave problems at high speeds. Important U.S. and Soviet aircraft developed shortly after the war, such as the Lockheed Sabre and MiG-15, had swept-back wings, and others adopted delta-wing layouts. Research and development in aerodynamics, structural engineering, materials science, and related fields led to the development of fighters and fighter–bombers with improved performance characteristics. In the 1950s and 1960s there was an emphasis on speed.

The best supersonic jet fighters from the late 1950s onward could generally fly at twice the speed of sound, had fast rates of climb, great maneuverability and heavy firepower. The prototype of the French Dassault–Breguet Mirage III flew in November 1956 and was capable of high- or low-level interception in all weathers, tactical reconnaissance, and could also carry nuclear armament. Apart from this it exhibited experimental vertical take-off and landing (VTOL) ability. The U.S. company McDonnell started the McDonnell F-4 Phantom II project in 1953 as a response to the U.S. Navy's request for a supersonic twin jet all-weather assault fighter. The F-4 went into service in 1961, and during the 1960s and 1970s, was considered the best fighter in the world.

In the 1960s several European nations and aircraft producers joined in aircraft development programs because cost proved too high for a single nation to proceed by itself. In 1965 British and French manufacturers set up the Sepecat Consortium to build the Jaguar, a tactical support aircraft; four years later, in 1969; British, Federal German, and Italian firms established the Panavia Consortium to produce the multirole combat aircraft (MRCA) Tornado. One of the advantages of a multirole aircraft is economical: an MRCA can perform different operational tasks ranging from ground attack to reconnaissance and interception. For most of the flight, the Tornado can be flown automatically by its avionics. One of its most important features is the adoption of the Turbo Union RB 199 three-spool turbofan, in which part of the thrust is obtained from a large diameter fan, which makes for high performance and improved economy.

Similar to the Tornado, the U.S. General Dynamics F-16 (Fighting Falcon) is a multirole combat fighter and attack aircraft. Chosen by the U.S. Air Force in 1975 it employs the concept of relaxed static stability, which makes the fighter extremely maneuverable. The F-16 is equipped with high performance flight controls signaled by electronic fly-by-wire systems; it can be regarded as probably the best, but definitely the most cost-effective, combat aircraft of the 1980s.

Besides large military aircraft producers like the U.S., the Soviet Union or some supranational European consortia, Sweden is an interesting example of a relatively small, neutral state which, on the basis of excellent technical know-how and production facilities, developed advanced military aircraft of its own. The best known are the Saab-Scania J-35 Draken, an all-weather interceptor with maximum speed exceeding Mach 2; the Saab-37 Viggen; and recently, the JAS-39 Gripen. With the Viggen the designers employed a canard configuration (a canard is a fixed or moveable foreplane located ahead of the main wing thus making the aircraft virtually a supersonic biplane), which renders the plane highly maneuverable and enables it to operate out of short airstrips. The Gripen, while not superior in performance to the Draken or Viggen, is relatively small and comparatively easy to fly and maintain.

The "Eurofighter" is an aircraft designed to replace fighters and fighter–bombers in service in Europe and elsewhere. Although it is highly agile and has other first-class performance characteristics, it is also very expensive. But it is still cheaper than the US F-22 Rapier, a multimission air superiority combat aircraft developed jointly by Lockheed and Boeing. Its main features are advanced stealth technology and a thrust-vectoring system that yields phenomenal agility, although the Russian fighters SU-27 and SU-37 are in the same league. The French Dassault Rafale is equal to the Eurofighter and the F-22. The U.S. JSF (Joint Strike Fighter) currently being developed will also incorporate short take-off and landing ability.

See also **Aircraft Design; Warplanes, Bombers; Warplanes, Reconnaissance**

HANS-JOACHIM BRAUN

Further Reading

Angelucci, E. *The Rand McNally Encyclopedia of Military Aircraft 1914 to the Present*. Rand McNally, Chicago, 1981.

Batchelor, J. *Fighter: A History of Fighter Aircraft*. McDonald, London, 1973.

Bilstein, R.E. *Flight in America: From the Wrights to the Astronauts*, 3rd edn. Johns Hopkins University Press, Baltimore, 2001.

Bättig, T. On Target: *Moderne Kampfflugzeuge, deren Erkennungsmerkmale, Bewaffnung und Einsatzweise*. E.S. Mittler & Sohn, Hamburg, 1998.

Gunston, B. and Taylor, M.J. *Jane's Encyclopedia of Aviation*. Crescent, New York, 1993.

Gunston, B. *An Illustrated Guide to Modern Fighters and Attack Aircraft*. Prentice Hall, New York, 1987.

Munson, K. *The Blandford Book of Warplanes*. Blandford Press, Poole, Dorset, 1981.

Richardson, D. *Modern Warplanes: A Technical Survey of World's Most Significant Combat Aircraft in Service*. Salamander, London, 1982.

Stanley, W.L. *Measuring Technological Change in Jet Fighter Aircraft*. Rand Corporation, report 2249, Santa Monica, 1979.

Warplanes, Reconnaissance

The Montgolfier brothers' balloon flights in 1783 took the first step toward the development of aerial reconnaissance. Military application of the new technology came quickly with France's war with the Austrians in the last decade of the eighteenth century, though the small balloon corps established as part of the Napoleonic army lasted only a few years. Both the Union and the Confederacy used tethered balloons to spot for their artillery during the American Civil War and by the end of the nineteenth century, European armies were adding hydrogen balloons to their inventories. Design of the Parseval–Sigisfeld balloon, colloquially known as the "Drachen" (dragon), departed from the spherical type used earlier in the U.S. by the addition of a wind sock to the bottom of the envelope to point the main bag into the wind for greater stability. This innovation reduced the bobbing that had nauseated the crews of round balloons and prevented them from staying aloft for meaningful periods of time. Albert Caquot, director of the technical division of the French Aviation Militáire during World War I, further refined balloon design with the addition of side fins. Balloon use during World War I offered a technological advantage over heavier-than-air craft in that the balloon could remain aloft indefinitely and could be linked by telephone directly to the artillery batteries for which it was spotting.

Airplanes offered their own advantages. The Blériot XIs, BE-2s, and Albatros-B and -C types that made up the squadrons that mobilized at the outset of the Great War were able to roam over the battlefield, which allowed them to take intelligence photographs or to range targets beyond the sight of balloon observers. Airplanes had a downside, however. Their pilots and observers were unable to carry on a real-time conversation with anyone on the ground. Until development of the airborne radio receiver, which would have to wait for the next war, they were forced to fire flares, signal with lamps, fly to headquarters or the nearest artillery battery to land and deliver their message, or drop a note in a weighted bag. Eventually transmitters became available, but their use required deployment of an aerial consisting of a lead weight at the end of a long wire, which made the aircraft as long and unwieldy as the light planes of today which tow advertising banners over sports stadiums. If the mission came under attack while the antenna was out, the chance of outmaneuvering the enemy and surviving the combat was not good.

The science of photography also took some time to develop. Early wartime photographs offered either perpendicular or oblique views and were taken with handheld cameras by crewmembers leaning out of their cockpits. The results were shaky and the odds were against getting a series of exposures that overlapped into a complete view of the enemy area—one that could be made into a useful map. That changed with the invention of the serial camera. The serial camera could be mounted on the floor of the cockpit with its lens poking through a hole in the floor. The camera took a continuous series of photos. All that was necessary for the production of a good map was for the pilot to fly in a straight line at a constant altitude, and while that was easier said than done in the presence of enemy aircraft and antiaircraft guns—the determined reconnaissance crew sometimes being required to break off several times to fight and then return to finish the mission—it freed the crew to concentrate on defending the aircraft and represented a considerable advance over earlier practice.

After World War I and during World War II, technological effort was aimed at putting the camera at higher altitudes, theoretically out of the ability of the enemy to reach and destroy it, and to further increase its operational effectiveness by allowing it to operate in the dark. This led to development of electrical heating apparatus that prevented camera shutters from being adversely affected by the cold at high altitudes and to the slit camera that adjusted the speed at which film was fed through the camera to the speed of the aircraft, an advance that improved the production of maps of enemy territory. Nighttime operations were aided by aerial flash equipment designed by Harold Edgerton of the Massachusetts Institute of Technology, which provided not only well-lit scenery, but also frozen imagery of the target, an asset vital to effective bomb sighting.

With the opening of the Cold War, the mindset that kept pushing reconnaissance to increasingly high altitudes and greater speeds took on a new importance as it not only kept the camera out of the enemy's physical reach, but out of his legal and political reach as well. The Royal Air Force's first jet bomber, the Canberra, counted on both speed and altitude to keep it away from enemy fighters. These advantages were to prove useful to reconnaissance as well and the Canberra still serves in the RAF inventory. The quest for speed and height led ultimately to the two best-known Cold War reconnaissance aircraft, the U-2 and the SR-71. Capable of cruising at 740 kilometers per hour (km/h), with a range of 3540 kilometers, and a ceiling of 17,000 meters. (21,000 meters and above in the later models), the U-2 represented the cutting edge in aerial intelligence gathering until it was superceded by the faster and higher flying SR-71. The Blackbird pushed the altitude envelope to over 26,000 meters and was able to maintain speeds of Mach 3.2.

In the wake of such performance, the next logical step was to put cameras out of the Earth's atmosphere and to take man out of the cockpit altogether. Today's Earth-orbiting satellites are able to photograph objects the size of a pack of cigarettes and maintain ongoing surveillance of areas of interest every hour-and-a-half as they continually circle the globe. The price of exclusive reliance on such high-tech intelligence may have revealed itself in the wake of the 11 September 2001 attacks on the World Trade Center and the Pentagon, however, as interest is beginning to focus on technology that combines the advantages of the satellite with systems more immediately controllable by and useful to the soldier on the battlefield. The Predator, an unmanned reconnaissance aircraft that sends imagery through receiving systems portable enough to be contained in a trailer but sufficiently sophisticated enough to be capable of being linked by satellite to military and intelligence offices all over the world is a twenty-first century system that combines high performance with relatively low operational costs. It is well attuned to the needs of a modern military that emphasizes the combination of high performance and low risk.

See also **Aircraft Design; Warplanes, Fighters and Fighter Bombers**

JAMES STRECKFUSS

Further Reading

Angelucci, E. *The Rand McNally Encyclopedia of Military Aircraft 1914 to the Present*. Rand McNally, Chicago, 1981.

Bilstein, R.E. *Flight in America: From the Wrights to the Astronauts*, 3rd edn. Johns Hopkins University Press, Baltimore, 2001.

Gunston, B. and Taylor, M.J. *Jane's Encyclopedia of Aviation*. Crescent, New York, 1993.

Munson, K. *The Blandford Book of Warplanes*. Blandford Press, Poole, Dorset, 1981.

Waste Processing, *see* **Nuclear Waste Processing and Storage**

Wind Power Generation

Wind is essentially the movement of substantial air masses from regions of high pressure to regions of low pressure induced by the differential heating of

Figure 4. At the USDA-ARS Conservation and Production Research Laboratory in Bushland, Texas, wind turbines generate power for submersible electric water pumps that are far more efficient than traditional windmills (background).
[*Photo by Scott Bauer, ARS/USDA*].

the Earth's surface. This simplistic view belies the complexity of atmospheric weather systems but serves to indicate the origin of climatic airflow.

The first attempts to harness wind power for electricity production date back to the 1930s. In Germany, Honnef planned a monstrous five turbine, 20 megawatt (MW), wind tower, several hundred meters high, a far cry from the sleek aerospace wind turbine generators (WTGs) of today. The design of a large scale WTG is limited to one of two formats realistically held to have good prospects. First, the horizontal axis type descended from those encountered by Don Quixote and common until recently in the flat lands of Europe; and second, the vertical axis machines of which the Darrieus rotor is perhaps the most common. Of the two, horizontal axis machines predominate, although the vertical axis type has many positive attributes, not the least of these being simplicity. The advantages and disadvantages of the two types are summarized in Table 1.

Horizontal Axis Wind Turbines

To gain some idea of the basic parameters that are taken into consideration in windmill design, we must resort to some basic fluid mechanics. A windmill of the horizontal axis variety is similar to a propeller, but it takes energy from the fluid instead of imparting energy to it. The flow pattern for the windmill is the opposite of that for the propeller: slipstream widening as it passes the disc.

Table 1 Relative advantages and disadvantages of horizontal axis and vertical axis wind turbine generators.

Type of generator	Advantages	Disadvantages
Horizontal axis	Rotor adjustments through pitch angle control relatively simple Optimal aerodynamic blade design	Complex Large structure
Vertical axis	Independence of wind direction Simple construction	Most designs have no blade control to regulate output Poor efficiency Problematical dynamic behavior

Since

$$\text{Kinetic energy} = \tfrac{1}{2}\,\text{mass (m)} \times \text{velocity (v)}^2$$

then

$$\text{Power available} = \tfrac{1}{2}\,\rho A v \times v^2$$

where ρ = air density, A = area swept by rotor and v = wind velocity.

In practice the actual power gained from the wind is far less than this. To account for this a nondimensional coefficient Cp expressed as

$$\text{Cp} = (\text{Force on rotor}) = (\tfrac{1}{2}\,Av^2)$$

is introduced and termed the power coefficient. Thus,

$$\text{Power delivered} = \text{Cp} \times \tfrac{1}{2}\,Av^3$$

The power coefficient is a measure of the efficiency of a wind turbine, and from theoretical considerations it is possible to show that it cannot be greater than 0.593. Actual efficiencies for the traditional windmill with a small number of sail-like blades are commonly of the order of about 5 percent, although much higher efficiencies of the order of 35 percent are achievable today.

Perhaps the single most important component of a horizontal axis wind turbine is the rotor. The most significant design parameters for the rotor are:

- Number of blades.
- Rotor blade construction.
- Hub design.

The number of blades is essentially a compromise between conversion efficiency and cost. In general practice, on large machines it has been found that because of diminishing returns the extra expense of incorporating more than two blades is not justified by the gain in output.

The rotor must be designed for a life of 20 to 30 years and must be able to withstand the complex oscillatory stress patterns created by asymmetrical wind loading, shadow effects, and the not inconsiderable bending loads due to the weight of the blades (sometimes as much as 50 meters long). Various construction techniques have been adopted for the rotor blades ranging from laminated wood and other composites through to steel; the only untried accepted technique is that employed by the aircraft industry of riveted aluminum, which is expensive.

The major components of the system for converting mechanical energy into electricity are step-up gearbox, generator, current collectors, connecting shafts, couplings and locking brake. These components are usually housed in a nacelle (wood or metal enclosure), although some WTGs transfer the mechanical power to the base of the tower via a shaft or hydraulic pump/motor system, thus allowing the weighty generator to be ground based.

Apart from special cases, for relatively large WTGs only the normal three-phase generators of the synchronous or asynchronous type can be considered. Synchronization with the public grid is a problem because of the variability of the wind, so although synchronous generators are available relatively inexpensively and are in the main fitted to WTGs, they are by no means trouble free.

The use of asynchronous generator would circumvent the synchronization problem by allowing the generator to take its lead from the system. However, there is a limit to the number of asynchronous generators that may be connected to a particular grid system, and the efficiency of such generators is lower than for the synchronous variety.

Vertical Axis Wind Turbines

Vertical axis wind turbines (VAWTGs) have not been completely eclipsed by the development of large-scale horizontal axis machines. Much interest has been stirred by the arrival of the Musgrove variable geometry VAWTG, which combines the advantages of a vertical format with the ability to control the power output via the rotor blades.

The Darrieus rotor, first used in 1925, has for the greater part been limited to small- and medium-size test plants. The concept involved two, three, or four airfoils fixed at top and bottom to a vertical rotor shaft and bowed out at the center. Restraint of the rotor speed to design limits is achieved by means of brakes both in the form of spoilers at the center of the airfoils and disk brakes on the output shaft.

The Musgrove design, while retaining all the advantages of the vertical axis format, eliminates the complex blades of the Darrieus machine and uses conventional airfoils instead. The rotors were designed making extensive use of carbon fiber-reinforced plastic with titanium fittings and have an expected life of 40 years. Increasing wind speed causes the variable geometry rotors gradually to collapse vertically effectively reefing the blades and controlling the speed of rotation.

To put wind power in perspective, it must be considered that although the gross U.K. potential for electricity production from this source is as much as 20 percent of the total current consumption at a load factor varying between 10 and 45 percent. Thus wind power might be more usefully envisaged as an energy displacer for more expensive plants.

See also **Electrical Energy Generation and Supply, Large Scale; Energy and Power**

IAN BURDON

Further Reading

Gipe, P. *Wind Energy Comes of Age*. Wiley, New York, 1995.

Leggett, J. *The Carbon War: Global Warming at the End of the Oil Era*. Penguin, London, 1999.

MacNeill, J.R., Ed. *Something New Under the Sun: An Environmental History of the Twentieth Century*. Penguin, London, 2001.

Righter, R.W. *Wind Energy in America: A History*. University of Oklahoma Press, Norman, 1996.

Useful Website

http://telosnet.com/win/20th.html

World Wide Web

The World Wide Web (Web) is a "finite but unbounded" collection of media-rich digital resources that are connected through high-speed digital networks. It relies upon an Internet protocol suite that supports cross-platform transmission and makes available a wide variety of media types (i.e., multimedia). The cross-platform delivery environment represents an important departure from more traditional network communications protocols such as e-mail, telnet, and file transfer protocols (FTP) because it is content-centric. It is also to be distinguished from earlier document acquisition systems such as Gopher, which was designed in 1991, originally as a mainframe program but quickly implemented over networks, and wide area information systems (WAIS), also released in 1991. WAIS accommodated a narrower range of media formats and failed to include hyperlinks within their navigation protocols. Following the success of Gopher on the Internet, the Web quickly extended and enriched the metaphor of integrated browsing and navigation. This made it possible to navigate and peruse a wide variety of media types effortlessly on the Web, which in turn led to the Web's hegemony as an Internet protocol.

While earlier network protocols were special purpose in terms of both function and media formats, the Web is highly versatile. It became the first convenient form of digital communication to have sufficient rendering and browsing utilities to allow any person or group with network access to share media-rich information with their peers. It also became the standard for hyperlinking cyberspace multimedia, or cybermedia, connecting concept to source in manifold directions and identifying them primarily by uniform resource locators (URLs), which became network-wide addresses.

In a formal sense, the Web is a client-server model for packet-switched networked computer systems defined by the protocol pair hypertext transfer protocol (HTTP) and hypertext markup language (HTML). HTTP is the primary transport protocol of the Web, while HTML defines the organization and structure of the Web documents to be exchanged via hyperlinks. HTTP and HTML are higher-order Internet protocols specifically created for the Web. In addition, the Web must also utilize the lower-level Internet protocols, Internet Protocol (IP) and Transmission Control Protocol (TCP). The basic Internet protocol suite is thus designated TCP/IP. IP determines how datagrams will be exchanged via packet-switched networks while TCP builds upon IP by adding control and reliability checking.

The Web can be thought of as an extension of the digital computer network technology that began in the 1960s. Localized, platform-dependent, low-performance networks became prevalent in the 1970s. These local area networks (LANs) were largely independent of, and incompatible with, each other. In a quest for technology that could integrate these LANs, the U.S. Department of Defense, through its Advanced Research Projects Agency (ARPA) funded research in internetworking (i.e., interconnecting LANs via a wide area network (WAN) which resulted in the first national network, ARPANET. For most of the 1970s and 1980s ARPANET served as the primary network backbone for interconnecting LANs for both the research community and the U.S. government. The open architecture and being built upon a robust, highly versatile and enormously popular protocol suite, TCP/IP, resulted in rapid growth and the gradual evolution into the Internet.

The Web was conceived by Tim Berners-Lee and his colleagues at CERN (now called European Laboratory for Particle Physics) in 1989 as a shared information space that could support collaborative work. Berners-Lee defined HTTP and HTML at that time and as a proof-of-concept prototype developed the first Web client navigator–browser in 1990 for the NeXTStep platform. Nicola Pellow developed the first cross-platform Web browser in 1991 while Berners-Lee and Bernd Pollerman developed the first server application—a phone book database. The Web began as a text-only interface, but the NCSA Mosaic browser added a graphic interface by the early 1990s.

Despite the original design goal of supporting collaborative work, Web use has become highly variegated. The Web has extended into a wide range of products and services offered by individuals and organizations—for commerce, education, entertainment, "edutainment," and even propaganda. Most Web resources are still set up for noninteractive multimedia downloads with the dominant Web theme being static HTML documents and noninteractive animations.

Web technologies have evolved beyond the original concept. The support of the common gateway interface (CGI) within HTTP in 1993 added interactive computing capability to the Web. An important use of CGI has been the processing of CGI forms, which enable input from the Web user-client to be passed to the server for processing. Forms were added to HTML around 1994 and allowed users to give feedback through standard graphic user interface (GUI) objects (e.g., text boxes, check boxes, buttons). Another technological advance began in 1994 with "helper apps:" extensions of the network browser metaphor, which diminished the browser-centricity by supporting multimedia through separate, special-purpose "players." In this way, a wider range of multimedia could be rendered than could be economically and practically built into the browser itself. Web browsers now include generic launchpads that could spawn prespecified multimedia players based on the file type/file extent. This generic, browser-independent approach would be challenged twice in 1996, first by "plug-ins" and then by "executable content." Plug-ins, as the name implies, are external applications which extend the browser's built-in capability for rendering multimedia files. Unlike helper apps which rendered the multimedia in an external window, plug-ins render the media within the browser's window in the case of video, or with simultaneous presentation in the case of audio. In this way the functionality of the plug-in is seamlessly integrated with the operation of the browser and as a result often proprietary and browser-specific because of this tight integration. While plug-ins proved to be a useful notion for creating extendable browsers, plug-in developers had to write and compile code for each target platform. This was eliminated through the notion of executable content, which maintained the tight integration between the multimedia peruser and browser. The enabled browser will download the executable files that render the multimedia and execute them as well, all within the browser's own workspace on the client. This added a high level of animated media rendering and interactive content on the client side. There are several competing paradigms for Web-oriented executable content (e.g., scripting languages). However, executing foreign programs downloaded across the networks is not without security risk.

Several methods for dynamically creating web-page content have evolved; for example, dynamic HTML (DHTML) and server-side includes (SSI). Both server-push and client-pull technologies provide data downloads without user intervention. Server-push has been used to produce multiple-cell animations, slide shows, "ticker tapes," automatic pass-through of splash pages, and so on. A multitude of media-rich (if not content-rich) channels are available for use with this technology.

The World Wide Web represents the closest technology to the ideal of a completely distributed network environment for multiform communication. Perhaps the most significant impact of the Web will occur when it becomes a fully interactive, participatory, and immersive medium by default. Security and privacy issues will continue to emerge as new methods are integrated into current Web technologies. The secure socket layer is a security protocol that sits on top of TCP/IP to prevent eavesdropping, tampering, or message forgery over the Internet. Secure HTTP is a secure protocol over HTTP for identification when entering a server. However, it currently takes extra effort for Web users and servers to insure secure communication so needed by commonplace commerce. Although "cookies" were introduced to make the Web experience more useful by recording information about individual network transactions, they allow tracking of a user, seen by many as a loss of privacy.

The world will continue to become a smaller place as cultures continue to be only a click away. The Web promises to have one of the largest impacts on general society of any technology thus far created.

See also **Computer Networks; Electronic Communications; Internet; Packet Switching**

HAL BERGHEL AND YONINA COOPER

Further Reading

Berghel, H. Cyberprivacy for the new millennium. *IEEE Computer*, 34:1, 132–134, 2001.

Berghel, H. The World Wide Web, in *Encyclopedia of Computer Science*, 4th edn, Ralston, A., Reilly, E. and Hemmendinger, D., Eds. Nature Publishing Group, 2000, pp. 1867–1874.

Berghel, H. The client side of the Web, in *Encyclopedia of Library and Information Science*, Kent, A. and Hall, C.M., Eds. Marcel Dekker, New York, 1999.

Berghel, H. and Blank, D. The World Wide Web, in *Advance in Computers*, Zelkowitz, M.V. Academic Press, New York, 1999.

Berners-Lee, Tim. WWW: Past, present and future. *Computer*, 29,10, 1996, 69–77.

Comer, Douglas E. and Droms, R. E. *Computer Networks and Internets,* 2nd edn. Prentice Hall, Upper Saddle River, NJ, 1999.

Comer, D. *The Internet Book: Everything You Need To Know About Computer Networking And How The Internet Works*, 2nd edn., Prentice Hall, Upper Saddle River, NJ, 1997.

Comer, D. and Stevens, D. *Internetworking with TCP/IP,* 3rd edn. Prentice Hall, Upper Saddle River, NJ, 1996.

Hall, E. *Internet Core Protocols.* O'Reilly & Associates, Sebastopol, 2000.

Stevens, W. R. *et al.* . *TCP/IP Illustrated Volumes 1–3.* Addison Wesley, Boston, 1994.

Wright Flyers

By 1905 Wilbur and Orville Wright had developed, made, and flown the first practical airplane. This, their Flyer III, had a structure that withstood many take-offs and landings; it could bank, turn, fly figures of eight, and circle; and it was reliable enough to remain airborne for half an hour or more. The Wrights, despite having no formal training in engineering, had invested some six years of work in the production of this machine, and had combined observations of nature, experiments with wind tunnels, flying kites and gliders to construct their first Flyer. Each part of the Flyers—there were three in all—displayed the approach and originality of the Wrights, and their progression from early ideas to the eminently practical Flyer III can be marked out in four stages:

1. From 1896 to 1899 they studied the work of others and observed the flight of birds.
2. From 1899 to August 1901 they experimented with kites and gliders and validated their control systems, but found serious problems with existing knowledge.
3. From September 1901 to August 1902 they undertook intensive research on the form of airfoils, and reworking some aerodynamic problems allowed them to obtain a mastery over glider control.
4. From late 1902 to the winter of 1903 they concentrated on Flyer I, which carried an

Figure 5. Close up of Wright brothers' airplane, including the pilot and passenger seats, 1911. [*Library of Congress, Prints & Photographs Division, LC-DIG-ppprs-00690*].

engine of their own design that drove two highly efficient propellers, again of their own design. It was this airplane that is credited as the first heavier-than-air machine to fly under its own power when Orville piloted it for 12 seconds on 17 December 1903.

The brothers quickly realized that a good system of control was essential in order to achieve flight, and they concluded that methods which relied on shifting the center of gravity of a machine were unsatisfactory. Any effective airplane must be controllable, or be stable, about the three axes of roll, pitch, and yaw; and it was the flight of birds that provided them with their ideas. Bird flight is in fact very complicated, but it is clear that the main control elements are the wings and the tail. Birds use combinations of movements in their flying, but the Wrights decided that the two most important elements were a twisting of the wings, and a raising and lowering of the tail. To test the effectiveness of wing twisting, or wing warping as they were to call it, a biplane kite equipped with a fixed elevator was built and flown in August 1899. The tests were successful, and wing warping was incorporated into a full-sized glider (No. 1) which they flew in October 1900—usually unmanned. The fixed tail-plane was replaced with a front-mounted elevator. Initially the glider's wings were given upward slopes (dihedral), as this configuration confers a degree of stability in roll, but in fact the glider was difficult to control, and the Wrights made little use of dihedral in later designs. Their second glider (No. 2) incorporated lessons they had learned with No. 1. It was larger and well able to carry a pilot, and its wings were given downward slopes (anhedral). It flew reasonably well, but showed a tendency when banked to side slip, slew round and crash. The Wrights were led to doubt data on airfoils and control they had acquired from Otto Lilienthal in Germany and Octave Chanute in the U.S., and they decided to carry out and rely on their own investigations.

To find the best form of airfoils they used wind tunnels and horizontal force balances carried on the handlebars of a bicycle. Extensive series of tests led them to believe that they would be able to calculate the performance of an airplane in advance of its construction, for they could relate the geometry of the wing to airspeed and lift. The major aerodynamic problem to be solved was to find a means of counteracting the tendency of the planes to slew round, side slip, and crash in turns or when hit by gusts. A rear-mounted vertical fixed rudder was seen to be a solution, but this acted as a lever and aggravated the slewing tendency. Coupling the rudder with the wing-warping controls was a partial solution, and in fact it was found that as a result of overcompensating for slewing it was possible to make smoothly banked turns. Their glider No. 3 was the vehicle for these changes and enabled the Wrights to demonstrate the efficiency of their control system. They were almost ready to build a powered machine.

Two formidable problems had to be solved before a powered machine could be built—the provision of a suitable engine, and efficient propellers. To obtain both, the brothers relied on their own resources—they designed and built their own internal combustion, four-cylinder, 12-horsepower engine, and used their work on airfoils to design propellers. Flyer 1 was a biplane and incorporated all the lessons learnt during their experiments with gliders. Its propellers were contrarotating to counteract gyroscopic effects and were chain driven. The pilot lay flat on the lower wing, and operated an independently operated elevator, and a double rudder whose action was effected by the hip-cradle with which the pilot worked the wing-warping controls. Orville's 12-second, 37-meter flight took place at Kill Devils Hill, North Carolina, at 10:35 a.m. on December 17, 1903. Three more flights were made that morning; the last, made by Wilbur at noon, lasted for 59 seconds and covered 862 meters over the ground. These flights were the first in which a piloted machine had taken off under its own power, had flown under full control, and had landed on ground level with its take-off point. However, there were problems with control, and Flyer I was not a fully practical machine. Flyer II, similar to I, was used to learn more of the techniques of flying and to iron out some minor problems, but it tended to stall in tight turns. This difficulty was overcome in Flyer III, which first flew in June 1905. In this machine the Wrights had a practical airplane. Their solution was to decouple the rudder from the warp controls, enabling rotation about all three axes to be controlled independently, a situation that persists to this day. Like all their flying machines Flyer III was essentially unstable and had to be flown by the pilot, yet so good were their machines that they remained unchallenged by others until around the end of the Edwardian period in 1910.

See also **Aircraft Design; Internal Combustion Piston Engine**

COLIN HEMPSTEAD

Further Reading

Combs, H. *Kill Devil Hill: Discovering the Secret of the Wright Brothers*. Boston, 1979.
Gibbs-Smith, C. H. *The Wright Brothers: A Brief Account of their Work, 1899–1911*. London, 1963.
Hallion, R. *Wright Brothers, Heirs of Prometheus*. Washington, 1978.
Jakab, P.L. *Visions of a Flying Machine*. Shrewsbury, 1990.
Kirk, S. *First in Flight: The Wright Brothers in North Carolina*. Winston-Salem, NC, 1995.
Scott, P. *The Pioneers of Flight: A Documentary History*. Princeton, NJ, 1999.
Westcott, L. *Wind and Sand: The Story of the Wright Brothers at Kitty Hawk*. New York, 1983.
Unitt, P. Charlie Taylor and the Wrights' 1909 Military Flyer. *Am. Aviat. Hist. Soc.*, 43, 228–35, 1998.

X-Ray Crystallography

X-ray crystallography is a technique allowing for the determination of crystal structure. Throughout the twentieth century it has been used to determine increasingly complex structures, from the inorganic, through organic, to the biological and, most famously, the determination of the structure of DNA. It has been of tremendous utility in the materials sciences and in molecular biology. The knowledge of three-dimensional molecular structures can determine the unknown chemical formula of a compound, and it is critical for understanding biological processes such as how molecules interact, how enzymes catalyze reactions, and how drugs act. It is also a prerequisite for new drug design.

The technique relies on the fact that x-rays are waves with a wavelength of about 10^{-10} meters or 10 Å, which is comparable to intermolecular distances. X-rays can thus be diffracted by the periodic arrangement or lattice of molecules or atoms in a crystal. Following the discovery of x-rays by William Conrad Roentgen in 1895, the diffraction of x-rays by a crystal was discovered in 1912 by Max von Laue, for which he was awarded the Nobel Prize for physics in 1914. In 1913, the British physicist, William Lawrence Bragg, analyzed the diffraction pattern created on photographic film by simple crystals such as rock salt, reasoning what the three-dimensional crystal structure must be in order to create just that pattern. Using the x-ray spectrometer developed by his father, William Henry Bragg, he made a series of such determinations, from which he provided a general conceptualization of the relationship between crystal structure and the diffraction patterns created upon irradiation with x-rays of varying wavelengths. Bragg and his father received a Nobel Prize in 1915 for the analysis of crystal structure by means of x-rays. Since then, many improvements have been made both in the technique for the generation and detection of diffraction patterns and also in the reasoning from diffraction pattern to crystal structure.

During the first decade, progress was made in the basics of the technique, partly in the understanding of what intensities of x-rays led to what degree of blackening of the diffraction spots, and partly in the control of monochromatic and multichromatic x-rays. The development of x-ray tubes with higher intensities, especially of the hot-cathode variety, helped crystallography along. As in radiology, a problem continued to be posed by the heating of the x-ray tube's target, and many solutions for cooling were tried. In this early period it was also realized that one could analyze powders in addition to the single crystals initially used. Finally, the relationship between Fourier analysis and diffraction was being probed; the eventual conceptualization being that each single instance of diffraction corresponds to a Fourier component of the crystal's charge density. A Fourier transform is a mathematical operation that can be thought of as operating on the charge density: the observed diffraction pattern is the outcome. Taking the result and running it through an inverse Fourier transform gets back the original charge density.

In the 1920s this was taken up and developed into two-dimensional Fourier analysis. In the same decade, x-ray intensity was being standardized for the burgeoning discipline of radiotherapy, also yielding the possibility of absolute intensities for diffraction work. X-ray crystallography was now mature enough to yield information about organic crystals—structural information agreeing with the structures determined by organic chemistry.

In the 1930s, much of the theory of x-ray crystallography was consolidated in concepts useful for the increasingly routine deductions. Fourier projections demand a large number of measurements.

In the 1930s, the technique of rotating the photograph replaced the use of the spectrometer because Fourier projections require a large number of reflection measurements. J. D. Bernal at the University of London provided the interpretation of rotation photographs. With this technique the structure of several organic crystals were determined, along with their bond length and bond angle. Linus Pauling at the California Institute of Technology prominently contributed to the general understanding of the nature of the chemical bond. Fourier analysis was imaginatively manipulated and became so pliable as to replace the original method: guessing structures and then working out the resultant theoretical diffraction pattern until it matched the one measured. Phase indeterminacy remained a general problem (both the amplitude *and* the phase of the diffracted waves are needed to compute the inverse Fourier transform), but techniques were worked out for individual cases. Important concepts of x-ray crystallography came into general use in this period, such as reciprocal lattices and Brillouin zones.

During World War II the development of x-ray crystallography was generally interrupted, but after 1945, it accelerated due to increased funding for science. The community of crystallographers was consolidated with the establishment of the International Union of Crystallography and the founding of the journal *Acta Crystallographica* in 1948. The determination of inorganic compounds was by now fairly routine and the race with inorganic chemists was on to determine ever more complex structures, for example strychnine. Crystallographers themselves found that they broke through a barrier in this period by surpassing organic chemists' knowledge with the determination of crystal structure and chemical formula of vitamin B12 in 1956 by Dorothy Hodgkin, who won the Nobel Prize for her crystallography work. Vitamin B12 was by far the largest molecular structure ever solved by x-ray crystallography at that time, and this discovery led the way to the synthetic development of B12.

In the second half of the twentieth century, x-ray crystallography has developed into the most prominent tool in structural biology. In the 1950s, Pauling and his colleagues determined the α-helix in protein structure, Watson and Crick (at the University of Cambridge, U.K.) used the measurements of Rosalind Franklin (working with Bernal in London) and others to determine the double-helix structure of DNA, and John Kendrew and Max Perutz (also at Cambridge) determined the structure of myoglobin and hemoglobin. The initial placing of heavy atoms within large molecules happened to solve the phase problem, and this was developed into a central technique. The experimental set-up was increasingly automated, yielding more and more data, and the so-called direct approach rendered the calculations amenable to electronic computers. Human conceptualizations remained important, but increasingly tasks in the inference from diffraction pattern to crystal structure were algorithmized.

While the demands of structural biology for x-ray crystallography have continued to grow dramatically for the last half century, the most important technical development has been the shift from x-ray tubes to synchrotron radiation (i.e., radiation that is produced by charged particles moving at relativistic speeds in a magnetic field). High-energy physics experiments required storage rings for electrons; and as these particles are accelerated centripetally through bending magnets, x-rays are produced.

Demand and supply have interacted closely; for example, in the early stages of the European Molecular Biology Laboratory (EMBL). Diffraction experiments at the Deutscher Elektron-Synchrotron (DESY), in Hamburg, Germany, on insect flight muscle in 1971 not only made an immediate impact on studies in structural biology but also provided a practical example of how an international molecular biology facility could foster novelty. John Kendrew, EMBL's first director, used this synchrotron research to bolster the demands for large-scale funding.

Initially, the interest focused on greater intensity of x-ray fluxes but as time went on, other properties of synchrotron x-rays (continuous spectrum and definite time structure) have also been utilized. New equipment and beam lines were added at many facilities during the late 1970s and early 1980s, providing increasing access to the radiation for small-angle scattering, high-resolution x-ray spectroscopy, and protein crystallography experiments. Computer specialists helped to improve technology and analysis. They worked to construct data acquisition systems and analysis packages for structural problems, most notably for fast time-resolved synchrotron radiation experiments. They also developed early interactive computer graphics for x-ray crystallography and molecular modeling, a mini-revolution that replaced balsa wood and mechanical models. The demand for access to synchrotron radiation by visiting scientists intensified considerably. The demand led to the development of two-dimensional detectors and online imaging plate scanners for protein crystallography.

The experiments at DESY were parasitic, in the sense that they had to adapt a beamline for structural biology at a synchrotron that had been designed for high-energy physics experiments, such as particle collisions. X-rays produced at these physics facilities had originally been regarded as waste. In the 1980s, second-generation facilities were dedicated specifically to the production of synchrotron radiation. Area detector systems were developed for the new facilities, replacing photographic film. Several improvements relate to the increase in data processing power: for example, graphical display, such as the fitting of models onto electron density maps; and stereochemically restrained refinement, generally adopted through computer programs. The introduction of recombinant DNA technology was very significant for x-ray crystallography in this period because it enabled the production of samples of arbitrary molecules large enough to generate diffraction patterns.

In the second-generation facilities, it was discovered that magnetic array devices, termed wigglers and undulators, markedly enhanced the x-ray beams. The design of synchrotrons incorporating wigglers and undulators has led to a third generation of synchrotrons in the 1990s. Other important developments in the 1990s include charge-coupled device detectors, cryopreservation, multiwavelength anomalous diffraction (MAD) phasing, selenomethionyl proteins, and structure-solving automation. Since the turn of the century, MAD has become the method of choice.

See also **Cancer, Radiation Therapy; X-Rays in Diagnostic Medicine**

ARNE HESSENBRUCH

Further Reading

Bragg, W. L. *The History of X-Ray Analysis*. Longman, Greens & Company, London, 1943.

Ewald, P.P., Ed. *Fifty Years of X-Ray Diffraction*. Oosthoek's Uitgevermaatschappij, Utrecht, 1962.

Hendrickson, W.A. Synchrotron crystallography. *Trends Biochem. Sci.*, 25, 637–643, 2000.

Hessenbruch, A. A brief history of x-rays. *Endeavour*, 26, 137–141, 2002,

Michette, A. and Pfauntsch, S. *X-Rays: The First Hundred Years*. Wiley, New York, 1996.

Olby, R. *The Path to the Double Helix*. Macmillan, London, 1974.

X-Rays in Diagnostic Medicine

William Conrad Roentgen, a German physicist, discovered x-rays in November 1895. X-rays are a form of electromagnetic radiation, ranging in wavelength from approximately 10^{-9} to 10^{-12} meters, or 0.01 to 100 angstrom (Å). Unlike visible light, x-rays have sufficient energy to penetrate human tissues, and it is this property that makes them so useful in diagnostic imaging. Roentgen discovered x-rays while investigating discharge from an induction coil through a partially evacuated glass tube. The tube was covered in black paper and the room was in darkness, yet he noticed that a fluorescent screen in the room became illuminated. Other scientists had encountered similar phenomenon while experimenting with electrical discharges through gas filled tubes. Sir William Crookes, a leading scientist in the field, found photographic plates in his laboratory fogged but did not realize it was as a result of his experiments, and he returned them to the manufacturer. Roentgen, however, was the first to realize the significance of the effect, and he received the first Nobel Prize for physics in 1901.

X-rays, as opposed to visible light rays, penetrate matter because of their higher energy. This is a result of high-energy electrons generated in the x-ray tube; the energy of electrons incident on the target sets a maximum limit on the energy of the x-rays produced. The incident electrons gain energy from the high electrical potentials across the tubes. Up to the 1920s most x-ray tubes used induction coils (Figure 1). The inner coil is connected to the battery through a switch and an interrupter device. Once the switch is closed, current flows in the circuit, and the iron core becomes magnetized. The interrupter consists of a screw and sprung switch. The iron head of the sprung switch is attracted by the magnet and disconnects from the contact screw breaking the current path. The core becomes demagnetized, the sprung switch springs back to make contact with the screw, which once more allows current to pass, magnetizing the core. This process repeats at high speed causing a varying current in the inner coil, which induces alternating current of high potential in the secondary coil. This places a potential across the terminals T1 and T2, which are in parallel with the x-ray tube. The potential is sufficient for a spark to cross the gap between the two terminals. If the terminals are further separated, after a certain point the tube's

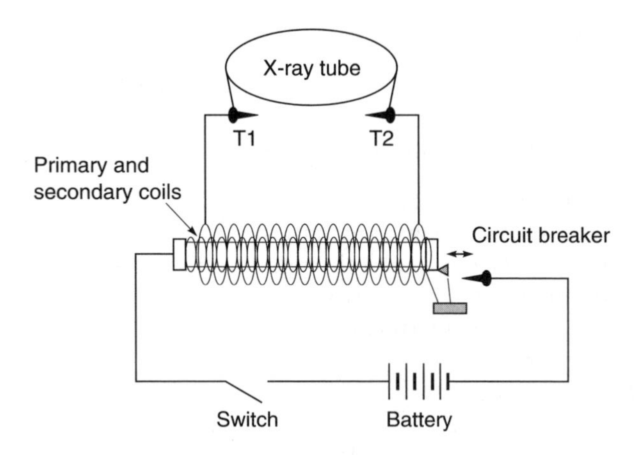

Figure 1. Induction coil.

resistance is less than the resistance between the terminals and the potential is available across the tube. The oldest method of specifying penetrating power of x-rays was by determining the length of the spark gap (the kilovoltage is equal to the equivalent spark gap in air).

The military were among the first to take advantage of x-ray technology to locate bullets and shrapnel in wounded soldiers. Power sources were a particular problem for x-ray systems in the field, however, and solutions ranged from gasoline engine-driven dynamos, to biplane engines in World War I, with even pedal power from a tandem bicycle used in the Sudan in 1898 (with the bike used for charging storage batteries). High-voltage transformers were introduced around 1919 and became the basis of high voltage generators in x-ray tubes.

In gas-filled tubes the potential across the tube causes ionization of the gas and electrons were attracted to the positive side of the tube. When the stream of electrons hit the target (originally the end of the glass tube), heat and x-rays were produced. Most of the energy was converted to heat, which placed a limit on the amount of time x-rays could be produced because the target would melt. The two most important early improvements in x-ray gas tube design were by Campbell Swinton, who introduced a sheet of platinum as a metal target, and by Herbert Jackson, who used concave cathodes to focus electrons onto a small area of the metal target producing more sharply defined images. The gas tube was unreliable, as x-ray production depended on the variable factor of the gas content. Richardson's discovery of thermionic emission in 1902 formed the basis for a major advance in x-ray tube design. W. D. Coolidge of the General Electric Company developed a thermionic x-ray tube (the Coolidge tube) in 1913 that used an almost perfect vacuum. Electrons are boiled off the cathode (thermionic emission) when current passed through the filament circuit. These electrons are accelerated across the tube producing x-rays when they strike the anode target. By 1915 Coolidge had developed rotating anodes, achieving target rotation of 750 revolutions per second, which effectively increased the area of the anode target and increased the anode's ability to dissipate heat. Modern x-ray tubes have most of the design features developed in the first 40 years of research; however, tubes generally have a selection of cathode filaments, anodes rotate at much higher speeds, the target is usually tungsten (which has twice the melting point of platinum), and improved designs for high-voltage generators and regulators have allowed a greater degree of control over x-ray energy (Figure 2).

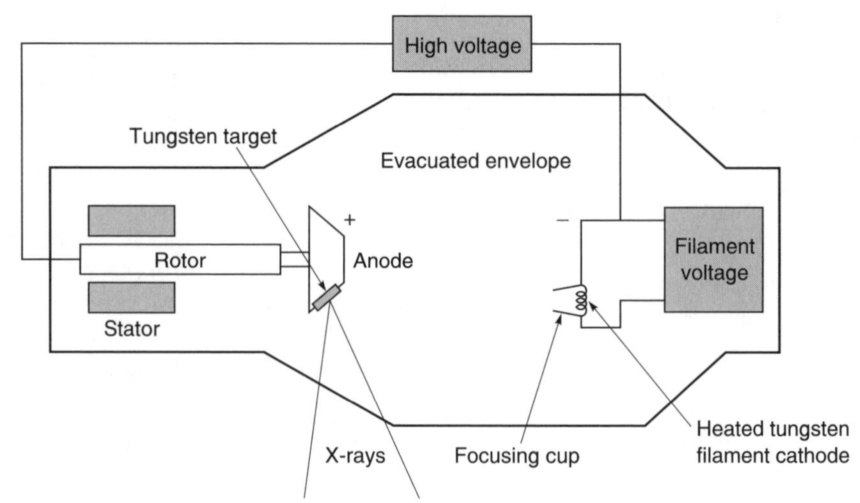

Figure 2. X-ray tube.

X-rays have a number of properties that allow their detection, and they affect a photographic emulsion in the same way as light. Films are placed behind the patient to capture the emerging x-rays (Figure 3). Screens, which reduce exposure requirements in both dose and time, have been used since the early days of x-rays. When x-rays interact with matter they may be scattered at different angles, and scatter disturbs the correspondence between points in the image and in the patient, resulting in reduced image contrast. Moving grids for scatter rejection were introduced in the 1920s. The grid blocked photons emerging from the patient at oblique angles to the detector. The grid was moved during exposures to prevent the grid lines from appearing on the image. Digital radiographic techniques, developed in the 1980s by the Fuji Corporation, introduced a photo-stimulatible phosphor screen as a primary image receptor. Use of digital detectors for radiography occurred in the 1980s and 1990s, although film remained the dominant detector type.

Another property exploited for the detection of x-rays is that they cause fluorescence in certain materials. By 1896, Thomas Edison had examined 1800 chemicals to detect and compare their x-ray fluorescent properties. His skiascope used a platino–barium cyanide fluoroscopic plate installed in a visor, which was held up to the eyes to allow observation of x-ray fluorescence. Fluoroscopic images are dynamic, with the illumination pattern on the screen responding to changes in the object being imaged. Fluorescence produced very low levels of illumination, and in many applications the eye had to be dark-adapted before images could be properly viewed. Image intensifiers, which were developed in the 1960s, greatly improved fluoroscopic imaging. These intensifiers convert the fluorescent light into electrons and accelerate them across to an output plate, which amplifies the signal. The output plate, being smaller, produces an additional geometric gain. The output screen is linked to a camera that allows display of dynamic images on a monitor.

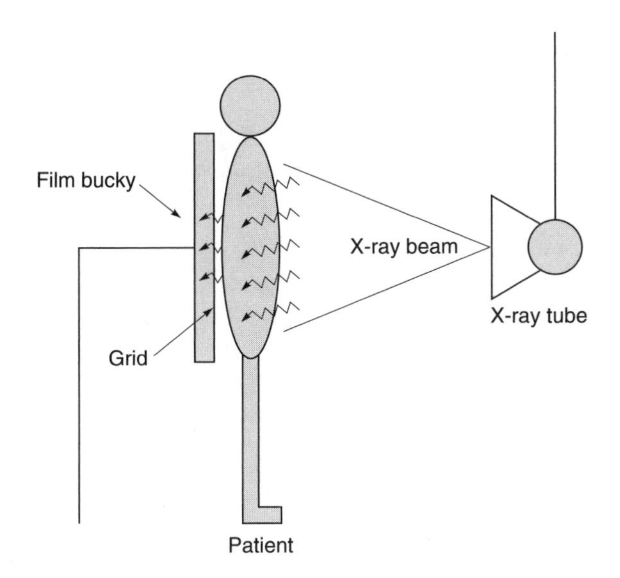

Figure 3. Basic chest x-ray radiography system.

Digital technology has revolutionized x-ray imaging. Digital subtraction angiography (DSA), which exploits the use of a contrast agent and digital imaging processing for use in vascular imaging, was developed in the early 1980s. The contrast, a dye with high x-ray attenuation properties, is injected into the blood stream. In DSA, images of a region before and after the dye is injected can be digitally subtracted. The resultant image clearly shows the vein (or area where contrast was injected), while the overlying bone structures, which might inhibit visualization, are subtracted.

In the first few years after the discovery of x-rays, it became apparent that there were hazards associated with their use. Accounts of the early years contain numerous reports of injuries and even the death of radiation workers. In 1904 Beck described three levels of Roentgen ray burns varying from itching symptoms to painful blisters and skin discoloration and ulcers. In 1915 the Roentgen society in London adopted radiation protection recommendations, and the first dose limits were proposed in 1925. The International Committee on X-ray and Radium Protection (ICXRP, later the ICRP) was formed in 1928. In 1954 the U.S. National Committee on Radiation Protection (NCRP) put forward the concept of ALARA, an acronym for "as low as reasonably achievable." Because x-rays are a form of ionizing radiation, all examinations must be clinically justified and doses kept "as low as reasonably achievable." The ALARA concept was stated in the first ICRP publication in 1959. Radiation protection recommendations continued to be reviewed and updated throughout the twentieth century in an effort to optimize safety in the use of diagnostic x-rays.

X-ray technology applications routinely used in hospitals include film screen systems, fluoroscopy systems that can provide dynamic image information, and computed tomography (CT), which provides high-contrast image slices through the patient. The many clinical applications include examination of broken bones, angiography pro-

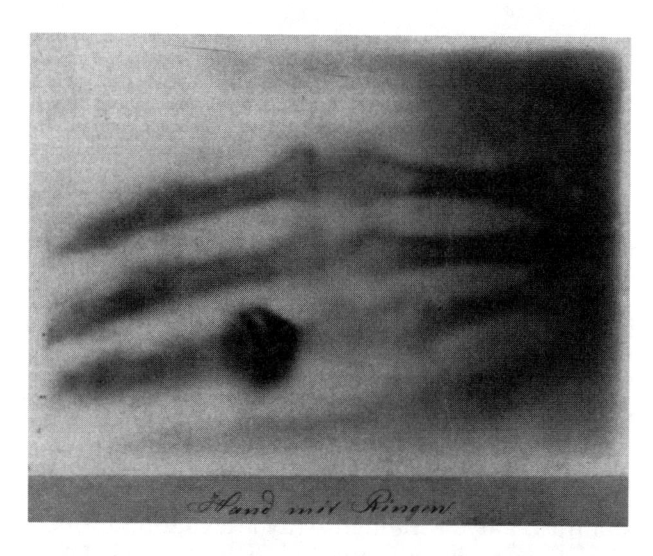

Figure 4. Hand of Mrs. Wilhelm Roentgen: the first x-ray image, 1895.

cedures, and identification of tumors, such as in mammography. In interventional procedures, x-ray technology is used to assist radiologists and surgeons in removing blockages in blood vessels and inserting pacemakers. High-energy x-ray systems may also be used therapeutically to provide lethal radiation doses to cancerous tumors. Future developments are likely to concentrate on digital systems, and many hospitals are implementing wholly digital departments. Picture Archive Communications Systems (PACS) allow images to be stored centrally and transmitted to display devices around the hospital. Instead of looking at x-ray films on light boxes, clinicians examine films displayed on high-quality monitors.

See also **Angiography; Cancer; Radiation Therapy; Particle Accelerators, Linear; Tomography in Medicine**

COLIN WALSH

Further Reading

Boone, J.M. X-ray production, interaction, and detection in diagnostic imaging, in *Handbook of Medical Imaging, Volume 1: Physics and Psychophysics*. Beutel, J., Kundel, H.L. and Van Metter, R.L., Eds. SPIE Press, U.S., 2000.

Brown, B.H., Smallwood, R.H., Barber, D.C., Lawford, P.V. and Hose D.R. *Medical Physics and Biomedical Engineering*. Institute of Physics Publishing, U.K., 1999.

Bushberg, J., Seibert, J.A., Leidholdt, E.M. and Boone, J.M. *The Essential Physics of Medical Imaging*, 2nd edn. Lippincott, Williams & Wilkins, Philadelphia, 2002.

Curry, T.S., Dowdey, J.E. and Murry, R.C. *Christensen's Physics of Diagnostic Radiology*, 4th edn. Lea & Febiger, Philadelphia, 1990.

Dance, D.R. Diagnostic radiology with x-rays, in *The Physics of Medical Imaging*, Webb, S., Ed. Institute of Physics Publishing, U.K., 1988.

Davidovits, P. *Physics in Biology and Medicine*, 2nd edn. Harcourt/Academic Press, U.S., 2001.

Mould, R.F.*A Century of X-Rays and Radioactivity in Medicine*. Institute of Physics Publishing, U.K., 1993.

Smith, F.A. *A Primer in Applied Radiation Physics*. World Scientific Publishing, Singapore, 2000.

Index

B

C

D

E

F

G

H

I

J

K

L

M

N

Q

R

S

T

U

V

W

X

Y

Z